D0031902

BIOMETRY

The principles and practice of statistics in biological research

A SERIES OF BOOKS IN BIOLOGY

BIOMETRY

The principles and practice of statistics in biological research

Robert R. Sokal
and
F. James Rohlf

State University of New York at Stony Brook

W. H. FREEMAN AND COMPANY
San Francisco

Printed in the United States of America
Library of Congress Catalog Card Number: 68-16819

To our parents

Klara and Siegfried Sokal
Harriet and Gilbert Rohlf

PREFACE

Apologia

Anyone venturing to bring out yet another textbook of statistics at this time must be imbued with a healthy dose of optimism. Few subjects have experienced such a veritable bibliographic explosion as statistics. There are generally several textbooks for every conceivable application of statistics to a field of knowledge and at the more elementary level there are a good many more. Under these circumstances aspiring authors should furnish adequate reasons for adding to this deluge with a volume of their own.

We have written this book because we feel there is a need for an up-to-date text aimed primarily at the academic biologist—a text that develops the subject from an elementary introduction up to the advanced methods necessary nowadays for biological research and for an understanding of the published literature. Most available texts represent the outlook and interests of agricultural experiment stations. This is quite proper in view of the great application of statistics in this field; in fact, modern statistics originated at such institutions. However, personal inclination and the nature of the institution at which we teach make us address ourselves to general zoologists, botanists, microbiologists, geneticists, physiologists, and other biologists, working largely on nonapplied subjects in universities, research institutes, and museums. Considerable overlap exists between the needs of these two, somewhat artificially contrasted groups and of necessity some of our presentation will deal with agricultural experimentation. Since it is a well-known pedagogical dictum that people learn best by familiar examples, we have endeavored to make our examples as pertinent as possible for our intended audience.

Much of agricultural work is by its very nature experimental. This book, while furnishing ample directions for the analysis of experimental work, also stresses descriptive and analytical statistical study of biological phenomena. The existence of these powerful methods is often overlooked and sometimes, by implication, the validity of nonexperimental biological work is put in doubt. We feel that descriptive, analytical, and experimental approaches are all of value, and we have tried to strike a balance among them.

The readers whom we hope to interest are graduate students in biological departments who require a knowledge of biometry as part of their professional training and professional biologists in universities, museums, and research positions who for one reason or another did not obtain biometric training during their student careers; these persons need to acquire knowledge largely in an autodidactic way. This book has been designed to serve not only as a text to accompany a lecture course, but also to be as complete as possible for self-study.

This book is the outgrowth of eighteen years of experience by both authors in teaching biometry to classes of graduate students from the departments of anatomy, bacteriology, biochemistry, botany, entomology, geology, pharmacology, physiology, radiation biophysics, and zoology at the University of Kansas and also at the University of California, Santa Barbara. These students generally had had no previous experience in statistics or mathematics beyond college algebra, although in recent years the incidence of new graduate students who have had mathematics through calculus and a beginning course in statistics as undergraduates has increased noticeably. However, for the foreseeable future a course such as the one at Kansas must continue to assume no prior knowledge of statistics on the part of its participants.

Much of the approach of the present text relies directly on the form of the Kansas biometry course developed by the senior author as a result of considerable experimentation. The emphasis in this course is largely on "doing": in the laboratory students are taught efficient operation of a calculating machine, and by a series of assignments they are then instructed how to solve biological problems statistically, both by thinking through the nature of the problem and by being able to carry out the computations. Experience in teaching statistics to students has shown us that simple numerical problems that are easy to do (hence not time-consuming) are not adequate to give students the facility needed to solve problems of their own when these arise during their research. *Students have to be taught the solution of real biological problems involving complicated and copious data, similar to those that would be obtained in an actual research project.* In biometry, as in most other fields, there is no substitute for experience. Teaching biometry is in many ways similar to the teaching of biological microtechnique. Theoretical knowledge of the procedures does not equip a person to prepare sections and mount these on slides.

We aim to instill in our students an ability to think through biological research problems in such a way as to grasp the essentials of the experimental or analytical setup, to know which types of statistical tests to apply in a given case, and to carry out the computations on a variety of computational equipment. A modicum of statistical theory has to be presented to introduce the subject matter, but once this is done further knowledge of statistical theory is acquired as part of the learning process in solving statistical problems, rather than as a preamble to their solution. Because of limitations of time and the limited mathematical preparation of most students, discussion of theory must be curtailed and some methods are presented almost entirely in "cook book" style.

In the course at Kansas we have found that students with very limited mathematical backgrounds are able to do excellently in biometry. There appears remarkably little correlation between innate mathematical ability and capability to understand the ordinary biometric methodologies. By contrast, in the experience of the authors a very high correlation exists between success in the biometry course and success in the chosen field of biological specialization. Teachers should encourage students who feel they have no mathematical aptitude, as this does not apply to the study of applied statistics.

We must make our apologies to the mathematical statisticians who may chance upon this volume. Compromises had to be made on occasion between precise definitions as required by mathematical statistics and those formulations of theory which teaching experience has shown to be easiest understood by beginning classes. In judging our effort, allowance should be made for such changes and circumscriptions.

To make our book more useful to the autodidactic reader we have employed a more direct and personal style than is customary in textbooks of this sort. No book can replace a dedicated and able teacher, but we have tried to make our written approach as similar as we can to our classroom manner.

The arrangement of this book

Since this book deviates in several important ways from the customary textbook of statistics, we would like to point these out to the prospective reader so that he may derive the maximum benefit from its use. Subject matter is arranged in chapters and sections, numbered by the conventional decimal numbering system; the number before the decimal point refers to the chapter, the number after the point to the section. Topics can be referred to in the Table of Contents as well as in the Index. The tables are numbered with the same decimal system, the first number denoting the chapter, the second the number of the table within the chapter. Certain special tables are entitled "boxes" and are numbered as such, again by means of the decimal system. These boxes serve a double function. They illustrate computational methods for solving various types of biometric problems and can therefore be used as convenient patterns for computation by persons using the book. They contain within them usually all the steps necessary—from the initial setup of the problem to the final result; hence to students familiar with the book the boxes can serve as quick summary reminders of the technique. A second important use of the boxes relates to their origin as mimeographed sheets handed out in the biometry course at the University of Kansas. In the lecture time at the disposal of the instructor it would have been impossible to convey even as much as half the subject matter covered in the course (or in this book) if the material contained in the sheets had to be put on the blackboard. By asking the students to refer in class to their biometry sheets (now the boxes and some of the tables in this book) much time was saved and, incidentally, students were able to devote their attention to understanding the contents of these sheets rather than to copying them. Instructors who employ this book as a text may wish to avail themselves of the boxes in a similar manner.

Figures are numbered by the same decimal system as tables and boxes. A similar numbering system is also used for (sampling) Experiments, (mathematical) Expressions, and (homework) Exercises. Numbers for appendixes carry the prefix letter A.

Since the emphasis in this book is on practical applications of statistics to biology, discussions of statistical theory are deliberately kept to a minimum. However, we feel that some exposure to statistical theory is salutary for any biologist and we have provided some in various places in the book. Derivations are given for a number of formulas but these are consigned to Appendix A1, where they should be studied and reworked by the student.

The book presents what we consider the minimum required knowledge in statistics for a Ph.D. in the biological sciences at the present time. The most important topics treated are the simple distributions (binomial, Poisson, and normal), simple statistical tests (including nonparametric tests), analysis of variance through factorial analysis, analysis of frequencies, regression, and correlation. This is more material than is covered in most elementary texts, yet even this much is usually not sufficient to treat even the common problems of everyday biological research. To enable the reader to become acquainted with some advanced methods we have therefore introduced at appropriate places throughout the book so-called signpost paragraphs, set off by the symbols

Such paragraphs will explain some of these methods and their application and refer the reader to books and papers that we have found to be most useful in explaining a given technique in simple and instructive fashion. We hope that these signposted discussions will open up for the reader the wider field of statistics and permit him at the very least to be an intelligent consulter of professional statisticians.

Our interest in and concern for proper computational techniques should be evident to anyone reading the book. Many a biologist learning statistics has become discouraged from applying his newly won knowledge because he had not been taught efficient computational procedures that permitted him to obtain correct results with a minimum of effort. Time and again the authors have run into students and colleagues attempting to solve statistical problems by methods appropriate to the computational facilities of the 1920's. We feel that the emphasis given of old to paper and pencil methods is no longer warranted at a time when access to a digital computer is becoming quite general for most scientists who wish to avail themselves of one. Access to a desk calculator should be a matter of course to research biologists even at smaller and less well-endowed institutions.

A general discussion of desk calculator computation is presented in Chapter 3 and machine practice exercises are given in Appendix A2. Suggestions for efficient solutions of various statistical formulas are scattered throughout the book, following mention of a given formula. Information about special procedures for desk calculators is generally set off by the symbols

□□ □□

Chapter 3 also discusses (we believe for the first time in a biological statistics book) the application of digital computers to the solution of the types of problems discussed in this book. Even though most research biologists will not choose to become intimately acquainted with programming a computer, it will be to their considerable advantage to understand what types of problems can profitably be handled by computer methods and what others had best be retained for the desk calculator. To help those trying to run statistical problems on a computer we have given in Appendix A3 some simple programs useful in computing most common statistics. These programs are written in the widely employed FORTRAN language.

In Appendix A4 we present a tabular guide to the types of problems frequently encountered in biological research and to the statistical methods appropriate for their solutions. It is deferred to the Appendix since it cannot be understood in its entirety without mastery of the statistical subject matter to which it refers. This is not a new idea and it must have occurred to many others. The senior author discussed it with several of his fellow graduate students at the University of Chicago as early as 1950. Siegel (1956) presented a useful key to nonparametric statistics on the inside covers of his text. We are awaiting with interest the reaction of the users of this book to our guide and will greatly appreciate suggestions for changes and improvements.

Homework exercises for students reading this book in connection with a biometry course (or studying on their own) are given at the end of each chapter. In keeping with our own convictions, these are largely real research problems. Some of them, therefore, require a fair amount of computation for their solution.

Persons familiar with textbooks of statistics will note a major departure from the customary sequence of presentation. Though we progress in conventional fashion from descriptive statistics to the fundamental distributions and into the testing of elementary statistical hypotheses, we then proceed immediately to the analysis of variance. The familiar and time-honored t-test is treated merely as a special case of the analysis of variance and is relegated to several sections in various appropriate chapters of the book. Although this may disturb traditionalists, we have taken this step deliberately for two reasons. (1) It is urgent that students become acquainted with analysis of variance at the earliest possible moment, since a thorough foundation in analysis of variance is essential nowadays to every biologist. There will be some, of course, who will study this book but not proceed beyond a certain point. By introducing analysis of variance as early as possible we hope to include it in even a cursory study of the subject matter. (2) If analysis of variance is presented and understood early, the necessity for an employment of the t-distribution is materially reduced, except for the setting of confidence limits and in a few other special situations. All t-tests can be carried out as analyses of variance and many are more informative when carried out as such. The amount of computation is generally equivalent. We trust therefore that the novice who has heard of the t-test (as just about every biologist has) is not going to be disappointed at not finding it expounded at great length in this book. The methods presented here will enable him to do all that can be done by the various t-tests, and much more besides.

In other topics we have also been concerned with introducing new and improved techniques, stressing these over earlier methods that we feel are less suitable. Notable instances of such innovations are the adoption of the simultaneous test procedure for a posteriori multiple comparisons tests and the employment of the G-statistic for the analysis of frequencies in lieu of the traditional chi-square tests, which are therefore less extensively discussed than might otherwise have been the case. We believe this to be the first beginning text to cover these new topics for biologists.

We have been dissatisfied with the customary placement of statistical tables at the end of textbooks of biometry and statistics. Users of these books and tables are constantly inconvenienced by having to turn back and forth between the text material on a certain method and the table necessary for the test of significance or for some other computational step. Occasionally, the tables are interspersed throughout a textbook at sites of their initial application; they are then difficult to locate and turning back and forth in the book is not avoided. Constant users of statistics, therefore, generally use one or more sets of statistical tables, not only because such handbooks usually contain more complete and diverse statistical tables than do the textbooks, but also to avoid the constant turning of pages in the latter.

When we first planned to write this textbook, we thought to eliminate tables altogether, asking readers to furnish their own statistical tables from those available. However, for pedagogical reasons, it was found desirable to refer to a standard set of tables and we consequently undertook to furnish such tables to be bound separately from the text.

The statistical tables (*Statistical Tables,* by Rohlf and Sokal) are identified by boldface letters, such as Table **A**. Table **Z** is followed by Tables **AA** through **GG**.

Lawrence, Kansas
September 1968

ROBERT R. SOKAL
F. JAMES ROHLF

ACKNOWLEDGMENTS

In preparing this book we have had the benefit of constructive criticism from several colleagues. Professors K. R. Gabriel (Hebrew University) and R. C. Lewontin (University of Chicago) read the entire manuscript and commented in detail on all aspects of the book. Many of their suggestions have been incorporated and have contributed significantly to any merit that the book possesses. Professors F. J. Sonleitner (University of Oklahoma), Theodore J. Crovello (University of Notre Dame), and Albert J. Rowell (University of Kansas) also furnished many detailed comments on the text. Numerous students in several biometry classes also contributed comments on the mimeographed version of the text. Since not all their suggestions were accepted, the reviewers named here cannot be held responsible for errors and other failings in the book.

We are indebted to Edwin Bryant, David Fisher, Koichi Fujii, John Kishpaugh, and David Wool for carefully checking the numerical accuracy of tables and boxes. Mr. Koichi Fujii helped prepare some of the computer programs and made a final check of the accuracy of the computations on a desk computer.

We are obligated to the many generations of biometry students who have enabled us to try out our ideas for teaching biometry and whose responses—positive or adversely critical—have guided us in developing our course. It has been a never-ending source of satisfaction to see these students appropriate biometric techniques as their own and employ them actively in their researches.

The project continued long enough to involve several successions of secretaries. Mrs. Dolores Vandermeer, Mrs. Kaye de Gutierrez, Mrs. Joetta Weaver, Mrs. Barbara Maycock, and especially Mrs. Connie Meister, who did the greatest part of the work, helped prepare and organize the manuscript. We are greatly in their debt.

To our wives, Julie and Pat, we owe a continuing debt of gratitude for their forebearance during the long period of gestation of this book.

ROBERT R. SOKAL
F. JAMES ROHLF

CONTENTS

1 INTRODUCTION

Dear Reader:

If you are starting your study of this book at this point and have skipped over our Preface, we would urge you respectfully, but firmly, to turn back and read it. We realize that prefaces are often uninspiring reading; you may not, for example, be thrilled to learn through whose kind permission we obtained the use of this or that statistical table. However, the Preface in this book contains important information for the user. We have put into it our reasons for writing this book and also have described the audience to which we address ourselves. More importantly, however, we explain our philosophy of teaching statistics and how we expect the arrangement of this book to meet our objectives. Unless the system of the book is understood, we fear that you are not likely to receive full benefit from reading it. Thank you.

The Authors

This introductory chapter is concerned with setting the stage for your study of biometry. We shall first of all define the field itself (Section 1.1). We shall then cast a necessarily brief glance at its historical development (Section 1.2). Section 1.3 concludes the chapter with a discussion of the attitudes which the person trained in statistics brings to biological research. The application of statistics to biology has revolutionized not only the methodology of research but the very interpretation of the phenomena under study. The effects which this philosophical attitude has had and is having on biology are briefly outlined. We hope that the reader of this book will absorb and appropriate some of these points of view.

1.1 Some definitions

Scientific etiquette demands that a field be defined before its study is begun. Therefore, before we begin our study of *biometry*, how is the field to be defined? The Greek roots are *bios* (life) and *metron* (measure); hence biometry means the measurement of life, which is really not a bad definition even though it is not a very useful one. For the purposes of this book biometry is conceived in a very broad sense. We shall define it as *the application of statistical methods to the solution of biological problems.* As will be seen in Section 1.2, the original meaning of biometry was much narrower and implied a special field related to the study of evolution and natural selection. However, the wider definition is customary now. Biometry is also called *biological statistics* or simply *biostatistics.*

Our definition tells us that biometry is the application of statistics to biological problems, but the definition leaves us somewhat up in the air—"statistics" has not been defined. *Statistics* is a science which by name at least is well known even to the layman (and quite often unjustly abused by him). The number of definitions you can find for it is only limited by the number of books you wish to consult. In its modern sense we might define statistics as *the scientific study of numerical data based on natural phenomena.* All parts of this definition are important and deserve emphasis.

Scientific study: We are concerned with the commonly accepted criteria of validity of scientific evidence. Objectivity in presentation and in evaluation of data and the general ethical code of scientific methodology must constantly be in evidence if the old canard that "figures never lie, only statisticians do" is not to be revived.

Data: Statistics generally deals with populations or groups of individuals; hence it deals with *quantities* of information, not with a single *datum.* Thus the measurement of a single animal or the response from a single biochemical test will generally not be of interest; unless a sample of animals is measured or several such tests are performed, statistics ordinarily can play no role. (We are old-fashioned enough to insist that the word *data* be used only in its plural sense, despite the latest dictionaries.)

Numerical: Unless the data of a study can be quantified in one way or the other, they would not be amenable to statistical analysis. Numerical data can be measurements—the length or width of a structure or the amount of a chemical in a body fluid—or counts—the number of bristles or teeth. The different kinds of variables will be discussed in greater detail in Chapter 2.

Natural phenomena: We use this term in a wide sense, including all those events that happen in animate and inanimate nature not under the control of man, plus those evoked by the scientist and partly under his control, as in an experiment. Different biologists will concern themselves with different levels of natural phenomena; other kinds of scientists, with yet different ones.

But all scientists would agree that the chirping of crickets, the number of peas in a pod, and the age at maturity in a chicken are natural phenomena. The heartbeat of rats in response to adrenalin or the mutation rate in maize after irradiation may still be considered natural, even though man has interfered with the phenomenon through his experiment. However, the average biologist would not consider the number of stereo hi-fi sets bought by persons in different states in a given year to be a natural phenomenon, although sociologists or human ecologists might so consider it and deem it worthy of study. The qualification "natural phenomena" is included in the definition of statistics largely to make certain that the phenomena studied are not arbitrary ones that are entirely under the will and control of the researcher, such as the number of animals employed in an experiment.

The word statistics is also used in another, though related, meaning. It is used as the plural of the noun *statistic,* which refers to any one of many computed or estimated statistical quantities, such as the mean, the standard deviation, or the correlation coefficient. Each one of these is a statistic. (By extension, the term statistic has also been used to indicate figures or pieces of data—that is, the individuals or items of data to be defined. This is the way the word statistic is used in the familiar warning to the careless driver not to become a statistic. This is not, however, a recommended use of the word.)

1.2 The development of biometry

The nature of this book and the space at our disposal do not permit us to indulge in an extended historical review. The origin of modern statistics can be traced back into the 17th century, in which it derived from two sources. The first of these related to political science and developed as a quantitative description of the various aspects of the affairs of a government or state (hence the term statistics). This subject also became known as political arithmetic. Taxes and insurance caused people to become interested in problems of censuses, longevity, and mortality. Such considerations assumed increasing importance, especially in England, as the country prospered during the development of its empire. John Graunt (1620–1674) and William Petty (1623–1687) were early students of vital statistics and others followed in their footsteps.

At about the same time there developed the second root of modern statistics: the mathematical theory of probability engendered by the interest in games of chance among the leisure classes of the time. Important contributions to this theory were made by Blaise Pascal (1623–1662) and Pierre de Fermat (1601–1665), both Frenchmen. Jacques Bernoulli (1654–1705), a Swiss, laid the foundation of modern probability theory in *Ars Conjectandi,* published posthumously. Abraham de Moivre (1667–1754), a Frenchman living in England, was the first to combine the statistics of his day with probability theory in working out annuity values. De Moivre also was the first to approximate the important normal distribution through the expansion of the binomial.

A later stimulus for the development of statistics came from the science of astronomy, in which many individual observations had to be digested into a coherent theory. Many of the famous astronomers and mathematicians of the 18th century such as Pierre Simon Laplace (1749–1827) in France and Karl Friedrich Gauss (1777–1855) in Germany were among the leaders in this field. The latter's lasting contribution to statistics is the development of the method of least squares, which will be encountered in later chapters of this book in a variety of forms. Gauss already realized the importance of his ideas to any situation in which errors in observations occur.

Perhaps the earliest important figure in biometric thought was Adolphe Quetelet (1796–1874), a Belgian astronomer and mathematician, who in his work combined the theory and practical methods of statistics and applied them to problems of biology, medicine, and sociology. He introduced the concept of the "average man" and was the first to recognize the significance of the constancy of large numbers—an idea of prime importance in modern statistical work.

Much progress was made in the theory of statistics by mathematicians in the 19th century, but this development is of less interest to us than the work of Francis Galton (1822–1911), a cousin of Charles Darwin. Galton has been called the father of biometry and eugenics, two subjects that he studied interrelatedly. The publication of Darwin's work and the inadequacy of Darwin's genetic theories stimulated Galton to try to solve the problems of heredity. Galton's major contribution to biology is his application of statistical methodology to the analysis of biological variation, such as the analysis of variability and his study of regression and correlation in biological measurements. His hope of unraveling the laws of genetics through these procedures was in vain. He started with the most difficult material and with the wrong assumptions. However, his methodology has become the foundation for the application of statistics to biology. Karl Pearson (1857–1936) at University College, London, became interested in the application of statistical methods to biology, particularly in the demonstration of natural selection, through the influence of W. F. R. Weldon (1860–1906), a zoologist at the same institution. Weldon, incidentally, is credited with coining the term biometry for the type of studies pursued by him. Pearson continued in the tradition of Galton and laid the foundation for much of descriptive and correlational statistics. The dominant figure in statistics and biometry in this century has been Ronald A. Fisher (1890–1962). His many contributions to statistical theory will become obvious even to the cursory reader of this book.

Statistics today is a broad and extremely active field, whose applications touch almost every science and even the humanities. New applications for statistics are constantly being found, and no one can predict from what branch of statistics new applications to biology will be made. The prospective student of biological statistics need not let the formidable size of the field overwhelm him, but he should be aware that statistics is a constantly changing and growing science and in this respect is quite different from a subject such as elementary algebra or trigonometry. Methods of teaching these subjects may have changed in the last fifty years, but the basic principles and theorems in these fields have changed not at all. A goodly proportion of the methods presented in this book were only published in the technical literature ten or fifteen years ago or were then even entirely unknown. Time-honored methods such as tests for normality, statistics of skewness and kurtosis, and the t-test are possibly becoming obsolete for a variety of reasons. Anyone would be rash to predict what biometry will look like thirty years from now. However, whatever its nature, it is more than likely that it will occupy an even more commanding position in biology at the end of this century than it does at present.

1.3 The statistical frame of mind

The ever-increasing importance and application of statistics to biological data is evident even on cursory inspection. We show in Figure 1.1 the results of a survey of eight volumes of *The American Naturalist*; because of its wide coverage, this journal can serve as a general indicator of trends in biological research. Papers in the decennial volumes were catalogued into those containing no numerical results, those containing numbers but no statistical computations, those containing simple statistics, and those having a major emphasis on mathematics or statistics. The dramatic increase of the percentage of papers involving some quantitative method is clearly shown in the graph in spite of the fluctuation to be expected in a sample of this sort. We believe that this general trend can be duplicated for almost any biological journal.

Why has there been such a marked increase in the use of statistics in biology? It has apparently come with the realization that in biology the interplay of causal and response variables obeys laws that are not in the classic mold of 19th century physical science. In that century biologists such as Robert Mayer, Helmholtz, and others, in trying to demonstrate that biological processes were nothing but physicochemical phenomena, helped create the impression that the experimental methods and natural philosophy that had led to such dramatic progress in the physical sciences should be imitated fully in biology. Regrettably, opposition to this point of view was confounded with the vitalistic movement, which led to unproductive theorizing.

Thus, many biologists even to this day have retained the tradition of strictly mechanistic and deterministic concepts of thinking, while physicists, as their science became more refined and came to deal with ever more "elementary" particles, began to resort to statistical approaches. In biology most phenomena are affected by many causal factors, uncontrollable in their variation and often unidentifiable. Statistics is needed to measure such variable phenomena with a predictable error and to ascertain the reality of minute but important differences. Whether biological phenomena are in fact

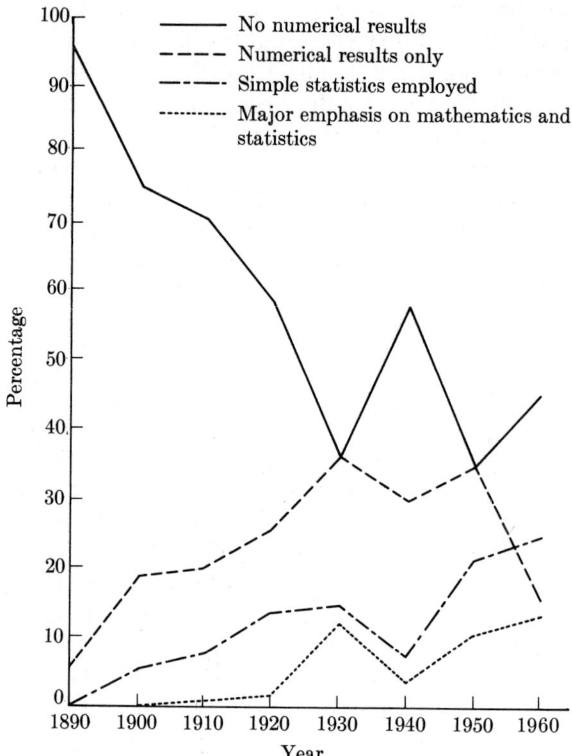

FIGURE 1.1 Percentage of articles involving numerical and statistical work in decennial issues of the *American Naturalist.*

fundamentally deterministic and only the variety of causal variables and our inability to control these make these phenomena appear probabilistic, or whether biological processes are truly probabilistic, as postulated in quantum mechanics for elementary particles, is a deep philosophical question beyond the scope of this book and the competence of the authors. The fact remains that, as recorded by the observer, biological phenomena can only be discussed within a probabilistic framework.

A misunderstanding of these principles and relationships has given rise to the attitude of some biologists that if differences induced by an experiment, or observed in nature, are not such as to be clear on plain inspection (and therefore not in need of statistical analysis), they are not worth investigating. This attitude is also related to the search for all-or-none type responses, often looked for in fields such as experimental embryology (one of the least touched by the biometric approach, somewhat to its disadvantage, we would think). There are still a few legitimate fields of inquiry in which, from the nature of the phenomena studied, statistical investigation is unnecessary. A notable instance in recent years has been the elegant work of Karl von Frisch (1950) on the sensory physiology of bees. This is an example *par excellence* of the type of problem in which the results are obvious without statistical analysis. However, as soon as the details and refinements in this work are being investigated, as is now being done by his students and associates and critics, statistical analysis becomes necessary.

It should be stressed that statistical thinking is not really different in kind from ordinary disciplined scientific thinking, in which we try to quantify our observations. In statistics we express our degree of belief or disbelief as a probability rather than as a vague, general statement. For example, the statements that species A is larger than species B or that females are more often found sitting on tree M rather than on tree N are of a kind commonly made by biological scientists. Such statements can and should be more precisely expressed in quantitative form. In many ways the human mind is a remarkable statistical machine, absorbing many facts from the outside world, digesting these, and regurgitating them in simple summary form. From our experience we know certain events to occur frequently, others rarely. "Man smoking cigarette" is a frequently observed event, "Man slipping on banana peel," rare. We know from experience that Japanese are on the average shorter than Englishmen and that Egyptians are on the average darker than Swedes. We associate thunder with lightning, almost always, flies with garbage cans frequently in the summer, but snow with the southern Californian desert extremely rarely. All such knowledge comes to us as a result of our lifetime experience, both directly and through that of others by direct communication or through reading. All these facts have been processed by that remarkable computer, the human brain, which furnishes an abstract. This summary is constantly under revision, and though occasionally faulty and biased, it is on the whole astonishingly sound; it is our knowledge of the moment.

Although statistics arose to satisfy the needs of scientific research, the development of its methodology in turn affected the sciences in which statistics is applied. Thus, in a positive feedback type of operation, statistics, created to serve the needs of natural science, has itself affected the philosophy of the biological sciences. Two examples may be given. First, analysis of variance has had a tremendous effect in influencing the types of experiments researchers carry out; the whole field of quantitative genetics, one of whose problems is the separation of environmental from genetic effects, depends upon the analysis of variance for its realization, and many of the concepts of quantitative genetics have been directly built around the analysis of variance. Second, factor analysis, an important tool of psychometric research, has by itself initiated numerous studies in psychology and in recent years also in biology.

2 DATA IN BIOLOGY

In Section 2.1 we explain the statistical meaning of the terms "sample" and "population," which we shall be using throughout this book. Then we come to the types of observations that we obtain from biological research material and with which we shall perform the various computations in the rest of this book (Section 2.2). In obtaining data we shall run into the problem of the degree of accuracy necessary for recording the data. This problem and the procedure for rounding off figures are discussed in Section 2.3. We shall then be ready to consider in Section 2.4 certain kinds of derived data, such as ratios and indices, frequently used in biological science, which present peculiar problems with relation to their accuracy and distribution. Knowing how to arrange data in frequency distributions is important, because such arrangements permit us to get an overall impression of their general appearance and also to set them up for further computational procedures. Frequency distributions as well as the presentation of numerical data are discussed in the last section (2.5) of this chapter.

2.1 Samples and populations

We shall now define a number of important terms necessary for an understanding of biological data. The data in a biometric study are generally based on *individual observations. They are observations or measurements taken on the smallest sampling unit.* These smallest sampling units frequently, but not necessarily, are also individuals in the ordinary biological sense. If we measure weight in 100 rats, then the weight of each rat is an individual observation; the hundred rat weights together represent the *sample of observations,* defined as *a collection of individual observations selected by a specified procedure.* In this instance, one individual observation is based on one individual in a biological sense—that is, one rat. However, if we had studied weight in a single rat over a period of time, the sample of individual observations would be the weights recorded on one rat at successive times. If we wish to measure temperature in a study of ant colonies, where each colony is a basic sampling unit, each temperature reading for one colony is an individual observation, and the sample of observations is the temperatures for all the colonies considered. If we can estimate the DNA content of a single mammalian sperm cell, we shall consider such an estimate to be an individual observation, and the sample of observations may be the estimates of DNA content of all the sperm cells studied in one individual mammal. A synonym for individual observation is *item*.

We have carefully avoided so far specifying what particular variable was being studied, because the terms "individual observation" and "sample of observations" as used above define only the structure but not the nature of the data in a study. *The actual property measured by the individual observations is the character or variable.* The more common term employed in general statistics is *variable*. However, in biology the word *character* is frequently used synonymously. (Character also has other meanings in biology, especially in taxonomy, but we shall not be concerned with these here.) More than one character can be measured on each smallest sampling unit. Thus, in a group of 25 mice we might measure the blood *p*H and the erythrocyte count. Each of the 25 mice (a biological individual) is the smallest sampling unit; blood *p*H and red cell count would be the two characters studied; the *p*H readings and cell counts are individual observations, and two samples of 25 observations on *p*H and erythrocyte count would result. Or we may speak of a *bivariate sample* of 25 observations, each referring to a *p*H reading paired with an erythrocyte count.

Next we define *population*. The biological definition of this term is well known. It refers to all the individuals of a given species (perhaps of a given life history stage or sex) found in a circumscribed area at a given time. In statistics, population always means *the totality of individual observations about which inferences are to be made, existing anywhere in the world or at least within a definitely specified sampling area limited in space and time.* If you take five men and study the number of leucocytes in their peripheral blood and you are prepared to draw conclusions about all men from this sample of five, then the population from which the sample has been drawn represents the leucocyte counts of all mankind—that is, all extant members of the species *Homo sapiens*. If, on the other hand, you restrict yourself to a more narrowly specified sample, such as five male Chinese, aged 20, and you are restricting your conclusions to this particular group, then the population from which you are sampling will be leucocyte numbers of all Chinese males of age 20. The population in this statistical sense is sometimes referred to as the *universe*. A population may refer to variables of a concrete collection of objects or creatures, such as the tail lengths of all the white mice in the world.

the leucocyte counts of all the Chinese men in the world of age 20, the DNA content of all the hamster sperm cells in existence; or it may refer to the outcomes of experiments, such as all the heartbeat frequencies produced in guinea pigs by injections of adrenalin. In the first cases the population is generally finite. Although in practice it would be impossible to collect, count, and examine all hamster sperm cells, all Chinese men of age 20, or all white mice in the world, these populations are in fact finite. Certain smaller populations, such as all the whooping cranes in North America or all the pocket gophers in a given colony, may well lie within reach of a total census. By contrast, an experiment can be repeated an infinite number of times (at least in theory). A given experiment such as the administration of adrenalin to guinea pigs could be repeated as long as the experimenter could obtain material and his health and patience held out. The sample of experiments actually performed is a sample from an infinite number that *could* be performed. Some of the statistical methods to be developed later make a distinction between sampling from finite and from infinite populations. However, though populations are theoretically finite in most applications in biology, they are generally so much larger than samples drawn from them, that they can be considered as *de facto* infinitely sized populations.

2.2 Variables in biology

To try to enumerate all the possible kinds of variables that could be studied in biological research would be a hopeless task. Each discipline has its own set of variables, which may include conventional morphological measurements, concentrations of chemicals in body fluids, rates of certain biological processes, frequencies of certain events as in genetics and radiation biology, physical readings of optical or electronic machinery used in biological research, and many more. We assume that persons reading this book more often than not will already have special interests in biology and will have become acquainted with the methodologies of research in their areas of interest, so that the variables commonly studied in their fields are at least to some degree familiar. In any case, the problems for measurement and enumeration must suggest themselves to the researcher; biometry will not, in general, contribute to the discovery and definition of such variables.

Some exception must be made to the above statement. Once a variable has been chosen, a statistical analysis of it may demonstrate it to be unreliable. If a number of variables are studied, certain rather elaborate procedures of multivariate analysis can assign weights to these variables, indicating in this manner their value for a given procedure. For example, in taxonomy the method of discriminant functions can decide what combination of a series of variables can best be used to distinguish between two groups (Section 14.14). Another multivariate technique, factor analysis, has been employed to define those characters best representing distinct geographic variation patterns (see Sokal, 1962). However, as a general rule, and particularly within the framework of this book, choice of a variable as well as definition of the problem to be solved is primarily the responsibility of the biological researcher.

We have already referred to biological variables in a general way, but we have not yet defined them. We shall define a *variable as a property with respect to which individuals in a sample differ in some ascertainable way.* If the property does not differ within a sample at hand or at least among the samples being studied, it cannot be of statistical interest. Being entirely uniform, such a property would also not be a "variable" from the etymological point of view and should not be so called. Length, height, weight, number of teeth, vitamin C content, and genotypes are examples of variables in ordinary, genetically and phenotypically diverse, groups of organisms. Warm-bloodedness in a group of mammals is not, since they are all alike in this regard, although body temperature of individual mammals would, of course, be a variable. Also, if we had a heterogeneous group of animals some of which were homeothermic and others were not, then the character body temperature (with its two states or forms of expression "warm-blooded" and "cold-blooded") would be a variable.

We can divide biological variables as follows:

Variables
- Measurement variables
  - Continuous variables
  - Discontinuous variables
- Ranked variables
- Attributes

Measurement variables are *all those whose differing states can be expressed in a numerically ordered fashion.* They are divisible into two kinds. The first of these are *continuous variables,* which at least theoretically can assume an infinite number of values between any two fixed points. For example, between the two length measurements 1.5 and 1.6 cm there is an infinite number of lengths that could be measured if one were so inclined and had a precise enough method of calibration to obtain such measurements. Any given reading of a continuous variable, such as a length of 1.57 mm, is therefore an approximation to the exact reading, which in practice is unknowable. However, for purposes of computation these approximations are usually sufficient and, as will be seen below, may even be made more approximate by rounding off. Many of the variables studied in biology are continuous variables. Examples are lengths, areas, volumes, weights, angles, temperatures, periods of time, percentages, rates.

Contrasted with continuous variables are the *discontinuous variables,* also known as *meristic* or *discrete variables.* These are variables that have only certain fixed numerical values, with no intermediate values possible in between. Thus the number of segments in a certain insect appendage may be

4 or 5 or 6 but never $5\frac{1}{2}$ or 4.3. Examples of discontinuous variables are numbers of a certain structure (such as segments, bristles, teeth, or glands), the numbers of offspring, the numbers of colonies of microorganisms or animals, or the numbers of plants in a given quadrat.

A word of caution: not all variables restricted to integral numerical values are meristic. An example will illustrate this point. If an animal behaviorist were to code the reactions of animals in a series of experiments as 1. very aggressive, 2. aggressive, 3. neutral, 4. submissive, and 5. very submissive, we might be tempted to believe that these five different states of the variable were meristic because they assume integral values. However, they are clearly only arbitrary points (class marks, see Section 2.5) along a continuum of aggressiveness; the only reason why no values such as 1.5 occur is because the experimenter did not wish to subdivide his behavior classes too finely, either for reasons of convenience or because he felt unable to determine more than five subdivisions of this spectrum of behavior with accuracy. Thus, this variable is clearly continuous rather than meristic, as it might have appeared at first sight.

Some variables cannot be measured but at least can be ordered or ranked by their magnitude. Thus, in an experiment one might record the rank order of emergence of ten pupae without specifying the exact time at which each pupa emerged. In such cases we code the data as a *ranked variable*, the order of emergence. Special methods for dealing with such variables have been developed and several are furnished in this book. By expressing a variable as a series of ranks, such as 1, 2, 3, 4, 5, we do not imply that the difference in magnitude between, say, ranks 1 and 2 is identical to or even proportional to the difference between 2 and 3. Such an assumption is made for the measurement variables, discussed above.

Variables that cannot be measured but must be expressed qualitatively are called *attributes*. These are all properties, such as black or white, pregnant or not pregnant, dead or alive, male or female. When such attributes are combined with frequencies, they can be treated statistically. Of 80 mice, we may, for instance, state that four were black and the rest gray. When attributes are combined with frequencies into tables suitable for statistical analysis, they are referred to as *enumeration data*. Thus the enumeration data on color in mice just mentioned would be arranged as follows:

Color	*Frequency*
Black	4
Gray	76
Total number of mice	80

In some cases attributes can be changed into variables if this is desired. Thus colors can be changed into wavelengths or color chart values, which are measurement variables. Certain other attributes that can be ranked or ordered can be coded to become ranked variables. For example, three attributes referring to a structure as "poorly developed," "well developed," and "hypertrophied" could conveniently be coded 1, 2, and 3. These values imply the rank order of development, but not the relative magnitudes of these attribute states.

A term that has not yet been explained is *variate*. In this book we shall use it as a single reading, score, or observation of a given variable. Thus, if we have measurements of the length of the tails of five mice, tail length will be a continuous variable, and each of the five readings of length will be a variate. In this text we identify variables by capital letters, the most common symbol being Y. Thus Y may stand for tail length of mice. A variate will refer to a given length measurement; Y_i is the measurement of tail length of the ith mouse, and Y_4 is the measurement of tail length of the fourth mouse in our sample. The use of the terms variable and variate differs somewhat from author to author; frequently the two terms are used synonymously. We are accustomed to the usage presented here and in searching through statistical literature find ample precedent for it. We must confess, however, that to be consistent we should refer to univariable and multivariable statistics instead of univariate and multivariate statistics, which we have retained. Happily, this is not a point on which consistency or uniformity are especially important.

2.3 Accuracy and precision of data

Scientists are generally aware of the importance of accuracy and precision in their work, and it is self-evident that accuracy must extend to the numerical results of their work and to the processing of these data. Owing to the great diversity of approaches employed in different disciplines of biology, it would seem futile to attempt to furnish specific rules for obtaining data accurately. However, we should make certain that, whatever the method of obtaining data, it is consistent, so that, given the same observational or experimental setup, different numerical readings of the same structure or event would be identical or within acceptable and predictable limits of each other.

"Accuracy" and "precision" are used synonymously in everyday speech, but in statistics we define them more rigorously. *Accuracy* is the closeness of a measured or computed value to its true value; *precision* is the closeness of repeated measurements of the same quantity. A biased but sensitive scale might yield inaccurate but precise weight. By chance an insensitive scale might result in an accurate reading, which would however be imprecise, since a repeated weighing would be unlikely to yield an equally accurate weight. Unless there is bias in a measuring instrument, precision will lead to accuracy. We need therefore mainly be concerned with the former.

Precise variates are usually but not necessarily whole numbers. Thus, when we count four eggs in a nest, there is no doubt about the exact number of eggs in the nest if we have counted correctly; it is four, not three nor five, and clearly it could not be four plus or minus a fractional part. Meristic variables are generally measured as exact numbers. Seemingly, continuous variables derived from meristic ones can under certain conditions also be exact numbers. For instance, ratios between exact numbers are themselves also exact. If in a colony of animals there are 18 females and 12 males, the ratio of females to males is 1.5, a continuous variate but also an exact number.

Most continuous variables, however, are approximate. We mean by this that the exact value of the single measurement, the variate, is unknown and probably unknowable. The last digit of the measurement stated should imply precision, that is the limits on the measurement scale between which we believe the true measurement to lie. Thus a length measurement of 12.3 mm implies that the true length of the structure lies somewhere between 12.25 and 12.35 mm. Exactly where between these *implied limits* the real length is we do not know. Some might object to defining the limits of the number as 12.25 to 12.35 mm. This is because the adjacent measurement of 12.2 would imply limits of 12.15 to 12.25. Where then, it is asked, would a true measurement of 12.25 fall? Would it not equally likely fall in either of the two classes 12.2 and 12.3, clearly an unsatisfactory state of affairs? For this reason some authors define the implied limits of 12.3 as 12.25 to 12.34999 . . . . Such an argument is correct, but when we record a number as either 12.2 or 12.3 we imply that the decision whether to put it into the higher or lower class has already been taken. This decision was not taken arbitrarily, but presumably was based on the best available measurement. If the scale of measurement is so precise that a value of 12.25 would clearly have been recognized, then the measurement should have been recorded originally to four significant figures. Implied limits therefore always carry one more figure beyond the last significant one measured by the observer.

Hence it follows that if we record the measurement as 12.32, we are implying that the true value lies between 12.315 and 12.325. Unless this is what we mean, there would be no point in adding another decimal figure to our original measurements. If we do add another figure we must imply an increase in precision. We see therefore that accuracy and precision in numbers is not an absolute concept, but is relative. Assuming there is no bias, a number becomes increasingly more accurate as we are able to write more significant figures for it (increase its precision). To illustrate this concept of the relativity of accuracy you are asked to look at the following three numbers.

	Implied limits
193	192.5 –193.5
192.8	192.75 –192.85
192.76	192.755–192.765

We may imagine these numbers to be the recorded measurements of the same structure. Let us assume that we had extramundane knowledge of the true length of the given structure as being 192.758 units. If that were so, the three measurements increase in accuracy from the top down. You will note that the implied limits of the topmost one are wider than those of the measurement below it, which in turn are wider than those of the third measurement.

Meristic variates, though ordinarily exact, may be recorded approximately when large numbers are involved. Thus when counts are reported to the nearest thousand, a count of 36,000 insects in a cubic meter of soil implies that the true number varied somewhere from 35,500 to 36,500 insects.

To how many significant figures should we record measurements? If we array the sample by order of magnitude from the smallest individual to the largest one, an easy rule to remember is that *the number of unit steps from the smallest to the largest measurement in an array should be between 30 and 300.* Thus, if we are measuring a series of shells to the nearest millimeter and the largest is 8 mm and the smallest is 4 mm wide, there are only four unit steps between the largest and the smallest measurement. Hence we should have measured our shells to one more significant decimal place. Then the two extreme measurements might have been 8.2 mm and 4.1 mm, with 41 unit steps between them (counting the last significant digit as the unit); this would have been an adequate number of unit steps. The reason for such a rule is that an error of 1 in the last significant digit of a reading of 4 mm would constitute an inadmissible error of 25%, but an error of 1 in the last digit of 4.1 is less than 2.5%. Similarly, if we had measured the height of the tallest of a series of plants as 173.2 cm and that of the shortest of these plants as 26.6 cm, the difference between these limits would comprise 1466 unit steps (of 0.1 cm), which are far too many. It would therefore have been advisable to record the heights to the nearest centimeter, as follows: 173 cm for the tallest and 27 cm for the shortest. This would yield 146 unit steps. Using the rule stated above, we shall record two or three digits for most measurements. However, on occasion more digits will be necessary when the leading digits fulfill the function of a conventional code, as in the following example. In a representative series of *p*H readings measured with great precision, the highest value might be 7.456 and the lowest value 7.434. The first two digits of such a measurement are in effect a code that is conventional for *p*H readings of the material under study. The variation observed in the sample all centers on the second and third decimal place. It is therefore necessary to use that many places in order to obtain an adequate amount of variation for analysis. The perceptive reader may have noticed that the number of unit steps from lowest to highest variate in the *p*H example was only 22, or less than the 30 given as the lower limit for the desired number of steps. This rule, as are all such "rules of thumb," is to be taken with a grain of statistical salt. These rules are general guidelines but not proscriptions whose mild infraction would immediately invalidate all further work. With some experience an approximate adherence to these rules becomes second nature to the biologist engaging in statistical analysis.

The last digit of an approximate number should always be significant; that is, it should imply a range for the true measurement of from half a "unit step" below to half a "unit step" above the recorded score, as illustrated earlier. This applies to all digits, zero included. Zeros should therefore not be written at the end of approximate numbers to the right of the decimal point unless they are meant to be significant digits. Thus 7.80 must imply the limits 7.795 to 7.805. If 7.75 to 7.85 is implied, the measurement should be recorded as 7.8.

When the number of significant digits is to be reduced, we carry out the process of *rounding off* numbers. The rules for rounding off are very simple. A digit to be rounded off is not changed if it is followed by a digit less than 5. If the digit to be rounded off is followed by a digit greater than that 5 or by 5 followed by other nonzero digits, it is increased by one. When the digit to be rounded off is followed by a 5 standing alone or followed by zeros, it is unchanged if it is even but increased by one if it is odd. The reason for this last rule is that when such numbers are summed in a long series we should have as many digits raised as are being lowered on the average; these changes would therefore balance out. Practice the above rules by rounding off the following numbers to the indicated number of significant digits.

Number	*Significant digits desired*	*Answer*
26.58	2	27
133.7137	5	133.71
0.03725	3	0.0372
0.03715	3	0.0372
18,316	2	18,000
17.3476	3	17.3

2.4 Derived variables

The majority of variables in biometric work are observations recorded as direct measurements or counts of biological material or as readings that are the output of various types of instruments. However, there is an important class of variables in biological research which we may call the *derived* or *computed variables* and which are generally based on two or more independently measured variables whose relations are expressed in a certain way. We are referring to ratios, percentages, indices, rates, and the like.

A *ratio* expresses as a single value the relation which two variables have one to the other. In its simplest form it is expressed as in 64:24, which may represent the number of wild type versus mutant individuals or the number of males versus females, or the proportion of parasitized individuals versus those not parasitized and so on. The above examples implied ratios based on counts; a ratio based on a continuous variable might be similarly expressed as 1.2:1.8, which may represent the ratio of width to length in a sclerite of an insect or the ratio between the concentrations of two minerals contained in water or soil. Ratios may also be expressed as fractions; thus the two ratios above could be expressed as $\frac{64}{24}$ and $\frac{1.2}{1.8}$. However, for computational purposes it is most useful to express the ratio as a quotient. The two ratios cited above would therefore be 2.666 . . . and 0.666 . . . , respectively. These are pure numbers, not expressed in measurement units of any kind. It is this form for ratios that we shall consider further below. *Percentages* are also a type of ratio. Ratios and percentages are basic quantities in much biological research, widely used and generally familiar.

Not all derived variables are ratios or percentages. The term *index* is used in a general sense for derived variables, although some would limit it to the ratio of one anatomic variable divided by a larger, so-called standard one. For instance, in a study of the cranial dimensions of cats, Haltenorth (1937) divided all such measures by the basal length of the skull. Thus each measurement was in fact a percentage of the basal length of the skull of the cat being measured. Another example of an index in this sense is the well-known cephalic index in physical anthropology. Conceived in the wide sense an index could be the average of two measurements—either simply, such as $\frac{1}{2}$ (length of A + length of B), or in weighted fashion, such as $\frac{1}{3}$ [(2 × length of A) + length of B]. An index may refer to the summation of a series of numerically scored properties. Thus, if an animal is given six behavioral tests in which its score can range from 0 to 4, an index of its behavior might be the sum of the scores of the six tests. Similar indices have been described for determining the degree of hybridity in organisms (the hybrid indices of various authors).

Rates will be important in many experimental fields of biology. The amount of a substance liberated per unit weight or volume of biological material, weight gain per unit time, reproductive rates per unit population size and time (birth rates), and death rates would fall in this category.

As we shall see below there are some serious drawbacks to the use of ratios and percentages in statistical work. In spite of these the use of these variables is deeply ingrained in scientific thought processes and is not likely to be abandoned. Furthermore, we should emphasize that ratios may be the only meaningful way to interpret and understand certain types of biological problems. If the biological process being investigated operates on the ratio of the variables studied, one must examine this ratio to understand the process. Thus, Sinnott and Hammond (1935) found that inheritance of the shapes of squashes, *Cucurbita pepo,* could be interpreted by a form index based on a length-width ratio, but not in terms of the independent dimensions of shape. Similarly, the evolution of shape in many burrowing animals is a function of the cross-sectional profile of the animal rather than of a single dimension. Selection affecting body proportions must be found to exist in the evolution of almost any organism when properly investigated.

The disadvantages of using ratios are several: first is their relative inaccuracy. Let us return to the ratio $\frac{1.2}{1.8}$ mentioned above and recall from the previous section that a measurement of 1.2 indicates a true range of measurement of the variable from 1.15 to 1.25; similarly, a measurement of 1.8 implies a range from 1.75 to 1.85. We realize therefore that the true ratio may vary anywhere from $\frac{1.15}{1.85}$ to $\frac{1.25}{1.75}$, or 0.622 and 0.714, respectively. We note a possible maximal error of 4.2% if 1.2 were an original measurement: [(1.25 − 1.2)/1.2]; the corresponding maximal error for the ratio is 7.0%: [(0.714 − 0.667)/0.667]. Furthermore, the best estimate of a ratio is not usually the midpoint between its possible ranges. Thus in our example the midpoint between the implied limits is 0.668 and the true ratio is 0.666 . . . , only a slight difference, which may, however, be greater in other instances. In many cases, therefore, ratios will not be as accurate as measurements obtained directly. This liability can be overcome by empirical determinations of their variability through measures of their variance (described in later sections of this book).

A second drawback to ratios and percentages is that their distributions may be rather unusual and they may therefore not be more or less normally distributed (see Chapter 6) as required by many statistical tests. This difficulty can frequently be overcome by transformation of the variable (as discussed in Chapter 13). Another disadvantage of ratios is that they do not provide information on the relationship between the two variables whose ratio is being taken. Often more may be learned by studying the variables singly and their relationships to each other (bivariate and multivariate analysis). For instance, in the cranial measurements of cats discussed above there will clearly be correlation due to general size among the variables making up the indices. Thus a large cat will have a large head; hence the basal length of the skull and the separate bones constituting the cranium will also be large. Through methods of multivariate analysis this general factor can be separated out, and we may find dimensions independent of general size representing the basal length of the skull and certain bones such as the width of the zygomatic arch. Under those circumstances it may be much more interesting to study variation and correlation of the various variables independent of general size rather than their ratios.

Finally, we should point out that when ratios involve enumeration data or meristic variables they may occasionally give rise to curious distributions. By way of an example we may cite percentages obtained from an experiment performed some years ago by Sokal. In this experiment ten *Drosophila* eggs were put in a vial and the positions of the pupae noted after pupation. Some pupae pupated at the margin or the wall of the vials; these were called "peripheral." Others pupated away from the wall; these were called "central." Ideally there should have been ten pupae—if all the eggs had hatched and there had been no larval mortality whatsoever. In fact, however, because of natural mortality and of errors in preparing the vials, ten pupae were not always found. The logical minimum number of survivors on which a result could be reported was one pupa. The proportion of peripheral pupae was then calculated by dividing the number of such pupae in a vial by the total number of pupae found in that vial. Although ratios thus obtained are a continuous variable in appearance, they are not so in fact, because certain values can never be obtained. For instance, by limiting the maximum number of pupae to 10 we cannot obtain percentages of peripheral pupation between 0 and 10% or between 90 and 100% (see Table 2.1). Also, we see that some

TABLE 2.1

Possible percentages of peripheral pupation in vials seeded with ten *Drosophila* eggs. Experimental results show number of peripheral pupae in the numerator, total number of pupae in the denominator.

Percent	*Possible experimental results yielding this percentage*	Percent	*Possible experimental results yielding this percentage*
00	0/10, 0/9, 0/8, 0/7, 0/6, 0/5, 0/4, 0/3, 0/2, 0/1	56	5/9
10	1/10	57	4/7
11	1/9	60	6/10, 3/5
12	1/8	62	5/8
14	1/7	67	6/9, 4/6, 2/3
17	1/6	70	7/10
20	2/10, 1/5	71	5/7
22	2/9	75	6/8, 3/4
25	2/8, 1/4	78	7/9
29	2/7	80	8/10, 4/5
30	3/10	83	5/6
33	3/9, 2/6, 1/3	86	6/7
38	3/8	88	7/8
40	4/10, 2/5	89	8/9
43	3/7	90	9/10
44	4/9	100	10/10, 9/9, 8/8, 7/7, 6/6, 5/5, 4/4, 3/3, 2/2, 1/1
50	5/10, 4/8, 3/6, 2/4, 1/2		

Source: From a study by Sokal (1966).

percentages are given by several ratios: 33% is obtained either by $\frac{3}{9}$ or $\frac{2}{6}$ or $\frac{1}{3}$, but 57% can only be obtained by having 4 out of 7 pupae peripheral. By examining Table 2.1 it can be seen which values are more likely to be encountered if we make the simplest assumption—that there is an equal probability for any given experimental outcome. In using ratios of meristic variates or of counts, such discontinuities and peculiarities of distribution must be taken into account. The problem in this instance was solved by transforming the data to probits, a procedure discussed in Section 14.12.

2.5 Frequency distributions

If we were to sample, for instance, a population of birth weights of infants, we could represent each sampled measurement by a point along an axis denoting magnitude of birth weights. This is illustrated in Figure 2.1, A, for a sample of 25 birth weights. If we sample repeatedly from the population and obtain 100 birth weights, we shall probably have to place some of these

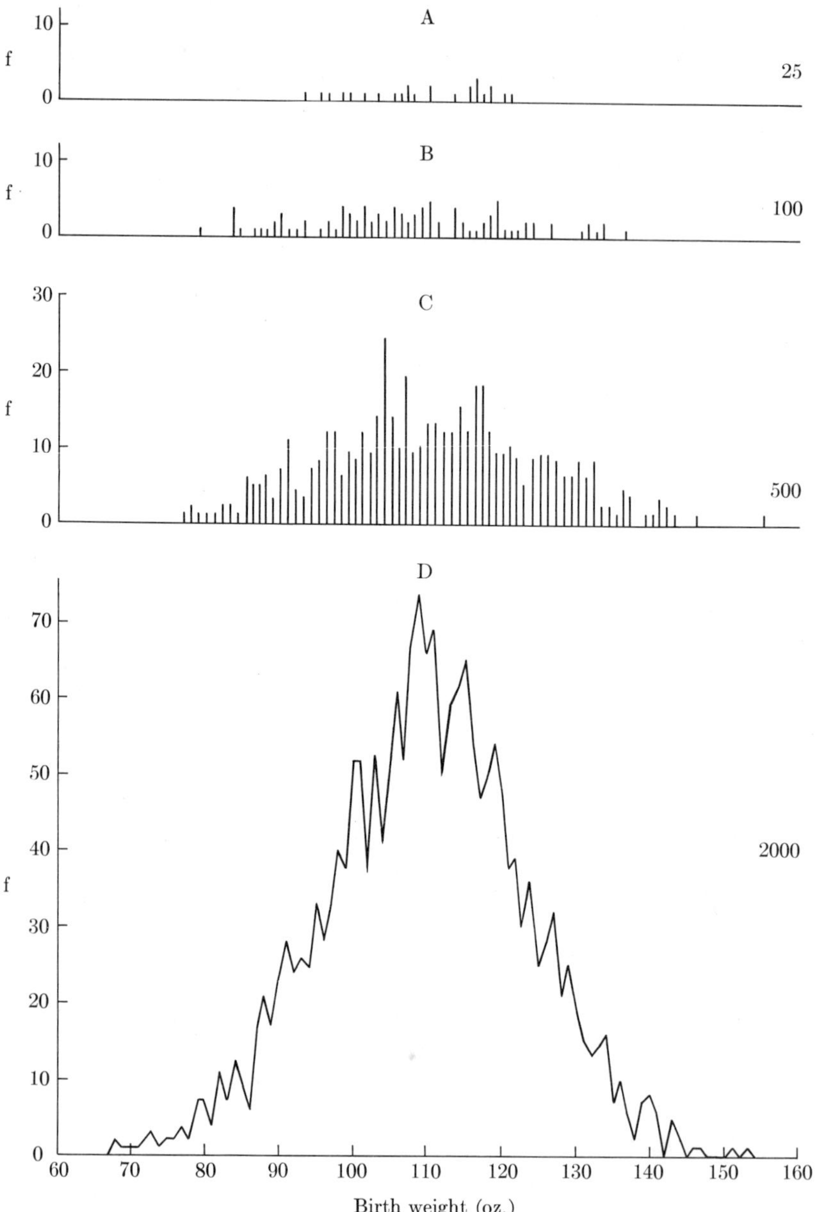

FIGURE 2.1 Sampling from a population of birth weights of infants (a continuous variable). A. A sample of 25. B. A sample of 100. C. A sample of 500. D. A sample of 2000.

points on top of other points in order to record them all correctly (Figure 2.1, B). As we continue sampling additional hundreds and thousands of birth weights (Figures 2.1, C and D), the assemblage of points will continue to increase in size but will assume a fairly definite shape. The curve tracing the outline of the mound of points approximates the distribution of the variable. Remember that a continuous variable such as birth weight can assume an infinity of values between any two points on the abscissa. The refinement of our measurements will determine how fine the number of recorded divisions between any two points along the axis will be.

The distribution of a variable is of considerable biological interest. If we find that the distribution is asymmetrical and drawn out in one direction, it tells us that there is, perhaps, selection for or against organisms falling in one of the tails of the distribution, or possibly that the scale of measurement chosen is such as to bring about a distortion of the distribution. If, in a sample of immature insects, we discover that the measurements are bimodally distributed (with two peaks), this would indicate that the population is dimorphic; different species or races may have become intermingled in our sample, or the dimorphism could have arisen from the presence of both sexes or of different instars. There are several characteristic shapes of frequency distributions, the most common of which is the symmetrical bell-shaped distribution (approximated by bottom graph in Figure 2.1), the normal frequency distribution discussed in Chapter 6. There are also skewed distributions (drawn out more at one tail than the other), L-shaped distributions as in Figure 2.2, U-shaped distributions, and others, which impart significant information about certain types of relationships. We shall have more to say about the implications of various types of distributions in later chapters and sections.

After data have been obtained in a given study they have to be arranged in a form suitable for computation and interpretation. We may assume that

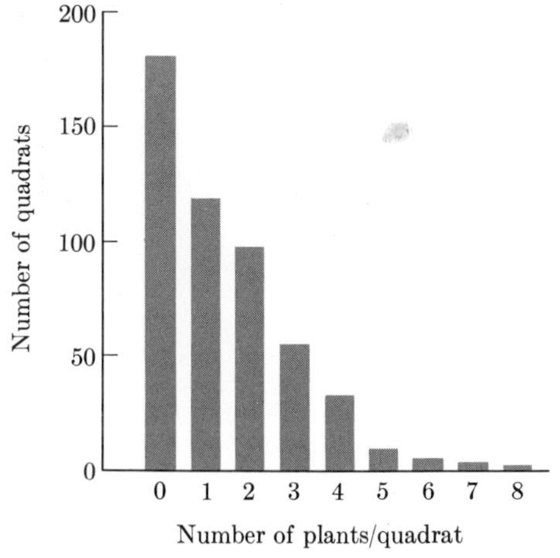

FIGURE 2.2 Bar diagram. Frequency of the sedge *Carex flacca* in 500 quadrats. Data from Table 2.3; originally from Archibald (1950).

variates are randomly ordered initially or are in the order in which the measurements had been taken. A simple arrangement would be an *array* of the data by order of magnitude. Thus, for example, the variates 7, 6, 5, 7, 8, 9, 6, 7, 4, 6, 7 could be arrayed in order of decreasing magnitude as follows: 9, 8, 7, 7, 7, 7, 6, 6, 6, 5, 4. Where there are some variates of the same value, such as the 6's and 7's in this fictitious example, a time-saving device might immediately have occurred to you—namely, to list a frequency for each of the recurring variates, thus: 9, 8, 7(4 ×), 6(3 ×), 5, 4. Such a shorthand notation is one way to represent a *frequency distribution,* which is simply an arrangement of the classes of variates with the frequencies of each class indicated. Conventionally, a frequency distribution is stated in tabular form, as follows for the above example.

Variable Y	*Frequency* f
9	1
8	1
7	4
6	3
5	1
4	1

The above is an example of a *quantitative frequency distribution,* since Y is clearly a measurement variable. However, arrays and frequency distributions need not be limited to such variables. We can make frequency distributions of attributes, called *qualitative frequency distributions.* In these, the

TABLE 2.2

A qualitative frequency distribution. Number of individuals of Heteroptera tabulated by family in total samples from a summer foliage insect community.

Family	f
Alydidae	2
Anthocoridae	37
Coreidae	2
Lygaeidae	318
Miridae	373
Nabidae	3
Neididae	5
Pentatomidae	25
Piesmidae	1
Reduviidae	3
Rhopalidae	2
Saldidae	1
Thyreocoridae	10
Tingidae	69
Total Heteroptera	851

Source: Data from Whittaker (1952).

various classes are listed in some logical or arbitrary order. For example, in genetics we might have a qualitative frequency distribution as follows:

	f
$A-$	86
aa	32

This tells us that there are two classes of individuals, those identified by the $A-$ phenotype, of which 86 were found, and the homozygote recessive aa, of which 32 were seen in the sample. In ecology it is quite common to have a species list of inhabitants of a sampled ecological area. The arrangement of such tables is usually alphabetical. An example of an ecological list constituting a qualitative frequency distribution of families is shown in Table 2.2. In this instance the families of insects referred to are listed alphabetically; in other cases the sequence may be by convention, as for the families of flowering plants in botany.

A quantitative frequency distribution based on meristic variates is shown in Table 2.3. This is an example from plant ecology: the number of plants

TABLE 2.3

A meristic frequency distribution. Number of plants of the sedge *Carex flacca* found in 500 quadrats.

No. of plants per quadrat Y	*Observed frequency* f
0	181
1	118
2	97
3	54
4	32
5	9
6	5
7	3
8	1
Total	500

Source: Data from Archibald (1950).

per quadrat sampled are listed at the left in the variable column; the observed frequency is shown at the right.

Quantitative frequency distributions based on a continuous variable are the most commonly employed frequency distributions and you should become thoroughly familiar with them. An example is shown in Box 2.1. It is based on 25 femur lengths of stem mothers (one of the life history stages) of a species of aphids. The 25 readings (measured in coded micrometer units) are shown at the top of Box 2.1 in the order in which they were obtained as measurements. They could have been arrayed according to their magnitude

BOX 2.1

Preparation of frequency distribution and grouping into fewer classes with wider class intervals.

Twenty-five femur lengths of stem mothers of the aphid *Pemphigus populi-transversus*. Measurements are in *mm* $\times 10^{-1}$.

Original measurements				
3.8	3.6	4.3	3.5	4.3
3.3	4.3	3.9	4.3	3.8
3.9	4.4	3.8	4.7	3.6
4.1	4.4	4.5	3.6	3.8
4.4	4.1	3.6	4.2	3.9

<table>
<tr><th colspan="4">Original frequency distribution</th><th colspan="5">Grouping into 8 classes of interval 0.2</th><th colspan="5">Grouping into 5 classes of interval 0.3</th></tr>
<tr><th>Implied limits</th><th>Y</th><th>Tally marks</th><th>f</th><th>Practical limits</th><th>Implied limits</th><th>Class mark</th><th>Tally marks</th><th>f</th><th>Practical limits</th><th>Implied limits</th><th>Class mark</th><th>Tally marks</th><th>f</th></tr>
<tr><td>3.25–3.35</td><td>3.3</td><td>|</td><td>1</td><td>3.3–3.4</td><td>3.25–3.45</td><td>3.35</td><td>|</td><td>1</td><td>3.3–3.5</td><td>3.25–3.55</td><td>3.4</td><td>||</td><td>2</td></tr>
<tr><td>3.35–3.45</td><td>3.4</td><td></td><td>0</td><td></td><td></td><td></td><td></td><td></td><td></td><td></td><td></td><td></td><td></td></tr>
<tr><td>3.45–3.55</td><td>3.5</td><td>|</td><td>1</td><td>3.5–3.6</td><td>3.45–3.65</td><td>3.55</td><td>卌</td><td>5</td><td></td><td></td><td></td><td></td><td></td></tr>
<tr><td>3.55–3.65</td><td>3.6</td><td>||||</td><td>4</td><td></td><td></td><td></td><td></td><td></td><td>3.6–3.8</td><td>3.55–3.85</td><td>3.7</td><td>卌 |||</td><td>8</td></tr>
<tr><td>3.65–3.75</td><td>3.7</td><td></td><td>0</td><td>3.7–3.8</td><td>3.65–3.85</td><td>3.75</td><td>||||</td><td>4</td><td></td><td></td><td></td><td></td><td></td></tr>
<tr><td>3.75–3.85</td><td>3.8</td><td>||||</td><td>4</td><td></td><td></td><td></td><td></td><td></td><td></td><td></td><td></td><td></td><td></td></tr>
<tr><td>3.85–3.95</td><td>3.9</td><td>|||</td><td>3</td><td>3.9–4.0</td><td>3.85–4.05</td><td>3.95</td><td>|||</td><td>3</td><td>3.9–4.1</td><td>3.85–4.15</td><td>4.0</td><td>卌</td><td>5</td></tr>
<tr><td>3.95–4.05</td><td>4.0</td><td></td><td>0</td><td></td><td></td><td></td><td></td><td></td><td></td><td></td><td></td><td></td><td></td></tr>
<tr><td>4.05–4.15</td><td>4.1</td><td>||</td><td>2</td><td>4.1–4.2</td><td>4.05–4.25</td><td>4.15</td><td>|||</td><td>3</td><td></td><td></td><td></td><td></td><td></td></tr>
<tr><td>4.15–4.25</td><td>4.2</td><td>|</td><td>1</td><td></td><td></td><td></td><td></td><td></td><td>4.2–4.4</td><td>4.15–4.45</td><td>4.3</td><td>卌 |||</td><td>8</td></tr>
</table>

4.25–4.35	4.3	||||	4	4.3–4.4	4.25–4.45	4.35	𝍸 ||	7					
4.35–4.45	4.4	|||	3										
4.45–4.55	4.5	|	1	4.5–4.6	4.45–4.65	4.55	|	1	4.5–4.7	4.45–4.75	4.6	||	2
4.55–4.65	4.6		0										
4.65–4.75	4.7	|	1	4.7–4.8	4.65–4.85	4.75	|	1					
Σf			25					25					25

Source: Data from R. R. Sokal.

Histogram of the original frequency distribution shown above and of the grouped distribution with 5 classes. Line below abscissa shows class marks for the grouped frequency distribution. Hatched bars represent original frequency distribution; hollow bars represent grouped distribution.

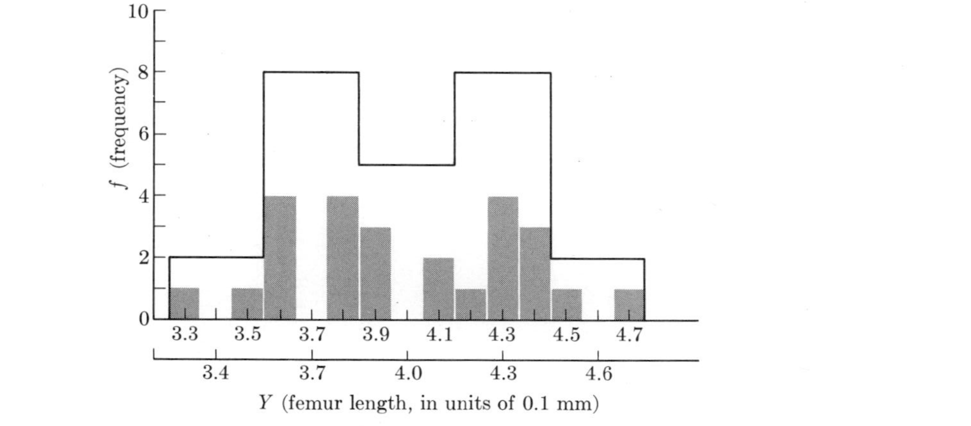

For a detailed account of the process of grouping, see Section 2.5

if we had so desired, but it is not necessary or customary to take this intermediate step in order to proceed from there to a frequency distribution. We have therefore not shown an array in Box 2.1, but have immediately set up the data in a frequency distribution. The variates increase in magnitude by unit steps of 0.1. The frequency distribution is prepared by entering each variate in turn on the scale and indicating a count by a conventional tally mark. When all of the items have been tallied in the corresponding class, the tallies are converted into numerals indicating frequencies in the next column. Their sum is indicated by $\sum f$.

Let us look at what we have achieved in summarizing our data. The original 25 variates are now represented by only 15 classes. We find that variates 3.6, 3.8, and 4.3 have the highest frequencies. However, we also note that there are several classes, such as 3.4 or 3.7, which are not represented by a single aphid. This gives the entire frequency distribution a rather drawn-out and scattered appearance. The reason for this is that we have only 25 aphids, rather too few to put into a frequency distribution with 15 classes. To obtain a more cohesive and smooth-looking distribution we have to condense our data into fewer classes. This process is known as *grouping of classes* of frequency distributions; it is illustrated in Box 2.1 and described in the following paragraphs.

As we embark on grouping we should realize that what we are doing when we group individual variates into classes of wider range is only an extension of the very same process that took place when we obtained the initial measurement. Thus, as we have seen in Section 2.3, when we measure an aphid and record its femur length as 3.3 units, we imply thereby that the true measurement lies between the range from 3.25 to 3.35 units, but that we were unable to measure to the second decimal place. In recording the measurement initially as 3.3 units, we estimated that it fell within this range. Had we estimated that it exceeded the value of 3.35, for example, we would have given it the next higher score, 3.4. Therefore all the measurements between 3.25 and 3.35 were in fact grouped into the class identified by the *class mark* 3.3. Our *class interval* was 0.1 units. If we now wish to make wider class intervals, we are doing nothing but extending the range within which measurements are placed into one class.

Reference to Box 2.1 will make this process clear. We group the data twice in order to impress upon the reader the flexibility of the process. In the first example of grouping, the class interval has been doubled in width; that is, it was made to equal 0.2 units. If we start at the lower end, the implied class limits will now be from 3.25 to 3.45, the limits for the next class from 3.45 to 3.65, and so forth. As discussed in Section 2.3, we prefer this formulation of the limits to the often encountered 3.25 to 3.44 and 3.45 to 3.64, or 3.25 to 3.4499 . . . and 3.45 to 3.6499 . . . , since in practice the last significant digit indicated in the implied limits is always one more decimal place than in the original measurements. We find that presenting the implied limits in this way facilitates their comprehension by students and also simplifies the computation of class marks.

Our next task is to find these class marks. This was quite simple in the frequency distribution shown at the left side of Box 2.1, in which the original measurements had been used as class marks. However, now we are using a class interval twice as wide as before, and the class marks are calculated by taking the midpoint of the new class intervals. Thus to find the class mark of the first class we take the midpoint between 3.25 and 3.45, which turns out to be 3.35. We note that the class mark has one more decimal place than the original measurements. We should not now be led to believe that we have suddenly achieved greater accuracy. Whenever we designate a class interval whose last significant digit is even (0.2 in this case), the class mark will carry one more decimal place than the original measurements. On the right side of the table in Box 2.1 the data are grouped once again, using a class interval of 0.3. Because of the odd end digit of the interval, the class mark now shows as many decimals as the original variates, the midpoint between 3.25 and 3.55 being 3.4.

Once the implied class limits and the class mark for the first class have been correctly found, the others can be written down by inspection without any special computation. Simply add the class interval repeatedly to each of the values. Thus, starting with the lower limit 3.25, by adding 0.2 we obtain 3.45, 3.65, 3.85, and so forth; similarly for the class marks, we obtain 3.35, 3.55, 3.75, and so forth. It should be obvious that the wider the class intervals, the more compact the data become but also the less accurate. However, looking at the frequency distribution of aphid femur lengths in Box 2.1, we notice that the initial rather chaotic structure is being simplified by grouping. When we group the frequency distribution into five classes with a class interval of 0.3 units, it becomes notably bimodal (that is, it possesses two peaks of frequencies).

Also provided in Box 2.1 are columns labeled *practical limits*. These are the limits you will sometimes find listed for grouping procedures in other textbooks of statistics. We find that beginners are likely to make errors on class intervals and class marks when using such limits, and therefore you should always write down the implied limits in order to be quite clear what the class limits and class marks of a frequency distribution really are. The practical limits have a function, however. They do tell you what the limits are with respect to the scale and accuracy of measurements which you originally employed for your variables. Thus in Box 2.1, with a class interval of 0.2 units the implied limits of the first class are 3.25 to 3.45, but the practical limits are 3.3 to 3.4 for the first class and 3.5 to 3.6 for the second class. We mean by this that all those aphids scored 3.3 and 3.4 should go into the first class, those 3.5 and 3.6 into the second class. Similarly, grouping them by class intervals of 0.3 at the right side of Box 2.1, the practical limits are 3.3 to 3.5 and 3.6 to 3.8 for the first two classes. These limits would tell immediately where a given original measurement should go. We find practical limits very convenient in preparing tally sheets, because one does not have to introduce the extra decimal place. Whenever an assistant with little theoretical insight into statistics is asked to prepare the tally sheets, the practical limits are of considerable usefulness. However, you must never imagine the practical limits to be the implied limits. Class marks and class intervals have to be computed on the basis of implied limits, not on practical limits.

In setting up frequency distributions, from 12 to 20, classes should be established. This rule need not be slavishly adhered to, but should be employed with some of the common sense that comes from experience in handling statistical data. The number of classes depends largely on the size of the sample studied. Samples of less than 40 or 50 should rarely be given as many as 12 classes, since that would provide too few frequencies per class. On the other hand, samples of several thousand may profitably be grouped into more than 20 classes. If the aphid data of Box 2.1 need to be grouped, they should probably not be grouped into more than 6 classes.

If the original data provide us with fewer classes than we think we should have, then nothing can be done if the variable is meristic, since this is the nature of the data in question. However, with a continuous variable a scarcity of classes would indicate that we probably have not made our measurements with sufficient accuracy. If we have followed the rules on accuracy in measurement stated in Section 2.3, this could not have happened. Often, regrettably, measurements have already been obtained before statistical advice is sought. Nothing can then be done if there are too few classes.

Whenever there are more than the desired number of classes, grouping should be undertaken. When the data are meristic, the implied limits of continuous variables are meaningless. Yet with many meristic variates, such as a bristle number varying from a low of 13 to a high of 81, it would probably be wise to group them into classes, each containing several counts. This can best be done by using an odd number as a class interval so that the class mark representing the data will be a whole rather than a fractional number. Thus if we were to group the bristle numbers 13, 14, 15, and 16 into one class, the class mark would have to be 14.5, a meaningless value in terms of bristle number. It would therefore be better to use a class ranging over 3 bristles or 5 bristles, giving the integral values 14 or 15 as a class mark.

Before we get carried away by our new ability to group data in frequency distributions, we should be quite clear under what conditions we wish to do so. In earlier biometric work, placing data in frequency distributions was *de rigeur*. Given the circumstances of the time, this was quite reasonable. When data are to be handled by paper and pencil methods and more than 20 or 25 observations are at hand, grouping them in frequency distributions is an absolute necessity, since computations would not be practicable otherwise. However, since paper and pencil methods have been essentially abandoned, frequency distributions are no longer necessary for that purpose. If a calculating machine is available and there are more than 100 or 150 observations, it is probably still worthwhile to set up the data in a frequency distribution before carrying out the statistical computations. Such a decision is based on the fact that the computational time saved when a frequency distribution is employed has to be balanced against the time it takes to set up a frequency distribution. At the lower frequencies (that is, <100) it would take longer to set up such a distribution than to compute from raw observations. If computer facilities are available the above rule is, of course, also invalid. Since a computer can handle hundreds and even thousands of observations in a very short time, it would generally not be necessary to group them unless we are dealing with extremely large samples, much in excess of 1000 observations per sample. In the latter case you have to take into account the type of computer available. On some extremely fast computers (both as regards input and computation speeds) it would still be more economic to process the data without grouping them previously. On other computers it might be of advantage to group the data first on a card-sorting machine and then to employ a computer program for frequency distributions. Computers are relatively inefficient for grouping data in frequency distributions and these operations can often be done faster and certainly more economically on card sorters. If only mean and variance are required, it is probably not worthwhile to employ a computer. Such values could be computed quite rapidly on a desk calculator from the frequency distribution prepared by the sorter (see the next chapter for more detail on methods of computation).

One further important reason for obtaining frequency distributions arises if the form of the distribution is a specific point of interest. Discovering in a sample of immature insects that some measurements are bimodally distributed (with two peaks) would indicate that the population is dimorphic. However, in certain cases, where repeated samples have been taken of a particular biological phenomenon or process and where the shape of the distribution is well known, we would not wish to set up a frequency distribution each time unless we needed to do so for computational reasons.

If the shape of a frequency distribution is of particular interest, we may often wish to present the distribution in graphic form when discussing the results. This is generally done by means of frequency diagrams, of which there are two common types. For a distribution of meristic data we employ a *bar diagram*, as shown in Figure 2.2 for the sedge data of Table 2.3. The abscissa represents the variable (in our case the number of plants per quadrat), and the ordinate represents the frequencies. The important point about such a diagram is that the bars do not touch each other, which indicates that the variable is not continuous. By contrast, continuous variables, such as the frequency distribution of the femur lengths of aphid stem mothers, are graphed as a *histogram*, in which the width of each bar along the abscissa represents a class interval of the frequency distribution and the bars touch each other

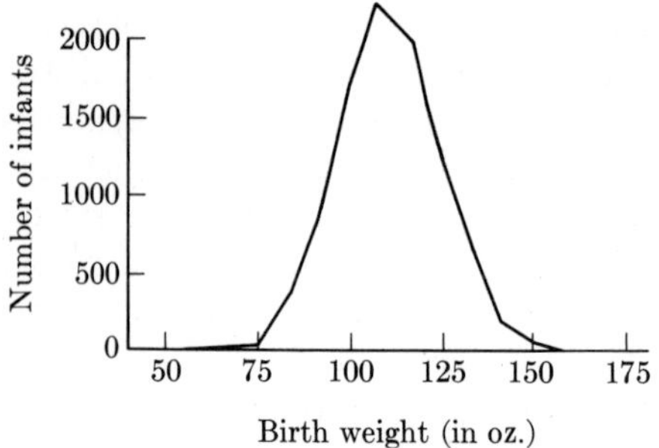

FIGURE 2.3 Frequency polygon. Birth weights of 9465 male infants. Chinese third-class patients in Singapore, 1950 and 1951. Data from Millis and Seng (1954).

to show that the actual limits of the classes are contiguous. The midpoint of the bar corresponds to the class mark. At the bottom of Box 2.1 are shown histograms of the frequency distribution of the aphid data, ungrouped and grouped. The height of the bars represents the frequency of each class. To illustrate that histograms are appropriate approximations to the continuous distributions found in nature, we may take a histogram and make the class intervals more narrow, producing more classes. The histogram would then clearly fit more closely to a continuous distribution. We can continue this process until the class intervals approach the limit of infinitesimal width. At this point the histogram becomes the continuous distribution of the variable. Occasionally the class intervals of a grouped continuous frequency distribution are unequal. For instance, in a frequency distribution of ages we might have more detail on the different stages of young individuals and less accurate identification of the ages of old individuals. In such cases, the class intervals for the older age groups would be wider, those for the younger age groups, narrower. In representing such data, the bars of the histogram are drawn with different widths. Figure 2.3 shows another graphical mode of representation of a frequency distribution of a continuous variable—birth weight in infants. As we shall see later the shapes of distributions as seen in such frequency diagrams can reveal much about the biological situations affecting a given variable.

Exercises 2

2.1 Differentiate between the following pairs of terms and give an example of each. (a) Statistical and biological populations. (b) Variate and individual. (c) Accuracy and precision (repeatability). (e) Class interval and class mark. (f) Bar diagram and histogram. (g) Abscissa and ordinate.

2.2 Round the following numbers to three significant figures: 106.55, 0.06819, 3.0495, 7815.01, 2.9149, and 20.1500. What are the implied limits before and after rounding? Round these same numbers to one decimal place.

2.3 Given 200 measurements ranging from 1.32 mm to 2.95 mm, how would you group them into a frequency distribution? Give class limits as well as class marks.

2.4 Group the following 40 measurements of interorbital width of a sample of domestic pigeons into a frequency distribution and draw its histogram (data from Olson and Miller, 1958).

12.2	12.9	11.8	11.9	11.6	11.1	12.3	12.2	11.8	11.8
10.7	11.5	11.3	11.2	11.6	11.9	13.3	11.2	10.5	11.1
12.1	11.9	10.4	10.7	10.8	11.0	11.9	10.2	10.9	11.6
10.8	11.6	10.4	10.7	12.0	12.4	11.7	11.8	11.3	11.1

2.5 How accurately should you measure the wing length of a species of mosquitoes in a study of geographic variation, if the smallest specimen has a length of about 2.8 mm and the largest a length of about 3.5 mm?

2.6 Look up the common logarithms (using Table **C**) of the 40 measurements given in Exercise 2.4 and make a frequency distribution of these transformed variates. Comment on the resulting change in the pattern of the frequency distribution from that found before.

3 THE HANDLING OF DATA

We have already stressed in the Preface that the skillful and expeditious handling of data is essential to the successful practice of statistics. Since the emphasis of this book is to a large extent on practical statistics, it is necessary to acquaint readers with the various techniques available for carrying out statistical computations. We discuss desk calculators in Section 3.1, electronic calculators in Section 3.2, and electronic computers in Section 3.3. It is very important to know which computational devices are most appropriate in terms of efficiency and economy for any given problem. It might appear initially that if access to a digital computer is available all computations should be done on such a computer. However, there are only certain types of statistical problems for which digital computer processing is most suited; they are discussed in Section 3.4.

Lacking any mechanical computational aids, we are reduced to carrying out statistical computations by the so-called pencil and paper methods. Textbooks of statistics until rather recently carried fairly extensive sections dealing with the adaptation of computational procedures for pencil and paper methods to insure the most expeditious handling of the computations. In this day and age, computations by pencil and paper methods can no longer be justified from an economic point of view, since both desk calculators and computers repay their investment many times over in terms of man-hours saved for carrying out the computations. In laboratories and museums a desk calculator is well on its way to becoming as indispensable a research tool as a microscope.

Thus pencil and paper methods are largely ignored in this book and we have assumed that a fully automatic calculator is available to the student. If, through one circumstance or another, you find yourself without such computational tools, pencil and paper methods may have to be resorted to.

As will be seen in the next chapter, data need to be coded in order to be rapidly processed, and the type of coding necessary for pencil and paper methods differs from that for machine computation. Pointers on how coding for pencil and paper methods would be carried out are given after coding has been explained (Section 4.9).

Two other developments for the simplification of computation may be mentioned here. The first is the tendency to develop tables providing answers for certain standard statistical problems without any computation whatsoever. An example of this is Finney's table for the test of independence in a 2×2 contingency table containing small frequencies (Pearson and Hartley, 1958; Table 38). For certain problems this table replaces Fisher's method of finding exact probabilities, which is very tedious (see Section 16.4).

The second development is of some newer statistical techniques, which are so easy to carry out that no mechanical aids are needed. Some of these techniques are inherently simple (for example, some of the nonparametric methods, such as the sign test, Section 13.11); others are approximate and consequently less than fully efficient methods which can serve as a first summary of the desired results. An example of a less than fully efficient method is the employment of the median (Section 4.3) in lieu of the mean (Section 4.1).

Accuracy in computation will depend on the number of figures carried during computation. Different computing devices vary in their capacity for retaining the number of digits; therefore, results of solving the same problem on a desk calculator and a computer, for example, may differ. Even different computers will not necessarily give exactly the same answers.

3.1 Desk calculators

We should distinguish among three types of mechanical calculating machines. *Adding machines* are familiar to most persons from their extensive use in stores and offices. They consist of a keyboard for entering numbers into the machines and certain operational keys for adding, subtracting, subtotaling, totaling, correcting entries, and for several other functions. The results are printed on a roll of paper tape attached to the machine. An adding machine by itself is not especially useful in statistics, since multiplication and division are frequently required. However, in an institution where a considerable amount of computing is done, an adding machine is an important auxiliary tool because it permits the checking of the addition of long columns of numbers and provides a visual record of these computations. Thus some of the computational steps carried out on automatic desk calculators need not be repeated, but can be checked from adding-machine tapes.

A modified version of an adding machine is the so-called *printing calculator* now marketed by several firms. These machines are lineal descendants of electric adding machines, to which automatic multiplication and division of numbers have been added. The results of these operations and some intermediate steps are printed on a paper tape. These machines can, of course, also be used as simple adding machines. Until recently, printing calculators were not especially useful in statistical work, being too slow and generally providing scant operational flexibility. However, in the last few years new models have appeared that are competitive with conventional rotary desk calculators in the speed and complexity of operations that can be carried out. They also provide a printed record of computations, making special checking operations unnecessary. The design of these machines is quite different from rotary calculators so that although almost all the conventional statistical computations can be performed on them, the flow of operations is appreciably different. The instructions given in Appendix A2 relate principally to rotary desk calculators and may need some revisions before being applicable to printing calculators. It is suggested that you consult the instruction booklet for your machine if you plan to work the problems in this text on a printing calculator.

A conventional *desk calculator* (also called rotary desk calculator) does not print the results of its calculations on paper tape but makes them visible to the operator in dials located in the carriage of the machine. Hand-operated calculating machines can still occasionally be seen. These basically perform four operations: addition of numbers, subtraction of numbers, shifting of the carriage with relation to the keyboard, and clearing of the dials. Multiplication is performed by means of repeated addition; division is accomplished by means of repeated subtraction of the divisor from the dividend. Both multiplication and division are greatly simplified by shifting the carriage the appropriate number of decimal places. The modern, complex-looking, fully automatic calculating machines are nothing but elaborate versions of these earlier, simple prototypes. In Appendix A2 (Operation of Desk Calculators), we discuss only the fully automatic calculators, which were first introduced in the 1930's and which have undergone numerous improvements and refinements since then.

Three brands of fully automatic calculators are most frequently seen in the United States. These are the Friden, Marchant (now SCM), and Monroe calculators. We shall not discuss in detail the separate features and differences of these machines, since they vary in different models of each brand. Instead, we shall refer in Appendix A2 to a composite of the representative characteristics of present-day fully automatic calculators.

Persons planning to buy a desk calculator should try to obtain as many of the features listed in Appendix A2 as can be obtained with the funds at their disposal. It is quite important to get maximal decimal capacity; that is, the keyboard should permit the entry of ten digits and the long dials should have at least a 21-digit capacity. Different operators have varied personal preferences among the variable features. Some prefer a divided keyboard (with a separate multiplier keyboard); others swear by a single keyboard.

Since even the two authors of this book cannot entirely agree on the ideal features of a machine, we hesitate to furnish further recommendations.

3.2 Electronic desk calculators

The most recent arrivals among the calculating machines are electronic desk calculators, which carry out the computations by means of tubes, transistors, and circuitry, as in an electronic computer. These machines combine the convenience of accessibility and a relatively low price with the instantaneous speeds of computation available by electronic rather than mechanical means. Figures are entered by means of a keyboard and results are displayed on a screen or printed on a paper tape. Several of the models are programmable to some degree, permitting repetitive operations to be performed automatically as soon as a number is entered into the machine. Some of these models permit fairly sophisticated programs to be written (up to several hundred instructions), and these programs are recorded on tapes or cards that can be inserted in the machine, making it immediately ready to carry out a given set of computations.

These calculators are still in the early stages of their evolution. Most models seem slower than they need to be and several are not capable of handling a number of digits adequate for statistical computations. If these calculators are not programmable, they are of limited use for statistical work. The accumulation of squares and cross products on such a machine would be more cumbersome than on a conventional desk calculator. Although the actual computation would be faster, more keys would have to be depressed and more operator decisions taken. There seems little justification for the purchase of an electronic calculator that is not to some degree programmable. The development of remote computer terminals (see next section) may be rapid enough to inhibit the further development of electronic desk calculators.

3.3 Computers

Electronic machines, since the middle 40's, have revolutionized the handling and processing of data. These machines are generally known as computers and are of two basic kinds: analog and digital. An *analog computer* generally does not work with discrete numbers but simulates continuous phenomena by analogous physical processes. Although analog computers are finding many uses in biology, they are generally of little application in statistical work, since their numerical accuracy is only between 95% and 98%. We shall therefore be concerned only with the digital computers, which work with discrete numbers.

A *digital computer* has the same basic arithmetical abilities as the desk calculators mentioned in the previous section (addition, subtraction, multiplication, and division), but it also has the following distinctive features.

Data are entered and the results of computations displayed automatically under the control of a previously prepared set of detailed instructions stored within the computer. These instructions also control the exact sequence of arithmetic computations to be performed. Computers also have the ability to execute different sets of instructions, depending on whether a given quantity is negative, zero, or positive. Thus it is possible to develop sophisticated programs that can bring into play different commands, as a function of the outcomes of previous computations. The program may even change some of its own instructions because of intermediate results obtained during a computation.

A typical digital computer consists of three main components: the *memory*, which stores both the data and the instructions; the *processor*, which actually performs the arithmetic computations; and the *peripheral devices*, which handle the input, output, and intermediate storage of data and instructions. Computers differ greatly in memory capacity and in the speed at which arithmetic computations can be performed. Small computers may hold around 2000 instructions and/or data items in memory, but some large systems can hold up to 512,000 such items. The slower computers add two 10-digit numbers in about 0.002 seconds, and faster computers perform this task in about 0.00001 seconds. Generally speaking, the largest and fastest computers are most efficient and economical for carrying out a given task but may not necessarily be economically justifiable for any but the largest research centers. It is primarily their great speed that makes computers important. Tasks that previously took hundreds of man-years to complete are now processed routinely by computers in a matter of minutes.

The instructions to the machine are known as the *program*, and the persons who specialize in the writing of computer programs are known as programmers. Writing instructions in the generally unique language for each model of computer is a difficult procedure, requiring specialized knowledge and experience. Fortunately *compilers* have been developed, which are programs that translate instructions written in a simple, problem-oriented *language* into the machine language for a particular computer. The best-known compiler is FORTRAN, in which operations to be performed are stated in a language close to algebra and ordinary English. FORTRAN is convenient for many scientific applications, especially those involving mathematics or statistics. Most persons can be taught in a very few hours to write programs in FORTRAN, and numerous graduate students in biology are already familiar with this language. The obvious advantages of knowing how to program will make ever-increasing numbers of students (and their seniors) acquire the art. Manuals for FORTRAN include McCracken (1965), Golden (1965), and Farina (1966).

The material presented in this book consists of relatively standard statistical computations, most of which have quite likely already been programmed in any given computer installation. Some standard programs covering topics in this book and written in FORTRAN IV are given in Appendix A3. They have been checked out for use on the GE 625 computer system, but they were designed to make their adaptation to other computers as simple as possible. For example, special features of certain "dialects" of FORTRAN were avoided, and only those features available at nearly all computer centers were used. To make use of these programs, they must be carefully punched onto tabulating cards and checked out with test data, as shown in Appendix A3.

An outline of the usual procedures required for carrying out a calculation by computer may be helpful.

Step 1. Before embarking on writing a new program, check for the availability of a previously prepared program to perform the desired computation. Most computer centers have a library of available programs. The steps that follow assume availability of a working program for the desired computation.

Step 2. The data must be keypunched onto tabulating cards (occasionally paper tape) in a format compatible with the program to be used. Detailed descriptions of this format should be provided in program write-ups. In addition to the actual data, information must usually be included on the number of observations to be read in and possibly a series of codes added to specify certain options to be taken in the computations (parameter cards). After the data have been keypunched and carefully proofed, they are assembled along with cards containing the program itself and special control cards containing information required for accounting purposes by the computer center. When using a previously prepared standard program (a *library program*), often one need not submit the program itself each time, but can simply insert a special control card requesting the computer to obtain the program from its library.

Step 3. The cards are turned over to the personnel of the computer center who carry out the actual operations of entering information into the computer and collecting the output. Usually jobs are submitted to the computer in batches, and one must normally wait two to twelve hours (and occasionally longer) before obtaining results of the computation.

Step 4. Output from the computer center should be checked to make certain that the data cards were entered in the proper sequence and that the proper program had been used. Unless using a well-tested library program, one should also verify that the desired computations were actually performed. This can be done with a small set of test data (with known results) included along with the actual data.

One must be very careful and precise in the preparation of a set of instructions. A single misplaced comma can make a very large difference in the execution of the computations. As a result, it is very unusual for a program to run the first time it is used; rather, one must repeatedly test the program, make corrections ("debug"), retest, and so on, until the program processes the test data correctly.

The delay involved in having to prepare data cards, submitting them to the computer center, and awaiting the returned output is an inconvenience. This is especially so during preparation of programs or when one wishes to watch the computations as they are being performed so that one may, if need be, stop the computations, change some of the parameters, or otherwise manually modify the computations in progress. Such needs have resulted in the development of remote computer terminals, which are typewriter-like or cathode-ray tube devices for the input and output of information to and from a computer at a remote site. They may be located in laboratories or offices and are connected to a computer that may be hundreds or thousands of miles away. The centralized computer is programmed in such a way that many terminals can be attached to a single computer. The input from each terminal can be processed instantaneously and simultaneously for all the users. Immediate responses to the user's inquiry establishes a new phase of man-computer interaction, which permits continuous interrogation and manipulation of evidence. Such a *conversational mode* of computer use was previously impossible.

3.4 Efficiency and economy in data processing

From the foregoing descriptions of machinery available for data processing it may seem to the reader that computers, when available, should always be preferred over desk calculators because of their obvious advantages. This would only be so under ideal conditions, when each research worker would have a computer or at least a remote terminal to a computer in his office and when input into this computer and call-up of a program could be made as efficient as input into a desk calculator. Whenever the amount of time required to process a problem on the computer exceeds the time it would take to do the problem on a desk calculator, the use of the computer is not justified. Punching data on cards or tapes, checking these data, taking them to the computation center, having them run there, printing up the results, and bringing them back to the office, all these steps entail a considerable amount of time. Thus one should not immediately turn with any trifling problem to the computer, because this may waste both time and money.

The types of problems most efficiently done on a computer might be summarized as follows. (1) Simple computational operations for large sets of data. An example would be the computation of a mean and standard deviation of a large sample of variates. It has, however, already been stated that when there are too many variates it becomes more efficient to sort them mechanically on a sorter, set up a frequency distribution, and then compute the required statistics either on a desk calculator or with a computer program for a frequency distribution. (2) Simple operations on small samples, which must be repeated many times on different sets of data. Thus, for example, no one would go to a computer to test a single Mendelian ratio. However, if in an extensive series of experiments many progenies are obtained and all of them are to be tested, computer processing becomes worthwhile. Thus some years ago one of us (RRS) was able to process tests on progeny data of some 50 separate experiments in housefly genetics in less than 1 minute on the IBM 650, the "model T" among computers. (3) Finally, the more complex the computations, the more useful a computer becomes. Thus for estimates of skewness and kurtosis (g_1's and g_2's; see Section 6.6) a very tedious procedure on a desk calculator, prone to numerous possibilities of error, computation by digital computer is both faster and much more reliable. Other computations, such as the fitting of probit regression lines (Section 14.12), the computation of discriminant functions, or factor analysis (Section 15.8), are all so involved and time-consuming that only with the development of digital computers have breakthroughs in their application taken place.

Certain types of computations that vary considerably from problem to problem, such as many kinds of analysis of variance (Chapters 8–13), are relatively difficult to program because of the flexibility that would have to be inherent in such programs. In spite of this, some general analysis of variance programs have been written (for example, programs A3.4 and A3.5), but with the rapidly proliferating knowledge of FORTRAN among biologists many will simply write a specialized program whenever they need it.

4 DESCRIPTIVE STATISTICS

An early and fundamental stage in any science is the descriptive stage. Until the facts as they are can be accurately described, an analysis of their causes is premature. The question "What?" comes before "How?" The application of statistics to biology has followed these general trends. Before Francis Galton could begin to think about the relations between the heights of fathers and those of their sons he had to have adequate tools for measuring and describing heights in a population. Similarly, unless we know something about the usual distribution of the sugar content of blood in a population of guinea pigs, as well as its fluctuations from day to day and within days, we shall be unable to ascertain the effect of a given dose of a drug upon this variable. In a sizable sample it would obviously be tedious to obtain knowledge of the material by contemplating all the individual observations. We need some form of summary to permit us to deal with the data in manageable form, as well as to be able to share our findings with others in scientific talks and publications. A histogram or bar diagram of the frequency distribution would be one type of summary. However, for most purposes, a numerical summary is needed to describe concisely, yet accurately, the properties of the observed frequency distribution. Quantities providing such a summary may be called *descriptive statistics*. This chapter will introduce you to some of them and show how they are computed.

Two kinds of descriptive statistics will be discussed in this chapter: statistics of location and statistics of dispersion. The *statistics of location* describe the position of a sample along a given dimension representing a variable. Thus, when measuring the length of certain animals, we would like to know whether the sample measurements lie in the vicinity of 2 cm or

20 cm. A statistic of location, therefore, must yield a representative value for the sample of observations. However, such a statistic (sometimes also known as a measure of central tendency) will not describe the shape of a frequency distribution. This may be long or very narrow; it may be humped or U-shaped, it may contain two humps, or it may be markedly assymmetrical. Quantitative measures of such aspects of frequency distributions are required. To this end we need to define and study *statistics of dispersion*.

The arithmetic mean described in Section 4.1 is undoubtedly the most important single statistic of location, but others (the geometric mean, the harmonic mean, the median, and the mode) are briefly mentioned in Sections 4.2, 4.3, and 4.4. Two simple statistics of dispersion (the range and the quartile deviation) are briefly noted in Section 4.5, and the standard deviation, the most common statistic for describing dispersion, is explained in Section 4.6. Our first encounter with contrasts between sample statistics and population parameters occurs in Section 4.7, in connection with statistics of location and dispersion. In Section 4.8 there is a description of methods of coding data in order to simplify them for the machine computation of the mean and standard deviation, which is discussed in Section 4.9. The coefficient of variation (a statistic that permits us to compare the relative amount of dispersion in different samples) is explained in the last section (4.10).

The techniques that will be at your disposal after you have mastered this chapter will not be very powerful in solving biological problems, but they will be indispensable tools for any further work in biometry. Other descriptive statistics, of both location and dispersion, will be taken up in later chapters.

An important note: We shall first encounter the use of logarithms in this chapter. To avoid confusion here and in subsequent chapters, common logarithms have been consistently abbreviated as log, and natural logarithms as ln. Thus, log x means $\log_{10} x$ and ln x means $\log_e x$.

4.1 The arithmetic mean

The most common statistic of location is familiar to everyone. It is the *arithmetic mean*, commonly called the *mean* or *average*. The mean is calculated by summing all the individual observations or items of a sample and dividing this sum by the number of items in the sample. For instance, as the result of a gas analysis in a respirometer an investigator obtains the following four readings of oxygen percentages:

	14.9
	10.8
	12.3
	23.3
Sum =	61.3

He calculates the mean oxygen percentage as the sum of the four items divided by the number of items—here, by four. Thus the average oxygen percentage is

$$\text{Mean} = \frac{61.3}{4} = 15.325\%$$

Calculating a mean presents us with the opportunity for learning statistical symbolism. We have already seen (Section 2.2) that an individual observation is symbolized by Y_i, which stands for the ith observation in the sample. Four observations could be written symbolically as follows:

$$Y_1, Y_2, Y_3, Y_4$$

We shall define n, the *sample size*, as the number of items in a sample. In this particular instance, the sample size n is 4. Thus, in a large sample, we can symbolize the array from the first to the nth item as follows:

$$Y_1, Y_2, \ldots, Y_n$$

When we wish to sum items, we use the following notation:

$$\sum_{i=1}^{i=n} Y_i = Y_1 + Y_2 + \cdots + Y_n$$

Don't let the impressive looking symbol on the left side of the identity scare you. The capital Greek sigma, $\sum$, simply means the sum of the items indicated. The $i = 1$ means that the items should be summed, starting with the first one, and ending with the nth one as indicated by the $i = n$ above the Σ. The subscript and superscript are necessary to indicate how many items should be summed. The "$i =$" in the superscript is usually omitted as superfluous. For instance, if we had wished to sum only the first three items, we would have written $\sum_{i=1}^{3} Y_i$. On the other hand, had we wished to sum all of them except the first one, we would have written $\sum_{i=2}^{n} Y_i$.

We call $\sum$ an operator symbol. It tells you what to do with the variables that follow it. Summation signs with subscripts and superscripts give an explicit mathematical formulation of the summation operation. However, with some exceptions (which will appear in later chapters), it is desirable to omit subscripts and superscripts, especially for the nonmathematical readers of this book. Subscripts generally add to the apparent complexity of the formula and, when they are unnecessary, distract the student's attention from the important relations expressed by the formula. Below are seen increasing simplifications of the complete summation notation shown at the extreme left:

$$\sum_{i=1}^{i=n} Y_i = \sum_{i=1}^{n} Y_i = \sum_{i} Y_i = \sum^{n} Y = \sum Y$$

The third of the symbols might be interpreted as meaning: sum the Y_i's over all available values of i. This is a frequently used notation, although we shall not employ it in this book. The next notation, with n as a superscript, tells us to sum n items of Y; take note that the i subscript of the Y has been dropped as unnecessary. Finally, the simplest notation is shown at the right. It merely says sum the Y's. This will be the form we shall use most frequently, and if a summation sign precedes a variable the summation will be understood to be over n items (all the items in the sample) unless subscripts or superscripts specifically tell us otherwise.

We shall use the symbol $\bar{Y}$ for the arithmetic mean of the variable Y. Some textbooks use $\bar{y}$. The reader should be forewarned that many other textbooks use the symbol X for a variable and, consequently, use $\bar{X}$ or $\bar{x}$ for the mean. In agreement with several modern texts we have switched to Y, for reasons that will be explained in Chapter 14. We would, however, like to take this occasion to mention that statistical symbolism is, regrettably, still far from uniform, although laudable progress toward uniformity has been made in recent years. We shall try to point out from time to time common alternative symbolisms to those presented in the text in order to prepare you for encounters with the statistical writings of others.

Our newly acquired knowledge of symbols enables us to write down the formula for the arithmetic mean:

$$\bar{Y} = \frac{\sum Y}{n} = \frac{1}{n} \sum Y \tag{4.1}$$

Both forms shown are acceptable; the form at the right, with the term $\frac{1}{n}$ preceding the $\sum$, is probably slightly neater looking. In effect, this formula tells us: sum all the (n) items and divide the sum by n. We shall postpone until Section 4.9 a discussion of the mechanics of efficiently computing a mean on a desk calculator.

The mean of a sample represents the center of the observations in the sample. If you were to draw a histogram of an observed frequency distribution on a sheet of cardboard, then cut out the histogram and lay it flat against a blackboard, supporting it with a pencil beneath, chances are that it would be out of balance, toppling to either the left or the right. If you moved the supporting pencil point to a position about which the histogram would exactly balance, this point of balance would be the arithmetic mean. In fact, this would be an empirical method of finding the arithmetic mean of a frequency distribution.

4.2 Other means

We shall see in Chapters 13 and 14 that variables are sometimes transformed into their logarithms or reciprocals. If we calculate the means of such transformed variables and then change them back into the original scale, these means will not be the same as if we had computed the arithmetic means of the original variables. The resulting means have received special names in statistics. The retransformed mean of the logarithmically transformed variables is called the *geometric mean*. It is computed as

$$G.M._Y = \text{antilog} \frac{1}{n} \sum \log Y \tag{4.2}$$

which indicates that the geometric mean $G.M._Y$ is the antilogarithm of the mean of the logarithms of variable Y. Since addition of logarithms is equivalent to multiplication of their antilogarithms, another way of representing this quantity is

$$G.M._Y = \sqrt[n]{Y_1 Y_2 Y_3 \cdots Y_n} \tag{4.3}$$

The geometric mean permits us to become familiar with another operator symbol: capital pi, Π, which may be read as product. Just as $\sum$ symbolizes summation of the items that follow it, so Π symbolizes the multiplication of the items that follow it. The subscripts and superscripts have exactly the same meaning as in the summation case. Thus formula (4.3) for the geometric mean can be rewritten more compactly as follows:

$$G.M._Y = \sqrt[n]{\prod_{i=1}^{n} Y_i} \tag{4.3a}$$

Without a digital computer, the computation of the geometric mean by Expression (4.3a) would become quite tedious. Samples of $n > 10$ would exceed the capacity of most desk calculators, and in practice the geometric mean has to be computed by transforming the variates into logarithms.

The reciprocal of the arithmetic mean of reciprocals is called the *harmonic mean*. If we symbolize it by H_Y, the formula for the harmonic mean can be written in concise form (without subscripts and superscripts) as

$$\frac{1}{H_Y} = \frac{1}{n} \sum \frac{1}{Y} \tag{4.4}$$

The reader may wish to convince himself that the geometric mean and the harmonic mean of the four oxygen percentages are 14.65% and 14.09%, respectively. Unless the individual items do not vary, the geometric mean is always less than the arithmetic mean, and the harmonic mean is always less than the geometric mean.

Some beginners in statistics have difficulty in accepting the fact that measures of location or central tendency other than the arithmetic mean are permissible or even desirable. They feel that the arithmetic mean is the "logical" average, and that any other mean would distort the data. This whole problem relates to the proper scale of measurement for representing data; this scale is not always the linear scale familiar to everyone, but is sometimes by preference a logarithmic or reciprocal scale. If you have doubts about this question, we shall try to allay these in Chapter 13, where we discuss the reasons for transforming variables.

4.3 The median

The *median* M is a statistic of location occasionally useful in biological research. It is defined as that value of the variable (in an ordered array) that has an equal number of items on either side of it. Thus, the median divides a frequency distribution into two halves. In the following sample of five measurements,

14, 15, 16, 19, 23

$M = 16$, since the third observation has an equal number of observations on both sides of it. We can visualize the median easily if we think of an array from largest to smallest—for example, a row of men lined up by their heights. The median individual will then be that person having an equal number of men on his right and left sides. His height will be the median height of the sample considered. This quantity is easily evaluated from a sample array with an odd number of individuals. When the number in the sample is even, the median is conventionally calculated as the midpoint between the $(n/2)$th and the $[(n/2) + 1]$th variate. Thus, for the sample of four measurements

14, 15, 16, 19

the median would be the midpoint between the second and third items, or 15.5.

Whenever any one value of a variate occurs more than once, problems may develop in locating the median. Computation of the median item becomes more involved, because all the members of a given class in which the median item is located will have the same class mark. The computation of the median of a frequency distribution either with the class marks of the original variate or with variates grouped into wider classes is illustrated and explained in Box 4.1.

The median is just one of a family of statistics dividing a frequency distribution into equal areas. It divides the distribution into two halves. The three *quartiles* cut the distribution at the 25, 50, and 75% points—that is, at points dividing the distribution into first, second, third, and fourth quarters by area (and frequencies). The second quartile is, of course, the median. (There are also quintiles, deciles, and percentiles, dividing the distribution into 5, 10, and 100 equal portions, respectively.)

From the point of view of applying it in later, more advanced statistical work, the median is not a useful statistic (except for "nonparametric" methods; see Chapter 13). However, in certain special cases it is a more representative measure of location than the arithmetic mean. Such instances almost always involve asymmetric distributions. An often quoted example from economics would be a suitable measure of location for the "typical" salary of an employee of a corporation. The very high salaries of the few senior executives would shift the arithmetic mean, the center of gravity, toward a completely unrepresentative value. The median, on the other hand, would be

BOX 4.1

Example of computation of the median of a frequency distribution.

Birth weights of male Chinese in ounces.

(1)	(2)	(3)	(4)
Y	f	*Cumulative frequency* F	*Reverse cumulative frequency* F
59.5	2	2	9465
67.5	6	8	9463
75.5	39	47	9457
83.5	385	432	9418
91.5	888	1320	9033
99.5	1729	3049	8145
107.5	2240	5289	6416
115.5	2007	7296	4176
123.5	1233	8529	2169
131.5	641	9170	936
139.5	201	9371	295
147.5	74	9445	94
155.5	14	9459	20
163.5	5	9464	6
171.5	1	9465	1
	9465		

Source: Millis and Seng (1954).

The data above represent a frequency distribution (obviously necessary for computation because of the large number of observations in the sample) but it also shows a new feature, namely a cumulative frequency column. This column is produced by successive addition of the frequencies for each class from the column to the left of it, accumulating the values either as the variable increases or as it decreases (reverse cumulative frequencies). The first 2 in column (3) is simply a copy of the 2 to its left, the 8 is the sum of 2 and 6, the 47 is $8 + 39$, the 432 is $47 + 385$, and so forth. The cumulative frequencies represent the total number of items below the upper class limit of their class (or above the lower class limit in the case of a reverse cumulative frequency distribution). Thus there are 432 individuals below 87.5 oz., the upper limit of the 83.5 oz. class. Obviously, the final value in the cumulative frequency column is the sum of all the frequencies in the sample.

The median for ranked data in an array is the $(n + 1)/2$th variate. For data in a frequency distribution such as the birth weights in this box, the median is the $n/2$th variate

$$\frac{n}{2} = \frac{9465}{2} = 4732.5$$

The 4732.5th item is located in the weight class 107.5 oz., i.e., somewhere be-

BOX 4.1 continued

tween the implied limits of 103.5 and 111.5. In this class there are 2240 items (5289 − 3049 in the cumulative frequency column) and item 4732.5 is the 1683.5th item in it. Assuming equal distribution of items in the class, we consider item 4732.5 to be

$$\frac{1683.5}{2240} = 0.7516 \text{ of the total class interval,}$$

or 75.16% of the distance from the lower class limit to the upper class limit. Since each class interval is 8.0 oz., the median item is

$$(0.7516) \times (8.0) = 6.013 \text{ oz.}$$

above the lower class limit (103.5 oz.); that is, the median of this frequency distribution is located at 103.5 + 6.013 = 109.513 oz. The median could also have been computed by going

$$\frac{556.5}{2240} = 0.2484$$

of a class interval below the upper class limit (111.5). For this approach we would use the reverse cumulative frequency column given above to compute 4732.5 − 4176 = 556.5.

little affected by a few high salaries; it would give the particular point on the salary scale above which lie 50% of the salaries in the corporation, the other half being lower than this figure.

In biology an example of the preferred application of a median over the arithmetic mean may be in populations showing skewed distribution, such as weights. Thus a median weight of American males 50 years old may be a more meaningful statistic than the average weight. The median is also of importance in cases where it may be difficult or impossible to obtain and measure all the items of a sample necessary to obtain a mean. An example will clarify this situation. An animal behaviorist is studying the time it takes for a sample of animals to perform a certain behavioral step. The variable which he is measuring is the time from the beginning of the experiment until each individual has performed. What he wants to obtain is an average time of performance. Such an average time, however, could only be calculated after records have been obtained on all the individuals. It may take a long time for the slowest animals to complete their performance, longer than the observer wishes to spend looking at them. Moreover, some of them may never respond appropriately, making the computation of a mean impossible. Therefore, a convenient statistic of location to describe these animals may be the median time of performance, or a related statistic such as the 75th or 90th percentile. Thus, so long as the observer knows what the total sample size is, he need not have measurements for the right-hand tail of his distribution. Similar examples would be the responses to a drug or poison in a group of individuals (the median lethal or effective dose, LD_{50} or ED_{50}) or the median time for a mutation to appear in a number of lines of a species.

4.4 The mode

The *mode* refers to the most "fashionable" value of the variable in a frequency distribution, or the value represented by the greatest number of individuals. When seen on a frequency distribution it is the value of the variable at which the curve peaks. In grouped frequency distributions the mode as a point has not much meaning. It usually suffices to identify the modal class. In biology, the mode does not have many applications.

Distributions having two peaks (equal or unequal in height) are called *bimodal*, those with more than two peaks are *multimodal*. In those rare distributions that are U-shaped, we refer to the low point at the middle of the distribution as an *antimode*.

In evaluating the relative merits of the arithmetic mean, the median, and the mode, a number of considerations have to be kept in mind. The mean

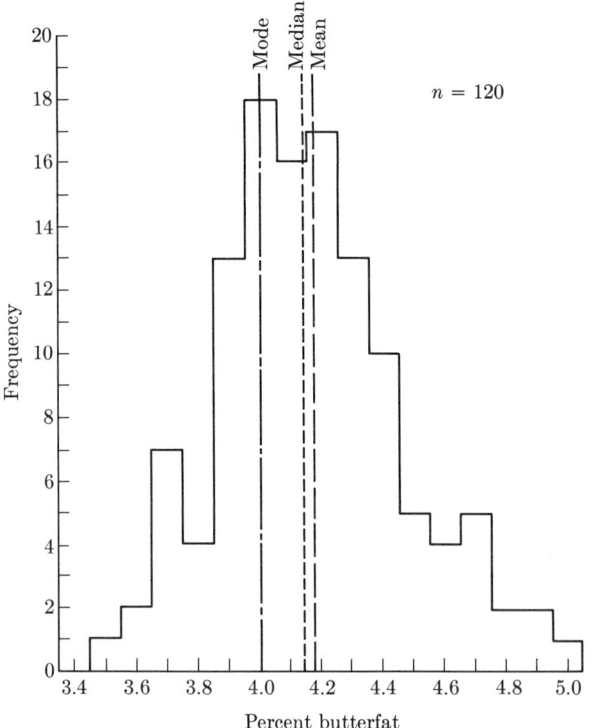

FIGURE 4.1 An asymmetrical frequency distribution (skewed to the right) showing location of the mean, median, and mode. Percent butterfat in 120 samples of milk (from a Canadian cattle breeders' record book).

is generally preferred in statistics since it has a smaller standard error than other statistics of location (explained in Section 7.2), it is easier to work with mathematically, and it has an additional desirable property (explained in Section 7.1): it will tend to be normally distributed even if the original data are not. The mean is markedly affected by outlying observations, the median and mode are not. The mean is generally more sensitive to changes in the shape of a frequency distribution and, if it is desired to have a statistic reflecting such changes, the mean may be preferred. In unimodal, symmetrical distributions the mean, the median, and the mode are all identical. A prime example of this is the well-known normal distribution of Chapter 6. In a typical asymmetrical distribution such as the one shown in Figure 4.1, the relative positions of the mode, median, and mean are generally these: the mean is closest to the drawn-out tail of the distribution, the mode is farthest, and the median is in between these. An easy way to remember this sequence is to recall that they occur in alphabetical order from the longer tail of the distribution.

4.5 Simple statistics of dispersion

We now turn to measures of dispersion. Figure 4.2 demonstrates that radically different looking distributions may possess the same mean median and mode. It is therefore obvious that other ways of characterizing distributions must be found.

One simple measure of dispersion is the *range*. It is the difference between the largest and the smallest items in a sample. Thus the range of the four oxygen percentages listed earlier (Section 4.1) is

$$\text{Range} = 23.3 - 10.8 = 12.5\%$$

and the range of the birth weights (Box 4.1) is

$$\text{Range} = 171.5 - 59.5 = 112.0 \text{ oz.}$$

Since the range is a measure of the span of the variates along the scale of the variable, it is in the same units as the original measurements. The range is clearly affected by even a single outlying value and for this reason is only a rough estimate of the dispersion of all the items in the sample.

Another measure of dispersion, now largely of historical interest, is the *quartile deviation*. This statistic, also known as the *semi-interquartile range*, is computed as

$$Q.D. = \tfrac{1}{2}(Q_3 - Q_1)$$

where Q_3 and Q_1 are the third and first quartiles of the distribution, respectively. Although this statistic gives an adequate description of the dispersion in the central half of the distribution, it obviously does not take the tails of the distribution into consideration and thus it suffers from defects opposite to those of the range; that is, while the range is determined by outliers and

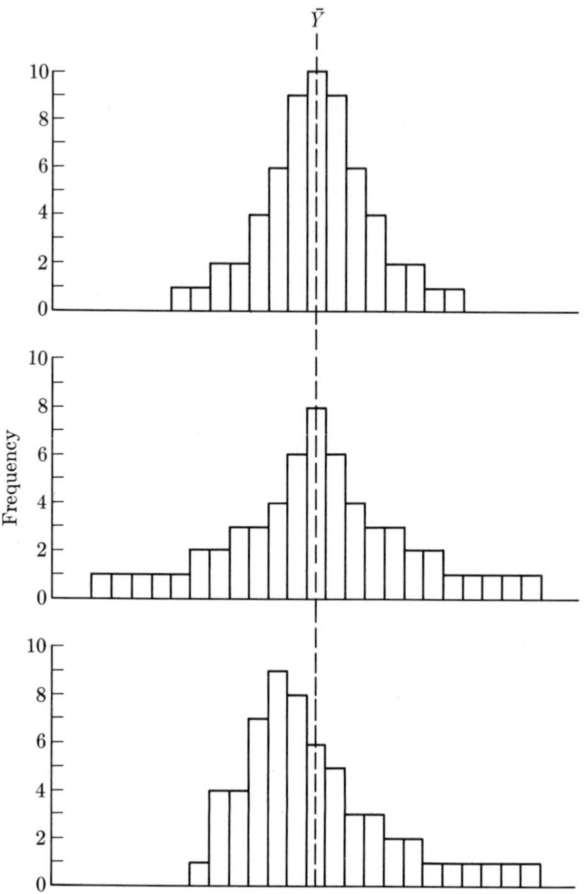

FIGURE 4.2 Three frequency distributions with identical means and sample sizes but differing in dispersion pattern.

ignores the distribution of items close to the mean, *Q.D.* is not affected by items lying beyond Q_1 and Q_3.

4.6 The standard deviation

A desirable measure of dispersion will take all items of a distribution into consideration, weighting each item by its distance from the center of the distribution. We shall now try to construct such a statistic. In Table 4.1 we show the grouped frequency distribution of the femur lengths of aphid stem mothers from Box 2.1. The first two columns show the class marks and the frequencies. The third column is needed for the computation of the mean of the frequency distribution (see Section 4.9 below). It is the class mark Y

multiplied by the frequency f. The computation of the mean is shown below the table. The mean femur length turns out to be 4.0 units.

The distance of each class mark from the mean is computed as the following deviation:

$$y = Y - \bar{Y}$$

Each individual deviation or *deviate* is by convention computed as the individual observation minus the mean, $Y - \bar{Y}$, rather than the reverse, $\bar{Y} - Y$. Deviates are symbolized by lowercase letters corresponding to the capital letters of the variables. Column (4) in Table 4.1 gives the deviates computed

TABLE 4.1

Deviations from the mean. Femur lengths of aphids (from Box 2.1).

(1)	(2)	(3)	(4)		(5)		(6)	(7)
Y	f	fY	$y = Y - \bar{Y}$		fy		$\|y\|$	$f\|y\|$
3.4	2	6.8	−0.6		−1.2		0.6	1.2
3.7	8	29.6	−0.3		−2.4		0.3	2.4
4.0	5	20.0	0.0		0.0		0.0	0.0
4.3	8	34.4		0.3		2.4	0.3	2.4
4.6	2	9.2		0.6		1.2	0.6	1.2
Total	25	100.0			−3.6	3.6		7.2

$$\bar{Y} = \frac{\Sigma fY}{\Sigma f} = \frac{100.0}{25} = 4.0 \qquad A.D. = \frac{\Sigma f|y|}{\Sigma f} = \frac{7.2}{25} = 0.29$$

in this manner. The column has been split into negative and positive deviates for convenience in computation. The deviates now have to be multiplied by their respective frequencies, so that those deviates that occur frequently contribute more to our measure of dispersion than those present only rarely. We therefore multiply the deviates by their frequencies f. The results of these computations are shown in column (5), which retains the separation between negative and positive deviates.

We now propose to calculate an average deviation that would sum all the deviates and divide them by the number of deviates in the sample. However, when we sum our deviates we note that the negative and positive deviates cancel out, as is shown by the sums at the bottom of column (5). This is always true of the sum of deviations from the arithmetic mean and is related to the fact that the mean is the center of gravity. Consequently the average deviation would also always equal zero. You are urged to study Appendix A1.1, which demonstrates that the sum of deviations around the mean of a sample is always equal to zero. Do not be frightened by what is indeed a very simple proof; it will give you good experience in the algebraic handling of statistical symbols.

Thus it is obvious that simply averaging the deviations from the mean will never provide us with a useful measure of dispersion. Statisticians have overcome this problem by employing absolute deviations. "Absolute" is a mathematical term meaning that a quantity is considered to be positive regardless of whether its actual sign is positive or negative. An absolute quantity is symbolized by being placed between two vertical lines, as shown at the head of column (6) in Table 4.1. In column (7) these deviates are again multiplied by their frequencies. The sum of these deviates is shown at the foot of the column; it equals 7.2. The *mean deviation* or *average deviation* (generally abbreviated *A.D.*) is then computed as shown at the bottom of Table 4.1. Its value is 0.29 in the original units of measurement. This statistic, popular in the early decades of this century, has now been almost entirely abandoned in favor of the standard deviation, discussed below. The only reason we discuss the average deviation at all is a pedagogical one. Having seen how it is calculated and having understood the reasons for trying to find a deviation of this sort, we are in a good position to comprehend the development of the standard deviation.

Squaring the deviates also overcomes the fact that the sum of the deviations around the mean is always zero and results in other desirable mathematical properties, which we shall take up in a later section. In Table 4.2 the

TABLE 4.2

The standard deviation. Long method, not recommended for actual computations but shown here to illustrate meaning of standard deviation. Same data as in Table 4.1.

(1)	*(2)*	*(3)*	*(4)*	*(5)*
Y	f	$y = Y - \bar{Y}$	y^2	fy^2
3.4	2	−0.6	0.36	0.72
3.7	8	−0.3	0.09	0.72
4.0	5	0.0	0.00	0.00
4.3	8	0.3	0.09	0.72
4.6	2	0.6	0.36	0.72
Total	25			2.88

$\bar{Y} = 4.0$ $\Sigma fy^2 = 2.88$

$$\text{Variance} = \frac{\Sigma fy^2}{n} = \frac{2.88}{25} = 0.1152 \qquad \text{Standard deviation} = \sqrt{0.1152} = 0.3394$$

data on femur length of aphids are again presented. Columns (1), (2), and (3) are the class marks, frequencies, and deviates as given previously. The deviates are now not separated into negative and positive columns. Column (4) lists their squares, all of which are, of course, positive. Finally, column (5) shows the squared deviates multiplied by their frequencies that is, column (4) multiplied by column (2). The sum of these squared deviates is 2.88. This is a very important quantity in statistics, known for short as the *sum of squares* and identified symbolically as $\sum y^2$. In Table 4.2 the sum of squares is symbolized as $\sum fy^2$, but the f is usually omitted since the $\sum$ indicates summation over all the items possible. Another common symbol for the sum of squares is SS.

The next step is to obtain the average of the n squared deviations. The resulting quantity is known as the *variance* or the *mean square:*

$$\text{Variance} = \frac{\sum y^2}{n} = \frac{2.88}{25} = 0.1152$$

The variance is a measure of fundamental importance in statistics and we shall employ it throughout this book. At the moment we need only remember that because of the squaring of the deviations the variance is expressed in squared units. To undo the effect of the squaring we now take the positive square root of the variance and obtain the *standard deviation:*

$$\text{Standard deviation} = +\sqrt{\frac{\sum y^2}{n}} = 0.3394$$

A standard deviation is again expressed in the original units of measurement, since it is a square root of the squared units of the variance.

An important note: Do not use the technique just learned and illustrated in Table 4.2 for an actual computation of a variance and standard deviation. This technique is far too tedious and furthermore is likely to introduce serious rounding errors because the mean in most cases will not be an integral number, as in this example. We have shown this approach here for purely pedagogical and heuristic reasons.

The observant reader may have noticed that we have avoided assigning any symbol to either the variance or the standard deviation. We shall explain why in the next section.

4.7 Sample statistics and parameters

Up to now we have calculated statistics from samples without giving too much thought to what these statistics represent. When correctly calculated, a mean and standard deviation will always be absolutely true measures of location and dispersion for the samples on which they are based. Thus the true mean of the four oxygen percentage readings in Section 4.1 is in fact 15.325%. The standard deviation of the 25 femur lengths is 0.3394 units when the items are grouped as shown. However, rarely in biology (or in statistics in general for that matter) are we interested in measures of location and dispersion only as descriptive summaries of the samples we have studied. Almost always we are interested in the *populations* from which the samples were taken. We therefore would like to know not the mean of the particular four oxygen percentages but rather the true oxygen percentage of the universe of readings from which the four readings had been sampled. Similarly, we would like to know the true mean femur length of the population of aphid stem mothers, not merely the mean of the 25 individuals we measured. When studying dispersion we generally wish to learn the true standard deviations of the populations and not those of the samples. These population statistics, however, are unknown and (generally speaking) are unknowable. Who would be able to collect all the stem mothers of this particular aphid population and measure them? Thus we need to use *sample statistics* as estimators of *population statistics* or *parameters*.

It is conventional in statistics to use Greek letters for population parameters and Roman letters for sample statistics. Thus the sample mean $\bar{Y}$ estimates μ, the parametric mean of the population. Similarly, a sample variance, symbolized by s^2, estimates a parametric variance, symbolized by σ^2. Such estimators should be *unbiased*. By this we mean that samples (regardless of the sample size), taken from a population with a known parameter, should give sample statistics which, when averaged, will give the parametric value. An estimator that does not do so is called *biased*. The sample mean $\bar{Y}$ is an unbiased estimator of the parametric mean μ. However, the sample variance as computed above (in Section 4.6) is not unbiased. On the average it will underestimate the magnitude of the population variance σ^2. In order to overcome this bias, mathematical statisticians have shown that when sums of squares are divided by $n - 1$ rather than by n the resulting sample variances will be unbiased estimators of the population variance. For this reason it is customary to compute variances by dividing the sum of squares by $n - 1$. The formula for the standard deviation is therefore customarily given as follows:

$$s = +\sqrt{\frac{\Sigma y^2}{n - 1}} \tag{4.5}$$

In the case of the aphid stem mother data the standard deviation would thus be computed as

$$s = \sqrt{\frac{2.88}{24}} = 0.3464$$

We note that this value is only slightly larger than our previous estimate of 0.3394. Of course, the greater the sample size, the less difference there will be between division by n and by $n - 1$. However, regardless of sample size, it is good practice to divide a sum of squares by $n - 1$ when computing a variance or standard deviation. It may be assumed that when the symbol s^2 is encountered, it refers to a variance obtained by division of the sum of squares by the *degrees of freedom*, as the quantity $n - 1$ is generally referred to. The only time when division of the sum of squares by n is appropriate is when the interest of the investigator is truly limited to the sample at hand and to its variance and standard deviation as descriptive statistics of the sample, in contrast to using these as estimates of the population parameters. There are also the rare cases in which the investigator possesses data on the entire population; in such cases division by n is perfectly justified, because then he is not estimating a parameter but is in fact evaluating it. Thus the variance of the wing lengths of all adult whooping cranes would be a parametric value; similarly, if the I.Q.'s of all winners of the Nobel Prize in physics had been measured, their variance would be a parameter since it is based on the entire population.

4.8 Coding of data before computation

Before we can discuss how to compute means and standard deviations in a practical manner on a desk calculator, one more important procedure must be learned—the coding of the original data. By *coding* we mean the addition or subtraction of a constant number to the original data, and/or the multiplication or division of these data by a constant. Data frequently need to be coded because they are originally expressed in too many digits or are very large numbers that may cause difficulties and errors during data handling. The importance of correct coding of data cannot be overestimated. Successful statistical computation—obtaining the correct result and not becoming bogged down in a morass of figure—is often facilitated by correct coding of the original data.

We shall call *additive coding* the addition or subtraction of a constant (since subtraction is only addition of a negative number). We shall similarly call *multiplicative coding* the multiplication or division by a constant (since division is multiplication by the reciprocal of the divisor). We shall call *combination coding* the application of both additive and multiplicative coding to the same set of data.

In Appendix A1.2 we examine the consequences of the three types of coding in the computation of means, variances, and standard deviations. The results *for means* can be summarized as follows: when a constant has been added to each variate, the coded mean is decoded by subtracting the constant.

When variates have been coded by being multiplied by a constant, the mean can be decoded by being divided by the same constant; similarly, when variates have been coded by division, the mean is decoded by multiplication by this same constant.

Means coded by combination coding can be decoded by performing *the inverse operation, in reverse sequence*. This can be made clear by a little mnemonic diagram which, incidentally, works equally well for simple multiplicative or additive coding. Thus if we have coded a set of variates by subtracting 5 and multiplying by 10, we decode the mean by first dividing by 10, then adding 5:

$$\text{code} \xrightarrow{-5 \times 10}$$
$$\xleftarrow[+5 \div 10]{} \text{decode}$$

Similarly, a coded mean based on variates that had been divided by 100 and to which 1.2 had then been added could be decoded by first subtracting 1.2, then multiplying $\times$ 100:

$$\text{code} \xrightarrow{\div 100 + 1.2}$$
$$\xleftarrow[\times 100 - 1.2]{} \text{decode}$$

The following diagrams illustrate simple coding:

$$\text{code} \xrightarrow{+10} \qquad\qquad \text{or} \qquad\qquad \text{code} \xrightarrow{\times 4}$$
$$\xleftarrow[-10]{} \text{decode} \qquad\qquad\qquad\qquad \xleftarrow[\div 4]{} \text{decode}$$

On considering the effects of coding variates on the values of *variances and standard deviations*, we find first of all that additive codes have no effect on the sums of squares, variances, or standard deviations. The mathematical proof is given in Appendix A1.2, but this can be seen intuitively because an additive code has no effect on the distance of an item from its mean. The distance from an item of 15 to its mean of 10 would be 5. If we were to code the variates by subtracting a constant of 10, the item would now be 5 and the mean zero. The difference between them would still be 5. Thus if only additive coding is employed, the only statistic in need of decoding is the mean. But multiplicative coding does have an effect on sums of squares, variances, and standard deviations. The standard deviations have to be divided by the multiplicative code, just as had to be done for the mean. However, the sums of squares or variances have to be divided by the multiplicative codes squared because they are squared terms, and the multiplicative factor became squared during the operations. In combination coding the additive code can be ignored. The following mnemonic diagrams can be used to summarize decoding for standard deviations:

$$\text{code} \xrightarrow{+5 \times 10}$$
$$\xleftarrow[\div 10]{} \text{decode}$$

and for sums of squares or variances:

$$\text{code} \xrightarrow{\div 100 + 1.2}$$
$$\xleftarrow[\times 100^2]{} \text{decode}$$

Examples of coding and decoding data are shown in Boxes 4.2 and 4.3, which are discussed in the next section.

4.9 Machine methods for computing mean and standard deviation

The procedure used to compute the sum of squares in Section 4.6 can be expressed by the following formula:

$$\sum y^2 = \sum (Y - \bar{Y})^2 \tag{4.6}$$

This formulation explains most clearly the nature of the sum of squares, but for computational purposes it is the most awkward one; if it were the only one available it would seriously discourage people from doing statistical work. To carry out the computation of $\sum y^2$ by Formula (4.6), we would first have to compute the mean, then calculate the deviation of each item from the mean. Next, each deviation would have to be squared and finally these squares accumulated. The most tiresome procedure would be the calculation and squaring of the deviations, which on most desk calculators cannot be done automatically without transcribing the deviations or at least entering them manually. In our experience many students once having learned some particular technique will repeat it by rote. Therefore, having learned to calculate a sum of squares as shown in Table 4.2 they might tend to repeat this particular procedure. We cannot therefore emphasize strongly enough that Table 4.2 is presented for pedagogic reasons only; the sum of squares is *never* ordinarily calculated as shown there. Let us now develop a computational formula for the standard deviation. In review, there are three steps necessary for computing this statistic: (1) find $\sum y^2$, the sum of squares, (2) divide by $n - 1$ to give the variance, (3) take the square root of the variance to obtain the standard deviation. Steps (2) and (3) both are single operations and no time saving is possible for them. Our efforts at simplifying the computation must center on step (1), the computation of the sum of squares. The customary computational formula for this quantity is

$$\sum y^2 = \sum Y^2 - \frac{(\sum Y)^2}{n} \tag{4.7}$$

Let us be clear what this formula represents. The first term on the right side of the equation, $\sum Y^2$, is the sum of all the Y's squared as follows:

$$\sum Y^2 = Y_1^2 + Y_2^2 + Y_3^2 + \cdots + Y_n^2$$

When referred to by name, $\sum Y^2$ should be called the "sum of Y squared" and should be carefully distinguished from $\sum y^2$, "the sum of squares of Y." These names are unfortunate, but they are too well established to think of amending them. The other quantity in Expression (4.7) is $(\sum Y)^2/n$. It is often called the *correction term, CT*. The numerator of this term is the square of the sum of the Y's; that is, all the Y's are first summed and this sum is then squared. In general, this quantity is different from $\sum Y^2$, which first squares the Y's and then sums them. These two terms are identical only if all the Y's are equal. If you are not certain about this, you can convince yourself of this fact by calculating these two quantities for a few numbers.

Why is Expression (4.7) identical with Expression (4.6)? The proof of this identity is very simple and is given in Appendix A1.3. You are urged to work through it to build up your confidence in handling statistical symbols and formulae.

□□ From Expressions (4.1) and (4.7) we see that it is necessary to have three quantities—n, $\sum Y$, and $\sum Y^2$—for computing the mean and standard deviation of a sample. The actual desk calculator operations depend on whether the data are unordered or arrayed in a frequency distribution. We have recommended leaving the data unordered when the number of variates is less than 150 because the time needed to set up a frequency distribution would be about equivalent to the time saved during computation if the data were in a frequency distribution form. When the data are unordered the computation proceeds as in Box 4.2, which is based on the unordered aphid femur

BOX 4.2

Machine calculation of $\bar{Y}$ and s from unordered data.

Aphid femur length data, unordered as shown at the head of Box 2.1.

The data were coded by multiplying by 10 to avoid decimal points during the computation. Coded variables and statistics are identified by the subscript c.

Computation	*Coding and decoding*
$n = 25$	Code: $Y_c = 10Y$
$\sum Y_c = 1001$	
$\bar{Y}_c = 40.04$	To decode $\bar{Y}_c$: $\bar{Y} = \dfrac{\bar{Y}_c}{10} = \dfrac{40.04}{10} = 4.004$
$\sum Y_c^2 = 40{,}401$	
$\sum y_c^2 = \sum Y_c^2 - \dfrac{(\sum Y_c)^2}{n}$	
$= 40{,}401 - \dfrac{(1001)^2}{25}$	
$= 320.960$	
$s_c^2 = \dfrac{\sum y_c^2}{n-1} = \dfrac{320.960}{24}$	
$= 13.37$	To decode s_c^2: $s^2 = \dfrac{s_c^2}{10^2} = \dfrac{13.37}{100} = 0.1337$
$s_c = \sqrt{13.37} = 3.656$	To decode s_c: $s = \dfrac{s_c}{10} = \dfrac{3.656}{10} = 0.3656$

length data shown at the head of Box 2.1. The data were coded for convenience by multiplying variates $\times$ 10 in order to remove the decimal point during the computation. With unordered data and a fully automatic rotary desk calculator, the strategy of obtaining n, $\sum Y$, and $\sum Y^2$ will depend on whether the calculator has an automatic squaring device. With such a device, it is probably fastest to square the figures accumulatively, locking both long and short dials. At the end of this operation the $\sum Y$ should be in the short dials and $\sum Y^2$ should be in the long dials. It is very important that all statistical computations be checked independently by a second operator or by the same worker—using a slightly different technique, if at all possible, or by using one of the error-proof computational checks as one proceeds. In this case it is usually sufficient to obtain $\sum Y$ independently by adding the numbers on an adding machine. If this total checks with the total shown on the short dial of the squaring operation, one may assume that $\sum Y^2$ is also correct (on some machines depressed keys may disengage accidentally during the squaring operation and thus introduce error).

On machines that do not have automatic squaring, the recommended procedure is the so-called *1, 5-zeros method.* By this method one puts a 1, followed by five zeros and the Y value into the multiplier dial and multiplies this times Y in the keyboard. Such an operation will result in showing $100000Y$ in the short dial and $Y00 \ldots Y^2$ in the long dial. The two Y quantities represent the double entry of Y, once in the multiplier dial and once in the main keyboard. By locking the dials or using accumulative multiplication the Y's become $\sum Y$, the Y^2's become $\sum Y^2$, and the 1's accumulate to n. So long as the two $\sum Y$ values in the long and short dials agree, the machine operator can be assured that the computation is still correct and that the $\sum Y^2$ is also correct. This is true unless a double compensating error has occurred, which is most unlikely. During operations of this sort it is not wise to process too many variates at once. The inexperienced machine operator would be wise to stop after every thirty items or so, write down the intermediate result and then proceed. Thus when it is discovered that the two $\sum Y$ quantities do not check the operator can return to an earlier correct point rather than having to start from the beginning. With the 1, 5-zeros method no further check is necessary. There are numerous modifications of these techniques and with the hints already given you may well develop your own technique, most suitable for your particular machine.

When the data are arrayed in a frequency distribution the computations are considerably simpler. An example is shown in Box 4.3. These are the birth weights of male Chinese children, first encountered in Box 4.1. First of all they have to be coded, to remove the rather awkward class marks. This is done by subtracting 59.5, the lowest class mark of the array. The resulting class marks are values such as 0, 8, 16, 24, 32, etc. They are then divided by 8, which changes them to 0, 1, 2, 3, 4, etc., which is the desired format. The computation is now carried out quite simply and elegantly. We put the f value into the multiplier dial, enter the Y and the Y^2 value into the keyboard at its left and right margins, respectively, and start to multiply accumula-

BOX 4.3

Machine calculation of $\overline{Y}$, s, and CV from a frequency distribution.

Birth weights of male Chinese in ounces.

(1) Class mark Y	(2) f	(3) Coded class mark Y_c
59.5	2	0
67.5	6	1
75.5	39	2
83.5	385	3
91.5	888	4
99.5	1729	5
107.5	2240	6
115.5	2007	7
123.5	1233	8
131.5	641	9
139.5	201	10
147.5	74	11
155.5	14	12
163.5	5	13
171.5	1	14
	$9465 = n$	

Source: Millis and Seng (1954).

Computation

$$\sum fY_c = 59{,}629$$

$$\overline{Y}_c = 6.300$$

$$\sum fY_c^2 = 402{,}987$$

$$CT = \frac{(\sum fY_c)^2}{n} = 375{,}659.550$$

$$\sum fy_c^2 = \sum fY_c^2 - CT = 27{,}327.450$$

$$s_c^2 = \frac{\sum fy_c^2}{n-1} = 2.888$$

$$s_c = 1.6991$$

Coding and decoding

$$\text{Code: } Y_c = \frac{Y - 59.5}{8}$$

$$\text{To decode } \overline{Y}_c\text{: } \overline{Y} = 8\overline{Y}_c + 59.5 = 50.4 + 59.5 = 109.9 \text{ oz.}$$

$$\text{To decode } s_c\text{: } s = 8s_c = 13.593 \text{ oz.}$$

$$CV = \frac{s}{\overline{Y}} \times 100 = \frac{13.593}{109.9} \times 100 = 12.369\%$$

tively for each class mark. Thus for the first line we would multiply 2 times 0 ($= Y$) and 0 ($= Y^2$). For the second line 6 times 1 and 1, for the third line 39 times 2 and $4 (= 2^2)$, etc. Most people remember the squares of numbers up to 15 by heart and can therefore enter them easily. Rarely will there be more than 15 classes. Even if there are more, it is probably preferable to look up the squares in a table than to use a code that would produce some negative class marks (as would occur by subtraction of code 107.5, the modal class mark, in the present example). The handling of negative numbers on most desk calculators is rather tricky and may give rise to errors. There is no automatic check for the procedure just outlined, and it is probably advisable to repeat the operation from the bottom up. Another method of checking is to recode the data in a slightly different way and then recalculate the mean and the sum of squares. □□

For computation by pencil and paper methods, data should be arrayed in frequency distributions even for very small sample sizes. The class marks should be coded as shown in Box 4.3, but the codes should be such that the median or modal class receives a coded class mark of zero. Although this will cause all lower class marks to be negative, handling simple negative numbers is not difficult by paper and pencil methods. The main advantage of such a scheme is that none of the class marks becomes a large number. In general, two-digit numbers should be avoided, since handling their squares would cause difficulties without a calculating machine. Thus the example in Box 4.3 might be coded for paper and pencil methods with a class mark of 0 for the 107.5-oz. class, which would yield -6 and $+8$ for the 59.5 and 171.5-oz. classes, respectively. The arithmetic involved in computing the sum of squares under this system of coding is not too heavy to be carried out with paper and pencil, if necessary.

Computer programs that compute the basic statistics, $\overline{Y}$, s^2, s, (and others which we have not yet discussed) are given in Appendix A3. Program A3.1 assumes the raw unordered observations as input, in contrast to Program A3.2, which requires data to be presented in the form of a frequency distribution.

A common rule of thumb for estimating the mean is to average the largest and smallest observation to obtain the so-called *midrange*. For the aphid stem mother data of Box 2.1, this value is $(4.7 + 3.3)/2 = 4.0$, which happens to fall exactly on the computed sample mean. Standard deviations can be estimated from ranges by appropriate division of the range.

For samples of .	*divide the range by*
10	3
30	4
100	5
500	6
1000	$6\frac{1}{2}$

The range of the aphid data is 1.4. When this value is divided by 4 we get an estimate of the standard deviation of 0.35, which compares not too badly with the calculated value of 0.3656 in Box 4.2. These approximate methods are very useful for detecting gross errors in computation.

A more accurate procedure is by employment of Table **Z** in the *Statistical Tables*, accompanying this volume. This table furnishes the mean range for different sample sizes of a normal distribution (see Chapter 6) with a variance of one. When we divide the range of a sample by a mean range from Table **Z**, we obtain an estimate of the standard deviation of the population from which the sample was taken. Thus, for the aphid data we look up $n = 25$ in Table **Z** and obtain 3.931. We estimate $s = 1.4/3.931 = 0.356$, a value closer to the sample standard deviation than that obtained by the rougher method discussed above, which, however, is based on the same principle and assumptions.

4.10 The coefficient of variation

Having obtained the standard deviation as a measure of the amount of variation in the data, you may be led to ask, "Now what?" At this stage in our comprehension of statistical theory, nothing really useful comes of the computations we have carried out, although the skills just learned are basic to all later statistical work. So far, the only use that we might have for the standard deviation is as an estimate of the amount of variation in a population. Thus we may wish to compare the magnitudes of the standard deviations of similar populations and see whether population A is more or less variable than population B. However, when populations differ appreciably in their means, the comparison of their variances or standard deviations would be quite risky. For instance, the standard deviation of the tail lengths of elephants is obviously much greater than the entire tail length of a mouse. In order to compare the amount of variation in populations having different means, the *coefficient of variation* has been developed. This is simply the standard deviation expressed as a percentage of the mean. Its formula is

$$CV = \frac{s \times 100}{\bar{Y}} \tag{4.8}$$

For example, the coefficient of variation of the birth weights in Box 4.3 is 12.37%, as shown at the bottom of that box. The coefficient of variation is independent of the unit of measurement and is expressed as a percentage.

Coefficients of variation are extensively used when one wishes to compare the variation of two populations independent of the magnitude of their means. It is probably of little interest to discover whether the birth weights of the Chinese children are more or less variable than the femur lengths of the aphid stem mothers. We can calculate the latter as $0.3656 \times 100/4.004 = 9.13\%$, which would suggest that the birth weights are more variable. More commonly, we shall wish to test whether a given biological sample is more variable for one character than for another. Thus, for a sample of rats, is body weight more variable than blood sugar content? A second, frequent type of comparison, especially in systematics, is among different populations for the same character. Thus, we may have measured wing length in samples of birds from several localities. We wish to know whether any one of these populations is more variable than the others. An answer to this question can be obtained by examining the coefficients of variation of wing length in these samples. At one time systematists took great stock by the coefficients of variation and even based some classificatory decisions on the magnitude of these coefficients. However, there is little, if any, foundation for such actions. An extensive discussion of the coefficient of variation, especially as it relates to systematics, can be found in Simpson, Roe, and Lewontin (1960).

Exercises 4

4.1 Find the mean, standard deviation, and coefficient of variation for the pigeon data given in Exercise 2.4. Group the data into ten classes, recompute $\bar{Y}$ and s, and compare them with the results obtained from the ungrouped data. Compute the median for the grouped data.

4.2 Find $\bar{Y}$, s, CV, and the median for the following data. (mg of glycine per mg of creatinine in the urine of 37 chimpanzees; from Gartler, Firschein, and Dobzhansky, 1956.) ANS.: $\bar{Y} = 0.115$, $s = 0.10404$.

.008	.018	.056	.055	.135	.052	.077	.026	.440	.300
.025	.036	.043	.100	.120	.110	.100	.350	.100	.300
.011	.060	.070	.050	.080	.110	.110	.120	.133	.100
.100	.155	.370	.019	.100	.100	.116			

4.3 The following are percentages of butterfat from 120 registered three-year-old Ayrshire cows selected at random from a Canadian stock record book.

(a) Calculate $\bar{Y}$, s, and CV directly from the data. You will find it advantageous to calculate $\sum Y$ and $\sum Y^2$ separately for each of the four columns, since in case of an error you will have to recompute only one column instead of all four. Save the data on each column. They will be used later.

(b) Group the data in a frequency distribution and again calculate $\bar{Y}$, s, and CV. Compare the results with those of (a). How much accuracy has been lost by grouping? Also calculate the median.

4.32	4.24	4.29	4.00
3.96	4.48	3.89	4.02
3.74	4.42	4.20	3.87
4.10	4.00	4.33	3.81
4.33	4.16	3.88	4.81
4.23	4.67	3.74	4.25
4.28	4.03	4.42	4.09
4.15	4.29	4.27	4.38
4.49	4.05	3.97	4.32
4.67	4.11	4.24	5.00
4.60	4.38	3.72	3.99
4.00	4.46	4.82	3.91

(columns continued on next page.)

4.71	3.96	3.66	4.10
4.38	4.16	3.77	4.40
4.06	4.08	3.66	4.70
3.97	3.97	4.20	4.41
4.31	3.70	3.83	4.24
4.30	4.17	3.97	4.20
4.51	3.86	4.36	4.18
4.24	4.05	4.05	3.56
3.94	3.89	4.58	3.99
4.17	3.82	3.70	4.33
4.06	3.89	4.07	3.58
3.93	4.20	3.89	4.60
4.38	4.14	4.66	3.97
4.22	3.47	3.92	4.91
3.95	4.38	4.12	4.52
4.35	3.91	4.10	4.09
4.09	4.34	4.09	4.88
4.28	3.98	3.86	4.58

4.4 What effect would adding a constant 5.2 to all observations have upon the numerical values of the following statistics? $\overline{Y}$, s, CV, average deviation, median, mode, range. What would be the effect of adding 5.2 and then multiplying the sums by 8.0? Would it make any difference in the above statistics if we multiplied by 8.0 first and then added 5.2?

4.5 Show that the equation for the variance can also be written as

$$s^2 = \frac{\sum Y^2 - n\overline{Y}^2}{n - 1}$$

(In actual practice this expression is not used much because it may lead to serious rounding errors unless $\overline{Y}$ is calculated to many significant figures.)

4.6 Estimate μ and σ using the midrange and the range (see Section 4.9) for the data in Exercises 4.1, 4.2, and 4.3. How well do these estimates agree with the estimates given by $\overline{Y}$ and s? ANS.: Estimates of μ and σ for Exercise 4.2 are 0.224 and 0.1014.

5 INTRODUCTION TO PROBABILITY DISTRIBUTIONS: BINOMIAL AND POISSON

In Section 2.5 we first encountered frequency distributions. For example, Table 2.3 shows a distribution for a meristic, or discrete (discontinuous) variable, the number of sedge plants per quadrat. Examples of distributions for continuous variables are the femur lengths of aphid stem mothers in Box 2.1 or the human birth weights in Box 4.3. Each of these distributions informs us about the absolute frequency of any given class and permits computation of the relative frequencies of any class of variable. Thus, most of the quadrats contained either no sedges, or one or two plants. In the 139.5-oz. class of birth weights we find only 201 out of the total of 9465 babies recorded; that is, approximately only 2.1% of the infants are in that birth weight class. We realize, of course, that these frequency distributions are only samples from given populations. The birth weights represent a population of male Chinese infants from a given geographical area. But if we knew our sample to be representative of that population, we could make all sorts of predictions based upon the sample frequency distribution. For instance, we could say that approximately 2.1% of male Chinese babies born in this population should weigh between 135.5 and 143.5 oz. at birth. Similarly, we might say that the probability of any one baby in this population weighing 139.5 oz. at birth is quite low. If each of the 9465 weights were given a registration number, the numbers mixed up in a hat, and a single one pulled out, the probability that we would pull out one of the 201 in the 139.5-oz. class would be very low indeed—only 2.1%. It would be much more probable that we would sample an infant of 107.5 or 115.5 oz., since the infants in these classes are represented by frequencies 2240 and 2007, respectively. Finally, if we were to sample from an unknown population of babies and find that the very first individual sampled had a birth weight of 170 oz., we would probably reject any hypothesis that the unknown population was the same as that sampled in Box 4.3. We would arrive at this conclusion because in the distribution in Box 4.3 only one out of almost 10,000 infants had a birth weight that high. Though it is possible that we could have sampled from the population of male Chinese babies and obtained a birth weight of 170 oz., the probability that the first individual sampled would have such a value is very low indeed. It seems much more reasonable to suppose that the unknown population from which we are sampling is different in mean and possibly in variance from the one in Box 4.3.

We have used this empirical frequency distribution to make certain predictions (with what frequency a given event will occur) or to make judgments and decisions (is it likely that an infant of a given birth weight belongs to this population?). In many cases in biology, however, we shall make such predictions not from empirical distributions, but on the basis of theoretical considerations which in our judgment are pertinent. We may feel that the data should be distributed in a certain way because of basic assumptions about the nature of the forces acting on the example at hand. If our actually observed data do not conform to the values expected on the basis of these assumptions, we shall have serious doubts about our assumptions. This is a common use of frequency distributions in biology. The assumptions being tested generally lead to a theoretical frequency distribution known also as a *probability distribution.* This may be a simple two-valued distribution such as the 3:1 ratio in a Mendelian cross, or it may be a more complicated function trying to predict the number of plants in a quadrat. If we find that the observed data do not fit the expectations on the basis of theory, we are often led to the discovery of some biological mechanism causing this deviation from expectation. The phenomena of linkage in genetics, of preferential mating between different phenotypes in animal behavior, of congregation of animals at certain favored places or, conversely, their territorial dispersion are cases in point. We shall thus make use of probability theory to test our assumptions about the laws of occurrence of certain biological phenomena. We should point out to the reader, however, that probability theory underlies the entire structure of statistics, since, owing to the nonmathematical orientation of this book, this may not be entirely obvious.

In the sections that follow we shall first of all present a sketchy discussion of probability (Section 5.1), limited to the amount necessary for comprehension of the sections that follow at the intended level of mathematical sophistication. Next we shall take up the binomial frequency distribution (Section 5.2), which not only is important in certain types of studies, such as genetics, but is fundamental to an understanding of the various kinds of probability distributions to be discussed in this book.

The Poisson distribution, which follows in Section 5.3, is of wide applicability in biology, especially for tests of randomness of occurrence of certain events. Both the binomial and Poisson distributions are discrete probability distributions. Some other discrete distributions are briefly mentioned in

Section 5.4. The most common continuous probability distribution is the normal frequency distribution, discussed in the following chapter.

5.1 Some simple considerations of probability, random sampling, and hypothesis testing

We shall start this discussion with an example that is not biometrical or biological in the strict sense. We have often found it pedagogically effective to introduce new concepts through situations thoroughly familiar to the student, even if the example is not relevant to the general subject matter of biometry.

Let us betake ourselves to Matchless University, a state institution somewhere between the Appalachians and the Rockies. Looking at its enrollment figures we notice that 4% of the student body comes from abroad and 96% are Americans. In much of our work we shall use proportions rather than percentages as a useful convention. Thus the enrollment consists of 0.04 foreign students and 0.96 American students. The total student body, corresponding to 100%, is therefore represented by the figure 1.0.

If we were to assemble all the students and sample 100 of them at random, we would intuitively expect that, on the average, 4 will be foreign students. The actual outcome might vary. There might not be a single foreign student among the 100 sampled or there may be quite a few more than 4. The ratio of the number of foreign students sampled divided by the total number of students sampled might therefore vary from zero to considerably greater than 0.04. If we increased our sample size to 500 or 1000, it is less likely that the ratio will fluctuate widely around 0.04. The greater the sample taken, the closer the ratio of foreign students sampled to the total students sampled will approach 0.04. In fact, the *probability* of sampling a foreign student is defined as the limit reached by the ratio of foreign students to the total number of students sampled, as sample size keeps increasing. Thus, we may formally summarize the situation by stating that the probability of a student at Matchless University being a foreign student is $p_F = 0.04$. Similarly, the probability of sampling an American student is $q_A = 0.96$. Since $p_F + q_A = 1.0$, $q_A = 1 - p_F$.

Now let us imagine the following experiment: we try to sample a student at random from among the student body at Matchless University. This is not as easy a task as might be imagined. If we wish to do this operation physically, we would have to set up a collection or trapping station somewhere on campus. And to make certain that the sample is truly random with respect to the entire student population, we would have to know the ecology of students on campus very thoroughly. We should try to locate our trap at some station where each student has an equal probability of passing. Few, if any, such places can be found in a university. The student union facilities are likely to be frequented more by independent and foreign students, less by those living in organized houses and dormitories. Fewer foreign students might be found along fraternity row. Clearly we would not wish to place our trap near the International Club or House because our probability of sampling a foreign student would be greatly enhanced. In front of the bursar's window we might sample students paying tuition. But those on scholarships might not be found there. We do not know whether the proportion of scholarships among foreign students is the same or different from that among the American students. Athletic events, political rallies, dances, and the like would all draw a differential spectrum of the student body; indeed, no easy solution seems in sight. The time of sampling is equally important, in the seasonal as well as the diurnal cycle.

Those among the readers who are interested in sampling of organisms from nature will already have perceived parallel problems in their work. If we were to sample only students wearing turbans or saris, their probability of being foreign students would be almost 1. We could no longer speak of a random sample. In the familiar ecosystem of the university these violations of proper sampling procedure are obvious to all of us, but they are not nearly so obvious in real biological instances where we are unfamiliar with the true nature of the environment. How should we proceed to obtain a random sample of leaves from a tree, of insects from a field, or of mutations in a culture? In sampling at random we are attempting to permit the frequencies of various events occurring in nature to be reproduced unalteredly in our records; that is, we hope that on the average the frequencies of these events in our sample will be the same as they are in the natural situation. Another way of saying this is that in a random sample every individual in the population being sampled has an equal probability of being included in the sample.

We might go about obtaining a random sample by using records representing the student body, such as the student directory, selecting a page from it at random and a name at random from the page. Or we could assign an arbitrary number to each student, write each on a chip or disk, put these in a large container, stir well, and then pull out a number.

Imagine now that we sample a single student physically by the trapping method, after carefully planning the placement of the trap in such a way as hopefully to make sampling random. The single student we have drawn from the student body of Matchless University turns out to be a foreign student. What can we conclude from this? By chance alone, this result would happen 0.04 or 4% of the time, or not very frequently. The assumption that we have sampled at random should probably be rejected, since if we accept the hypothesis of random sampling the outcome of the experiment is improbable. Please note that we said *improbable*, not *impossible*. It is obvious that we could have chanced upon a foreign student as the very first one to be sampled. However, it is not very likely. The probability is 0.96 that a single student sampled would be an American student. If we could be certain that our sampling method was random (as when drawing student numbers out of a container), we would of course have to decide that an improbable event has occurred. The decisions of this paragraph are all based on our definite knowledge that the proportion of foreign students at Matchless University is indeed 0.04. If we were uncertain about this, we would be led to assume a higher proportion of foreign students as a consequence of the outcome of our sampling experiment.

We shall now extend our experiment and sample two students rather than just one. What are the possible outcomes of this sampling experiment? Clearly we could obtain two American students, one American student and one foreign student, or two foreign students. What would be the expected probabilities of these outcomes? We know the expected outcomes for sampling one foreign student, but what would be the probability of sampling a second foreign student? Now the nature of the sampling procedure becomes quite important. We may sample with or without *replacement;* that is, we may return the first student sampled to the population or may keep him out of the pool of the individuals to be sampled. If we do not replace the first individual sampled, the probability of sampling a foreign student will no longer be exactly 0.04. This is easily seen. Let us assume that Matchless University has 10,000 students. Then, since there are 4% of foreign students, there must be 400 foreign students at the university. After sampling a foreign student first, this number is reduced to 399 out of 9999 students. Consequently, the probability of sampling a foreign student now becomes $399/9999 = 0.0399$, a slightly lower probability than the value of 0.04 for sampling the first foreign student. If, on the other hand, we return the original foreign student to the student population and make certain that the population is thoroughly randomized before being sampled again (that is, give him a chance to lose himself among the campus crowd or, in the case of drawing student numbers out of a container, mixing up the disks with the numbers on them), the probability of sampling a second student is the same as before, 0.04. In fact, if we keep on replacing the sampled individuals in the original population, we can sample from it as though it were an infinitely sized population.

Biological populations are, of course, finite, but they are frequently so large that for purposes of sampling experiments we can consider them effectively infinite whether we replace sampled individuals or not. After all, even in this relatively small population of 10,000 students, the probability of sampling a second foreign student (without replacement) is only minutely different from 0.04. For the rest of this section we shall consider sampling to be with replacement, so that the probability level of obtaining a foreign student does not change. The distribution resulting from sampling without replacement is discussed in Section 5.4.

There is a second potential source of difficulty in this design. We not only have to assume that the probability of sampling a second foreign student is equal to that of the first, but also that it is *independent* of it. By independence of events we mean that if one event occurs with a certain probability, a second such event will occur with the same probability, not affected by whether the first event has already occurred or not. In the case of the students, having sampled one foreign student, is it more or less likely that a second student sampled in the same manner will also be a foreign student? Independence of the events may depend on where we sample the students or on the method of sampling. If we sampled students on campus it is quite likely that the events are not independent; that is, having sampled one foreign student, the probability that the second student is foreign is increased, since foreign students tend to congregate. Thus, at Matchless University the probability that a student walking with a foreign student is also foreign would be greater than 0.04.

If the sampling probabilities for the second student are independent of the type of student sampled first, we can compute the probabilities of the outcomes simply as the product of the independent probabilities. Thus the probability of obtaining two foreign students is

$$p_F \times p_F = p_F^2 = (0.04) \times (0.04) = 0.0016$$

and the probability of obtaining two American students is

$$q_A \times q_A = q_A^2 = (0.96) \times (0.96) = 0.9216$$

According to this rule the probability of obtaining one of each type of student in the sample should be the product 0.04×0.96. However, it is in fact twice that probability. It is easy to see why. There is only one way of obtaining two Americans, namely by sampling first one American and then again an-an American. Similarly, there is only one way to sample two foreign students. However, sampling one of each type student can be done by first sampling other American student followed by a foreign student or by first sampling a foreign student followed by an American student. Thus the probability is

$$2p_Fq_A = 2 \times (0.04) \times (0.96) = 0.0768$$

If we conduct such an experiment and obtain a sample of two foreign students we would be led to the following conclusions. Only 0.0016 of the samples (16/100th of 1% or 16 out of 10,000 cases) would be expected to consist of two foreign students. It is quite improbable to obtain such a result by chance alone. Given $p_F = 0.04$ as a fact, we would therefore suspect that sampling was not random or that the events were not independent (or that both assumptions—random sampling and independence of events—were incorrect).

Random sampling is sometimes confused with randomness in nature. The former is the faithful representation in the sample of the distribution of the events in nature, the latter is the independence of the events in nature. The first of these generally is or should be under the control of the experimenter and is related to the strategy of good sampling. The second generally describes an innate property of the objects being sampled and thus is of greater biological interest. The confusion between random sampling and independence of events arises because lack of either can yield observed frequencies of events differing from expectation. We have already seen how lack of independence in samples of foreign students can be interpreted from both points of view in our illustrative example from Matchless University.

The above account of probability is adequate for our present purposes but far too sketchy to convey an understanding of the field. Readers interested in extending their knowledge of the subject are referred to the simple accounts in Alder and Roessler (1964) and Gnedenko and Khinchin (1961).

5.2 The binomial distribution

We have seen how expected frequencies for samples of two students can be obtained. If we were to sample three students, the possible outcomes are three foreign students, two foreign and one American, one foreign and two American, or three American students. Samples of three foreign or three American students can again be obtained in only one way, and their probabilities are p_F^3 and q_A^3, respectively. However, in samples of three there are three ways of obtaining two students of one kind and one student of the other. As before, if A stands for American and F stands for foreign, then the sampling sequence could be AFF, FAF, and FFA for two foreign students and one American. Thus the probability of this outcome will be $3p_F^2q_A$. Similarly, the probability for two Americans and one foreign student is $3p_Fq_A^2$.

A convenient way to summarize these results is by means of the binomial expansion, which is applicable to samples of any size from populations in which objects occur only in two classes—students who may be foreign or American, individuals who may be dead or alive, male or female, black or white, rough or smooth, and so forth. This is accomplished by expanding the binomial term $(p + q)^k$, where k equals sample size, p equals the probability of occurrence of the first class, and q equals the probability of occurrence of the second class. By definition, $p + q = 1$; hence q is a function of p: $q = 1 - p$. We shall expand the expression for samples of k from 1 to 3:

For samples of 1 $(p + q)^1 = p + q$
For samples of 2 $(p + q)^2 = p^2 + 2pq + q^2$
For samples of 3 $(p + q)^3 = p^3 + 3p^2q + 3pq^2 + q^3$

It will be seen that these expressions yield the same outcomes discussed previously. The coefficients (the numbers before the powers of p and q) express the number of ways a particular outcome is obtained. An easy method for evaluating the coefficients of the expanded terms of the binomial expression is through the use of Pascal's triangle, which is shown below.

k	
1	1 1
2	1 2 1
3	1 3 3 1
4	1 4 6 4 1
5	1 5 10 10 5 1
. . .	. . . .

Pascal's triangle provides the coefficients of the binomial expression—that is, the number of possible outcomes of the various combinations of events. For $k = 1$ the coefficients are 1, 1 respectively; for the second line ($k = 2$), write 1 at the left-hand margin of the line. The 2 in the middle of this line is the sum of the values to the left and right of it in the line above. The line is concluded with a 1. Similarly, the values at the beginning and end of the third line are 1, the other numbers are sums of the values to their left and right in the line above; thus 3 is the sum of 1 and 2. This principle continues for every line. You can work out the coefficients for any size sample in this manner. The line for $k = 6$ would consist of the following coefficients: 1, 6, 15, 20, 15, 6, 1. The p and q values receive powers in a consistent pattern, which should be easy to imitate for any value of k. We give it here for $k = 4$:

$$p^4q^0 + p^3q^1 + p^2q^2 + p^1q^3 + p^0q^4$$

The power of p decreases from 4 to 0 (k to 0 in the general case) as the power of q increases from 0 to 4 (0 to k in the general case). Since any value to the power 0 is 1 and any term to the power 1 is simply itself, we can simplify this expression as shown below and at the same time provide it with the coefficients from Pascal's triangle for the case $k = 4$:

$$p^4 + 4p^3q + 6p^2q^2 + 4pq^3 + q^4$$

Thus we are able to write down almost by inspection the expansion of the binomial to any reasonable power. When the value of k is considerable (> 15), Pascal's triangle is tedious to elaborate (unless you have already worked it out once, in which case you should save it). For such cases the coefficients have been tabled in Miller (1954). Let us now practice our newly learned ability to expand the binomial.

Suppose we have a population of insects, exactly 40% of which are infected with a given virus X. If we take samples of $k = 5$ insects each and examine each insect separately for presence of virus, what distribution of samples could we expect if the probability of infection of each insect in a sample were independent from that of other insects in the sample? In this case $p = 0.4$, the proportion infected, and $q = 0.6$, the proportion not infected. It is assumed that the population is so large that the question of whether sampling is with or without replacement is irrelevant for practical purposes. The expected frequencies would be the expansion of the binomial:

$$(p + q)^k = (0.4 + 0.6)^5$$

With the aid of Pascal's triangle this expansion is

$$p^5 + 5p^4q + 10p^3q^2 + 10p^2q^3 + 5pq^4 + q^5$$

or

$$(0.4)^5 + 5(0.4)^4(0.6) + 10(0.4)^3(0.6)^2 + 10(0.4)^2(0.6)^3 + 5(0.4)(0.6)^4 + (0.6)^5$$

representing the expected proportions of samples of five infected insects, four infected and one noninfected insects, three infected and two noninfected insects, and so on. The reader has probably realized by now that the terms of the binomial expansion actually yield a type of frequency distribution for these different outcomes. Associated with each outcome, such as "five infected insects," there is a probability of occurrence—in this case $(0.04)^5 = 0.01024$. This is a theoretical frequency distribution or *probability distribution* of events that can occur in two classes. It describes the expected distribution of outcomes in random samples of five insects, 40% of which are infected. The probability distribution described here is known as the *binomial distribution,* and the binomial expansion yields the expected frequencies of the classes of the binomial distribution.

A convenient layout for presentation and also computation of a binomial distribution is shown in Table 5.1. The first column lists the number of infected insects per sample, the second column shows decreasing powers of p from p^5 to p^0, and the third column shows increasing powers of q from q^0 to q^5. The binomial coefficients from Pascal's triangle are shown in column (4). The *relative expected frequencies,* which are the probabilities of the various outcomes, are shown in column (5). We label such expected frequencies $\hat{f}_{\mathrm{rel}}$. They are simply the product of columns (2), (3), and (4). Their sum is equal to 1.0, since the events listed in column (1) exhaust the possible outcomes. We see from column (5) in Table 5.1 that only about 1% of samples are expected to consist of 5 infected insects, and 25.9% are expected to contain 1 infected and 4 noninfected insects. We shall test whether these predictions hold in an actual experiment.

Experiment 5.1. Simulate the case of the infected insects by using a table of random numbers such as Table **O**. These are randomly chosen one-digit numbers in which each digit 0 through 9 has an equal probability of appearing. The numbers are grouped in blocks of twenty for convenience. Since there is an equal probability for any one digit to appear, you can let any four digits (say 0, 1, 2, 3) stand for the infected insects and the remaining digits (4, 5, 6, 7, 8, 9) stand for the noninfected insects. The probability of any one digit selected from the table representing an infected insect (that is, being a 0, 1, 2, or 3) is therefore 40% or 0.4, since these are four of the ten possible digits. Also, successive digits are assumed to be independent of the values of previous digits. Thus the assumptions of the binomial distribution should be met in this experiment. Enter the table of random numbers at an arbitrary point (not always at the beginning!) and look at successive groups of five digits, noting in each group how many of the digits were 0, 1, 2, or 3. Take as many groups of five as you can find time to do, but no less than 100 groups.

Column (7) in Table 5.1 shows the results of one such experiment during one year by a biometry class. A total of 2423 samples of five numbers were obtained from the table of random numbers and the distribution of the four digits simulating the percentage of infection is shown in this column. The observed frequencies are labeled f. In order to calculate the expected frequencies for this actual example we multiplied the relative expected frequencies $\hat{f}_{\mathrm{rel}}$ of column (5) times $n = 2423$, the number of samples taken. This

TABLE 5.1

Expected frequencies of infected insects in samples of 5 insects sampled from an infinitely large population with an assumed infection rate of 40 percent.

(1) *Number of infected insects per sample* Y	(2) *Powers of* $p = 0.4$	(3) *Powers of* $q = 0.6$	(4) *Binomial coefficients*	(5) *Relative expected frequencies* $\hat{f}_{rel}$	(6) *Absolute expected frequencies* $\hat{f}$	(7) *Observed frequencies* f
5	0.01024	1.00000	1	0.01024	24.8	29
4	0.02560	.60000	5	0.07680	186.1	197
3	0.06400	.36000	10	0.23040	558.3	535
2	0.16000	.21600	10	0.34560	837.4	817
1	0.40000	.12960	5	0.25920	628.0	643
0	1.00000	.07776	1	0.07776	188.4	202
			$\Sigma\hat{f}$ or Σf ($= n$)	1.00000	2423.0	2423
			ΣY	2.00000	4846.1	4815
			Mean	2.00000	2.00004	1.98721
			Standard deviation	1.09545	1.09543	1.11934

results in *absolute expected frequencies*, labeled $\hat{f}$, shown in column (6). When we compare the observed frequencies in column (7) with the expected frequencies in column (6) we note general agreement between the two columns of figures. The two distributions are also illustrated in Figure 5.1. If the observed frequencies would not fit expected frequencies, we might believe that the lack of fit was due to chance alone. Or we might be led to reject one or more of the following hypotheses: one, that the true proportion of digits 0, 1, 2, and 3 is 0.4 (rejection of this hypothesis would normally not be reasonable, for we may rely on the fact that the proportion of digits

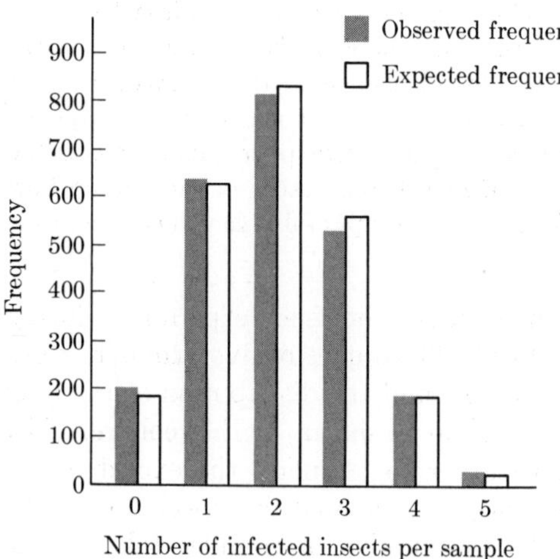

FIGURE 5.1 Bar diagram of observed and expected frequencies given in Table 5.1.

0, 1, 2, and 3 in a table of random numbers *is* 0.4 or very close to it); two, that sampling was at random; and three, that events are independent.

These statements can be reinterpreted in terms of the original infection model with which we started this discussion. If, instead of a sampling experiment of digits by a biometry class, this had been a real sampling experiment of insects, we would conclude that the insects had indeed been randomly sampled and that we had no evidence to reject the hypothesis that the proportion of infected insects was 40%. If the observed frequencies had not fitted the expected frequencies, the lack of fit might be attributed to chance, or to the conclusion that the true proportion of infection is not 0.4, or we would have to reject one or both of the following assumptions: (1) that sampling was at random, and (2) that the occurrence of infected insects in these samples was independent.

Experiment 5.1 was designed to yield random samples and independent events. How could we simulate a sampling procedure in which the occurrences of the digits 0, 1, 2, and 3 were not independent? We could, for example, instruct the sampler to sample as indicated previously, but every time he found a 3, to search through the succeeding digits until he found another one of the four digits standing for infected individuals and to incorporate this in the sample. Thus, once a 3 was found, the probability would be 1.0 that another one of the indicated digits would be included in the sample. After repeated samples, this would result in higher frequencies of classes of two or more indicated digits and in lower frequencies than expected (on the basis of the binomial distribution) of classes of one event. A variety of such different sampling schemes could be devised. It should be quite clear to the reader that the probability of the second event occurring would be different from that of the first and dependent on it.

How would we interpret a large departure of the observed frequencies from expectation in another example? We have not as yet learned techniques for testing whether observed frequencies differ from those expected by more than can be attributed to chance alone. This will be taken up in Chapter 16. Assume that such a test has been carried out and that it has shown us that our observed frequencies are significantly different from expectation. Two main types of departure from expectation are likely: (1) *clumping* and (2) *repulsion*, shown in fictitious examples in Table 5.2. In actual examples we would have no a priori notions about the magnitude of p, the probability of one of the two possible outcomes. In such cases it is customary to obtain p from the observed sample and to calculate the expected frequencies, using the sample p. This would mean that the hypothesis that p is a given value cannot be tested, since by design the expected frequencies will have the same p value as the observed frequencies. Therefore the hypotheses tested are whether the samples are random and the events independent.

The clumped frequencies in Table 5.2 have an excess of observations at the tails of the frequency distribution and consequently a shortage of observations at the center. Such a distribution is also called *contagious*. Remember that the total number of items must be the same in both observed and expected frequencies in order to make them comparable. In the repulsed frequency distribution there are more observations than expected at the center of the distribution and fewer at the tails. These discrepancies are easiest seen in columns (4) and (6) of Table 5.2, where the deviations of observed from

TABLE 5.2

Artificial distributions to illustrate clumping and repulsion. Expected frequencies from Table 5.1.

(1) Number of infected insects per sample Y	*(2)* Absolute expected frequencies $\hat{f}$	*(3)* Clumped (contagious) frequencies f	*(4)* Deviation from expectation	*(5)* Repulsed frequencies f	*(6)* Deviation from expectation
5	24.8	47	+	14	−
4	186.1	227	+	157	−
3	558.3	558	0	548	−
2	837.4	663	−	943	+
1	628.0	703	+	618	−
0	188.4	225	+	143	−
$\Sigma\hat{f}$ or n	2423.0	2423		2423.0	
ΣY	4846.1	4846		4846	
Mean	2.00004	2.00000		2.00000	
Standard deviation	1.09543	1.20074		1.01435	

expected frequencies are shown as plus or minus signs. (These two types of distributions are also called overdispersed and underdispersed, but there has been some confusion in the literature about the meaning of these terms and we shall not use them here.)

What do these phenomena imply? In the clumped frequencies more samples were entirely infected (or largely infected) and similarly more samples were entirely noninfected (or largely noninfected) than you would expect if probabilities of infection were independent. This could be due to poor sampling design. If, for example, the investigator in collecting his samples of five insects always tended to pick out like ones—that is, infected ones or noninfected ones—then such a result would likely appear. But if the sampling design is sound, the results become more interesting. Clumping would then mean that the samples of five are in some way related, so that if one insect is infected, others in the same sample are more likely to be infected. This could be true if they come from adjacent locations in a situation in which neighbors are easily infected. Or they could be siblings jointly exposed to a source of infection. Or possibly the infection might spread among members of a sample between the time that the insects have been sampled and the time they are examined.

The opposite phenomenon, repulsion, is more difficult to interpret biologically. There are fewer homogeneous groups and more mixed groups in such a distribution. This involves the idea of a compensatory phenomenon: if some of the insects in a sample are infected, the others in the sample are less likely to be. If the infected insects in the sample could in some way transmit immunity to their associates in the sample, such a situation could arise logically, but it is biologically improbable. A more reasonable interpretation of such a finding is that for each sampling unit there were only a limited number of pathogens available and that once several of the insects have become infected, the others go free of infection simply because there is no more infectious agent. This is an unlikely situation in microbial infections, but in situations in which a limited number of parasites enter the body of the host, repulsion may be more reasonable.

From the expected and observed frequencies in Table 5.1, we may calculate the mean and standard deviation of the number of infected insects per sample. These values are given at the bottom of columns (5), (6), and (7) in Table 5.1. We note that the means and standard deviations in columns (5) and (6) are almost identical and differ only trivially because of rounding errors. Column (7), being a sample from a population whose parameters are the same as those of the expected frequency distribution in columns (5) or (6), differs somewhat. The mean is slightly smaller and the standard deviation is slightly greater than in the expected frequencies. If we wish to know the mean and standard deviation of expected binomial frequency distributions we need not go through the computations shown in Table 5.1. The mean and standard deviation of a binomial frequency distribution are, respectively,

$$\mu = kp, \qquad \sigma = \sqrt{kpq}$$

Substituting the values $k = 5$, $p = 0.4$, and $q = 0.6$ of the above example, we obtain $\mu = 2.0$ and $\sigma = 1.09545$, which are identical to the values computed from column (5) in Table 5.1. Note that we use the Greek parametric notation here because μ and σ are parameters of an expected frequency distribution, not sample statistics, as are the mean and standard deviation in column (7). The proportions p and q are parametric values also and strictly speaking should be distinguished from sample proportions. In fact, in later chapters we resort to $\hat{p}$ and $\hat{q}$ for parametric proportions (rather than π which conventionally is used as the ratio of the circumference to the diameter of a circle). Here, however, we prefer to keep our notation simple. It is interesting to look at the standard deviations of the clumped and repulsed frequency distributions of Table 5.2. We note that the clumped distribution has a standard deviation greater than expected, and that of the repulsed one is less than expected. Comparison of sample standard deviations with their expected values is a useful measure of dispersion in such instances. If we wish to express our variable as a proportion rather than as a count—that is, to indicate mean incidence of infection in the insects as 0.4, rather than as 2 per sample of 5—we can use other formulas for the mean and standard deviation in a binomial distribution:

$$\mu = p, \qquad \sigma = \sqrt{pq/k}$$

We shall now employ the binomial distribution to solve a biological problem. On the basis of our knowledge of the cytology and biology of species A, we expect the sex ratio among its offspring to be 1:1. The study of a litter in nature reveals that of 17 offspring 14 were females and 3 were males. What conclusions can we draw from this evidence? Assuming that $p_{♀}$ (the probability of being a female offspring) $= 0.5$ and that this probability is independent among the members of the sample, the pertinent probability distribution is the binomial for sample size $k = 17$. Expanding the binomial to the power 17 is a formidable task which, as we shall see, fortunately need not be done in its entirety. However, we must have the binomial coefficients, which can be obtained either from an expansion of Pascal's triangle (fairly tedious unless once obtained and stored for future use) or by working out the expected frequencies for any given class of Y from the general formula for any term of the binomial distribution

$$C(k, Y)p^{k-Y}q^{Y} \tag{5.1}$$

The expression $C(k, Y)$ stands for the number of combinations that can be formed from k items taken Y at a time. This can be evaluated as $k!/[Y!(k - Y)!]$, where ! means "factorial." In mathematics k factorial is the product of all the integers from 1 up to and including k. Thus: $5! = 1 \times 2 \times 3 \times 4 \times 5 = 120$. By convention, $0! = 1$. In working out fractions containing factorials, note that any factorial will always cancel against a higher factorial. Thus $5!/3! = (5 \times 4 \times 3!)/3! = 5 \times 4$. For example, the binomial coefficient for the expected frequency of samples of 5 items containing 2 infected insects is $C(5, 2) = 5!/2!3! = 5 \times 4/2 = 10$.

The setup of the example is shown in Table 5.3. The first probability in column (3) is $p_{♀}^{17}$, which can be computed directly on a desk calculator with transfer multiplication or obtained by looking up the antilogarithm of 17 times the logarithm of $p_{♀}$. Decreasing powers of $p_{♀}$ may be similarly computed. Successive powers of $q_{♂}$ are computed (from power 0 to power 4). Note that for the purposes of our problem we need not proceed beyond the term for 13 females and 4 males. Calculating the relative expected frequencies in column (6), we note that the probability of 14 females and 3 males is 0.005,188,40, a very small value. If we add to this value all "worse" outcomes—that is, all outcomes that are even more unlikely than 14 females and 3 males on the assumption of a 1:1 hypothesis—we obtain a probability of 0.006,363,42, still a very small value. It is a common practice in statistics

TABLE 5.3

Some expected frequencies of males and females for samples of 17 offspring on the assumption that the sex ratio is 1:1 [$p_♀ = 0.5$, $q_♂ = 0.5$; $(p_♀ + q_♂)^k = (0.5 + 0.5)^{17}$].

(1) ♀♀	(2) ♂♂	(3) $p_♀$	(4) $q_♂$	(5) *Binomial coefficients*	(6) *Relative expected frequencies* $\hat{f}_{rel}$	
17	—	0.000,007,63	1	1	0.000,007,63	0.006,363,42 (sum of rows 17–14)
16	1	0.000,015,26	.5	17	0.000,129,71	
15	2	0.000,030,52	.25	136	0.001,037,68	
14	3	0.000,061,04	.125	680	0.005,188,40	
13	4	0.000,122,07	.0625	2380	0.018,157,91	

to calculate the probability of a deviation as large or larger than a given value.

On the basis of these findings one or more of the following assumptions is unlikely: (1) that the true sex ratio in species A is 1:1; (2) that we have sampled at random in the sense of obtaining an unbiased sample; or (3) that the sexes of the offspring are independent of one another. Lack of independence of events may mean that although the average sex ratio is 1:1, the individual sibships or litters are largely unisexual, so that the offspring from a given mating would tend to be all (or largely) females or all (or largely) males. To confirm this hypothesis we would need to have more samples and then examine the distribution of samples for clumping, which would indicate a tendency for unisexual sibships.

We must be very precise about the questions we ask of our data. There are really two questions we could ask about the sex ratio. One, are the sexes unequal in frequency so that females will appear more often than males? Two, are the sexes unequal in frequency? We may be only concerned with the first of these questions, since we know from past experience that in this particular group of organisms the males are never more frequent than females. In such a case the reasoning followed above is appropriate. However, if we know very little about this group of organisms and if our question is simply whether the sexes among the offspring are unequal in frequency, then we have to consider both tails of the binomial frequency distribution; departures from the 1:1 ratio could occur in either direction. We should then not only consider the probabilities of samples with 14 females and 3 males (and all worse cases) but also the probability of samples of 14 males and 3 females (and all worse cases in that direction). Since this probability distribution is symmetrical (because $p_♀ = q_♂ = 0.5$), we simply double the cumulative probability of 0.006,363,42 obtained previously, which results in 0.012,726,84. This new value is still very small, making it quite unlikely that the true sex ratio is 1:1. This is your first experience with one of the most important applications of statistics—hypothesis testing. A formal introduction to this field will be deferred until Section 7.8. We may simply point out here that the two approaches followed above are known appropriately as *one-tailed tests* and *two-tailed tests,* respectively. Students sometimes have difficulty knowing which of the two tests to apply. In future examples we shall try to point out in each case why a one-tailed or a two-tailed test is being used.

We have said that a tendency for unisexual sibships would result in a clumped distribution of observed frequencies. An actual case of this nature is a classic in the literature, the sex ratio data obtained by Geissler (1889) from hospital records in Saxony. Table 5.4 reproduces sex ratios of 6115

TABLE 5.4

Sex ratios in 6115 sibships of twelve in Saxony.

(1)	*(2)*	*(3)*	*(4)*	*(5)*	*(6)*	*(7)*	*(8)*	*(9)*
♂♂	♀♀	$p_{♂}$	$q_{♀}$	*Binomial coefficients*	*Relative expected frequencies* $\hat{f}_{rel}$	*Absolute expected frequencies* $\hat{f}$	*Observed frequencies* f	*Deviation from expectation* $(f - \hat{f})$
12	—	0.000,384	1	1	0.000,384	2.3	7	+
11	1	0.000,739	0.480,785	12	0.004 264	26.1	45	+
10	2	0.001,424	0.231,154	66	0.021,725	132.8	181	+
9	3	0.002,742	0.111,135	220	0.067,041	410.0	478	+
8	4	0.005,282	0.053,432	495	0.139,703	854.3	829	−
7	5	0.010,173	0.025,689	792	0.206,973	1265.6	1112	−
6	6	0.019,592	0.012,351	924	0.223,590	1367.3	1343	−
5	7	0.037,734	0.005,938	792	0.177,459	1085.2	1033	−
4	8	0.072,676	0.002,855	495	0.102,708	628.1	670	+
3	9	0.139,972	0.001,373	220	0.042,280	258.5	286	+
2	10	0.269,584	0.000,660	66	0.011,743	71.8	104	+
1	11	0.519,215	0.000,317	12	0.001,975	12.1	24	+
—	12	1	0.000,153	1	0.000,153	0.9	3	+
Total					0.999,998	6115.0	6115	
			$\bar{Y}$	6.23058	s^2	3.48985		

Source: Geissler (1889).

sibships of 12 children each from the more extensive study by Geissler. All columns of the table should by now be familiar. The expected frequencies were not calculated on the basis of a 1:1 hypothesis, since it is known that in human populations the sex ratio at birth is not 1:1. As the sex ratio varies in different human populations, the best estimate of it for the population in Saxony was simply obtained using the mean proportion of males in these data. This can be obtained by calculating the average number of males per sibship ($\bar{Y} = 6.23058$) for the 6,115 sibships and converting this into a proportion. This value turns out to be 0.519,215. Consequently the proportion of females = 0.480,785. In the deviations of the observed frequencies from the absolute expected frequencies shown in column (9) of Table 5.4, we notice considerable clumping. There are many more instances of families with all male or all female children (or nearly so) than independent probabilities would indicate. The genetic basis for this is not clear, but it is evident that there are some families which "run to girls" and similarly those which "run to boys." Evidence of clumping can also be seen from the fact that s^2 is much larger than we would expect on the basis of the binomial distribution [$\sigma^2 = kpq = 12(0.519215)0.480785 = 2.99557$].

There is a distinct contrast between the data in Table 5.1 and those in Table 5.4. In the insect infection data of Table 5.1 we had a hypothetical proportion of infection based on outside knowledge. In the sex ratio data of Table 5.4 we had no such knowledge; we used an *empirical value of p obtained from the data,* rather than a *hypothetical value external to the data.* This is a distinction whose importance will become apparent later. In the sex ratio data of Table 5.3, as in much work in Mendelian genetics, a hypothetical value of p is used.

To round out this section on the binomial distribution, we show you an efficient method for calculating expected binomial frequencies on a calculating machine. This method is given in Box 5.1 and the instructions are self-explanatory. Since this is a step-by-step procedure in which the correctness of a given result depends on the correctness of the previous step, careful calculation with a sufficient number of decimal places is especially important. On calculating machines with a transfer device from the long dials to the multiplier dials, this operation can be especially efficiently carried out. There are tables of relative expected frequencies of the binomial. Table 37 of the Biometrika Tables (Pearson and Hartley, 1958) is useful for some probabilities and sample sizes, and the Tables of the Binomial Probability Distribution published by the U.S. National Bureau of Standards, provide a more extensive range of sample sizes $k \leq 49$. A computer program for expected binomial frequencies is given in Appendix A3.3.

5.3 The Poisson distribution

In the typical application of the binomial we had relatively small samples (2 students, 5 insects, 17 offspring, 12 siblings) in which two alternative states occurred at varying frequencies (American and foreign, infected and noninfected, male and female). Quite frequently, however, we study cases in which sample size k is very large. These would present a considerable computational problem. We have seen that the expansion of the binomial $(p + q)^k$ is quite tiresome when k is large. Suppose you had to expand the expression $(0.001 + 0.999)^{1000}$. Note that not only is the sample size large in this expression, but one of the events (represented by probability q) is very much more

BOX 5.1

Machine method for computing expected binomial frequencies. Adapted from Bliss and Calhoun (1954).

Insect data of Table 5.1: p (probability of being infected) $= 0.4$; q (probability of being noninfected) $= 0.6$; k (sample size) $= 5$.

(1) Number of infected insects per sample of 5 insects	*(2)* Number of noninfected insects per sample of 5 insects	*(3)* Relative expected frequencies $\hat{f}_{rel}$
5	0	0.01024
4	1	0.07680
3	2	0.23040
2	3	0.34560
1	4	0.25920
0	5	0.07776

Computational steps

□□ **1.** Calculate a quotient $Q = q/p$ to five significant decimal places. $Q = 0.6/0.4 = 1.50000$.

2. Calculate p^k using logarithms or for small k use transfer multiplication feature if available.

$$\log p^k = k \log p = 5 \log(0.4)$$
$$= 5(9.60206 - 10)$$
$$= 48.01030 - 50$$
$$= 8.01030 - 10$$
$$\text{antilog}\,(8.01030 - 10) = 0.01024$$

The first relative expected frequency, $\hat{f}_{rel_1} = 0.01024$.

2A. When the data are set up so that the first $\hat{f}$-value required is q^k rather than p^k, the quotient is calculated as $Q = p/q$, and $\hat{f}_{rel_1}$ is computed as q^k similar to step **2** above.

3. Multiply $\hat{f}_{rel_1}$ by n (number of samples in the study) if you want absolute expected frequencies and enter the product as $\hat{f}_1$, the first value in the $\hat{f}$ column. (In our case, since we want relative expected frequencies, we multiply by 1 and enter 0.01024 in the $\hat{f}_{rel}$ column.)

4. The subsequent expected frequencies $\hat{f}_i$ from $\hat{f}_2$ to $\hat{f}_{k+1}$ can be computed as

$$f_i = \hat{f}_{i-1} Q \left(\frac{k - i + 2}{i - 1}\right)$$

BOX 5.1 continued

Thus, in this case

$\hat{f}_2 = \hat{f}_1 Q(5/1) = (0.01024)(1.5)(5) \quad = 0.07680$

$\hat{f}_3 = \hat{f}_2 Q(4/2) = (0.07680)(1.5)(2) \quad = 0.23040$

$\hat{f}_4 = \hat{f}_3 Q(3/3) = (0.23040)(1.5)(1) \quad = 0.34560$

$\hat{f}_5 = \hat{f}_4 Q(2/4) = (0.34560)(1.5)(0.5) = 0.25920$

$\hat{f}_6 = \hat{f}_5 Q(1/5) = (0.25920)(1.5)(0.2) = 0.07776$

Note the regular progression of the third factor from $k/1$ to $1/k$, the numerator and denominator being decreased and increased, respectively, by unity at each step. Each expected frequency is computed from the previous one, so it is especially important to be accurate in your figures and to carry a sufficient number of decimal places. On a machine with a transfer multiplication device the method of this box is especially simple to carry out.

5. As a final check, sum the expected frequencies to obtain 1 or n. □□

frequent than the other (represented by probability p). Expanding this binomial by the methods you have learned so far requires a very accurate table of logarithms (logarithms to 10 decimal places were used in Table 5.8). Yet such expressions are often of great biological importance and are commonly encountered. In such cases we are generally interested in one tail of the distribution only. This is the tail represented by the terms

$$p^0q^k,\ C(k, 1)p^1q^{k-1},\ C(k, 2)p^2q^{k-2},\ C(k, 3)p^3q^{k-3}, \ldots$$

The first term represents no rare events and k frequent events in a sample of k events, the second term represents one rare event and $k - 1$ frequent events, the third term 2 rare events and $k - 2$ frequent events, and so forth. The expressions of the form $C(k, i)$ are the binomial coefficients, represented by the combinatorial terms discussed in the previous section. Although this expression would permit the computation of the desired tail of the curve, it would still be a very complicated procedure in view of the magnitude of k. The reader might try computing one or more of these terms for $(0.001 + 0.999)^{1000}$ to convince himself of this. Fortunately, it is much easier to compute another distribution, the Poisson distribution, which closely approximates the desired results.

The *Poisson distribution* is also a discrete frequency distribution of the number of times a rare event occurs. But, in contrast to the binomial distribution, the number of times that an event does not occur is infinitely large. For purposes of our treatment here, a Poisson variable will be studied in either a spatial or temporal sample. An example of the first would be the number of moss plants in a sampling quadrat on a hillside or the number of parasites on an individual host; an example of a temporal sample is the number of mutations occurring in a genetic strain in the time interval of one month or the reported cases of influenza in one town during one week. The Poisson variable, Y, will be the number of events per sample. It can assume discrete values from 0 on up. To be distributed in Poisson fashion the variable must have two properties. (1) Its mean must be small relative to the maximum possible number of events per sampling unit. Thus the event should be "rare." For example, a quadrat in which moss plants are counted must be large enough that a substantial number of moss plants could occur there physically if the biological conditions were such as to favor the development of numerous moss plants in the quadrat. A quadrat of 1-cm square would be far too small for mosses to be distributed in Poisson fashion. Similarly, a time span of 1 minute would be unrealistic for reporting new influenza cases in a town, but within 1 week a great many such cases could occur. (2) An occurrence of the event must be independent of prior occurrences within the sampling unit. Thus, the presence of one moss plant in a quadrat must not enhance or diminish the probability of other moss plants developing in the quadrat. Similarly, the fact that one influenza case has been reported must not affect the probability of reporting subsequent influenza cases. Events that meet these conditions ("rare and random events") should be distributed in Poisson fashion.

The purpose of fitting a Poisson distribution to numbers of rare events in nature is to test whether the events occur independently with respect to each other. If they do, they will follow the Poisson distribution. If the occurrence of one event enhances the probability of a second such event, we obtain a clumped or contagious distribution. If the occurrence of one event impedes that of a second such event in the sampling unit, we obtain a repulsed or spatially uniform distribution. The Poisson can be used as a test for randomness or independence of distribution not only spatially but also in time, as some examples below will show.

The distribution is named after the French mathematician Poisson, who described it in 1837. It is an infinite series whose terms add to one (as must be true for any probability distribution). The series can be represented as

$$\frac{1}{e^{\mu}}, \frac{\mu}{1!e^{\mu}}, \frac{\mu^2}{2!e^{\mu}}, \frac{\mu^3}{3!e^{\mu}}, \frac{\mu^4}{4!e^{\mu}}, \ldots, \frac{\mu^r}{r!e^{\mu}}, \ldots \tag{5.2}$$

which are the relative expected frequencies corresponding to the following counts of the rare event Y:

$$0, \quad 1, \quad 2, \quad 3, \quad 4, \quad \ldots, \quad r, \quad \ldots$$

Thus, the first of these terms represents the relative expected frequency of samples containing no rare event; the second term, one rare event; the third term, two rare events; and so on. The denominator of each term contains e^{μ}, where e is the base of the natural or Naperian logarithm, a constant whose value, accurate to 5 decimal places, is 2.71828. We recognize μ as the parametric mean of the distribution; it is a constant for any given problem. The exclamation mark after the coefficient in the denominator means "factorial" and has been explained in the previous section.

One way to learn more about the Poisson distribution is to apply it to an actual case. At the top of Box 5.2 is a well-known result from the early statistical literature. It tests the distribution of yeast cells in 400 squares of a hemacytometer, a counting chamber such as is used in making counts of blood cells and other microscopic structures suspended in liquid. Column (1) lists the number of yeast cells observed in each hemacytometer square and column (2) gives the observed frequency—the number of squares containing a given number of yeast cells. We note that 75 squares contained no yeast cells, but that most squares held either 1 or 2 cells. Only 17 squares contained 5 or more yeast cells.

Why would we expect this frequency distribution to be distributed in Poisson fashion? We have here a relatively rare event, the frequency of yeast cells per hemacytometer square, the mean of which has been calculated and found to be 1.8. That is, on the average there are 1.8 cells per square. Relative to the amount of space provided in each square and the number of cells that could have come to rest in any one square, the actual number found is low indeed. We might also expect that the occurrence of individual yeast cells in a square is independent of the occurrence of other yeast cells. This is a commonly encountered class of application of the Poisson distribution.

The mean of the rare event is the only quantity that we need to know to calculate the relative expected frequencies of a Poisson distribution. Since we do not know the parametric mean of the yeast cells in this problem, we employ an estimate (the sample mean) and calculate expected frequencies of a Poisson distribution whose μ equals the sample mean of the observed frequency distribution of Box 5.2. It is convenient for the purpose of computation to rewrite Expression (5.2) as

$$\frac{1}{e^{\bar{Y}}}, \frac{1}{e^{\bar{Y}}}\left(\frac{\bar{Y}}{1}\right), \frac{\bar{Y}}{e^{\bar{Y}}}\left(\frac{\bar{Y}}{2}\right), \frac{\bar{Y}^2}{2e^{\bar{Y}}}\left(\frac{\bar{Y}}{3}\right), \frac{\bar{Y}^3}{2 \times 3e^{\bar{Y}}}\left(\frac{\bar{Y}}{4}\right), \ldots \tag{5.3}$$

Note first of all that the parametric mean μ has been replaced by the sample mean $\bar{Y}$. Each term is mathematically exactly the same as its corresponding term in Expression (5.2), but it has been factored into a convenient form for computation. After the first term in Expression (5.3), all subsequent terms consist of the previous term multiplied by the mean over an integer that increases by 1 for each succeeding term. Thus, we need only compute the expression $1/e^{\bar{Y}}$ once to obtain the frequency of the first term, multiply this by $\bar{Y}/1$ to get the second term, multiply the second term by $\bar{Y}/2$ for the third term, and so forth. It is important to make no computational error, since in such a chain multiplication the correctness of each term depends on the accuracy of the term before it. Expression (5.3) yields relative expected frequencies. If, as is more usual, absolute expected frequencies are desired,

BOX 5.2

Calculation of expected Poisson frequencies.

Yeast cells in 400 squares of a hemacytometer: $\bar{Y} = 1.8$ cells per square; $n = 400$ squares sampled.

(1) *Number of cells per square*	(2) *Observed frequencies* f		(3) *Absolute expected frequencies* $\hat{f}$		(4) *Deviation from expectation* $f - \hat{f}$	
0	75		66.1		+	
1	103		119.0		−	
2	121		107.1		+	
3	54		64.3		−	
4	30		28.9		+	
5	13		10.4		+	
6	2		3.1		−	
7	1	17	0.8	14.5	+	+
8	0		0.2		−	
9	1		0.0		+	
	400		399.9			

Source: "Student" (1907).

Computational steps

Flow of computation based on Expression (5.3) multiplied by n, since we wish to obtain absolute expected frequencies, $\hat{f}$.

1. Calculate $e^{\bar{Y}}$. If a table of exponentials is available $e^{\bar{Y}}$ may be looked up directly. Thus, consulting Table **E** of e^{-x} we look up the function for 1.8 and obtain 0.16530. The reciprocal of this value, 6.0496, equals $e^{\bar{Y}}$. If no such table is available, the following procedure may be used.

$$\log e^{\bar{Y}} = \bar{Y} \log e = \bar{Y}(0.43429) = (1.80)(0.43429) = 0.78172$$

$$\text{antilog } (0.78172) = 6.0495$$

2. $\hat{f}_0 = \dfrac{n}{e^{\bar{Y}}} \qquad = \dfrac{400}{6.0495} \qquad = 66.12$

3. $\hat{f}_1 = \left(\dfrac{n}{e^{\bar{Y}}}\right)\bar{Y} \qquad = 66.12(1.8) \qquad = 119.02$

4. $\hat{f}_2 = \left(\dfrac{n\bar{Y}}{e^{\bar{Y}}}\right)\dfrac{\bar{Y}}{2} \qquad = 119.02\left(\dfrac{1.8}{2}\right) = 107.12$

5. $\hat{f}_3 = \left(\dfrac{n\bar{Y}^2}{2e^{\bar{Y}}}\right)\dfrac{\bar{Y}}{3} \qquad = 107.12\left(\dfrac{1.8}{3}\right) = 64.27$

BOX 5.2 continued

6. $\hat{f}_4 = \left(\frac{n\bar{Y}^3}{2 \times 3e^{\bar{Y}}}\right)\frac{\bar{Y}}{4} \qquad = 64.27\left(\frac{1.8}{4}\right) = 28.92$

7. $\hat{f}_5 = \left(\frac{n\bar{Y}^4}{2 \times 3 \times 4e^{\bar{Y}}}\right)\frac{\bar{Y}}{5} \qquad = 28.92\left(\frac{1.8}{5}\right) = 10.41$

8. $\hat{f}_6 = \left(\frac{n\bar{Y}^5}{2 \times 3 \times 4 \times 5e^{\bar{Y}}}\right)\frac{\bar{Y}}{6} \qquad = 10.41\left(\frac{1.8}{6}\right) = 3.12$

9. $\hat{f}_7 = \left(\frac{n\bar{Y}^6}{2 \times 3 \times 4 \times 5 \times 6e^{\bar{Y}}}\right)\frac{\bar{Y}}{7} \qquad = 3.12\left(\frac{1.8}{7}\right) = 0.80$

10. $\hat{f}_8 = \left(\frac{n\bar{Y}^7}{2 \times 3 \times 4 \times 5 \times 6 \times 7e^{\bar{Y}}}\right)\frac{\bar{Y}}{8} = 0.80\left(\frac{1.8}{8}\right) = 0.018$

Total 399.80

$\hat{f}_9$ and beyond 0.2

□□ At step **3** enter $\bar{Y}$ into the multiplier and lock it in so it can be used as a constant multiplier. Then multiply it by $n/e^{\bar{Y}}$ from step **2.** At each subsequent step multiply by the result of the previous step and then divide by the appropriate integer (or multiply by its reciprocal on machines with a transfer device). □□

simply multiply the first term by n, the number of samples, and then proceed with the computational steps as before. By this process of chain multiplication the n continues as a factor in every term. The actual computation is illustrated in Box 5.2 and the expected frequencies so obtained are listed in column (3) of the frequency distribution.

What have we learned from this computation? When we compare the observed with the expected frequencies we notice quite a good fit of our observed frequencies to a Poisson distribution of mean 1.8, although we have not as yet learned a statistical test for goodness of fit (Chapter 16). No clear pattern of deviations from expectation is shown. We cannot test a hypothesis about the mean because the mean of the expected distribution was taken from the sample mean of the observed variates. As in the binomial distribution, clumping or aggregation would mean that the probability that a second yeast cell will be found in a square is not independent of the presence of the first one, but is higher. This would result in a clumping of the items in the classes at the tails of the distribution so that there would be some squares with large numbers of cells.

The biological interpretation of the dispersion pattern varies with the problem. The yeast cells seem to be randomly distributed in the counting chamber, indicating thorough mixing of the suspension. However, unless the proper suspension fluid is used, red blood cells will often stick together because of an electrical charge. This so-called rouleaux effect would be indicated by clumping of the observed frequencies.

Note that in Box 5.2, as in the subsequent tables giving examples of the application of the Poisson distribution, we group the low frequencies at one tail of the curve, uniting them by means of a bracket. This tends to simplify the patterns of distribution somewhat. However, the main reason for this grouping is related to the chi-square test for goodness of fit (of observed to expected frequencies), which is discussed in Section 16.2. For purposes of this test, no expected frequency $\hat{f}$ should be less than 5.

Before we turn to other examples we need to learn a few more facts about the Poisson distribution. You probably noticed that in computing expected frequencies we needed to know only one parameter—the mean of the distribution. By comparison, in the binomial we needed two parameters, p and k. Thus the mean completely defines the shape of a given Poisson distribution. From this it follows that the variance is some function of the mean and in fact in a Poisson distribution we have a very simple relationship between the two: $\mu = \sigma^2$, the variance being equal to the mean. The variance of the number of yeast cells per square based on the observed frequencies in Box 5.2 equals 1.965, not much larger than the mean of 1.80, indicating again that the yeast cells are distributed in Poisson fashion, hence randomly. This relationship between variance and mean suggests a rapid test of whether an observed frequency distribution is distributed in Poisson fashion even without fitting expected frequencies to the data. We simply compute a *coefficient of dispersion*

$$C.D. = \frac{s^2}{\bar{Y}}$$

This value will be near 1 in distributions that are essentially Poisson, will be >1 in clumped samples, and <1 in cases of repulsion. In the yeast cell example, $C.D. = 1.092$.

Although the computation of expected Poisson frequencies by the method of Box 5.2 is not very difficult or time consuming, this can be done even faster by means of tables of the Poisson distribution. For example, Table **X** furnishes the relative expected frequencies of Poisson distributions to six decimal places from values of $\mu = 0.1$ to $\mu = 11.0$ in increments of 0.1. The Biometrika Tables for Statisticians (Pearson and Hartley, 1958) provide in Table 39 relative expected frequencies up to a mean of $\mu = 15$. For a more extensive table, including some of the very low means, see Molina (1942). It may be necessary to interpolate in these tables between two adjacent means. Linear interpolation (see Introduction to the set of statistical tables accompanying this book) is sufficient in this instance. It may be faster to compute expected frequency on a desk calculator equipped with a transfer key for chain multiplication than to interpolate in Poisson tables. A computer program for expected Poisson frequencies is given in Appendix A3.3.

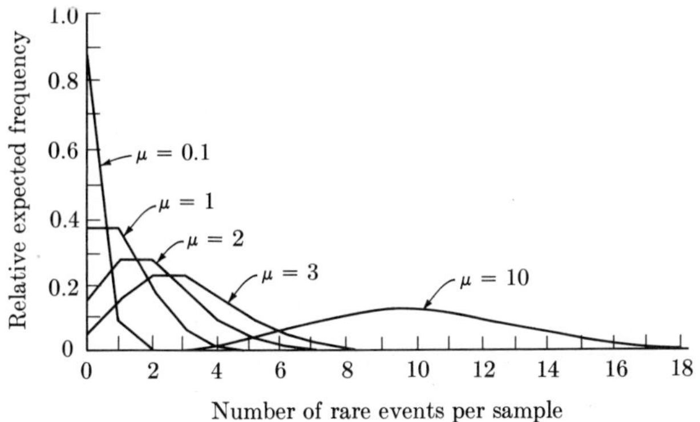

FIGURE 5.2 Frequency polygons of the Poisson distribution for various values of the mean.

From Poisson tables we can obtain an idea of the shapes of the Poisson distributions of different means. Figure 5.2 gives frequency polygons (line connecting the midpoint of a bar diagram) for five Poisson distributions. We notice that for the low value of $\mu = 0.1$ the frequency polygon is extremely L-shaped, but with an increase in the value of μ the distributions become humped and eventually nearly symmetrical.

We conclude our study of the Poisson distribution with a consideration of several examples from a variety of fields. The first example (Table 5.5) is quite analogous to that of the yeast cells. It is from an ecological study of mosses of the species *Hypnum schreberi*, invading mica residues of china clay.

TABLE 5.5

Number of moss shoots (*Hypnum schreberi*) per quadrat on china clay residues (mica).

(1) *Number of moss shoots per quadrat* Y	(2) *Observed frequencies* f		(3) *Absolute expected frequencies* $\hat{f}$		(4) *Deviation from expectation* $f - \hat{f}$	
0	100		77.7		+	
1	9		37.6		−	
2	6		9.1		−	
3	8		1.5		+	
4	1	17	0.2	10.8	+	+
5	0		0.0		0	
6	2		0.0		+	
Total	126		126.1			

$\bar{Y} = 0.4841$ $s^2 = 1.308$ $C.D. = 2.702$

Source: Barnes and Stanbury (1951).

These residues occur on deposited "dams" (often 5,000 sq. yds. in area), on which the ecologists laid out 126 quadrats. In each quadrat they counted the number of moss shoots. Expected frequencies are again computed, using the mean number of moss shoots, $\bar{Y} = 0.4841$, as an estimate of μ. There are many more quadrats than expected at the two tails of the distribution than at its center. Thus, we would expect only approximately 78 quadrats without a moss plant, but in fact we find 100. Similarly, while there are 11 quadrats containing 3 or more moss shoots, the Poisson distribution predicted only 1.7 such. By way of compensation the central classes are shy of expectation. Instead of the 38 expected quadrats with one moss plant each, only 9 were found, and there is a slight deficiency also in the 2 mosses per quadrat class. This is another illustration of the phenomenon of clumping, first encountered in the binomial distribution. The sample variance $s^2 = 1.308$, and is much larger than $\bar{Y} = 0.4841$, yielding a coefficient of dispersion $C.D. = 2.702$.

Searching for a biological explanation of the clumping, the investigators found that the protonema or spores of the moss were carried in by the water and deposited at random, but that each protonema gave rise to a number of upright shoots, so that counts of the latter would indicate a clumped distribution. In fact, when the clumps instead of individual shoots were used as variates, the investigators found that the clumps followed a Poisson distribution—that is, were randomly distributed. There are problems in directly applying this approach, since the size of the sampling units used can influence the results obtained. This would happen, for example, if the plants produced

TABLE 5.6

Mites (*Arrenurus* sp.) infesting 589 chironomid flies (*Calopsectra akrina*).

(1) *Number of mites per fly*	(2) *Observed frequencies* f		(3) *Poisson expected frequencies* $\hat{f}$		(4) *Deviation from expectation* $f - \hat{f}$	
0	442		380.7		+	
1	91		166.1		−	
2	29		36.2		−	
3	14	27	5.3	5.7	+	+
4	4		0.6		+	
5	6		0.1		+	
6	2		0.0		+	
7	0		0.0		0	
8	1		0.0		+	
Total	589		589.0			

$\bar{Y} = 0.4363$ $s^2 = 0.9709$ $C.D. = 2.225$

Source: Data by F. J. Rohlf.

substances that prevented other plants from growing very close to them (repulsed distribution), but on a larger scale the plants were clumped in favorable regions (clumped distribution). Organisms moving under their own power, when distributed in clumped fashion, may have responded to various social instincts or may have accumulated in clumps as a result of, or in response to, environmental forces.

The second example shows the distribution of a water mite on adults of a chironomid fly (Table 5.6). This example is similar to the case of the mosses except that here the sampling unit is a fly instead of a quadrat. The rare event is a mite infesting the fly. The coefficient of dispersion is 2.225 and this is clearly reflected by the observed frequencies, which are greater than expected in the tails and less than expected in the center. This relationship is easily seen in the last column of the frequency distribution, which shows the signs of the deviations (observed minus expected frequencies) and shows a characteristic clumped pattern. A possible explanation is that the density of mites in the several ponds from which the chironomid flies emerged differed considerably. Those chironomids that emerged from ponds with many mites would be parasitized by more than one mite, but those from ponds where mites were scarce would show little or no infestation.

The third example (Table 5.7) tests the randomness of the distribution of weed seeds in samples of grass seed. Here the total number of seeds in a quarter ounce sample could be counted and we could therefore get an estimate of k (which is several thousand) and also of q, which represents the large

TABLE 5.7

***Potentilla* (weed) seeds in 98 quarter-ounce samples of grass seeds (*Phleum pratense*).**

(1) Number of weed seeds per sample of seeds	(2) Observed frequencies f		(3) Poisson expected frequencies $\hat{f}$		(4) Deviation from expectation $f - \hat{f}$	
0	37		31.3		+	
1	32		35.7		−	
2	16		20.4		−	
3	9	13	7.8	10.6	+	+
4	2		2.2		−	
5	0		0.5		−	
6	1		0.1		+	
7	1		0.0		+	
Total	98		98.0			
$\bar{Y} = 1.1429$	$s^2 = 1.711$		$C.D. = 1.497$			

Source: Modified from Leggatt (1935).

proportion of grass seed as contrasted with p, the small proportion of weed seed. The data are therefore structured as in a binomial distribution with alternative states, weed seed and grass seed. However, for purposes of computation only the number of weed seeds need be considered. This is an example of a case mentioned at the beginning of this section, a binomial in which the frequency of one outcome is very much smaller than that of the other and the sample size is large. We can use the Poisson as a useful approximation to the expected binomial frequencies for the tail of the distribution. We use the average number of weed seeds per sample of seeds as our estimate of the mean and calculate expected Poisson frequencies for this mean. Thus for purposes of computation only the number of weed seeds need be considered. Although the pattern of the deviations and the coefficient of dispersion again indicate clumping, this tendency is not marked and the methods of Chapter 16 would indicate insufficient evidence for the conclusion that the distribution is not a Poisson. We conclude that the weed seeds are randomly distributed through the samples. If clumping had been significant it might have been because the weed seeds stuck together, possibly because of interlocking hairs, sticky envelopes, or the like.

Perhaps you need to be convinced that the binomial under these conditions does in fact approach the Poisson distribution. The mathematical proof is too involved for this text but we can demonstrate the fact by an empirical example shown in Table 5.8. Here the expected binomial frequencies for the expression $(0.001 + 0.999)^{1000}$ are given in column (2), and in column (3)

TABLE 5.8

Expected frequencies compared for binomial and Poisson distributions. (For explanation see text.)

(1) Y	*(2)* *Expected binomial frequencies* $p = 0.001$ $k = 1000$ $\hat{f}_{rel}$	*(3)* *Expected frequencies approximated by Poisson* $\mu = 1$ $\hat{f}_{rel}$
0	0.367695	0.367879
1	0.368063	0.367879
2	0.184032	0.183940
3	0.061283	0.061313
4	0.015290	0.015328
5	0.003049	0.003066
6	0.000506	0.000511
7	0.000072	0.000073
8	0.000009	0.000009
9	0.000001	0.000001
Total	1.000000	0.999999

these frequencies are approximated by a Poisson distribution with mean equal to 1.0, since the expected value for p is 0.001, which is one event for a sample size of $k = 1000$. We note that the two columns of expected frequencies are extremely close.

The next distribution (Table 5.9) is extracted from an experimental study of the effects of different densities of parents of the Azuki bean weevil.

TABLE 5.9

Azuki bean weevils (*Callosobruchus chinensis*) emerging from 112 azuki beans (*Phaseolus radiatus*).

(1) Number of weevils emerging per bean Y	(2) Observed frequencies f		(3) Poisson expected frequencies $\hat{f}$		(4) Deviation from expectation $f - \hat{f}$	
0	61		70.4		−	
1	50		32.7		+	
2	1	1	7.6	8.9	−	−
3	0		1.2		−	
4	0		0.1		−	
Total	112		112.0			

$\bar{Y} = 0.4643$ $s^2 = 0.269$ $C.D. = 0.579$

Source: Utida (1943).

We are studying the number of weevils emerging per bean. Larvae of these weevils enter the beans, feed and pupate inside them, and then emerge through an emergence hole. Thus the number of holes per bean is a good measure of the number of adults that have emerged. The rare event in this case is the weevil present in the bean. We note that the distribution is strongly repulsed. There are many more beans containing one weevil than the Poisson distribution would predict. A statistical finding of this sort leads to the biological question, "Why?" In this case it was found that the adult female weevils tended to deposit their eggs evenly rather than randomly over the available beans. This prevented too many eggs being placed on any one bean and precluded heavy competition among the developing larvae on any one bean. A contributing factor was competition between remaining larvae feeding on the same bean, resulting in all but one generally being killed or driven out. Thus it is easily understood how the above biological phenomena would give rise to a repulsed distribution.

We do not wish you to miss two of the classical tragicomic applications of the Poisson distribution. Table 5.10 is a frequency distribution of men killed by being kicked to death by a horse in ten Prussian army corps over

TABLE 5.10

Men killed by being kicked by a horse in 10 Prussian army corps in the course of 20 years.

(1) *Number of men killed per year per army corps*	(2) *Observed frequencies* f		(3) *Poisson expected frequencies* $\hat{f}$		(4) *Deviation from expectation* $f - \hat{f}$	
0	109		108.7		+	
1	65		66.3		−	
2	22		20.2		+	
3	3		4.1		−	
4	1	4	0.6	4.8	+	−
5	0		0.1		−	
Total	200		200.0			
$\bar{Y} = 0.610$	$s^2 = 0.611$		$C.D. = 1.002$			

Source: Bortkiewicz (1898).

twenty years. The basic sampling unit is temporal in this case, one army corps per year. We are not certain exactly how many men are involved, but surely a considerable number. The 0.610 men killed per army corps per year is the rare event. If we knew the number of men in each army corps we could calculate the probability of not being killed by a horse in one year. This would then give us a binomial that could be approximated by the Poisson. However, knowing that the sample size (the number of men in an army corps) is large, we need not concern ourselves with values of p and k and can consider the example simply from the Poisson model, using the observed mean number of men killed per army corps per year as an estimate of μ. The example is an almost perfect fit of expected frequencies to observed ones. What would clumping have meant in such an example? An unusual number of deaths in a certain army corps in a given year could have been due to poor discipline in the particular corps or to a particularly vicious horse that killed several men before the corps got rid of it. Repulsion might mean that one death per year per corps occurred more frequently than expected. This might have been so if men in each corps were careless each year until someone had been killed, after which time they all became more careful for a while.

Finally, Table 5.11 presents an accident record of 647 women working in a munitions plant during a five-week period. The sampling unit is one woman during this period. The rare event is the number of accidents that occurred to one woman in this period. The model assumes, of course, that the accidents are not fatal or very serious, removing the individual from further exposure. The noticeable clumping in these data probably arises from accident proneness or because some women have more dangerous jobs than others.

TABLE 5.11

Accidents to 647 women working on high explosive shells in 5 weeks.

(1) *Number of accidents per woman*	(2) *Observed frequencies* f	(3) *Poisson expected frequencies* $\hat{f}$	(4) *Deviation from expectation* $f - f$
0	447	406.3	+
1	132	189.0	−
2	42	44.0	−
3	21 }	6.8 }	+ }
4	3 } 26	0.8 } 7.7	+ } +
5	2 }	0.1 }	+ }
Total	647	647.0	

$\bar{Y} = 0.4652$ $s^2 = 0.692$ $C.D. = 1.488$

Source: Greenwood and Yule (1920).

↛ Occasionally samples of items that might possibly be distributed in Poisson fashion are taken in such a way that the zero class is missing. For example, an entomologist is studying the number of mites per leaf of a tree. He may not have sampled leaves at random from the tree, but collected only those leaves containing mites, excluding those without these animals. Estimates of parameters and calculation of expected frequencies for these so-called *truncated Poisson distributions* can be made by special methods. Cohen (1960) gives tables and graphs for rapid estimation to ease the otherwise heavy computation. ↚

5.4 Some other discrete probability distributions

The binomial and Poisson distributions are both examples of discrete probability distributions. Thus there is either no moss plant per quadrat or one moss plant, two plants, three plants, and so forth, but no values in between these. Numerous other discrete distributions are known in probability theory and in recent years many have been suggested as suitable for one or the other application. The beginner need not concern himself with these, but it might be well for you to be familiar by name at least with those briefly discussed below.

The *hypergeometric distribution* is the distribution equivalent to the binomial case but sampled from a *finite population without replacement.* In Section 5.1 we estimated the probability of finding a second foreign student in Matchless University. We pointed out that if there are 10,000 students at Matchless University, 400 of whom are foreign, then the probability of sampling one foreigner is indeed 0.04. However, once a foreign student has been sampled, then the probability of sampling another foreign student is no longer 0.04 (even when independent) but is 399/9999, which equals 0.0399. It would be 0.04 only *with replacement*—that is, if we returned the first foreign student sampled to the University population before we sample again. The binomial distribution is entirely correct only in cases of sampling with replacement or with infinite population size, which amounts to the same thing. For practical purposes, when small samples are taken from large populations, as in the case of Matchless University, these populations can be considered essentially infinite. But if you have a population of 100 animals, 4% of which carry a mutation, a sample of one mutant reduces the population to 3 out of 99, or from 4% to 3.03%. Thus repeated samples of 5 from this population will follow not the binomial distribution but a different distribution, the hypergeometric distribution. The individual terms of the hypergeometric distribution are given by the expression

$$\frac{C(pN, r)C(qN, k - r)}{C(N, k)}$$

which gives the probability of sampling r items of the type represented by probability p out of a sample of k items from a population of size N. The mean of a hypergeometric distribution is kp and the variance of this distribution is $kpq(N - k)/(N - 1)$.

You will note that the mean is the same as that of the binomial distribution and its variance is that of the binomial multiplied by $(N - k)/(N - 1)$. When N is very large as compared with k, this term is approximately 1, showing that the hypergeometric distribution approximates the binomial. The expected frequencies for any sizable distribution are tedious to compute and you need a table of factorials for rapid calculation. We suggest using a digital computer with facilities for double precision arithmetic to evaluate expected hypergeometric frequencies for a sizable distribution.

In biology, sampling of small samples from a finite distribution occurs in certain problems of evolutionary genetics. Another application is in mark-recapture studies, in which a certain proportion of a population is caught, marked, released, and subsequently recaptured, leading to estimates of the number of the entire population.

A number of probability distributions have been employed as underlying mathematical models for cases of contagious distribution. The difficulty of fitting these varies with the distribution. Regrettably, no general, simply phrased account exists to which we can refer the reader. Greig-Smith (1964) gives a nontechnical account of the application of these distributions to ecology and provides references that may lead the interested reader deeper into the subject. Numerous ecological examples of the application of contagious distributions are given in Williams (1964).

Of these contagious distributions, two are mentioned here. The *negative binomial distribution* has probably been used more frequently than any other contagious distribution. The theoretical conditions that would give rise to a negative binomial distribution are discussed by Bliss and Calhoun (1954), who also give methods of calculating expected frequencies. The account in Bliss and Fisher (1953), though somewhat more rigorous, may be more easily located in a library. Volume 3 of Statistics in Biology by C. I. Bliss (in preparation) will undoubtedly contain a comprehensive account.

The *logarithmic distribution* (or logarithmic series) has been used extensively in studying the distribution of taxonomic units in faunal samples. This distribution has been frequently employed by C. B. Williams and can be looked up in his recent book (Williams, 1964).

Exercises 5

5.1 In man the sex ratio of newborn infants is about 100 ♀♀ : 105 ♂♂. Were we to take 10,000 random samples of 6 newborn infants from the total population of such infants for one year, what would be the expected frequency of groups of 6 males, 5 males, 4 males, and so on?

5.2 Show algebraically why the machine method of Box 5.1 works.

5.3 The two columns below give fertility of eggs of the CP strain of *Drosophila melanogaster* raised in 100 vials of 10 eggs each (data from R. R. Sokal). Find the expected frequencies on the assumption of independence of mortality for each egg in a vial. Use the observed mean. Calculate the expected variance and compare it with the observed variance. Interpret results, knowing that the eggs of each vial are siblings and that the different vials contain descendants from different parent pairs. ANS.: $\sigma^2 = 2.417$, $s^2 = 6.636$.

Number of eggs hatched Y	*Number of vials* f
0	1
1	3
2	8
3	10
4	6
5	15
6	14
7	12
8	13
9	9
10	9

5.4 Calculate Poisson expected frequencies for the frequency distribution given in Table 2.3 (number of plants of the sedge *Carex flacca* found in 500 quadrats).

5.5 The Army Medical Corps is concerned over the intestinal disease X. From previous experience they know that soldiers suffering from the disease invariably harbor the pathogenic organism in their feces and that to all practical purposes every diseased stool specimen contains these organisms. However, the organisms are never abundant and thus only 20% of all slides prepared by the standard procedure will contain some (we assume that if an organism is present on a slide it will be seen). How many slides should they direct their laboratory technicians to prepare and examine per stool specimen, so that in case a specimen is positive it will be erroneously diagnosed negative in less than 1% of the cases (on the average)? On the basis of your answer would you recommend that the Corps attempt to improve their diagnostic methods? ANS.: 21 slides.

5.6 A cross is made in a genetic experiment in drosophila in which it is expected that $\frac{1}{4}$ of the progeny will have white eyes and $\frac{1}{2}$ will have the trait called "singed bristles." Assume that the two gene loci segregate independently. (a) What proportion of the progeny should exhibit both traits simultaneously? (b) If 4 flies are sampled at random, what is the probability that they would all be white-eyed? (c) What is the probability that none of the 4 flies would have either white eyes or "singed bristles"? (d) If two flies are sampled, what is the probability that at least one of the flies has either white eyes or "singed bristles" or both traits?

5.7 Those readers who have had a semester or two of calculus may wish to try to prove that Expression (5.1) tends to Expression (5.2) as k becomes indefinitely large (and p becomes infinitesimal, so that $\mu = kp$ remains constant).
HINT: $\left(1 - \frac{x}{n}\right)^n \longrightarrow e^{-x}$ as $n \longrightarrow \infty$.

5.8 Summarize and compare the assumptions and parameters on which the binomial and Poisson distributions are based.

6 THE NORMAL PROBABILITY DISTRIBUTION

The theoretical frequency distributions in the last chapter were all discrete. Their variables assumed values that changed in integral steps (meristic variables). Thus the number of infected insects per sample was 0 or 1 or 2 but could not assume an intermediate value between these. Similarly, the number of yeast cells per hemacytometer square is a meristic variable and requires a discrete probability function to describe it. However, most variables encountered in biology are continuous (such as the infant birth weights or the aphid femur lengths used as examples in Chapters 2 and 4). This chapter deals more extensively with the distributions of continuous variables.

The first section (6.1) introduces frequency distributions of continuous variables. In Section 6.2 we show one way of deriving the most common such distribution, the normal probability distribution, and we examine its properties in Section 6.3. A few applications of the normal distribution are illustrated in Section 6.4. The technique of fitting a normal distribution to observed data is described in Section 6.5, and the following section (6.6) contains a discussion of some of the reasons for departure from normality in observed frequency distributions and a description of techniques for measuring the nature and amount of such departures. Various graphic techniques for pointing out departures from normality and for estimating mean and standard deviation in approximately normal distributions are given in Section 6.7. The final section (6.8) makes some brief comments about other types of theoretical frequency distributions of continuous variables.

6.1 Frequency distributions of continuous variables

For continuous variables the theoretical probability distribution or *probability density function* can be represented by a continuous curve, as shown in Fig-

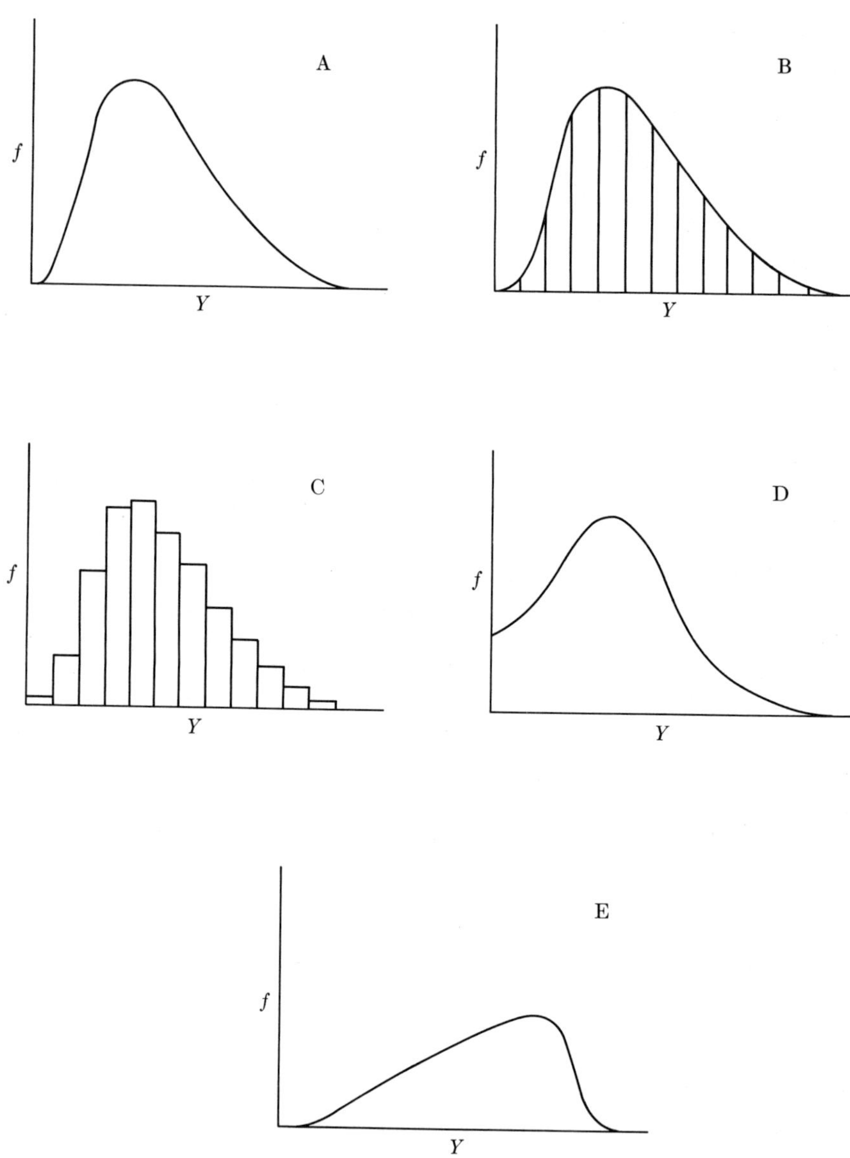

FIGURE 6.1 Examples of some probability distributions of continuous variables. (See text for explanation.)

ure 6.1A. The height of the curve gives the density for a given value of the variable. By *density* we mean the relative concentration of variates along the Y-axis (as indicated in Figure 2.1). Note that the Y-axis is the abscissa in the figure, with the frequency f or the density being the ordinate. In order to compare the theoretical with the observed frequency distribution it is necessary to divide the two into corresponding classes, as shown by the vertical lines in Figure 6.1B. Probability density functions are defined so that the expected frequency of observations between two class limits (vertical lines) is represented by the area between these limits under the curve. The total area under the curve is therefore equal to the sum of the expected frequencies (1.0 or n, depending on whether relative or absolute expected frequencies were calculated).

When forming a frequency distribution of observations of a continuous variable, the choice of class limits is arbitrary because all values of a variable are theoretically possible. In a continuous distribution one cannot evaluate the probability of the variable being exactly equal to a given value such as 3 or 3.5. One can only estimate the frequency of observations falling between two limits. This is so because the area of the curve corresponding to any point along the curve is an infinitesimal.

The usual way to graph continuous distributions is in the form of a histogram—that is, to make each of the segments in Figure 6.1B into rectangles of the same area, as shown in Figure 6.1C. So long as each class interval is of the same width, the height of each rectangle is directly proportional to its area and the frequency of each class may be read off the ordinate directly.

Thus in order to calculate expected frequencies for a continuous distribution we shall have to calculate the proportions of the area under the curve between the class limits. We shall see how this is done in the case of the normal frequency distribution in Sections 6.3 and 6.4.

Continuous frequency distributions may start and terminate at finite points along the Y-axis, as shown in Figure 6.1A, or one or both ends of the curve may extend indefinitely, as in Figures 6.1D and 6.1E. The idea of an area under a curve, when one or both ends go to infinity, may trouble those of you not acquainted with the calculus. Fortunately, however, this is not a great conceptual stumbling block, since in all the cases that we shall encounter the tail of the curve will approach the Y-axis rapidly enough, so that the portion of the area beyond a certain point is for all practical purposes zero and the frequencies it represents are infinitesimal.

We may fit continuous frequency distributions to some sets of meristic data as, for example, the number of teeth in an organism. In such cases we have reason to believe that underlying biological variables causing differences in numbers of the structure are really continuous, even though expressed as a discrete variable.

We shall now proceed to discuss the most important probability density function in statistics, the normal frequency distribution.

6.2 Derivation of the normal distribution

There are several ways of deriving the normal frequency distribution from elementary assumptions. Most of these require more mathematics than we expect of our readers. We shall therefore use a largely intuitive approach, which we have found of heuristic value.

Let us consider a binomial distribution of the familiar form $(p + q)^k$ in which k becomes indefinitely large. What type of biological situation could give rise to such a binomial distribution? An example might be one in which many factors cooperate additively in producing a biological result. The following hypothetical case is possibly not too far removed from reality. The intensity of skin pigmentation in an animal will be due to the summation of many factors, some genetic, others environmental. As a simplifying assumption let us state that every factor can occur in two states only: present or absent. When the factor is present it contributes one unit of pigmentation to skin color, but it contributes nothing to pigmentation when it is absent. Each factor, regardless of its nature or origin, has the identical effect and the effects are additive; if three out of five possible factors are present in an individual, the pigmentation intensity would be three units, being the sum of three contributions of one unit each. One final assumption: each factor has an equal probability of being present or absent in a given individual. Thus, $p_F = 0.5$, the probability of the factor being present, while $q_f = 0.5$, the probability of the factor being absent.

With only one factor ($k = 1$), expansion of the binomial $(p_F + q_f)^1$ would yield two pigmentation classes among the animals as follows:

F	f	pigmentation classes
0.5	0.5	expected frequency
1	0	pigmentation intensity

Half the animals would have intensity 1, the other half 0. With $k = 2$ factors present in the population (the factors are assumed to occur independently of each other), the distribution of pigmentation intensities would be represented by the expansion of the binomial $(p_F + q_f)^2$:

FF	Ff	ff	pigmentation classes
0.25	0.50	0.25	expected frequency
2	1	0	pigmentation intensity

One-fourth of the individuals would have pigmentation intensity 2, half 1, and the remaining fourth 0.

The expected frequencies for four, six, and ten factors are shown in Table 6.1. We note that the number of classes in the binomial increases with the number of factors. The frequency distributions are symmetrical and the expected frequencies at the tails become progressively less as k increases. When these binomial distributions are graphed as histograms in

TABLE 6.1

Demonstration of how the expanded binomial $(p + q)^k$ approaches the normal curve as k approaches infinity. (For explanation see text.)

	F	f	*Pigmentation intensity*	f_{rel}
With four factors	4	—	4	.0625
	3	1	3	.25
	2	2	2	.375
	1	3	1	.25
	—	4	0	.0625
With six factors	6	—	6	.0156
	5	1	5	.09375
	4	2	4	.2344
	3	3	3	.3125
	2	4	2	.2344
	1	5	1	.09375
	—	6	0	.0156
With ten factors	10	—	10	.001
	9	1	9	.010
	8	2	8	.044
	7	3	7	.117
	6	4	6	.205
	5	5	5	.246
	4	6	4	.205
	3	7	3	.117
	2	8	2	.044
	1	9	1	.010
	0	10	0	.001

Figure 6.2 (rather than as bar diagrams as they should be drawn), we note that they approach the familiar bell-shaped outline of the normal frequency distribution (Figures 6.3 and 6.4). Were we to expand the expression for $k = 20$, our histogram would be so close to a normal frequency distribution that we could not show the difference between them on a graph the size of this page. Of course, just stating that the expected frequencies of the binomial approach the shape of a normal probability distribution is not acceptable mathematical proof. The mathematically inclined reader will find a relatively simple proof of this proposition in Section 8.15 of Yule and Kendall (1950). Readers not so inclined must take it on faith that the model postulated above, the result of many factors acting independently and additively, will approach normality. However, at the beginning of this procedure we made a number of severe limiting assumptions for the sake of simplicity. Let us now return to these and see if we can remove them.

1. The two states of the factors are present in equal frequency, or $p = q$. When $p \neq q$, the distribution also approaches normality as k approaches

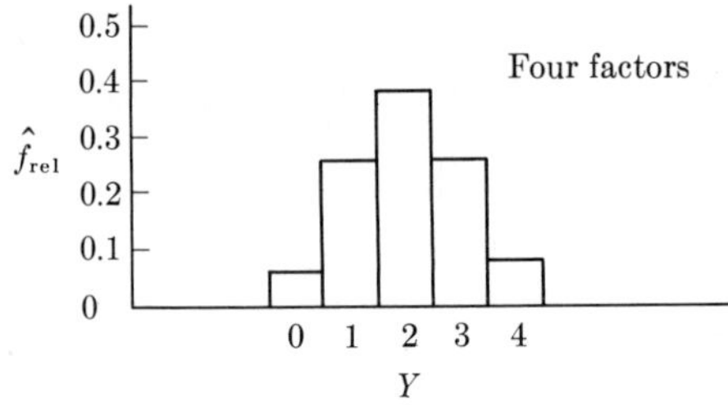

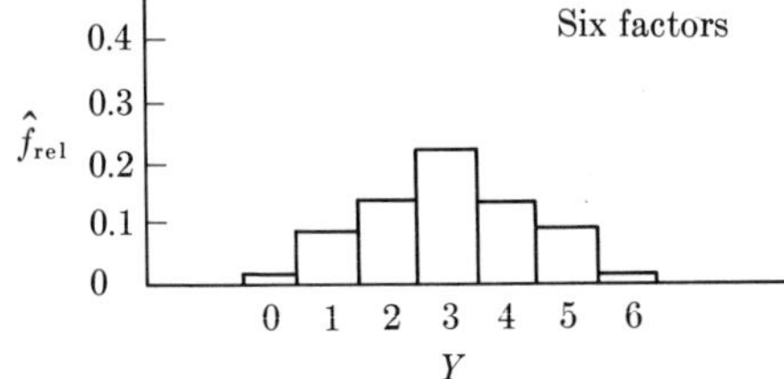

FIGURE 6.2 Histograms based on relative expected frequencies of Table 6.1.

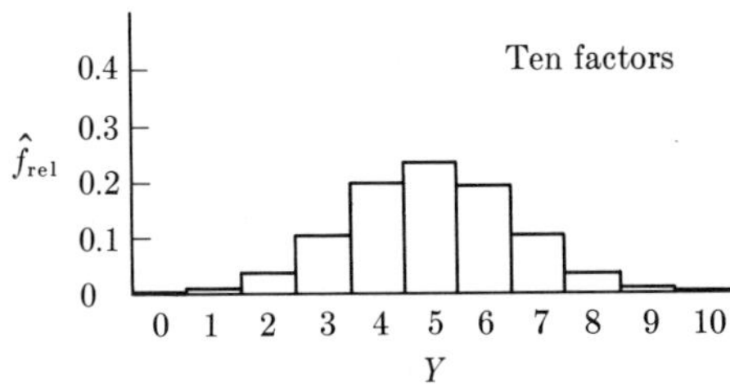

infinity. This is intuitively difficult to see because when $p \neq q$ the histogram is at first asymmetrical. However, it can be shown mathematically that when k, p, and q are such that $kpq \geq 3$, the normal distribution will be closely approximated.

2. We stated that factors occur in two states only—present or absent. In a more realistic situation factors would be permitted to occur in several states, one state making a large contribution, a second state a smaller contribution, and so forth. However, it can also be shown that the multinomial $(p + q + r + \cdots + z)^k$ approaches the normal frequency distribution as k approaches infinity.

3. Different factors may be present in different frequencies and also may have different quantitative effects. As long as these are additive and independent, normality is still approached when k approaches infinity.

Lifting these restrictions makes the assumptions leading to a normal distribution compatible with innumerable biological situations. It is therefore not surprising that so many biological variables are approximately normally distributed.

Let us review the conditions that would tend to produce normal frequency distributions: (1) if there are many factors that are single or composite;

(2) if these factors are independent in occurrence; (3) if the factors are independent in effect—that is, if their effects are additive; and (4) if they make equal contributions to the variance. The fourth condition we are not yet in a position to discuss and mention here only for completeness. It will be discussed in Chapter 8.

6.3 Properties of the normal distribution

Formally the *normal probability density function* can be representd by the expression

$$Z = \frac{1}{\sigma\sqrt{2\pi}} e^{-(Y-\mu)^2/2\sigma^2} \tag{6.1}$$

Here Z indicates the height of the ordinate of the curve, which represents the density of the items. It is the dependent variable in the expression, being a function of the variable Y. There are two constants in the equation: π, well known to be approximately 3.14159, making $1/\sqrt{2\pi}$ equal 0.39894, and e, the base of the Naperian or natural logarithms, whose value approximates 2.71828. There are two parameters in a normal probability density function. These are the parametric mean μ and the parametric standard deviation σ, which determine the location and shape of the distribution. Thus there is not just one normal distribution, as might appear to the uninitiated who keep encountering the same bell-shaped image in elementary textbooks; rather, there are an infinity of such curves, since these parameters can assume an infinity of values. This is illustrated by the three normal curves in Figure 6.3, representing the same total frequencies. Curves A and B differ in their means, and hence are at different locations. Curves B and C have identical means but different standard deviations. Since the standard deviation of

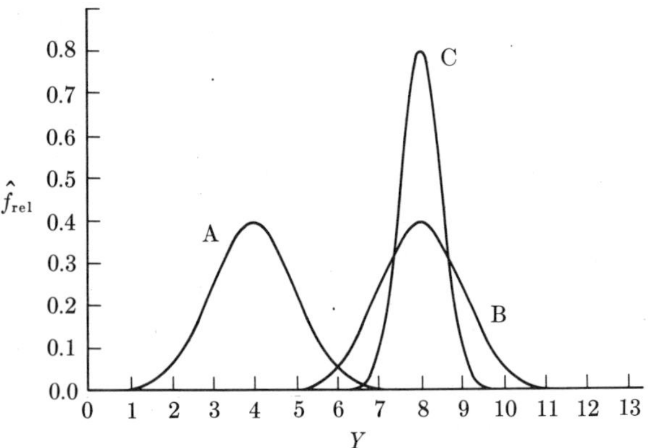

FIGURE 6.3 Illustration of how changes in the two parameters of the normal distribution affect the shape and position of curves. A. $\mu = 4$, $\sigma = 1$; B. $\mu = 8$, $\sigma = 1$; C. $\mu = 8$, $\sigma = 0.5$.

curve C is only half that of curve B, it presents a much narrower appearance.

In theory a normal frequency distribution extends from negative infinity to positive infinity along the axis of the variable (labeled Y, although it is frequently the abscissa). This means that a normally distributed variable can assume any value, however large or small, although values farther from the mean than plus or minus three standard deviations are quite improbable, their relative expected frequencies being very rare. This can be seen from Expression (6.1). When Y is very large or very small, the term $(Y - \mu)^2/2\sigma^2$ will necessarily become very large. Hence e raised to the negative power of that term will be very small and Z will therefore be very small. The ordinate Z for a normal distribution is tabled in many books (see Pearson and Hartley, 1958, Table 1) as a function of distance from the mean in standard deviation units.

↛ Fitting a normal distribution as a curve superimposed upon observed data is only rarely done. If needed, this is carried out by means of the table of ordinates of the normal curve. The procedure is described in Croxton, Cowden, and Klein (1967, Chapter 23, Table 23.2). The usual technique is to calculate expected normal frequencies for the classes of a sampled distribution, as shown in Section 6.5. ↚

The curve is symmetrical around the mean. Therefore the mean, median, and mode of the normal distribution are all at the same point. The following percentages of items in a normal frequency distribution lie within the indicated limits:

$\mu \pm \sigma$ contains 68.26% of the items
$\mu \pm 2\sigma$ contains 95.46% of the items
$\mu \pm 3\sigma$ contains 99.73% of the items

Conversely,

50% of the items fall between $\mu \pm 0.674\sigma$
95% of the items fall between $\mu \pm 1.960\sigma$
99% of the items fall between $\mu \pm 2.576\sigma$

These relations are shown in Figure 6.4. How have these percentages been calculated? The direct calculation of any portion of the area under the normal curve requires an integration of the function shown as Expression (6.1). Fortunately, for those of you who do not know calculus (and even for those of you who do) the integration has already been carried out in an alternative form of the normal distribution: the *normal distribution function* (the theoretical *cumulative* distribution function of the normal probability density function), shown in Figure 6.4. It gives the total frequency from negative infinity up to any point along the abscissa. We can therefore look up directly the probability that an observation will be less than a specified value of Y.

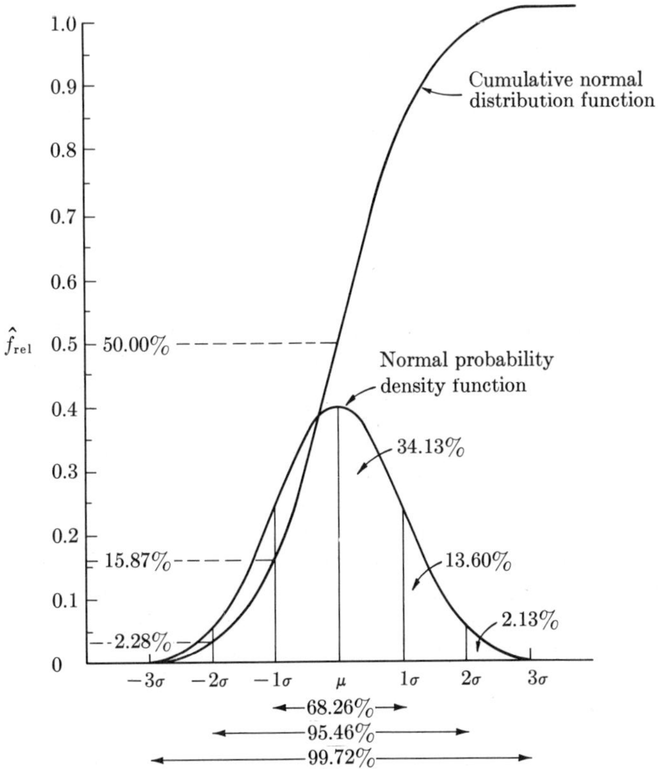

FIGURE 6.4 Areas under the normal probability density function and the cumulative normal distribution function.

For example, Figure 6.4 shows that the total frequency up to the mean is 50.00% and the frequency up to a point one standard deviation below the mean is 15.87%. These frequencies are found, graphically, by raising a vertical line from a point, such as $-\sigma$, until it intersects the cumulative distribution curve, and then reading the frequency (15.87%) off the ordinate. The probability of an observation falling between two arbitrary points can be found by subtracting the probability of an observation falling below the lower point from the probability of an observation falling below the upper point. For example, we can see from Figure 6.4 that the probability of an observation falling between the mean and a point one standard deviation below the mean is $0.5000 - 0.1587 = 0.3413$.

The normal distribution function is tabulated in Table **P**, Areas of the Normal Curve, and, for convenience in later calculations, 0.5 has been subtracted from all of the entries. This table therefore lists the proportion of the area between the mean and any point a given number of standard deviations above it. Thus, for example, the area between the mean and the point 0.50 standard deviations above the mean is 0.1915 of the total area of the curve. Similarly, the area between the mean and the point 2.64 standard deviations above the mean is 0.4959 of the curve. A point 4.0 standard deviations from the mean includes 0.499968 of the area between it and the mean. However, since the normal distribution extends from negative to positive infinity, one needs to go an infinite distance from the mean to reach fully half of the area under the curve. The use of the table of areas of the normal curve will be illustrated in the next section.

A sampling experiment will give you a "feel" for the distribution of items under a normal curve.

Experiment 6.1. You are asked to sample from two populations. The first one is an approximately normal frequency distribution of 100 wing lengths of houseflies. The second population deviates strongly from normality. It is a frequency distribution of the total annual milk yield of 100 Jersey cows. Both populations are shown in Table 6.2. You are asked to sample from them repeatedly in order to simulate sampling from an infinite population. Obtain samples of 35 items from each of the two populations. This is done by obtaining two sets of 35 two digit random numbers from the table of random numbers (Table **O**) with which you became familiar in Experiment 5.1. Write down the random numbers in blocks of five, and copy next to them the value of Y (for either wing length or milk yield) corresponding to the random number. An example of such a block of five numbers and the computations required for it are shown below, using the housefly wing lengths as an example.

Random number		*Wing length* Y
16		41
59		46
99		54
36		44
21		42
	$\sum Y =$	227
	$\sum Y^2 =$	10,413
	$\bar{Y} =$	45.4

These samples and the computations carried out for each sample will be used in subsequent chapters. Therefore, preserve your data carefully!

In this experiment consider the 35 variates for each variable as a single sample, without breaking them down into groups of five. Since the true mean and standard deviation (μ and σ) of the two distributions are known, you can calculate the expression $(Y_i - \mu)/\sigma$ for each variate Y_i. Thus for the first housefly wing length sampled above, you compute

$$\frac{(41 - 45.5)}{3.90} = -1.1538$$

This means that the first wing length is 1.1538 standard deviations below the true mean of the population. The deviation from the mean measured in standard deviation units is called a *standardized deviate* or *standard deviate.* The arguments of Table **P**, expressing distance from the mean in units of σ are called *standard normal deviates.* Group all 35 variates in a frequency dis-

TABLE 6.2

Populations of wing lengths and milk yields. *Column 1.* Rank number. *Column 2.* Lengths (in mm $\times 10^{-1}$) of 100 wings of houseflies arrayed in order of magnitude; $\mu = 45.5$, $\sigma^2 = 15.21$, $\sigma = 3.90$; distribution approximately normal. *Column 3.* Total annual milk yield (in hundreds of pounds) of 100 two-year-old registered Jersey cows arrayed in order of magnitude; $\mu = 66.61$, $\sigma^2 = 124.4779$, $\sigma = 11.1597$; distribution departs strongly from normality.

(1)	*(2)*	*(3)*	*(1)*	*(2)*	*(3)*	*(1)*	*(2)*	*(3)*	*(1)*	*(2)*	*(3)*	*(1)*	*(2)*	*(3)*
01	36	51	21	42	58	41	45	61	61	47	67	81	49	76
02	37	51	22	42	58	42	45	61	62	47	67	82	49	76
03	38	51	23	42	58	43	45	61	63	47	68	83	49	79
04	38	53	24	43	58	44	45	61	64	47	68	84	49	80
05	39	53	25	43	58	45	45	61	65	47	69	85	50	80
06	39	53	26	43	58	46	45	62	66	47	69	86	50	81
07	40	54	27	43	58	47	45	62	67	47	69	87	50	82
08	40	55	28	43	58	48	45	62	68	47	69	88	50	82
09	40	55	29	43	58	49	45	62	69	47	69	89	50	82
10	40	56	30	43	58	50	45	63	70	48	69	90	50	82
11	41	56	31	43	58	51	46	63	71	48	70	91	51	83
12	41	56	32	44	59	52	46	63	72	48	72	92	51	85
13	41	57	33	44	59	53	46	64	73	48	73	93	51	87
14	41	57	34	44	59	54	46	65	74	48	73	94	51	88
15	41	57	35	44	60	55	46	65	75	48	74	95	52	88
16	41	57	36	44	60	56	46	65	76	48	74	96	52	89
17	42	57	37	44	60	57	46	65	77	48	74	97	53	93
18	42	57	38	44	60	58	46	65	78	49	74	98	53	94
19	42	57	39	44	60	59	46	67	79	49	75	99	54	96
20	42	57	40	44	61	60	46	67	80	49	76	00	55	98

Source: Column 2—Data adapted from Sokal and Hunter (1955). Column 3—Data from Canadian government records.

tribution, then do the same for milk yields. Since you know the parametric mean and standard deviation, you need not compute each deviate separately, but can simply write down class limits in terms of the actual variable as well as in standard deviation form. The class limits for such a frequency distribution are shown in Table 6.3. Combine the results of your sampling with those of your classmates and study the percentage of the items in the distribution one, two, and three standard deviations to each side of the mean. Note the marked differences in distribution between the housefly wing lengths and the milk yields.

TABLE 6.3

Table for recording frequency distributions of standard deviates $(Y_i - \mu)/\sigma$ for samples of Experiment 6.1.

	Wing lengths			*Milk yields*	
	Practical class limits	f		*Practical class limits*	f
$-\infty$			$-\infty$		
-3σ			-3σ		
$-2\frac{1}{2}\sigma$			$-2\frac{1}{2}\sigma$		
-2σ	36, 37		-2σ		
$-1\frac{1}{2}\sigma$	38, 39		$-1\frac{1}{2}\sigma$		
$-\sigma$	40, 41		$-\sigma$	51–55	
$-\frac{1}{2}\sigma$	42, 43		$-\frac{1}{2}\sigma$	56–61	
$\mu = 45.5$	44, 45		$\mu = 66.61$	62–66	
$\frac{1}{2}\sigma$	46, 47		$\frac{1}{2}\sigma$	67–72	
σ	48, 49		σ	73–77	
$1\frac{1}{2}\sigma$	50, 51		$1\frac{1}{2}\sigma$	77–83	
2σ	52, 53		2σ	84–88	
$2\frac{1}{2}\sigma$	54, 55		$2\frac{1}{2}\sigma$	89–94	
3σ			3σ	95–98	
$+\infty$			$+\infty$		

6.4 Applications of the normal distribution

The normal frequency distribution is the most widely used distribution in statistics, and time and again we shall have recourse to it in a variety of situations. For the moment we may subdivide its applications as follows.

1. We sometimes have to know whether a given sample is normally distributed before we can apply a certain test to it. To test whether a given sample is normally distributed we have to calculate expected frequencies for a normal curve of the same mean and standard deviation. The next section will teach you how to do this and Section 6.7 provides some shortcuts.

2. Knowing whether a sample is normally distributed may confirm or

reject certain underlying hypotheses about the nature of the factors affecting the phenomenon studied. This is related to the conditions making for normality in a frequency distribution, discussed in Section 6.2. Thus if we find a given variable to be normally distributed we have no reason for rejecting the hypothesis that the causal factors affecting the variable are additive and independent and of equal variance. On the other hand, when we find departure from normality, this may indicate certain forces, such as selection, affecting the variable under study. For instance, bimodality may indicate a mixture of samples from two populations. Skewness of milk yield data may reflect the fact that these were records of selected cows and substandard milk cows were not included in the record.

3. If we assume a given distribution to be normal, we may make predictions and tests of given hypotheses based upon this assumption. An example of such an application is shown below.

You will recall the birth weights of male Chinese children, last illustrated in Box. 4.3. The mean of this sample of 9,465 birth weights is 109.9 oz. and its standard deviation is 13. 593oz. Sampling at random from the birth records of this population, what is the chance of obtaining a birth weight of 151 oz. or heavier? Such a birth weight is considerably above the mean of our sample, the difference being $151 - 109.9 = 41.1$ oz. However, we cannot consult the table of areas of the normal curve with a difference in ounces. We must *standardize* the difference—that is, divide it by the standard deviation to convert it into a standard deviate. When we divide the difference by the standard deviation we obtain $41.1/13.593 = 3.02$. This means that a birth weight of 151 oz. is 3.02 standard deviation units greater than the mean. Assuming that the birth weights are normally distributed we may consult the table of areas of the normal curve (Table **P**), where we find a value of 0.4987 for 3.02 standard deviations. This means that 49.87% of the area of the curve lies between the mean and a point 3.02 standard deviations from it. Conversely, 0.0013 or 0.13% of the area lies beyond 3.02 standard deviation units above the mean. Thus, assuming a normal distribution of birth weights, only 0.13% or 13 out of 10,000 of the infants would have a birth weight of 151 oz. *or farther* from the mean. It is quite improbable that a single sampled item from that population would deviate by so much from the mean, and if a random sample of one weight was obtained from the records of an unspecified population we might therefore be justified in doubting whether the observation did in fact come from the population known to us.

The above probability was calculated from one tail of the distribution. We found the probability of an individual being 3.02 standard deviations *greater* than the mean. If we have no prior knowledge that the individual will be either heavier or lighter than the mean but are merely concerned with how different it is from the population mean, an appropriate question would be: assuming that the individual belongs to the population, what is the probability of observing a birth weight of an individual as deviant from the mean in either direction? That probability must be computed by using both tails of the distribution. The previous probability can be simply doubled, since the normal curve is symmetrical. Thus, $2 \times 0.0013 = 0.0026$. This, too, is so small that we would conclude that a birth weight as deviant as 151 oz. is unlikely to have come from the population represented by our sample of male Chinese children.

We can learn one more important point from this example. Our assumption has been that the birth weights are normally distributed. Inspection of the frequency distribution in Box 4.3, however, shows clearly that the distribution is asymmetrical, tapering to the right. Though there are eight classes above the mean class, there are only six classes below the mean class. In view of this asymmetry, conclusions about one tail of the distribution would not necessarily pertain to the second tail. We calculated that 0.13% of the items would be found beyond 3.02 standard deviations above the mean, which corresponds to 151 oz. In fact our sample contains 20 items $(14 + 5 + 1)$ beyond the 147.5-oz. class, the upper limit of which is 151.5 oz., almost the same as the single birth weight. However, 20 items of the 9465 of the sample is approximately 0.21%, more than the 0.13% expected from the normal frequency distribution. Although it would still be improbable to find a single birth weight as heavy as 151 oz. in the sample, conclusions based on the assumption of normality might be in error if the exact probability were critical for a given test. Our statistical conclusions are only as valid as our assumptions about the data.

6.5 Fitting a normal distribution to observed data

One way to test whether an observed frequency distribution is normally distributed is to calculate expected frequencies for a normal distribution of the same mean and standard deviation as the observed sample. This is the same approach which we used for calculating expected binomial or Poisson frequencies. The procedure is shown in Box 6.1 for the frequency distribution of birth weights of male Chinese children (last shown in Box 4.3). As with the binomial and Poisson distribution, we have not as yet learned a technique for testing the significance of the departures from expectation. This is done by the methods of Chapter 16. From general inspection the data do not fit a normal distribution too well, but seem skewed to the right (see next section).

6.6 Skewness and kurtosis

In the previous section we saw how to calculate expected frequencies for a normal distribution. In many cases an observed frequency distribution will depart obviously from normality, and it is useful to have a statistic that measures the nature and amount of departure. We shall emphasize two types of departure from normality. One is *skewness*, which is another name for asymmetry; skewness means that one tail of the curve is drawn out more than the other. In such curves the mean and the median will not coincide. Curves are called skewed to the right or left, depending upon whether the right or left tails are drawn out. The other type of departure from normality is *kurtosis*, or "peakedness" of a curve. A *leptokurtic* curve has more items near the mean and at the tails, with fewer items in the intermediate regions relative to a normal distribution with the same mean and variance. A *platykurtic* curve has fewer items at the mean and at the tails than the normal curve but has more items in intermediate regions. A bimodal distribution is an extreme platykurtic distribution.

The sample statistics for measuring skewness and kurtosis are called g_1 and g_2, representing population parameters γ_1 and γ_2. These measures of nonnormality were quite popular in the early decades of this century when much stress was laid on the form of distributions of biological variables. Three factors combined to bring the computation of g_1 and g_2 into disfavor. First, the computation is rather complex and time consuming. Second, easier graphic methods became available that give approximate ideas of the shape of frequency distributions, although they do not allow the actual testing of departures from normality. Finally, less emphasis has in general been placed on the form of distributions in recent years. However, most recently, the advent of the digital computer has again put a different light on this situation. In many institutions it is now far simpler to obtain precise estimates of g_1 and g_2 on a digital computer than to carry out a graphic estimate. It is therefore likely that the computation of g_1 and g_2 will again become more fashionable. In certain genetic studies one of us (*RRS*) was able to obtain estimates of g_1 and g_2 for hundreds of frequency distributions in a matter of minutes, an absolutely stupendous task in precomputer days.

A computer program for g_1 and g_2 is shown in Appendix A3.1 for ungrouped data and in Appendix A3.2 for frequency distributions. For those who do not yet have access to a computer the following notes for desk calculator operations are given. The formulas for g_1 and g_2 are so-called *moment statistics*. A central moment in statistics, as in physics, is $(1/n) \sum^n (Y - \bar{Y})^r$, the average of the deviations of all items from the mean, each raised to the power r. The first central moment, $(1/n) \sum (Y - \bar{Y})$, always equals zero, as we have seen in Section 4.6. The second moment, $(1/n) \sum (Y - \bar{Y})^2$, is the variance. The statistic g_1 is the third central moment divided by the cube of the standard deviation, $(1/ns^3) \sum (Y - \bar{Y})^3$, and g_2 is 3 less than the fourth central moment divided by the fourth power of the standard deviation, $(1/ns^4) \sum (Y - \bar{Y})^4 - 3$. The computation of these two statistics is outlined in Box 6.2 for the birth weights of Chinese males. Just as the sample variance had to be corrected for bias by dividing $\sum (Y - \bar{Y})^2$ by $n - 1$ rather than n, so sample g_1 and g_2 need to be corrected to allow for similar bias.

BOX 6.1

Calculation of expected normal frequencies

Birth weights of male Chinese in ounces, from Box 4.3: $n = 9465$; $\overline{Y} = 109.9$; $s = 13.593$.

	(1)	(2)	(3)	(4)	(5)	(6)	(7)	(8)	(9)
	Y	*Lower implied limit of class*	d (*lower limit* $-\overline{Y}$)	$\frac{\lvert d\rvert}{s}$	*Area between* $\frac{\lvert d\rvert}{s}$ *and* $\overline{Y}$	*Relative expected frequencies* $\hat{f}_{rel}$	*Absolute expected frequencies* $\hat{f}$	*Observed frequencies* f	*Deviation from expectation* $(f - \hat{f})$
					0.5000				
	59.5	55.5	−54.4	4.00	0.49997	0.0003	2.8	2	−
	67.5	63.5	−46.4	3.41	0.4997	0.0021	19.9	6	−
	75.5	71.5	−38.4	2.82	0.4976	0.0101	95.6	39	−
	83.5	79.5	−30.4	2.24	0.4875	0.0370	350.2	385	+
	91.5	87.5	−22.4	1.65	0.4505	0.0951	900.1	888	−
	99.5	95.5	−14.4	1.06	0.3554	0.1746	1652.6	1729	+
Mean class	107.5	103.5	− 6.4	0.47	0.1808	0.2286	2163.7	2240	+
	115.5	111.5	1.6	0.12	0.0478	0.2134	2019.8	2007	−
	123.5	119.5	9.6	0.71	0.2612	0.1403	1327.9	1233	−
	131.5	127.5	17.6	1.29	0.4015	0.0684	647.4	641	+
	139.5	135.5	25.6	1.88	0.4699	0.0233	220.5	201	−
	147.5	143.5	33.6	2.47	0.4932	0.0057	54.0	74	+
	155.5	151.5	41.6	3.06	0.4989	0.0010	9.5	14	+
	163.5	159.5	49.6	3.65	0.4999	0.0001	0.9	5	+
	171.5	167.5	57.6	4.24	0.5000	0.0000	0.0	1	+
	179.5	175.5	65.6	4.83	0.5000				
						1.0000	9464.9	9465	

Computational steps

1. In column (1) record class marks in increasing order of magnitude of Y. Add one class mark beyond those observed. Thus in this case add the class mark 179.5.

2. In column (2) record the lower implied limit of the class represented by the class mark in column (1). Once the first lower limit is found correctly (55.5 in this case), the others can be obtained by successive addition of the class interval (in this case 8).

3. Subtract $\bar{Y}$, the mean of the distribution, from each of the lower limits in column (2) and record as d in column (3).

4. Divide the absolute magnitude of d (that is, with sign neglected) by the calculated standard deviation s of the sample and record the result in column (4). The deviations of the lower class limits from the mean are now expressed in standard deviation units (standard deviates).

5. Consult the table of areas of the normal curve (Table **P**) and enter in column (5) the area corresponding to the argument in column (4). Thus in the first row 4.00 standard deviations give you 0.49997 of the area, and 3.41 standard deviations give you 0.4997.

6. For all of the class intervals except the one containing the mean (this is always the last class to have a negative d), the area within a class interval is the difference *with sign neglected* between the area in its line and the area in the line just below it. Thus, to obtain the relative expected frequency 0.0003 in column (6), we compute 0.49997 − 0.4997. The value of $\hat{f}_{rel} = 0.0233$ for the 139.5-oz. class results from subtracting 0.4699 from 0.4932. To obtain the relative expected frequency of the class containing the mean, sum the areas in column (5) for the class containing the mean and for the class in the line beneath it. This yields $0.1808 + 0.0478 = 0.2286$. The arrows connecting items in columns (5) and (6) show these operations graphically. Each value in column (6) is the resultant of the operations indicated by the arrows, these being additions or subtractions of the values in column (5). Arrows are shown only at boundaries where the pattern of computation changes. Should the mean fall exactly on a class limit, the + − arrow pattern changes directly to a − + pattern without a + + intermediary class.

7. The sum of the relative expected frequencies is exactly 1.0000, which is a good check on our computations. If the sum is not close to 1.0, check for computational errors. If the sum is correct and less than 1.0, you may need to extend the frequency distribution by one or more classes at each tail.

8. The absolute expected frequencies in column (7) are obtained by multiplying the relative expected frequencies by $n = 9465$. Their sum is very close to n, which again furnishes a check on the procedure.

9. In column (8) we show the observed frequencies and in column (9) the signs of the deviations from expectation. We note that classes above the mean generally show an excess and those below the mean show a shortage of observed frequencies with respect to expectation. This is what you would expect in a distribution skewed to the right.

These corrections are rather involved and are not shown here, but are included in the instructions furnished in Box 6.2, in which computational formulas are given (analogous to Expression (4.7),

$$\frac{1}{n-1}\left[\sum Y^2 - \frac{(\sum Y)^2}{n}\right]$$

BOX 6.2

Computation of g_1 and g_2 from a frequency distribution.

Birth weights of male Chinese in ounces, from Box 6.1.

(1) Class mark Y	*(2)* Coded class mark Y_c	*(3)* f	*(4)* Y_c^2	*(5)* Y_c^3	*(6)* Y_c^4
59.5	−7	2	49	−343	2401
67.5	−6	6	36	−216	1296
75.5	−5	39	25	−125	625
83.5	−4	385	16	− 64	256
91.5	−3	888	9	− 27	81
99.5	−2	1729	4	− 8	16
107.5	−1	2240	1	− 1	1
115.5	0	2007	0	0	0
123.5	1	1233	1	1	1
131.5	2	641	4	8	16
139.5	3	201	9	27	81
147.5	4	74	16	64	256
155.5	5	14	25	125	625
163.5	6	5	36	216	1296
171.5	7	1	49	343	2401
		9465			

Basic sums

$n = \Sigma f = 9465$ $\quad \Sigma f Y_c = -6626$ $\quad \Sigma f Y_c^2 = 31{,}966$

$\Sigma f Y_c^3 = -51{,}848$ $\quad f Y_c^4 = 301{,}690$

Computational steps

1. Set up a frequency distribution (from low to high) and code the class marks so that the coded middle class mark is zero. The purpose of this is to make the coded class marks as small as possible, because they have to be raised to their third and fourth powers. In this case we coded $Y_c = (Y - 115.5)/8$.
2. It is most convenient for computation to prepare a table of Y_c, Y_c^2, Y_c^3, and Y_c^4 from $Y_c = 1$ to the highest coded class mark, as shown above. Once such a table is prepared it should be saved for future problems.

BOX 6.2 continued

3. Calculate the sums $\sum f$, $\sum fY_c$, $\sum fY_c^2$, $\sum fY_c^3$, and $\sum fY_c^4$.

a. First compute $\sum f$, $\sum fY_c$, and $\sum fY_c^2$ as follows:

i. Enter the first f into the multiplier dials at the extreme right.
ii. Then enter the corresponding coded Y_c^2 at the left end of the keyboard and Y_c at the right (ignore the negative sign before the Y_c).
iii. Multiply.
iv. The f will appear in the short (counter) dials, fY_c^2 will appear in the middle of the long (accumulating) dials and $|fY_c|$ will appear at the right end of the long dials.

If steps **i** through **iii** are now repeated (with the dials locked), the quantities $\sum f$, $\sum fY_c^2$, and $\sum |fY_c|$ will be accumulated. Since some of the Y_c's are negative, one additional step is necessary in order to calculate $\sum fY_c$ correctly. As soon as the class corresponding to a coded class mark of zero is reached, the current $\sum |fY_c|$ (10,147 in the present example) must be written down. After the entire column has been summed, two times this intermediate $\sum |fY_c|$ must be subtracted from the final $\sum |fY_c|$ [in this case $\sum fY_c = 13{,}668 - 2(10{,}147) = -6626$].

b. Next calculate $\sum f$, $\sum fY_c^2$, and $\sum fY_c^3$ in the same manner as above. Enter f into the multiplier dials and $|Y_c^3|$ and Y_c^2 into the keyboard. $\sum |fY_c^3|$ must be recorded after the sign of Y_c changes and the total $\sum fY_c^3$ must be corrected as was $\sum fY_c$. Note that $\sum f$ and $\sum fY_c^2$ should agree with values calculated previously and thus provide a check on the accuracy of the computations.

c. Calculate $\sum f$, $\sum fY_c^3$, and $\sum fY_c^4$, checking $\sum fY_c^3$ in the process.

d. Finally, calculate $\sum f$, $\sum fY_c$, and $\sum fY_c^4$ as a check on the computation of $\sum fY_c$ and $\sum fY_c^4$.

4. Calculate the following statistics from the coded data.

$$\bar{Y}_c = \frac{\sum fY_c}{n} = \frac{-6626}{9465}$$

$$= -0.70005$$

$$s_c^2 = \frac{[\sum fY_c^2 - (\sum fY_c)^2/n]}{(n-1)} = \frac{[31{,}966 - (-6626)^2/9465]}{(9464)} = \frac{27{,}327.44998}{9464}$$

$$= 2.88752$$

$$s_c = 1.69927$$

$$g_1 = \frac{[n\sum fY_c^3 - 3(\sum fY_c)(\sum fY_c^2) + 2(\sum fY_c)^3/n]}{(n-1)(n-2)s_c^3}$$

$$= \frac{[9465(-51{,}848) - 3(-6626)(31{,}966) + 2(-6626)^3/9465]}{(9464)(9463)(2.88752)(1.69927)}$$

$$= 0.18934$$

$$g_2 = \frac{(n+1)\{n\sum fY_c^4 - 4(\sum fY_c)\sum fY_c^3 + [6(\sum fY_c)^2\sum fY_c^2/n] - 3(\sum fY_c)^4/n^2\}}{(n-1)(n-2)(n-3)s_c^4} - \frac{3(n-1)^2}{(n-2)(n-3)}$$

BOX 6.2 continued

$$= \frac{9466\{9465(301{,}690) - 4(-6626)(-51{,}848) + [6(-6626)^2\,31{,}966/9465] - 3(-6626)^4/9465^2\}}{(9464)(9463)(9462)(2.88752)(2.88752)} - \frac{3(9464)^2}{(9463)(9462)}$$

$$= 0.08912$$

5. Decode the mean and standard deviation as follows:

$\bar{Y} = D\bar{Y}_c + C$ where D and C are the multiplicative and additive codes employed to code column (2) in the manner of Appendix A1.2

$$= 8.0(-0.70005) + 115.5$$

$$= 109.8996$$

$$s^2 = D^2 s_c^2$$

$$= (8.0)^2(2.88752)$$

$$= 184.80128$$

$$s = Ds_c$$

$$= 8.0(1.69927)$$

$$= 13.59416$$

The g_1 and g_2 do not require decoding.

6. In a normal frequency distribution γ_1 and γ_2 are both zero. A negative g_1 indicates skewness to the left; a positive g_1, skewness to the right. A negative g_2 indicates platykurtosis, and a positive g_2 shows leptokurtosis.

Compare these results with the graphic analysis in Box 6.3.

for the variance). Although these statistics can also be computed for ungrouped data (computer program A3.1 works this way), it is usually more convenient to group even small samples. The systematic procedure outlined in Box 6.2 is less likely to lead to error than raising uncoded data to the third and fourth power.

In a normal frequency distribution both γ_1 and γ_2 are zero. A negative g_1 indicates skewness to the left, a positive g_1, skewness to the right. A negative g_2 indicates platykurtosis, while a positive g_2 shows leptokurtosis. Thus a repulsed distribution would have a positive g_2, but a clumped distribution should have a negative g_2. Absolute magnitude of g_1 and g_2 does not mean much; these statistics have to be tested for "significance" by methods not yet learned (see Section 7.9 and Box 7.4). It would appear that the birth weights in Box 6.2 are skewed to the right (weights generally are skewed in this manner) and that a very light tendency to leptokurtosis is also present.

6.7 Graphic methods

Computation of expected normal frequencies for comparison with an observed distribution and estimation of skewness and kurtosis in a frequency distribution are tedious procedures without a digital computer. This has encouraged the development of graphic methods that examine the shape of an observed distribution for departures from normality. These methods also permit estimates of the mean and standard deviation of the distribution without computation. Until almost instantaneous access to computers becomes generally available, such methods will continue to fulfill a useful role.

The graphic methods are based on a cumulative frequency distribution. We first met cumulative frequencies in Box 4.1 when computing the median. In Figure 6.4 we saw that a normal frequency distribution graphed in cumulative fashion describes an S-shaped curve, called a sigmoid curve. In Figure 6.5 the ordinate of the sigmoid curve is given as relative frequencies expressed as percentages. The slope of the cumulative curve reflects changes in height of the frequency distribution on which it is based. Thus the steep middle segment of the cumulative normal curve corresponds to the relatively greater height of the normal curve around its mean. The ordinate in Fig-

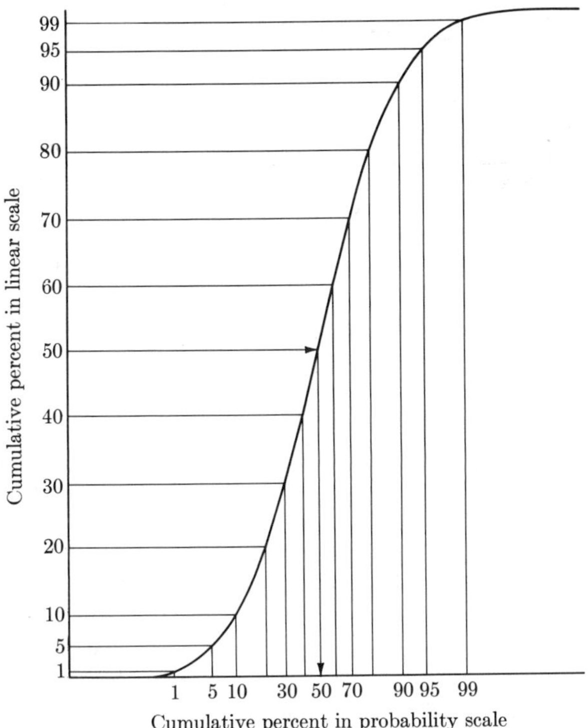

FIGURE 6.5 Transformation of cumulative percentages into normal probability scale.

ures 6.4 and 6.5 is in linear scale. Another possible scale is the *normal probability scale* (often simply called *probability scale*), which can be generated by dropping perpendiculars from the cumulative normal curve, corresponding to given percentages on the ordinate, to the abscissa (as shown in Figure 6.5). The scale represented by the abscissa compensates for the nonlinearity of the cumulative normal curve. It contracts the scale around the median

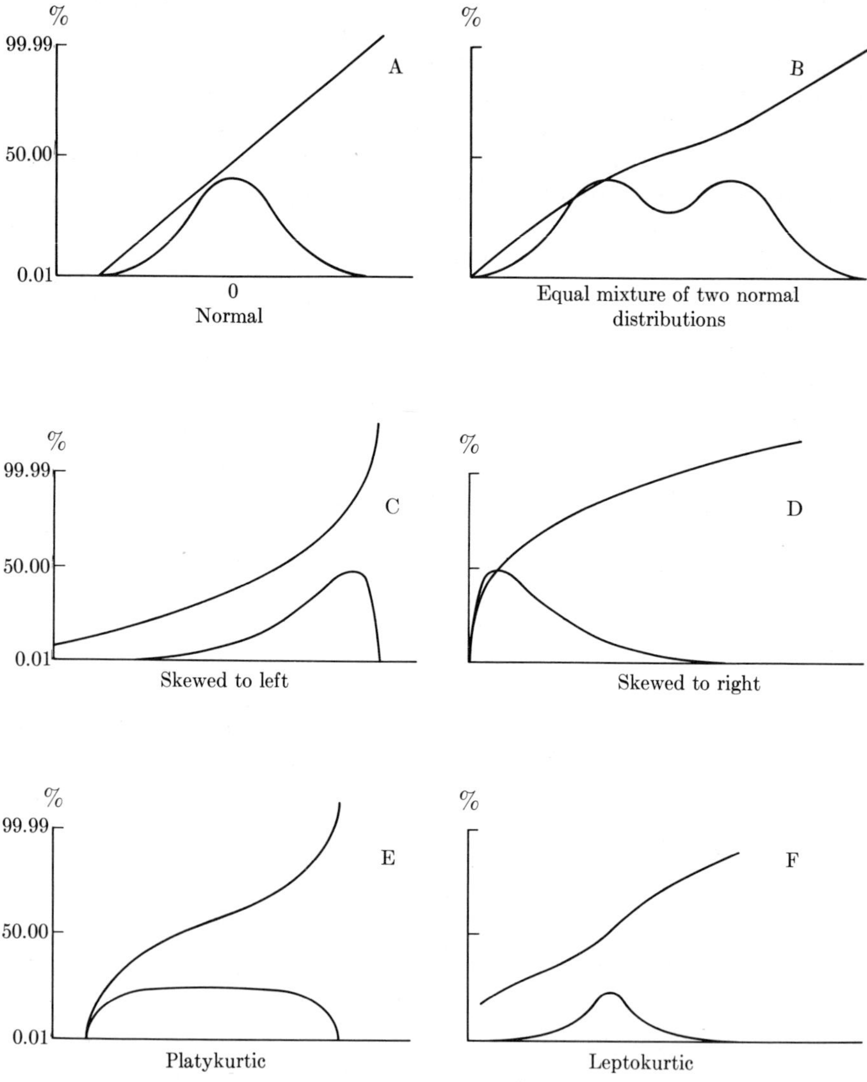

FIGURE 6.6 Examples of some frequency distributions with their cumulative distributions plotted with the ordinate in normal probability scale. (See Boxes 6.3 and 6.4 for explanation.)

and expands it at the low and high cumulative percentages. This scale can be found on *arithmetic* or *normal probability graph paper* (or simply *probability graph paper*), which is generally available. Such paper generally has the long edge graduated in probability scale, while the short edge is in linear scale. Note that there are no 0% or 100% points on the ordinate. This is not possible since the normal frequency distribution extends from negative to positive infinity, and however long we made our line we would never reach the limiting values of 0% and 100%. If we graph a cumulative normal distribution with the ordinate in normal probability scale, it will lie exactly on a straight line. Figure 6.6A shows such a graph drawn on probability paper. Box 6.3 shows you how to use probability paper to examine a frequency distribution for normality and to obtain graphic estimates of its mean and standard deviation.

Figure 6.6 shows a series of frequency distributions variously departing from normality. They are graphed as ordinary frequency distributions with density on a linear scale (ordinate not shown) and as cumulative distributions on probability paper. They are useful as guide lines when examining the distributions of data on probability paper.

The method described in Box 6.3 works best for fairly large samples ($n > 50$). In smaller samples a difference of one item per class would make a substantial difference in the cumulative percentage in the tails. For samples below 50 we use the method of *ranked normal deviates* or *rankits*. These can be explained as follows. Suppose we have a normal distribution whose $\mu = 0$ and $\sigma = \sigma^2 = 1$. Such a distribution is called a *standard normal distribution*. Items in this frequency distribution will be both positive and negative, scattering around the mean of 0, and will be in standard deviation units; that is, their magnitude will at the same time be their distance from the mean in standard deviations. If this is not immediately obvious, consider a normal distribution of mean 10.0 and standard deviation 2.0. If we subtract 10 from each item the new mean of the population will be 0. If we furthermore divide each item by the standard deviation we obtain a so-called standardized frequency distribution with a mean of 0 and a standard deviation (and variance) of 1, similar to the distribution postulated above. The magnitude of each variate will be in standard deviation units.

Supposing now that we take a sample of 5 variates from such a standard normal distribution, order them from lowest to highest variates, and record their scores. We might get a sample as follows: -1.8, -0.6, -0.1, $+0.9$, $+2.1$. If we repeat this process a great number of times we can calculate the average score of the first item in the ordered sample, and similarly for the second, and so forth. Table **AA** shows these average standardized scores or *rankits*. We find the rankits for a sample of 5 to be -1.163, -0.495, 0.0, 0.495, 1.163. These rankits have a variety of uses; one is for testing normality of a frequency distribution, as shown below. When rankits are plotted against

BOX 6.3

Graphic test for normality of a frequency distribution and estimate of mean and standard deviation. Use of arithmetic probability paper for large samples.

Birth weights of male Chinese in ounces, from Box 6.2.

(1) Class mark Y	(2) Upper class limit	(3) f	(4) Cumulative frequencies F	(5) Percent cumulative frequencies
59.5	63.5	2	2	0.02
67.5	71.5	6	8	0.08
75.5	79.5	39	47	0.50
83.5	87.5	385	432	4.6
91.5	95.5	888	1320	13.9
99.5	103.5	1729	3049	32.2
107.5	111.5	2240	5289	55.9
115.5	119.5	2007	7296	77.1
123.5	127.5	1233	8529	90.1
131.5	135.5	641	9170	96.9
139.5	143.5	201	9371	99.0
147.5	151.5	74	9445	99.79
155.5	159.5	14	9459	99.94
163.5	167.5	5	9464	99.99
171.5	175.5	1	9465	100.0
		9465		

Computational steps

1. Prepare a frequency distribution as shown in columns (1), (2), and (3).

2. Form a cumulative frequency distribution as shown in column (4). It is obtained by successive summation of the frequency values. In column (5) express the cumulative frequencies as percentages of total sample size n, which is 9465 in this example. Thus they are the values of column (4) divided by 9465.

3. Graph the upper class limit of each class along the abscissa (in linear scale) against percent cumulative frequency along the ordinate (in probability scale) on normal probability paper (see Figure 6.7). A straight line is fitted to the points by eye, preferably using a transparent plastic ruler, which permits all the points to be seen as the line is drawn. In drawing the line, most weight should be given to the points between cumulative frequencies of 25% to 75%. This is because a difference of a single item may make appreciable changes in the percentages at the tails. We notice that the upper frequencies deviate to the right of the straight line. This is typical of data that are skewed to the right (see Figure 6.6D).

4. Such a graph permits the rapid estimation of the mean and standard deviation

BOX 6.3 continued

of a sample. The mean is approximated by a graphic estimation of the median. The more normal the distribution is, the closer will the mean be to the median. The median is estimated by dropping a perpendicular from the intersection of the 50% point on the ordinate and the cumulative frequency curve to the abscissa (see Figure 6.7). The estimate of the mean of 110.7 oz. is quite close to the computed mean of 109.9 oz.

5. An estimate of the standard deviation is obtained by dropping similar perpendiculars from the intersections of the 15.9% and the 84.1% points with the cumulative curve, respectively. These points enclose the portion of a normal curve represented by $\mu \pm \sigma$. By measuring the difference between these perpendiculars and dividing this by 2, we obtain an estimate of one standard deviation. In this instance the estimate is $s = 13.6$, since the difference is 27.2 oz. divided by 2. This is a close approximation to the computed value of 13.59 oz.

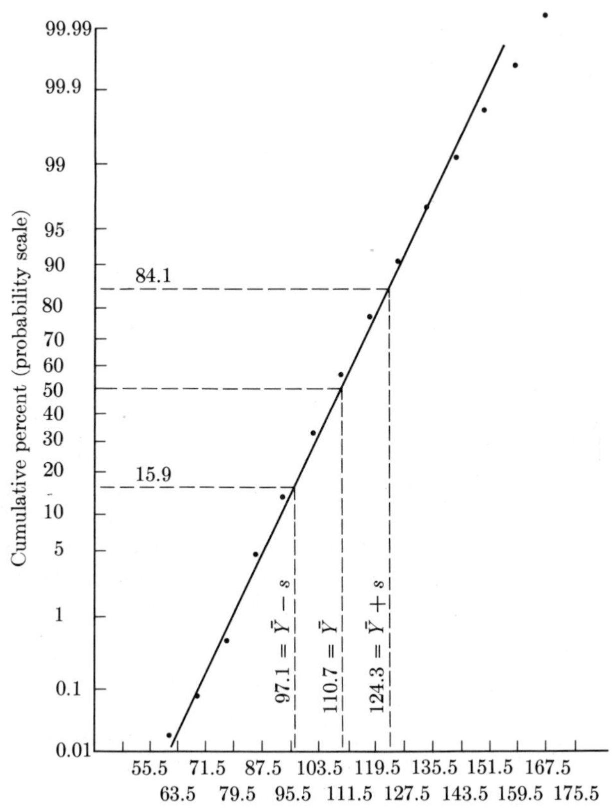

FIGURE 6.7 Graphic analysis of data from Box 6.3.

a ranked array of a normally distributed variable on ordinary graph paper, the points will again lie in a straight line. An illustration of the application of rankits to the testing for normality and for obtaining estimates of μ and σ is shown in Box 6.4 and Figure 6.8. Figure 6.6 may again be used to interpret

BOX 6.4

Graphic test for normality of a frequency distribution and estimate of mean and standard deviation. Rankit method for small samples.

Femur lengths of aphid stem mothers from Box 2.1: $n = 25$.

(1)	*(2)*	*(3)*	*(1)*	*(2)*	*(3)*
Y	*Rankits from Table* **AA**	*Rankits allowing for ties*	Y	*Rankits from Table* **AA**	*Rankits allowing for ties*
3.3	−1.97	−1.97	4.1	0.10	0.15
3.5	−1.52	−1.52	4.1	0.20	0.15
3.6	−1.26	−1.00	4.2	0.30	0.30
3.6	−1.07	−1.00	4.3	0.41	0.58
3.6	−0.91	−1.00	4.3	0.52	0.58
3.6	−0.76	−1.00	4.3	0.64	0.58
3.8	−0.64	−0.47	4.3	0.76	0.58
3.8	−0.52	−0.47	4.4	0.91	1.08
3.8	−0.41	−0.47	4.4	1.07	1.08
3.8	−0.30	−0.47	4.4	1.26	1.08
3.9	−0.20	−0.10	4.5	1.52	1.52
3.9	−0.10	−0.10	4.7	1.97	1.97
3.9	0.00	−0.10			

Computational steps

1. Write down sample arrayed in increasing order of magnitude in column (1). In column (2) put corresponding rankits from Table **AA**. The table gives only the rankits for the half of each distribution greater than the median for any sample size. The other half is the same but is negative in sign. All odd-size samples (such as this one) have 0 as the median value. The rankits for this example are looked up under sample size $n = 25$.

2. A special problem illustrated in this example is the case of ties, or variates of identical magnitude. In such a case we sum the rankit values for the corresponding ranks and find their mean. Thus the −1.00 occupying lines 3 to 6 in column (3) is the average of rankits −1.26, −1.07, −0.91, and −0.76 in column (2), which are the third, fourth, fifth, and sixth rankits for a sample of 25, respectively. Care must be taken when ties involve the median, because then we must sum positive as well as negative rankits. An example would be four rankits of values −0.24, −0.14, −0.05, and +0.05, respectively, which would give an average rankit of −0.095.

BOX 6.4 continued

3. The rankit values (ordinate) are plotted on ordinary graph paper against the variable (abscissa) as shown in Figure 6.8, and a straight line is fitted by eye. The values do not lie exactly on a straight line and there is some suggestion of bimodality in the data, as is also shown in the figure at the bottom of Box 2.1.
4. Mean and standard deviation can be computed as with probability paper (see Box 6.3). For the mean we use the intersection of rankit 0 with the fitted line, and for the standard deviation we use rankits +1.0 and −1.0, representing one standard deviation to each side of the median.

the meaning of the resulting curves. Not too much reliance should be placed on single samples, but repeated trends in different samples are likely to be meaningful.

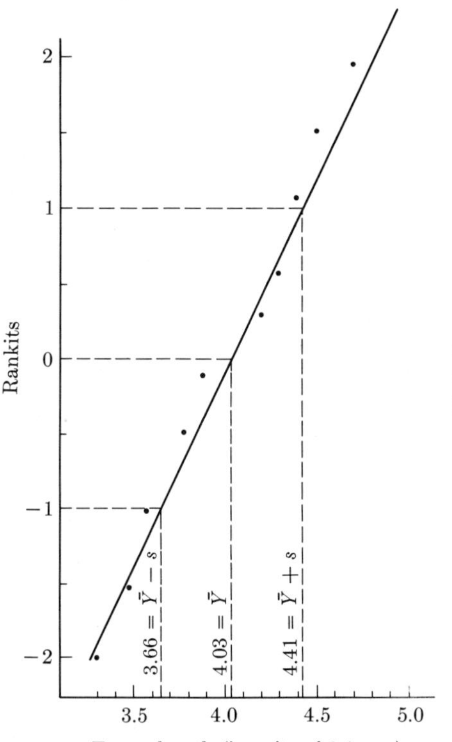

FIGURE 6.8 Graphic analysis of aphid femur length data using rankits. (See Box 6.4 for explanation.)

6.8 Other continuous distributions

Statisticians have described numerous continuous probability density functions and studied their properties. Some observed nonnormal distributions can be made normal by transformation to a suitable scale (discussed in Chapter 13). Even when a variable is normally distributed, various statistics obtained from it (such as the variance) are not normally distributed. We shall study several of these distributions and their applications in subsequent chapters.

Exercises 6

6.1 Perform the following operations on the data of Exercise 2.4. (a) If you have not already done so, make a frequency distribution from the data and graph the results in the form of a histogram. (b) Compute the expected frequencies for each of the classes based on a normal distribution with $\mu = \bar{Y}$ and $\sigma = s$. (c) Graph the expected frequencies in the form of a histogram and compare them with the observed frequencies. (d) Comment on the degree of agreement between observed and expected frequencies.

6.2 Carry out the operations listed in Exercise 6.1 on the transformed data generated in Exercise 2.6.

6.3 Perform a graphic analysis on the data of Exercises 2.4 and 2.6; examine for normality and estimate the mean and standard deviation. Use a table of rankits (Table **AA**).

6.4 Assume you know that the petal length of a population of plants of species X is normally distributed with a mean of $\mu = 3.2$ cm and a standard deviation of $\sigma = 1.8$. What proportion of the population would be expected to have a petal length (a) greater than 4.5 cm? (b) greater than 1.78 cm? (c) between 2.9 and 3.6 cm? ANS. (a) = 0.2353, (b) = 0.7845, and (c) = 0.154.

6.5 Compute g_1 and g_2 for the data given in Exercises 2.4 and 2.6. Do your results agree with what you would expect on the basis of the graphic analyses performed in Exercise 6.3? ANS. g_1 and g_2 for Exercise 2.4 are 0.2981 and 0.0512.

6.6 Perform a graphic analysis of the butterfat data given in Exercise 4.3, using probability paper. In addition, plot the data on probability paper with the abscissa in logarithmic units. Compare the results of the two analyses.

7 ESTIMATION AND HYPOTHESIS TESTING

You are now ready for two very important steps toward learning statistics. In this chapter we provide answers to two fundamental statistical questions that every biologist must answer repeatedly in the course of his work: (1) how reliable are the results I have obtained? and (2) how probable is it that the differences between observed results and those expected on the basis of an hypothesis have been produced by chance alone? The first question, about reliability, is answered through the setting of confidence limits to sample statistics. The second question leads into hypothesis testing. Both subjects belong to the field of statistical inference. The subject matter in this chapter is fundamental to an understanding of any of the subsequent chapters. We therefore urge you to study the material until it has been thoroughly mastered.

In Section 7.1 we shall first of all consider the form of the distribution of means and their variance. In the next section (7.2) we examine the distributions and variances of statistics other than the mean. This brings us to the general subject of standard errors, which are statistics measuring the reliability of an estimate. Confidence limits provide bounds to our estimates of population parameters. We develop the idea of a confidence limit in Section 7.3 and show its application to samples where the true standard deviation is known. However, one usually deals with small, more or less normally distributed samples with unknown standard deviations, in which case the t-distribution must be used. We shall introduce the t-distribution in Section 7.4. The application of t to the computation of confidence limits for statistics of small samples with unknown population standard deviations is shown in Section 7.5. Another important distribution (chi-square) is explained in Section 7.6 and then applied to setting confidence limits for the variance (Section 7.7). The theory of hypothesis testing is introduced in Section 7.8 and applied to a variety of cases exhibiting the normal or t-distributions

(Section 7.9). Finally, Section 7.10 illustrates hypothesis testing for variances by means of the chi-square distribution.

7.1 Distribution and variance of means

We commence our study of the distribution and variance of means with a sampling experiment.

Experiment 7.1. You were asked to retain from Experiment 6.1 the means of the seven samples of 5 housefly wing lengths and the seven similar means of milk yields. We can collect these means from every student in a class, possibly adding them to the sampling results of previous classes, and construct a frequency distribution of these means. For each variable we can also obtain the mean of the seven means, which is a mean of a sample of 35 items. Here again we shall make a frequency distribution of these means, although it takes a considerable number of samplers to accumulate a sufficient number of samples of 35 items for a meaningful frequency distribution.

In Table 7.1 we show a frequency distribution of 1400 means of samples of 5 housefly wing lengths. Consider only columns (1) and (3) for the time being. Actually these samples were not obtained by biometry classes but by

TABLE 7.1

Frequency distribution of means of 1400 random samples of 5 housefly wing lengths. (Data from Table 6.2.) Class marks chosen to give intervals of $\frac{1}{2}\sigma_{\bar{Y}}$ to each side of the parametric mean μ.

	(1) Class mark $\bar{Y}$ (in mm × 10^{-1})	(2) Class mark (in $\sigma_{\bar{Y}}$ units)	(3) f
	39.832	$-3\frac{1}{4}$	1
	40.704	$-2\frac{3}{4}$	11
	41.576	$-2\frac{1}{4}$	19
	42.448	$-1\frac{3}{4}$	64
	43.320	$-1\frac{1}{4}$	128
	44.192	$-\frac{3}{4}$	247
	45.064	$-\frac{1}{4}$	226
$\mu = 45.5 \rightarrow$	45.936	$\frac{1}{4}$	259
	46.808	$\frac{3}{4}$	231
	47.680	$1\frac{1}{4}$	121
	48.552	$1\frac{3}{4}$	61
	49.424	$2\frac{1}{4}$	23
	50.296	$2\frac{3}{4}$	6
	51.168	$3\frac{1}{4}$	3
			1400
	$\bar{\bar{Y}} = 45.480$	$s = 1.778$	$\sigma_{\bar{Y}} = 1.744$

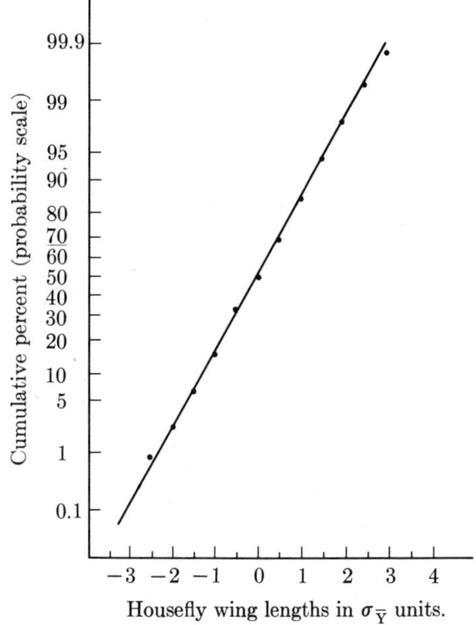

FIGURE 7.1 Graphic analysis of data from Table 7.1.

a digital computer, enabling us to collect these values in a very short time. Their mean and standard deviation are given at the foot of the table. These

TABLE 7.2

Frequency distribution of means of 200 random samples of 35 housefly wing lengths. (Data from Table 6.2.) Class marks chosen to give intervals of $\frac{1}{2}\sigma_{\bar{Y}}$ to each side of the parametric mean μ.

	(1) Class mark Y (in mm $\times 10^{-1}$)	(2) Class mark (in $\sigma_{\bar{Y}}$ units)	(3) f
	44.017	$-2\frac{1}{4}$	2
	44.347	$-1\frac{3}{4}$	7
	44.676	$-1\frac{1}{4}$	19
	45.006	$-\frac{3}{4}$	31
$\mu = 45.5 \rightarrow$	45.335	$-\frac{1}{4}$	46
	45.665	$\frac{1}{4}$	41
	45.994	$\frac{3}{4}$	31
	46.324	$1\frac{1}{4}$	16
	46.653	$1\frac{3}{4}$	4
	46.983	$2\frac{1}{4}$	2
	47.312	$2\frac{3}{4}$	1
			200
	$\bar{Y} = 45.475$	$s = 0.584$	$\sigma_{\bar{Y}} = 0.659$

values are plotted on probability paper in Figure 7.1. Note that the distribution appears quite normal, as does that of the means based on 200 samples of 35 wing lengths shown in Table 7.2 and Figure 7.2. This illustrates an important theorem. *The means of samples from a normally distributed population are themselves normally distributed regardless of sample size n.* Thus we note that the means of samples from the normally distributed housefly wing lengths are normally distributed whether they are based on 5 or 35 individual readings.

Both distributions of means of the heavily skewed milk yields (Tables 7.3 and 7.4, and Figures 7.3 and 7.4) appear to be close to normal distributions. However, the means based on five milk yields do not agree with the normal nearly as well as do the means of 35 items. This illustrates another theorem of fundamental importance in statistics. *As sample size increases, the means of samples drawn from a population of any distribution will approach the normal distribution.* This theorem, when rigorously stated (about sampling from populations with finite variances), is known as the *central limit theorem.* It is illustrated by the near normality of the milk yield means based on 35 items (Table 7.4 and Figure 7.4) as compared with the considerable skewness of the means based on 5 items (Table 7.3 and Figure 7.3). The importance of this theorem is that it permits us to use the normal distribution to make statistical inferences about means of populations in which the items are not at all normally distributed.

The next fact of importance is that the range of the means is considerably less than that of the original items. Thus the wing length means range from

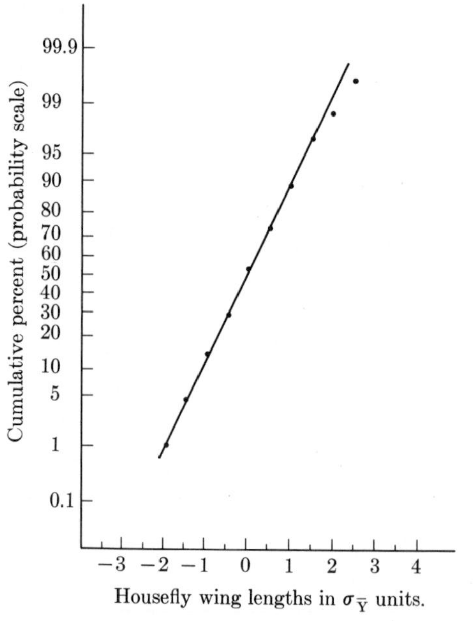

FIGURE 7.2 Graphic analysis of data from Table 7.2.

TABLE 7.3

Frequency distribution of means of 1400 random samples of 5 milk yields. (Data from Table 6.2.) Class marks chosen to give intervals of $\frac{1}{2}\sigma_{\bar{Y}}$ to each side of the parametric mean μ.

	(1) Class mark Y (*in hundreds of lbs.*)	(2) Class mark (*in $\sigma_{\bar{Y}}$ units*)	(3) f
	55.380	$-2\frac{1}{4}$	13
	57.876	$-1\frac{3}{4}$	51
	60.371	$-1\frac{1}{4}$	193
	62.867	$-\frac{3}{4}$	221
	65.362	$-\frac{1}{4}$	273
$\mu = 66.61 \rightarrow$	67.858	$\frac{1}{4}$	235
	70.353	$\frac{3}{4}$	195
	72.849	$1\frac{1}{4}$	124
	75.344	$1\frac{3}{4}$	53
	77.840	$2\frac{1}{4}$	27
	80.335	$2\frac{3}{4}$	8
	82.831	$3\frac{1}{4}$	5
	85.326	$3\frac{3}{4}$	1
	87.822	$4\frac{1}{4}$	1
			1400
	$\bar{Y} = 66.489$	$s = 5.040$	$\sigma_{\bar{Y}} = 4.991$

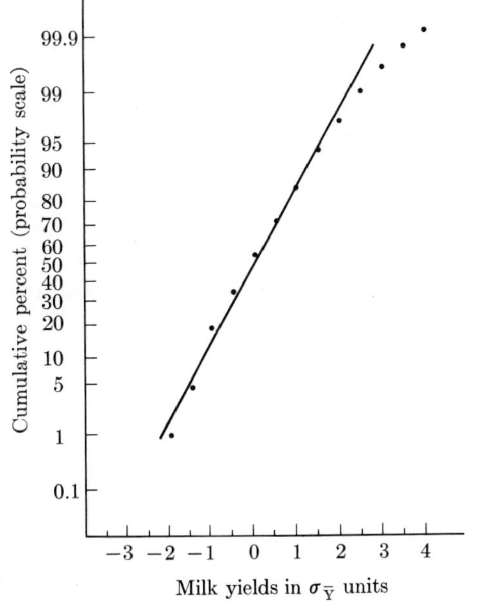

FIGURE 7.3 Graphic analysis of data from Table 7.3.

TABLE 7.4

Frequency distribution of means of 200 random samples of 35 milk yields. (Data from Table 6.2.) Class marks chosen to give intervals of $\frac{1}{2}\sigma_{\bar{Y}}$ to each side of the parametric mean μ.

	(1) Class mark Y (in hundreds of lbs.)	(2) Class mark (in $\sigma_{\bar{Y}}$ units)	(3) f
	62.366	$-2\frac{1}{4}$	5
	63.310	$-1\frac{3}{4}$	9
	64.252	$-1\frac{1}{4}$	16
	65.196	$-\frac{3}{4}$	31
	66.138	$-\frac{1}{4}$	35
$\mu = 66.61 \rightarrow$			
	67.082	$\frac{1}{4}$	42
	68.024	$\frac{3}{4}$	33
	68.968	$1\frac{1}{4}$	19
	69.910	$1\frac{3}{4}$	9
	70.854	$2\frac{1}{4}$	1
			200
	$\bar{Y} = 66.591$	$s = 1.799$	$\sigma_{\bar{Y}} = 1.886$

39.4 to 51.6 in samples of 5 and from 43.9 to 47.4 in samples of 35, but the individual wing lengths range from 36 to 55. The milk yield means range from 54.2 to 89.0 in samples of 5 and from 61.9 to 71.3 in samples of 35, but

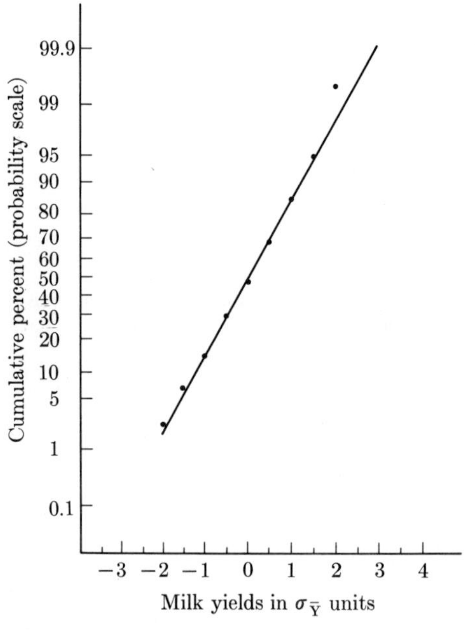

FIGURE 7.4 Graphic analysis of data from Table 7.4.

the individual milk yields range from 51 to 98. Not only do means show less scatter than the items upon which they are based (an easily understood phenomenon if you give some thought to it), but the range of the distribution of the means diminishes as the sample size upon which the means are based increases.

The differences in ranges are reflected in differences in the standard deviation of these distributions. If we calculate the standard deviations of the means in the four distributions in Tables 7.1 through 7.4, we obtain the following values:

	Observed standard deviations of distributions of means	
	$n = 5$	$n = 35$
Wing lengths	1.778	0.584
Milk yields	5.041	1.799

Note that the standard deviations of the sample means based on 35 items are considerably less than those based on 5 items. This is also intuitively obvious. Means based on large samples should be close to the parametric mean and means based on large samples will not vary as much as will means based on small samples. The variance of means is therefore partly a function of the sample size on which the means are based. It is also a function of the variance of the items in the samples. Thus, in the text table above, the means of milk yields have a much greater standard deviation than means of wing lengths based on comparable sample size simply because the standard deviation of the individual milk yields (11.1597) is considerably greater than that of individual wing lengths (3.90).

Mathematical statisticians have worked out the expected value of the variance of sample means. By *expected value* we mean the average value to be obtained by infinitely repeated sampling. Thus if we were to take samples of a means of n items repeatedly and were to calculate the variance of these a means each time, the average of these variances would be the expected value. The expected value of the variance of means based on n items is

$$\sigma^2_{\bar{Y}} = \frac{\sigma^2}{n} \tag{7.1}$$

Consequently the expected standard deviation of means is

$$\sigma_{\bar{Y}} = \frac{\sigma}{\sqrt{n}} \tag{7.2}$$

From this formula it is clear that the standard deviation of means is a function of the standard deviation of items as well as of sample size of means. The greater the sample size, the smaller will be the standard deviation of means. In fact, as sample size increases to a very large number, the standard deviation of means becomes vanishingly small. This makes good sense. Very large sample sizes, averaging many observations, should yield estimates of means less variable than those based on a few items.

When working with samples from a population we do not, of course, know its parametric standard deviation σ, but can only obtain a sample estimate, s, of the latter. Also, we would be unlikely to have numerous samples of size n from which to compute the standard deviation of means directly. Customarily, we therefore have to estimate the standard deviation of means from a single sample by using Expression (7.2), substituting s for σ:

$$s_{\bar{Y}} = \frac{s}{\sqrt{n}} \tag{7.3}$$

Thus we obtain, from the standard deviation of a single sample, an estimate of the standard deviation of means we would expect were we to obtain a collection of means based on equal-sized samples of n items from the same population. As we shall see, this estimate of the standard deviation of a mean is a very important and frequently used statistic.

Table 7.5 illustrates some estimates of the standard deviations of means that might be obtained from random samples of the two populations that we have been discussing. The means of 5 samples of wing lengths based on 5 individuals ranged from 43.6 to 46.8, their standard deviations from 1.095 to 4.827, and the estimate of standard deviation of the means from 0.548 to 2.414. Ranges for the other categories of samples in Table 7.5 similarly included the parametric values of these statistics. The estimates of the standard deviations of the means of the milk yields cluster around the expected value, since they are not dependent on normality of the variates. However, in a particular sample in which by chance the sample standard deviation is a poor estimate of the population standard deviation (as in the second sample of 5 milk yields), the estimate of the standard deviation of means is equally wide of the mark.

There is one point of difference between the standard deviation of items and the standard deviation of sample means. If we estimate a population standard deviation through the standard deviation of a sample, the magnitude of the estimate will not change as we increase our sample size. We may expect that the estimate will improve and will approach the true standard deviation of the population. However, its order of magnitude will be the same, whether the sample is based on 3, 30, or 3000 individuals. This can be clearly seen in Table 7.5. The values of s are closer to σ in the samples based on $n = 35$ than in samples of $n = 5$. Yet the general magnitude is the same in both instances. The standard deviation of means, however, decreases as sample size increases, as is obvious from Expression (7.3). Thus, means based on 3000 items will have a standard deviation only one-tenth that of means based on 30 items. This is obvious from

$$\frac{s}{\sqrt{3000}} = \frac{s}{\sqrt{30} \times \sqrt{100}} = \frac{s}{\sqrt{30} \times 10}$$

TABLE 7.5

Means, standard deviations, and standard deviations of means (standard errors) of five random samples of 5 and 35 housefly wing lengths and Jersey cow milk yields, respectively. (Data from Table 6.2.) Parametric values for the statistics are given in the sixth line of each category.

	(1) $\bar{Y}$	(2) s	(3) $s_{\bar{Y}}$
		Wing lengths	
	45.8	1.095	0.490
	45.6	3.209	1.435
$n = 5$	43.6	4.827	2.159
	44.8	4.764	2.131
	46.8	1.095	0.490
	$\mu = 45.5$	$\sigma = 3.90$	$\sigma_{\bar{Y}} = 1.744$
	45.37	3.812	0.644
	45.00	3.850	0.651
$n = 35$	45.74	3.576	0.604
	45.29	4.198	0.710
	45.91	3.958	0.669
	$\mu = 45.5$	$\sigma = 3.90$	$\sigma_{\bar{Y}} = 0.659$
		Milk yields	
	66.0	6.205	2.775
	61.6	4.278	1.913
$n = 5$	67.6	16.072	7.188
	65.0	14.195	6.348
	62.2	5.215	2.332
	$\mu = 66.61$	$\sigma = 11.160$	$\sigma_{\bar{Y}} = 4.991$
	65.429	11.003	1.860
	64.971	11.221	1.897
$n = 35$	66.543	9.978	1.687
	64.400	9.001	1.521
	68.914	12.415	2.099
	$\mu = 66.61$	$\sigma = 11.160$	$\sigma_{\bar{Y}} = 1.886$

Since our estimates of the sample variances show considerable variation (see Table 7.5), the estimates of the standard deviations of the means also vary accordingly. A poor estimate of σ will yield a poor estimate of $\sigma_{\bar{Y}}$.

7.2 Distribution and variance of other statistics

Just as we obtained a mean and a standard deviation from each sample of the wing lengths and milk yields, so we could also have obtained other statistics from each sample, such as a variance, a median, a coefficient of variation, or a g_1. After repeated sampling and computation we would have frequency distributions for these statistics and would be able to compute their standard deviations just as we did for the frequency distribution of means. In many cases the statistics are normally distributed, as was true for the means. In others the statistics will be distributed normally only if they are based on samples from a normally distributed population, or if they are based on large samples, or if both these conditions hold. In some instances, as in the case of variances, their distribution is never normal. This is illustrated by Figure 7.5, which shows a frequency distribution of the variances from the 1400 samples of housefly wing lengths. We notice that the distribution is strongly skewed to the right, which is characteristic of the distribution of variances.

Standard deviations of various statistics are generally known as *standard errors*. Beginners sometimes get confused by an imagined distinction between standard deviations and standard errors. The standard error of a statistic such as the mean (or g_1) is the standard deviation of a distribution of means (or g_1's) for samples of a given sample size n. Thus the terms standard error and standard deviation are used synonymously with the following exception: it is not customary to use standard error as a synonym of standard deviation of items in a sample or population. Standard error or standard deviation has to be qualified by referring to a given statistic, such as the standard deviation of g_1, which is the same as the standard error of g_1. Used without any qualification the term "standard error" conventionally implies the standard error of the mean. "Standard deviation" used without qualification generally means standard deviation of items in a sample or population. Thus when you read that means, standard deviations, standard errors, and coefficients of

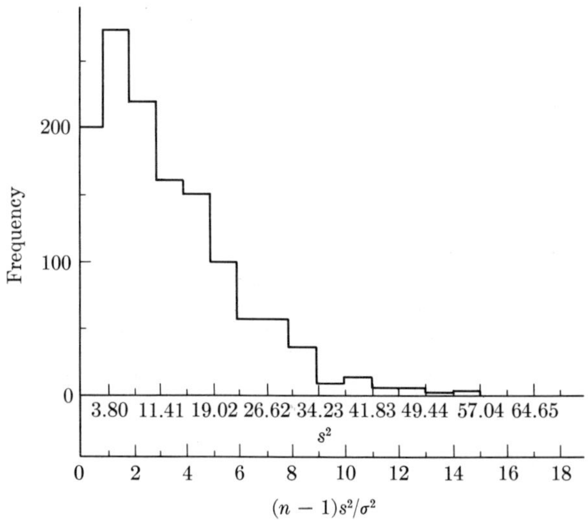

FIGURE 7.5 Histogram of variances based on 1400 samples of 5 housefly wing lengths from Table 6.2. Abscissa is given in terms of s^2 and $(n-1)s^2/\sigma^2$.

BOX 7.1

Standard errors for common statistics.

	(1) Statistic	(2) Estimate of standard error	(3) df	(4) Comments on applicability
1	$\bar{Y}$	$s_{\bar{Y}} = \frac{s}{\sqrt{n}} = \frac{s_Y}{\sqrt{n}} = \sqrt{\frac{s^2}{n}}$	$n-1$	True for any population with finite variance
2	Median	$s_{\text{med}} = (1.2533)s_{\bar{Y}}$	$n-1$	Large samples from normal populations
3	Average deviation (A.D.)	$s_{\text{A.D.}} = \frac{s}{n}\sqrt{\frac{2(n-1)}{\pi}\left[\frac{\pi}{2} + \sqrt{n(n-1)} - n - \arcsin\left\{\frac{1}{(n-1)}\right\}\right]}$	$n-1$	Samples from normal populations
		$\approx 0.6028\frac{s}{\sqrt{n}}$	$n-1$	For large $n(>100)$
4	s	$s_s = (0.7071068)\frac{s}{\sqrt{n}}$	$n-1$	Samples from normal populations $(n > 15)$
5	CV	$s_{CV} = \frac{CV}{\sqrt{2n}}\sqrt{1 + 2\left(\frac{CV}{100}\right)^2}$	$n-1$	Samples from normal populations
		$s_{CV} \approx \frac{CV}{\sqrt{2n}}$	$n-1$	Used when $CV < 15$
6	g_1	$s_{g_1} = \sqrt{\frac{6n(n-1)}{(n-2)(n+1)(n+3)}}$	∞	Samples from normal populations
		$\approx \sqrt{\frac{6}{n}}$	∞	For large $n(>100)$
7	g_2	$s_{g_2} = \sqrt{\frac{24n(n-1)^2}{(n-3)(n-2)(n+3)(n+5)}}$	∞	Samples from normal populations
		$\approx \sqrt{\frac{24}{n}}$	∞	For large $n(>100)$

variation are shown in a table, this signifies that arithmetic means, standard deviations of items in samples, standard deviations of their means (= standard errors of means), as well as coefficients of variation, are displayed. The following summary of terms may be helpful:

Standard deviation $= s = \sqrt{\sum y^2/(n-1)}$
Standard deviation of a statistic St
$=$ standard error of a statistic $St = s_{St}$
Standard error $=$ standard error of a mean
$=$ standard deviation of a mean $= s_{\bar{Y}}$

Standard errors are usually not obtained from a frequency distribution by repeated sampling but are estimated from only a single sample and represent the expected standard deviation of the statistic in case a large number of such samples had been obtained. You will remember that we estimated the standard error of a distribution of means from a single sample in this manner in the previous section.

Box 7.1 lists the standard errors of common statistics. Column (1) lists the statistic whose standard error is described; column (2) shows the formula for the estimated standard error; column (3) gives the degrees of freedom on which the standard error is based (their use is explained in Section 7.5); column (4) provides comments on the range of application of the standard error. The uses of these standard errors will be illustrated in subsequent sections.

7.3 Introduction to confidence limits

The various sample statistics we have been obtaining, such as means, standard deviations, or g_1's, are estimates of population parameters μ, σ, or γ_1, respectively. So far we have not discussed the reliability of these estimates. We first of all wish to know whether the sample statistics are *unbiased estimators* of the population parameters, as discussed in Section 4.7.

But knowing, for example, that $\bar{Y}$ is an unbiased estimate of μ is not enough. We would like to find out how reliable a measure of μ it is. Of course, what we really wish to know is the magnitude of μ, σ^2, and other parameters. However, this is an impossible task unless we sample exhaustively. Thus the true values of the parameters will almost always remain unknown and we commonly estimate reliability of a sample statistic by setting confidence limits to it.

To begin our discussion of this topic, let us start with the unusual case of a population whose parametric mean and standard deviation are known to be μ and σ, respectively. The mean of a sample of n items is symbolized by $\bar{Y}$. The expected standard error of the mean is $\sigma/\sqrt{n}$. As we have seen, the sample means will be normally distributed. Therefore, from Section 6.3, the region from $1.96\sigma/\sqrt{n}$ below μ to $1.96\sigma/\sqrt{n}$ above μ includes 95% of the sample means of size n. Another way of stating this is to consider the ratio $(\bar{Y} - \mu)/(\sigma/\sqrt{n})$. This is the standard deviate of a sample mean from the parametric mean. Since they are normally distributed, 95% of such standard deviates will lie between -1.96 and $+1.96$. We can express this statement symbolically as follows:

$$P\left\{-1.96 \leq \frac{\bar{Y} - \mu}{\sigma/\sqrt{n}} \leq +1.96\right\} = 0.95$$

This means that the probability P that the sample means $\bar{Y}$ will differ by more than 1.96 standard errors $\sigma/\sqrt{n}$ from the parametric mean μ equals 0.95. The expression between the brackets is an inequality, all terms of which can be multiplied by $\sigma/\sqrt{n}$ to yield

$$\{-1.96\sigma/\sqrt{n} \leq (\bar{Y} - \mu) \leq +1.96\sigma/\sqrt{n}\}$$

We can rewrite this expression as

$$\{-1.96\sigma/\sqrt{n} \leq (\mu - \bar{Y}) \leq +1.96\sigma/\sqrt{n}\}$$

because $-a \leq b \leq a$ implies $a \geq -b \geq -a$ which can be written as $-a \leq -b \leq a$. And finally we can transfer $-\bar{Y}$ across the inequality signs, just as in an equation it could be transferred across the equal sign. This yields the final desired expression:

$$P\left\{\bar{Y} - \frac{1.96\sigma}{\sqrt{n}} \leq \mu \leq \bar{Y} + \frac{1.96\sigma}{\sqrt{n}}\right\} = 0.95 \tag{7.4}$$

or

$$P\{\bar{Y} - 1.96\sigma_{\bar{Y}} \leq \mu \leq \bar{Y} + 1.96\sigma_{\bar{Y}}\} = 0.95 \tag{7.4a}$$

This means that the probability, P, that the term $\bar{Y} - 1.96\sigma_{\bar{Y}}$ is less than or equal to the parametric mean μ and that the term $\bar{Y} + 1.96\sigma_{\bar{Y}}$ is greater than or equal to μ is 0.95. We shall call the two terms, $\bar{Y} - 1.96\sigma_{\bar{Y}}$ and $\bar{Y} + 1.96\sigma_{\bar{Y}}$, L_1 and L_2, respectively, the lower and upper 95% *confidence limits* of the mean.

Another way of stating the relationship implied by Expression (7.4a) is that if we repeatedly obtained samples of size n from the population and constructed these limits for each, we could expect 95% of the intervals between these limits to contain the true mean, and only 5% of the intervals would miss μ. The interval from L_1 to L_2 is called a *confidence interval.*

If you were not satisfied to have the confidence interval contain the true mean only 95 times out of 100, you might employ 2.576 as a coefficient in place of 1.960. You may remember that 99% of the area of the normal curve lies between $\mu \pm 2.576\sigma$. Thus, to calculate 99% confidence limits, compute the two quantities $L_1 = \bar{Y} - 2.576\sigma/\sqrt{n}$ and $L_2 = \bar{Y} + 2.576\sigma/\sqrt{n}$ as lower and upper confidence limits, respectively. In this case 99 out of 100 confidence intervals obtained in repeated sampling would contain the true mean. The new confidence interval is wider than the 95% interval (since we have multiplied by a greater coefficient). If you were still not satisfied with the reliability of the confidence limit you could increase it, multiplying the standard error of the mean by 3.291 in order to obtain 99.9% confidence limits. This value (or 3.890 for 99.99% limits) could be found by inverse interpolation in an extensive table of areas of the normal curve (for example, Table 1 in Pearson and Hartley, 1956). The new coefficient would widen the interval further. Notice that you can construct confidence intervals that will be expected to contain μ an increasingly greater percentage of the time. First you would expect to be right 95 times out of 100, then 99 times out of 100, finally 999 times out of 1000. But as your confidence increases your statement becomes vaguer and vaguer, since the confidence interval lengthens. Let us examine this by way of an actual sample.

We obtain a sample of 35 housefly wing lengths from the population of Table 6.2 with known mean ($\mu = 45.5$) and standard deviation ($\sigma = 3.90$). Let us assume that the sample mean is 44.8. We can expect the standard deviation of means based on samples of 35 items to be $\sigma_{\bar{Y}} = \sigma/\sqrt{n} = 3.90/\sqrt{35} = 0.6592$. We compute confidence limits as follows:

$$\text{the lower limit is } L_1 = 44.8 - (1.960)(0.6592) = 43.51$$
$$\text{the upper limit is } L_2 = 44.8 + (1.960)(0.6592) = 46.09$$

Remember that this is an unusual case in which we happen to know the true mean of the population ($\mu = 45.5$) and hence we know that the confidence limits enclose the mean. We expect 95% of such confidence intervals obtained in repeated sampling to include the parametric mean. We could increase the reliability of these limits by going to 99% confidence intervals, replacing 1.960 in the above expression by 2.576 and obtaining $L_1 = 43.10$ and $L_2 = 46.50$. We could have greater confidence that our interval covers the mean, but we could be much less certain about the true value of the mean because of the wider limits. By increasing the degree of confidence still further, say to 99.99%, we could be virtually certain that our confidence limits ($L_1 = 42.24$, $L_2 = 47.36$) contain the population mean, but the bounds enclosing the mean are now so wide as to make our prediction far less accurate than previously.

Experiment 7.2. For the seven samples of 5 housefly wing lengths and the seven similar samples of milk yields last worked with in Experiment 7.1 (Section 7.1), compute 95% confidence limits to the parametric mean for each sample and for the total sample based on 35 items. Base the standard errors of the means on the parametric standard deviations of these populations (housefly wing lengths $\sigma = 3.90$, milk yields $\sigma = 11.1597$). Record how many in each of the four classes of confidence limits (wing lengths and milk yields, $n = 5$ and $n = 35$) were correct—that is, contained the parametric mean of the population. Pool your results with those of other class members.

We tried the experiment on a computer for the 200 samples of 35 wing lengths each, computing confidence limits of the parametric mean employing the parametric standard error of the mean, $\sigma_{\bar{Y}} = 0.6592$. Figure 7.6 shows these 200 confidence intervals plotted parallel to the ordinate. Of these, 194 (97.0%) cross the parametric mean of the population. This should give you a concrete idea of the reliability of confidence limits.

To reduce the width of the confidence interval we have to reduce the standard error of the mean. Since $\sigma_{\bar{Y}} = \sigma/\sqrt{n}$, this can be done only by reducing either the standard deviation of the items or increasing the sample size. The first of these alternatives is frequently not available. If we are sampling from a population in nature we ordinarily have no way of reducing its standard deviation. However, in many experimental procedures we may be able to reduce the variance of the data. For example, if we are studying heart weight in rats and find that its variance is rather large, we might be able to reduce this variance by taking rats of only one age group, in which the variation of heart weight would be considerably less. Thus, by controlling one of the variables of the experiment, the variance of the response variable, heart weight, is reduced. Similarly, by keeping temperature or other environmental variables constant in a procedure, we can frequently reduce the variance of our response variable and hence obtain more precise estimates of population parameters.

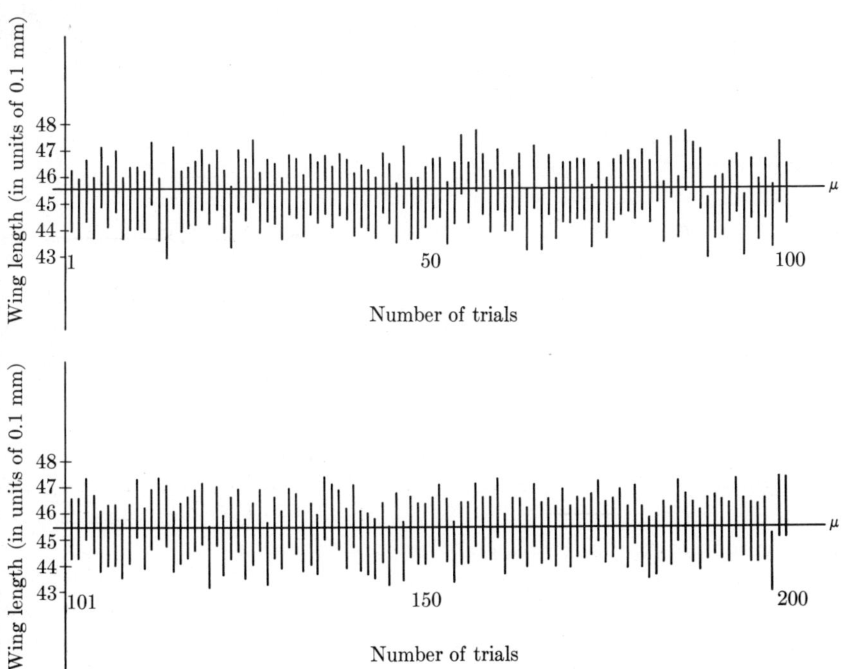

FIGURE 7.6 Ninety-five percent confidence intervals of means of 200 samples of 35 housefly wing lengths, based on parametric standard errors, $\sigma_{\bar{Y}}$. The heavy horizontal line is the parametric mean μ. The ordinate represents the variable.

A more common way to reduce the standard error is by increasing sample size. It is obvious from Expression (7.2) that as n increases the standard error decreases; hence as n approaches infinity, the standard error and the lengths of confidence intervals approach zero. This ties in with what we have learned before: in samples whose size approaches infinity, the sample mean would approach the parametric mean.

We must guard against a common mistake in expressing the meaning of the confidence limits of a statistic. When we have set lower and upper limits (L_1 and L_2, respectively) to a statistic, we imply that the probability of this interval covering the mean is 0.95 or, expressed in another way, that on the average 95 out of 100 confidence intervals similarly obtained would cover the mean. We *cannot state* that there is a probability of 0.95 that the true mean is contained within any particular observed confidence limits, although this may seem to be saying the same thing. The latter statement is incorrect because the true mean is a parameter; hence it is a fixed value and it is therefore either inside the interval or outside it. It cannot be inside a particular interval 95% of the time. It is important, therefore, to learn the correct statement and meaning of confidence limits.

Readers may occasionally encounter the term *fiducial limits*. Frequently this term is used synonymously for confidence limits. However, this is not correct. The two kinds of limits are based on different theoretical foundations (whose details are too intricate to be entered into here). We have already seen the nature of the statement made when applying confidence limits. Saying that the fiducial limits of a parameter lie between L_1 and L_2 at probability $P = 0.05$ means that the probability is no greater than 0.05 that the parameter could have a value other than those covered by the fiducial interval and yield the sample that was in fact obtained. For most ordinarily encountered applications, confidence limits and fiducial limits give the same numerical results, and hence they are frequently confused. But in certain special cases they would give different results. In this book we are entirely concerned with confidence limits.

So far we have only considered means based on normally distributed samples with known parametric standard deviations. We can, however, extend the methods just learned to samples from populations with unknown standard deviations but where the population is known to be normally distributed and the samples are large, say $n \geq 100$. In such cases we use the sample standard deviation for computing the standard error of the mean.

However, when the samples are small ($n < 100$) and we lack knowledge of the parametric standard deviation, we must take into consideration the reliability of our sample standard deviation. To do so we must make use of the so-called t or Student's distribution. We shall learn how to set confidence limits employing the t-distribution in Section 7.5. Before that, however, we shall have to become familiar with this distribution in the next section.

7.4 Student's *t*-distribution

The deviations $\bar{Y} - \mu$ of sample means from the parametric mean of a normal distribution are themselves normally distributed. If these deviations are divided by the parametric standard deviation, $(\bar{Y} - \mu)/\sigma_{\bar{Y}}$, they are still normally distributed, with $\mu = 0$ and $\sigma = 1$. In fact, they form a standard normal distribution (Section 6.7). Subtracting the constant μ from every $\bar{Y}_i$ is simply an additive code (Section 4.8) and will not change the form of the distribution of sample means, which is normal (Section 7.1). Dividing each deviation by the constant $\sigma_{\bar{Y}}$ reduces the variance to unity, but proportionately so for the entire distribution, so that its shape is not altered and a previously normal distribution remains so. We can demonstrate the validity of these statements by the fact that the frequency distributions of housefly wing length means shown in Tables 7.1 and 7.2 are normally distributed.

If, on the other hand, we had calculated the variance s_i^2 of each of the samples and calculated the deviation for each mean $\bar{Y}_i$ as $(\bar{Y}_i - \mu)/s_{\bar{Y}_i}$, where $s_{\bar{Y}_i}$ stands for the estimate of the standard error of the mean of the ith sample, we would have found the distribution of the deviations to be wider and flatter than the normal distribution. This is illustrated in Figure 7.7, which shows the ratio $(\bar{Y}_i - \mu)/s_{\bar{Y}_i}$ for the 1400 samples of five housefly wing lengths of

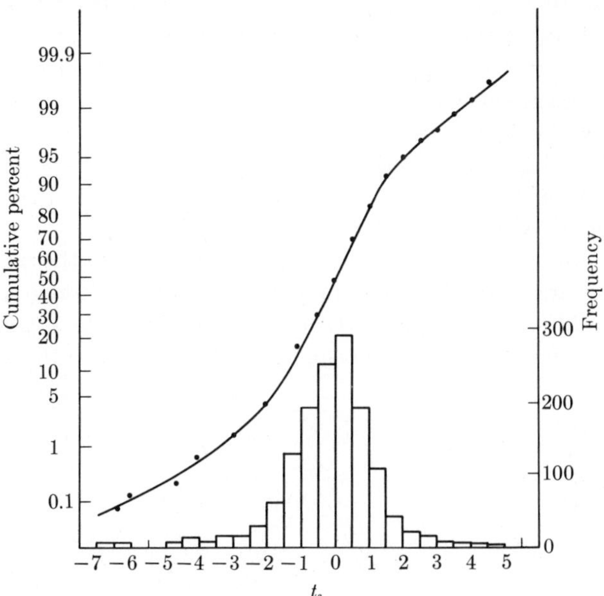

FIGURE 7.7 Distribution of quantity $t_s = (\bar{Y} - \mu)/s_{\bar{Y}}$ along abscissa computed for 1400 samples of 5 housefly wing lengths presented as a histogram and as a cumulative frequency distribution. Right-hand ordinate represents frequencies for the histogram, left-hand ordinate is cumulative frequency in probability scale.

Table 7.1. The new distribution ranges wider than the corresponding normal distribution, because the denominator is the sample standard error rather than the parametric standard error and will sometimes be smaller, at other times greater than expected. This increased variation will be reflected in the greater variance of the ratio $(\bar{Y} - \mu)/s_{\bar{Y}}$. The expected distribution of this ratio is called the *t*-distribution, also known as "Student's" distribution, named after W. S. Gossett who first described it, publishing under the pseudonym "Student." The *t*-distribution is a function with a complicated mathematical formula that need not be presented here.

The *t*-distribution shares with the normal the properties of being symmetric and of extending from negative to positive infinity. However, it differs from the normal in that it assumes different shapes depending on the number of degrees of freedom. By degrees of freedom is meant the quantity $n - 1$, where n is the sample size upon which a variance has been based. It will be remembered that this quantity $n - 1$ is the divisor in obtaining an unbiased estimate of the variance from a sum of squares. The number of degrees of freedom pertinent to a given Student's distribution are the same as the number of degrees of freedom of the standard deviation in the ratio $(\bar{Y} - \mu)/s_{\bar{Y}}$. Degrees of freedom (abbreviated *df* or sometimes ν) can range from 1 to infinity. A *t*-distribution for $df = 1$ deviates most markedly from the normal. As the number of degrees of freedom increases, Student's distribution approaches the shape of the standard normal distribution ($\mu = 0$, $\sigma = 1$) ever more closely, and in a graph the size of this page a *t*-distribution of $df = 30$ is essentially indistinguishable from a normal distribution. At $df = \infty$ the *t*-distribution *is* the normal distribution. Thus we can think of the *t*-distribution as the general case, considering the normal as a special case of Student's distribution with $df = \infty$. Figure 7.8 shows *t*-distributions for 1 and 2 degrees of freedom compared with a normal frequency distribution.

We were able to employ a single table for the areas of the normal curve by coding the argument in standard deviation units. However, since the *t*-distributions differ in shape for differing degrees of freedom, it will be necessary to have a separate *t*-table, corresponding in structure to the table of the

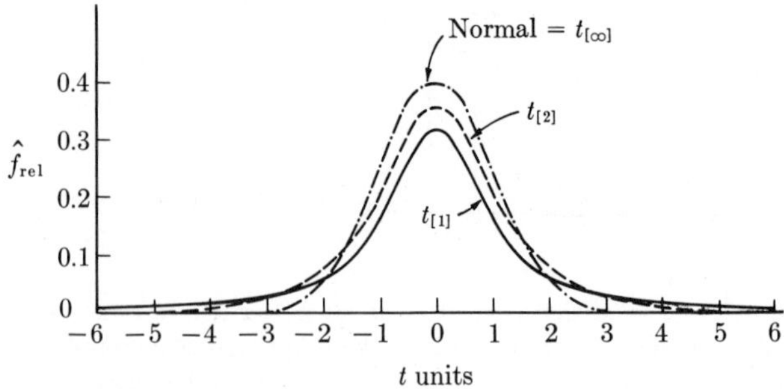

FIGURE 7.8 Frequency curves of *t*-distributions for 1 and 2 degrees of freedom compared with the normal distribution.

areas of the normal curve, for each value of df. This would make for very cumbersome and elaborate sets of tables. Some values of such cumulative t-distributions for $df = 1$ to 24, 30, 40, 60, 120, and ∞ are given in Table 9 of the *Biometrika Tables* (Pearson and Hartley, 1958). If you have that table available, you might try looking up a given value of t, say $t = 1.0$, for $df = 5$. The cumulative area includes 0.81839 of the area of the t-distribution for $df = 5$. The corresponding value for the normal, from a table of the areas of the normal curve, is 0.84134. The latter figure can also be found in the table of the cumulative t-distribution under $df = \infty$. Since a greater proportion of the area is drawn out into the tails, less of the area is found between one standard deviation above and below the mean in the t-distribution than in the normal distribution.

However, the conventional t-tables are differently arranged. Table **Q** shows degrees of freedom and probability as arguments and the corresponding values of t as functions. The probabilities indicate the percent of the area in both tails of the curve (to the right and left of the mean) beyond the indicated value of t. Thus, looking up the *critical value* of t at probability $P = 0.05$ and $df = 5$, we find $t = 2.571$ in Table **Q**. Since this is a two-tailed table, the probability of 0.05 means that there will be 0.025 of the area in each tail beyond a t-value of 2.571. You will recall that the corresponding value for infinite degrees of freedom (for the normal curve) is 1.960. Only those probabilities generally used are shown in Table **Q**. You should become very familiar with looking up t-values in this table. This is one of the most important tables to be consulted. A fairly conventional symbolism is $t_{\alpha[\nu]}$, meaning the tabled t-value for ν degrees of freedom and proportion α in both tails ($\alpha/2$ in each tail), which is equivalent to the t-value for the cumulative probability of $1 - (\alpha/2)$. Try looking up some of these values to become familiar with the table. For example, convince yourself that $t_{.05[7]}$, $t_{.01[3]}$, $t_{.02[10]}$, and $t_{.05[\infty]}$ correspond to 2.365, 5.841, 2.764, and 1.960, respectively.

We shall now employ the t-distribution for the setting of confidence limits to means of small samples.

7.5 Confidence limits based on sample statistics

Armed with a knowledge of the t-distribution we are now able to set confidence limits to the means of samples from a normal frequency distribution whose parametric standard deviation is unknown. The limits are computed as $L_1 = \bar{Y} - t_{\alpha[n-1]}s_{\bar{Y}}$ and $L_2 = \bar{Y} + t_{\alpha[n-1]}s_{\bar{Y}}$ for confidence limits of probability $P = 1 - \alpha$. Thus for 95% confidence limits we use values of $t_{.05[n-1]}$. We can rewrite Expression (7.4a) as

$$P\{L_1 < \mu < L_2\} = P\{\bar{Y} - t_{\alpha[n-1]}s_{\bar{Y}} \leq \mu \leq \bar{Y} + t_{\alpha[n-1]}s_{\bar{Y}}\} = 1 - \alpha \quad (7.5)$$

An example of the application of this expression is shown in Box 7.2. We can convince ourselves of the appropriateness of the t-distribution for setting

BOX 7.2

Confidence limits for μ.

Aphid stem mother femur lengths from Boxes 2.1 and 4.2: $\bar{Y} = 4.004$; $s = 0.366$; $n = 25$.

Values for $t_{\alpha[n-1]}$ from a two-tailed t-table (Table **Q**), where $1 - \alpha$ is the proportion expressing confidence and $n - 1$ are the degrees of freedom:

$$t_{.05[24]} = 2.064 \qquad t_{.01[24]} = 2.797$$

The 95% confidence limits for the population mean, μ, are given by the equations

$$L_1 \text{ (lower limit)} = \bar{Y} - t_{.05[n-1]} \frac{s}{\sqrt{n}}$$

$$= 4.004 - \left(2.064 \frac{0.366}{\sqrt{25}}\right) = 4.004 - 0.151$$

$$= 3.853$$

$$L_2 \text{ (upper limit)} = \bar{Y} + t_{.05[n-1]} \frac{s}{\sqrt{n}}$$

$$= 4.004 + 0.151$$

$$= 4.155$$

The 99% confidence limits are

$$L_1 = \bar{Y} - t_{.01[24]} \frac{s}{\sqrt{n}}$$

$$= 4.004 - \left(2.797 \frac{0.366}{\sqrt{25}}\right) = 4.004 - 0.205$$

$$= 3.799$$

$$L_2 = \bar{Y} + t_{.01[24]} \frac{s}{\sqrt{n}}$$

$$= 4.004 + 0.205$$

$$= 4.209$$

confidence limits to means of samples from a normally distributed population with unknown σ through a sampling experiment.

Experiment 7.3. Repeat the computations and procedures of Experiment 7.2 (Section 7.3), but base standard errors of the means on the standard deviations computed for each sample and use the appropriate t-value in place of a standard normal deviate.

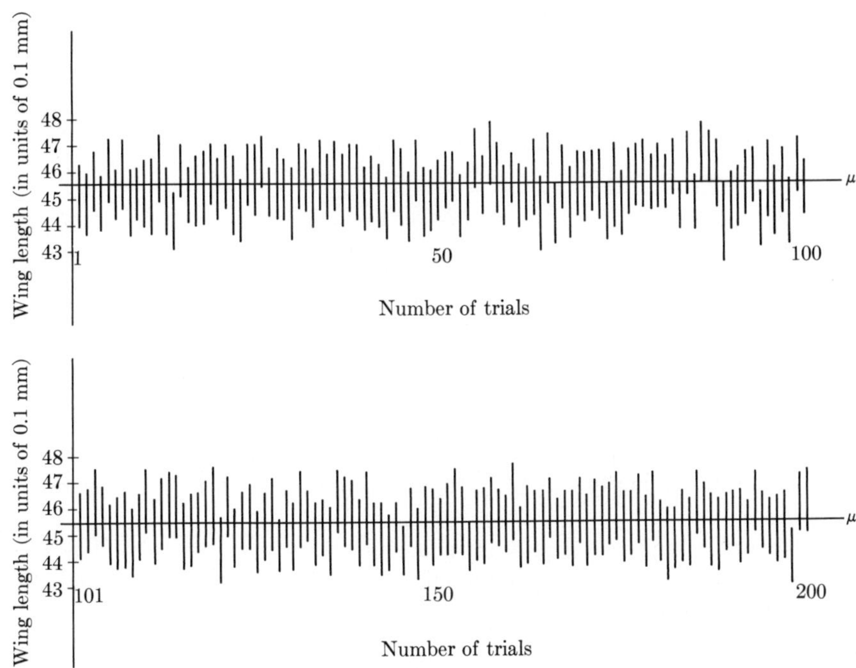

FIGURE 7.9 Ninety-five percent confidence intervals of means of 200 samples of 35 housefly wing lengths, based on sample standard errors, $s_{\bar{Y}}$. The heavy horizontal line is the parametric mean μ. The ordinate represents the variable.

Figure 7.9 shows 95% confidence limits of 200 sampled means of 35 housefly wing lengths, computed with t and $s_{\bar{Y}}$ rather than with the normal curve and $\sigma_{\bar{Y}}$ as in Figure 7.6. We note that 191 (95.5%) of the 200 confidence intervals cross the parametric mean.

We can use the same technique for setting confidence limits to any given statistic as long as it follows the normal distribution. This will apply in an approximate way to all the statistics of Box 7.1. Thus, for example, we may set confidence limits to the coefficient of variation of the aphid femur lengths of Boxes 2.1 and 4.2. These are computed as

$$P\{CV - t_{\alpha[n-1]}s_{CV} \leq CV_P \leq CV + t_{\alpha[n-1]}s_{CV}\} = 1 - \alpha$$

where CV_P stands for the parametric value of the coefficient of variation. Since the standard error of the coefficient of variation approximately equals $s_{CV} = CV/\sqrt{2n}$, we proceed as follows:

$$CV = \frac{100s}{\bar{Y}} = \frac{100(0.3705)}{4.004} = 9.25$$

$$s_{CV} = \frac{9.25}{\sqrt{2 \times 25}} = \frac{9.25}{7.0711} = 1.31$$

$$\begin{aligned} L_1 &= CV - t_{.05[24]}s_{CV} \\ &= 9.25 - (2.064)(1.31) \\ &= 9.25 - 2.70 \\ &= 6.55 \\ L_2 &= CV + t_{.05[24]}s_{CV} \\ &= 9.25 + 2.70 \\ &= 11.95 \end{aligned}$$

When sample size is very large or when σ is known, the distribution is effectively normal. However, rather than turn to the table of areas of the normal curve, we usually simply use $t_{\alpha[\infty]}$, the t-distribution with infinite degrees of freedom.

Although confidence limits are a useful measure of the reliability of a sample statistic, they are not commonly given in scientific publications, the statistic $\pm$ its standard error being cited in their place. Thus, you will frequently see column headings such as "Mean $\pm$ S.E." This indicates that the reader is free to use the standard error to set confidence limits if he is interested. It should be obvious to you from your study of the t-distribution that you could not set confidence limits to a statistic without knowing the sample size on which it is based, n being necessary to compute the correct degrees of freedom. Thus, the occasional citation of means and standard errors without also stating sample size n is to be strongly deplored.

You must also be careful in interpreting such tables of results. Often they are cited as "Mean and Standard Deviation." The meaning of standard deviation is ambiguous here. Is it the standard deviation of the items or the standard deviation of their mean? Desirable headings for $\bar{Y} \pm s_{\bar{Y}}$ are "Means and Their Standard Errors" or "Means and Their Standard Deviations." If the heading says "Means and Standard Deviations" one would ordinarily assume s, not $s_{\bar{Y}}$. Also, if for any reason you wish to cite confidence limits, do not give them as "Statistic $\pm \frac{1}{2}$(Confidence Interval)" because readers would confuse this with the conventional "Mean $\pm$ Standard Error"; state limits as L_1 and L_2, respectively. In literature prior to the mid-1930's a $\pm$ statistic following the mean could quite likely be the probable error, P.E., fashionable at the time but now quite abandoned. In a normal curve one P.E. = 0.6745 S.E. The region bounded by $\bar{Y} \pm$ P.E. therefore constitutes a 50% confidence interval when normality can be assumed, as in a large sample.

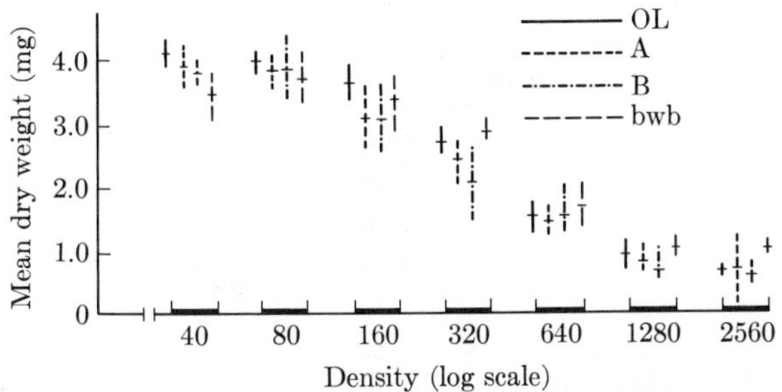

FIGURE 7.10 Weights of four strains of houseflies in pure culture at varying densities. Means and their 95% confidence limits are shown. Ordinate: dry weight in mg; abscissa: density in logarithmic scale, refers to number of eggs per 36 g of medium. The four strains are indicated by different patterns of lines as shown at upper right. (Data from Sullivan and Sokal, 1965.)

It is important to state a statistic and its standard error to a sufficient number of decimal places. The following rule of thumb helps. Divide the standard error by three, then note the decimal place of the first nonzero digit of the quotient; give the statistic significant to that decimal place and provide one further decimal for the standard error. This rule is quite simple, as an example will illustrate. If the mean and standard error of a sample are computed as 2.354 ± 0.363, we divide 0.363 by 3 which yields 0.121. Therefore the mean should be reported accurate to one decimal place, and the standard error should be reported accurate to two decimal places. Thus we report this result as 2.4 ± 0.36. If, on the other hand, the same mean had a standard error of 0.243, dividing this standard error by 3 would have yielded 0.081 and the first nonzero digit would have been in the second decimal place. Thus the mean should have been reported as 2.35 ± 0.243.

When representing the means of samples graphically, confidence limits are frequently indicated. This can be done simply as shown in Figure 7.10. In taxonomic work more elaborate graphs, the so-called Dice-grams, are popular. For a discussion of these, see Sokal (1965). They show for each sample the range of the items (by a line parallel to one of the two axes of the graph representing the variable), the location of the mean of the sample (by a small triangular wedge or line perpendicular to the line representing the range) and the confidence limits of the mean (by a bar resting on the range line around the mean point). An example is shown in Figure 7.11. These limits are sometimes stated as strict confidence limits; more commonly, they are simply two times the standard error of the mean. (This would approximate 95% confidence limits only for sample sizes greater than 30.) Some Dice-grams feature means, total range, standard deviation, confidence limits of the mean, sample variance, and confidence limits of the variance. This elaboration is largely self-defeating since such graphs, although informative, are not easy to digest.

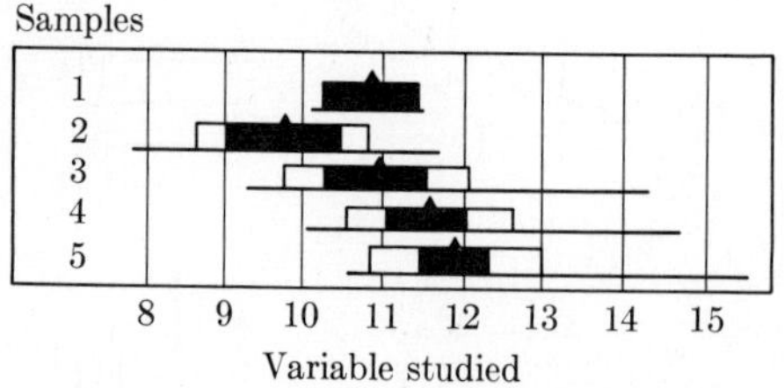

FIGURE 7.11 A "Dice-gram" as modified by Hubbs and Hubbs (1953). The sample mean is shown by the small triangle indicating a location along the abscissa (variable studied). The total range of variation is shown by the heavy horizontal line. The black portion of the bar represents two estimated standard errors to each side of the mean ($2s_{\bar{Y}}$). One-half of each black bar plus the white bar at either end indicates one sample standard deviation (s) on either side of the mean. The samples are arranged vertically in some logical order. (Modified from Pimentel, 1958.)

7.6 The chi-square distribution

Another continuous distribution of great importance in statistics is the distribution of χ^2 (read *chi-square*). We need to learn it now in connection with the distribution and confidence limits of variances.

The chi-square distribution is a probability density function whose values range from zero to positive infinity. Thus, unlike the normal distribution or t, the function approaches the abscissa asymptotically only at the right-hand tail of the curve, not at both tails. The function describing the χ^2-distribution is complicated and will not be given here. As in the case of t, there is not merely one χ^2-distribution, but there is one distribution for each number of degrees of freedom. Therefore, χ^2 is a function of ν, the number of degrees of freedom. Figure 7.12 shows probability density functions for the χ^2-distributions for 1, 2, 3, and 6 degrees of freedom. Notice that the curves are strongly skewed to the right, L-shaped at first, but more or less approaching symmetry for higher degrees of freedom.

We can generate a χ^2-distribution from a population of standard normal deviates. You will recall that we standardize a variable Y_i by subjecting it to the operation $(Y_i - \mu)/\sigma$. Let us symbolize a standardized variable as $Y'_i = (Y_i - \mu)/\sigma$. Now imagine repeated samples of n variates Y_i from a normal population with mean μ and standard deviation σ. For each sample we transform every variate Y_i to Y'_i, as defined above. The quantities $\sum^n Y'^2_i$ computed for each sample will be distributed as a χ^2-distribution with n degrees of freedom. Using the definition of Y'_i, we can rewrite $\sum^n Y'^2_i$ as

$$\sum^n \frac{(Y_i - \mu)^2}{\sigma^2} = \frac{1}{\sigma^2}\sum^n (Y_i - \mu)^2 \tag{7.6}$$

When we change the parametric mean μ to a sample mean in this expression, it becomes

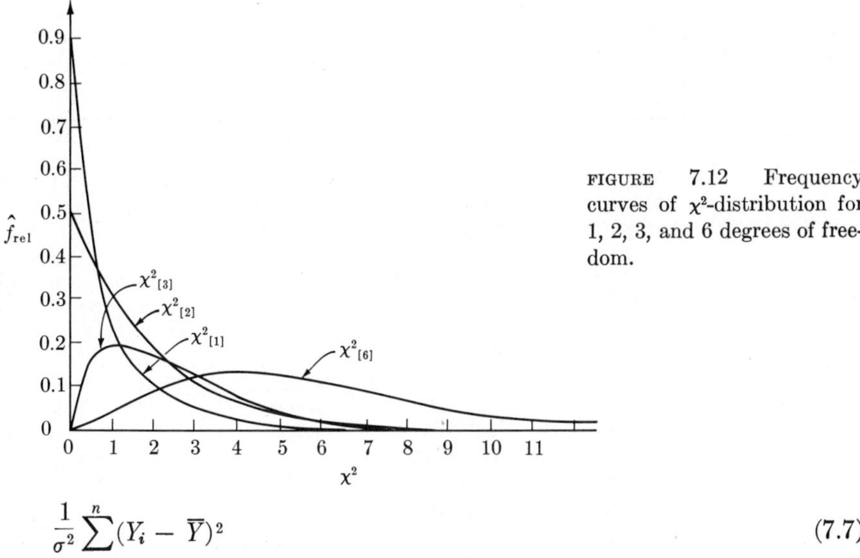

FIGURE 7.12 Frequency curves of χ^2-distribution for 1, 2, 3, and 6 degrees of freedom.

$$\frac{1}{\sigma^2}\sum^{n}(Y_i - \bar{Y})^2 \tag{7.7}$$

which is simply the sum of squares of the variable divided by a constant, the parametric variance. Another common way of stating this expression is

$$\frac{(n-1)s^2}{\sigma^2} \tag{7.8}$$

which simply replaced the numerator of Expression (7.7) with $n - 1$ times the sample variance, which is of course the sum of squares.

If we were to sample repeatedly n items from a normally distributed population, Expression (7.8) computed for each sample would yield a χ^2-distribution with $n - 1$ degrees of freedom. Notice that, although we have samples of n items, we have lost a degree of freedom because we are now employing a sample mean rather than the parametric mean. Figure 7.5, a sample distribution of variances, has a second scale along the abscissa, which is the first scale multiplied by the constant $(n - 1)/\sigma^2$. This scale converts the sample variances s^2 of the first scale into Expression (7.8). Since the second scale is proportional to s^2, the distribution of the sample variance will serve to illustrate a sample distribution approximating χ^2. The distribution is strongly skewed to the right, as would be expected in a χ^2-distribution.

Table 7.6 gives the data for the second abscissa of Figure 7.5, or Expression (7.8) computed for the 1400 random samples of 5 housefly wing lengths. Column (1) shows the lower class limits, column (2) absolute observed frequencies, column (3) relative expected frequencies of a χ^2-distribution for 4 degrees of freedom, and column (4) the absolute expected frequencies. The observed and expected frequencies in columns (2) and (4) do not agree too well. There are more observations near the low values of χ^2. We believe that

TABLE 7.6

Frequency distribution of the ratio $(n - 1)s^2/\sigma^2$ for 1400 samples of 5 housefly wing lengths. (Data from Table 6.2.) Columns 3 and 4 give relative and absolute expected frequencies based on a χ^2-distribution with four degrees of freedom.

(1) $(n-1)s^2/\sigma^2$ *(lower class limit)*	*(2)* *Observed frequencies* f	*(3)* *Relative expected frequencies* $\hat{f}_{rel}$	*(4)* *Absolute expected frequencies* $\hat{f}$
0	135	0.09020	126.3
1	196	0.17404	243.7
2	266	0.17793	249.1
3	215	0.15182	212.5
4	154	0.11871	166.2
5	148	0.08815	123.4
6	97	0.06326	88.6
7	57	0.04431	62.0
8	57	0.03048	42.7
9	36	0.02067	28.9
10	10	0.01387	19.4
11	13	0.00921	12.9
12	7	0.00607	8.5
13	6	0.00398	5.6
14	1	0.00260	3.6
15	2	0.00168	2.4
>15	—	0.00302	4.2
Total	1400	1.00000	1400.0

is so because the housefly wing length data of Table 6.2 are not quite normal. They do not extend from $-\infty$ to $+\infty$ but only from -2.44σ to $+2.44\sigma$. Means sampled from these data apparently were approximately normally distributed, but these ratios seem more sensitive to a departure from normality in the original population. In order to bring the population closer to normality we would have had to furnish a base population $n = 1000$ or more, rather than the mere 100 wing lengths of Table 6.2. The expected frequencies are obtained from tables of the cumulative frequencies of the χ^2-distribution. In such a table [for example, Table 7 in the *Biometrika Tables* (Pearson and Hartley, 1958)] values of χ^2 and ν (degrees of freedom) serve as arguments and the portion of the area of the curve to the right of a given value of χ^2 and for the given *df* is the function. Extensive tables of this sort are too cumbersome and also do not furnish the exact probability levels customarily required. Therefore, the conventional χ^2-tables as shown in Table **R** give these common probabilities and degrees of freedom as arguments and list the χ^2 corresponding to the probability and the *df* as the functions. Each chi-square in Table **R** is the value of χ^2 beyond which the area under the χ^2-distribution for ν degrees of freedom represents the indicated probability. Just as we used subscripts to indicate the cumulative proportion of the area as well as the degrees of freedom represented by a given value of t, we shall subscript χ^2 as follows: $\chi^2_{\alpha[\nu]}$ indicates the χ^2-value to the right of which is found proportion α of the area under a χ^2-distribution for ν degrees of freedom.

Let us learn how to use Table **R**. Looking at the distribution of $\chi^2_{[2]}$ we note that 90% of all values of $\chi^2_{[2]}$ would be to the right of 0.211, but only 5% of all values of $\chi^2_{[2]}$ would be greater than 5.991. Mathematical statisticians have shown that the expected value of $\chi^2_{[\nu]}$ (the mean of a χ^2-distribution) equals its degrees of freedom ν. Thus the expected value of a $\chi^2_{[5]}$-distribution is 5. When we examine 50% values (the medians) in the χ^2-table we notice that they are generally lower than the expected value (the means). Thus for $\chi^2_{[5]}$ the 50% point is 4.351. This illustrates the asymmetry of the χ^2-distribution, the mean being to the right of the median. Our first application of the χ^2-distribution will be in the next section. However, its most extensive use will be in connection with Chapter 16.

7.7 Confidence limits for variances

We saw in the last section that the ratio $(n - 1)s^2/\sigma^2$ is distributed as χ^2 with $n - 1$ degrees of freedom. We take advantage of this fact in setting confidence limits to variances.

First of all we can make the following statement about the ratio $(n - 1)s^2/\sigma^2$:

$$P\left\{\chi^2_{(1-(\alpha/2))[n-1]} \leq \frac{(n-1)s^2}{\sigma^2} \leq \chi^2_{(\alpha/2)[n-1]}\right\} = 1 - \alpha$$

This expression is similar to those encountered in Section 7.3 and implies that the probability P that this ratio will be within the indicated boundary values of $\chi^2_{[n-1]}$ is 0.95. Simple algebraic manipulation of the quantities in the inequality within brackets yields

$$P\{(n-1)s^2/\chi^2_{(\alpha/2)[n-1]} \leq \sigma^2 \leq (n-1)s^2/\chi^2_{(1-(\alpha/2))[n-1]}\} = 1 - \alpha \qquad (7.9)$$

Since $(n - 1)s^2 = \sum y^2$, we can simplify Expression (7.9) to

$$P\{\textstyle\sum y^2/\chi^2_{(\alpha/2)[n-1]} \leq \sigma^2 \leq \sum y^2/\chi^2_{(1-(\alpha/2))[n-1]}\} = 1 - \alpha \qquad (7.10)$$

This still looks like a formidable expression but simply means that if we divide the sum of squares $\sum y^2$ by the two values of $\chi^2_{[n-1]}$ bounding $1 - \alpha$ of the area of the $\chi^2_{[n-1]}$-distribution, the two quotients will enclose the true value of the variance σ^2 with a probability of $P = 1 - \alpha$.

An actual numerical example will make this clear. Suppose we have a sample of 5 housefly wing lengths with a sample variance of $s^2 = 13.52$. If we wish to set 95% confidence limits to the parametric variance we evaluate Expression (7.10) for the sample variance s^2. We first calculate the sum of squares for this sample: $4 \times 13.52 = 54.08$. Then we look up the values for

$\chi^2_{.025[4]}$ and $\chi^2_{.975[4]}$. Since 95% confidence limits are required, α in this case is equal to 0.05. These χ^2-values span between them 95% of the area under the χ^2-curve. They correspond to 11.143 and 0.484, respectively, and the limits in Expression (7.10) then become

$$L_1 = 54.08/11.143 \qquad \text{and} \qquad L_2 = 54.08/0.484$$

or

$$L_1 = 4.85 \qquad \text{and} \qquad L_2 = 111.74$$

This confidence interval is very wide but we must not forget that the sample variance is, after all, based on only 5 individuals. Note also that the interval is asymmetrical around 13.52, the sample variance. This is in contrast to the confidence intervals encountered earlier, which were symmetrical around the sample statistic.

The method described above is called the equal tails method because an equal amount of probability is placed in each tail (for example, $2\frac{1}{2}\%$). It can be shown that in view of the skewness of the distribution of variances this method does not yield the shortest possible confidence intervals. One may wish the confidence interval to be "shortest" in the sense that the ratio L_2/L_1 be as small as possible. Box 7.3 shows how to obtain these shortest unbiased

BOX 7.3

Confidence limits for σ^2. Method of shortest unbiased confidence intervals.

Aphid stem mother femur lengths from Boxes 2.1 and 4.2: $n = 25$; $s^2 = 0.1337$.

The factors from Table **V** for $\nu = n - 1 = 24$ *df* and confidence coefficient $(1 - \alpha) = 0.95$ are

$$f_1 = 0.5943 \qquad f_2 = 1.8763$$

and for a confidence coefficient of 0.99 they are

$$f_1 = 0.5139 \qquad f_2 = 2.3513$$

The 95% confidence limits for the population variance, σ^2, are given by the equations

$$L_1 = (\text{lower limit}) = f_1 s^2 = 0.5943(0.1337) = 0.07946$$

$$L_2 = (\text{upper limit}) = f_2 s^2 = 1.8763(0.1337) = 0.2509$$

The 99% confidence limits are

$$L_1 = f_1 s^2 = 0.5139(0.1337) = 0.06871$$

$$L_2 = f_2 s^2 = 2.3513(0.1337) = 0.3144$$

confidence intervals for s^2 using Table **V**, based on the method of Tate and Klett (1959). With the help of Table **V**, which gives $(n - 1)/\chi^2_{p[n-1]}$, where p is an adjusted value of $\alpha/2$ or $1 - (\alpha/2)$ designed to yield the shortest unbiased confidence intervals. The computation is very simple and much faster than calculating confidence limits by means of χ^2.

7.8 Introduction to hypothesis testing

The most frequent application of statistics in biological research is to test some scientific hypothesis. Statistical methods are important in biology because results of experiments are usually not clearcut and therefore need statistical tests to support decisions between alternative hypotheses. A statistical test examines a set of sample data and, on the basis of an expected distribution of the data, leads to a decision on whether to accept the hypothesis underlying the expected distribution or whether to reject that hypothesis and accept an alternative one. The nature of the tests varies with the data and the hypothesis, but the same general philosophy of hypothesis testing is common to all tests and will be discussed in this section. Study the material below very carefully because it is fundamental to an understanding of every subsequent chapter in this book!

We would like to refresh your memory on the sample of 17 animals of species A, 14 of which were females and 3 of which were males. These data were examined for their fit to the binomial frequency distribution presented in Section 5.2 and their analysis is shown in Table 5.3. We concluded from Table 5.3 that if the sex ratio in the population was 1:1 ($p_♀ = q_♂ = 0.5$), the probability of obtaining a sample with 14 males and 3 females is 0.005,188, making it very unlikely that such a result could be obtained by chance alone. We learned that it is conventional to include all "worse" outcomes—that is, all those that deviate even more from the outcome expected on the hypothesis $p_♀ = q_♂ = 0.5$. Including all worse outcomes the probability is 0.006,363, still a very small value. The above computation is based on the idea of a one-tailed test, in which we are only interested in departures from the 1:1 sex ratio that show a preponderance of females. If we have no preconception about the direction of the departures from expectation, we must calculate the probability of obtaining a sample as deviant as 14 females and 3 males *in either direction* from expectation. This requires the probability either of obtaining a sample of 3 females and 14 males (and all worse samples) or of obtaining 14 females and 3 males (and all worse samples). Such a test is two-tailed and, since the distribution is symmetrical, we double the previously discussed probability to yield 0.012,726.

What does this probability mean? It is our hypothesis that $p_♀ = q_♂ = 0.5$. Let us call this hypothesis H_0, the *null hypothesis*, which is the hypothesis under test. It is called the null hypothesis because it assumes that there is no real difference between the true value of p in the population from which we sampled and the hypothesized value of $\hat{p} = 0.5$; for example, in the present example we believe that the only reason our sample does not exhibit a 1:1 sex ratio is because of sampling error. If the null hypothesis $p_♀ = q_♂ =$

0.5 is true, then approximately 13 samples out of 1000 will be as deviant or more deviant than this one in either direction *by chance alone*. Thus, it is quite *possible* to have arrived at a sample of 14 females and 3 males by chance, but it is not very *probable*, since so deviant an event would occur only about 13 out of 1000 times or 1.3% of the time. If we actually obtain such a sample, we may make one of two decisions. We may decide that the null hypothesis is in fact true (that is, the sex ratio is 1:1) and that the sample obtained by us just happened to be one of those in the tail of the distribution, or we may decide that so deviant a sample is too improbable an event to justify acceptance of the null hypothesis. We may therefore decide that the hypothesis about the sex ratio being 1:1 is not true. Either of these decisions may be correct, depending upon the truth of the matter. If in fact the 1:1 hypothesis is correct, then the first decision (to accept the null hypothesis) will be correct. If we decide to reject the hypothesis under these circumstances, we commit an error. *The rejection of a true null hypothesis is called a type I error.* On the other hand, if in fact the true sex ratio of the population is other than 1:1, the first decision (to accept the 1:1 hypothesis) is an error, a so-called *type II error, which is the acceptance of a false null hypothesis.* Finally, if the 1:1 hypothesis is not true and we do decide to reject it, then we again make the correct decision. Thus, there are two kinds of correct decisions, accepting a true null hypothesis and rejecting a false null hypothesis, and two kinds of errors, type I, rejecting a true null hypothesis, and type II, accepting a false null hypothesis.

Before we carry out a test, we have to decide what magnitude of type I error (rejection of true hypothesis) we are going to allow. Even when we sample from a population of known parameters, there will always be some samples which by chance are very deviant. The most deviant of these are likely to mislead us into believing our hypothesis H_0 to be untrue. If we permit 5% of samples to lead us into a type I error, then we shall reject 5 out of 100 samples from the population, deciding that these are not samples from the given population. In the distribution under study, this means that we would reject all samples of 17 animals containing 13 of one sex plus 4 of the other sex. This can be seen by referring to column (3) of Table 7.7, where the expected frequencies of the various outcomes on the hypothesis $p_{♀} = q_{♂} = 0.5$ are shown. This table is an extension of the earlier Table 5.3, which showed only a tail of this distribution. Actually, you would obtain a type I error slightly less than 5% if we sum relative expected frequencies for both tails starting with the class of 13 of one sex and 4 of the other. From Table 7.7 it can be seen that the relative expected frequency in the two tails will be $2 \times 0.0245209 = 0.0490418$. In a discrete frequency distribution, such as the binomial, we cannot calculate errors of exactly 5% as we can in a continuous frequency distribution, where we can measure off exactly 5% of the area. If we decide on an approximate 1% error, we would reject the hypothesis $p_{♀} = q_{♂}$ for all samples of 17 animals having 14

TABLE 7.7

Relative expected frequencies for samples of 17 animals under two hypotheses. Binomial distribution.

(1)	*(2)*	*(3)*	*(4)*
♀♀	♂♂	$H_0: p_♀ = q_♂ = \frac{1}{2}$ $\hat{f}_{rel}$	$H_1: p_♀ = 2q_♂ = \frac{2}{3}$ $\hat{f}_{rel}$
17	0	0.0000076	0.0010150
16	1	0.0001297	0.0086272
15	2	0.0010376	0.0345086
14	3	0.0051880	0.0862715
13	4	0.0181580	0.1509752
12	5	0.0472107	0.1962677
11	6	0.0944214	0.1962677
10	7	0.1483765	0.1542104
9	8	0.1854706	0.0963815
8	9	0.1854706	0.0481907
7	10	0.1483765	0.0192763
6	11	0.0944214	0.0061334
5	12	0.0472107	0.0015333
4	13	0.0181580	0.0002949
3	14	0.0051880	0.0000421
2	15	0.0010376	0.0000042
1	16	0.0001297	0.0000002
0	17	0.0000076	0.0000000
	Total	1.0000002	0.9999999

or more of one sex (from Table 7.7 we find the $\hat{f}_{rel}$ in the tails equals $2 \times 0.0063629 = 0.0127258$). Thus, the smaller the type I error we are prepared to accept, the more deviant a sample has to be for us to reject the null hypothesis H_0. Your natural inclination might well be to have as little error as possible. You may decide to work with an extremely small type I error, such as 0.1% or even 0.01%, accepting the null hypothesis unless the sample is extremely deviant. The difficulty with such an approach is that although guarding against an error of the first kind, you might be falling into an error of the second kind (type II) of accepting the null hypothesis when in fact it is not true and an alternative hypothesis H_1 is true. Presently we shall show how this comes about.

First let us learn some more terminology. Type I error is most frequently expressed as a probability and is symbolized by α. When expressed as a percentage it is also known as the *significance level.* Thus a type I error of $\alpha = 0.05$ corresponds to a significance level of 5% for a given test. When we cut off on a frequency distribution areas proportional to α, the type I error, the portion of the abscissa under the area that has been cut off is called the *rejection region* or *critical region* of a test, and the portion of the abscissa that would lead to acceptance of the null hypotheses is called the *acceptance region*. Figure 7.13A is a bar diagram showing the expected distribution of outcomes in the sex ratio example, given H_0. The dotted lines separate approximate 1% rejection regions from the 99% acceptance region.

Now let us take a closer look at the type II error. This is the probability of accepting the null hypothesis when in fact it is false. If you try to evaluate the probability of type II error, you immediately run into a problem. If the null hypothesis H_0 is false, some other hypothesis H_1 must be true. But unless you can specify H_1 you are not in a position to calculate type II error. An example will make this clear immediately. Suppose in our sex ratio case we have only two reasonable possibilities—(1) our old hypothesis $H_0: p_{♀} = q_{♂}$, or (2) an alternative hypothesis $H_1: p_{♀} = 2q_{♂}$, which states that the sex ratio is 2:1 in favor of females so that $p_{♀} = \frac{2}{3}$ and $q_{♂} = \frac{1}{3}$. We now have to calculate expected frequencies for the binomial distribution $(p_{♀} + q_{♂})^k = (\frac{2}{3} + \frac{1}{3})^{17}$ to find the probabilities of the various outcomes under this hypothesis. These are shown graphically in Figure 7.13B and are tabulated and compared with expected frequencies of the earlier distribution in Table 7.7.

Suppose we had decided on a type I error of $\alpha \approx 0.01$ ($\approx$ means "approximately equal to") as shown in Figure 7.13A. At this significance level we would accept the H_0 for all samples of 17 having 13 or fewer animals of one sex. Approximately 99% of all samples will fall into this category. However, what if H_0 is not true and H_1 is true? Clearly, from the population represented by hypothesis H_1 we could also obtain outcomes in which one sex was represented 13 or fewer times in samples of 17. We have to calculate what proportion of the curve representing hypothesis H_1 will overlap the acceptance region of the distribution representing hypothesis H_0. In this case we find that 0.8634 of the distribution representing H_1 overlaps the acceptance region of H_0 (see Figure 7.13B). Thus, if H_1 is really true (and H_0 correspondingly false), we would erroneously accept the null hypothesis 86.34% of the time. This percentage corresponds to the proportion of samples from H_1 that fall within the limits of the acceptance regions of H_0. This proportion is called β, the type II error expressed as a proportion. In this example β is quite large. Clearly a sample of 17 animals is unsatisfactory to discriminate between the two hypotheses. Though 99% of the samples under H_0 would fall in the acceptance region, fully 86% would do so under H_1. A single sample that falls in the acceptance region would not enable us to reach a decision between the hypotheses with a high degree of reliability. If the sample had 14 or more females, we would conclude that H_1 was correct. If it had 3 or less females we might conclude that neither H_0 nor H_1 was true. As H_1 approaches H_0 (as in $H_1: p_{♀} = 0.55$, for example), the two distributions would overlap more and more and the magnitude of β would increase, making discrimination between the hypotheses even less likely. Conversely, if H_1 represented $p_{♀} = 0.9$, the distributions would be much farther apart and type II error β would be reduced. Clearly, then, the magnitude of β depends, among

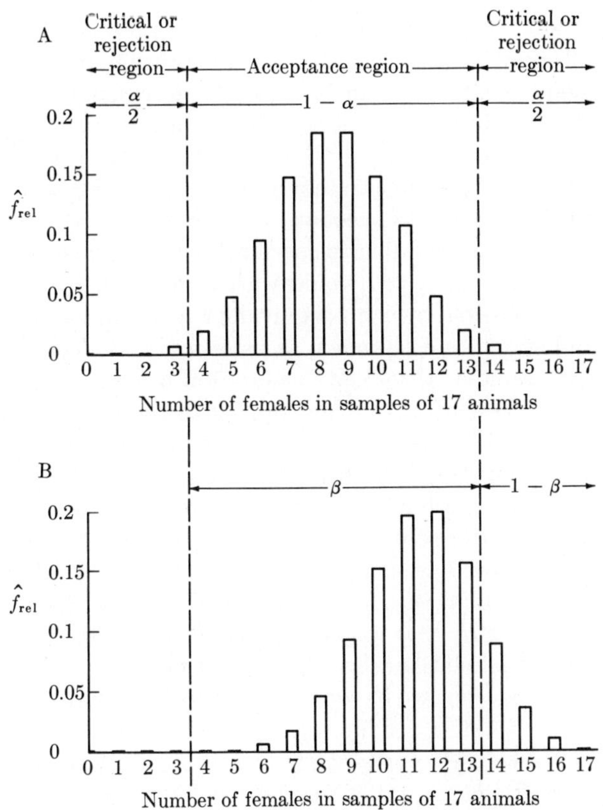

FIGURE 7.13 Expected distributions of outcomes when sampling 17 animals from two hypothetical populations. A. $H_0 : p_♀ = q_♂ = \frac{1}{2}$. B. $H_1 : p_♀ = 2q_♂ = \frac{2}{3}$. Dotted lines separate critical regions from acceptance region of the distribution of Figure A. Type I error α equals approximately 0.01.

other things, on the parameters of the alternative hypothesis H_1 and cannot be specified without knowledge of the latter.

When the alternative hypothesis is fixed as in the previous example ($H_1 : p_♀ = 2q_♂$), the magnitude of the type I error α we are prepared to tolerate will determine the magnitude of the type II error β. The smaller the rejection region α in the distribution under H_0, the greater will be the acceptance region $1 - \alpha$ in this distribution. The greater $1 - \alpha$, however, the greater will be its overlap with the distribution representing H_1, and hence the greater will be β. Convince yourself of this in Figure 7.13. By moving the dotted lines outward we are reducing the critical regions representing type I error α in diagram A. But as the dotted lines move outward, more of the distribution of H_1 in diagram B will lie in the acceptance region of the null hypothesis. Thus, by decreasing α we are increasing β and in a sense defeating our own purposes. In most applications, scientists would wish to keep both of these errors small, since they do not wish to reject a null hypothesis when it is true, nor do they wish to accept it when another hypothesis is correct. We shall see below what steps can be taken to decrease β while holding α constant at a preset level. However, we should note that there are special applications, often nonscientific, in which one type of error would be less serious than the other and our strategy of testing or method of procedure would obviously take this into account. Thus, if you were a manufacturer producing a certain item according to specifications (which correspond to the null hypothesis in this case), you would wish to maximize your profits and to reject as few as possible, which is equivalent to making α small. That is why in industrial statistics α is known as *producers' risk*. You might not be as concerned with samples that really came from a population specified by the alternative hypothesis H_1 but which appear to conform with H_0 because they fall within its acceptance range. Such products could conceivably be marketed as conforming to specification H_0. On the other hand, as a consumer you would not mind so much a large value of α, representing a large proportion of rejects in the manufacturing process. You would, however, be greatly concerned about keeping β as small as possible, since you would not wish to accept items as conforming to H_0, which in reality were samples from the population specified by H_1 that might be of inferior quality. For this reason β is known in industrial statistics as *consumers' risk*.

Let us summarize what we have learned up to this point. When we have to carry out a statistical test, we first specify a null hypothesis H_0 and establish a significance level that corresponds to a probability of α for a type I error. In the case of the sex ratios we defined $H_0: p_♀ = q_♂$ and $\alpha \approx 0.01$. Having done this, we take a sample and test whether the sample statistic is within the acceptance region of the null hypothesis. Our sample turned out to be 14 females and 3 males. This falls beyond the acceptance region; hence we reject the null hypothesis and conclude that this sample came from a population in which $p_♀ \neq q_♂$. If we can specify an alternative hypothesis we can calculate the probability of type II error. In this case $H_1: p_♀ = 2q_♂$ and $\beta = 0.8634$. This is the probability of accepting the null hypothesis when, in fact, the alternative hypothesis is true. In certain special situations in which the alternative hypotheses can be clearly specified, as in genetics and in this example, one might then test the previous alternative hypothesis, changing it into the null hypothesis. Thus, we might now wish to test whether the true sex ratio is $2♀♀ : 1♂$. From Figure 7.13B it is obvious that the probability of 14 males and 3 females is very small indeed and can be ignored. The probability of obtaining 14 or more females under the new null hypothesis is $1 - \beta$ of the old alternative hypothesis, as illustrated in Figure 7.13b. Hence it would be 0.1366 and, if we accept $\alpha = 0.05$, we cannot reject the new null hypothesis.

Significance levels can be varied at will by the investigator. However, an investigator is limited because for many tests cumulative probabilities of the appropriate distributions have not been tabulated and he has therefore to make use of published probability levels. These are commonly 0.05, 0.01, and 0.001, although several others are occasionally encountered. When a null hypothesis has been rejected at a specified level of α, we say that the sample is *significantly different* from the parametric or hypothetical population at probability $P \leq \alpha$. Generally, values of α greater than 0.05 are not considered to be *statistically significant*. A significance level of 5% ($P = 0.05$) corresponds to one type I error in 20 trials, a level of 1% ($P = 0.01$) to one error in 100 trials. Significance levels less than 1% ($P \leq 0.01$) are nearly always adjudged significant; those between 5% and 1% may be considered significant at the discretion of the investigator. Since statistical significance has a special technical meaning (H_0 rejected at $P \leq \alpha$), we shall use the adjective significant only in this sense; its use in scientific papers and reports, unless such a technical meaning is clearly implied, should be discouraged. For general descriptive purposes synonyms such as important, meaningful, marked, noticeable, and others can serve to underscore differences and effects.

A brief remark on null hypotheses represented by asymmetrical probability distributions is in order here. Suppose our null hypothesis in the sex ratio case had been $H_0: p_♀ = \frac{2}{3}$, as discussed above. The distribution of samples of 17 offspring from such a population is shown in Figure 7.13B. It is clearly asymmetrical and for this reason the critical regions have to be defined independently. For a given two-tailed test we can either double the probability P of a deviation in the direction of the closer tail and compare $2P$ with α, the conventional level of significance; or, we can compare P with $\alpha/2$, half the conventional level of significance. In this latter case 0.025 is the maximal value of P conventionally considered significant.

We shall review what we have learned by means of a second example, this time involving a continuous frequency distribution—the normally distributed housefly wing lengths—of parametric mean $\mu = 45.5$ and variance $\sigma^2 = 15.21$. Means based on 5 items sampled from these will also be normally distributed as was demonstrated in Table 7.1 and Figure 7.1. Let us assume that someone presents you with a single sample of 5 housefly wing lengths and you wish to test whether they could belong to the specified population. Your null hypothesis will be $H_0: \mu = 45.5$ or $H_0: \mu = \mu_0$, where μ is the true mean of the population from which you sampled and μ_0 stands for the hypothetical parametric mean of 45.5. We shall assume for the moment that we have no evidence that the variance of our sample is very much greater or smaller than the parametric variance of the housefly wing lengths. If it were, it would be unreasonable to assume that our sample comes from the specified population. There is a critical test of the assumption about the sample variance, which we shall take up later. The curve at the center of Figure 7.14 represents the expected distribution of means of samples of 5 housefly wing lengths from the specified population. Acceptance and rejection regions for a type I error $\alpha = 0.05$ are delimited along the abscissa. The boundaries of

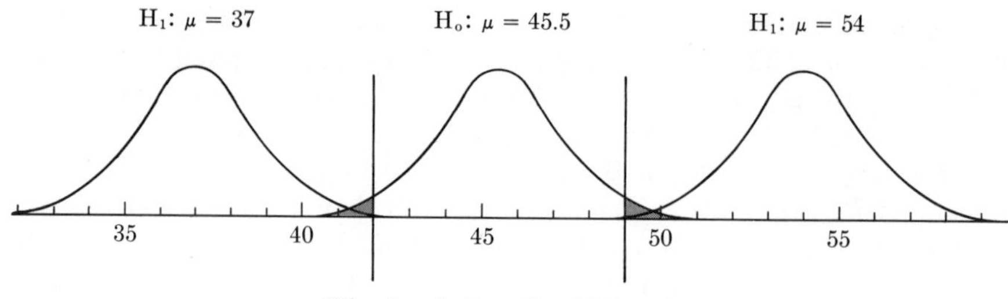

FIGURE 7.14 Expected distribution of means of samples of 5 housefly wing lengths from normal populations specified by μ as shown above curves and $\sigma_{\bar{Y}} = 1.744$. Center curve represents null hypothesis, $H_0{:}\mu = 45.5$, curves at sides represent alternative hypotheses, $\mu = 37$ or $\mu = 54$. Vertical lines delimit 5% rejection regions for the null hypothesis ($2\frac{1}{2}\%$ in each tail, shaded).

the critical regions are computed as follows (remember that $t_{[\infty]}$ is equivalent to the normal distribution):

$$L_1 = \mu_0 - t_{.05[\infty]}\sigma_{\bar{Y}} = 45.5 - (1.96)(1.744) = 42.08$$

and

$$L_2 = \mu_0 + t_{.05[\infty]}\sigma_{\bar{Y}} = 45.5 + (1.96)(1.744) = 48.92$$

Thus we would consider it improbable for means less than 42.08 or greater than 48.92 to have been sampled from this population. For such sample means we would therefore reject the null hypothesis. The test we are proposing is two-tailed because we have no a priori assumption about the possible alternatives to our null hypothesis. If we could assume that the true mean of the population from which the sample was taken could only be equal to or greater than 45.5, the test would be one-tailed.

Now let us examine various alternative hypotheses. One alternative hypothesis might be that the true mean of the population from which our sample stems is 54.0, but that the variance is the same as before. We can express this assumption as $H_1{:}\mu = 54.0$ or $H_1{:}\mu = \mu_1$, where μ_1 stands for the alternative parametric mean 54.0. From the table of the areas of the normal curve and our knowledge of the variance of the means, we can calculate the proportion of the distribution implied by H_1 that would overlap the acceptance region implied by H_0. We find that 54.0 is 5.08 measurement units from 48.92, the upper boundary of the acceptance region of H_0. This corresponds to $5.08/1.744 = 2.91\sigma_{\bar{Y}}$ units. From the table of areas of the normal curve (Table **P**) we find that 0.0018 of the area will lie beyond 2.91σ at one tail of the curve. Thus under this alternative hypothesis 0.0018 of the distribution of H_1 will overlap the acceptance region of H_0. This is β, the type II error under this alternative hypothesis. Actually this is not entirely correct. Since the left tail of the H_1 distribution goes all the way to negative infinity, it will leave the acceptance region and cross over into the left-hand rejection region of H_0. However, this represents only an infinitesimal amount of the area of H_1 (the lower critical boundary of H_0, 42.08, is $6.83\sigma_{\bar{Y}}$ units from $\mu_1 = 54.0$) and can be ignored.

Our alternative hypothesis H_1 specified that μ_1 is 8.5 units greater than μ_0. However, as said before, we may have no a priori reason to believe that the true mean of our sample is either greater or less than μ. Therefore we may simply assume that it is 8.5 measurement units away from 45.5. In such a case we must similarly calculate β for the alternative hypothesis that $\mu_1 = \mu_0 - 8.5$. Thus the alternative hypothesis becomes $H_1{:}\mu = 54.0$ or 37.0, or $H_1{:}\mu = \mu_1$, where μ_1 represents either 54.0 or 37.0, the alternative parametric means. Since the distributions are symmetrical, β is the same for both alternative hypotheses. Type II error for hypothesis H_1 is therefore 0.0018, regardless of which of the two alternative hypotheses is correct. If H_1 is really true, 18 out of 10,000 samples would lead to an incorrect acceptance of H_0, a very low proportion of error. These relations are shown in Figure 7.14.

You may rightly ask what reason we have to believe that the alternative parametric value for the mean is 8.5 measurement units to either side of $\mu_0 = 45.5$. It would be quite unusual if we had any justification for such a belief. As a matter of fact the true mean may just as well be 7.5 or 6.0 or any number of units to either side of μ_0. If we draw curves for $H_1{:}\mu = \mu_0 \pm 7.5$, we find that β has increased considerably, the curves for H_0 and H_1 now being closer together. Thus the magnitude of β will depend on how far the alternative parametric mean is from the parametric mean of the null hypothesis. As the alternative mean approaches the parametric mean, β increases up to a maximum value of $1 - \alpha$, which is the area of the acceptance region under the null hypothesis. At this maximum the two distributions would be superimposed upon each other. Figure 7.15 illustrates the increase in β as μ_1 approaches μ, starting with the test illustrated in Figure 7.14. To simplify the graph the alternative distributions are shown for one tail only. Thus we clearly see that β is not a fixed value but varies with the nature of the alternative hypothesis.

An important concept in connection with hypothesis testing is the *power* of a test. It is $1 - \beta$, the complement of β, and is the probability of rejecting the null hypothesis when in fact it is false and the alternative hypothesis is correct. Obviously, for any given test we would like to have the quantity $1 - \beta$ be as large as possible and the quantity β as small as possible. Since we generally cannot specify a given alternative hypothesis, we have to describe β or $1 - \beta$ for a continuum of alternative values. When $1 - \beta$ is graphed in this manner the result is called a *power curve* for the test under consideration. Figure 7.16 shows the power curve for the housefly wing length example just discussed. This figure can be compared with Figure 7.15, from which it is directly derived. Figure 7.15 emphasizes the type II error β, and Figure 7.16 graphs the complement of this value, $1 - \beta$. We note that the power of the test falls off sharply as the alternative hypothesis approaches

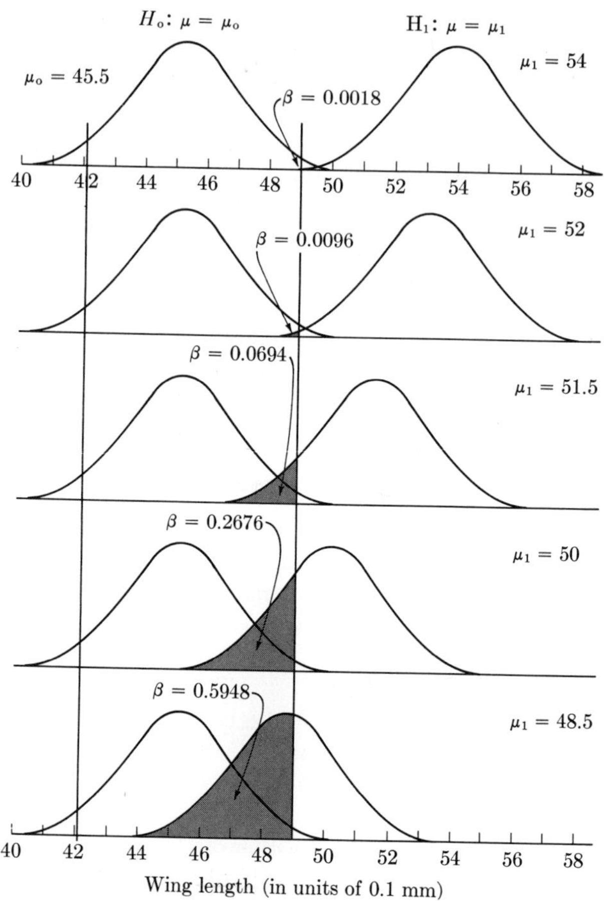

FIGURE 7.15 Diagram to illustrate increases in type II error, β, as alternative hypothesis, H_1, approaches null hypothesis H_0—that is, μ_1 approaches μ. Shading represents β. Vertical lines mark off 5% critical regions ($2\frac{1}{2}$% in each tail) for the null hypothesis. To simplify the graph the alternative distributions are shown for one tail only. Data identical to those in Figure 7.14.

the null hypothesis. Common sense confirms these conclusions: we can make clear and firm decisions about whether our sample comes from a population of mean 45.5 or 60.0. The power is essentially 1. But if the alternative hypothesis is that $\mu_1 = 45.6$, differing only by 0.1 from the value assumed under the null hypothesis, it will be difficult to decide which of these hypotheses is true and the power will be very low.

To improve the power of a given test (or decrease β) while keeping α constant for a stated null hypothesis, we must increase sample size. If instead of sampling 5 wing lengths we had sampled 35, the distribution of means would be much narrower. Thus rejection regions for the identical type I error would now commence at 44.21 and 46.79. Although the accept-ance and rejection regions have remained the same proportionately, the acceptance region has become much narrower in absolute value. Previously we could not, with confidence, reject the null hypothesis for a sample mean of 48.0. Now, when based on 35 individuals, a mean as deviant as 48.0 would occur only 15 times out of 100,000 and the hypothesis would, therefore, be rejected. What has happened to type II error? Since the distribution curves are not as wide as before, there is less overlap between them; if the alternative hypothesis $H_1: \mu = 54.0$ or 37.0 is true, the probability that the null hypothesis could be accepted by mistake (type II error) is infinitesimally small. If we let μ_1 approach μ_0, β will increase, of course, but it will always be smaller than the corresponding value for sample size $n = 5$. This comparison is shown in Figure 7.16, where the power for the test with $n = 35$ is much higher than that for $n = 5$. If we were to increase our sample size to 100 or 1000, the power would be still further increased. Thus we reach an important conclusion: if a given test is not sensitive enough we can increase its sensitivity (= power) by increasing sample size.

There is yet another way of increasing the power of a test. If we cannot increase sample size, the power may be raised by changing the nature of the test. Different statistical techniques testing roughly the same hypothesis may differ substantially in both the actual magnitude as well as in the slopes of their power curves. Tests that maintain higher power levels over substantial ranges of alternative hypotheses are clearly to be preferred. We have already mentioned in several places the various nonparametric tests which in recent years have become increasingly popular and have begun to replace time-honored statistical tests such as the t-test and others. Their popularity is due not only to the fact that these tests are simple to execute, but in many cases also because it has been shown that their power curves are less affected by failure of assumptions than are those of the parametric methods. However, it is also true that nonparametric tests have lower over-all power than parametric ones, when all the assumptions of the parametric test are met.

We might briefly look at a one-tailed test. The null hypothesis is $H_0: \mu_0 = 45.5$ as before. However, the alternative hypothesis assumes that we have reason to believe that the parametric mean of the population from

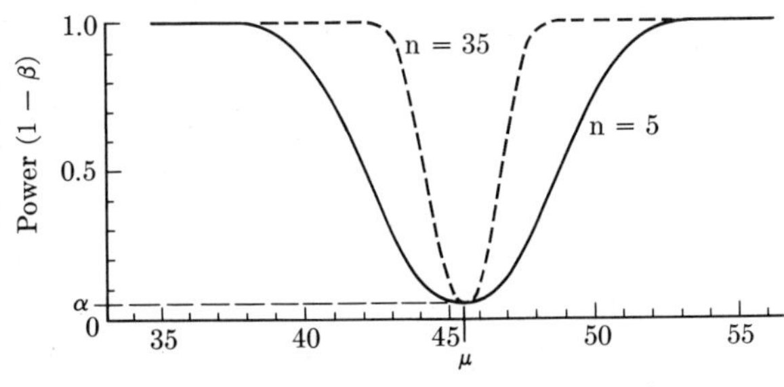

FIGURE 7.16 Power curves for testing $H_0: \mu = 45.5$, $H_1: \mu \neq 45.5$ for $n = 5$ (as in Figures 7.14 and 7.15) and for $n = 35$. (For explanation see text.)

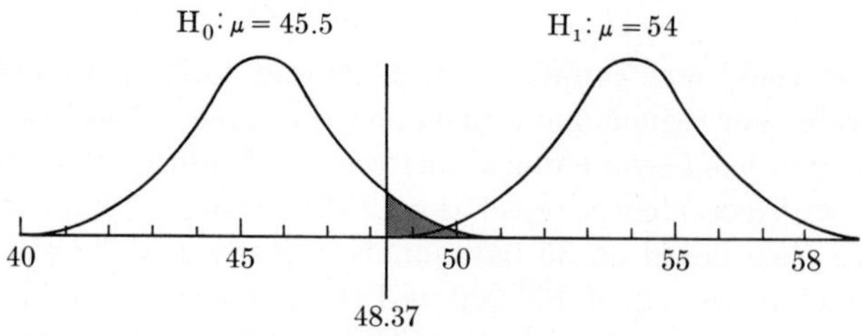

FIGURE 7.17 One-tailed significance test for the distribution of Figure 7.14. Vertical line now cuts off 5% rejection region from one tail of the distribution (corresponding area of curve has been shaded).

which our sample has been taken could not possibly be less than $\mu_0 = 45.5$. If it is different from that value it could only be greater than 45.5. We might have two grounds for such an hypothesis. First, we might have some biological reason for such a belief. Our parametric flies might be a dwarf population and any other population from which our sample could have come must be bigger. A second reason might be that we are interested in only one deviation of difference. For example, we may test the effect of a chemical in the larval food intended to increase the size of the sample of flies. Therefore, we would expect that $\mu_1 \geq \mu_0$, and we are not interested in testing for any μ_1 that is less than μ, because such an effect is the exact opposite of what we anticipate. Similarly, if we are investigating the effect of a certain drug as a cure for cancer, we might wish to compare the untreated population that has a mean fatality rate θ (from cancer) with the treated population, whose rate is θ_1. Our alternative hypotheses will be $H_1{:}\theta_1 < \theta$. That is, we are not interested in any θ_1 that is greater than θ, because if our drug will increase mortality from cancer it certainly is not much of a prospect for a cure.

When such a one-tailed test is performed, the rejection region along the abscissa is only under one tail of the curve representing the null hypothesis. Thus, for our housefly data (distribution of means of sample size $n = 5$) the rejection region will be in one tail of the curve only and for a 5% type I error will appear as shown in Figure 7.17. We compute the critical boundary as $45.5 + (1.645)(1.744) = 48.37$. The 1.645 is $t_{.10[\infty]}$, which corresponds to the 5% value for a one-tailed test. Compare this rejection region, which rejects the null hypothesis for all means greater than 48.37, with the two rejection regions in Figure 7.15, which reject the null hypothesis for means lower than 42.08 and greater than 48.92. The alternative hypothesis is considered for one tail of the distribution only, and the power curve of the test is not symmetrical but is drawn out with respect to one side of the distribution only.

7.9 Tests of simple hypotheses employing the normal and *t*-distributions

We shall proceed to apply our newly won knowledge of hypothesis testing to some simple examples involving the normal and *t*-distributions.

The first test to be considered is fairly uncommon. Given a single variate Y_1 or a sample mean $\bar{Y}$, we wish to know whether it differs significantly from a parametric mean μ—that is, whether it could have been sampled from a population with parameter μ for the character in question. Such a test assumes that we know the parametric mean of a population, which is not usual. It is more likely that μ represents a generally accepted standard or one promulgated by a government control or inspection agency. Examples might be the mean percentage of butterfat in a type of milk, or the mean number of microorganisms permitted in a grade of milk or in a vaccine. In systematics, a certain mean may represent a species or subspecies population and we might test whether a given sample can be considered as belonging to that population. If we know the variance of the standard or parametric population, the test is quite straightforward. From what has been learned before, it is obvious that either

$$\frac{Y_1 - \mu_0}{\sigma} \tag{7.11}$$

or

$$\frac{\bar{Y} - \mu_0}{\sigma_{\bar{Y}}} \tag{7.12}$$

would be normally distributed. As an example we shall test whether a single housefly with wing length of 43 units belongs to the population we are familiar with ($\mu = 45.5$, $\sigma = 3.90$). We compute Expression (7.11):

$$\frac{43 - 45.5}{3.90} = \frac{-2.5}{3.90} = -0.641$$

We ignore the negative sign of the numerator, since this is a two-tailed test and it is immaterial in which direction the item deviates from the mean. This is a two-tailed test because the question posed is whether the sample could belong to the population ($H_0\colon \mu = 45.5$). The alternative hypothesis is that the sample belongs to a population whose mean is either greater than or less than 45.5 ($H_1\colon \mu \neq 45.5$). Since the individual reading is 0.641 standard deviations from the mean, the table of areas of the normal curve (Table **P**, with linear interpolation) informs us that 0.2392 of the items under the normal curve lie between μ_0 and 0.641 standard deviations, and hence 0.2608 of the items lie beyond 0.641 standard deviations on one side of the mean; thus 0.5216 of the items will be 0.641 or more standard deviations from the mean on both sides of the distribution. Thus, by any of the conventional levels of significance (5% or 1%), we would accept the null hypothesis and conclude that the housefly wing length of 43 units belonged to the specified population. Ordinarily we would simply use the t-table (Table **Q**) and look up $df = \infty$. We would note that $t_{.8[\infty]} = 0.253$ and $t_{.5[\infty]} = 0.674$. Therefore the probability of such a deviation is between 0.50 and 0.80, but closer to 0.50, as was found more accurately from the table of the areas of the normal curve.

Similarly, a sample of 100 milk yields that gave a mean of 70.05 could be tested whether it belonged to the population of Table 6.2 ($\mu = 66.61$, $\sigma = 11.1597$). We compute Expression (7.12),

$$\frac{70.05 - 66.61}{11.1597/\sqrt{100}} = \frac{3.44}{1.11597} = 3.08$$

and find that the sample mean is 3.08 standard deviations above the parametric mean. Only 0.0011 of the area of the normal curve (Table **P**) extends beyond 3.08 standard deviations, and even when doubled for the two-tailed test appropriate in this case (0.0022 or 0.22% of the area of the curve) the percentage is much smaller than permitted at the conventional level of significance. We are led to reject the null hypothesis in this case. The mean based on the sample of 100 items does not seem to belong to the parametric population known to us.

Now let us proceed to a test that is similar except that we do not know the parametric standard deviation. Government regulations prescribe that the standard dosage in a certain biological preparation should be 600 activity units per cubic centimeter. We prepare 10 samples of this preparation and test each for potency. We find that the mean number of activity units per sample is 592.5 units per *cc* and the standard deviation of the samples is 11.2. Does our sample conform to the government standard? Stated more precisely, our null hypothesis is $H_0: \mu = \mu_0$. The alternative hypothesis is that the dosage is not equal to 600, or $H_1: \mu \neq \mu_0$. We proceed to calculate the significance of the deviation $\bar{Y} - \mu_0$ expressed in standard deviation units. The appropriate standard deviation is that of means (the standard error of the mean), *not* the standard deviation of items because the deviation is that of a sample mean around a parametric mean. We therefore calculate $s_{\bar{Y}} = s/\sqrt{n} = 11.2/\sqrt{10} = 3.542$. We next test the deviation $(\bar{Y} - \mu_0)/s_{\bar{Y}}$. We have seen earlier in Section 7.4 that a deviation divided by an estimated standard deviation will be distributed according to the *t*-distribution with $n - 1$ degrees of freedom. We therefore write

$$t_s = \frac{\bar{Y} - \mu_0}{s_{\bar{Y}}} \tag{7.13}$$

This indicates that we would expect this deviation to be distributed as a *t*-variate. Note that in Expression (7.13) we wrote t_s. In most textbooks you will find this ratio simply identified as *t*, but in fact the *t*-distribution is a parametric and theoretical distribution that generally is only approached, but never equaled, by observed, sampled data. This may seem a minor distinction, but readers should be quite clear that in any hypothesis testing of samples we are only *assuming* that the distributions of the tested variables follow certain theoretical probability distributions. To conform with general statistical practice, the *t*-distribution should really have a Greek letter (such as τ), with *t* serving as the sample statistic. Since this would violate long-standing practice, we prefer to use the subscript s to indicate the sample value.

The actual test is very simple. We calculate Expression (7.13),

$$t_s = \frac{592.5 - 600}{3.542} = \frac{-7.5}{3.542} = -2.12, \qquad df = n - 1 = 9,$$

and compare it with the expected values for t at 9 degrees of freedom. Since the t-distribution is symmetrical, we shall ignore the sign of t_s and always enter it in Table **Q** under its positive value. The two values on either side of t_s are $t_{.05[9]} = 2.26$ and $t_{.10[9]} = 1.83$. These are t-values for two-tailed tests, appropriate in this instance because the alternative hypothesis is that $\mu \neq 600$; that is, it can be smaller or greater. It appears that the significance level of our value of t_s is between 5% and 10%; if the null hypothesis is actually true, the probability of obtaining a deviation as great or greater than 7.5 is somewhere between 0.05 and 0.10. By conventional levels of significance this is insufficient for declaring the sample mean significantly different from the standard. We consequently accept the null hypothesis. In conventional language we would report the results of the statistical analysis as follows. "The sample mean is not significantly different from the accepted standard." Such a statement in a scientific report should always be backed up by a probability value, and the proper way of presenting this is to write $0.10 > P > 0.05$, which means that the probability of such a deviation is between 0.05 and 0.10. Another way of saying this is that the value of t_s is *not significant* (frequently abbreviated as *ns*).

A convention often encountered is the use of asterisks after the computed value of the significance test, $t_s = 2.86^{**}$. The symbols generally represent the following probability ranges:

$$* = 0.05 > P > 0.01, \qquad ** = 0.01 > P > 0.001, \qquad *** = P < 0.001$$

However, since some authors occasionally imply other ranges by these asterisks, the meaning of the symbols has to be specified in each scientific report.

It might be argued that in the case of the biological preparation the concern of the tester should not be whether the sample differs significantly from a standard, but whether it is significantly *below* the standard. This may be one of those biological preparations in which an excess of the active component is of no harm but a shortage would make the preparation ineffective at the conventional dosage. Then the test becomes one-tailed, performed in exactly the same manner except that the critical values of t for a one-tailed test are at half the probabilities of the two-tailed test. Thus $t_{.025[9]} = 2.26$, the former 0.05-value, and $t_{.05[9]} = 1.83$, the former 0.10-value, making our observed t_s-value of 2.12 "significant at the 5% level" or, more precisely stated, significant at $0.05 > P > 0.025$. If we are prepared to accept a 5% significance level, we would consider the preparation significantly below the standard.

You may be surprised that the same example, employing the same data and significance tests, should lead to two different conclusions, and you may begin to wonder whether some of the things you hear about statistics and statisticians are not, after all, correct. The explanation lies in the fact that the two results are answers to different questions. If we test whether our sample is significantly different from the standard in either direction, we must conclude that it is not different enough for us to reject the null hypothesis. If, on the other hand, we exclude from consideration the fact that the true sample mean μ could be greater than the established standard μ_0, the difference as found by us is clearly significant. It is obvious from this example that in any statistical test one must clearly state whether a one-tailed or a two-tailed test has been performed if the nature of the example is such that there could be any doubt about the matter. We should also point out that such a difference in the outcome of the results is not necessarily typical. It is only because the outcome in this case is in a borderline area between clear significance and nonsignificance. Had the difference between sample and standard been 10.5 activity units, the sample would have been unquestionably significantly different from the standard by the one-tailed or the two-tailed test.

The promulgation of a standard mean is generally insufficient for the establishment of a rigid standard for a product. If the variance among the samples is sufficiently large, it will never be possible to establish a significant difference between the standard and the sample mean. This is an important point that should be quite clear to you. Remember that the standard error can be increased in two ways—by lowering sample size or by increasing the standard deviation of the replicates. Both of these are undesirable aspects of any experimental setup. Yet, in the relatively rare cases where the purpose of the investigator is to show *no* difference rather than a significant difference, this can simply be obtained superficially by having a large enough error due to small sample size or to poor technique. One of us once had to review a research article, the main point of which was that the unpleasant side effects of drug B (newly developed) were no different from those of drug A (which was well established). The writers of the article used the appropriate statistical test but used it on only very few individuals. They apparently failed to understand that by using few enough individuals you could never refute the null hypothesis (that effect of drug B equals effect of drug A). One needs to calculate beforehand the sample size necessary to establish differences of a given magnitude at a certain level of significance, as discussed in detail in Section 9.8.

The test described above for the biological preparation leads us to a general test for the significance of any statistic—that is, for the significance of a deviation of any statistic from a parameter, which is illustrated in Box 7.4. Such a test applies whenever the statistics are expected to be normally distributed. When the standard error is estimated from the sample, the

BOX 7.4

Testing the significance of a statistic—that is, the significance of a deviation from a parameter. For normally distributed statistics.

Computational steps

1. Compute t_s as the following ratio,

$$t_s = \frac{St - St_p}{s_{St}}$$

where St is a sample statistic, St_p is the parametric value against which the sample statistic is to be tested, and s_{St} is its estimated standard error, obtained from Box 7.1.

2. The pertinent hypotheses are

$$H_0\colon St = St_p \qquad H_1\colon St \neq St_p$$

for a two-tailed test.

$$H_0\colon St = St_p \qquad H_1\colon St > St_p$$

or

$$H_0\colon St = St_p \qquad H_1\colon St < St_p$$

for a one-tailed test.

3. In the two-tailed test look up the critical value of $t_{\alpha[\nu]}$, where α is the type I error agreed upon and ν is the degrees of freedom pertinent to the standard error employed (see Box 7.1). In the one-tailed test look up the critical value of $t_{2\alpha[\nu]}$ for a significance level of α.

4. Accept or reject the appropriate hypothesis in **2** on the basis of the t_s value in **1** compared with critical values of t in **3**.

Example of computation

Test significance of g_1 from Box 6.2 ($g_1 = 0.18934$, $n = 9465$). Note from Box 7.1 that standard error of g_1 is $\sqrt{6/n}$ for samples of $n > 100$.

1. $$t_s = \frac{(g_1 - \gamma_1)}{s_{g_1}}$$

$$= \frac{0.18934 - 0}{\sqrt{6/9465}} = \frac{0.18934}{0.025178}$$

$$= 7.52$$

2. The null hypothesis is that there is no skewness—that is, that $\gamma_1 = 0$. The pertinent test is clearly two-tailed, since g_1 can be either negative (skewness to left) or positive (skewness to right). We wish to test whether there is *any* skewness, that is, the alternative hypothesis is that $\gamma_1 \neq 0$. Thus

$$H_0\colon \gamma_1 = 0 \qquad H_1\colon \gamma_1 \neq 0$$

3. The appropriate degrees of freedom employed with s_{g_1} are ∞ (see Box 7.1). We therefore employ critical values of t with degrees of freedom $\nu = \infty$:

$$t_{.05[\infty]} = 1.960 \qquad t_{.01[\infty]} = 2.576 \qquad t_{.001[\infty]} = 3.291$$

BOX 7.4 continued

4. $t_s = 7.52^{***}$ $(P < 0.001)$.

The above is a terse symbolism for stating that the observed g_1 is significant at $P < 0.001$. This means that the probability is much less than one in a thousand that a g_1 as observed can be sampled from a population in which γ_1 equals zero (no skewness). We accept the alternative hypothesis and conclude that $\gamma_1 \neq 0$. Since g_1 is positive, this indicates that $\gamma_1 > 0$; that is, the birth weights are skewed to the right. Thus the distribution of the birth weights is asymmetrical, drawn out at the right tail.

---

t-distribution is used. However, since the normal distribution is just a special case, $t_{[\infty]}$, of the t-distribution, most statisticians uniformly apply the t-distribution with the appropriate degrees of freedom from 1 to infinity.

In Box 7.4, as an example, we are testing the significance of g_1. By this we mean that we test whether the deviation of the observed value of g_1 is significantly different from the expected value of γ_1 for a normal distribution, which is zero. A number of coefficients such as g_1, g_2, regression coefficients, and correlation coefficients have their significance tested in this manner, by the appropriate standard error.

In Box 7.4 the conclusions are carefully spelled out in statistical shorthand, in the conventional way in which such results are reported in the scientific literature and in terms of the actual detailed implications of the test, which must be clearly understood by the student before he proceeds.

7.10 Testing the hypothesis $H_0{:}\sigma^2 = \sigma_0^2$

The method of Box 7.4 can be used only if the statistic is normally distributed. In the case of the variance this is not so. As we have seen in Section 7.7, sums of squares divided by σ^2 follow the χ^2-distribution. Therefore, for testing the hypothesis that a sample variance is different from a parametric variance we must employ the χ^2-distribution.

Let us use the biological preparation of the last section as an example. We were told that the standard deviation was 11.2 based on 10 samples. Therefore the variance must have been 125.44. Suppose the government postulates that the variance of samples from the preparation should be no greater than 100.0. Is our sample variance significantly above 100.0? Remembering from Expression (7.8) that $(n - 1)s^2/\sigma^2$ is distributed as $\chi^2_{[n-1]}$, we proceed as follows. We first calculate

$$\begin{aligned} X^2 &= (n - 1)s^2/\sigma^2 \\ &= (9)125.44/100 \\ &= 11.290 \end{aligned}$$

Note that we compute the quantity X^2 rather than χ^2, again to emphasize that we are obtaining a sample statistic that we shall compare to the parametric distribution. The use of X^2 to denote the sample statistic approximating a χ^2-distribution has become quite widely established. Following the general outline of Box 7.4 we next establish our null and alternative hypotheses, which are $H_0: \sigma^2 \leq \sigma_0^2$ and $H_1: \sigma^2 > \sigma_0^2$; that is, we are to perform a one-tailed test. The critical value of χ^2 is found next as $\chi^2_{\alpha[\nu]}$, where α indicates the type I error and ν the pertinent degrees of freedom. Quantity α represents the proportion of the χ^2-distribution to the right of the given value as described in Section 7.6, and you see now why we used the symbol α for that portion of the area; it corresponds with the type I error. For 9 degrees of freedom we find in Table **R** that

$$\chi^2_{.05[9]} = 16.919, \qquad \chi^2_{.10[9]} = 14.684, \qquad \chi^2_{.50[9]} = 8.343$$

We notice that the probability of getting a χ^2 as large as 11.290 is therefore higher than 0.10 but less than 0.50, assuming that the null hypothesis is true. Thus X^2 is not significant at the 5% level; we have no basis for rejecting the null hypothesis, and we must conclude that the variance of the 10 samples of the biological preparation may be no greater than the standard permitted by the government. If we had decided to test whether the variance is different from the standard, permitting it to deviate in either direction, the hypotheses for this two-tailed test would have been $H_0: \sigma^2 = \sigma_0^2$ and $H_1: \sigma^2 \neq \sigma_0^2$, and a 5% type I error would have yielded the following critical values for the two-tailed test:

$$\chi^2_{.975[9]} = 2.700, \qquad \chi^2_{.025[9]} = 19.023$$

The values represent chi-squares at points cutting off $2\frac{1}{2}\%$ rejection regions at each tail of the χ^2-distribution. A value of $X^2 < 2.700$ or >19.023 would have been evidence that the sample variance did not belong to this population. Our value of $X^2 = 11.290$ would again have led to an acceptance of the null hypothesis.

In the next chapter we shall see that there is another significance test available to test the hypotheses about variances of the present section. This is the mathematically equivalent F-test, which is, however, a more general test, allowing us to test the hypothesis that two sample variances come from populations with equal variances.

Exercises 7

7.1 Differentiate between type I and type II errors. What do we mean by the power of a statistical test?

7.2 Since it is possible to test a statistical hypothesis with any size sample, why does one prefer larger sample sizes?

7.3 The 95% confidence limits for μ as obtained in a given sample were 4.91 and 5.67 g. Is it correct to say that 95 times out of 100 the population mean, μ, falls inside the interval from 4.91 to 5.67 g? If not, what would the correct statement be?

7.4 Set 99% confidence limits to the mean, median, coefficient of variation, variance, and g_2 for the birth weight data given in Boxes 4.1, 4.3, and 6.2. ANS. The lower limits are 109.540, 109.060, 12.136, 178.698, and −0.0405, respectively.

7.5 Set 95% confidence limits to the means listed in Table 7.5. Are these limits all correct? (That is, do they contain μ?)

7.6 In a study of mating calls in the tree toad *Hyla ewingi*, Littlejohn (1965) found the note duration of the call in a sample of 39 observations from Tasmania to have a mean of 189 msec and a standard deviation of 32 msec. Set 95% confidence intervals to the mean and to the variance.

7.7 In Section 5.3 the coefficient of dispersion was given as an index of whether or not data agreed with a Poisson distribution. Since in a true Poisson distribution the mean, μ, equals the parametric variance, σ^2, the coefficient of dispersion is analogous to Expression (7.8). Using the weed seed data from Table 5.7, test the hypothesis that the true variance is equal to the sample mean—in other words, that we have sampled from a Poisson distribution (in which the coefficient of dispersion should equal unity).

7.8 Using the method described in Exercise 7.7, test the agreement of the observed distributions to Poisson distributions by testing the hypothesis that the true coefficient of dispersion equals unity for the data of Tables 5.5, 5.6, 5.9, 5.10, and 5.11. Note that in these examples the chi-square table is not adequate, so that approximate critical values must be computed using the method given with Table **R**. In Section 8.3 an alternative significance test that avoids this problem will be presented. ANS. For the data in Table 5.6, $(n-1) \times \text{C.D.} = 1308.30$, $\chi^2_{.05[588]} \approx 645.708$.

7.9 In direct klinokinetic behavior relating to temperature, animals turn more often in the warm end of a gradient and less often in the colder end, the direction of turning being at random, however. In a computer simulation of such behavior, the following results were found. The mean position along a temperature gradient was found to be −1.352. The standard deviation was 12.267 and n equaled 500 individuals. The gradient was marked off in units: zero corresponded to the middle of the gradient, the initial starting point of the animals; minus corresponded to the cold end; and plus corresponded to the warmer end. Test the hypothesis that direct klinokinetic behavior did not result in a tendency toward aggregation in either the warmer or colder end; that is, test the hypothesis that μ, the mean position along the gradient, was zero.

7.10 In a study of bill measurements of the dusky flycatcher, Johnson (1966) found that the bill length for the males had a mean of 8.14 ± 0.021 and a coefficient of variation of 4.67%. On the basis of this information, infer how many specimens must have been used? ANS. $n = 328$.

8 INTRODUCTION TO ANALYSIS OF VARIANCE

We now proceed to a study of the analysis of variance. This method, developed by R. A. Fisher, is fundamental to much of the application of statistics in biology and especially to experimental design. One way to approach the analysis of variance is to consider it as a test of whether two or more sample means could have been obtained from populations with the same parametric mean with respect to a given variable. Alternatively, we could conclude that these means differ from each other to such an extent that we must assume they were sampled from different populations. Where only two samples are involved, the t-distribution has been traditionally used to test significant differences between means. However, the analysis of variance is a more general test, which permits testing two samples as well as many, and we are therefore introducing it at this early stage in order to equip the reader with this most powerful weapon for his statistical arsenal. We shall discuss the t-test for two samples as a special case later on in Section 9.4.

The authors are very enthusiastic about analysis of variance. A knowledge of this subject is indispensable to any modern biologist and, after you have mastered it, you will undoubtedly use it numerous times to test scientific hypotheses. However, the analysis of variance is more than a technique for statistical analysis. Once understood, analysis of variance provides an insight into the nature of variation of natural events, into Nature in short, which is possibly of even greater value than the knowledge of the method as such. If one can speak of beauty in a statistical method, analysis of variance possesses it more than any other.

In Section 8.1 we shall approach the subject via familiar ground, the sampling experiment of the housefly wing lengths. From these samples we shall obtain two independent estimates of the population variance. We digress in Section 8.2 to introduce yet another continuous distribution, the F-distribution, needed for the significance test in analysis of variance. Section 8.3 is another digression in which we show how the newly learned F-distribution can be used to test whether two samples have the same variance. We are now ready for Section 8.4, in which we examine the effects of subjecting the samples to different treatments. The next section (8.5) describes the partitioning of sums of squares and of degrees of freedom, the actual *analysis* of variance. The last two sections (8.6 and 8.7) take up in a more formal way the two scientific models for which the analysis of variance is appropriate, the so-called fixed treatment effects model (Model I) and the variance component model (Model II).

Except for Section 8.3, the entire chapter is largely theoretical. We shall postpone the practical details of computation to Chapter 9. However, a thorough understanding of the material in Chapter 8 is necessary for working out actual examples of analysis of variance in Chapter 9.

One final comment: Professor J. W. Tukey of Princeton University has contributed many a well-turned phrase or term (in addition to many important theoretical concepts) to the field of statistics. The abbreviation "anova" for analysis of variance is one of them and it has quickly become quite generally used, if not yet respectable enough to enter the standard dictionary of statistical terms (Kendall and Buckland, 1960). We shall use anova interchangeably with analysis of variance throughout the text.

8.1 The variances of samples and their means

We shall approach analysis of variance through the familiar sampling experiment of housefly wing lengths (Experiment 6.1 and Table 6.2), in which we combined seven samples of 5 wing lengths to form samples of 35. We have reproduced one such sample in Table 8.1. The seven samples of 5, here called groups, are listed vertically in the upper half of the table. Before we proceed to explain Table 8.1 further, we must become familiar with added terminology and symbolism for dealing with this kind of problem. We call our samples *groups;* they are sometimes called *classes* and by yet other terms we shall learn later. In any analysis of variance problem we shall have two or more such samples or groups and we shall use the symbol a for the number of groups. Thus in the present example $a = 7$. Each group or sample is based on n items as before; in Table 8.1, $n = 5$. The total number of items in the table is a times n, which in this case equals 7×5 or 35.

The sums of the items in each group are shown in the row underneath the horizontal dividing line. In an anova, summation signs can no longer be as simple as heretofore. We can sum either the items of one group only or the items of the entire table. We therefore have to use superscripts with the summation symbol. In line with our policy of using the simplest possible notation, whenever this is not likely to lead to misunderstanding, we shall use $\sum^{n} Y$ to indicate the sum of the items of a group and $\sum^{an} Y$ to indicate the sum

TABLE 8.1

Seven samples (groups) of 5 wing lengths of houseflies randomly selected. (Data from Experiment 6.1 and Table 6.2.) Parametric mean, $\mu = 45.5$; variance, $\sigma^2 = 15.21$.

	a groups ($a = 7$)							*Computation of sum of squares of means*	*Computation of total sum of squares*
	1	*2*	*3*	*4*	*5*	*6*	*7*		
	41	48	40	40	49	40	41		
	44	49	50	39	41	48	46		
n individuals per group ($n = 5$)	48	49	44	46	50	51	54		
	43	49	48	46	39	47	44		
	42	45	50	41	42	51	42		
$\sum^{n} Y$	218	240	232	212	221	237	227	$\sum^{a} \overline{Y} = 317.4$	$\sum^{an} Y = 1587$
$\overline{Y}$	43.6	48.0	46.4	42.4	44.2	47.4	45.4	$\overline{\overline{Y}} = 45.34$	$\overline{Y} = 45.34$
$\sum^{n} Y^2$	9534	11,532	10,840	9034	9867	11,315	10,413	$\sum^{a} \overline{Y}^2 = 14{,}417.24$	$\sum^{an} Y^2 = 72535$
$\sum^{n} y^2$	29.2	12.0	75.2	45.2	98.8	81.2	107.2	$\sum^{a} (\overline{Y} - \overline{\overline{Y}})^2 = 25.417$	$\sum^{an} y^2 = 575.886$

of all the items in the table. The sum of the items of each group is shown in the first row under the horizontal line. The mean of each group, symbolized by $\bar{Y}$, is in the next row, and is computed simply as $\sum^{n} Y/n$. The remaining two rows in that portion of Table 8.1 list $\sum^{n} Y^2$ and $\sum^{n} y^2$, separately for each group. These are the familiar quantities, sum of the squared Y's and the sum of squares of Y.

From the sum of squares for each group we can obtain an estimate of the population variance of housefly wing length. Thus in the first group $\sum^{n} y^2 = 29.2$. Therefore our estimate of the population variance is

$$s^2 = \frac{\sum^{n} y^2}{(n - 1)} = 29.2/4 = 7.3$$

a rather low estimate. Since we have a sum of squares for each group, we could obtain an estimate of the population variance from each of these. However, it stands to reason that we would get a better estimate if we averaged these separate variance estimates in some way. This is done by computing a *weighted average*. Actually, in this instance a simple average would suffice, since each estimate of the variance is based on samples of the same size. However, we prefer to give the general formula, which works equally well for this case as well as for instances of unequal sample sizes, where the weighted average is necessary. A general formula for calculating a weighted average of any statistic St is as follows:

$$\overline{St} = \frac{\sum_{i=1}^{i=m} w_i St_i}{\sum_{i=1}^{i=m} w_i} \tag{8.1}$$

where m statistics, St_i, each weighted by factor w_i, are being averaged. In this case each sample variance s_i^2 is weighted by its degrees of freedom, $w_i = n_i - 1$, resulting in a sum of squares $(\sum y^2)_i$, since $(n_i - 1)s_i^2 = \sum y_i^2$. Thus the numerator of Expression (8.1) is the sum of the sums of squares. The denominator is $\sum^{a} (n_i - 1) = 7 \times 4$, the sum of the degrees of freedom of each group. The average variance therefore is

$$s^2 = \frac{29.2 + 12.0 + 75.2 + 45.2 + 98.8 + 81.2 + 107.2}{28} = \frac{448.8}{28} = 16.029$$

This quantity is an estimate of 15.21, the parametric variance of housefly wing lengths. We might call this estimate, based on 7 independent estimates of variances of groups, the *average variance within groups* or simply *variance within groups*. Note that we use the expression *within* groups, although in previous chapters we used the term variance *of* groups. The reason we do this is that the variance estimates used for computing the average variance have so far all come from sums of squares measuring the variation within one column. As we shall see below, one can also compute variances among groups, cutting across group boundaries.

To obtain a second estimate of the population variance we treat the seven group means, $\bar{Y}$, as though they were a sample of seven observations. The resulting statistics are shown in the lower right part of Table 8.1, headed Computation of sum of squares of means. There are seven means in this example; in the general case there will be a means. We first compute $\sum^{a} \bar{Y}$, the sum of the means. Note that this is rather sloppy symbolism. To be entirely proper, we should identify this quantity as $\sum_{i=1}^{i=a} \bar{Y}_i$, summing the means of group 1 through group a. The next quantity computed is $\bar{\bar{Y}}$, the grand mean of the group means, computed as $\bar{\bar{Y}} = \sum^{a} \bar{Y}/a$. The sum of the seven means is $\sum^{a} \bar{Y} = 317.4$ and the grand mean is $\bar{\bar{Y}} = 45.34$, a fairly close approximation to the parametric mean $\mu = 45.5$. The sum of squares represents the deviations of the group means from the grand mean, $\sum^{a}(\bar{Y} - \bar{\bar{Y}})^2$. For this we need, first of all, the quantity $\sum^{a} \bar{Y}^2$, which equals 14,417.24. The customary computational formula for sum of squares applied to these means is $\sum^{a} \bar{Y}^2 - [(\sum^{a} \bar{Y})^2/a] = 25.417$. From the sum of squares of the means we obtain a *variance among the means* in the conventional way as follows: $\sum^{a} (\bar{Y} - \bar{\bar{Y}})^2/(a - 1)$. We divide by $a - 1$ rather than $n - 1$ because the sum of squares was based on a items (means). Thus variance of the means $s_{\bar{Y}}^2 = 25.417/6 = 4.2362$. We learned in Chapter 7, Expression (7.1), that when randomly sampling from a single population

$$\sigma_{\bar{Y}}^2 = \frac{\sigma^2}{n}$$

and hence

$$\sigma^2 = n\sigma_{\bar{Y}}^2$$

Thus, we can estimate a variance of items by multiplying the variance of means by the sample size on which the means are based (assuming we have sampled from a single population). When we do this for our present example we obtain $s^2 = 5 \times 4.2362 = 21.181$. This is a second estimate of the parametric variance 15.21. It is not as close to the true value as the previous estimate based on the average variance within groups, but this is to be expected since it is only based on 7 "observations." We need a name describing this variance to distinguish it from the variance of means from which it was computed, as well as from the variance within groups with which it will be compared. We shall call it the *variance among groups;* it is n times the variance of means and is an independent estimate of the parametric variance σ^2 of the housefly wing lengths. It may not be clear at this stage why the two estimates of σ^2 which we have obtained, the variance within groups and the variance among groups, are independent. We ask you to take on faith that they are. Although this is in no way a proof of independence, note that in this example the two estimates are indeed different, being 16.029 for the variance within groups and 21.181 for the variance among groups.

Let us review what we have done so far by expressing in a more formal way the data table for an anova and the computations carried out up to now. Table 8.2 represents a generalized table for data such as the samples of house-

TABLE 8.2

Data arranged for simple analysis of variance, single classification, completely randomized.

		a groups							
		1	*2*	*3*	. . .	*i*	. . .	*a*
n items	1	Y_{11}	Y_{21}	Y_{31}	. . .	Y_{i1}	. . .	Y_{a1}
	2	Y_{12}	Y_{22}	Y_{32}	. . .	Y_{i2}	. . .	Y_{a2}
	3	Y_{13}	Y_{23}	Y_{33}	. . .	Y_{i3}	. . .	Y_{a3}
	⋮	⋮	⋮	⋮	.	⋮	.	⋮
	j	Y_{1j}	Y_{2j}	Y_{3j}	. . .	Y_{ij}	. . .	Y_{aj}
	⋮	⋮	⋮	⋮	.	⋮	.	⋮
	n	Y_{1n}	Y_{2n}	Y_{3n}	. . .	Y_{in}	. . .	Y_{an}
Sums		$\sum^n Y_1$	$\sum^n Y_2$	$\sum^n Y_3$	. . .	$\sum^n Y_i$	. . .	$\sum^n Y_a$
Means		$\bar{Y}_1$	$\bar{Y}_2$	$\bar{Y}_3$	. . .	$\bar{Y}_i$	. . .	$\bar{Y}_a$

(The "Sums" and "Means" rows also carry an additional column, headed by no group label, with $\sum^n Y$ and $\bar{Y}$ respectively.)

fly wing lengths. Do not be put off by the unfamiliar notation! Each individual wing length is represented by Y, subscripted to indicate its position in the data table. The wing length of the jth fly from the ith sample or group is given by Y_{ij}. Thus you will notice that the first subscript changes with each column representing a group in the table, and the second subscript changes with each row representing an individual item. (Those of you familiar with matrix algebra may wonder why the subscript notation here is the reverse of the one in that field. It is simply a well-established usage in anova, and we prefer not to change it.) Using this notation we can compute the variance of sample 1 as

$$\frac{1}{n-1}\sum_{j=1}^{j=n}(Y_{1j}-\bar{Y}_1)^2$$

The variance within groups, which is the average variance of the samples, is computed as

$$\frac{1}{a(n-1)}\sum_{i=1}^{i=a}\sum_{j=1}^{j=n}(Y_{ij}-\bar{Y}_i)^2$$

Note the double summation. It means that we first set $i = 1$ (i being the index of the outer $\sum$), when we start with the first group. We sum the squared deviations of all items from the mean of the first group, changing index j of the inner $\sum$ from 1 to n in the process. We then return to the outer summation, set $i = 2$, and sum the squared deviations for group 2 from $j = 1$ to $j = n$. This process is continued until i, the index of the outer $\sum$, is set to a. In other words, we sum all the squared deviations within one group first and add this sum to similar sums from all the other groups. The variance among means is computed as

$$\frac{1}{a-1}\sum_{i=1}^{i=a}(\bar{Y}_i - \bar{\bar{Y}})^2$$

Now that we have two independent estimates of the population variance, what shall we do with them? We might wish to find out whether they do in fact estimate the same parameter. To test this hypothesis we need a statistical test that will evaluate the probability that the two sample variances are from the same population. Such a test employs the F-distribution, which is taken up in the next section.

8.2 The *F*-distribution

Let us devise yet another sampling experiment. This is quite a tedious one, so we won't ask you to carry it out. Assume that you are sampling at random from a normally distributed population, such as the housefly wing lengths with mean μ and variance σ^2. The sampling procedure consists of first sampling n_1 items and calculating their variance s_1^2, followed by sampling n_2 items and calculating their variance s_2^2. Sample sizes n_1 and n_2 may or may not be equal to each other but are fixed for any one sampling experiment. Thus, for example, we might always sample 8 wing lengths for the first sample (n_1) and 6 wing lengths for the second sample (n_2). After each pair of values (s_1^2 and s_2^2) has been obtained, we calculate

$$F_s = \frac{s_1^2}{s_2^2}$$

This will be a ratio near 1, because these variances are estimates of the same quantity. Its actual value will depend on the relative magnitudes of variances s_1^2 and s_2^2. If we repeatedly take samples of sizes n_1 and n_2, calculating the ratios F_s of their variances, the average of these ratios will in fact approach $n_2/(n_2 - 2)$, which is close to 1.0 when n_2 is appreciably large. Statisticians have worked out the expected distribution of this statistic, which is called the *F-distribution* in honor of R. A. Fisher. This is another distribution described by a complicated mathematical function that need not concern us here. Unlike the t- and χ^2-distributions, the shape of the F-distribution is determined by *two* values for degrees of freedom, ν_1 and ν_2. Thus for every possible combination of values ν_1, ν_2, each ν ranging from 1 to infinity, there exists a separate F-distribution. Remember that the F-distribution is a theoretical probability distribution, like the t-distribution and the χ^2-distribution. Variance ratios based on sample variances, s_1^2/s_2^2, are sample statistics that may or may not follow the F-distribution. We have, therefore, distinguished the sample variance ratio by calling it F_s, conforming to our convention of separate symbols for sample statistics as distinct from probability distributions (such as t_s and X^2 for t and χ^2).

We have discussed the generation of an F-distribution by repeatedly taking two samples from the same normal distribution. We could also have generated it by sampling from two separate normal distributions differing in their mean but identical in their parametric variances—that is, with $\mu_1 \neq \mu_2$ but $\sigma_1^2 = \sigma_2^2$). Thus we obtain an F-distribution whether the samples come from the same normal population or from different ones, so long as their variances are identical.

Figure 8.1 shows several representative F-distributions. For very low degrees of freedom the distribution is L-shaped but becomes humped and strongly skewed to the right as both degrees of freedom increase. Table **S** shows the cumulative probability distribution of F for several selected probability values. The values in the table represent $F_{\alpha[\nu_1,\nu_2]}$, where α is the proportion of the F-distribution to the right of the given F-value (in one tail) and ν_1, ν_2 are the degrees of freedom pertaining to the numerator and the denominator of the variance ratio, respectively. The table is arranged so that across the top one reads ν_1, the degrees of freedom pertaining to the upper (numerator) variance, and along the left margin one reads ν_2, the degrees of freedom

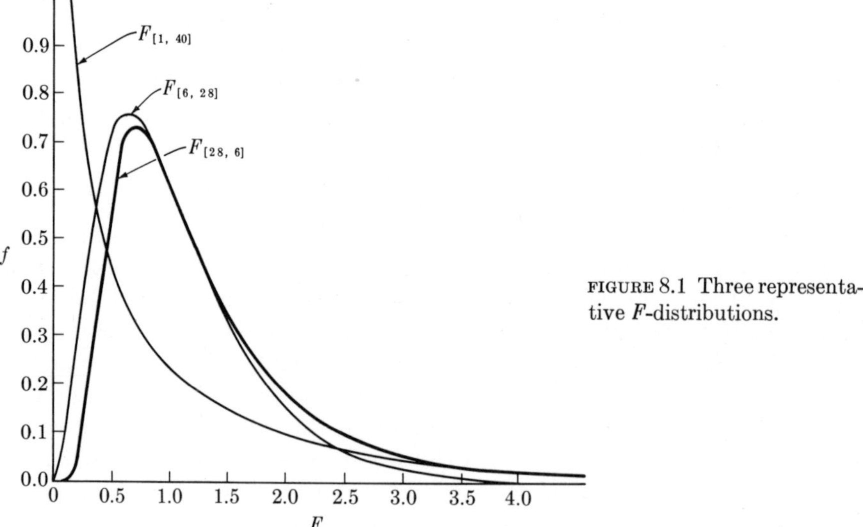

FIGURE 8.1 Three representative F-distributions.

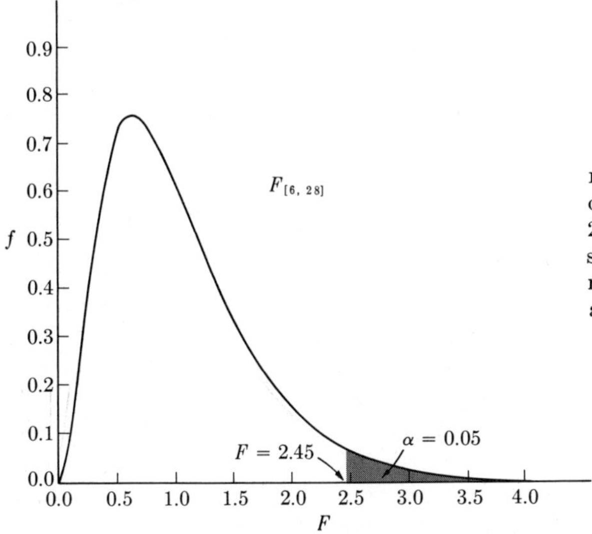

FIGURE 8.2 Frequency curve of the *F*-distribution for 6 and 28 degrees of freedom, respectively. A one-tailed 5% rejection region is marked off at $F = 2.45$.

pertaining to the lower (denominator) variance. At each intersection of degree of freedom values we list several values of F for decreasing magnitudes of α. The first value furnished is $\alpha = 0.75$. Note that for $\alpha = 0.5$, F is not $n_2/(n_2 - 2)$, as you might expect, because $\alpha = 0.5$ represents the median and not the mean of the asymmetrical *F*-distribution. Since in tests of hypotheses we are usually interested in small values of α, the right tail of the curve is especially emphasized in most *F*-tables. For example, an *F*-distribution with $\nu_1 = 6$, $\nu_2 = 28$ is 2.45 at $\alpha = 0.05$. By that we mean that 0.05 of the area under the curve lies to the right of $F = 2.45$. Only 0.01 of the area under the curve lies to the right of $F = 3.53$. Figure 8.2 illustrates this. Thus if we have a null hypothesis $H_0\colon \sigma_1^2 = \sigma_2^2$, with the alternative hypothesis $H_1\colon \sigma_1^2 > \sigma_2^2$, we would use a one-tailed *F*-test, as illustrated by Figure 8.2.

We can now test the two variances obtained in the sampling experiment of Section 8.1 and Table 8.1. The variance among groups based on 7 means was 21.180 and the variance within 7 groups of 5 individuals was 16.029. Our null hypothesis is that the two variances estimate the same parametric variance; the alternative hypothesis in an anova is always that the parametric variance estimated by the variance among groups is greater than that estimated by the variance within groups. The reason for this restrictive alternative hypothesis, which leads to a one-tailed test, will be explained in Section 8.4. We calculate the variance ratio $F_s = s_1^2/s_2^2 = 21.181/16.029 = 1.32$. Before we can inspect the *F*-table we have to know the appropriate degrees of freedom for this variance ratio. We shall learn simple formulas for degrees of freedom in an anova later, but at the moment let us reason it out for ourselves. The upper variance (among groups) was based on the variance of 7 means; hence it should have 6 degrees of freedom. The lower variance was based on an average of 7 variances, each of them based on 5 individuals yielding 4 degrees of freedom per variance; $7 \times 4 = 28$ degrees of freedom. Thus the upper variance has 6, the lower variance 28 degrees of freedom. If we check Table **S** for $\nu_1 = 6$, $\nu_2 = 28$ we find that $F_{.05[6,28]} = 2.45$. For $F = 1.32$, corresponding to the F_s-value actually obtained, α is between 0.5 and 0.25. Thus, we may expect between half and one-quarter of all variance ratios of samples based on 6 and 28 degrees of freedom, respectively, to have F_s-values greater than 1.32. We have no evidence to reject the null hypothesis and conclude that the two sample variances estimate the same parametric variance. This corresponds, of course, to what we knew anyway from our sampling experiment. Since the seven samples were taken from the same population, the variance of their means is expected to give us merely another estimate of the parametric variance of housefly wing length.

Whenever the alternative hypothesis is more generally that the two parametric variances are unequal (rather then the restrictive hypothesis $H_1: \sigma_1^2 > \sigma_2^2$), the sample variance s_1^2 could be smaller as well as greater than s_2^2. This leads to a two-tailed test and in such cases a 5% type I error means that rejection regions of $2\frac{1}{2}\%$ will occur at each tail of the curve. Because the F-distribution is asymmetrical, these rejection regions will not look identical, as they do in the t-distribution. Figure 8.3 shows the 5% two-tailed rejection regions for the same F-distribution illustrated in Figure 8.2. Since usually only the right half of the F-distribution is tabulated (our Table **S** shows only $\alpha = 0.75$ in the left half of the curve), when faced with a two-tailed test we deliberately divide the greater variance by the lesser variance and for a type I error of α look up the F-value for $\alpha/2$ in the F-table. Computational details of this method are illustrated in Section 8.3.

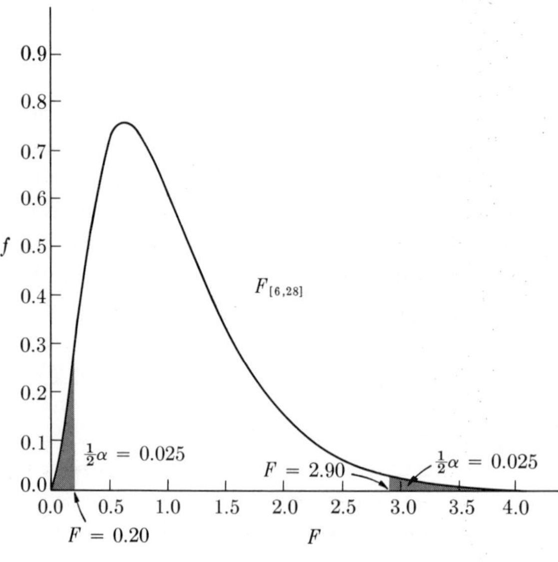

FIGURE 8.3 Frequency curve of the F-distribution for 6 and 28 degrees of freedom, respectively. A two-tailed 5% rejection region ($2\frac{1}{2}\%$ in each tail) is marked off at $F = 0.20$ and $F = 2.90$, respectively.

It is sometimes necessary to obtain F-values for $\alpha > 0.5$ (that is, in the left half of the F-distribution). Since, as we have just learned, these values are rarely tabulated, they can be obtained by a simple rule:

$$F_{\alpha[\nu_1,\nu_2]} = \frac{1}{F_{(1-\alpha)[\nu_2,\nu_1]}} \tag{8.2}$$

For example, $F_{.05[5,24]} = 2.62$. If we wish to obtain $F_{.95[5,24]}$ (the F-value for 5 and 24 degrees of freedom, respectively, to the right of which lies 95% of the area of the F-distribution), we first have to find $F_{.05[24,5]} = 4.53$. Then $F_{.95[5,24]}$ is the reciprocal of 4.53, which equals 0.221. Thus 95% of an F-distribution for 5 and 24 degrees of freedom lies to the right of 0.221.

There is an important relationship between the F-distribution and the χ^2-distribution. You may remember that the ratio $X^2 = \sum y^2/\sigma^2$ was distributed as a χ^2 with $n - 1$ degrees of freedom. If you divide the numerator of this expression by $n - 1$, you obtain the ratio $F_s = s^2/\sigma^2$, which is a variance ratio with an expected distribution of $F_{[n-1,\,\infty]}$. The upper degrees of freedom are $n - 1$, the degrees of freedom of the sum of squares or sample variance; the lower degrees of freedom are considered as infinity because only on the basis of an infinite number of items can we obtain the true, parametric variance of a population. Therefore, by dividing a value of X^2 by $n - 1$ degrees of freedom, we obtain an F_s-value with $n - 1$ and ∞ df, respectively. In general, $\chi^2_{[\nu]}/\nu = F_{[\nu,\infty]}$. We can convince ourselves of this by inspecting the F- and χ^2-tables. From the χ^2-table (Table **R**) we find that $\chi^2_{.05[10]} = 18.307$. Dividing this value by 10 df we obtain 1.8307. From the F-table (Table **S**) we find, for $\nu_1 = 10$, $\nu_2 = \infty$, that $F_{.05[10,\,\infty]} = 1.83$. Thus the two statistics of significance are closely related and, lacking a χ^2-table, we could make do with an F-table alone, using the values of $\nu F_{[\nu,\infty]}$ in place of $\chi^2_{[\nu]}$.

Before we return to analysis of variance we shall quickly apply our newly won knowledge of the F-distribution to testing a hypothesis about two sample variances.

8.3 The hypothesis H_0: $\sigma_1^2 = \sigma_2^2$

A test of the null hypothesis that two normal populations represented by two samples have the same variance is illustrated in Box 8.1. As will be seen later, the acceptance of this null hypothesis is a prerequisite for some tests leading to a decision about whether the means of two samples come from the same population. However, this test is of interest in its own right. Repeatedly it will be necessary to test whether two samples have the same variance. In genetics we may need to know whether an offspring generation is more variable for a character than the parent generation. In systematics we would like to find out whether two local populations are equally variable. In experimental biology we may wish to demonstrate under which of two experimental setups the readings will be more variable. In general, the less variable setup

BOX 8.1

Testing the significance of differences between two variances.

Survival in days of the cockroach *Blattella vaga* when kept without food or water.

Females	$n_1 = 10$	$\bar{Y}_1 = 8.5$ days	$s_1^2 = 3.6$
Males	$n_2 = 10$	$\bar{Y}_2 = 4.8$ days	$s_2^2 = 0.9$

$H_0: \sigma_1^2 = \sigma_2^2 \qquad H_1: \sigma_1^2 \neq \sigma_2^2$

Source: Data modified from Willis and Lewis (1957).

The alternative hypothesis is that the two variances are unequal. We have no reason to suppose that one sex should be more variable than the other. In view of the alternative hypothesis this is a two-tailed test. Since only the right tail of the F-distribution is tabled extensively in Table **S** and in most other tables, we calculate F_s as the ratio of the greater variance over the lesser one:

$$F_s = \frac{s_1^2}{s_2^2} = \frac{3.6}{0.9} = 4.00$$

Because the test is two-tailed, we look up the critical value $F_{\alpha/2[\nu_1,\nu_2]}$, where α is the type I error accepted, $\nu_1 = n_1 - 1$ and $\nu_2 = n_2 - 1$, the degrees of freedom for the upper and lower variance, respectively. Whether we look up $F_{\alpha/2[\nu_1,\nu_2]}$ or $F_{\alpha/2[\nu_2,\nu_1]}$ depends on whether sample 1 or sample 2 has the greater variance and was placed in the numerator.

From Table **S** we find $F_{.025[9,9]} = 4.03$, $F_{.05[9,9]} = 3.18$, and $F_{.10[9,9]} = 2.44$. Because this is a two-tailed test we double these probabilities. Thus, the F-value of 4.03 represents a probability of $\alpha = 0.05$, since the right-hand tail area of $\alpha = 0.025$ is matched by a similar left-hand area to the left of $F_{.975[9,9]} = 1/F_{.025[9,9]} = 0.248$. Therefore, assuming the null hypothesis is true, the probability of observing an F-value greater than 4.00 and smaller than $1/4.00 = 0.25$ is $0.10 > P > 0.05$. Strictly speaking, the two sample variances are not significantly different—the two sexes were equally variable in their duration of survival. However, the outcome is close enough to the 5% significance level to make us suspicious that possibly the variances were in fact different. It would be desirable to repeat this experiment in the hope that more decisive results would emerge.

would be preferred; if both setups were equally variable the experimenter would pursue the one that is simpler or less costly to undertake.

The test employed is a two-tailed test, since the alternative hypothesis is generally $H_1: \sigma_1^2 \neq \sigma_2^2$.

8.4 Heterogeneity among sample means

We shall now modify the data of Table 8.1, discussed in Section 8.1. Suppose these seven groups of houseflies did not represent random samples from the same population but resulted from the following experiment. Each sample had been reared in a separate culture jar and the medium in each of the culture jars had been prepared in a different way. Some had more water added, others more sugar, yet others more solid matter. Let us assume that sample 7 represents the standard medium against which we propose to compare the other samples. The various changes in the medium affect the sizes of the flies that emerge from it; this in turn affects the wing lengths we have been measuring.

We shall assume the following effects resulting from treatment of the medium:

Medium 1—decreases average wing length of a sample by 5 units
2—decreases average wing length of a sample by 2 units
3—does not change average wing length of a sample
4—increases average wing length of a sample by 1 unit
5—increases average wing length of a sample by 1 unit
6—increases average wing length of a sample by 5 units
7—(control) does not change average wing length of a sample

We shall symbolize the effect of treatment i as α_i. (Please note that this use of α is not related to its use as a symbol for type I error. There are, regrettably, more symbols needed in statistics than there are letters in the Greek and Roman alphabets.) Thus α_i assumes the following values for the above treatment effects:

$$\alpha_1 = -5 \qquad \alpha_4 = 1$$
$$\alpha_2 = -2 \qquad \alpha_5 = 1$$
$$\alpha_3 = 0 \qquad \alpha_6 = 5$$
$$\alpha_7 = 0$$

Note that $\sum^{a}\alpha_i = 0$; that is, the sum of the effects cancels out. This is a convenient property that is generally postulated but it is unnecessary for our argument. We can now modify Table 8.1 by adding the appropriate values of α_i to each sample. In sample 1 the value of α_1 is -5; therefore the first wing length, which was 41 (see Table 8.1), now becomes 36, the second wing length, formerly 44, becomes 39, and so on. For the second sample α_2 is -2, changing the first wing length from 48 to 46. Where α_i is 0, the wing lengths do not change; where α_i is positive, they are increased by the magnitude indicated. The changed values can be inspected in Table 8.3, which is arranged identically to Table 8.1.

We now repeat our previous computations. We first calculate the sum of squares of the first sample and find it to be 29.2. If you compare this value with the sum of squares of the first sample in Table 8.1, you find the two values to be identical. Similarly, all other values of $\sum^{n} y^2$, the sum of squares of each group, are identical to their previous values. Why is this so? The

TABLE 8.3

Data of Table 8.1 with fixed-treatment effects α_i or random effects A_i added to each sample.

	a groups ($a = 7$)							Computation of sum of squares of means	Computation of total sum of squares
	1	*2*	*3*	*4*	*5*	*6*	*7*		
α_i or A_i	−5	−2	0	+1	+1	+5	0		
n individuals per group ($n = 5$)	36	46	40	41	50	45	41		
	39	47	50	40	42	53	46		
	43	47	44	47	51	56	54		
	38	47	48	47	40	52	44		
	37	43	50	42	43	56	42		
$\sum^{n} Y$	193	230	232	217	226	262	227	$\sum^{a} \bar{Y} = 317.4$	$\sum^{an} Y = 1587$
$\bar{Y}$	38.6	46.0	46.4	43.4	45.2	52.4	45.4	$\bar{\bar{Y}} = 45.34$	$\bar{Y} = 45.34$
	$\binom{43.6}{-5}$	$\binom{48.0}{-2}$	$\binom{46.4}{+0}$	$\binom{42.4}{+1}$	$\binom{44.2}{+1}$	$\binom{47.4}{+5}$	$\binom{45.4}{+0}$	$\left(45.34 + \frac{-5 - 2 + 1 + 1 + 5}{7}\right)$	
$\sum^{n} Y^2$	7479	10,592	10,840	9463	10,314	13,810	10,413	$\sum^{a} \bar{Y}^2 = 14{,}492.44$	$\sum^{an} Y^2 = 72{,}911$
$\sum^{n} y^2$	29.2	12.0	75.2	45.2	98.8	81.2	107.2	$\sum^{a} (\bar{Y} - \bar{\bar{Y}})^2 = 100.617$	$\sum^{an} y^2 = 951.886$

effect of adding α_i to each group is simply that of an additive code, since α_i is constant for any one group. We saw in Appendix A1.2 that additive codes do not affect sums of squares or variances. Therefore, not only is each separate sum of squares the same as before, but the average variance within groups is still 16.029. Now let us compute the variance of the means. It is $100.617/6 = 16.770$, which is a value much higher than the variance of means found before, 4.236. When we multiply times $n = 5$ to get an estimate of σ^2 we obtain the variance of groups, which now is 83.848 and is no longer even close to an estimate of σ^2. We repeat the F-test with the new variances and find that $F_s = 83.848/16.029 = 5.23$, which is much greater than the critical value of $F_{.05[6,28]} = 2.45$. In fact, the observed value of F_s is greater than $F_{.005[6,28]} = 4.02$. Clearly, the upper variance, representing the variance among groups, has become significantly larger. The two variances are most unlikely to represent the same parametric variance.

What has happened? We can easily explain it by means of Table 8.4, which represents Table 8.3 symbolically in the manner that Table 8.2 represented Table 8.1. We note that each group has a constant α_i added and that

TABLE 8.4

Data of Table 8.3 arranged in the manner of Table 8.2.

		a groups						
		1	*2*	*3*	...	*i*	...	*a*
n items	1	$Y_{11} + \alpha_1$	$Y_{21} + \alpha_2$	$Y_{31} + \alpha_3$	...	$Y_{i1} + \alpha_i$	...	$Y_{a1} + \alpha_a$
	2	$Y_{12} + \alpha_1$	$Y_{22} + \alpha_2$	$Y_{32} + \alpha_3$	...	$Y_{i2} + \alpha_i$	...	$Y_{a2} + \alpha_a$
	3	$Y_{13} + \alpha_1$	$Y_{23} + \alpha_2$	$Y_{33} + \alpha_3$	...	$Y_{i3} + \alpha_i$	...	$Y_{a3} + \alpha_a$
	⋮	⋮	⋮	⋮	·	⋮	·	⋮
	j	$Y_{1j} + \alpha_1$	$Y_{2j} + \alpha_2$	$Y_{3j} + \alpha_3$	...	$Y_{ij} + \alpha_i$	...	$Y_{aj} + \alpha_a$
	⋮	⋮	⋮	⋮	·	⋮	·	⋮
	n	$Y_{1n} + \alpha_1$	$Y_{2n} + \alpha_2$	$Y_{3n} + \alpha_3$	...	$Y_{in} + \alpha_i$	...	$Y_{an} + \alpha_a$
Sums		$\sum^{n} Y_1 + n\alpha_1$	$\sum^{n} Y_2 + n\alpha_2$	$\sum^{n} Y_3 + n\alpha_3$	...	$\sum^{n} Y_i + n\alpha_i$	...	$\sum^{n} Y_a + n\alpha_a$
Means		$\bar{Y}_1 + \alpha_1$	$\bar{Y}_2 + \alpha_2$	$\bar{Y}_3 + \alpha_3$	...	$\bar{Y}_i + \alpha_i$	...	$\bar{Y}_a + \alpha_a$

this constant changes the sums of the groups by $n\alpha_i$ and the means of these groups by α_i. In Section 8.1 we computed the variance within groups as

$$\frac{1}{a(n-1)} \sum_{i=1}^{i=a} \sum_{j=1}^{j=n} (Y_{ij} - \bar{Y}_i)^2$$

If we try to repeat this, our formula becomes more complicated because to each Y_{ij} and each $\bar{Y}_i$ there has now been added α_i. We therefore write

$$\frac{1}{a(n-1)} \sum_{i=1}^{i=a} \sum_{j=1}^{j=n} [(Y_{ij} + \alpha_i) - (\bar{Y}_i + \alpha_i)]^2$$

When we open the parentheses inside the square brackets, the second α_i

changes sign and the α_i's cancel out, leaving the expression exactly as before, substantiating our earlier observation that the variance within groups does not change despite the treatment effects.

The variance of means was previously calculated by the formula

$$\frac{1}{a-1}\sum_{i=1}^{i=a}(\bar{Y}_i - \bar{Y})^2$$

However, from Table 8.4 we see that the new grand mean equals

$$\frac{1}{a}\sum_{i=1}^{i=a}(\bar{Y}_i + \alpha_i) = \frac{1}{a}\sum_{i=1}^{i=a}\bar{Y}_i + \frac{1}{a}\sum_{i=1}^{i=a}\alpha_i = \bar{Y} + \bar{\alpha}$$

When we substitute the new values for the group means and the grand mean, the formula appears as

$$\frac{1}{a-1}\sum_{i=1}^{i=a}[(\bar{Y}_i + \alpha_i) - (Y + \bar{\alpha})]^2$$

which in turn yields

$$\frac{1}{a-1}\sum_{i=1}^{i=a}[(\bar{Y}_i - \bar{Y}) + (\alpha_i - \bar{\alpha})]^2$$

Squaring the expression in the square brackets, we obtain the terms

$$\frac{1}{a-1}\sum_{i=1}^{i=a}(\bar{Y}_i - \bar{\bar{Y}})^2 + \frac{1}{a-1}\sum_{i=1}^{i=a}(\alpha_i - \bar{\alpha})^2 + \frac{2}{a-1}\sum_{i=1}^{i=a}(\bar{Y}_i - \bar{\bar{Y}})(\alpha_i - \bar{\alpha})$$

The first of these terms we immediately recognize as the previous variance of the means, $s_{\bar{Y}}^2$. The second is a new quantity but is familiar by general appearance. It clearly is a variance or at least a quantity akin to a variance. The third expression is a new type. It is a so-called covariance, which we have not yet encountered. We shall not be concerned with it at this stage except to say that in cases such as the present one, where the magnitude of treatment effects α_i is independent of the $\bar{Y}_i$ to which they are added, the expected value of this quantity is zero; hence it would not contribute to the new variance of means.

The independence of the treatments effects and the sample means is an important concept that we must clearly understand. If we had not applied different treatments to the medium jars, but simply treated all jars as controls, we would still have obtained differences among the wing length means. Those are the differences found in Table 8.1 with random sampling from the same population. By chance some of these means are greater, some are smaller. In our planning of the experiment we had no way of predicting which sample means would be small and which would be large. Therefore, in planning our treatments we had no way of matching up large treatment effect, such as that of medium 6, with the mean which by chance would be the greatest, not as for sample 2. Also, the smallest sample mean (sample 4) is not associated with the smallest treatment effect. Only if the magnitude of the treatment effects were deliberately correlated with the sample means (this would be difficult to do in the experiment designed here), would the third term in the expression, the covariance, have an expected value other than zero.

We have seen that the first quantity is simply the variance of means as described before. The second quantity is clearly added as a result of the treatment effects. It is analogous to a variance but cannot be called such since it is not based on a random variable but rather on deliberately chosen treatments largely under our control. By changing the magnitude and nature of the treatments we can more or less alter the variance-like quantity at will. We shall, therefore, call it the *added component due to treatment effects*. Since the α_i's are arranged so that $\bar{\alpha} = 0$, we can rewrite the middle term as

$$\frac{1}{a-1}\sum_{i=1}^{i=a}(\alpha_i - \bar{\alpha})^2 = \frac{1}{a-1}\sum_{i=1}^{i=a}\alpha_i^2 = \frac{1}{a-1}\sum^{a}\alpha^2.$$

In analysis of variance we multiply the variance of the means by n in order to estimate the parametric variance of the items. As you know, we call the quantity so obtained the variance of groups. When we do this for the case in which treatment effects are present we obtain

$$n\left(s_{\bar{Y}}^2 + \frac{1}{a-1}\sum^{a}\alpha^2\right) = s^2 + \frac{n}{a-1}\sum^{a}\alpha^2$$

Thus we see that the estimate of the parametric variance of the population is increased by the quantity

$$\frac{n}{a-1}\sum^{a}\alpha^2$$

which is n times the added component due to treatment effects. We found the variance ratio F_s to be significantly greater than could be reconciled with the null hypothesis. It is now obvious why this is so. We were testing the variance ratio expecting to find F approximately equal to $\sigma^2/\sigma^2 = 1$. In fact, however, we have

$$F \approx \frac{\sigma^2 + \dfrac{n}{a-1}\sum^{a}\alpha^2}{\sigma^2}$$

It is clear from this formula, deliberately displayed in this lopsided manner, that the F-test is sensitive to the presence of the added component due to treatment effects.

At this point you should have your first insight into the purpose of analysis of variance. It permits us to test whether there are added treatment effects—that is, whether a group of means can simply be considered random samples from the same population or whether treatments which have affected each group separately have resulted in shifting these means sufficiently so that they can no longer be considered samples from the same population. If that is so, an added component due to treatment effects will be present and may be detected by an F-test in the significance test of the analysis of variance. In such a study we are generally not interested in the magnitude of

$$\frac{n}{a-1} \sum^{a} \alpha^2$$

but we are interested in the magnitude of the separate values of α_i. In our example these are the effects of different formulations of the medium on wing length. If instead of housefly wing length we were measuring blood pressure in samples of rats, and the different groups had been subjected to different drugs or different doses of the same drug, the quantities α_i would represent the effects of drugs on the blood pressure, which is clearly the issue of interest to the investigator. We may also be interested in studying differences of the type $\alpha_1 - \alpha_2$, leading us to the question of the significance of the differences between the effects of any two types of medium or any two drugs. But we are a little ahead of our story.

When analysis of variance involves treatment effects of the type just studied, we call it a Model I anova. Later in this chapter (Section 8.6), Model I will be defined precisely. There is another model, called a Model II anova, in which the added effects for each group are not fixed treatments but are random effects. By this we mean that we have not deliberately planned or fixed the treatment for any one group but that the actual effects on each group are random and only partly under our control. Suppose that the seven samples of houseflies in Table 8.3 represented the offspring of seven randomly selected females from a population reared on a uniform medium. There would be genetic differences among these females and their seven broods would reflect this. The exact nature of these differences are unclear and unpredictable. Before actually measuring them, we have no way of knowing whether brood 1 would have longer wings than brood 2, nor have we any way of controlling this experiment so that brood 1 did in fact grow longer wings. So far as we can ascertain, the genetic factors for wing length are distributed in an unknown manner in the population of houseflies (we might hope that they are normally distributed), and our sample of seven is a random sample of these factors. Another example for a Model II might be that instead of making up our seven cultures from a single batch of medium, we have prepared seven batches separately, one right after the other, and are now analyzing the variation among the batches. We would not be interested in the exact differences from batch to batch. Even if these were measured we would not be in a position to interpret them. Not having deliberately varied batch 3, we have no idea why it should produce longer wings than batch 2, for example. We would, however, be interested in the magnitude of the variance of the added effects. Thus, if we used seven jars of medium derived from one batch, we could expect the variance of the jar means to be $\sigma^2/5$, since there were 5 flies per jar. But when based on different batches of medium the variance could be expected to be greater, because all the imponderable accidents of formulation and environmental differences during medium preparation that make one batch of medium different from another would come into play. Interest would focus on the added variance component arising from differences among batches. Similarly, in the other example we would be interested in the added variance component arising from genetic differences among the females.

We shall now take a rapid look at the algebraic formulation of the anova in the case of Model II. In Table 8.3 the second row at the head of the data columns shows not only α_i but also A_i, which is the symbol we shall use for a random group effect. We use a capital letter to indicate that the effect is a variable. We shall also consider the random effects to sum to zero, or

$$\frac{1}{a}\sum^{a} A_i = \bar{A} = 0$$

The algebra of calculating the two estimates of the population variance is the same as in Model I, except that in place of α_i we imagine A_i substituted in Table 8.4. The estimate of the variance among means now represents the quantity

$$\frac{1}{a-1}\sum_{i=1}^{i=a}(\bar{Y}_i - \bar{\bar{Y}})^2 + \frac{1}{a-1}\sum_{i=1}^{i=a}(A_i - \bar{A})^2 + \frac{2}{a-1}\sum_{i=1}^{i=a}(\bar{Y}_i - \bar{\bar{Y}})(A_i - \bar{A})$$

The first term is the variance of means $s_{\bar{Y}}^2$ as before, and the last term is the covariance between the group means and the random effects A_i, the expected value of which is zero (as before) because the random effects are independent of the magnitude of the means. The middle term is a true variance since A_i is a random variable. We symbolize it by s_A^2 and call it the *added variance component among groups*. It would represent the added variance component among females or among medium batches, depending on which of the designs discussed above we think of. The existence of this added variance component is demonstrated by the F-test. If the groups are random samples we may expect F to approximate $\sigma^2/\sigma^2 = 1$, but with an added variance component the expected ratio is

$$F \approx \frac{\sigma^2 + n\sigma_A^2}{\sigma^2}$$

Note that σ_A^2, the parametric value of s_A^2, is multiplied by n, since we have to multiply the variance of means by n to obtain an independent estimate of the variance of the population. In a Model II we are not interested in the magnitude of any A_i or in differences such as $A_1 - A_2$, but we are interested in the magnitude of σ_A^2 and its relative magnitude with respect to σ^2, which is generally expressed as the percentage $100s_A^2/(s^2 + s_A^2)$. Since the variance among groups estimates $\sigma^2 + n\sigma_A^2$, we can calculate s_A^2 as

$$\frac{1}{n}\text{(variance among groups} - \text{variance within groups)}$$

$$= \frac{1}{n}[(s^2 + ns_A^2) - s^2] = \frac{1}{n}(ns_A^2) = s_A^2$$

For the present example $s_A^2 = \frac{1}{5}(83.848 - 16.029) = 13.56$. This added variance component among groups is

$$\frac{100 \times 13.56}{16.029 + 13.56} = \frac{1356}{29.589} = 45.83\%$$

of the sum of the variances among and within groups. Model II will be formally discussed at the end of this chapter (Section 8.7); the methods of estimating variance components are treated in detail in the next chapter.

8.5 Partitioning the total sum of squares and degrees of freedom

You may be discouraged to find more algebraic formulas ahead in this section. But these describe only various simple properties and, given a little good will, can be mastered by anyone. We present them simply because these quantities will be useful in computational methods for analysis of variance.

So far we have ignored one other variance that can be computed from the data in Table 8.1. If we remove the classification into groups, we can consider the housefly data to be a single sample of $an = 35$ wing lengths and calculate the mean and variance of these items in the conventional manner. The various quantities necessary for this computation are shown in the last column at the right in Tables 8.1 and 8.3, headed Computation of total sum of squares. We obtain a mean of $\overline{Y} = 45.34$ for the sample in Table 8.1, which is, of course, the same as the quantity $\overline{\overline{Y}}$ computed previously from the seven group means. The sum of squares of the 35 items is 575.886, which gives a variance of 16.938 when divided by 34 degrees of freedom. Repeating these computations for the data in Table 8.3 we obtain $\overline{Y} = 45.34$ (the same as in Table 8.1 because $\sum^{a}\alpha_i = 0$) and $s^2 = 27.997$, which is considerably greater than the corresponding variance from Table 8.1. The total variance computed from all an items is another estimate of σ^2. It is a good estimate in the first case, but in the second sample (Table 8.3), where added components due to treatment effects or added variance components are present, it is a poor estimate of the population variance.

However, the purpose of calculating the total variance in an anova is not for using it as yet another estimate of σ^2 but for ease of computation. This is best seen when we arrange our results in a conventional *analysis of variance table* as shown in Table 8.5. Such a table is divided into four columns. The first identifies the source of variation as among groups, within groups,

TABLE 8.5

Anova table for data in Table 8.1.

	(1) Source of variation	(2) df	(3) Sums of squares SS	(4) Mean squares MS
$\bar{Y} - \bar{\bar{Y}}$	Among groups	6	127.086	21.181
$Y - \bar{Y}$	Within groups	28	448.800	16.029
$Y - \bar{\bar{Y}}$	Total	34	575.886	16.938

and total (groups amalgamated to form a single sample). The column headed df gives the degrees of freedom by which the sums of squares pertinent to that source of variation must be divided in order to yield the variances. The degrees of freedom for variation among groups is $a - 1$, that for variation within groups is $a(n - 1)$, and that for the total variation is $an - 1$. The next two columns show sums of squares and variances, respectively. Notice that the sums of squares entered in the anova table are the sum of squares among groups, the sum of squares within groups, and the sum of squares of the total sample of an items. You will note that variances are not called such in anova, but are generally called *mean squares*, since, in a Model I anova, they do not estimate a population variance. These quantities are not true *mean* squares because the sums of squares are divided by the degrees of freedom rather than sample size. The sums of squares and mean squares are frequently abbreviated SS and MS, respectively.

The sums of squares and mean squares in Table 8.5 are the same as those obtained previously, except for minute rounding errors. Note, however, an important property of the sums of squares. They were obtained independently of each other, but when we add the SS among groups to the SS within groups we obtain the total SS. The sums of squares are additive! Another way of saying this is that we can decompose the total sum of squares into a portion due to variation among groups and another portion due to variation within groups. Observe that the degrees of freedom are also additive and that the total of 34 df can be decomposed into 6 df among groups and 28 df within groups. Thus, if we know any two of the sums of squares (and their appropriate degrees of freedom), we can compute the third and complete our analysis of variance. Note that the mean squares are not additive. This is obvious since generally $(a + b)/(c + d) \neq a/c + b/d$.

We shall use the computational formula for sum of squares [Expression (4.7)] to demonstrate why these sums of squares are additive. Although it is an algebraic derivation, it is placed here rather than in the Appendix because these formulas will also lead us to the actual computational formulas for analysis of variance. As you may have suspected, the method we used to obtain the sums of squares is not the most rapid computational procedure;

we shall employ a far simpler method for routine computation. The sum of squares of means in simplified notation is

$$SS_{\text{means}} = \sum^{a}(\bar{Y} - \bar{\bar{Y}})^2 = \sum^{a}\bar{Y}^2 - \frac{\left(\sum^{a}\bar{Y}\right)^2}{a}$$

$$= \sum^{a}\left(\frac{1}{n}\sum^{n}Y\right)^2 - \frac{1}{a}\left[\sum^{a}\left(\frac{1}{n}\sum^{n}Y\right)\right]^2$$

$$= \frac{1}{n^2}\sum^{a}\left(\sum^{n}Y\right)^2 - \frac{1}{an^2}\left(\sum^{a}\sum^{n}Y\right)^2$$

Note that the deviation of means from the grand mean is first rearranged to fit the computational formula [Expression (4.7)] and then each mean is written in terms of its constituent variates. Collection of denominators outside the summation signs yields the final desired form. To obtain the sum of squares of groups, we multiply SS_{means} by n, as before. This yields

$$SS_{\text{groups}} = n \times SS_{\text{means}} = \frac{1}{n}\sum^{a}\left(\sum^{n}Y\right)^2 - \frac{1}{an}\left(\sum^{a}\sum^{n}Y\right)^2$$

Next we evaluate the sum of squares within groups:

$$SS_{\text{within}} = \sum^{a}\sum^{n}(Y - \bar{Y})^2 = \sum^{a}\left[\sum^{n}Y^2 - \frac{1}{n}\left(\sum^{n}Y\right)^2\right]$$

$$= \sum^{a}\sum^{n}Y^2 - \frac{1}{n}\sum^{a}\left(\sum^{n}Y\right)^2$$

The total sum of squares represents

$$SS_{\text{total}} = \sum^{a}\sum^{n}(Y - \bar{\bar{Y}})^2$$

$$= \sum^{a}\sum^{n}Y^2 - \frac{1}{an}\left(\sum^{a}\sum^{n}Y\right)^2$$

We now copy the formulas for these sums of squares, slightly rearranged as follows:

$$\begin{aligned}
SS_{\text{groups}} &= \frac{1}{n}\sum^{a}\left(\sum^{n}Y\right)^2 & & - \frac{1}{an}\left(\sum^{a}\sum^{n}Y\right)^2 \\
SS_{\text{within}} &= -\frac{1}{n}\sum^{a}\left(\sum^{n}Y\right)^2 & + \sum^{a}\sum^{n}Y^2 & \\
\hline
SS_{\text{total}} &= & \sum^{a}\sum^{n}Y^2 & - \frac{1}{an}\left(\sum^{a}\sum^{n}Y\right)^2
\end{aligned}$$

Adding the expression for SS_{groups} to that for SS_{within} we obtain a quantity that is identical to the one we just developed as SS_{total}. This demonstration explains why the sums of squares are additive.

We shall not go through any derivation but simply state that the degrees of freedom pertaining to the sums of squares are also additive. The total degrees of freedom are split up into the degrees of freedom corresponding to variation among groups and those of variation of items within groups.

Before we continue, let us review the meaning of the three mean squares in the anova. The total MS is a statistic of dispersion of the 35 (an) items around their mean, the grand mean 45.34. It describes the variance in the entire sample due to all and sundry causes and estimates σ^2 when there are no added treatment effects or variance components among groups. The within-group MS, also known as the *individual* or *intragroup* or *error mean square*, gives the average dispersion of the 5 (n) items in each group around the group means. If the a groups are random samples from a common homogeneous population, the within-group MS should estimate σ^2. The MS among groups is based on the variance of group means, which describes the dispersion of the 7 (a) group means around the grand mean. If the groups are random samples from a homogeneous population, the expected variance of their mean will be σ^2/n. Therefore, in order to have all three variances of the same order of magnitude, we multiply the variance of means by n to obtain the variance among groups. If there are no added treatment effects or variance components the MS among groups is an estimate of σ^2. Otherwise it is an estimate of

$$\sigma^2 + \frac{n}{a-1}\sum^{a}\alpha^2 \qquad \text{or} \qquad \sigma^2 + n\sigma_A^2$$

depending on whether the anova at hand is Model I or II.

The additivity relations we have just learned are independent of the presence of added treatment or random effects. We could show this algebraically, but it is simpler to inspect Table 8.6, which summarizes the anova of Table 8.3 in which α_i or A_i are added to each sample. The additivity relation

TABLE 8.6

Anova table for data in Table 8.3.

	(1) Source of variation	*(2)* df	*(3)* Sums of squares SS	*(4)* Mean squares MS
$\bar{Y} - \bar{\bar{Y}}$	Among groups	6	503.086	83.848
$Y - \bar{Y}$	Within groups	28	448.800	16.029
$Y - \bar{\bar{Y}}$	Total	34	951.886	27.997

still holds, although the values for group SS and total SS are different from those of Table 8.5.

Another way of looking at the partitioning of the variation is to study the deviation from means in a particular case. Referring to Table 8.1 we can look at the wing length of the first individual in the 7th group, which happens to be 41. Its deviation from its group mean is

$$Y_{71} - \bar{Y}_7 = 41 - 45.4 = -4.4$$

The deviation of the group mean from the grand mean is

$$\bar{Y}_7 - \bar{\bar{Y}} = 45.4 - 45.34 = 0.06$$

and the deviation of the individual wing length from the grand mean is

$$Y_{71} - \bar{\bar{Y}} = 41 - 45.34 = -4.34$$

Note that these deviations are additive. The deviation of the item from the group mean and that of the group mean from the grand mean add to the total deviation of the item from the grand mean. These deviations are stated algebraically as $(Y - \bar{Y}) + (\bar{Y} - \bar{\bar{Y}}) = (Y - \bar{\bar{Y}})$. Squaring and summing these deviations for an items will result in

$$\sum^{a} \sum^{n} (Y - \bar{Y})^2 + n \sum^{a} (\bar{Y} - \bar{\bar{Y}})^2 = \sum^{an} (Y - \bar{\bar{Y}})^2$$

Before squaring, the deviations were in the relationship $a + b = c$. After squaring we would expect them in the form $a^2 + b^2 + 2ab = c^2$. What happened to the cross-product term corresponding to $2ab$? This is

$$2 \sum^{an} (Y - \bar{Y})(\bar{Y} - \bar{\bar{Y}}) = 2 \sum^{a} \left[(\bar{Y} - \bar{\bar{Y}}) \sum^{n} (Y - \bar{Y})\right]$$

a covariance type term that is always zero, since $\sum^{n}(Y - \bar{Y}) = 0$ for each of the a groups (proof in Appendix A1.1).

We identify the deviations represented by each level of variation at the left margins of the tables giving the analysis of variance results (Tables 8.5 and 8.6). Note that the deviations add up correctly: the deviation among groups plus the deviation within groups equals to the total deviation of items in the analysis of variance, $(\bar{Y} - \bar{\bar{Y}}) + (Y - \bar{Y}) = (Y - \bar{\bar{Y}})$.

8.6 Model I anova

We have already mentioned the two different models in analysis of variance. An important point to remember is that the basic setup of data, as well as the actual computation and significance test, in most cases is the same for both models. The purposes of analysis of variance differ for the two models, as do some of the supplementary tests and computations following the initial significance test. The two models were first defined by Eisenhart (1947). Although analysis of variance dates back to the 1930's, the distinction between the models had not been made in earlier work.

Let us now try to resolve the variation found in an analysis of variance case. This will not only lead us to a more formal interpretation of anova but will also give us a deeper understanding of the nature of variation itself. For purposes of discussion we shall refer back to the housefly wing lengths of Table 8.3. We ask the question, "What makes any given housefly wing length assume the value it does?" The third wing length of the first sample of flies is recorded as 43 units. How can we explain such a reading?

First of all, knowing nothing else about this individual housefly, our best guess as to its wing length is the grand mean of the population, which we know to be $\mu = 45.5$. However, we have additional information about this fly. It is a member of group 1, which has undergone a treatment shifting the mean of the group downward by 5 units. Therefore, $\alpha_1 = -5$ and we would expect our individual Y_{13} (the third individual of group 1) to measure $45.5 - 5 = 40.5$ units. In fact, however, it is 43 units, which is 2.5 units above this latest expectation. To what can we ascribe this deviation? It is individual variation of the flies within a group because of the variance of individuals in the population ($\sigma^2 = 15.21$). All the genetic and environmental effects that make one housefly different from another housefly come into play to produce this variance. By means of carefully designed experiments we might learn something about the causation of this variance and attribute it to certain specific genetic or environmental factors. We might also be able to eliminate some of the variance. For instance, by using only full sibs (brothers and sisters) in any one culture jar we would decrease the genetic variation in individuals, and undoubtedly the variance within groups would be smaller. However, it is hopeless to try to eliminate all variance completely. Even if we could remove all genetic variance, there would still be environmental variance, and even in the most improbable case in which we could remove both there would remain measurement error, so that we would never obtain exactly the same reading even on the same individual fly. The within groups MS always remains as a residual, greater or smaller from experiment to experiment,—part of the nature of things. This is why the within groups variance is also called the error variance or error mean square. It is not an error in the sense of someone having made a mistake, but in the sense of providing you with a measure of the variation you have to contend with when trying to estimate significant differences among the groups. The error variance is composed of individual deviations for each individual, symbolized by ϵ_{ij}, the random component of the jth individual variate in the ith group. In our case, $\epsilon_{13} = 2.5$, since the actual observed value is 2.5 units above its expectation of 40.5.

We shall now state this relationship more formally. In a Model I analysis of variance we assume that the differences among group means, if any, are due to the fixed treatment effects determined by the experimenter. The purpose of the analysis of variance is to estimate the true differences among the group means. Any single variate can be decomposed as follows:

$$Y_{ij} = \mu + \alpha_i + \epsilon_{ij} \tag{8.3}$$

where $i = 1, \ldots, a$, $j = 1, \ldots, n$, ϵ_{ij} represents an independent, normally distributed variable with mean $\bar{\epsilon}_{ij} = 0$, and variance $\sigma_\epsilon^2 = \sigma^2$. Therefore a given reading is composed of the grand mean μ of the population, a fixed deviation α_i of the mean of group i from the grand mean μ, and a random deviation ϵ_{ij} of the jth individual of group i from its expectation, which is $(\mu + \alpha_i)$. Remember that both α_i and ϵ_{ij} can be positive as well as negative. The expected value (mean) of the ϵ_{ij}'s is zero and their variance is the parametric variance of the population, σ^2. For all the assumptions of the analysis of variance to hold, the distribution of ϵ_{ij} must be normal.

In a Model I anova we test for differences of the type $\alpha_1 - \alpha_2$ among the group means by testing for the presence of an added component due to treatments. If we find that such a component is present, we reject the null hypothesis that the groups come from the same population and accept the alternative hypothesis that at least some of the group means are different from each other, which indicates that at least some of the α_i's are unequal in magnitude. Next we generally wish to test which α_i's are different from each other. This is done by significance tests, with alternative hypotheses such as $H_1{:}\alpha_1 > \alpha_2$ or $H_1{:}\frac{1}{2}(\alpha_1 + \alpha_2) > \alpha_3$. In words, these test whether the mean of group 1 is significantly greater than the mean of group 2, or whether the mean of group 3 is smaller than the average of the means of groups 1 and 2.

Some examples of Model I analyses of variance in various biological disciplines follow. An experiment in which we try the effects of different drugs on batches of animals results in a Model I anova. We are interested in the results of the treatments and the differences between them. The treatments are fixed and determined by the experimenter. This is true also when we test the effects of different doses of a given factor—a chemical or the amount of light to which a plant has been exposed or temperatures at which culture bottles of insects have been reared. The treatment does not have to be entirely understood and manipulated by the experimenter; so long as it is fixed and repeatable, Model I will apply. If we had wanted to compare the birth weights of the Chinese children in the hospital in Malaya with weights of Chinese children born in a hospital on the Chinese mainland, it would also have been a Model I anova. The treatment effects then would be "mainland versus Malaya," which sums up a whole series of different factors, genetic and environmental—some known to us but most of them not understood. However, this is a definite treatment we can describe and also repeat; namely, we can if we wish, again sample birth weights of infants in Malaya as well as on the Chinese mainland.

Another example of Model I anova would be a study of body weights for several age groups of animals. The treatments would be the ages, which are fixed. If we find that there is a significant difference in weight among the ages, we might follow this up with the question of whether there is a difference from age 2 to age 3 or only from age 1 to age 2. To a very large extent Model I anovas are the result of an experiment and of deliberate manipulation of factors by the experimenter. However, the study of differences such as the comparison of birth weights from two countries, while not an experiment proper, also falls into this category.

Let us review once more the idea of decomposing an individual reading, using the birth weights as an example. An individual birth weight of an infant in Malaya can be thought of as consisting of the parametric mean μ of the birth weights of all Chinese children plus the treatment effects of being in Malaya α_1, plus a random component ϵ_{1j}, which represents the genetic and environmental differences making this individual infant deviate from its population mean.

8.7 Model II anova

The structure of variation in a Model II anova is quite similar to that in Model I:

$$Y_{ij} = \mu + A_i + \epsilon_{ij} \tag{8.4}$$

where $i = 1, \ldots, a$, $j = 1, \ldots, n$, ϵ_{ij} represents an independent, normally distributed variable with mean $\bar{\epsilon}_{ij} = 0$ and variance $\sigma_\epsilon^2 = \sigma^2$, and A_i represents a normally distributed variable, independent of all ϵ's, with mean $\bar{A}_i = 0$ and variance σ_A^2. The main distinction is that in place of fixed treatment effects α_i we now consider random effects A_i, differing from group to group. Since the effects are random, it is futile to estimate the magnitude of these random effects for any one group, or the differences from group to group, but we can estimate their variance, the added variance component among groups σ_A^2. We test for its presence and estimate its magnitude s_A^2, as well as its percentage contribution to the variation in a Model II analysis of variance.

Some examples will illustrate the applications of Model II anova. Suppose we wish to determine the *DNA* content of rat liver cells. We take five rats and make three preparations from each of the five livers obtained. The assay readings will be for $a = 5$ groups with $n = 3$ readings per group. The five rats presumably are sampled at random from the colony available to the experimenter. They must be different in various ways, genetically and environmentally, but we have no definite information about the nature of these differences. Thus, if we learn that rat 2 has slightly more *DNA* in its liver cells than rat 3, we can do little with this information, because we are unlikely to have any basis for following up this problem. We will, however, be interested in estimating the variance of the three replicates within any one liver and the variance among the five rats; that is, is there variance σ_A^2 among rats in addition to the variance σ^2 expected on the basis of the three replicates? The variance among the three preparations presumably arises only from differences in technique and possibly from differences in *DNA* content in different parts of the liver (unlikely in a homogenate). Added variance among rats, if it existed, might be due to differences in ploidy or related phenomena. The relative amounts of variation among rats and "within" rats (= among preparations) would guide us in designing further studies of this sort. If there was little variance among the preparations and relatively more variation among the rats, we would need fewer preparations and more rats. On the other hand, if the variance among rats is proportionately smaller, we would use fewer rats and more preparations per rat. This whole subject of efficient planning of experiments will be treated in Section 10.4.

In a study of the amount of variation in skin pigment in human populations we might wish to study different families within a homogeneous racial group and brothers and sisters within each family. The variance within families would be the error mean square and we would test for an added variance component among families. We would expect an added variance component σ_A^2 because there are genetic differences among families that determine amount of skin pigmentation. We would be especially interested in the relative proportions of the two variances σ^2 and σ_A^2 because they would provide us with important genetic information. From our knowledge of genetic theory we would expect the variance among families to be greater than the variance among brothers and sisters within a family.

The above examples illustrate the two types of problems involving Model II analysis of variance that are most likely to arise in biological work. One is concerned with the general problem of the design of an experiment and the magnitude of the experimental error at different levels of replication, such as error among replicates within rat livers and among rats, error among batches, experiments, and so forth. The other relates to variation among and within families, among and within females, among and within populations, and so forth, being concerned with the general problem of the relation between genetic and phenotypic variation.

In the next chapter we shall learn computational formulas for carrying out an analysis of variance and shall practice these on several different examples. In these examples we shall have an opportunity to distinguish cases of Model I and Model II anovas and to learn the proper way of analyzing each model and of presenting the results of such analyses.

Exercises 8

8.1 In a study comparing the chemical composition of the urine of chimpanzees and gorillas (Gartler, Firschein, and Dobzhansky, 1956), the following results were obtained. For 37 chimpanzees the variance for the amount of glutamic acid in milligrams per milligram of creatinine was 0.01069. A similar study based on six gorillas yielded a variance of 0.12442. Is there a significant difference between the variability in chimpanzees and gorillas? ANS. $F_s = 11.639$, $F_{.025[5,36]} \approx 2.90$.

8.2 The following data are from an experiment by Sewall Wright. He crossed Polish and Flemish giant rabbits and obtained 27 F_1 rabbits. These were inbred and 112 F_2 rabbits were obtained. We have extracted the following data on femur length of these rabbits.

	n	$\bar{Y}$	s
F_1	27	83.39	1.65
F_2	112	80.5	3.81

Is there a significantly greater amount of variability in femur lengths among the F_2 than among the F_1 rabbits? What well-known genetic phenomenon is illustrated by these data?

8.3 Show that it is possible to represent the value of an individual variate as follows: $Y_{ij} = (\bar{\bar{Y}}) + (\bar{Y}_i - \bar{\bar{Y}}) + (Y - \bar{Y}_i)$. What do each of the terms in parentheses estimate in a Model I anova and in a Model II anova?

8.4 For the data in Table 8.3, make tables to represent partitioning of the value of each variate into its three components, $\bar{\bar{Y}}$, $(\bar{Y}_i - \bar{\bar{Y}})$, $(Y - \bar{Y}_i)$. The first table would then consist of 35 values, all equal to the grand mean. In the second table all entries in a given column would be equal to the difference between the mean of that column and the grand mean. And the last table will consist of the deviations of each individual variate from its column mean. These tables represent one's estimates of the individual components of Expression (8.3). Compute the mean and sums of squares for each table.

9 SINGLE CLASSIFICATION ANALYSIS OF VARIANCE

We are now ready to study actual cases of analysis of variance in a variety of applications and designs. The present chapter deals with the simplest kind of anova, *single classification analysis of variance.* By this is meant that the groups of samples are classified by only a single criterion. Both interpretations of the seven samples of housefly wing lengths studied in the last chapter, different medium formulations (Model I), and progenies of different females (Model II) would represent a single criterion for classification. Other examples would be different temperatures at which groups of animals were raised or different soils in which samples of plants have been grown.

We shall start out in Section 9.1 by stating the basic computational formulas for analysis of variance, based on the topics covered in the previous chapter. Section 9.2 gives an example of the general case with unequal sample sizes, since all groups in the anova need not necessarily have the same sample size. We shall illustrate this case by means of a Model II anova. Some computations special to a Model II anova are also shown—estimating variance components and setting confidence limits to them. Since the basic computations for the analysis of variance are the same in either model, it is not necessary to repeat the illustration with a Model I. The latter model is featured in Section 9.3, which shows the computational simplification resulting from equal sample sizes. Formulas become especially simple for the two-sample case (Section 9.4). In this case a t-test can be applied as well, and we shall show that it is mathematically equivalent to an analysis of variance F-test when there are only two groups. Another special case, the comparison of a single specimen with a sample, is illustrated in Section 9.5.

When a Model I analysis of variance has been found to be significant, leading to the conclusion that the means are not from the same population, we wish to test the means in a variety of ways to discover which pairs of means are different from each other, and whether the means can be divided into groups that are significantly different from each other. These so-called multiple comparisons tests are covered in Sections 9.6 and 9.7. The earlier section deals with so-called planned comparisons designed before the test is run, the latter section with a posteriori tests that suggest themselves to the experimenter as a result of his analysis.

Experimenters frequently ask statisticians what sample sizes are required for a given study. This topic is taken up in Section 9.8.

In contrast to some previous chapters dealing largely with theory, much of the present discussion will center around boxes illustrating these various applications. Computational details will be discussed in the boxes; general implications and additional comments will be found in the text.

9.1 Computational formulas

We saw in Section 8.5 that the total sum of squares and degrees of freedom can be additively partitioned into those pertaining to variation among groups and those of variation within groups. For the analysis of variance proper we need only the sum of squares among groups and the sum of squares within groups. However, on a desk calculator it is simpler to calculate the total sum of squares and the sum of squares among groups, leaving the sum of squares within groups to be obtained by the subtraction $SS_{\text{total}} - SS_{\text{groups}}$. This rule does not apply to digital computers; there the tedium of computing sums of squares within groups is of no consequence. In Section 8.5 we arrived at the following computational formulas for the total and among groups sums of squares:

$$SS_{\text{total}} = \sum^{a}\sum^{n} Y^2 - \frac{1}{an}\left(\sum^{a}\sum^{n} Y\right)^2$$

$$SS_{\text{groups}} = \frac{1}{n}\sum^{a}\left(\sum^{n} Y\right)^2 - \frac{1}{an}\left(\sum^{a}\sum^{n} Y\right)^2$$

These formulas assume equal sample size n for each group and will be rewritten in the next section to make them generally applicable for either equal or unequal sample sizes. However, they suffice in their present form to illustrate some general points about computational procedures in analysis of variance.

We note first of all that the second, subtracted term in each sum of squares is identical. This term represents the sum of all the variates in the anova (the grand total), quantity squared and divided by the total number of variates. It is comparable to the second term in the computational formula for the ordinary sum of squares [Expression (4.7)]. This term is often called the *correction term* (abbreviated CT).

The first term for the total sum of squares is simple. It is the sum of all squared variates in the anova table. Thus the total sum of squares, which describes the variation of a single unstructured sample of an items, is simply the familiar sum of squares formula of Expression (4.7).

The first term of the sum of squares among groups is obtained by squaring the sum of the items of each group, dividing each square by its sample size and summing the quotients from this operation for each group. Since the sample size of each group is equal in the above formulas, we can first sum all the squares of the group sums and then divide their sum by the constant n.

From the formula for the sum of squares among groups emerges an important computational rule of analysis of variance. *To find the sum of squares among any set of groups, square the sum of each group and divide it by its sample size; sum the quotients of these operations and subtract from this sum a correction term. To find this correction term, sum all the items in the set, square this sum, and divide it by the number of items on which this sum is based.*

9.2 General case: unequal n

Our first computational example of a single classification analysis of variance has unequal sample sizes. We shall develop general formulas for carrying out such an analysis and shall see in the next section that the more usual anova with equal sample sizes is only a special case, which is simpler to compute. For our example we shall use a Model II analysis of variance. Up to and including the F-test for significance, the computations are exactly the same whether the anova is Model I or Model II. We shall point out the stage in the computations at which there would be a divergence of operations depending on the model.

The example is shown in Box 9.1. It concerns a series of morphological measurements of the width of the scutum (dorsal shield) of samples of tick larvae obtained from four different host individuals of the cottontail rabbit. These four hosts were obtained at random from one locality. We know nothing about their origins or their genetic constitution. They represent a random sample of the population of host individuals from the given locality. We would not be in a position to interpret differences between larvae from any given host, since we know nothing of the origins of the individual rabbits. Population biologists are nevertheless interested in such analyses because they provide an answer to the following question. Are the variances of means of larval characters among hosts greater than expected on the basis of variances of the characters within hosts? We can calculate the average variance of width of larval scutum on a host. This will be our "error" term in the analysis of variance. We then test the observed mean square among groups and see if it contains an added component of variance. What would such an added component of variance represent? The mean square within host individuals (that is, of larvae on any one host) represents genetic differences among larvae and differences in environmental experiences of these larvae. Added variance among hosts demonstrates significant differentiation among the larvae possibly due to differences among the hosts affecting the larvae. It also may be due to genetic differences among the larvae should each host carry a family of ticks, or at least a population whose individuals are more related to each other than they are to tick larvae on other host individuals. The emphasis in this example is on the magnitudes of the variances. In view of the random choice of hosts this is a clear case of a Model II anova.

The computation is illustrated in Box 9.1. After quantities **1** through **7** have been evaluated they are entered into an analysis of variance table, as shown in the box. General formulas for such a table are shown first, followed by a table filled in for the specific example. We note 3 degrees of freedom among groups, there being four host individuals, and 33 *df* within groups, these being the sum of 7, 9, 12, and 5 degrees of freedom. We note that the mean square among groups is considerably greater than the error mean square, giving rise to a suspicion that an added variance component is present. If the MS_{groups} is equal to or less than the MS_{within}, we do not bother going on with the analysis, for we would not have evidence for the presence of an added variance component. You may wonder how it could be possible for the MS_{groups} to be less than the MS_{within}. You must remember that these two are independent estimates. If there is no added variance component among groups, the estimate of the variance among groups is as likely to be less as it is to be greater than the variance within groups.

Expressions for the expected values of the mean squares are also shown in the first anova table of Box 9.1. They are those you learned in the previous section for a Model II anova, with one exception. There is n_0 in place of n. Since sample size n_i differs among groups in this example, it is obvious that no single value of n would be appropriate in the formula. We therefore use an average n; this, however, is not simply $\bar{n}$, the arithmetic mean of the n_i's, but is

$$n_0 = \frac{1}{a-1}\left(\sum^{a} n_i - \frac{\sum^{a} n_i^2}{\sum^{a} n_i}\right) \tag{9.1}$$

which is an average usually close to, but always less than $\bar{n}$, unless sample sizes are equal, when $n_0 = \bar{n}$. So, as a computational check when computing n_0, also compute $\bar{n}$ and make certain that $n_0 < \bar{n}$. Mathematical statisticians have determined that the appropriate mean value of n for the expected mean square among groups is n_0 as defined above.

The critical 5%, 1%, and 0.1% values of F are shown below the second anova table in Box 9.1 (2.89, 4.44, and 6.89, respectively). You should confirm them for yourself in Table **S**. Note that the argument $\nu_2 = 33$ is not given. You therefore have to interpolate between arguments representing 30 and 40 degrees of freedom, respectively. The values shown in Box 9.1 we computed using harmonic interpolation (see section on interpolation in the introduction to the statistical tables). However, it is frequently not necessary

BOX 9.1

Single classification anova with unequal sample sizes.

Width of scutum (dorsal shield) of larvae of the tick *Haemaphysalis leporispalustris* in samples from 4 cottontail rabbits. Measurements in microns. This is a Model II anova.

	Hosts (a = 4)			
	1	*2*	*3*	*4*
	380	350	354	376
	376	356	360	344
	360	358	362	342
	368	376	352	372
	372	338	366	374
	366	342	372	360
	374	366	362	
	382	350	344	
		344	342	
		364	358	
			351	
			348	
			348	
$\sum^{n_i} Y$	2978	3544	4619	2168
n_i	8	10	13	6
$\sum^{n_i} Y^2$	1,108,940	1,257,272	1,642,121	784,536

Source: Data by P. A. Thomas.

Preliminary computations

1. Grand total $= \sum_{i=1}^{a} \sum_{j=1}^{n_i} Y_{ij} = 2978 + 3544 + \cdots + 2168 = 13{,}309$

Note: For simplicity we shall abbreviate our notation in the rest of this box as follows: $\sum^{a} = \sum_{i=1}^{a}, \quad \sum^{n} = \sum_{j=1}^{n_i}, \quad Y = Y_{ij}$

2. Sum of the squared observations $= \sum^{a} \sum^{n} Y^2$

$$= 1{,}108{,}940 + 1{,}257{,}272 + \cdots + 784{,}536 = 4{,}792{,}869$$

3. Sum of the squared group totals, each divided by its sample size

$$= \sum^{a} \frac{\left(\sum^{n} Y\right)^2}{n_i} = \frac{(2978)^2}{8} + \frac{(3544)^2}{10} + \cdots + \frac{(2168)^2}{6} = 4{,}789{,}091$$

4. Grand total squared and divided by total sample size = correction term

$$CT = \frac{\left(\sum^{a} \sum^{n} Y\right)^2}{\sum^{a} n_i} = \frac{(13{,}309)^2}{37} = 4{,}787{,}283$$

BOX 9.1 continued

5. $SS_{\text{total}} = \sum^{a}\sum^{n} Y^2 - CT$

$= \text{quantity } \mathbf{2} - \text{quantity } \mathbf{4} = 4{,}792{,}869 - 4{,}787{,}283 = 5586$

6. $SS_{\text{groups}} = \sum^{a} \frac{\left(\sum^{n} Y\right)^2}{n_i} - CT$

$= \text{quantity } \mathbf{3} - \text{quantity } \mathbf{4} = 4{,}789{,}091 - 4{,}787{,}283 = 1808$

7. $SS_{\text{within}} = SS_{\text{total}} - SS_{\text{groups}}$

$= \text{quantity } \mathbf{5} - \text{quantity } \mathbf{6} = 5586 - 1808 = 3778$

□□ *Computational hints.*—Obtain sums and sums of squared observations separately for each group, using your customary technique, then calculate quantities **1** and **2** by summation. To compute quantity **3** we need a desk calculator technique for accumulating quotients as illustrated in Appendix A2.14. Some mathematical relations will be useful in checking the accuracy of your computations. Quantity **3** cannot be greater than quantity **2** and will almost always be less than **2**. Quantity **4** must be less than either **2** or **3**. Quantity **6** will be less than quantity **5**. □□

The anova table is constructed as follows.

Source of variation	*df*	*SS*	*MS*	F_s	*Expected MS*
$\bar{Y} - \bar{\bar{Y}}$ Among groups	$a - 1$	**6**	$\frac{\mathbf{6}}{(a-1)}$	$\frac{MS_{\text{groups}}}{MS_{\text{within}}}$	$\sigma^2 + n_0\sigma^2_A$
$Y - \bar{Y}$ Within groups	$\sum^{a} n_i - a$	**7**	$\frac{\mathbf{7}}{\left(\sum^{a} n_i - 1\right)}$		σ^2
$Y - \bar{\bar{Y}}$ Total	$\sum^{a} n_i - 1$	**5**			

Substituting the computed values into the above table we obtain the following.

Source of variation	*df*	*SS*	*MS*	F_s
$\bar{Y} - \bar{\bar{Y}}$ Among groups (among hosts)	3	1808	602.7	5.26**
$Y - \bar{Y}$ Within groups (error; among larvae on a host)	33	3778	114.5	
$Y - \bar{\bar{Y}}$ Total	36	5586		

$F_{.05[3,33]} = 2.89$ $F_{.01[3,33]} = 4.44$ $F_{.001[3,33]} = 6.89$

** = $P < 0.01$

Conclusion.—There is a significant ($P < 0.01$) added variance component among hosts for width of scutum in larval ticks.

to carry out such an interpolation. One can simply look up the "conservative" value of F, which is the next tabled F with less degrees of freedom; in our case $F_{\alpha[3,30]}$, which is 2.92 and 4.51 for $\alpha = 0.05$ and $\alpha = 0.01$, respectively. The observed value of F_s is 5.26, considerably above the interpolated as well as the conservative value of $F_{.01}$. We therefore reject the null hypothesis ($H_0: \sigma_A^2 = 0$) that there is no added variance component among groups and that the two mean squares estimate the same variance, allowing a type I error of less than 1%. We accept instead the alternative hypothesis of the existence of an added variance component σ_A^2.

What is the biological meaning of this conclusion? For some reason or other the ticks on different host individuals differ more from each other than do individual ticks on any one host. This may be due to some modifying influence of individual hosts on the ticks (biochemical differences in blood, differences in the skin, differences in the environment of the host individual; all of them rather unlikely in this case) or may be due to genetic differences among the ticks. Possibly the ticks on each host represent a sibship (the descendants of a single pair of parents), and the differences of the ticks among host individuals represent genetic differences among families; or selection has acted differently on the tick populations on each host; or the hosts have migrated to the collection locality from different geographic areas in which the ticks differ in width of scutum. Of these various possibilities, genetic differences among sibships seem most reasonable, in view of the biology of the organism.

The computations up to this point would have been identical in a Model I anova. If this had been Model I, the conclusion would have been that there is a significant treatment effect rather than an added variance component. Now, however, we must complete the computations appropriate to a Model II anova. These may include the estimation of the added variance component, computation of its confidence limits, and the calculation of percentage variation at the two levels.

The computation of the variance components for the example in Box 9.1 is given in Box 9.2. From the expressions for the expected mean squares shown in the anova table, it is obvious how the variance component among groups σ_A^2 and the error variance σ^2 are obtained. Of course the values that we obtain are simply estimates and therefore are written as s_A^2 and s^2. We are frequently not so much interested in the absolute values of these variance components but in their relative magnitudes. For this purpose we sum them and express each as a percentage of this sum. Thus $s^2 + s_A^2 = 114.5 + 54.190 = 168.690$ and s^2 and s_A^2 are 67.9% and 32.1% of this sum, respectively. Relatively more variation occurs within groups (larvae on a host) than among groups (hosts). The proportion of the variation among groups is also known as r_I, the *coefficient of intraclass correlation*. When this coefficient is high, it means that most of the variation in the sample is among groups. If it were unity it would mean that all of the variance in the sample is among groups and that there

BOX 9.2

Estimation of variance components in a single classification anova with unequal sample sizes.

Data from Box 9.1.

Anova table

Source of variation	*df*	*MS*	*Expected MS*
Among groups (among hosts)	3	602.7	$\sigma^2 + n_0\sigma^2{}_A$
Within groups (error; among larvae on a host)	33	114.5	σ^2

The sizes of the four samples are 8, 10, 13, and 6. Hence from Expression (9.1) the average sample size, n_0, is

$$n_0 = \frac{1}{a-1}\left(\sum^a n_i - \frac{\sum^a n_i^2}{\sum^a n_i}\right)$$

$$= \frac{1}{4-1}\left([8+10+13+6] - \frac{8^2+10^2+13^2+6^2}{8+10+13+6}\right) = 9.009$$

As a check we confirm that $n_0 < \bar{n}$ when sample sizes are unequal:

$$\bar{n} = \frac{\sum^a n_i}{a} = (8+10+13+6)/4 = 9.25$$

We may estimate the variance component as follows:

$$s_A^2 = \frac{(MS_{\text{groups}} - MS_{\text{within}})}{n_0} = \frac{(602.7 - 114.5)}{9.009} = 54.190$$

If one is interested only in the relative magnitudes of the variance component and the error variance, one often expresses these as a percentage of their sum, $s^2 + s_A^2 = 114.5 + 54.190 = 168.690$

Percent of variation among groups (hosts):

$$\frac{s_A^2}{s^2 + s_A^2} \times 100 = \frac{54.190}{114.5 + 54.190} \times 100 = \frac{54.190}{168.690} \times 100 = 32.1\%$$

Percent of variation within groups (among larvae on a host):

$$\frac{s^2}{s^2 + s_A^2} \times 100 = \frac{114.5}{168.690} \times 100 = 67.9\%$$

r_I, the coefficient of intraclass correlation

= proportion of variation among groups

= 0.321

is no variance at all within groups. In such a case, all the individuals within one group would be identical. This is a measure of the similarity or "correlation" among the individuals within a group relative to the amount of differences found among the groups. This is not the sense in which we use correlation nowadays, and the percentage variation among and within groups seems a more expressive way to represent this statistic.

Since a procedure for the computation of confidence intervals of variance components from an anova with unequal sample sizes has not been developed, Box 9.3 shows a computation of confidence limits of the variance components

BOX 9.3

Added variance components and their confidence limits in a Model II anova with equal sample sizes.

Fertility (number of larvae hatched out of 10 eggs deposited) for the offspring of 7 females of *Drosophila melanogaster*, PP strain (a = 7 females). There were n = 3 replicates for each female. Data by R. R. Sokal.

Anova table

Source of variation	*df*	*MS*	*Expected MS*	F_s
Among groups (among females)	6	9.111	$\sigma^2 + n\sigma^2{}_A$	2.20 *ns*
Within groups (replicates)	14	4.142	σ^2	
Total	20			

Since MS_{within} is an unbiased estimate of σ^2 and MS_{groups} is an unbiased estimate of $\sigma^2 + n\sigma_A^2$ (where σ_A^2 is the added variance component due to differences among females), σ_A^2 may be estimated as follows:

$$s_A^2 = \frac{1}{n}(MS_{\text{groups}} - MS_{\text{within}})$$

$$= \tfrac{1}{3}(9.111 - 4.142) = 1.656$$

It is also possible to calculate approximate confidence limits for σ_A^2 (the confidence limits for MS_{within}, σ^2, may be calculated using the methods described in Section 7.7).

1. For degree of freedom $\nu_1 = df_{\text{groups}} = 6$, $\nu_2 = df_{\text{within}} = 14$, α = desired level of significance = .05 (for 95% confidence intervals), look up the following tabulated F-values in Table **S**:

$$F_{\alpha/2[\nu_1,\nu_2]} = F_{.025[6,14]} = 3.50$$

$$F_{\alpha/2[\nu_2,\nu_1]} = F_{.025[14,6]} = 5.30 \qquad \text{(obtained by interpolation in Table S)}$$

$$F_{\alpha/2[\nu_1,\infty]} = F_{.025[6,\infty]} = 2.41$$

$$F_{\alpha/2[\infty,\nu_1]} = F_{.025[\infty,6]} = 4.85$$

BOX 9.3 continued

2. The lower confidence limit is

$$L_1 = \frac{MS_{\text{within}}}{n}\left[\frac{F_s}{F_{\alpha/2[\nu_1,\infty]}} - 1 - \left(\frac{F_{\alpha/2[\nu_1,\nu_2]}}{F_{\alpha/2[\nu_1,\infty]}} - 1\right)\frac{F_{\alpha/2[\nu_1,\nu_2]}}{F_s}\right]$$

$$= \frac{4.142}{3}\left[\frac{2.20}{2.41} - 1 - \left(\frac{3.50}{2.41} - 1\right)\frac{3.50}{2.20}\right]$$

$$= 1.38067\{0.91286 - 1 - [0.45228(1.59091)]\}$$

$$= 1.38067(-0.80668)$$

$$= -1.11376$$

The lower limit is negative because F_s was not significant. Conventionally the lower limit is set equal to zero in such cases, because a negative variance is nonsensical:

$$L_1 = 0$$

3. The upper confidence limit is

$$L_2 = \frac{MS_{\text{within}}}{n}\left[F_s F_{\alpha/2[\infty,\nu_1]} - 1 + \left(1 - \frac{F_{\alpha/2[\infty,\nu_1]}}{F_{\alpha/2[\nu_2,\nu_1]}}\right)\frac{1}{F_{\alpha/2[\nu_2,\nu_1]}}\right]$$

$$= \frac{4.142}{3}\left[(2.20)(4.85) - 1 + \left(1 - \frac{4.85}{5.30}\right)\frac{1}{5.30}\right]$$

$$= 1.38067\left[10.6700 - 1 + \frac{1 - 0.91509}{5.30}\right]$$

$$= 1.38067(9.68602)$$

$$= 13.373$$

of another example with equal sample sizes. The method shown here is that of Scheffé (1959). This method is claimed to give more accurate confidence limits than the methods previously used (see Anderson and Bancroft, 1952, for examples of other approximate confidence limits). In a Monte Carlo study (see Section 17.3 for an explanation of this approach) of the properties of variance components in a variety of examples, we found this claim to be valid. The example shown in Box 9.3 treats fertility expressed as number of larvae hatched out of 10 eggs laid by females of *Drosophila melanogaster*. From each of the 7 females, which had been randomly chosen from a certain genetic strain, three replicate egg samples (of 10 eggs each) were taken and tested for hatchability. We therefore have a basic data table (not shown) of 7 groups (females) and 3 replicates within groups (hatch counts). The anova table is shown in Box 9.3. We note that the F-value is not significant. Thus we cannot reject the null hypothesis for the variance component. However, we shall proceed to set confidence limits to it anyway in order to obtain experience with this computation. The lower confidence limit is negative. This was to be expected, since the variance component among groups is not significant. The computation is straightforward and is laid out in Box 9.3.

9.3 Special case: equal *n*

Equal sample sizes in a single classification analysis of variance result in slightly simpler computational formulas, although these are basically not different from the general case shown in the previous section. We shall illustrate a single classification anova with equal sample sizes by a Model I example. As we have stated before, the computation up to and including the first test of significance is identical for both models. Thus the computation of Box 9.4 could also serve for a Model II anova with equal sample size.

BOX 9.4

Single classification anova with equal sample sizes.

The effect of the addition of different sugars on length, in ocular units ($\times$ 0.114 = mm), of pea sections grown in tissue culture with auxin present: $n = 10$ (replications per group). This is a Model I anova.

	Treatments ($a = 5$)				
Observations, i.e., replications	*Control*	*2% glucose added*	*2% fructose added*	*1% glucose + 1% fructose added*	*2% sucrose added*
1	75	57	58	58	62
2	67	58	61	59	66
3	70	60	56	58	65
4	75	59	58	61	63
5	65	62	57	57	64
6	71	60	56	56	62
7	67	60	61	58	65
8	67	57	60	57	65
9	76	59	57	57	62
10	68	61	58	59	67
$\sum^{n} Y$	701	593	582	580	641
$\bar{Y}$	70.1	59.3	58.2	58.0	64.1

Source: Data by W. Purves.

Preliminary computations

1. Grand total $= \sum^{a}\sum^{n} Y = 701 + 593 + \cdots + 641 = 3097$

2. Sum of the squared observations $= \sum^{a}\sum^{n} Y^2$

$$= 75^2 + 67^2 + \cdots + 68^2 + 57^2 + \cdots + 67^2 = 193{,}151$$

3. Sum of the squared group totals divided by $n = \frac{1}{n}\sum^{a}\left(\sum^{n} Y\right)^2$

$$= \tfrac{1}{10}[701^2 + 593^2 + \cdots + 641^2] = \tfrac{1}{10}[1{,}929{,}055] = 192{,}905.50$$

BOX 9.4 continued

4. Grand total squared and divided by total sample size = correction term

$$CT = \frac{1}{an}\left(\sum^{a}\sum^{n} Y\right)^2 = \frac{(3097)^2}{5 \times 10} = \frac{9{,}591{,}409}{50} = 191{,}828.18$$

5. $SS_{\text{total}} = \sum^{a}\sum^{n} Y^2 - CT$

$= \text{quantity } \mathbf{2} - \text{quantity } \mathbf{4} = 193{,}151 - 191{,}828.18 = 1322.82$

6. $SS_{\text{groups}} = \frac{1}{n}\sum^{a}\left(\sum^{n} Y\right)^2 - CT$

$= \text{quantity } \mathbf{3} - \text{quantity } \mathbf{4} = 192{,}905.50 - 191{,}828.18 = 1077.32$

7. $SS_{\text{within}} = SS_{\text{total}} - SS_{\text{groups}}$

$= \text{quantity } \mathbf{5} - \text{quantity } \mathbf{6} = 1322.82 - 1077.32 = 245.50$

Anova table

Source of variation	*df*		*SS*		*MS*		*F₀*		*Expected MS*
	Formula	*Value*	*Symbol*	*Value*	*Formula*	*Value*	*Formula*	*Value*	
$\bar{Y} - \bar{\bar{Y}}$ among groups (among treatments)	$a - 1$	4	**6**	1077.32	$\frac{\mathbf{6}}{(a-1)}$	269.33	$\frac{MS_{\text{groups}}}{MS_{\text{within}}}$	49.33	$\sigma^2 + \frac{n}{a-1}\sum^{a}\alpha^2$
$Y - \bar{Y}$ within groups (error, replicates)	$a(n - 1)$	45	**7**	245.50	$\frac{\mathbf{7}}{a(n-1)}$	5.46			σ^2
$Y - \bar{\bar{Y}}$ Total	$an - 1$	49	**5**	1322.82					

$F_{.05[4,45]} = 2.58$ $\quad F_{.01[4,45]} = 3.77$ $\quad F_{.001[4,45]} = 5.57$

Conclusions.—There is a highly significant ($P \ll 0.001$) added component due to treatment effects in the mean square among groups (treatments). The different sugar treatments clearly have a significant effect on growth of the pea sections.

See Sections 9.6 and 9.7 and Boxes 9.8 and 9.9 for the completion of a Model I analysis of variance: that is, the method for determining which means are significantly different from each other.

The data are from an experiment in plant physiology. They record the length in coded units of pea sections grown in tissue culture with auxin present. The purpose of the experiment was to test the effects of the addition of various sugars on growth as measured by length. Four experimental groups, representing three different sugars and one mixture of sugars, were used, plus one control without sugar. Ten observations (replicates) were made for each treatment. The term "treatment" already implies a Model I anova. It is obvious that the five groups do not represent random samples from all possible experimental conditions but were deliberately designed to test the effects of certain sugars on the growth rate. We are interested in the effect of the sugars on length, and our null hypothesis will be that there is no added component due to treatment effects among the five groups; that is, the population means are all assumed to be equal. In the preliminary computations, steps **1** and **2** are the same as before, but the computation of quantity **3** can be simplified. You will remember that this represents the sum of the quotients obtained by squaring each group sum and dividing it by its own sample size. Since sample sizes are equal, we can simply sum the square of each group sum and divide once by the constant n. The denominator of quantity **4** is also simpler because of the constancy of n.

It may seem that we are carrying an unnecessary number of digits in the computations in Box 9.4. This is often necessary to ensure that the error sum of squares, quantity **7**, has sufficient accuracy.

Quantities **5**, **6**, and **7** are obtained as before. The analysis of variance table is similar to that shown in the general case (Box 9.1) except that there are simpler expressions for number of degrees of freedom.

Since ν_2 is relatively large, the critical values of F have been computed by harmonic interpolation in Table **S** (see section on interpolation in the introduction to the statistical tables). They have been given here only to present a complete record of the analysis. Ordinarily, when confronted with this example, you would not bother working out these values of F. Comparison of the observed variance ratio $F_s = 49.33$ with $F_{.001[4,40]} = 5.70$, the conservative critical value, would convince you to reject the null hypothesis. The probability that the five groups differ as much as they do by chance is almost infinitesimally small. Clearly the sugars produce an added treatment effect, apparently inhibiting growth and consequently reducing the length of the pea sections.

At this stage we are not in a position to say whether each treatment is different from every other treatment, or whether the sugars are different from the control but not different from each other. Such tests are necessary to complete a Model I analysis but we defer their discussion until Sections 9.6 and 9.7.

9.4 Special case: two groups

A frequent test in statistics is to establish the *significance of the difference between two means*. This can easily be done by means of an *analysis of variance for two groups*. Box 9.5 shows this procedure for a Model I anova, the more frequent model in the two sample case. However, if the assumptions of a Model II anova are met, it too can be carried out for two groups.

The example in Box 9.5 concerns the onset of reproductive maturity in water fleas, *Daphnia longispina*. This is measured as the average age (in days) at beginning of reproduction. Each variate in the table is in fact an average and a possible flaw in the analysis might be that these averages are not based on equal sample sizes. However, we are not given this information and have to proceed on the assumption that each reading in the table is an equally reliable variate. The two series represent different genetic crosses and the seven replicates in each series are clones derived from the same genetic cross. This example is clearly a Model I, since the question to be answered is whether series I differs from series II in average age at the beginning of reproduction. Inspection of the data shows that the mean age at beginning of reproduction is very similar for the two series. It would surprise us therefore to find that they are significantly different. However, we shall carry out a test anyway. As you realize by now, one cannot tell from the absolute magnitude of a difference whether it is significant. This depends on the magnitude of the error mean square, representing the variance within series.

The computations are shown in Box 9.5. This is an example with equal sample sizes and we follow the computational outline for Box 9.4. If the example had been one of unequal sample sizes in the two groups, we would have followed the outline of Box 9.1. With equal sample sizes and only two groups there is one further computational shortcut. Quantity **6**, SS_{groups}, can be computed by the very simple formula shown in Box 9.5. There is only 1 degree of freedom between the two groups. The critical value of $F_{.05[1,12]}$ is given underneath the anova table but it is really not necessary to consult it. Inspection of the mean squares in the anova shows that MS_{groups} is much smaller than MS_{within}; therefore the value of F_s is far below unity and there cannot possibly be an added component due to treatment effects between the series. In cases where $MS_{\text{groups}} \leq MS_{\text{within}}$, we do not usually bother to calculate F_s because the analysis of variance could not possibly be significant.

The present case is unusual in that the mean square between groups is very much lower than that within groups. You may wonder whether these two could really be estimates of the same parametric variance, as they are supposed to be according to theory? Let us test whether the within-group variance is significantly greater than the among-group variance. Our null hypothesis is that $H_0{:}\,\sigma^2_{\text{groups}} = \sigma^2_{\text{within}}$ and the alternative hypothesis is $H_1{:}\,\sigma^2_{\text{groups}} < \sigma^2_{\text{within}}$. In view of the alternative hypothesis it is a one-tailed test. We calculate $F_s = MS_{\text{within}}/MS_{\text{groups}} = 0.45714/0.00643 = 71.09$. Since $F_{.05[12,1]} = 244$, we accept H_0 and conclude that the two mean squares are estimates of the same parametric variance. It should not surprise us that there is such a great discrepancy between the variance estimates. After all, the mean square between groups is based on only 1 degree of freedom.

Some statisticians would object to the above procedure on the grounds that once a one-tailed test had been made with the alternative hypothesis $(H_1{:}\,\sigma^2_{\text{groups}} > \sigma^2_{\text{within}})$ and the null hypothesis accepted, it is not valid to retest the data with the converse alternative hypothesis $(H_1{:}\,\sigma^2_{\text{groups}} < \sigma^2_{\text{within}})$. The net effect of such a procedure is that of a two-tailed test, with twice the

BOX 9.5

Single classification anova (Model I) with two groups with equal sample sizes.

Average age (in days) at beginning of reproduction in *Daphnia longispina* (each variate is a mean based on approximately similar numbers of females). Two series derived from different genetic crosses and containing seven clones each are compared; $n = 7$ clones per series.

	Series ($a = 2$)	
	I	*II*
	7.2	8.8
	7.1	7.5
	9.1	7.7
	7.2	7.6
	7.3	7.4
	7.2	6.7
	7.5	7.2
$\sum^{n} Y$	52.6	52.9
$\bar{Y}$	7.5143	7.5571
$\sum^{n} Y^2$	398.28	402.23

Source: Data by Ordway from Banta (1939).

Preliminary computations

1. Grand total $= \sum^{a}\sum^{n} Y = 52.6 + 52.9 = 105.5$

2. $\sum^{a}\sum^{n} Y^2 = 398.28 + 402.23 = 800.51$

3. Sum of the squared group totals divided by $n = \frac{1}{n}\sum^{a}\left(\sum^{n} Y\right)^2$

$$= \frac{(52.6)^2 + (52.9)^2}{7} = 795.02429$$

4. Grand total squared and divided by total sample size = correction term

$$CT = \frac{1}{an}\left(\sum^{a}\sum^{n} Y\right)^2 = \frac{(105.5)^2}{14} = 795.01786$$

5. SS_{total} = quantity **2** − quantity **4** = 5.49214

6. SS_{groups} = quantity **3** − quantity **4** = 0.00643

Note: With only two equal-sized groups, the group SS can be directly computed by a simple formula, avoiding calculation of quantity **3**:

$$SS_{\text{groups}} = \frac{\left(\sum^{n} Y_1 - \sum^{n} Y_2\right)^2}{2n} = \frac{(52.6 - 52.9)^2}{14} = 0.00643$$

BOX 9.5 continued

7. $SS_{\text{within}} = SS_{\text{total}} - SS_{\text{groups}}$

$$= \text{quantity } \mathbf{5} - \text{quantity } \mathbf{6} = 5.49214 - 0.00643 = 5.48571$$

Anova table

Source of variation	*df*	*SS*	*MS*	F_s
$\bar{Y} - \bar{\bar{Y}}$ Between groups (series)	1	0.00643	0.00643	0.0141
$Y - \bar{Y}$ Within groups (error; clones within series)	12	5.48571	0.45714	
$Y - \bar{\bar{Y}}$ Total	13	5.49214		

$F_{.05[1,12]} = 4.75$

Conclusions.—Since $F_s \ll F_{.05[1,12]}$, the null hypothesis is accepted. The means of the two series are not significantly different; that is, the two series do not differ in average age at beginning of reproduction.

intended type I error. Reversing the alternative hypothesis should therefore be done only with great caution.

If the mean square among groups had been significantly lower than the error mean square, this would have meant that either an unusual event has occurred by chance (type I error) or that for some reason or another the group means are less variable than they should be on the basis of the variation of their items. This could occur whenever items are not sampled at random but are sampled with a type of compensatory bias. A grocer filling baskets of strawberries may not wish to fill them at random but may fill them so that the smaller and less desirable strawberries are at the bottom of each basket, which is topped by the larger and more attractive ones on view to the customer. Such a filling of strawberry baskets would result in considerable within-basket variance of strawberry size, but remarkable constancy in means for each basket (if the grocer has done a thorough job of biased filling of baskets), so that the mean square among groups (baskets) would be remarkably low. In theory at least MS_{groups} would be zero if the grocer is able to fill each basket with the exact same assortment of strawberries. In scientific work similar poor sampling may result from unrecognized biases by which the person doing the sampling unconsciously compensates for obvious deviations from what he believes to be the norm. When biased sampling can be ruled out, the underdispersion of sample means may be due to a compensation effect in the natural phenomenon under study. Whenever resources for a process are limited, some individuals may obtain a large part and others may have to be satisfied with only a small part. Thus, if there is a limited space in which to live, several large individuals may occupy most of it, leaving space for only a few small ones. This is related to the idea of repulsion encountered in the binomial distribution (Section 5.2). Situations such as these indicate that the basic assumptions of an anova are not met (see Chapter 13), and ordinarily one would not continue the anova.

There is another method of solving a Model I two-sample analysis of variance. This is a *t-test of the differences between two means*. This t-test is the traditional method of solving such a problem, and may already be familiar to you from previous acquaintance with statistical work. It has no real advantage in either ease of computation or understanding, and as you will see below it is mathematically equivalent to the anova that was just computed in Box 9.5. It is presented here mainly for the sake of completeness. It would seem too much of a break with tradition not to have the "t-test" in a biometry text.

In Section 7.4 we learned about the t-distribution and saw that a t-distribution of $n - 1$ *df* could be obtained from a distribution of the term $(\bar{Y}_i - \mu)/s_{\bar{Y}_i}$, where $s_{\bar{Y}_i}$ has $n - 1$ degrees of freedom and $\bar{Y}$ is normally distributed. The numerator of this term represents a deviation of a sample mean from a parametric mean and the denominator a standard error for such a deviation. We now learn that the expression

$$t_s = \frac{(\bar{Y}_1 - \bar{Y}_2) - (\mu_1 - \mu_2)}{\sqrt{\left[\dfrac{(n_1 - 1)s_1^2 + (n_2 - 1)s_2^2}{n_1 + n_2 - 2}\right]\left(\dfrac{n_1 + n_2}{n_1 n_2}\right)}} \tag{9.2}$$

is also distributed as t. Expression (9.2) looks complicated, but it really has the same structure as the simpler term for t. The numerator is a deviation, this time not between a single sample mean and the parametric mean, but between a single difference between two sample means, $\bar{Y}_1$ and $\bar{Y}_2$, and the true difference between the means of the populations represented by these means. In a test of this sort our null hypothesis is that the two samples come from the same population; that is, they must have the same parametric mean. Thus the difference $\mu_1 - \mu_2$ is expected to be zero. We therefore test the deviation of the difference $\bar{Y}_1 - \bar{Y}_2$ from zero. The denominator of Expression (9.2) is a standard error, the standard error of the difference between two means $s_{\bar{Y}_1 - \bar{Y}_2}$. The left portion of the expression, which is in square brackets, is a weighted average of the variances of the two samples, s_1^2 and s_2^2, computed in the manner of Section 8.1. The right term of the standard error is the computationally easier form of $(1/n_1) + (1/n_2)$, which is the factor by which the average variance within groups must be multiplied in order to convert it into a variance of the difference of means. The analogy with the multiplication of a sample variance s^2 by $1/n$ to transform it into a variance of a mean $s_{\bar{Y}}^2$ should be obvious.

The test as outlined here assumes equal variances in the two samples. This is also an assumption of the analyses of variance carried out so far, although we have not stressed this. If the variances of the two samples are significantly different from each other, other methods have to be applied. These are discussed in Section 13.3, where we also present the tests for checking whether several variances are equal. With only two variances, equality may be tested by the procedure in Box 8.1.

When sample sizes are equal in a two-sample test, Expression (9.2) simplifies to

$$t_s = \frac{(\bar{Y}_1 - \bar{Y}_2) - (\mu_1 - \mu_2)}{\sqrt{\frac{1}{n}(s_1^2 + s_2^2)}} \tag{9.3}$$

which is the formula applied in the present example in Box 9.6. When the sample sizes are unequal but rather large, so that the differences between n_i and $n_i - 1$ are relatively trivial, Expression (9.2) reduces to the simpler

$$t_s = \frac{(\bar{Y}_1 - \bar{Y}_2) - (\mu_1 - \mu_2)}{\sqrt{\frac{s_1^2}{n_2} + \frac{s_2^2}{n_1}}} \tag{9.4}$$

The derivation of Expressions (9.3) and (9.4) from Expression (9.2) is shown in Appendix A1.4. The pertinent degrees of freedom for Expressions (9.2) and (9.4) are $n_1 + n_2 - 2$ and for Expression (9.3) df is $2(n - 1)$.

The test of significance for differences between means using the t-test is shown in Box 9.6. This is a two-tailed test because our alternative hypothesis is $H_1: \mu_1 \neq \mu_2$. The results of this test are identical to those of the anova in Box 9.5: the two means are not significantly different. We mentioned earlier that these two results are in fact mathematically identical. We can show this simplest by squaring the obtained value for t_s, which should be identical to the F_s-value of the corresponding analysis of variance. Since $t_s = -0.1184$ in Box 9.6, $t_s^2 = 0.0140$. Within rounding error this is equal to the F_s obtained in the anova of Box 9.5 ($F_s = 0.0141$). Why is this so? We shall show two reasons for this. We learned that $t_{[\nu]} = (\bar{Y} - \mu)/s_{\bar{Y}}$, where ν is the degrees of freedom of the variance of the mean $s_{\bar{Y}}^2$; therefore $t_{[\nu]}^2 = (\bar{Y} - \mu)^2/s_{\bar{Y}}^2$. However, this expression can be regarded as a variance ratio. The denominator is clearly a variance with ν degrees of freedom. The numerator is also a variance. It is a single deviation squared, which represents a sum of squares possessing 1 rather than zero degrees of freedom (since it is a deviation from the true mean μ rather than a sample mean). A sum of squares based on 1 degree of freedom is at the same time a variance. Thus t^2 is a variance ratio; specifically, $t_{[\nu]}^2 = F_{[1,\nu]}$, as we have seen above. In Appendix A1.5 we demonstrate algebraically that the t_s^2 obtained in Box 9.6 and the F_s-value obtained in Box 9.5 are identical quantities.

We can also demonstrate the relationship between t and F from the statistical tables. In Table **S** we find that $F_{.05[1,10]} = 4.96$. This value is supposed to be equal to $t_{.05[10]}^2$. The square root of 4.96 is 2.227 and when we look up $t_{.05[10]}$ in Table **Q** we find it to be 2.228, which is within acceptable

BOX 9.6

A t-test of the hypothesis that two sample means come from a population with equal μ: also confidence limits of the difference between two means.

Average age at beginning of reproduction in *Daphnia*. Data from Box 9.5.

This test assumes that the variances in the populations from which the two samples were taken are identical. If in doubt about this hypothesis, test by method of Box 8.1, Section 8.3, and if variances of the two samples are highly significantly different do not employ this test but carry out the test shown in Section 13.3.

The appropriate formula for t_s is one of the following.

Expression (9.2), when sample sizes are unequal and n_1 or n_2 or both sample sizes are small (< 30): $df = n_1 + n_2 - 2$

Expression (9.3), when sample sizes are identical (regardless of size): $df = 2(n - 1)$

Expression (9.4), when both n_1 and n_2 are unequal but large (> 30): $df = n_1 + n_2 - 2$

For the present data, since sample sizes are equal, we choose Expression (9.3):

$$t_s = \frac{(\bar{Y}_1 - \bar{Y}_2) - (\mu_1 - \mu_2)}{\sqrt{\frac{1}{n}(s_1^2 + s_2^2)}}$$

We do not have variances computed, but we can modify the formula to one based on sums of squares, which are obtained one computational step earlier from the quantities furnished in Box 9.5:

$$t_s = \frac{\bar{Y}_1 - \bar{Y}_2 - (\mu_1 - \mu_2)}{\sqrt{\dfrac{\sum^n y_1^2 + \sum^n y_2^2}{n(n-1)}}}$$

We are testing the null hypothesis that $\mu_1 - \mu_2 = 0$. Therefore we replace this quantity by zero in this example and, drawing upon the data of Box 9.5, obtain $\sum y_1^2 = 3.029$ and $\sum y_2^2 = 2.457$. Then

$$t_s = \frac{7.5143 - 7.5571}{\sqrt{(3.029 + 2.457)/7 \times 6}} = \frac{-0.0428}{\sqrt{5.486/42}} = \frac{-0.0428}{0.3614} = -0.1184$$

The degrees of freedom for this example are $2(n - 1) = 2 \times 6 = 12$. The critical value of $t_{.05[12]} = 2.179$. Since the absolute value of our observed t_s is less than the critical t-value, the means are found to be not significantly different, which is the same result as was obtained by the anova of Box 9.5.

A comparison of the results of the present t-test with those of the F-test of Box 9.5 shows close correspondence. We have learned that $F_{[1,\nu_2]} = t^2_{[\nu_2]}$. Since $t_s = -0.1184$, $t_s^2 = 0.0140$, which is equal to the F_s-value of 0.0141 from Box 9.5, within rounding error.

BOX 9.6 continued

Confidence limits of the difference between two means

$$L_1 = (\bar{Y}_1 - \bar{Y}_2) - t_{\alpha[\nu]} s_{\bar{Y}_1 - \bar{Y}_2}$$

$$L_2 = (\bar{Y}_1 - \bar{Y}_2) + t_{\alpha[\nu]} s_{\bar{Y}_1 - \bar{Y}_2}$$

In this case $\bar{Y}_1 - \bar{Y}_2 = -0.0428$, $t_{.05[12]} = 2.18$, and $s_{\bar{Y}_1 - \bar{Y}_2} = 0.3614$ as computed earlier as the denominator of the t-test. Therefore

$$L_1 = -0.0428 - (2.18)(0.3614) = -0.8307$$

$$L_2 = -0.0428 + (2.18)(0.3614) = 0.7451$$

The 95% confidence limits contain the zero point (no difference), as was to be expected, since the difference $\bar{Y}_1 - \bar{Y}_2$ was found to be not significant.

---

rounding error of the previous value. Since t approaches the normal distribution as its degrees of freedom approach infinity, $F_{\alpha[1,\nu]}$ approaches the distribution of the square of the normal deviate as $\nu \longrightarrow \infty$.

The relation between t^2 and F leads to another relation. We have just learned that when $\nu_1 = 1$, $F_{[\nu_1,\nu_2]} = t^2_{[\nu_2]}$. We know from Section 8.2 that $\chi^2_{[\nu_1]}/\nu_1 = F_{[\nu_1,\infty]}$. Therefore, when $\nu_1 = 1$ and $\nu_2 = \infty$, $\chi^2_{[1]} = F_{[1,\infty]} = t^2_{[\infty]}$. This can be demonstrated from Tables **R**, **S**, and **Q**:

$$\chi^2_{.05[1]} = 3.841$$

$$F_{.05[1,\infty]} = 3.84$$

$$t_{.05[\infty]} = 1.960 \qquad t^2_{.05[\infty]} = 3.8416$$

The t-test for differences between two means is useful when we wish to set confidence limits to such a difference. Box 9.6 shows how to calculate 95% confidence limits to the difference between the series means in the *Daphnia* example. The appropriate standard error and degrees of freedom depend on whether Expression (9.2), (9.3), or (9.4) is chosen for t_s. It does not surprise us to find that the confidence limits of the difference in this case enclose the value of zero, ranging from -0.8307 to $+0.7451$. This must be so when a difference is found to be not significantly different from zero. We can interpret this by saying that we cannot exclude zero as the true value of the difference between the means of the two series. In fact, this confidence interval contains all parameter values for the difference between two means, for which hypotheses would not be rejected.

9.5 Special case: a single specimen compared with a sample

A type of significance test quite frequent in taxonomy, but also of use in other fields, is whether a single specimen sampled at random could belong to a given population. If the parameters and distribution of the population are known, this test presents no new features. Thus in the case of a normal distribution with mean μ and standard deviation σ, if we sample a single specimen and measure it for the variable in question, recording it as Y_1, we can test whether it belongs to the population by considering $(\bar{Y}_1 - \mu)/\sigma$ as a standard normal deviate (Section 6.3) and inspecting its magnitude in the table of areas of the normal curve (Table **P**).

However, if the population is known to us only through a sample, the way to deal with this problem is to consider it a two-sample case similar to that in Section 9.4 just discussed. The new aspect here is that one of the samples is represented by only a single variate, and thus it does not contribute to the degrees of freedom or to the estimate of the within group variance. An example of such a test is shown in Box 9.7. The appropriate formula is

$$t_s = \frac{Y_1 - \bar{Y}_2 - (\mu_1 - \mu_2)}{s_2 \sqrt{\dfrac{n_2 + 1}{n_2}}} \tag{9.5}$$

where Y_1 is the single specimen, $\bar{Y}_2$ is the sample mean, s_2 is the sample standard deviation, and n_2 is the sample size. Expression (9.5) is easily obtained from Expression (9.2) by substituting Y_1 for $\bar{Y}_1$, and 1 for n_1. The pertinent degrees of freedom are as before: $n_1 + n_2 - 2$, which reduces to $n_2 - 1$.

The problem in Box 9.7 is whether a single mosquito collected in Kansas

BOX 9.7

Comparison of a single observation with the mean of a sample.

Wing lengths (in mm) of a sample of mosquitos, *Aedes canadensis*, from Minnesota and a single specimen from Kansas. Data by F. J. Rohlf.

A special case of the method given in Box 9.6, in which one of the two samples contains only a single observation, so that the variance of that sample cannot be estimated.

The modified formula is Expression (9.5):

$$t_s = \frac{Y_1 - \bar{Y}_2 - (\mu_1 - \mu_2)}{s_2 \sqrt{\dfrac{n_2 + 1}{n_2}}}$$

$$df = n_2 - 1$$

Sample from Minnesota:

Y	4.02	3.88	3.34	3.87	3.18

$n = 5$ $\sum Y = 18.29$ $\bar{Y} = 3.658$ $\sum Y^2 = 67.4597$

$\sum y^2 = 0.5549$ $s^2 = 0.1387$ $s = 0.3725$

Single variate from Kansas:

Y_1 3.02

BOX 9.7 continued

$$t_s = \frac{3.02 - 3.658 - 0}{0.3725\sqrt{(5+1)/5}} = \frac{-0.64}{0.3725 \times 1.09544} = \frac{-0.64}{0.40805} = -1.5684$$

$df = 5 - 1 = 4$

$t_{.05[4]} = 2.776 \qquad t_{.10[4]} = 2.132 \qquad t_{.20[4]} = 1.533$

The probability of obtaining a variate as extreme as 3.02 mm is between 0.20 and 0.10 under the null hypothesis that both samples were from the same statistical population. We therefore accept the null hypothesis and conclude that the Kansas specimen could well have come from a population whose wing length did not differ from that of the Minnesota population. In view of the small sample size, the conclusion could also be that we do not have sufficient evidence to reject the hypothesis that the Kansas specimen came from a population with the same mean as the Minnesota population.

could belong to a population with the same mean wing length as the Minnesota mosquito population. This is clearly a two-tailed test because our alternative hypothesis is that the mean of the Kansas population from which the single specimen comes could be greater or less than that of the Minnesota population ($H_1: \mu_1 \neq \mu_2$).

The conclusions from Box 9.7 are that between 10 and 20% of wing lengths sampled from the Minnesota population would be as far from the Minnesota mean in either direction as is the Kansas specimen. We therefore do not have sufficient evidence to reject the null hypothesis, and we conclude that the Kansas population does not differ in wing length from the Minnesota population.

It is important to stress that such a test has few, if any, direct taxonomic implications. It does not prove or disprove the assertion that the Kansas specimen is in the same species, subspecies, or in any other taxon with the Minnesota population. The Kansas specimen could well be in another genus and still not be different in wing length. Conversely, there may be a statistical difference in wing length between the Kansas population and that in Minnesota, but, for a variety of reasons, it may not be wise to recognize it in any formal sense as a taxonomically distinct unit. In conventional taxonomy criteria for specific distinction are usually not statistical but "biological" (Simpson, 1961) and those for subspecific definition would need a considerably more detailed statistical analysis, involving numerous variables (Sokal, 1965).

Occasionally the comparison of single individuals with sample means is for individuals that are very deviant from their sample mean. This involves considerations of statistical theory that are beyond the present text. Readers confronted with such problems are advised to consult Dixon and Massey

(1957, Section 16-5) or Gumbel (1954), who has treated this problem in depth. ↚

9.6 Comparisons among means: a priori tests

We have seen that after the initial significance test a Model II analysis of variance is completed by estimation of the added variance components. We usually complete a Model I anova of more than two groups by examining the data in greater detail, testing which means are different from which other ones or which groups of means are different from other such groups or from single means. Let us refer back to the Model I anovas treated so far in this chapter. We can dispose right away of the two-sample case in Box 9.5, the average age of water fleas at beginning of reproduction. As you will recall, there was no significant difference in age between the two genetic series. But even if there had been such a difference, no further tests are possible. However, the data on length of pea sections given in Box 9.4 show a significant difference among the five treatments (based on 4 degrees of freedom). Although we know that the means are not all equal, we do not know which ones differ from which other ones. This leads us to the subject of tests among pairs and groups of means. Thus, for example, we might test the control against the 4 experimental treatments representing added sugars. The question to be tested would be, Does the addition of sugars have an effect on length of pea sections? We might also test for differences among the sugar treatments. A reasonable test might be pure sugars (glucose, fructose, and sucrose) versus the mixed sugar treatment (1% glucose, 1% fructose).

An important point about such tests is that they are designed and chosen independently of the results of the experiment. They should be planned *before* the experiment has been actually carried out and the results obtained. Such comparisons are called *planned* or *a priori comparisons*. Such tests are applied regardless of the results of the preliminary overall anova. By contrast, after the experiment has been carried out we might wish to compare certain means that we notice to be markedly different. For instance, sucrose with a mean of 64.1 appears to have had less of a growth-inhibiting effect than fructose with a mean of 58.2. We might therefore wish to test whether there is in fact a significant difference between the effects of fructose and sucrose. Such comparisons, which suggest themselves as a result of the completed experiment, are called *unplanned* or *a posteriori comparisons*. These tests are performed only if the preliminary overall anova is significant. They include tests of the comparisons between all possible pairs of means. When there are a means, there can, of course, be $a(a - 1)/2$ possible comparisons between the means. The reason we make this distinction between a priori and a posteriori comparisons is that the tests of significance appropriate for the two comparisons are different. A simple, analogous example will explain why this is so.

Let us assume we have sampled from an approximately normal population of heights of men. We have computed their mean and standard deviation. If we sample two men at a time from this population, we can predict the difference between them on the basis of ordinary statistical theory. Some men will be very similar, others relatively very different. Their differences will be distributed normally with a mean of 0 and an expected variance of $2\sigma^2$, for reasons that will be learned in a later chapter (Section 15.2). Thus, if we obtain a large difference between a randomly sampled pair of men, it will have to be a sufficient number of standard deviations greater than zero for us to reject our null hypothesis that the two men come from the specified population. If, on the other hand, we were to look at the heights of the people before sampling them and then take pairs that seem to be very different, it is obvious that we would repeatedly obtain differences between pairs of men that are several standard deviations apart. Such differences would be outliers in the expected frequency distribution of differences, and time and again we would reject our null hypothesis when in fact it is true. The men were sampled from the same population, but because they were not sampled at random but were inspected before being sampled, the probability distribution on which our hypothesis testing rested is no longer valid. It is obvious that the tails in a large sample from a normal distribution will be anywhere from 5 to 7 standard deviations apart, and that if we deliberately take individuals from each tail and compare them, they will appear to be highly significantly different from each other, even though they belong to the same population.

When we compare means differing greatly from each other as the result of some treatment in the analysis of variance, we are doing exactly the same thing as taking the tallest and the shortest men from the frequency distribution of heights. If we wish to know whether these are significantly different from each other, we cannot use the ordinary probability distribution on which the analysis of variance rested but we have to use special tests of significance. These a posteriori tests will be discussed in the next section. The present section concerns itself with the carrying out of a priori tests, those comparisons planned before the execution of the experiment.

There are two ways of going about a priori tests. We can subdivide the treatment sum of squares and the treatment degrees of freedom into separate comparisons, carrying these out as part of the analysis of variance, using F-tests as tests of significance. We can also carry out t-tests for all comparisons between two groups. However, the latter are mathematically equivalent to corresponding F-tests (for the same reason that F-tests and t-tests were equivalent in Section 9.4), and we prefer the F-tests because of their simplicity and elegance and because they tie in well with the overall anova. We shall use the t-tests in connection with the least significant difference, a special approach treated at the end of this section.

The general rule for making a planned comparison is extremely simple and relates back to the rule for obtaining the sum of squares for any set of groups (discussed at the end of Section 9.1). To compare k groups of any size n_i, take the sum of each group, square it, divide it by its sample size n_i, and sum the k quotients so obtained. From the sum of these quotients subtract a correction term, which will be the grand sum of all the groups in this comparison, quantity squared and divided by the number of items in this grand sum. If the comparison involves all the groups in the anova, the correction term will be the main CT of the study. If, however, the comparison involves only some of the groups of the anova, the CT will be different, being restricted only to these groups.

These rules can best be learned by means of an example. Table 9.1 lists

TABLE 9.1

Means, group sums, and sample sizes from the data in Box 9.4. Length of pea sections grown in tissue culture (in ocular units).

	Control	*2% glucose*	*2% fructose*	*1% glucose + 1% fructose*	*2% sucrose*	Σ
$\bar{Y}$	70.1	59.3	58.2	58.0	64.1	$(61.94 = \bar{\bar{Y}})$
$\sum^{n} Y$	701	593	582	580	641	3097
n	10	10	10	10	10	50

the means, group sums, and sample sizes of the experiment with the pea sections first discussed in Box 9.4. You will recall that there were highly significant differences among the groups. We now wish to test whether the controls are different from the four treatments representing addition of sugar. There will thus be two groups, one the control group and the other the "sugars" group, with a sum of 2396 and a sample size of 40. We therefore compute

SS (control versus sugars)

$$= \frac{(701)^2}{10} + \frac{(593 + 582 + 580 + 641)^2}{40} - \frac{(701 + 593 + 582 + 580 + 641)^2}{50}$$

$$= \frac{(701)^2}{10} + \frac{(2396)^2}{40} - \frac{(3097)^2}{50} = 832.32$$

In this case the correction term is the same as for the anova because it involves all the groups of the study. The result is a sum of squares for the comparison between these two groups. Since a comparison between two groups has only 1 degree of freedom, the sum of squares is at the same time a mean square. This mean square is tested over the error mean square of the anova to give the following comparison:

$$F_s = \frac{MS\text{ (control versus sugars)}}{MS_{\text{within}}} = \frac{832.32}{5.46} = 152.44$$

$$F_{.05[1,45]} = 4.05, \qquad F_{.01[1,45]} = 7.23, \qquad F_{.001[1,45]} = 12.4$$

This comparison is highly significant, showing that the additions of sugars significantly retarded the growth of the pea sections.

Next we test whether the mixture of sugars is significantly different from the pure sugars. Using the same technique we calculate

SS (mixed sugars versus pure sugars)

$$= \frac{(580)^2}{10} + \frac{(593 + 582 + 641)^2}{30} - \frac{(593 + 582 + 580 + 641)^2}{40}$$

$$= \frac{(580)^2}{10} + \frac{(1816)^2}{30} - \frac{(2396)^2}{40} = 48.13$$

Here the CT is different, being based on the sum of the sugars only. The appropriate test statistic is

$$F_s = \frac{MS\text{ (mixed sugars versus pure sugars)}}{MS_{\text{within}}} = \frac{48.13}{5.46} = 8.82$$

This is significant in view of the critical values of $F_{\alpha[1,45]}$ given above.

A final test is among the three sugars. This mean square has 2 degrees of freedom, since it is based on three means. Thus we compute

$$SS\text{ (among pure sugars)} = \frac{(593)^2}{10} + \frac{(582)^2}{10} + \frac{(641)^2}{10} - \frac{(1816)^2}{30} = 196.87$$

$$MS\text{ (among pure sugars)} = \frac{SS\text{ (among pure sugars)}}{df} = \frac{196.87}{2} = 98.435$$

$$F_s = \frac{MS\text{ (among pure sugars)}}{MS_{\text{within}}} = \frac{98.435}{5.46} = 18.06$$

This F_s is highly significant, since $F_{.001[2,45]} = 8.09$.

We conclude that the addition of the three sugars retards growth in the pea sections, that mixed sugars affect the sections differently from pure sugars, and that the pure sugars are significantly different among themselves, probably because of the sucrose, which has a far higher mean. We cannot test the sucrose against the other two, because that would be an a posteriori test, which suggested itself to us after we had looked at the results. To carry out such a test we need the methods of the next section.

However, our a priori tests might have been quite different, depending entirely on our initial hypotheses. Thus, we could have tested control versus sugars initially, followed by disaccharides (sucrose) versus monosaccharides (glucose, fructose, glucose + fructose), followed by mixed versus pure monosaccharides and finally by glucose versus fructose.

The pattern and number of a priori tests are determined by one's hypotheses about the data. However, there are certain restrictions. It would clearly not be proper to decide a priori that one wished to compare every mean against every other mean [$a(a-1)/2$ comparisons]. For a groups the sum of the degrees of freedom of the separate a priori tests should not exceed $a-1$. In addition, it is desirable to structure the tests in such a way that each test tests an independent relationship among the means (as was done in the example above). For example, we would prefer not to test if means one, two, and three differed if we had already found that mean one differed from mean three, since significance of the latter implies significance of the former. More definite rules for determining if a set of tests are independent will be given in Section 14.10.

Since these tests are independent, the three sums of squares we have so far obtained, based on 1, 1, and 2 *df*, respectively, together add up to the sum of squares among treatments of the original analysis of variance based on 4 degrees of freedom. We can convince ourselves of that by summing the *SS* of the comparisons:

			df
SS (control versus sugars)	=	832.32	1
SS (mixed versus pure sugars)	=	48.13	1
SS (among pure sugars)	=	196.87	2
SS (among treatments)	=	1077.32	4

This again illustrates the elegance of analysis of variance. The treatment sums of squares can be decomposed into separate parts that are sums of squares in their own right, with degrees of freedom pertaining to them. One sum of squares measures the difference between the controls and the sugars, the second that between the mixed sugars and the pure sugars, and the third the remaining variation among the three sugars. We can present all of these results as an anova table, as shown in Table 9.2.

Not all sets of comparisons yield sums of squares that add up to the treatment sum of squares. For example, another comparison that might

TABLE 9.2

Anova table from Box 9.4 with treatment sum of squares decomposed into planned comparisons.

Source of variation	*SS*	*df*	*MS*	F_s
Treatments	1077.32	4	269.33	49.33***
Control vs sugars	832.32	1	832.32	152.44***
Mixed vs pure sugars	48.13	1	48.13	8.82**
Among pure sugars	196.87	2	98.44	18.03***
Within	245.50	45	5.46	
Total	1322.82	49		

legitimately be made is glucose and fructose considered together against the mixture of glucose and fructose:

SS (glucose and fructose together versus glucose and fructose mixed)

$$= \frac{(593 + 582)^2}{20} + \frac{(580)^2}{10} - \frac{(593 + 582 + 580)^2}{30}$$

$$= \frac{(1175)^2}{20} + \frac{(580)^2}{10} - \frac{(1753)^2}{30} = 0.238$$

The appropriate test statistic is

$$F_s = \frac{MS \text{ (of above comparison)}}{MS_{\text{within}}} = \frac{0.238}{5.46} = 0.044$$

This is clearly not significant.

This one-*df* comparison, though leading to a legitimate test of significance, cuts across the comparisons just made and therefore is not independent of the tests we have already made. The comparison of mixed versus pure sugars tested nearly the same thing. It is also clear that we cannot decompose four degrees of freedom into comparisons summing to more than four degrees of freedom. This whole subject of decomposing the treatment sum of squares into individual degrees of freedom and the distinction between the two kinds of comparisons (called orthogonal when they can be added and nonorthogonal when they cannot) will be discussed in Sections 14.10 and 14.11.

Now let us apply these techniques to another example, as shown in Box 9.8. These are egg-laying records from 25 females of three genetic lines of *Drosophila*, two selected for resistance and susceptibility to *DDT*, the third a nonselected control line. The data in Box 9.8 show that average fecundity is considerably higher in the nonselected line. Since we are interested in differences among the lines, each of which was treated in a specific manner, this is clearly Model I. The computations are the same as for Box 9.4, which also was a Model I single classification anova with equal sample sizes. The various steps are shown less fully because it is assumed that you are by now familiar with the basic computations of the analysis of variance. The anova table shows a highly significant difference among lines. The three critical values are no longer given, but the three asterisks indicate a probability of $P < 0.001$ that the null hypothesis is correct. The significant outcome of the basic anova encourages us to continue further tests, although we would proceed to make a priori tests even if it were not significant. Occasionally, in spite of a nonsignificant overall analysis of variance, some a priori comparisons are significant.

Two a priori hypotheses occurred to the investigator. It was of interest to test whether the selected lines (R.S. and S.S.) had a significantly different fecundity from the nonselected line. The second test was whether the line

BOX 9.8

A priori comparison of means and groups of means in a single classification Model I anova.

Per diem fecundity (number of eggs laid per female per day for first 14 days of life) for 25 females each of three lines of *Drosophila melanogaster*. The R.S. and S.S. lines were selected for resistance and for susceptibility to DDT, respectively, and the N.S. line is a nonselected control strain; $n = 25$ females per line.

	Lines ($a = 3$)					
	Resistant (R.S.)		*Susceptible* (S.S.)		*Nonselected* (N.S.)	
	12.8	22.4	38.4	23.1	35.4	22.6
	21.6	27.5	32.9	29.4	27.4	40.4
	14.8	20.3	48.5	16.0	19.3	34.4
	23.1	38.7	20.9	20.1	41.8	30.4
	34.6	26.4	11.6	23.3	20.3	14.9
	19.7	23.7	22.3	22.9	37.6	51.8
	22.6	26.1	30.2	22.5	36.9	33.8
	29.6	29.5	33.4	15.1	37.3	37.9
	16.4	38.6	26.7	31.0	28.2	29.5
	20.3	44.4	39.0	16.9	23.4	42.4
	29.3	23.2	12.8	16.1	33.7	36.6
	14.9	23.6	14.6	10.8	29.2	47.4
	27.3		12.2		41.7	
$\sum^{n} Y$	631.4		590.7		834.3	
$\bar{Y}$	25.256		23.628		33.372	

Source: Data by R. R. Sokal.

A. *Anova*

1. $\sum^{a}\sum^{n} Y = 631.4 + 590.7 + 834.3 = 2056.4$

2. $\sum^{a}\sum^{n} Y^2 = 63{,}404.98$

3. $\dfrac{\sum^{a}\left(\sum^{n} Y\right)^2}{n} = \dfrac{(631.4)^2 + (590.7)^2 + (834.3)^2}{25} = 57{,}745.96$

4. $CT = \dfrac{(2056.4)^2}{75} = 56{,}383.75$

5. $SS_{\text{total}} = 7021.23$

6. $SS_{\text{groups}} = 1362.21$

7. $SS_{\text{within}} = 5659.02$

BOX 9.8 continued

Anova table

Source of variation	*df*	*SS*	*MS*	F_s
$\bar{Y} - \bar{\bar{Y}}$ Among groups (among genetic lines)	2	1362.21	681.11	8.67***
$Y - \bar{Y}$ Within groups	72	5659.02	78.60	
$Y - \bar{\bar{Y}}$ Total	74	7021.23		

B. *A priori tests among the means*

There were two interesting a priori hypotheses. Their rationale is discussed in the text.

1. H_0: $\mu_{(\text{R.S. and S.S.})} = \mu_{\text{N.S.}}$ $\qquad H_1$: $\mu_{(\text{R.S. and S.S.})} \neq \mu_{\text{N.S.}}$

SS[for (R.S. + S.S.) vs N.S.]

$$= \frac{(\sum Y_{\text{R.S.}} + \sum Y_{\text{S.S.}})^2}{n_{\text{R.S.}} + n_{\text{S.S.}}} + \frac{(\sum Y_{\text{N.S.}})^2}{n_{\text{N.S.}}} - \frac{(\sum Y_{\text{R.S.}} + \sum Y_{\text{S.S.}} + \sum Y_{\text{N.S.}})^2}{n_{\text{R.S.}} + n_{\text{S.S.}} + n_{\text{N.S.}}}$$

$$= \frac{(631.4 + 590.7)^2}{25 + 25} + \frac{(834.3)^2}{25} - \frac{(631.4 + 590.7 + 834.3)^2}{25 + 25 + 25}$$

$$= \frac{(1222.1)^2}{50} + \frac{(834.3)^2}{25} - \frac{(2056.4)^2}{75}$$

$$= 1329.08$$

There being one df for this comparison, $SS = MS$.

The significance test for this hypothesis is

$$F_s = \frac{MS}{MS_{\text{within}}} = \frac{1329.08}{78.60} = 16.91^{***}$$

Since $F_{.001[1,60]} = 12.0$, the comparison is highly significant. We can reject the null hypothesis far below the 0.1% level.

2. H_0: $\mu_{\text{R.S.}} = \mu_{\text{S.S.}}$ $\qquad H_1$: $\mu_{\text{R.S.}} \neq \mu_{\text{S.S.}}$

Since, in this case, the two groups to be compared have equal sample sizes, instead of testing them as

$$\frac{(\sum Y_{\text{R.S.}})^2}{n_{\text{R.S.}}} + \frac{(\sum Y_{\text{S.S.}})^2}{n_{\text{S.S.}}} - \frac{(\sum Y_{\text{R.S.}} + \sum Y_{\text{S.S.}})^2}{n_{\text{R.S.}} + n_{\text{S.S.}}}$$

we may use a simpler equation for the sum of squares first learned in Box 9.5:

$$SS(\text{R.S. vs S.S.}) = \frac{(\sum Y_{\text{R.S.}} - \sum Y_{\text{S.S.}})^2}{2n} = \frac{(631.4 - 590.7)^2}{50} = 33.13$$

There is one df for this comparison, and again $SS = MS$.

The significance test for this hypothesis is

$$F_s = \frac{MS}{MS_{\text{within}}} = \frac{33.13}{78.60} = 0.4215$$

which is not significant.

BOX 9.8 continued

Anova table

Source of variation	*df*	*SS*	*MS*
Among groups (among genetic lines)	2	1362.21	681.11***
Selected vs nonselected lines	1	1329.08	1329.08***
R.S. vs S.S.	1	33.13	33.13 *ns*
Within groups (error; females within genetic lines)	72	5659.02	78.60

Since each of these tests has a single *df* in its numerator, it may also be carried out using a *t*-test. For example, test **2**, R.S. vs S.S., is repeated, using the MS_{within} as a pooled estimate of σ^2. Since n is equal,

$$t_s = \frac{\bar{Y}_1 - \bar{Y}_2}{\sqrt{\dfrac{2}{n} MS_{\text{within}}}} = \frac{25.256 - 23.628}{\sqrt{\dfrac{2(78.60)}{25}}} = \frac{1.628}{2.508} = 0.649 \; ns$$

The applicable degrees of freedom are 72 because MS_{within} was based on 72 *df*. The difference between the lines is not significant, as was found by the *F*-test. We note that $t_s^2 = 0.4212 \approx F_s$.

selected for resistance had a different fecundity from that selected for susceptibility. These hypotheses are tested in the manner just learned. The results show that the selected lines have a significantly different (lower) fecundity from the nonselected line and that the two selected lines do not differ from each other in fecundity. These data are tabulated in a new anova table shown in Box 9.8 in which both one-degree of freedom tests are included. Again you can see that the sums of squares are additive, together yielding the sum of squares among lines. Note also that this anova table is given in more abbreviated form than earlier ones to show you how such tables are published in the literature. The F's are not given in this table nor is the total sum of squares. As you will see later, sometimes even the sums of squares are not stated, the remaining essential information being the degrees of freedom and the mean squares.

Since the two tests compared two groups (in the first case one of the groups was composite), they were single degree of freedom tests and consequently could have been tested by using a *t*-test. The second comparison (R.S.-line versus the S.S.-line) is repeated to illustrate how the same test could be carried out by means of t. The test is based on Expression (9.3), which is simplified by our assumption that the two population variances are identical. With this assumption the variance of Expression (9.3) (the denominator squared) changes to $(2/n)MS_{\text{within}}$. The error mean square, MS_{within}, is an average of all the sample variances, as shown in Section 8.1. The test is carried out as

$$t_s = \frac{(\bar{Y}_1 - \bar{Y}_2) - (\mu_1 - \mu_2)}{\sqrt{\frac{2}{n} MS_{\text{within}}}} \tag{9.6}$$

with the same number of degrees of freedom as the error mean square has. This is an important difference from previous t-tests. Although in the specific test carried out here only 50 flies are involved (25 R.S. flies and 25 S.S. flies), we use 72 *df*—the number of degrees of freedom on the basis of which the MS_{within} had been computed. As before, the result is not significant. For other tests with unequal n you could modify Expressions (9.2) and (9.4) to allow for the fact that s_1^2 and s_2^2 will be replaced by MS_{within}.

If we assume that $\mu_1 - \mu_2 = 0$ and set t_s equal to $t_{\alpha[\nu]}$, Expression (9.6) can be rearranged to read

$$\bar{Y}_1 - \bar{Y}_2 = t_{\alpha[\nu]} \sqrt{\frac{2}{n} MS_{\text{within}}} = LSD \tag{9.7}$$

It should be obvious that this quantity $\bar{Y}_1 - \bar{Y}_2$ is that difference of means that will just be significant between any pair of means in an equal sample size study at a probability α. Any two means that differ by more than this amount will, when put into Expression (9.6), give a significant value of t_s. This quantity, called the *least significant difference* and abbreviated *LSD*, has much appeal. It is a single value which, once computed, permits the investigator to look at any pair of means and tell whether they are significantly different. In the *Drosophila* example,

$$LSD = t_{.01[72]} \sqrt{\frac{2}{n} MS_{\text{within}}} = 2.65 \times 2.507 = 6.644$$

Thus any two means differing by 6.644 are significantly different from each other at $P \leq 0.01$. The *LSD* would indicate (correctly) that the nonselected line differs from either selected line in fecundity, but that the two selected lines are not different from each other.

However, we must be very cautious in our employment of the *LSD*. It is only legitimate to use it in cases of a priori planned comparisons. If we employ it freely to test the differences between any pair of means that strikes our fancy, especially between pairs of means that seem to us suspicious because of their different size means, the probability of a type I error will actually be much larger than that stated in the t-table. For such comparisons we must turn to the a posteriori methods of the next section.

9.7 Comparisons among means: a posteriori tests

The development of proper methods for testing a posteriori hypotheses in the analysis of variance has occupied the attention of statisticians considerably during the last fifteen years. Several different approaches have been suggested and the subject matter is still far from exhausted. One reason why there have been so many diverse approaches is that the problem of testing all of the "interesting" differences—those differences among means that loom large in the observed sample—is subject to a variety of interpretations. The exact probabilities of Type I and Type II errors have not yet been worked out for most of the methods, and it is difficult to say which method is "best." For this reason we shall present a few techniques that appear to be reasonable on the basis of current knowledge and should suffice for most problems.

If a single classification anova is significant it means, of course, that

$$\frac{MS_{\text{groups}}}{MS_{\text{within}}} \geq F_{\alpha[a-1,a(n-1)]} \tag{9.8}$$

Since $MS_{\text{groups}}/MS_{\text{within}} = SS_{\text{groups}}/[(a-1)\ MS_{\text{within}}]$, we can rewrite Expression (9.8) as

$$SS_{\text{groups}} \geq (a-1)\ MS_{\text{within}}\ F_{\alpha[a-1,a(n-1)]} \tag{9.9}$$

For example, in Box 9.4, where the anova is significant, $SS_{\text{groups}} = 1077.32$. Substituting into Expression (9.9) we obtain

$$1077.32 > (5-1)(5.46)(2.58) = 56.35$$

It is therefore possible to compute a critical SS value for a test of significance of an anova. Thus another way of calculating overall significance would be to see whether the SS_{groups} is greater than this critical SS. It is of interest to investigate why the SS_{groups} is as large as it is and to test for the significance of the various contributions made to this SS by differences among the sample means. This was discussed in the previous section, where separate sums of squares were computed based on comparisons among means planned before the data had been examined. A comparison was called significant if its F_s-ratio was $> F_{\alpha[k-1,a(n-1)]}$, where k is the number of means being compared. We can now also state this in terms of sums of squares. An SS is significant if it is greater than $(k-1)\ MS_{\text{within}} F_{\alpha[k-1,a(n-1)]}$.

The above tests were a priori comparisons. One procedure for testing a posteriori comparisons would be to set $k = a$ in the last formula, no matter how many means we compare. Thus the critical value of the SS will be larger than in the previous method, making it more difficult to demonstrate the significance of a sample SS. This is done to allow for the fact that we choose for testing those differences between group means that appear to be contributing substantially to the significance of the overall anova.

For an example, let us return to the effects of sugars on growth in pea sections (Box 9.4). We write down the means in ascending order of magnitude: 58.0 (glucose + fructose), 58.2 (fructose), 59.3 (glucose), 64.1 (sucrose), 70.1 (control). We notice that the first three treatments have quite similar means and suspect that they do not differ significantly among themselves and hence do not contribute substantially to the significance of the SS_{groups}.

To test this we compute the SS among these three means by the usual formula:

$$SS = \frac{593^2 + 582^2 + 580^2}{10} - \frac{(593 + 582 + 580)^2}{3(10)}$$
$$= 102{,}677.3 - 102{,}667.5 = 9.8$$

The differences among these means are not significant because this SS is less than the critical SS (56.35) calculated above.

The sucrose mean looks suspiciously different from the means of the other sugars. To test this we compute

$$SS = \frac{641^2}{10} + \frac{(593 + 582 + 580)^2}{30} - \frac{(641 + 593 + 582 + 580)^2}{10 + 30}$$
$$= 41{,}088.1 + 102{,}667.5 - 143{,}520.4 = 235.2$$

which is greater than the critical SS. We conclude, therefore, that sucrose retards growth significantly less than the other sugars tested. We may continue in this fashion, testing all the differences that look suspicious or even testing all possible sets of means, considering them 2, 3, 4, and 5 at a time. This latter approach may require a computer if there are more than 5 means to be compared, since there are very many possible tests that could be made. We furnish such a program in Appendix A3.6. This procedure was proposed by Gabriel (1964), who called it a *sum of squares simultaneous test procedure* (*SS-STP*).

A related way of carrying out a posteriori tests is to use the largest difference found among a set of means (their range) as a statistic in place of their SS. A sample range is then compared with

$$LSR = Q_{\alpha[k,\nu]} \sqrt{\frac{MS_{\text{within}}}{n}} \tag{9.10}$$

where $Q_{\alpha[k,\nu]}$ is the critical value of a special statistic for significance testing, called the *studentized range*, and shown in Table **U**. This table gives critical values for the distribution of Range/s for k normally distributed items, where s is an independent estimate of the standard deviation of the items. The table has two arguments: across the top is k, the number of items in the range, and along the left margin is ν, the degrees of freedom of the standard deviation. Expression (9.10) is called the *least significant range, LSR*. For the special case $k = 2$ (when comparing 2 means only), this is the same as the least significant difference mentioned in the previous section, because $Q_{\alpha[2,\nu]} = \sqrt{2}t_{\alpha[\nu]}$.

We shall again employ a uniform value of $k = a$ for all tests of any number of means; ν will be $a(n - 1)$. For the pea sections of Box 9.4,

$$LSR = Q_{.05[5,45]} \sqrt{\frac{5.46}{10}} = 4.02 \sqrt{0.546} = 2.970$$

Any sets of means with ranges greater than *LSR* are significantly heterogeneous. For example, the range among the three sugars tested above is $59.3 - 58.0 = 1.3$, which is not significant. To test whether sucrose differed from the other three by this method we have to test the homogeneity of the range of the four sugar means. This range is $64.1 - 58.0 = 6.1$, which is greater than the *LSR*, and hence the set of four means is heterogeneous. This procedure was developed by Tukey (1951) as "the method of allowances" and has also been called *Tukey's* w-*procedure*, or the *honestly significant difference* (hsd) procedure. We shall call it a *range simultaneous test procedure* (Range-*STP*) (Gabriel, 1964). If we wish to compare a pair of means that do not have the same sample size, the *LSR* must be modified slightly to take this into account by replacing n in Expression (9.10) by $2n_1n_2/(n_1 + n_2)$, which is an average of the two sample sizes. For testing the range of 3 or more means based upon unequal sample sizes, this test is no longer exact. It has been suggested that one simply use the average n given above, using the sample sizes of the largest and smallest means, but this is at best an approximate procedure. The *SS-STP* method described above is exact whether sample sizes are equal or not.

In both the *SS-STP* and the Range-*STP* (which usually give similar results) and in the original anova, the chance of making any type I error at all is α, the probability selected for the critical F or studentized range from the appropriate table. By "making any type I error at all" we mean making such an error in the overall test of significance of the anova and in any of the subsidiary comparisons among means or sets of means needed to complete the analysis of the experiment. This probability α is called the "experimentwise" error rate. It should be noted that though the probability of any error at all is α, the probability of error for any particular test of some subset, such as a test of the difference among 3 or between two means, is necessarily less than α. Thus for the test of each subset one is really using a significance level α', which may be much less than the experimentwise α, and if there are many means in the anova this actual error rate α' may be a 1/10th, 1/100th, or even a 1/1000th of the experimentwise α (Gabriel, 1964). For this reason the a posteriori tests discussed so far and the overall anova are not very sensitive to differences of individual means or differences within small subsets. Obviously not many differences are going to be considered significant if α' is minute. This is the price one pays for not planning one's comparisons before the data are examined. If a priori tests are made the error rate of each would still be α.

In an attempt to prevent the effective error rate α' from becoming too small when comparing differences among subsets of the original means, procedures have been proposed that use different critical values, depending upon the number of means being compared at one time. The purpose of these is to make it easier to call differences significant as one compares sets of fewer means. These procedures increase the probability of making a type I error, bringing α' closer to α. However, the exact values of α' are not known.

These methods are applied in a *stepwise* manner, testing differences among a set of means only if the set is contained within a larger set that was found to be significant. This is necessary because it would otherwise be possible that the set of means A, B, and C were not significantly different but that A and B were, clearly, an undesirable situation. The test procedures discussed earlier are called "simultaneous" because they do not require the prior testing of larger sets before testing significance of smaller sets. In these methods it is inherently impossible that a significant smaller set is contained within a nonsignificant larger set.

The *Student-Newman-Keuls (SNK) procedure* is an example of a stepwise method using the range as the statistic to measure differences among means. It is described in Boxes 9.9 and 9.10 for equal and unequal sample sizes, respectively. As in the procedures described above, the first step consists of calculating a pooled standard error of a group mean $= \sqrt{MS_{\text{within}}/n}$. Next we look up $Q_{\alpha[a,\nu]}$, $Q_{\alpha[a-1,\nu]}$, . . . , $Q_{\alpha[2,\nu]}$ in Table **U**, the critical studentized ranges at a significance level α for groups of a, $a - 1$, $a - 2$, . . . , 2 means and with ν = the *df* of the SS_{within}. Then we compute *LSR*'s for groups of a, $a - 1$, $a - 2$, . . . , 2 means by Expression (9.10). The means are arrayed in order of magnitude, either ascending or descending. They are thus not necessarily in the same order in which they were originally presented. We first test the total range from the largest to the smallest mean by comparing it with the *LSR* for a means. If this is not significant we stop testing; if it is, we test the range from the largest to the next to smallest mean and the range from the smallest to the next to largest mean by comparing the observed range with *LSR* for $a - 1$ means. If these are significant we test largest versus third smallest, second largest versus second smallest, and the third largest versus the smallest. We continue testing in a similar manner. If any range is not significant we do not test any ranges enclosed by it.

Boxes 9.9 and 9.10 show a consistent method of testing all possible comparisons between means and are self-explanatory. The final results are often represented diagrammatically by underlining those sets of means that are not significantly heterogeneous (illustrated at the ends of these boxes). Any two means not connected by a line are considered significantly different. The true probability levels of these statements (α') are not known. They cannot be greater than α, the intended error rate of the statistic of significance $Q_{\alpha[k,\nu]}$, but may be nearly as low as the α' values of the *STP* (Gabriel, 1964). A computer program for the SNK procedure is furnished in Appendix A3.7.

You may wonder how it is that we have analyzed by a posteriori tests the same experiment (effect of sugars on growth of pea sections) treated by a priori tests in the previous section. In that section we made several distinct a priori hypotheses because they seemed reasonable on grounds extraneous

BOX 9.9

Multiple comparisons among means based on equal sample sizes (a posteriori tests): the Student-Newman-Keuls test.

Lengths of pea sections grown in tissue culture. Data from Box 9.4.

First calculate the standard error of a group mean, using the error mean square (MS_{within}) from the anova as a pooled estimate of the variance among items within a group:

$$s_{\bar{Y}} = \sqrt{\frac{MS_{\text{within}}}{n}} = \sqrt{\frac{5.46}{10}} = \sqrt{0.546} = 0.7389$$

This standard error is based on $a(n - 1) = 45$ degrees of freedom (the *df* of the MS_{within})

Next calculate the least significant ranges *LSR* for 2, 3, . . . , *a* means, using the table of studentized ranges *Q* given in Table **U** (harmonic interpolation is necessary for $\nu = 45$; see section on interpolation in the introduction to the statistical tables):

$$LSR \text{ (for } k \text{ groups)} = Q_{\alpha[k,\nu]} \times s_{\bar{Y}}$$

For $\alpha = 0.05$ and $\nu = 45$ degrees of freedom, the *Q*'s and *LSR*'s are the following

	k			
	2	*3*	*4*	*5*
Q	2.848	3.428	3.772	4.018
LSR	2.10	2.53	2.79	2.97

The group means are then arrayed in order of magnitude:

Glucose + Fructose	*Fructose*	*Glucose*	*Sucrose*	*Control*
$\bar{Y}_1$	$\bar{Y}_2$	$\bar{Y}_3$	$\bar{Y}_4$	$\bar{Y}_5$
58.0	58.2	59.3	64.1	70.1

The sequence of steps for testing is as follows.

1. Compare highest mean $\bar{Y}_5$ minus lowest mean $\bar{Y}_1$ with *LSR* for $k = 5$. We find that $70.1 - 58.0 = 12.1$ is greater than $LSR_5 = 2.97$. Therefore this range is significant (at $P < 0.05$).

2. Compare range of four means with *LSR* for $k = 4$:

$$\bar{Y}_4 - \bar{Y}_1 = 64.1 - 58.0 = 6.1$$

$$\bar{Y}_5 - \bar{Y}_2 = 70.1 - 58.2 = 11.9$$

These are both significant because they are greater than $LSR_4 = 2.79$.

BOX 9.9 continued

3. Compare ranges of groups of three means with *LSR* for $k = 3$:

$$\bar{Y}_3 - \bar{Y}_1 = 59.3 - 58.0 = 1.3$$
$$\bar{Y}_4 - \bar{Y}_2 = 64.1 - 58.2 = 5.9$$
$$\bar{Y}_5 - \bar{Y}_3 = 70.1 - 59.3 = 10.8$$

The first range of 1.3 is less than $LSR_3 = 2.53$ and the others are greater. Hence the three means $\bar{Y}_1$, $\bar{Y}_2$, and $\bar{Y}_3$ are not significantly different from one another and this group of three means is considered homogeneous. Since the other two ranges are significant, further testing is necessary.

4. Compare ranges of pairs of means with *LSR* for $k = 2$. The only pairs that need be tested are

$$\bar{Y}_4 - \bar{Y}_3 = 64.1 - 59.3 = 4.8$$
$$\bar{Y}_5 - \bar{Y}_4 = 70.1 - 64.1 = 6.0$$

Both ranges are significant because they are greater than $LSR_2 = 2.10$.

Our conclusions can be summarized as $\mu_1 = \mu_2 = \mu_3 < \mu_4 < \mu_5$. This is often shown diagrammatically by underlining those means that are not significantly different:

Glucose + Fructose	*Fructose*	*Glucose*	*Sucrose*	*Control*
$\bar{Y}_1$	$\bar{Y}_2$	$\bar{Y}_3$	$\bar{Y}_4$	$\bar{Y}_5$
58.0	58.2	59.3	64.1	70.1
______	______	______		

to the actual results of the experiment. We thought that we should test all sugars against control, then mixed sugars against pure sugars, finally the pure sugars among themselves. You may recall that we also tested fructose and glucose considered together against the mixture of fructose and glucose. In Box 9.9, however, every single test was suggested not by the nature of the treatment (that is, whether it was a pure sugar or a mixed sugar or a control treatment), but simply by the magnitude of the response mean. Therein lies the essential difference between a priori and a posteriori tests. The results we obtained by the a priori test actually are consistent with the results we found by the a posteriori *SNK* test. However, this is not always so. When the decision is near the borderline for a given probability level or there are many means in the anova, the a priori test will always be more sensitive; it is more likely to result in significance than the a posteriori test. If two means are significantly different by the a posteriori test, we can be certain that they would have been significant by the a priori test.

The diagrammatic representation of the results of a multiple comparisons test will not always be as simple as those in Boxes 9.9 and 9.10. Often there are no clear boundaries (gaps) between sets of means not significantly different from each other, and such sets overlap. An example in point is the following data (left side of page), which represent the means of the character thorax width for the aphid *Pemphigus populi-transversus* studied in 23 localities in eastern North America. The means are arrayed by order of magnitude in coded units and the lines beside them represent nonsignificant sets of means. This symbolism was employed by Sokal and Rinkel (1963) at the margin of a geographic distribution map to provide an easy visual test of significance of any pair of means. When there are too many overlapping lines they are difficult to keep apart by eye and a different technique is preferable (Sokal and Thomas, 1965), illustrated in the middle of the page for the same data.

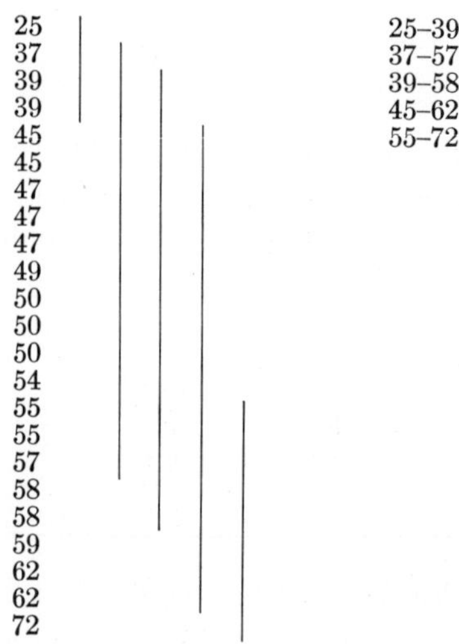

Any pair of means enclosed by the range of any one line in the second convention is not significantly different. Thus means 47 and 50 are not different from each other because they are enclosed in the range 39 to 58, but means 39 and 62 are different from each other because there is no range that encloses them.

When the means are not based on equal sample sizes, only approximate values of *LSR* can be found for the *SNK* test, as was the case with the Range-*STP*. An analysis of this type is shown in Box 9.10, which is a new example of a Model I analysis of variance with unequal sample sizes. This

BOX 9.10

Multiple comparisons among means based on unequal sample sizes (a posteriori tests): the Student-Newman-Keuls test.

Head widths (in mm) of a species of tiger beetle (*Cicindela circumpicta*) collected from eight localities.

BOX 9.10 continued

	Localities (ranked by magnitude of means)							
	Barnard, Kansas	*Roswell, New Mexico*	*Nebraska*	*Talmo, Kansas*	*Okeene, Oklahoma*	*Kackley, Kansas*	*Mayfield, Oklahoma*	*Stafford Co., Kansas*
$\bar{Y}$	3.5123	3.6940	3.7223	3.7229	3.7376	3.7479	3.7843	3.8451
n_i	20	20	20	14	10	15	11	20
s^2	0.0357	0.0445	0.0135	0.0284	0.0204	0.0184	0.0338	0.0157

Source: Selected data from a study by H. L. Willis.

The pooled variance = weighted average variance = MS_{within}

$$= \frac{\sum^{a} (n_i - 1)s_i^2}{\sum^{a} (n_i - 1)}$$

$$= \frac{(20 - 1)0.0357 + (20 - 1)0.0445 + \cdots + (20 - 1)0.0157}{19 + 19 + \cdots + 10 + 19}$$

$$= \frac{3.2270}{122}$$

$$= 0.02645$$

For ease in computation we code the data as follows: $Y_c = (Y - 3) \times 10$. This code is convenient since all of the means are between 3 and 4. Thus, the MS_{within} becomes $0.02645 \times 10^2 = 2.645$. Also, since there are many means, it is most efficient to make a two-way table of differences between the coded means so that the sequence of testing discussed below will be easier to follow. The differences are computed as row-mean − column-mean.

			Rank							
			1	*2*	*3*	*4*	*5*	*6*	*7*	*8*
		$\bar{Y}_c$	5.123	6.940	7.223	7.229	7.376	7.479	7.843	8.451
		n_i	20	20	20	14	10	15	11	20
Rank	$\bar{Y}_c$	n_i								
1	5.123	20	—							
2	6.940	20	1.817*	—						
3	7.223	20	2.100*	0.283	—					
4	7.229	14	2.106*	0.289	0.006	—				
5	7.376	10	2.253*	0.436	0.153	0.147	—			
6	7.479	15	2.356*	0.539	0.256	0.250	0.103	—		
7	7.843	11	2.720*	0.903	0.620	0.614	0.467	0.364	—	
8	8.451	20	3.328*	1.511*ns*	1.228	1.222	1.075	0.972	0.608	—

* Tested and found to be significant at $P = 0.05$.
ns = Tested and found to be not significant at $P = 0.05$. Unmarked differences were not tested because a larger range was found to be not significant.

BOX 9.10 continued

For a difference in the above table to be significant at the α level, it must be equal to or greater than

$$LSR = Q_{\alpha[k,\nu]}\sqrt{MS_{\text{within}}}\sqrt{\frac{n_1 + n_2}{2n_1n_2}}$$

where $k = 1 +$ difference in the ranks of two means. This equals the number of means in the set whose range we are testing. The degrees of freedom equal $\nu = df_{\text{within}}$ ($= 122$ in the present example). The sample sizes of the two means are n_1 and n_2, and $Q_{\alpha[k,\nu]}$ is a value from Table **U**, which gives the upper α points of the studentized range of k items when the estimate of the error variance has ν *df*. The following is a list of the $Q_{.05[k,120]}$-values we shall need for the present example for $k = 2, \ldots, 8$ (we shall not interpolate since Q for $\nu = 122$ is so close to the tabled values for $\nu = 120$):

2.800, 3.356, 3.685, 3.917, 4.096, 4.241, 4.363

When multiplied by $\sqrt{MS_{\text{within}}} = 1.626$ (for the coded data), these become

4.553, 5.457, 5.992, 6.369, 6.660, 6.896, 7.094

To test any particular range between two means we need only compare it with these quantities multiplied by $\sqrt{(n_1 + n_2)/2n_1n_2}$. If we had equal sample sizes we would need only to multiply these by the constant $1/\sqrt{n}$. The sequence of tests is carried out in a stepwise manner.

1. Test total range of the 8 means:

$$LSR = Q_{\alpha[8,\nu]}\sqrt{MS_{\text{within}}}\sqrt{\frac{n_1 + n_2}{2n_1n_2}}$$

$$= 7.094\sqrt{\frac{20 + 20}{2 \times 20 \times 20}} = 7.094\left(\frac{1}{\sqrt{20}}\right)$$

$$= 7.094(0.2236) = 1.586$$

Since the observed range (3.328) is greater than this, it is significant at the 5% level.

2. Test the smallest to the next to largest mean and the range from the next to smallest to the largest mean.

$$LSR = Q_{\alpha[7,\nu]}\sqrt{MS_{\text{within}}}\sqrt{\frac{n_1 + n_2}{2n_1n_2}}$$

Owing to the unequal sample sizes, we must compute separate *LSR*'s for these two tests:

$$LSR_{(1\rightarrow 7)} = 6.896\sqrt{\frac{20 + 11}{2 \times 20 \times 11}}$$

$$= 6.896\sqrt{\frac{31}{440}} = 6.896(0.265430)$$

$$= 1.830$$

BOX 9.10 continued

which is less than the observed range of 2.720, and

$$LSR_{(2\to8)} = 6.896 \frac{1}{\sqrt{20}}$$

$$= 6.896(0.2236) = 1.542$$

which is slightly greater than the observed range of 1.511. We now consider the range of means 2, 3, 4, 5, 6, 7, and 8 not significant and do not perform any further tests within this range. In terms of the table of ranges given above, we need not test any ranges listed above and to the right of this difference in row 8, column 2.

3. Since the first range (1 ⟶ 7) was significant, we now must test a range of 6 means in a similar fashion. We find range 1 ⟶ 6 significant. (Ranges 2 ⟶ 7 and 3 ⟶ 8 are not tested because they are within a nonsignificant range 2 ⟶ 8.)
4. Next we test ranges of 5 means, and find 1 ⟶ 5 significant.
5. We continue testing with ranges of 4, 3, and 2 means (in our case 1 ⟶ 4, 1 ⟶ 3, and 1 ⟶ 2), all of which are significant. The results can be summarized by underlining those ranges which were found to be not significant.

1   2   3   4   5   6   7   8

(underline spans 2 through 8)

is an analysis of head widths based upon samples of a species of tiger beetle from 8 localities. It is clearly Model I, because the 8 samples are not a random sample from some population but are from 8 localities that are of particular interest to the investigator. The SNK test is shown in detail in Box 9.10 and is self-explanatory.

Although all of the procedures described above lead to the same conclusions in the pea section data (Box 9.4 and 9.9), such agreement need not, in general, occur. When the tiger beetle data in Box 9.10 were analyzed by the SS-STP we found the following pattern of significances

Locality rank	*1*	*2*	*3*	*4*	*5*	*6*	*7*	*8*
$\bar{Y}$	3.5123	3.6940	3.7223	3.7229	3.7376	3.7479	3.7843	3.8451

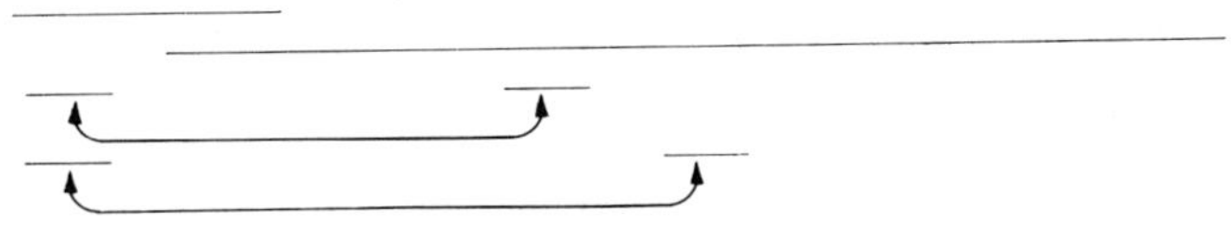

(Lines: 1–2 underlined; 2–8 underlined; double-headed arrows connect 1 with 4 and 1 with 5.)

which is not consistent with the results indicated by the Range-STP and the SNK test (Box 9.10), even though all these methods are testing the same null hypothesis of equality of means. Since sample sizes are unequal, means in nonsignificant sets are not necessarily adjacent, as indicated by the double-headed arrows.

A reason for this difference is that *SS* and range tests differ in their sensitivity to various alternative hypotheses. The range, of course, cannot take into account any differences in the distribution of means within it, as does a sum of squares. For example, means 3 and 4 are very close to each other in magnitude. Thus the range from 1 to 4 is very similar to the range from 1 to 3, and the sum of squares of 1, 3, and 4 is quite different from that of 1 and 4 alone (particularly since the sample sizes are not all equal).

Regrettably, we cannot at this time furnish firm recommendations on which methods seem appropriate for a posteriori testing. Choice of the method may have to be guided by computational facilities. Except with a very few means the *SS-STP* will require a digital computer and consequently workers limited to a desk calculator should employ a range method for numerous means, be this by *STP* or the *SNK* test. Arguments in support of either of these tests could be advanced, but these are not of such a nature as to make a firm recommendation possible.

↛ The subject of multiple comparisons is an active research field and new methods will undoubtedly be developed in the future. Among the methods not covered in this text, perhaps the best known is that of Duncan, which is another multiple range procedure and which is computationally identical with the *SNK* test except for the use of another table of critical values. Computational steps for Duncan's procedures are illustrated in Steel and Torrie (1960) in which this table of critical values is furnished. To compare all treatment means with a control in a Model I anova, a procedure has been worked out by Dunnett that is also furnished in Steel and Torrie (1960). ↚

9.8 Finding the sample size *n* required for a test

Statisticians are frequently asked how large a sample must be obtained for an experiment. Regrettably, there is usually no simple answer. Without an estimate of the variability of the items, no answer at all can be given. The statistician needs to know what the experimenter wishes to test. Does he wish to establish a significant difference between means, or to estimate the variance of the items to set confidence limits to a single mean? Given a precise statement of the problem, statisticians have provided a number of computational solutions. In the case of a posteriori tests there is again some disagreement about whether to obtain experimentwise error rates or single test error rates, but we will be concerned only with a priori tests here.

We shall show you a simple test that provides an answer to the most common question: How big a sample size must I obtain in order to be able to show that a true difference δ is significant at a significance level α with a probability P that the significance will be found if it exists? The formula for this is shown near the top of Box 9.11, where the meaning of the separate symbols is also explained. To use this formula one needs to specify the true

BOX 9.11

Number of replications needed to detect a given "true" difference between means.

From previous studies it is known that the coefficient of variation of wing length of a species of bird is about 6%. You plan to study four populations by analysis of variance. How many measurements need be made from each population to be 80% certain of detecting a 5% difference between two of the four means at the 1% level of significance? Since $a = 4$, the error MS will have $\nu = 4(n - 1)$ degrees of freedom.

The appropriate formula is

$$n \geq 2\left(\frac{\sigma}{\delta}\right)^2 \{t_{\alpha[\nu]} + t_{2(1-P)[\nu]}\}^2$$

where n = number of replications

σ = true standard deviation

δ = the smallest true difference that it is desired to detect. (NOTE: it is necessary to know only the ratio of σ to δ, not their actual values)

ν = degrees of freedom of the sample standard deviation ($\sqrt{MS_{\text{within}}}$) with a groups and n replications per group

α = significance level (such as 0.05)

P = desired probability that a difference will be found to be significant (if it is as small as δ)

$t_{\alpha[\nu]}$ and $t_{2(1-P)[\nu]}$ = values from a two tailed t-table with ν degrees of freedom and corresponding to probabilities of α and $2(1 - P)$, respectively

Iterative solution

We try an initial value of $n = 20$ as a reasonable guess. Then $\nu = 4(20 - 1) = 4 \times 19 = 76$. Since $CV = 6\%$, $s = 6\bar{Y}/100$. We wish δ to be 5% of the mean; that is, $\delta = 5\bar{Y}/100$. Using s as an estimate of σ we obtain $\sigma/\delta = (6\bar{Y}/100)/(5\bar{Y}/100) = 6/5$:

$$n \geq 2\left(\frac{6}{5}\right)^2 \{t_{.01[76]} + t_{2(1-0.80)[76]}\}^2$$

$$= 2\left(\frac{6}{5}\right)^2 [2.64 + 0.847]^2 = 2(1.44)12.16 = 35.0$$

Next we try $n = 35$, making $\nu = 4(35 - 1) = 4(34) = 136$:

$$n \geq 2(1.44)[2.61 + 0.845]^2 = 2.88(11.94) = 34.4$$

This is close to the previous solution, indicating that we have iterated to stability. It appears that 35 replications per population sample are necessary.

standard deviation σ, which can be estimated from a preliminary analysis. Thus it is always necessary to have some prior idea of how variable the data are. For example, assume that you are about to start a study of the geographic variation of birds, that wing length is one of the characters you would measure, and that you wish to know how many specimens of birds to study per locality. You must have studied at least a few localities previously by analysis of variance to estimate the within-locality variance of wing length for this particular species of bird. Similarly, for an experiment in which you are going to determine the glycogen content of rat livers, you must have several prior determinations to estimate their variance before you can calculate how many replicates you will need. Next you need to establish δ, the smallest true difference you desire to detect. It is obvious that if you are interested in picking up only a very crude difference (a large value of δ), you would need relatively few specimens; but for a fine difference (a small value of δ), a large sample size would be required. The significance level at which you are prepared to consider two means significantly different will be α; the desired probability that a significant difference will be found (if it exists and is as small as δ) is P. One property of the formula in Box 9.11 is that we do not need to have actual values for σ and δ but need to know only their ratio. The example in the box makes this clear. We only know the coefficient of variation and the percentage of difference that it is desired to detect. We can therefore write $(6 \times \bar{Y})/(5 \times \bar{Y}) = 6/5 = 1.2$, since this is the ratio of a 6% coefficient of variation over a 5% difference of means. The mean $\bar{Y}$ cancels out and a ratio of 1.2 results.

The example in Box 9.11 is solved in iterative fashion. This means that we enter a value for n into the formula and use it to obtain a better estimate of n. We need a first estimate of n because without it we would not know the degrees of freedom and could not look up t in Table **Q**. Once we have such an estimate, however far off, we can then get a better estimate by application of the formula, use it to look up a better value of t, and so on. Such an iterative procedure is frequently employed in mathematics when a direct solution for a term is difficult or impossible. If you are not familiar with iteration, an analogy due to Thomson (1951) may help. This is the problem of Robinson Crusoe wishing to make a wooden lathe. To do so he needs to have wheels and spindles, but he needs a lathe to make these. So Robinson Crusoe fashions a crude wheel by whittling with a knife, constructing a crude lathe with the wheel. He now can prepare a slightly better wheel from which he constructs a better lathe, which yields a yet better wheel, and so on. Such iterative procedures are often convergent; that is, they give increasingly better estimates. Even if you make a mistake in the computations along the way you will only be slowed down in finding the right solution. The correct procedure should always home in on the correct answer eventually.

To carry out the iteration efficiently on a desk calculator, we first compute the left portion of the expression, which is $2(6/5)^2 = 2(1.44) = 2.88$. This value will remain constant, while the expression in square brackets will change. For the first iteration we guess a sample size of $n = 20$, and the solution of the equation yields an estimate of $n = 35.0$. Substituting this value into the expression we arrive at $n = 34.4$ after the second iteration. This brings us right back to a sample size of 35. Thus we conclude that for a study of these bird wing lengths and for the percentage of difference required as well as for the other specifications of the problem (α and P), 35 birds per locality sample should be adequate.

A little experimentation with this equation will show that it is not too sensitive to changes in α and P but is very sensitive to change in the ratio of σ/δ, so that it takes very large sample sizes to detect small differences. It can also be seen that any refinement of experimental technique that reduces σ results in smaller required sample sizes or the possibility of detecting smaller differences.

Exercises 9

9.1 The following is an example with easy numbers to help you become familiar with the analysis of variance. A plant ecologist wishes to test the hypothesis that the height of plant species X depends on the type of soil it grows in. He measured the height of three plants in each of four plots representing different soil types, all four plots being contained in an area of two miles square. His results are tabulated below. (Height is given in centimeters.) Does your analysis support this hypothesis? ANS. $F_s = 6.951$, $F_{.05[3,8]} = 4.07$.

	Localities			
Observation number	*1*	*2*	*3*	*4*
1	15	25	17	10
2	9	21	23	13
3	14	19	20	16

9.2 The following are measurements (in coded micrometer units) of the thorax length of the aphid *Pemphigus populi-transversus*. The aphids were collected in 28 galls on the cottonwood, *Populus deltoides*. Four alate (winged) aphids were randomly selected from each gall and measured. The alate aphids of each gall are isogenic (identical twins), being descended parthenogenetically from one stem mother. Thus, any variance within galls can be due to environment only. Variance between different galls may be due to differences in genotype and also to environmental differences between galls. If this character, thorax length, is affected by genetic variation, significant intergall variance must be present. The converse is not necessarily true; significant variance between galls need not indicate genetic variation; it could as well be due to environmental differences between galls (data by Sokal, 1952). Analyze the variance of thorax length. Is there significant intergall variance present? Give estimates of the added component of intergall variance, if present. What percentage of the variance is controlled by intragall and what percentage by intergall factors? Discuss your results. (Remember to check your computations step by step; otherwise an error committed early in the calculation may ruin your entire effort. For computational purposes ignore the decimal point in the variates.)

Gall No.					*Gall No.*				
1.	6.1,	6.0,	5.7,	6.0	15.	6.3,	6.5,	6.1,	6.3
2.	6.2,	5.1,	6.1,	5.3	16.	5.9,	6.1,	6.1,	6.0
3.	6.2,	6.2,	5.3,	6.3	17.	5.8,	6.0,	5.9,	5.7
4.	5.1,	6.0,	5.8,	5.9	18.	6.5,	6.3,	6.5,	7.0
5.	4.4,	4.9,	4.7,	4.8	19.	5.9,	5.2,	5.7,	5.7
6.	5.7,	5.1,	5.8,	5.5	20.	5.2,	5.3,	5.4,	5.3
7.	6.3,	6.6,	6.4,	6.3	21.	5.4,	5.5,	5.2,	6.3
8.	4.5,	4.5,	4.0,	3.7	22.	4.3,	4.7,	4.5,	4.4
9.	6.3,	6.2,	5.9,	6.2	23.	6.0,	5.8,	5.7,	5.9
10.	5.4,	5.3,	5.0,	5.3	24.	5.5,	6.1,	5.5,	6.1
11.	5.9,	5.8,	6.3,	5.7	25.	4.0,	4.2,	4.3,	4.4
12.	5.9,	5.9,	5.5,	5.5	26.	5.8,	5.6,	5.6,	6.1
13.	5.8,	5.9,	5.4,	5.5	27.	4.3,	4.0,	4.4,	4.6
14.	5.6,	6.4,	6.4,	6.1	28.	6.1,	6.0,	5.6,	6.5

9.3 Millis and Seng (1954) published a study on the relation of birth order to the birth weights of infants. The data below on first-born and eighth-born infants are extracted from a table of birth weights of male infants of Chinese third-class patients at the Kandang Kerbau Maternity Hospital in Singapore in 1950 and 1951.

Birth weight	*Birth order*	
lb:oz	*1*	8
3:0–3:7	—	—
3:8–3:15	2	—
4:0–4:7	3	—
4:8–4:15	7	4
5:0–5:7	111	5
5:8–5:15	267	19
6:0–6:7	457	52
6:8–6:15	485	55
7:0–7:7	363	61
7:8–7:15	162	48
8:0–8:7	64	39
8:8–8:15	6	19
9:0–9:7	5	4
9:8–9:15	—	—
10:0–10:7	—	1
10:8–10:15	—	—
	1932	307

Which birth order appears to be accompanied by heavier infants? Is this difference significant? Can you conclude that birth order causes differences in birth weight? (Computational note: The variable should be coded as simply as possible.) Reanalyze, using the *t*-test, and verify that $t_s^2 = F_s$. ANS. $t_s = 11.016$.

9.4 The following cytochrome oxidase assessments of male *Periplaneta* roaches in cubic millimeters per ten minutes per milligram were taken from a larger study by Brown and Brown (1956):

	n	$\bar{Y}$	$s_{\bar{Y}}$
24 hours after methoxychlor injection	5	24.8	0.9
Control	3	19.7	1.4

Are the two means significantly different?

9.5 The following data are measurements of five random samples of domestic pigeons collected during the months of January, February, and March in Chicago in 1955. The variable is the length from the anterior end of the narial opening to the tip of the bony beak and is recorded in millimeters. Data from Olson and Miller (1958).

Samples				
1	*2*	*3*	*4*	*5*
5.4	5.2	5.5	5.1	5.1
5.3	5.1	4.7	4.6	5.5
5.2	4.7	4.8	5.4	5.9
4.5	5.0	4.9	5.5	6.1
5.0	5.9	5.9	5.2	5.2
5.4	5.3	5.2	5.0	5.0
3.8	6.0	4.8	4.8	5.9
5.9	5.2	4.9	5.1	5.0
5.4	6.6	6.4	4.4	4.9
5.1	5.6	5.1	6.5	5.3
5.4	5.1	5.1	4.8	5.3
4.1	5.7	4.5	4.9	5.1
5.2	5.1	5.3	6.0	4.9
4.8	4.7	4.8	4.8	5.8
4.6	6.5	5.3	5.7	5.0
5.7	5.1	5.4	5.5	5.6
5.9	5.4	4.9	5.8	6.1
5.8	5.8	4.7	5.6	5.1
5.0	5.8	4.8	5.5	4.8
5.0	5.9	5.0	5.0	4.9

Are the five samples homogeneous?

9.6 If a study such as that given in Exercise 9.5 on pigeons were to be repeated with only two samples (one a control and the other representing pigeons fed a special diet), what sample size should be used in order to be 80% certain of observing a true difference between two means as small as a tenth of a millimeter at the 5% level of significance? Assume that the variance in this new experiment would be the same as in the previous one.

9.7 P. E. Hunter (1959, detailed data unpublished) selected two strains of *Drosophila melanogaster*, one for short larval period (SL) and one for long larval periods (LL). A nonselected control strain (CS) was also maintained. At generation 42 the following data were obtained for the larval period (measured in hours). Analyze and interpret.

	Strain		
	SL	*CS*	*LL*
n_i	80	69	33
$\sum^{n_i} Y$	8070	7291	3640

$$\sum^{3} \sum^{n_i} Y^2 = 1{,}994{,}650$$

Note that part of the computation has already been performed for you. Perform a priori tests among the three means (short versus long larval periods and each against the control). Set 95% confidence limits to the observed differences of means for which these comparisons are made. ANS. $MS_{(\text{SL vs. LL})} = 2076.6697$.

9.8 The following data were taken from a study of the systematics of honeybees (DuPraw, 1965). Of the 15 variables reported in the original study, the statistics for wing length variable *A–O* (in mm × 83) are given below for fifteen localities. From the information given below (sample size, mean, and standard error) perform an analysis of variance and use the Student-Newman-Keuls test to separate the locality means into nonsignificant groups. Note that standard errors were not published for localities 7 and 11, so that the estimate of the within groups mean square must be based on the variation present in the other localities.

	Locality						
	1	*2*	*3*	*5*	*6*	*7*	*8*
n	20	20	20	16	12	11	16
$\bar{Y}$	379.9	384.1	376.8	362.9	374.1	360.4	341.8
$s_{\bar{Y}}$	5.6	3.2	3.3	3.8	2.0	—	7.3

	Locality							
	9	*10*	*11*	*13*	*14*	*15*	*16*	*17*
n	20	20	20	10	10	10	10	10
$\bar{Y}$	371.0	353.3	354.2	330.7	342.6	378.6	368.5	368.8
$s_{\bar{Y}}$	4.0	3.7	—	14.0	4.0	6.5	4.5	3.5

Locality codes:

1. Europe northwest of the Alps
2. Europe southeast of the Alps
3. Italy
5. Cyprus
6. Caucausus
7. Eastern Mediterranean
8. Egypt
9. North Africa
10. Central Africa
11. Cape of Good Hope
13. India and Pakistan
14. Japan
15. Mt. Kilimanjaro and vicinity
16. Yugoslavia
17. Yellow "oasis bee" from Sahara

Analyze and interpret. You may wish to compare your results with those in the original article based upon all fifteen measurements.

9.9 The following data were taken from a study of blood protein variations in deer (Cowan and Johnston, 1962). The variable is the mobility of serum protein fraction *II* expressed as 10^{-5} cm²/volt seconds.

	$\bar{Y}$	$s_{\bar{Y}}$
Sitka	2.8	0.07
California Blacktail	2.5	0.05
Vancouver Island Blacktail	2.9	0.05
Mule Deer	2.5	0.05
White Tail	2.8	0.07

$n = 12$ for each mean. Perform an analysis of variance and a multiple comparisons test, using the sums of squares *STP* procedure. ANS. $MS_{\text{within}} = 0.0416$, maximal nonsignificant sets (at $P = 0.05$) are samples 1, 3, 5 and 2, 4.

10 NESTED ANALYSIS OF VARIANCE

This chapter deals with slightly more complicated cases of single classification anova. Each class or group is subdivided into two or more randomly chosen subgroups. The reasons for employing such a design are explained in Section 10.1, and the computation for subsamples of equal size is shown in Section 10.2. Some complications ensue when subsamples are of unequal sizes; these are treated in Section 10.3. The last section, 10.4, introduces the concept of relative efficiency of designs. We consider in this section how best to plan an experiment to obtain a maximum amount of information for any given investment of effort and money.

10.1 Nested anova: design

The simple design of a single classification anova as learned in the last chapter is frequently insufficient to represent the complexity of a given experiment and to extract all the relevant information from it. The present chapter deals with cases in which each major grouping or class is subdivided into randomly chosen subgroups. Thus for each of the seven batches of medium from Section 8.4, we might prepare five jars and a sample of houseflies may be reared from each of these jars. Each group in such an analysis would represent a type of medium, each subgroup a jar for one medium, and the items in a subgroup would be the individual flies being measured from each jar.

What added information do we obtain from such a design? In an experiment with a single jar for each different medium formulation we can never be certain that observed differences among the flies in different jars are due to different formulations. Possibly ecological and other accidental differences among the jars result in morphological differences among the flies, which would have occurred even if the same formulation had been used in all of the jars. The only way to separate these two effects is to have two or more jars for each medium formulation. If we do not find any differences among the jars *within* medium formulations—beyond what we should expect on the basis of the amount of variation observed within a jar—we can ascribe differences among the media to their formulations. But even if we were to find that there was a significant amount of added variance among the jars within a formulation, we can still test to see if the variation among the media is greater than that to be expected on the basis of the observed variation among the jars that were treated alike. If the tested mean square is significant, we can conclude that the differences among medium formulations are significant above and beyond differences among jars.

We call the analysis of such a design a *nested analysis of variance,* because the subordinate classification is nested within the higher level of classification. Thus, in our example the jars are subordinate to the medium formulations. The design is also often called a *hierarchic analysis of variance.* A crucial point is that groups representing a subordinate level of classification must be randomly chosen. Thus the replicate jars within one medium formulation must not be deliberately picked to represent certain microenvironments. The subordinate level of a nested anova is always Model II. The highest level of classification in a nested anova may be Model I or Model II. If it is Model II, we speak of a *pure Model II* nested or hierarchic anova. If the highest level is Model I, we called it a *mixed model* nested anova.

There are two general classes of applications of nested anova. The first is similar to the hypothetical example just discussed. It usually serves to ascertain the magnitude of error at various stages of an experiment or an industrial process. Thus in the hypothetical housefly example we wish to know whether there is substantial variation among jars as well as among flies within jars, and we need to know the magnitude of this variation to enable us to test the magnitude of the added component due to treatments (= different formulations of medium). Nested anova is not limited to the two levels discussed so far. We can subdivide the subgroups into subsubgroups, and even further, as long as these are randomly chosen. Thus we might design an experiment in which we tested the effect of five drugs on the quantity of pigment in the skin of an animal. The five drugs and one control (six groups) are the major classification and clearly are fixed treatment effects (Model I). For each drug we might use five randomly selected rats. These would provide a measure of variance of rats within a drug class or of "rats within drugs," as one would say in statistical jargon. From each rat we might take three skin samples at random from its ventral side. This presents us with a new subordinate level of variation—skin samples within rats. Each skin sample is macerated and divided into two lots, which are hydrolyzed separately. This level is hydrolysates within skin samples. Finally, the quantity of pigment might be read as an optical density, and two replicate readings might be made of each hydrolysate. The basic error variance would be the variance of the replicate readings per hydrolysate, but we would also have estimates of the variance between hydrolysates within one skin sample, among skin samples within one rat, and among rats within one drug. These variance estimates are important in designing similar experiments because they point out to us at which level of the experiment most of our efforts should be concentrated. The most variable aspect of our experiment will need the greatest replication or will need better experimental control. Thus if we find that the two hydrolysates show the greatest proportion of variance, our method of hydrolysis clearly is not standardized enough and should be improved or, if this is impossible, we should divide each macerated skin sample into more lots to provide several more hydrolysates. This whole topic is taken up in greater detail in Section 10.4.

The second major use of a nested analysis of variance is in cases that are usually pure Model II. These are frequently from the field of quantitative genetics, where we wish to know the magnitude of the variance attributable to various levels of variation in a study. Such an approach is also used by systematists interested in discovering the sources of variation in natural populations. To illustrate such an application, we could have an experiment in which separate litters by the same mother have been sired by different fathers. If there are ten females (dams), each with five litters sired by five different males (sires), the variance can be subdivided into that among dams, that among sires within one dam, and that among offspring within one sire. Genetical theory (based on the statistical consequences of Mendelian genetics) can predict the relative magnitudes of such components on various assumptions. One of the aims of the science of quantitative genetics has been to verify such predictions and analyze departures from them. A similar approach can be used in geographic variation studies in which variances among samples within one locality and among localities can be estimated and may lead to important conclusions about the population distribution pattern of an organism. For further information on the application of analysis of variance to quantitative genetics, readers are referred to the very readable account by Falconer (1960).

Before we proceed to the computation of an actual example we should examine the model upon which a nested anova is based. In Expressions (8.3) and (8.4) we gave equations for decomposing variates in a single classification anova for Model I and Model II. For a two-level nested anova (Model II) this equation becomes

$$Y_{ijk} = \mu + A_i + B_{ij} + \epsilon_{ijk} \tag{10.1}$$

where Y_{ijk} is the kth observation in the jth subgroup of the ith group, μ is the parametric mean of the population, A_i is the random contribution for the ith group of (the higher) level A, and B_{ij} is the random contribution for the jth subgroup (level B) of the ith group; ϵ_{ijk} is the error term of the kth item in the jth subgroup of the ith group. We assume that A_i, B_{ij}, and ϵ_{ijk} are normally distributed with means of zero and variances of σ_A^2, $\sigma_{B \subset A}^2$, and σ^2, respectively. We use the symbol $\sigma_{B \subset A}^2$, rather than σ_B^2, to indicate that the variance is of level B *within* level A, rather than of level B as the highest level in the hierarchy. We shall use A, B, C, . . . , as subscripts for levels of variance components. Statistical workers frequently use letters that help them remember the meaning of variance components. Thus σ_G^2 might signify genetic variance, σ_D^2 variance among dogs, and so on. However, in a textbook a uniform and logical system is necessary.

When the model is mixed (that is, the highest level has fixed treatment effects) we decompose a variate as follows:

$$Y_{ijk} = \mu + \alpha_i + B_{ij} + \epsilon_{ijk} \tag{10.2}$$

This is the same expression as (10.1) except that α_i is a fixed treatment effect, unlike A_i in the earlier model. Expressions (10.1) and (10.2) can be expanded to further levels, which are identified by C, D, and so forth.

We now proceed to the computation of a nested analysis of variance.

10.2 Nested anova: computation

The example we have chosen to illustrate the computation of a nested analysis of variance is of the second type described in Section 10.1. It studies the percentage of phenotypic variation at two sampling levels. These data were obtained by Rohlf on rearing field-caught mosquito pupae (*Aedes intrudens*) in the laboratory. Twelve female pupae were brought into the laboratory and divided at random into three rearing cages, in which four pupae each were reared. The purpose of dividing them among different cages was to see whether the differences among cages would add variance to the overall variation among females that is due to their presumable genetic and ecological differences. The structure of the experiment and of the analysis, therefore, is three cages of four females each. Length of left wing was measured on each female and the measurement was repeated once, the mosquitoes being presented to the measurer in a random order so as not to bias the results. The measurements are shown at the top of Box 10.1.

To understand the nature of the computations it helps to rearrange these nested data in two ways, making them each time into a different single classification anova. In Table 10.1 we show first one of these arrangements in which we ignore the fact that the data are divided into three cages but simply treat each of the 12 females as one group in a single classification anova. Convince yourself that the data at the head of Table 10.1 are the same as those at the head of Box 10.1. Now we simply have $a = 12$ females, each of which was measured $n = 2$ times. The fact that these 12 females come from three cages is hidden by this arrangement of the data. The analysis of variance of this table is straightforward, and the quantities shown beneath it yield the ensuing anova table. The error mean square in the table represents variance within females, which estimates the average variance between the two measurements for each female. The upper mean square represents variance among females, but in this arrangement of the data it confounds two sources of variation: the differences among females within a cage as well as possible differences from cage to cage. We note that this mean square yields a highly significant $F_s = 166.88$, which would occur with an infinitesimal probability if the null hypothesis were correct. Obviously this is a mean square we shall eventually want to break down into its components, representing variation of females within a cage and that among cages. It is reassuring to find this mean square significant because this tells us that we shall find some significant differentiation at either or both of the upper levels of variation in the study. If there were no differences at all among the females, either because they did not differ inherently or because the different cages had not made them different, this would have shown up in the preliminary analysis we have just considered. This rule does not hold absolutely, for occasionally significant differences will be found even if the analysis is not significant, but in general a significant preliminary analysis indicates one or more significant levels in the nested analysis, and lack of significance in the preliminary analysis indicates the same for the nested analysis. Certainly when the mean square among groups is greater than the error mean square by as much as it is in Table 10.1, we can expect a substantial added variance component at some level of the nested anova.

The second rearrangement of the data of Box 10.1 is shown further down in Table 10.1. We can remove the subdivision into subgroups and simply treat our data as a single classification anova of items within the higher level of classification. Such a procedure would yield eight measurements in each of three cages. Sample sizes a and n now assume new values and should not be confused with the same symbols used earlier in Table 10.1. The analysis of variance shows significant variation among cages but only at the 5% level. The error term in this anova represents a mixture of variances among measurements and among females in a cage. One way of interpreting a nested classification is as a further partitioning of the within groups SS of a single classification anova. Thus, in this case we separate variation among measurements of one female from variation among females in a cage.

The group sum of squares of the first anova comprises a mixture of variation among cages and among females. When we subtract from it the sum of squares among cages we obtain a new sum of squares representing the variation of females within a cage, which is the quantity we are after in a nested analysis of variance. We therefore compute $2386.3534 - 665.6759 = 1720.6775$ with $11 - 2 = 9$ degrees of freedom. This SS is called the *sum of squares of subgroups within groups*. Three sums of squares are now arranged in an anova table, as shown in Box 10.1. They are (1) the SS among groups (of the original design—that is, the SS among cages computed in the second anova of Table 10.1), (2) the SS of subgroups within groups just mentioned,

TABLE 10.1

Preliminary analyses of data in Box 10.1.

	a females ($a = 12$)											
	1	*2*	*3*	*4*	*5*	*6*	*7*	*8*	*9*	*10*	*11*	*12*
n measurements ($n = 2$)	58.5	77.8	84.0	70.1	69.8	56.0	50.7	63.8	56.6	77.8	69.9	62.1
	59.5	80.9	83.6	68.3	69.8	54.5	49.3	65.8	57.5	79.2	69.2	64.5
$\sum^{n} Y$	118.0	158.7	167.6	138.4	139.6	110.5	100.0	129.6	114.1	157.0	139.1	126.6

Preliminary computations

1. Grand total $\sum^{a}\sum^{n} Y = 1599.2$

2. Sum of the squared observations $\sum^{a}\sum^{n} Y^2 = 108{,}962.00$

3. Sum of the squared group totals, divided by their sample size

$$\frac{\sum^{a}\left(\sum^{n} Y\right)^2}{n} = 108{,}946.3800$$

4. Correction term $\dfrac{\left(\sum^{a}\sum^{n} Y\right)^2}{na} = 106{,}560.0267$

5. SS_{total} = quantity **2** − quantity **4**
= 108,962.00 − 106.560.0267
= 2401.9733

6. SS_{groups} = quantity **3** − quantity **4**
= 108,946.38 − 106,560.0267
= 2386.3533

7. SS_{within} = quantity **5** − quantity 6
= 2401.9733 − 2386.3533
= 15.6200

Anova table

	Source of variation	*df*	*SS*	*MS*	F_s
$\bar{Y} - Y$	Among groups (females)	11	2386.3533	216.94	166.88**
$Y - \bar{Y}$	Within groups (between measurements on each female)	12	15.6200	1.30	

	a cages ($a = 3$)		
	1	*2*	*3*
n measurements ($n = 8$)	58.5	69.8	56.6
	59.5	69.8	57.5
	77.8	56.0	77.8
	80.9	54.5	79.2
	84.0	50.7	69.9
	83.6	49.3	69.2
	70.1	63.8	62.1
	68.3	65.8	64.5
$\sum^{n} Y$	582.7	479.7	536.8

Preliminary computations

1. Grand total $\sum^{a}\sum^{n} Y = 1599.2$
2. Sum of the squared observations $\sum^{a}\sum^{n} Y^2 = 108{,}962.00$
3. Sum of the squared group totals, divided by their sample size $\dfrac{\sum^{a}\left(\sum^{n} Y\right)^2}{n} = 107{,}225.7025$
4. Correction term $\dfrac{\left(\sum^{a}\sum^{n} Y\right)^2}{na} = 106{,}560.0267$
5. SS_{total} = quantity **2** − quantity **4** $= 108{,}962.00 - 106{,}560.0267 = 2401.9733$
6. SS_{groups} = quantity **3** − quantity **4** $= 107{,}225.7025 - 106{,}560.0267 = 665.6758$
7. SS_{within} = quantity **5** − quantity **6** $= 2401.9733 - 665.6758 = 1736.2975$

Anova table

	Source of variation	*df*	*SS*	*MS*	F_s
$\bar{Y} - Y$	Among groups (cages)	2	665.6758	332.84	4.026*
$Y - \bar{Y}$	Within groups (among measurements on all females in a cage)	21	1736.2975	82.68	

BOX 10.1

Two-level nested anova.

Two independent measurements of the left wings of each of 4 female mosquitoes (*Aedes intrudens*) reared in each of 3 cages; $b = 4$ females. The data are given in micrometer units. This is a Model II anova.

		Groups of females ($a = 3$)											
		Cage 1				*Cage 2*				*Cage 3*			
		1	*2*	*3*	*4*	*1*	*2*	*3*	*4*	*1*	*2*	*3*	*4*
Measurements	1	58.5	77.8	84.0	70.1	69.8	56.0	50.7	63.8	56.6	77.8	69.9	62.1
($n = 2$)	2	59.5	80.9	83.6	68.3	69.8	54.5	49.3	65.8	57.5	79.2	69.2	64.5
Subgroup sums	$\sum^{n} Y$	118.0	153.7	167.6	138.4	139.6	110.5	100.0	129.6	114.1	157.0	139.1	126.6
Group sums	$\sum^{b}\sum^{n} Y$	582.7				479.7				536.8			

Source: Data by F. J. Rohlf.

Preliminary computations

1. Grand total $= \sum^{a}\sum^{b}\sum^{n} Y = 582.7 + \cdots + 536.8 = 1599.2$

2. Sum of the squared observations $= \sum^{a}\sum^{b}\sum^{n} Y^2 = (58.5)^2 + \cdots + (64.5)^2 = 108{,}962.00$

3. Sum of the squared subgroup totals, divided by the sample size of the subgroups

$$= \frac{\sum^{a}\sum^{b}\left(\sum^{n} Y\right)^2}{n} = \frac{[(118.0)^2 + \cdots + (126.6)^2]}{2} = 108{,}946.3800$$

4. Sum of the squared group totals divided by the sample size of the groups

$$= \frac{\sum^{a}\left(\sum^{b}\sum^{n} Y\right)^2}{nb} = \frac{[(582.7)^2 + \cdots + (536.8)^2]}{(2 \times 4)} = 107{,}225.7025$$

5. Grand total squared and divided by the total sample size = correction term $CT = \dfrac{\left(\sum^{a}\sum^{b}\sum^{n} Y\right)^2}{nba}$

$$= \frac{(1599.2)^2}{(2 \times 4 \times 3)} = 106{,}560.0267$$

6. $SS_{\text{total}} = \sum^{a}\sum^{b}\sum^{n} Y^2 - CT =$ quantity **2** − quantity **5** = 108,962.0000 − 106,560.0267 = 2401.9733

7. $SS_{\text{groups}} = \left[\dfrac{\sum^{a}\left(\sum^{b}\sum^{n} Y\right)^2}{nb}\right] - \left[\dfrac{\left(\sum^{a}\sum^{b}\sum^{n} Y\right)^2}{nba}\right]$ = quantity **4** − quantity **5** = 107,225.7025 − 106,560.0267 = 665.6758

8. SS_{subgr} (subgroups within groups) $= \left[\dfrac{\sum^{a}\sum^{b}\left(\sum^{n} Y\right)^2}{n}\right] - \left[\dfrac{\sum^{a}\left(\sum^{b}\sum^{n} Y\right)^2}{nb}\right]$ = quantity **3** − quantity **4**

= 108,946.3800 − 107,225.7025 = 1720.6775

9. SS_{within} (within subgroups; error SS) $= \sum^{a}\sum^{b}\sum^{n} Y^2 - \left[\dfrac{\sum^{a}\sum^{b}\left(\sum^{n} Y\right)^2}{n}\right]$ = quantity **2** − quantity **3**

= 108,962.0000 − 108,946.3800 = 15.6200

□□ *Computational hints.*—The machine computation of these quantities is self-evident and is similar to that carried out for earlier anovas. The following relations for some of the above quantities must hold: **2** ≥ **3** ≥ **4** ≥ **5**. Note the regular alternation of quantities in steps **7** through **9**. □□

Step	*Source of variation*	*SS (expressed as difference of indicated quantities)*
7	Among groups	4–5
8	Among subgroups within groups	3–4
9	Within subgroups	2–3
6	Total	2–5

BOX 10.1 continued

Now fill in the anova table.

Anova table: formulas

Source of variation	*df*	*SS*	*MS*	F_s	*Expected MS*
$\bar{Y}_A - \bar{\bar{Y}}$ Among groups	$a - 1$	**7**	$\frac{\mathbf{7}}{a-1}$	$\frac{MS_{\text{groups}}}{MS_{\text{subgr}}}$	$\sigma^2 + n\sigma^2_{B\subset A} + nb\sigma^2_A$
$\bar{Y}_B - \bar{Y}_A$ Among subgroups within groups	$a(b - 1)$	**8**	$\frac{\mathbf{8}}{a(b-1)}$	$\frac{MS_{\text{subgr}}}{MS_{\text{within}}}$	$\sigma^2 + n\sigma^2_{B\subset A}$
$Y - \bar{Y}_B$ Within subgroups	$ab(n - 1)$	**9**	$\frac{\mathbf{9}}{ab(n-1)}$		σ^2
$Y - \bar{\bar{Y}}$ Total	$abn - 1$	**6**			

Completed anova

Source of variation	*df*	*SS*	*MS*	F_s
$\bar{Y}_A - \bar{\bar{Y}}$ Among groups (among cages)	2	665.6759	332.84	1.741 *ns*
$\bar{Y}_B - \bar{Y}_A$ Among subgroups within groups (among females within cages)	9	1720.6775	191.19	147.07**
$Y - \bar{Y}_B$ Within subgroups (error; between measurements on each female)	12	15.6200	1.30	
$Y - \bar{\bar{Y}}$ Total	23	2401.9734		

$F_{.05[2,9]} = 4.26 \qquad F_{.01[9,12]} = 4.39$

$^{**}P < 0.01$

Conclusions.—There is no evidence for a significant variance component among cages ($P > 0.05$) but there is a highly significant added variance component among females ($P \ll 0.01$) for wing length in these mosquitoes.

From the components of the expected mean squares shown in the anova table presenting general formulas, we can derive a procedure for estimating the variance components:

Within subgroups (error; between measurements on each female) $= s^2 = 1.30$

Among subgroups within groups (among females within cages) $= s^2_{B \subset A} = \dfrac{MS_{\text{subgr}} - MS_{\text{within}}}{n} = \dfrac{191.19 - 1.30}{2} = 94.94$

Among groups (among cages) $= s^2_A = \dfrac{MS_{\text{groups}} - MS_{\text{subgr}}}{nb} = \dfrac{332.84 - 191.19}{8} = 17.71$

Since we are frequently interested only in their relative magnitudes, the variance components can be expressed as percentages of the overall variance (their sum):

$$s^2 + s^2_{B \subset A} + s^2_A = 1.30 + 94.94 + 17.71 = 113.95$$

$$s^2 \text{ represents } \frac{100 \times 1.30}{113.95} = 1.14\%$$

$$s^2_{B \subset A} \text{ represents } \frac{100 \times 94.94}{113.95} = 83.32\%$$

$$s^2_A \text{ represents } \frac{100 \times 17.71}{113.95} = 15.54\%$$

and (3) the sum of squares within subgroups (the error SS of the first anova in Table 10.1). Note that the three sums of squares add to the total sum of squares; the three df-values similarly add to the total degrees of freedom.

Before we complete the analysis of this example we should point out that the actual computations of the sums of squares need not be as involved as we have shown them here for pedagogical reasons. Differences between sums of squares can be obtained directly from quantities (2) through (5) as shown in Box 10.1. After carrying out computational steps (1) through (9) in that box we are ready to fill in the anova table, general formulas for which are shown first, followed by the specific example. Notice especially the expected mean squares in a nested anova. Each level above the error variance contains within it the variation of all levels below. Thus the expected variance of subgroups within groups is $\sigma^2 + n\sigma^2_{B \subset A}$, which resembles the familiar expression from a single classification anova. The only new aspect is that σ^2_B has changed to $\sigma^2_{B \subset A}$ to symbolize that it is variance of level B *within* level A. The expected mean square among groups contains the terms below it plus $nb\sigma^2_A$. From these expected mean squares the tests of significance are obvious.

For reasons explained later in this section, we always test lower levels before upper levels. Thus we first test $MS_{\text{subgr}}/MS_{\text{within}}$ for the significance of $\sigma^2_{B \subset A}$ and then test $MS_{\text{groups}}/MS_{\text{subgr}}$ for the existence of σ^2_A. In our actual example we find an added variance component among females within cages, which is highly significant with an F_s value of 147.07, representing a probability of type I error far less than 1%. However, when we test the cages over the MS_{subgr} we find the F-ratio to be only 1.741, which is not significant. Thus we have no reason to reject the null hypothesis $H_0: \sigma^2_A = 0$. Although the females differed from each other for genetic reasons, or possibly because of different environmental experiences before being brought into the laboratory or during development, the fact that they were reared in different cages did not add significant variation to their wing lengths.

You may wonder why it is that the MS is significant among cages in the second anova of Table 10.1 but is not so now. The reason is that we tested it over a different error term. The error term in the second anova of Table 10.1 is a mixture of the mean squares from the two lower levels of the analysis of variance in Box 10.1. If you added the two sums of squares and their degrees of freedom, you would obtain 1736.2975, the SS_{within} of the second anova of Table 10.1, and the 21 error degrees of freedom. However, this error term confounds two separate sources of variation: the variation among measurements with 12 degrees of freedom and a sum of squares of 15.6200, and differences among females within cages with 9 degrees of freedom and a sum of squares of 1720.6775. Clearly, if each cage is represented by several females, the mean square among the latter must be included in the error term.

This Model II nested anova is completed by estimating the three variance components as shown in Box 10.1. The operations are simple extensions of the method of estimating variance components learned in Chapter 9. When we divide the overall variation into percentages attributable to each level, we note that the error variance (measurements within one female) represents only 1.14% of the total variation. The measurements evidently did not differ much from each other, and in future work it would probably suffice to measure each wing only once. The bulk of the variation (83.32%) represents variation among females (within cages), presumably largely genetic but possibly also environmental, as already discussed. You may question the 15.54% of variance among cages. Having decided earlier that there is no added variance component among cages, why did we estimate it at all? First of all, to provide a general outline of all the computations in the box, so that you can use the example as a model for others you will carry out in your research, but more importantly because even though we were not in a position to reject the null hypothesis, we have not really proven that there is not a small contribution to the variance from cage differences. Possibly if we had more degrees of freedom (more cages) we could have shown such a difference to be significant. Thus we might quite legitimately estimate the variance component and calculate its percentage contribution to the total variation, since the computation gives us our best estimate of what this variance component would be if it did indeed exist.

In the example in Box 10.1 the mean square for subgroups within groups was significant and hence we used it as a divisor to test the significance of the mean square among groups. However, what should we have done if the mean square among subgroups had not been significant? From the expected mean squares in Box 10.1 we see that in that case the MS_{subgr} estimates σ^2, as does MS_{within}. One might well pool these two estimates of σ^2 to obtain a better estimate thereof based on more degrees of freedom. Such pooling is done by summing the respective sums of squares and degrees of freedom and dividing the sum of the sums of squares by the sum of the degrees of freedom. Statisticians do not agree about the conditions under which mean squares should be pooled and about the desirability of pooling. The experimenter cannot go wrong if he does not pool. However, there are occasions when pooling seems desirable because it would increase the number of degrees of freedom of the denominator mean square. A conservative set of rules is illustrated in Box 10.2.

As a second illustration of nested analysis of variance we shall take a design that is perhaps the most common in the literature. It is a two-level mixed model shown in Table 10.2. The groups represent different strains of houseflies, some known to be DDT-resistant, others not. It was desired to test whether there were differences in various morphological characters among these strains and to see whether these differences could be related to DDT resistance. Since the morphology of flies is generally quite labile, depending substantially on their environmental influences as immatures, three jars of the same medium were prepared for each strain in order to separate out the

BOX 10.2

Rules for pooling mean squares in anova (when intermediate level *MS* is nonsignificant).

Notation

Levels of variation	*df*	*SS*	*MS*	*Variance estimated*
3	df_3	SS_3	MS_3	$\sigma^3{}_2$
2	df_2	SS_2	MS_2	$\sigma^2{}_2$
1	df_1	SS_1	MS_1	$\sigma^2{}_1$

The question often arises in anova whether to pool two mean squares, in different lines of an anova table, that appear to be estimates of the same quantity. Research workers would often like to do this since the pooled mean squares will have more degrees of freedom. But because the actual population parameters estimated by the mean squares are unknown, one does not know for certain when one should pool. The rules listed below have been proposed by Bancroft (1964).

Dichotomous key to the rules for pooling mean squares

1. (a) On the basis of prior knowledge (from related experiments), σ_2^2/σ_1^2 is thought to be small (between 1 and 2) 2
   (b) No prior knowledge of σ_2^2/σ_1^2, or ratio thought to be large (> 2). Set $\alpha_1 = 0.75$ (NOTE: If σ_2^2/σ_1^2 is actually small the significance levels of the subsequent tests will be affected. For this reason many statisticians would recommend that one never pool.) 3
2. (a) $df_3 \geq df_2$ and $df_1 \geq 5 \times df_2$. Set $\alpha_1 = 0.50$ 3
   (b) Relations not as stated above. Set $\alpha_1 = 0.25$ 3
3. (a) $F_s = MS_2/MS_1$ is significant at $P \leq \alpha_1$. Do not pool. Test MS_3 over MS_2 or MS_1 as indicated by anova procedure. Use conventional levels of significance.
   (b) $F_s = MS_2/MS_1$ is not significant at $P \leq \alpha_1$. Pool; that is, test MS_3 over $(SS_2 + SS_1)/(df_2 + df_1)$ at conventional levels of significance.

Some hypothetical examples will illustrate the above rules.

We shall assume the anova tables below to represent Model II nested analyses of variance so that MS_3 should be tested over MS_2. It is in such examples that pooling is most frequently applied when permissible, because df_2 is usually small, and by pooling MS_2 with MS_1 we can increase our degrees of freedom.

Example A

Level	*df*	*SS*	*MS*
3	1	50	50
2	10	250	25
1	20	400	20

BOX 10.2 continued

Assume no prior knowledge of σ_2^2/σ_1^2. Therefore, by 1(b) set $\alpha_1 = 0.75$. $F_s = MS_2/MS_1 = 25/20 = 1.25$ is greater than $F_{.75[10,20]} = 0.656$; hence by 3(a) do not pool and test MS_3 over MS_2. $F_s = 50/25 = 2$, which is less than $F_{.05[1,10]} = 4.96$ and therefore not significant.

Example B

Level	*df*	*SS*	*MS*
3	1	50	50
2	10	100	10
1	20	400	20

Assume no prior knowledge of σ_2^2/σ_1^2. By 1(b) set $\alpha_1 = 0.75$. $F_s = MS_2/MS_1 = 10/20 = 0.5$, which is less than $F_{.75[10,20]} = 0.656$. Therefore by 3(b) one may pool and test MS_3 over $(SS_2 + SS_1)/(df_2 + df_1) = (100 + 400)/(10 + 20) = 16.67$. $F_s = 50/16.67 = 3.00$, which is less than $F_{.05[1,30]} = 4.17$ and therefore not significant. Note that if we had not pooled we would have found $F_s = MS_3/MS_2 = 50/10 = 5.00$, greater than $F_{.05[1,10]} = 4.96$ and therefore significant. Although this test using a pooled error MS is therefore less powerful than the test one would perform without pooling, presumably the pooled error MS is a better estimate of the population variance σ^2.

Example C

Level	*df*	*SS*	*MS*
3	1	90	90
2	10	250	25
1	20	400	20

On the basis of prior knowledge we believe σ_2^2/σ_1^2 to be small. We therefore choose option 1(a) and proceed to decision 2. Since $df_3 = 1$ is not $\geq df_2 = 10$ and $df_1 = 20$ is not $\geq 5 \times df_2 = 5 \times 10 = 50$, we follow 2(b) and set $\alpha_1 = 0.25$. $F_s = 25/20 = 1.25$ is not significant compared with $F_{.25[10,20]} = 1.40$. Therefore by decision 3(b) we test MS_3 over the pooled error $MS = (250 + 400)/(10 + 20) = 650/30 = 21.67$. $F_s = MS_3/MS_{\text{pooled}} = 100/21.67 = 4.615 > F_{.05[1,30]} = 4.17$ and therefore significant. If we had not pooled, we would have found $F_s = 100/25 = 4.0$, less than $F_{.05[1,10]} = 4.96$.

These rules can also be used for pooling in orthogonal analyses of variance (see Chapter 11), in which the treatment mean square has to be tested over the error mean square. In such cases MS_3 is tested over MS_1, whose degrees of freedom might be increased by pooling it with MS_2 (the interaction mean square) if the above rules permit.

TABLE 10.2

Two-level nested anova (mixed model). Eight housefly strains differing significantly in DDT resistance were tested for morphological differences. Three jars of each strain were prepared and all jars were incubated together. Eight females were taken at random from each jar after all the adults had emerged. The character analyzed here is setae number on the third abdominal sternum. To conserve space original variates are not shown. We find only the ΣY for each jar: n (number of items per subgroup) $= 8$; b (number of subgroups per group) $= 3$; a (number of groups) $= 8$; abn (total sample size) $= 192$.

Strain	*Jar*	*Jar sum*	*Strain sum*	*Strain*	*Jar*	*Jar sum*	*Strain sum*
LDD	1	216		LC	1	228	
	2	222			2	215	
	3	213	651		3	216	659
OL	1	267		RH	1	236	
	2	305			2	243	
	3	250	822		3	226	705
NH	1	220		NKS	1	241	
	2	213			2	237	
	3	228	661		3	254	732
RKS	1	254		BS	1	223	
	2	254			2	205	
	3	282	790		3	220	648

Preliminary computations

1. $\sum^{a}\sum^{b}\sum^{n} Y = 5668$

2. $\sum^{a}\sum^{b}\sum^{n} Y^2 = 173{,}668$

3. $\dfrac{\sum^{a}\left(\sum^{b}\sum^{n} Y\right)^2}{n} = 169{,}004.75$

4. $\dfrac{\sum^{a}\sum^{b}\left(\sum^{n} Y\right)^2}{nb} = 168{,}647.50$

5. $CT = \dfrac{(\text{quantity } \mathbf{1})^2}{abn} = 167{,}324.08$

6. SS_{total} = quantity **2** − quantity **5** = 6343.92

7. SS_{groups} = quantity **4** − quantity **5** = 1323.42

8. SS_{subgr} = quantity **3** − quantity **4** = 357.25

9. SS_{within} = quantity **2** − quantity **3** = 4663.25

Anova table

	Source of variation	*df*	*SS*	*MS*	F_s	*Expected MS*
$\bar{Y}_A - \bar{\bar{Y}}$	Among strains (groups)	7	1323.42	189.06	8.47***	$\sigma^2 + n\sigma^2_{B\subset A} + nb\dfrac{\Sigma\alpha^2}{a-1}$
$\bar{Y}_B - \bar{Y}_A$	Among jars within strains (subgroups within groups)	16	357.25	22.33	*ns*	$\sigma^2 + n\sigma^2_{B\subset A}$
$Y - \bar{Y}_B$	Within jars (error)	168	4663.25	27.76		σ^2
$Y - \bar{\bar{Y}}$	Total	191	6343.92			

Source: Data from Sokal and Hunter (1955).

differences among strains from the variances among separate jars. Clearly, if each strain is reared in only one jar, we cannot separate differences among jars from differences among strains. The character reported here is setae number on the third abdominal sternum.

The data in Table 10.2 are presented in abbreviated form. Setae on eight females were counted per jar but only the sum of these eight counts for each jar is given in the table, the individual variates being omitted to conserve space. This is a nested anova because the jars were allocated at random, three to a strain. The quantities necessary for the computation are shown in the table, as is the completed anova. It is obvious from the table that there is no added variance component among the jars, the mean square among jars being less than that within jars. It appears therefore that differences in the microclimate of the jars do not affect setae number in these flies. However, this is not true of other characters. Sokal and Hunter (1955) found that many measurements of lengths of structures, such as wing length, were strongly affected by variation among the jars within strains. In this way we learn about the differential responses of various characters to environmental conditions. Should we pool MS_{subgr} with MS_{within} before testing the significance of MS_{groups}? Consulting Box 10.2, we make no assumption about the ratio of the two variances and set $\alpha_1 = 0.75$. $F_s = MS_{\text{subgr}}/MS_{\text{within}} = 0.804$, which is clearly significant at $\alpha_1 = 0.75$, since even $F_{.75[20,\infty]} = 0.773 < F_s$. Note that to avoid interpolation we use the F-value for the next higher tabled degrees of freedom in place of $F_{.75[16,168]}$, because F-values below 1 increase as their degrees of freedom increase while those above 1 show the reverse relationship. In view of the significance of the variance ratio we do not pool and test $MS_{\text{groups}}/MS_{\text{subgr}}$ at the conventional level of significance. We find the mean square among strains highly significant. Since the highest level of classification is Model I, we would ordinarily complete the analysis of these data with multiple comparisons in the manner of Sections 9.6 or 9.7.

We now take up a three-level nested anova. The example that we have chosen is a mixed model; the highest level of classification is Model I and the two lower levels of classification are randomly chosen subgroups and subsubgroups (Model II). Box 10.3 features data on glycogen content of rat livers. The measurement is in arbitrary units. Duplicate readings were made on each of three preparations of rat livers from each of two rats for three different treatments, as shown in the box. The setup of the data should by now be familiar. Note that we have a new symbol, c, the number of subsubgroups per subgroup, which in this case equals 3. If you were to have yet lower level classifications, you could introduce symbols d, e, ...

The preliminary computations shown in Box 10.3 follow the same pattern as those in Box 10.1. Note that we first find the grand total and the sum of the squared observations and follow this up with quantities **3**, **4**, and **5**, representing the squared subsubgroup totals, squared subgroup totals, and squared group totals each divided by their sample sizes. Quantity **6**, the

BOX 10.3

Three-level nested anova: Mixed Model.

Glycogen content of liver in arbitrary units. Duplicate readings on each of 3 preparations of rat livers from each of 2 rats for each of 3 treatments.

Treatments ($a = 3$)	*Control*						*Compound 217*						*Compound 217 plus sugar*					
Rats ($b = 2$)	*1*			*2*			*1*			*2*			*1*			*2*		
Preparations ($c = 3$)	*1*	*2*	*3*	*1*	*2*	*3*	*1*	*2*	*3*	*1*	*2*	*3*	*1*	*2*	*3*	*1*	*2*	*3*
Readings ($n = 2$)	131	131	136	150	140	160	157	154	147	151	147	162	134	138	135	138	139	134
	130	125	142	148	143	150	145	142	153	155	147	152	125	138	136	140	138	127
Preparation sums	261	256	278	298	283	310	302	296	300	306	294	314	259	276	271	278	277	261
Rat sums		795			891			898			914			806			816	
Treatment sums			1686						1812						1622			
Grand sum									5120									
Treatment means			140.5						151.0						135.2			

Preliminary computations

1. Grand total $= \sum^{a}\sum^{b}\sum^{c}\sum^{n} Y = 5120$

2. Sum of the squared observations $\sum^{a}\sum^{b}\sum^{c}\sum^{n} Y^2 = 131^2 + 130^2 + \cdots + 134^2 + 127^2 = 731{,}508$

3. Sum of the squared subsubgroup (preparation) totals divided by the sample size of the subsubgroups

$$= \frac{\sum^{a}\sum^{b}\sum^{c}\left(\sum^{n} Y\right)^2}{n} = \frac{(261^2 + 256^2 + \cdots + 277^2 + 261^2)}{2} = \frac{1{,}462{,}254}{2} = 731{,}127.0$$

4. Sum of the squared subgroup (rat) totals divided by the sample size of the subgroups

$$= \frac{\sum^{a}\sum^{b}\left(\sum^{c}\sum^{n} Y\right)^2}{cn} = \frac{(795^2 + 891^2 + \cdots + 806^2 + 816^2)}{(3 \times 2)} = \frac{4,383,198}{6} = 730,533.0$$

5. Sum of the squared group (treatment) totals divided by the sample size of the groups

$$= \frac{\sum^{a}\left(\sum^{b}\sum^{c}\sum^{n} Y\right)^2}{bcn} = \frac{(1686^2 + 1812^2 + 1622^2)}{(2 \times 3 \times 2)} = \frac{8,756,824}{12} = 729,735.3$$

6. Grand total squared and divided by the total sample size = correction term

$$CT = \frac{\left(\sum^{a}\sum^{b}\sum^{c}\sum^{n} Y\right)^2}{abcn} = \frac{5120^2}{(3 \times 2 \times 3 \times 2)} = \frac{26,214,400}{36} = 728,177.8$$

7. $SS_{\text{total}} = \sum^{a}\sum^{b}\sum^{c}\sum^{n} Y^2 - CT =$ quantity **2** − quantity **6** = 731,508.0 − 728,177.8 = 3330.2

8. $SS_{\text{groups}} = \frac{\sum^{a}\left(\sum^{b}\sum^{c}\sum^{n} Y\right)^2}{bcn} - CT =$ quantity **5** − quantity **6** = 729,735.3 − 728,177.8 = 1557.5

9. SS_{subgr} (subgroups within groups) $= \frac{\sum^{a}\sum^{b}\left(\sum^{c}\sum^{n} Y\right)^2}{cn} - \frac{\sum^{a}\left(\sum^{b}\sum^{c}\sum^{n} Y\right)^2}{bcn}$

= quantity **4** − quantity **5** = 730,533.0 − 729,735.3 = 797.7

10. SS_{subsubgr} (subsubgroups within subgroups) $= \frac{\sum^{a}\sum^{b}\sum^{c}\left(\sum^{n} Y\right)^2}{n} - \frac{\sum^{a}\sum^{b}\left(\sum^{c}\sum^{n} Y\right)^2}{cn}$

= quantity **3** − quantity **4** = 731,127.0 − 730,533.0 = 594.0

11. SS_{within} (within subsubgroups; error) $= \sum^{a}\sum^{b}\sum^{c}\sum^{n} Y^2 - \frac{\sum^{a}\sum^{b}\sum^{c}\left(\sum^{n} Y\right)^2}{n}$

= quantity **2** − quantity **3** = 731,508 − 731,127.0 = 381.0

Rules for the flow of these computations for any number of classificatory levels are given in the text, as are rules for working out the degrees of freedom.

BOX 10.3 continued

Anova table

Source of variation	*df*	*SS*	*MS*	F_s	Expected *MS*
$\bar{Y}_A - \bar{\bar{Y}}$ Among groups (treatments)	2	1557.5	778.75	2.93 *ns*	$\sigma^2 + n\sigma^2_{C\subset B} + nc\sigma^2_{B\subset A} + ncb\,\frac{\Sigma\alpha^2}{a-1}$
$\bar{Y}_B - \bar{Y}_A$ Among subgroups within groups (rats within treatments)	3	797.7	265.90	5.37*	$\sigma^2 + n\sigma^2_{C\subset B} + nc\sigma^2_{B\subset A}$
$\bar{Y}_C - \bar{Y}_B$ Among subsubgroups within subgroups (preparations within rats)	12	594.0	49.50	2.34*	$\sigma^2 + n\sigma^2_{C\subset B}$
$Y - \bar{Y}_C$ Within subsubgroups (error; readings within preparations)	18	381.0	21.10		σ^2
$Y - \bar{\bar{Y}}$ Total	35	3330.2			

$F_{.05[12,18]} = 2.34$ $\quad F_{.01[12,18]} = 3.37$ $\quad F_{.05[3,12]} = 3.49$ $\quad F_{.01[3,12]} = 5.95$ $\quad F_{.05[2,3]} = 9.55$

* $= 0.01 < P \leq 0.05$

Estimation of variance components

$s^2_{B\subset A} = (265.90 - 49.50)/6 = 36.1$ (rats within treatments)
$s^2_{C\subset B} = (49.50 - 21.17)/2 = 14.2$ (preparations within rats)
$s^2 = 21.1$ (readings within preparations)

These can also be expressed as percentages.

Since $s^2_{B\subset A} + s^2_{C\subset B} + s^2 = 71.5$
for $s^2_{B\subset A}$ (rats within treatments) $= 100 \times 36.1/71.5 = 50.5\%$
for $s^2_{C\subset B}$ (preparations within rats) $= 100 \times 14.2/71.5 = 19.9\%$
for s^2 (readings within preparations) $= 100 \times 21.17/71.5 = 29.6\%$

correction term, is followed by quantity **7**, SS_{total}, which is always quantity **2**, the sum of the squared observations, minus the correction term. The total sum of squares is followed by the sum of squares of groups, subgroups, subsubgroups, and the error, respectively, in that order. You may wonder how you would be able to construct such a flow chart for computations if yet another level of variation were added. There is a simple relationship evident in obtaining the sums of squares after the correction term has been computed. This relationship is true for the computations in Box 10.1, Table 10.2, and Box 10.3, as well as for other nested anovas with any number of hierarchic levels. First compute the total sum of squares, which is always quantity **2** − CT. Then the highest level sum of squares (SS groups) will be the quantity computed just before the correction term [quantity **5** in Box 10.3] minus the correction term. In general, an SS for a level E within a level D will be the sum of the squared E totals minus the sum of the squared D totals (in each case the sum will, of course, have been divided by the appropriate sample sizes). Thus for $SS_{\text{subgr}} = SS_{B \subset A}$ we compute quantity **4** minus quantity **5**, representing subgroups and groups or levels B and A, respectively. We have already noted the alternation of quantities resulting from these operations:

Groups	**5–6**
Subgroups	**4–5**
Subsubgroups	**3–4**
Within	**2–3**

Go through these operations for Boxes 10.1 and 10.3 to make sure that you have mastered this simple relationship.

The number of degrees of freedom could be provided by formula as in the earlier case of Box 10.1. However, we feel it is more profitable to reason it out for the individual case. There are $a = 3$ treatments, and hence $a - 1 = 2$ degrees of freedom among groups (treatments). There are $b = 2$ rats in each treatment corresponding to $b - 1 = 1$ degree of freedom for rats in each treatment. Since there are $a = 3$ treatments, we have $a(b - 1) = 3$ df for subgroups within groups (rats within treatments). From each rat were made $c = 3$ preparations, yielding $c - 1 = 2$ degrees of freedom for preparations within one rat. Since there were $ab = 6$ rats in the study, we have $ab(c - 1) = 12$ df among subsubgroups within subgroups (preparations within rats). Finally, $n = 2$ readings per preparation yield $n - 1 = 1$ df per preparation and, since there are $abc = 18$ preparations in the study, there are $abc(n - 1) = 18$ df within subsubgroups, corresponding to the error term (readings within preparations). These degrees of freedom, shown in the anova table in Box 10.3, total 35, as they should, there being $abcn = 36$ readings in all.

The anova table is shown in Box 10.3. We proceed to test each mean square over the one immediately beneath it, starting with the subsubgroups. Since the mean squares of subsubgroups and subgroups are significant, the question of pooling does not arise and the treatment mean square is tested over that of subgroups. However, the F-ratio for the latter test is not large enough to be significant. We conclude that though there is significant added variance among preparations within rats and among rats within treatments, we cannot establish significant differences among treatments. Note that this test is based on very few degrees of freedom. We might suspect that there really are differences among the treatments but that we cannot pick them up with so few degrees of freedom. We would be tempted to repeat this experiment using more rats per treatment and might well be able to demonstrate significant differences among the treatments. We shall return to this example in Section 10.4 and examine the efficiency of its design.

10.3 Nested anovas with unequal sample sizes

Regrettably, we do not always have equal sample sizes in an experiment or a study that would lend themselves to a nested anova. The complications arising from unequal replication are sufficiently annoying to urge us to design our work in such a way as to obtain equal sample sizes whenever possible. The basic principles of a nested anova are, of course, the same when sample sizes are unequal. However, we face complications on two counts.

First of all, our symbolism and notation become considerably more cumbersome. If we were to provide the fully correct mathematical symbolism, including subscripts for the variates and subscripts and superscripts for the summation signs, the formulas would look forbiddingly complex for what is essentially a simple computational setup. Let us look at an example of a two-level nested anova with unequal sample sizes to see what the notational and computational problems are. In Box 10.4 we record such a study carried out on blood pH of mice. This is an experiment in which 15 female mice (dams) have been successively mated over a period of time to either two or three males (sires). These sires were different for the 15 dams. Thus, in the present study a total of 37 sires was employed. The litters resulting from each mating were kept separate and the blood pH of female members of these litters was determined. The number of female offspring in these litters ranged from three to five. The data are arranged in the data table in Box 10.4, which shows individual readings as well as litter sums (the sums of the several readings for each litter), litter sizes, dam sums (the sums of the readings for the several litters of any one dam), the sample sizes on which these dam sums are based

BOX 10.4

Two-level nested anova with unequal sample sizes.

Blood pH for 3 to 5 female mice within litters resulting from matings of 2 or 3 different sires to each of 15 dams. The data have been coded by subtracting 7.0 and then multiplying by 100: $a = 15$ dams; $b_i = 2$ or 3 sires per dam where $i = 1, \ldots, a$; $n_{ij} = 3, \ldots, 5$ mice per litter (from one sire) where $i = 1, \ldots, a$, and $j = 1, \ldots, b_i$. This is a Model II anova.

BOX 10.4 continued

Dam number	*Sire number*	*Blood pH readings of individual mice*					*Litter sum*	*Litter size*	*Dam sum*	*Sample size of dam sum*
		1	*2*	*3*	*4*	*5*	$\sum_{k=1}^{n_{ij}} Y_{ijk}$	n_{ij}	$\sum_{j=1}^{b_i} \sum_{k=1}^{n_{ij}} Y_{ijk}$	n_i
1	1	48	48	52	54		202	4		
	2	48	53	43	39		183	4	385	8
2	1	45	43	49	40	40	217	5		
	2	50	45	43	36		174	4	391	9
3	1	40	45	42	48		175	4		
	2	45	33	40	46		164	4		
	3	40	47	40	47	47	221	5	560	13
4	1	38	48	46			132	3		
	2	37	31	45	41		154	4	286	7
5	1	44	51	49	51	52	247	5		
	2	49	49	49	50		197	4		
	3	48	59	59			166	3	610	12
6	1	54	36	36	40		166	4		
	2	44	47	48	48		187	4		
	3	43	52	50	46	39	230	5	583	13
7	1	41	42	36	47		166	4		
	2	47	36	43	38	41	205	5		
	3	53	40	44	40	45	222	5	593	14
8	1	52	53	48			153	3		
	2	40	48	50	40	51	229	5	382	8
9	1	40	34	37	45		156	4		
	2	42	37	46	40		165	4	321	8
10	1	39	31	30	41	48	189	5		
	2	50	44	40	45		179	4	368	9
11	1	52	54	52	56	53	267	5		
	2	56	39	52	49	48	244	5	511	10
12	1	50	45	43	44	49	231	5		
	2	52	43	38	33		166	4	397	9
13	1	39	37	33	43	42	194	5		
	2	43	38	44			125	3		
	3	46	44	37	54		181	4	500	12
14	1	50	53	51	43		197	4		
	2	44	45	39	52		180	4		
	3	42	48	45	51	48	234	5	611	13
15	1	47	49	45	43	42	226	5		
	2	45	42	52	51	32	222	5		
	3	51	51	53	45	51	251	5	699	15
Grand sum $\sum_{i=1}^{a} \sum_{j=1}^{b_i} \sum_{k=1}^{n_{ij}} Y_{ijk}$							7197			
Total sample size $\sum_{i=1}^{a} \sum_{j=1}^{b_i} n_{ij}$										160

The structure of the analysis is within litters, among litters due to different sires (sires within dams), and among dams.

BOX 10.4 continued

Preliminary computations

1. Grand total $= \sum^{a}\sum^{b_i}\sum^{n_{ij}} Y = 48 + 48 + \cdots + 45 + 51 = 7197$

2. Sum of the squared observations $= \sum^{a}\sum^{b_i}\sum^{n_{ij}} Y^2$

$$= 48^2 + 48^2 + \cdots + 45^2 + 51^2 = 329{,}353$$

3. Sum of squared subgroup (litter) totals, each divided by its sample size

$$= \sum^{a}\sum^{b_i}\frac{\left(\sum^{n_{ij}} Y\right)^2}{n_{ij}} = \frac{202^2}{4} + \frac{183^2}{4} + \cdots + \frac{222^2}{5} + \frac{251^2}{5} = 326{,}310.47$$

4. Sum of the squared group (dam) totals, each divided by its sample size

$$= \sum^{a}\frac{\left(\sum^{b_i}\sum^{n_{ij}} Y\right)^2}{\sum^{b_i} n_{ij}} = \frac{385^2}{8} + \frac{391^2}{9} + \cdots + \frac{611^2}{13} + \frac{699^2}{15} = 325{,}510.23$$

5. Grand total squared and divided by total sample size = correction term

$$CT = \frac{\left(\sum^{a}\sum^{b_i}\sum^{n_{ij}} Y\right)^2}{\sum^{a}\sum^{b_i} n_{ij}} = \frac{7197^2}{160} = 323{,}730.06$$

6. $SS_{\text{total}} = \sum^{a}\sum^{b_i}\sum^{n_{ij}} Y^2 - CT$

$$= \text{quantity } \mathbf{2} - \text{quantity } \mathbf{5} = 329{,}353.00 - 323{,}730.06 = 5622.94$$

7. $SS_{\text{groups}} = \sum^{a}\frac{\left(\sum^{b_i}\sum^{n_{ij}} Y\right)^2}{\sum^{b_i} n_{ij}} - CT$

$$= \text{quantity } \mathbf{4} - \text{quantity } \mathbf{5} = 325{,}510.22 - 323{,}730.06 = 1780.17$$

8. SS_{subgr} (subgroups within groups) $= \sum^{a}\sum^{b_i}\frac{\left(\sum^{n_{ij}} Y\right)^2}{n_{ij}} - \sum^{a}\frac{\left(\sum^{b_i}\sum^{n_{ij}} Y\right)^2}{\sum^{b_i} n_{ij}}$

$$= \text{quantity } \mathbf{3} - \text{quantity } \mathbf{4} = 326{,}310.47 - 325{,}510.22 = 800.24$$

9. SS_{within} (within subgroups; error SS) $= \sum^{a}\sum^{b_i}\sum^{n_{ij}} Y^2 - \sum^{a}\sum^{b_i}\frac{\left(\sum^{n_{ij}} Y\right)^2}{n_{ij}}$

$$= \text{quantity } \mathbf{2} - \text{quantity } \mathbf{3} = 329{,}353.00 - 326{,}310.47 = 3042.53$$

BOX 10.4 continued

Anova table

Source of variation	*df*	*SS*	*MS*	F_s	*Expected MS*
$\bar{Y}_A - \bar{\bar{Y}}$ Among groups (dams)	14	1780.16	127.17	3.470[a]	$\sigma^2 + n_0'\sigma^2_{B\subset A} + (nb)_0\sigma^2_A$
$\bar{Y}_B - \bar{Y}_A$ Among subgroups within groups (sires within dams)	22	800.24	36.37	1.470 *ns*	$\sigma^2 + n_0\sigma^2_{B\subset A}$
$Y - \bar{Y}_B$ Within subgroups (error; among mice of one litter—that is, from one sire)	123	3042.53	24.74		σ^2
$Y - \bar{\bar{Y}}$ Total	159	5622.94			

$F_{.05[24,\infty]} = 1.52$, therefore $F_s = MS_{\text{subgr}}/MS_{\text{within}}$ with 22 and 123 *df* cannot be significant. By the rules of Box 10.2, we do not pool these two mean squares, but test MS_{groups} over MS_{subgr}.

[a] Since the coefficient n_0 is not equal to n_0' there is no MS over which we can test MS_{groups} directly. See below how this value of F_s has been obtained.

The coefficients of the variance components in a two-level nested anova with unequal sample sizes are calculated as follows.

Preliminary computations

1. $\sum^a \sum^{b_i} n_{ij} = 4 + 4 + \cdots + 5 + 5 = 160$

2. $\sum^a \sum^{b_i} n_{ij}^2 = 4^2 + 4^2 + \cdots + 5^2 + 5^2 = 708$

3. $\sum^a \left(\sum^{b_i} n_{ij}\right)^2 = 8^2 + 9^2 + \cdots + 13^2 + 15^2 = 1800$

4. $\sum^a \left(\dfrac{\sum^{b_i} n_{ij}^2}{\sum^{b_i} n_{ij}}\right) = \dfrac{4^2 + 4^2}{8} + \dfrac{5^2 + 4^2}{9} + \cdots + \dfrac{5^2 + 5^2 + 5^2}{15} = 65.69$

$$n_0' = \frac{\sum^a \left(\dfrac{\sum^{b_i} n_{ij}^2}{\sum^{b_i} n_{ij}}\right) - \dfrac{\sum^a \sum^{b_i} n_{ij}^2}{\sum^a \sum^{b_i} n_{ij}}}{df_{\text{groups}}} = \frac{\text{quantity } \mathbf{4} - [\text{quantity } \mathbf{2}/\text{quantity } \mathbf{1}]}{df_{\text{groups}}}$$

$$= \frac{65.69 - (708/160)}{14} = 4.376$$

$$n_0 = \frac{\sum^a \sum^{b_i} n_{ij} - \sum^a \left(\dfrac{\sum^{b_i} n_{ij}^2}{\sum^{b_i} n_{ij}}\right)}{df_{\text{subgr}}} = \frac{\text{quantity } \mathbf{1} - \text{quantity } \mathbf{4}}{df_{\text{subgr}}}$$

$$= \frac{160 - 65.69}{22} = 4.287$$

BOX 10.4 continued

$$(nb)_0 = \frac{\sum^{a}\sum^{b_i} n_{ij} - \dfrac{\sum^{a}\left(\sum^{b_i} n_{ij}\right)^2}{\sum^{a}\sum^{b_i} n_{ij}}}{df_{\text{groups}}} = \frac{\text{quantity } \mathbf{1} - [\text{quantity } \mathbf{3}/\text{quantity } \mathbf{1}]}{df_{\text{groups}}}$$

$$= \frac{160 - (1800/160)}{14} = 10.625$$

Solving for estimates of the 3 unknown variance components:

$$MS_{\text{groups}} = 127.16 = s^2 + 4.376s^2_{B\subset A} + 10.625s^2_A$$

$$MS_{\text{subgr}} = 36.37 = s^2 + 4.287s^2_{B\subset A}$$

$$MS_{\text{within}} = 24.74 = s^2$$

Therefore

$$s^2 = MS_{\text{within}} = 24.74$$

$$s^2_{B\subset A} = \frac{MS_{\text{subgr}} - MS_{\text{within}}}{4.287} = \frac{36.37 - 24.74}{4.287} = 2.71$$

$$s^2_A = \frac{MS_{\text{group}} - MS_{\text{within}} - 4.376s^2_{B\subset A}}{10.625}$$

$$= \frac{127.16 - 24.74 - (4.376)(2.71)}{10.625} = 8.52$$

For an approximate test of significance we synthesize a new denominator mean square as $MS'_{\text{subgr}} = s^2 + n'_0 s^2_{B\subset A}$.

$$F_s = \frac{MS_{\text{groups}}}{MS'_{\text{subgr}}} = \frac{s^2 + 4.376s^2_{B\subset A} + 10.625s^2}{s^2 + 4.376s^2_{B\subset A}}$$

$$= \frac{24.74 + 4.376(2.71) + 10.625(8.52)}{24.74 + 4.376(2.71)}$$

$$= \frac{127.124}{36.599} = 3.470$$

Next we need to find a critical value of F against which to test F_s. For MS_{groups} in the numerator the degrees of freedom are unchanged at 14. However, the denominator is a new mean square with new degrees of freedom df'_{subgr}. To compute these we first express MS'_{subgr} in terms of the original mean squares.

$$MS'_{\text{subgr}} = s^2 + 4.376s^2_{B\subset A} = MS_{\text{within}} + 4.376\,\frac{(MS_{\text{subgr}} - MS_{\text{within}})}{4.287}$$

$$= MS_{\text{within}} + \frac{4.376}{4.287}MS_{\text{subgr}} - \frac{4.376}{4.287}MS_{\text{within}}$$

$$= \left(1 - \frac{4.376}{4.287}\right)MS_{\text{within}} + \frac{4.376}{4.287}MS_{\text{subgr}}$$

$$= -0.021MS_{\text{within}} + 1.021MS_{\text{subgr}}$$

$$= -0.021(24.74) + 1.021(36.37)$$

$$= 36.61$$

This checks within rounding error with the earlier estimate 36.599.

BOX 10.4 continued

A general formula for the df' of a reconstituted mean square MS' is

$$df' = \frac{(MS')^2}{\sum_i (w_i MS_i)^2/df_i}$$

where w_i are the coefficients of the mean squares (such as -0.021 and 1.021 above), df_i are the original degrees of freedom of mean squares MS_i, and $\sum_i$ indicates summation over all constituent mean squares MS_i (2 in this case).

$$df'_{\text{subgr}} = \frac{(36.61)^2}{[(-0.021 \times 24.74)^2/123] + [(1.021 \times 36.37)^2/22]}$$

$$= \frac{1340.2921}{0.2699/123 + 1379.6753/22}$$

$$= \frac{1340.2921}{0.0022 + 62.6780} = \frac{1340.2921}{62.6802}$$

$$= 21.38$$

$$\approx 21$$

Since $F_s = 3.470 > F_{.01[12,21]} = 3.17$ (conservative critical value employed to avoid interpolation for $F_{.01[14,21]}$ in Table **S**) there is a significant variance component among dams, but as we have seen there is no added variance among sires.

(the sum of litter sizes), and finally the grand sum of all readings as well as the total sample size of 160 readings. There are $a = 15$ groups (dams) in this study, each of them containing b_i subgroups (sires), where the i refers to the dam number. Thus b_7 is the number of sires of dam number 7, and this equals 3. The number of replicates (individual mice in a litter) must be symbolized with two subscripts, n_{ij}, where i refers to the dam number and j to the sire number of dam i. Thus $n_{7,2}$ is the sample size of the number of female offspring from the mating of dam 7 and sire 2, which equals 5. To express symbolically how many sires there are in the study we have to write $\sum^a b_i = 37$, and to find out how many mice were measured in the entire study we have to sum the number of mice in each litter for all sires and over all dams as follows $\sum^a \sum^{b_i} n_{ij} = 160$.

The computations are straightforward and are given in Box 10.4. The degrees of freedom can be worked out directly from the table. There are 15 dams, and hence 14 degrees of freedom among groups; there are 37 sires within dams, which lose one degree of freedom per dam, and therefore their $df = 37 - 15 = 22$. The within subgroups degrees of freedom are based on 160 mice belonging to 37 litters, and hence losing 37 degrees of freedom, which yields 123 df. By formula the degrees of freedom could be obtained as $a - 1$, $\sum^a b_i - a$, and $\sum^a \sum^{b_i} n_{ij} - \sum^a b_i$, respectively.

The first significance test is for the MS_{subgr}. We note that $F_s = 1.470 < F_{.05[24,\infty]} = 1.52$; therefore it must be $< F_{.05[22,123]}$, the correct critical value. Should we pool MS_{subgr} with MS_{within}? No assumption is made about the variance ratio (Box 10.2), and $F_s = 1.470 \gg F_{.75[24,\infty]} = 0.793$, the approximate critical value. We do not pool and plan to test MS_{groups} over MS_{subgr}.

We now encounter the second difficulty arising from unequal sample sizes. Not only must average sample sizes be computed (similar to the special formula learned in Box 9.2), but a glance at the expected mean squares of the anova table in Box 10.4 shows that the average coefficients of the same variance components do not correspond at different levels. We notice that $\sigma^2_{B\subset A}$ has n_o as a coefficient in the MS_{subgr}, but has another quantity n'_o as a coefficient in the MS_{groups}. The formulas for these coefficients are shown next in Box 10.4. In this example, the coefficients have similar values, being 4.287 for n_o and 4.376 for n'_o, but in other examples they might differ more. We next estimate the three unknown variance components, within subgroups, among subgroups within groups, and among groups by simple substitutions as shown in Box 10.4. The test of significance (usually referred to as Satterthwaite's approximation) can be carried out by synthesizing a new denominator mean square against which to test the mean square of groups. This denominator mean square has $s^2_{B\subset A}$ multiplied by $n'_o = 4.376$ instead of $n_o = 4.287$ as before. We now obtain a variance ratio $F_s = 3.470$ whose significance must be evaluated against a critical value of F. However, since our denominator mean square is an approximation, we have to calculate a corresponding approximate number of degrees of freedom. This is done by rewriting our new mean square in terms of the previous mean squares as shown in Box 10.4. From the coefficients of the old mean squares we calculate the new degrees of freedom df', which turn out to be 21.38. This is clearly only an approximation, because degrees of freedom should be integers. To be conservative we use 21 degrees of freedom for the denominator mean square and compare $F_s = 3.470$ with $F_{.01[14,21]}$. In Table **S** we find $F_{.01[12,21]} = 3.17$, showing the added variance component among dams to be highly significant. In this case the fairly elaborate computations to test the significance of the variance among dams seem unnecessary. Had we simply taken the variance ratios of the mean squares in the anova table we would have obtained a value of $F_s = 3.50$, which compared with $F_{[14,22]}$ is equally significant. However, as mentioned earlier, not always will the coefficients be as similar as in this example. The approximate method described here is necessary since the other test, comparing unequal quantities, is clearly not legitimate.

The biological interpretation of this example is interesting. Since sires within dams did not show an added variance component, there does not seem to be significant differentiation depending on the sires. Either the sires came from an inbred strain and were identical genetically, or the variance among dams, but not among sires, can be explained through maternal effects. In this way the analysis of variance can be of great help in pointing out further research in problems of genetics.

The next example is a three-level nested anova with unequal sample sizes. This is an analysis carried out not for a planned experiment but to analyze a series of data that had accumulated in the laboratory. The data are diameters of pollen grains of a species of tree that had been deposited in deep layers at the margin of a marsh. Samples of this pollen had been obtained and analyzed over a period of time. The variable measured was mean diameter of 100 pollen grains found on one slide. Thus the basic variate is a mean, but this need not concern us here. It was possible to group the various samples into three depths, which are considered the highest level category in this study and clearly are fixed treatment effects and hence Model I. For the first two depths two core samples each had been obtained, but three were available for the third depth. Varying numbers of preparations had been made from these core samples and varying numbers of slides examined from these preparations. The structure of the analysis is shown in the table in Box 10.5; to conserve space the data themselves are not given. The core samples within depths and the preparations within these samples can be considered to have been obtained at random and for this reason the analysis becomes a mixed model nested anova.

The actual computations are not shown in Box 10.5, but given the data

BOX 10.5

Three-level nested anova with unequal sample sizes.

Diameter of pollen grains of a species of tree from a paleobotanical study. All core samples were collected from the same locality at the margin of a marsh. Several core samples were obtained at each of three depths; varying numbers of preparations were made from each core sample and varying numbers of slides examined from each preparation. Variates are means of diameters of 100 pollen grains on a slide. Actual data are not shown here. This is a Mixed Model anova.

The structure of the analysis and the sample sizes.

Depths (a)	*1*				*2*					*3*					
Core samples (b_i)	*1*		*2*		*1*			*2*		*1*		*2*		*3*
Preparations (c_{ij})	*1*	*2*	*1*	*2*	*1*	*2*	*3*	*1*	*2*	*1*	*2*	*1*	*2*	*1*
Number of slides per preparation (n_{ijk})	4	5	4	4	5	4	4	5	5	3	4	2	2	5
Number of slides in core sample (n_{ij})	9		8		13			10		7		4		5
Number of slides per depth (n_i)	17				23					16				
Total sample size (n)					56									

BOX 10.5 continued

From the original data, the various sums of squares were computed following the procedures in Boxes 10.3 and 10.4. The results of these computations are given below.

Anova table

Source of variation	*df*	*SS*	*MS*	F_s
$\bar{Y}_A - \bar{\bar{Y}}$ Among groups (depths)	2	396.64	198.32	1.131[a] *ns*
$\bar{Y}_B - \bar{Y}_A$ Among subgroups within groups (core samples within depths)	4	640.44	160.11	4.366[a]*
$\bar{Y}_C - \bar{Y}_B$ Among subsubgroups within subgroups (preparations within core samples)	7	249.41	35.63	1.712 *ns*
$Y - \bar{Y}_C$ Within subsubgroups (error; within preparations)	42	874.02	20.81	
$Y - \bar{\bar{Y}}$ Total	55	2160.51		

* = $0.01 < P \leq 0.05$

[a] These F_s values are not computed as simple variance ratios of the mean squares in the table, because of unequal coefficients in the expected mean squares. Their computation is explained below.

As in Box 10.4, only the next to the lowest *MS* (MS_{subsubgr}) may be tested exactly. [In the present case it is not significant and, since we have no prior knowledge about the magnitude of the ratio of the variance components, we do not pool because $F_s = 1.712$ is greater than $F_{.75[7,60]} = 0.604$ which is a conservative approximation in Table **S** for $F_{.75[7,42]}$ (see Box 10.2).]

For the approximate tests of significance of the other mean squares we must calculate the coefficients of the variance components, estimate the components, and then synthesize new mean squares to use as the denominators of the F_s ratios.

The expected mean squares

Source of variation	*Expected MS for a Model II*
Among groups	$\sigma^2 + n_0''\sigma^2_{C\subset B} + (nc)_0'\sigma^2_{B\subset A} + (ncb)_0\sigma^2_A$
Subgroups within groups	$\sigma^2 + n_0'\sigma^2_{C\subset B} + (nc)_0\sigma^2_{B\subset A}$
Subsubgroups within subgroups	$\sigma^2 + n_0\sigma^2_{C\subset B}$
Within subsubgroups (error)	σ^2

In our case we have a mixed model. The highest level is differentiated by fixed treatment effects. The top line of the expected mean squares (MS_{groups}) should therefore read $\sigma^2 + n_0''\sigma^2_{C\subset B} + (nc)_0'\sigma^2_{B\subset A} + (ncb)_0 \dfrac{\sum \alpha^2}{a-1}$.

The basic quantities needed to compute the coefficients are the following.

1. $\sum^{a}\sum^{b_i}\sum^{c_{ij}} n_{ijk}^2 = 4^2 + 5^2 + \cdots + 2^2 + 5^2 = 238$

BOX 10.5 continued

2. $\sum^{a}\sum^{b_i} n_{ij}^2 = 9^2 + 8^2 + \cdots + 4^2 + 5^2 = 504$

3. $\sum^{a} n_i^2 = 17^2 + 23^2 + 16^2 = 1074$

4. $\sum^{a}\sum^{b_i}\sum^{c_{ij}} n_{ijk} = \sum^{a}\sum^{b_i} n_{ij} = \sum^{a} n_i = 56$

5. $$\sum^{a}\sum^{b_i}\left(\frac{\sum^{c_{ij}} n_{ijk}^2}{\sum^{c_{ij}} n_{ijk}}\right) = \frac{4^2+5^2}{9} + \frac{4^2+4^2}{8} + \cdots + \frac{2^2+2^2}{4} + \frac{5^2}{5} = 28.512$$

6. $$\sum^{a}\left(\frac{\sum^{b_i}\sum^{c_{ij}} n_{ijk}^2}{\sum^{b_i}\sum^{c_{ij}} n_{ijk}}\right)$$
$$= \frac{4^2+5^2+4^2+4^2}{17} + \cdots + \frac{3^2+4^2+2^2+2^2+5^2}{16} = 12.571$$

7. $$\sum^{a}\left(\frac{\sum^{b_i} n_{ij}^2}{\sum^{b_i} n_{ij}}\right) = \frac{9^2+8^2}{17} + \frac{13^2+10^2}{23} + \frac{7^2+4^2+5^2}{16} = 25.850$$

$$n_0'' = \frac{\sum^{a}\left(\frac{\sum^{b_i}\sum^{c_{ij}} n_{ijk}^2}{\sum^{b_i}\sum^{c_{ij}} n_{ijk}}\right) - \left(\frac{\sum^{a}\sum^{b_i}\sum^{c_{ij}} n_{ijk}^2}{\sum^{a}\sum^{b_i}\sum^{c_{ij}} n_{ijk}}\right)}{df_{\text{groups}}}$$
$$= \frac{\text{quantity } \mathbf{6} - [\text{quantity } \mathbf{1}/\text{quantity } \mathbf{4}]}{df_{\text{groups}}} = \frac{12.571 - (238/56)}{2} = 4.160$$

$$n_0' = \frac{\sum^{a}\sum^{b_i}\left(\frac{\sum^{c_{ij}} n_{ijk}^2}{\sum^{c_{ij}} n_{ijk}}\right) - \sum^{a}\left(\frac{\sum^{b_i}\sum^{c_{ij}} n_{ijk}^2}{\sum^{b_i}\sum^{c_{ij}} n_{ijk}}\right)}{df_{\text{subgr}}} = \frac{\text{quantity } \mathbf{5} - \text{quantity } \mathbf{6}}{df_{\text{subgr}}}$$
$$= \frac{(28.512 - 12.571)}{4} = 3.985$$

$$n_0 = \frac{\sum^{a}\sum^{b_i}\sum^{c_{ij}} n_{ijk} - \sum^{a}\sum^{b_i}\left(\frac{\sum^{c_{ij}} n_{ijk}^2}{\sum^{c_{ij}} n_{ijk}}\right)}{df_{\text{subsubgr}}} = \frac{\text{quantity } \mathbf{4} - \text{quantity } \mathbf{5}}{df_{\text{subsubgr}}}$$
$$= \frac{(56 - 28.512)}{7} = 3.927$$

$$(nc)_0' = \frac{\sum^{a}\left(\frac{\sum^{b_i} n_{ij}^2}{\sum^{b_i} n_{ij}}\right) - \left(\frac{\sum^{a}\sum^{b_i} n_{ji}^2}{\sum^{a}\sum^{b_i} n_{ij}}\right)}{df_{\text{groups}}}$$
$$= \frac{\text{quantity } \mathbf{7} - [\text{quantity } \mathbf{2}/\text{quantity } \mathbf{4}]}{df_{\text{groups}}} = \frac{25.850 - (504/56)}{2} = 8.425$$

BOX 10.5 continued

$$(nc)_0 = \frac{\sum^{a}\sum^{b_i} n_{ij} - \sum^{a}\left(\frac{\sum^{b_i} n_{ij}^2}{\sum^{b_i} n_{ij}}\right)}{df_{\text{subgr}}} = \frac{\text{quantity } \mathbf{4} - \text{quantity } \mathbf{7}}{df_{\text{subgr}}}$$

$$= \frac{(56 - 25.850)}{4} = 7.538$$

$$(ncb)_0 = \frac{\sum^{a} n_i - \left(\frac{\sum^{a} n_i^2}{\sum^{a} n_i}\right)}{df_{\text{groups}}} = \frac{\{\text{quantity } \mathbf{4} - [\text{quantity } \mathbf{3}/\text{quantity } \mathbf{4}]\}}{df_{\text{groups}}}$$

$$= \frac{56 - (1074/56)}{2} = 18.411$$

In order to solve for estimates of the 4 unknown variance components, we must solve the 4 simultaneous equations in which the 4 variances are the unknowns.

$$MS_{\text{groups}} = 198.32 = s^2 + 4.160s^2_{C\subset B} + 8.425s^2_{B\subset A} + 18.411s^2_A$$

$$MS_{\text{subgr}} = 160.11 = s^2 + 3.985s^2_{C\subset B} + 7.538s^2_{B\subset A}$$

$$MS_{\text{subsubgr}} = 35.63 = s^2 + 3.927s^2_{C\subset B}$$

$$MS_{\text{within}} = 20.81 = s^2$$

Therefore

$$s^2 = MS_{\text{within}} = 20.81$$

$$s^2_{C\subset B} = \frac{MS_{\text{subsubgr}} - MS_{\text{within}}}{3.927} = \frac{35.63 - 20.81}{3.927} = 3.774$$

$$s^2_{B\subset A} = \frac{MS_{\text{subgr}} - MS_{\text{within}} - 3.985s^2_{C\subset B}}{7.538}$$

$$= \frac{160.11 - 20.81 - [(3.985)(3.774)]}{7.538} = 16.485$$

Variance components may also be expressed as percentages.

Since $s^2_{B\subset A} + s^2_{C\subset B} + s^2 = 41.069$

$s^2_{B\subset A}$ (core samples within depths) $= 100 \times 16.485/41.069 = 40.1\%$

$s^2_{C\subset B}$ (preparations within core samples) $= 100 \times 3.774/41.069 = 9.2\%$

s^2 (slides within preparations) $= 100 \times 20.81/41.069 = 50.7\%$

In the present example there is no s^2_A, because the highest level classification is due to fixed treatment effects. However, we show the computation of s^2_A here to provide a complete pattern for all computations in a Model II three-level anova.

$$s^2_A = \frac{MS_{\text{groups}} - MS_{\text{within}} - 4.160s^2_{C\subset B} - 8.425s^2_{B\subset A}}{18.411}$$

$$= \frac{198.32 - 20.81 - [(4.160)(3.774)] - [(8.425)(16.485)]}{18.411} = 1.245$$

BOX 10.5 continued

Tests of significance

MS_{subsubgr} may be tested directly over MS_{within}:

$$F_s = \frac{35.63}{20.81} = 1.712\ ns$$

Following the rules in Box 10.2 we do not pool.

In order to test MS_{subgr} we synthesize a MS'_{subsubgr}, whose expectation is $\sigma^2 + n'_0\sigma^2_{C \subset B}$. We may construct such a mean square using our sample estimates.

$$MS'_{\text{subsubgr}} = s^2 + n'_0 s^2_{C \subset B} = 20.81 + (3.985)(3.774) = 35.849$$

$$F_s = \frac{MS_{\text{subgr}}}{MS'_{\text{subsubgr}}} = \frac{160.11}{35.849} = 4.466$$

To obtain df'_{subsubgr} we express MS'_{subsubgr} in terms of the original mean squares.

$$MS'_{\text{subsubgr}} = s^2 + 3.985 s^2_{C \subset B} = MS_{\text{within}} + 3.985\,\frac{MS_{\text{subsubgr}} - MS_{\text{within}}}{3.927}$$

$$= MS_{\text{within}} + \frac{3.985}{3.927} MS_{\text{subsubgr}} - \frac{3.985}{3.927} MS_{\text{within}}$$

$$= \left(1 - \frac{3.985}{3.927}\right) MS_{\text{within}} + \frac{3.985}{3.927} MS_{\text{subsubgr}}$$

$$= (-0.015) MS_{\text{within}} + 1.015 MS_{\text{subsubgr}}$$

$$= (-0.015)(20.81) + (1.015)(35.63)$$

$= 35.852$ (a close approximation to the more correct value of 35.849)

$$df'_{\text{subsubgr}} = \frac{(MS'_{\text{subsubgr}})^2}{\sum_i [(w_i MS_i)^2/df_i]}$$

$$= \frac{(35.849)^2}{[(-0.015 \times 20.81)^2/42] + [(1.015 \times 35.63)^2/7]}$$

$$= \frac{1285.1508}{186.8405} = 6.88 \approx 7$$

Therefore the critical value for the above variance ratio is $F_{.05[4,7]} = 4.12$, so that the subgroup MS (among core samples within a depth) is significant.

We proceed to test MS_{groups} over a synthesized MS'_{subgr}.

$$MS'_{\text{subgr}} = s^2 + n''_0 s^2_{C \subset B} + (nc)'_0 s^2_{B \subset A}$$

$$= 20.81 + (4.160)(3.774) + (8.425)(16.485) = 175.396$$

$$F_s = \frac{MS_{\text{groups}}}{MS'_{\text{subgr}}} = \frac{198.32}{175.396} = 1.131$$

It is improbable that such a low variance ratio will be significant in view of the few degrees of freedom available, but we shall proceed anyway to compute df'_{subgr} to furnish an outline of the necessary computations.

We first express MS'_{subgr} in terms of the original mean squares.

BOX 10.5 continued

$$MS'_{\text{subgr}} = s^2 + 4.160 s^2_{C \subset B} + 8.425 s^2_{B \subset A}$$

$$= MS_{\text{within}} + 4.160 \frac{MS_{\text{subsubgr}} - MS_{\text{within}}}{3.927} + 8.425 \frac{MS_{\text{subgr}} - MS_{\text{within}} - 3.985 s^2_{C \subset B}}{7.538}$$

$$= MS_{\text{within}} + \frac{4.160}{3.927}(MS_{\text{subsubgr}} - MS_{\text{within}}) + \frac{8.425}{7.538}(MS_{\text{subgr}} - MS_{\text{within}}) - \frac{(8.425)(3.985)(MS_{\text{subsubgr}} - MS_{\text{within}})}{(7.538)(3.927)}$$

$$= MS_{\text{within}} + 1.059(MS_{\text{subsubgr}} - MS_{\text{within}}) + 1.118(MS_{\text{subgr}} - MS_{\text{within}}) - 1.134(MS_{\text{subsubgr}} - MS_{\text{within}})$$

$$= (1 - 1.059 - 1.118 + 1.134) MS_{\text{within}} + (1.059 - 1.134) MS_{\text{subsubgr}} + (1.118) MS_{\text{subgr}}$$

$$= -0.043 MS_{\text{within}} - 0.075 MS_{\text{subsubgr}} + 1.118 MS_{\text{subgr}}$$

$$= (-0.043)(20.81) - (0.075)(35.63) + (1.118)(160.11)$$

$= 175.436$ (a close approximation to the correct value of 175.396)

$$df'_{\text{subgr}} = \frac{(MS'_{\text{subgr}})^2}{\sum_i (w_i MS_i)^2/df_i}$$

$$= \frac{(175.396)^2}{[(-0.043 \times 20.81)^2/42] + [(-0.075 \times 35.63)^2/7] + [(1.118 \times 160.11)^2/4]}$$

$$= \frac{30{,}763.7568}{8011.5559} = 3.840 \approx 4$$

Therefore the critical value of the above variance ratio is $F_{.05[2,4]} = 6.94$ and F_s is clearly nonsignificant.

they could easily be derived by a combination of Boxes 10.3 and 10.4. As a result of these computations the anova table shown in Box 10.5 was obtained. It is the purpose of Box 10.5 to show the analysis of this table because of the complexities of testing significance and estimating variance components in an example with unequal sample sizes. The first significance test is MS of subsubgroups over the error MS, which turns out to be not significant but large enough according to the rules in Box 10.2 to contraindicate pooling. The subsequent estimation of the variance components as well as approximate tests of significance for the subgroup and group mean squares are laid out in detail in Box 10.5. No new procedures are introduced; they simply become more involved because of the extra level of variation. Since this is a mixed model, the highest level is Model I, and we do not estimate a variance component for it but simply test the significance of added treatment effects. The box shows computation of a variance component for this level simply to complete the outline of all necessary computations for other examples that might be pure Model II. Some discrepancy in the several estimates of the synthetic mean squares are found, which are due to unavoidable rounding errors.

The conclusions from the analysis are that there is a significant added variance component among subgroups (core samples), while neither of the other two mean squares is significant. We therefore conclude that the diameter of the pollen grains shows added variance among core samples but not among preparations and does not differ for the depths recorded in this study. Thus there must be some local heterogeneity in this character, and different core samples appear to yield different pollen samples. This is reflected in the percentage of variation attributable to the several levels: 40.1% among core samples, 9.2% among preparations within core samples, and 50.7% among slides within preparations. The preparations seem to represent core samples with high replicability, but the techniques for making slides and measuring the pollen grains on these merit further examination and refinement in view of the high percentage of variation of that level.

↠ The general method of analysis of nested anovas is the same for any number of levels. To enable you to generate coefficients for variance components for cases with unequal samples sizes, other than those discussed in this section, general rules for finding such coefficients are shown in Federer (1955). ↞

The computer program A3.4 given in Appendix A3 can be used to carry out the computations described in Sections 10.2 and 10.3.

10.4 The optimal allocation of resources

In Box 9.11 we learned a method for estimating the sample size necessary to detect a given difference between means. In the designs implied by nested analyses of variance this is a more complex problem. Before we can know how many replications to use, we must decide how these replications should be structured. Take the experiment of Box 10.3 as an example. We wish to test for differences among treatments and each treatment is represented by one or more rats. However, how are these rats to be represented? You will recall that three preparations were made from each rat liver and two readings taken on each preparation. Would it have been wiser to take four readings per preparation but only one preparation per rat, or would it have been better to take five preparations from one rat liver but only one reading per preparation? Only after such decisions have been taken can we proceed to solve for required sample size (number of rats per treatment) in the manner of Box 9.11.

We approach the problem of optimal allocation of resources by asking what would be the expected variance of means of various designs. In a Model I single classification analysis of variance (as in Box 9.4) the anova is structured as follows:

Treatments (groups)	$\sigma^2 + n \dfrac{\sum \alpha^2}{a-1}$
Replicates (within groups)	σ^2

The error variance estimate is s^2, and the expected variance of a group mean will be $s_{\bar{Y}}^2 = s^2/n$, which is simply the familiar formula for the variance of the mean.

In a mixed model two-level nested anova (as in Table 10.2) the expected mean squares are structured as below:

Treatments (groups)	$\sigma^2 + n\sigma^2_{B \subset A} + nb \dfrac{\sum \alpha^2}{a-1}$
Subgroups within groups	$\sigma^2 + n\sigma^2_{B \subset A}$
Replicates within subgroups	σ^2

and the variance of group means can be estimated as MS_{subgr}/nb, which yields

$$s_{\bar{Y}}^2 = \frac{s^2}{nb} + \frac{s^2_{B \subset A}}{b} \tag{10.3}$$

where s^2 and $s^2_{B \subset A}$ are estimates of σ^2 and $\sigma^2_{B \subset A}$, respectively. Note that the error variance is now divided by nb, the total number of variates on which a subgroup mean is based, while the variance component of subgroups within groups is divided only by b, the number of subgroups constituting a mean.

In a mixed model three-level nested anova (as in Box 10.3), expected mean squares have the following structure:

Treatments (groups)	$\sigma^2 + n\sigma^2_{C \subset B} + nc\sigma^2_{B \subset A} + ncb \dfrac{\sum \alpha^2}{a-1}$
Subgroups within groups	$\sigma^2 + n\sigma^2_{C \subset B} + nc\sigma^2_{B \subset A}$
Subsubgroups within subgroups	$\sigma^2 + n\sigma^2_{C \subset B}$
Replicates within subsubgroups	σ^2

with s^2, $s^2_{C \subset B}$, and $s^2_{B \subset A}$ estimating variance components for increasing hierarchic levels. The expected variance of a group mean based on b subgroups, c subsubgroups, and n replicates within subsubgroups is MS_{subgr}/ncb, which yields

$$s_{\bar{Y}}^2 = \frac{s^2}{ncb} + \frac{s^2_{C \subset B}}{cb} + \frac{s^2_{B \subset A}}{b} \tag{10.4}$$

The variances of group means are expectations of the variances among means sampled from identical populations. They do not therefore include added treatment or random effects among groups. Thus we always divide the highest MS beneath the MS_{groups} by the number of replicates on which a group mean is based, yielding $s_{\bar{Y}}^2$ in its familiar form or as shown in Expressions (10.3) and (10.4).

A concrete example will make these formulas more meaningful to you. In a hypothetical pharmacological experiment testing the effects of various drugs, each drug is tested on several dogs. Several replicated readings of the chemical composition of each dog's blood were made. A mixed model two-level nested anova (among treatments, among dogs within treatments, readings within dogs) resulted in the following variance component estimates:

Among dogs $\quad s_{B \subset A}^2 = 25$

Readings within dogs $\quad s^2 = 100$

We shall now contrast two of many possible designs for further research with these drugs. In design (1) we propose to employ $b = 10$ dogs per treatment, making $n = 2$ readings on each dog. The expected variance of treatment means (if no added treatment effects are present) can be obtained from Expression (10.3):

$$s_{\bar{Y}}^2(1) = \frac{100}{2 \times 10} + \frac{25}{10} = 7.50$$

An alternative design (2) might consist of 2 dogs per treatment, but with 9 readings on each dog. Its expected variance would be

$$s_{\bar{Y}}^2(2) = \frac{100}{9 \times 2} + \frac{25}{2} = 18.06$$

Thus design (2) would have a considerably greater variance than design (1) and, all other considerations being equal, would be less desirable because it would be far less sensitive to picking up differences among treatment means than design (1).

To compare two designs we compute the *relative efficiency* (*RE*) of one design with respect to the other. *RE* is a ratio, usually expressed as a percentage, of the variances resulting from the two designs. The design whose variance is in the denominator is the one whose relative efficiency is being evaluated. Thus we measure the relative efficiency of design (2) with respect to design (1) by

$$RE = \frac{s_{\bar{Y}}^2(1)}{s_{\bar{Y}}^2(2)} \times 100 \tag{10.5}$$

In the previous example the relative efficiency of design (1) with respect to design (2) is

$$RE = \frac{100 s_{\bar{Y}}^2(2)}{s_{\bar{Y}}^2(1)} = \frac{100 \times 18.06}{7.50} = 240.8\%$$

The increase in efficiency on changing from design (2) to design (1) is

$$240.8\% - 100\% = 140.8\%$$

A simple meaning of relative efficiency is as follows. If we take the design whose relative efficiency is being measured and divide its replication by RE (expressed as a ratio), the resulting design would be as sensitive as the original design with which it was being compared. Thus in the pharmocological experiment if we take $10/2.408 = 4.153$ dogs per treatment (with 2 readings per dog), we would get a denominator MS for testing the treatment MS approximately equal to that of design (2), 2 dogs with 9 readings per dog. The equality is only approximate because the denominator mean squares would be based on different degrees of freedom which affect the sensitivity of the design. Also, we obviously cannot apportion 4.153 dogs per treatment, but would need to settle for 4 dogs.

However, the relative efficiency of one design with respect to another is not very meaningful unless the relative costs of obtaining the two designs are taken into consideration. Clearly, if one design is twice as efficient as another (that is, has half the other's variance) but at the same time is ten times as expensive to achieve, we would not settle on it. To introduce the idea of cost we write a so-called *cost function*, which tells us the cost of the entire experiment. For a two-level mixed model,

$$C = bc_{B \subset A} + nbc_{R \subset B} \tag{10.6}$$

where C is the cost of the entire experiment, $c_{B \subset A}$ is the cost of one subgroup unit, $c_{R \subset B}$ is the cost of one replicate reading within a subgroup, and b and n have their familiar meanings. Let us assume for the pharmacological experiment above that the cost of one subgroup unit (the cost of buying and feeding one dog) is $c_{B \subset A} = \$15$. Similarly, suppose the cost of one replicate within the subgroup (the price of one blood test on a dog) is $c_{R \subset B} = \$3$; then the total cost of design (1) by Expression (10.6) will be

$$C(1) = (10 \times \$15) + (10 \times 2 \times \$3) = \$210$$

while design (2) costs

$$C(2) = (2 \times \$15) + (2 \times 9 \times \$3) = \$84$$

Thus, although design (1) is more efficient than design (2), it also costs considerably more.

We shall now ask two questions. (a) For a given amount of money, such as \$210, what is the optimal (most efficient) design? (b) What is the least expensive design for obtaining a given variance of a treatment mean?

We need one more formula,

$$n = \sqrt{\frac{c_{B \subset A} s^2}{c_{R \subset B} s^2_{B \subset A}}} \tag{10.7}$$

which yields the optimal number of replicates n per subgroup unit—that is, a number that will result in minimal cost and minimal variance.

From Expression (10.7) we can see that the optimal number of replicates per subgroup is dependent upon the ratio of costs of the two levels as well as upon the ratio of their variance components. If the costs (= effort) of obtaining a replication at either hierarchic level are the same, Expression (10.7) becomes the ratio of the standard deviations corresponding to the variance components. Expression (10.7) may yield $n < 1$. In such a case we set $n = 1$, since obviously we must have at least 1 reading per subgroup. A dog will be of no use to us in the pharmacological experiment unless we take at least one blood sample from it! Cases in which the optimal number of replicates per subgroup is < 1 are clearly those in which variance among subgroups is great and the error variance is relatively small. In our problem,

$$n = \sqrt{\frac{15 \times 100}{3 \times 25}} = \sqrt{\frac{1500}{75}} = \sqrt{20} = 4.472$$

Thus our optimal number of readings per dog is 4, ignoring the decimal fraction. The answers to question (a) and (b) will then be as follows.

(a) Assume that we wish to spend the same amount of money as in design (1), or \$210. What is the optimal design at this cost? We substitute the known values in Expression (10.6) and solve for b:

$$C = bc_{B \subset A} + nbc_{R \subset B}$$

$$210 = (b \times 15) + (4 \times b \times 3) = 27b$$

$$b = 210/27 = 7.778 \approx 8$$

Thus 8 dogs with 4 readings per dog should yield the most efficient way of spending approximately \$210. Note that we said "approximately," for obviously we cannot use a fractional number of dogs or blood samples, which would be necessary to yield an exact cost of \$210. To check our results we calculate the new cost of the design, which is

$$C = (8 \times \$15) + (4 \times 8 \times \$3) = \$216$$

close to the figure of \$210 we aimed at. To prove that this is the most efficient design for the given cost we need calculus, but we can at least show that the new design is more efficient than the old one. We calculate the new variance of a treatment mean based on 8 dogs and 4 blood tests per dog by substituting in Expression (10.3):

$$s_{\bar{Y}}^2 = \frac{100}{8 \times 4} + \frac{25}{8} = 6.25$$

The new variance is smaller than that of design (1) at nearly the same cost. The relative efficiency,

$$RE = \frac{7.50 \times 100}{6.25} = 120\%$$

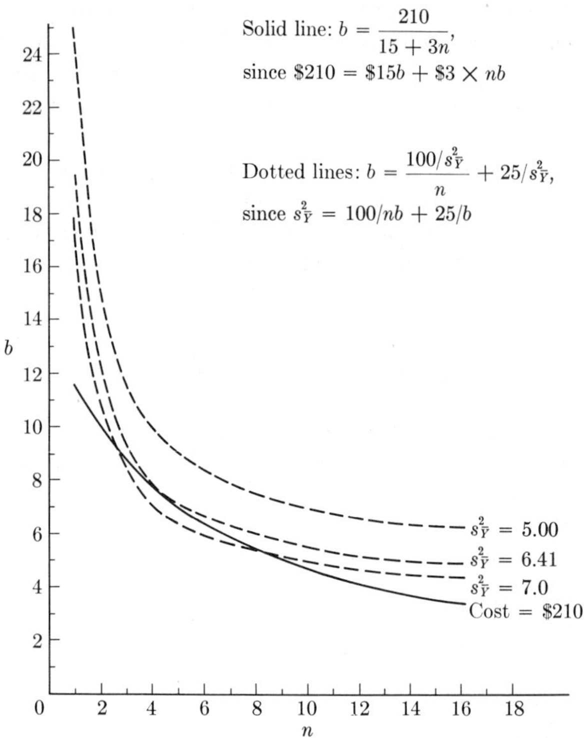

FIGURE 10.1 Equations for cost and estimated variance of a group mean in a two-level anova. (For explanation see text.)

and the increase in efficiency is 20%, but the increase in cost is only 2.9%. Clearly the new design is preferable, even though design (1) was relatively not a bad design to start with.

This process may also be shown graphically. The solid line in Figure 10.1 is a graph of the cost equation \$210 = \$15b + \$3nb and the line represents all combinations of n and b that would yield an experiment costing \$210. The dotted lines represent the equation $s_{\bar{Y}}^2 = 100/nb + 25/b$ for various values of $s_{\bar{Y}}^2$. All intersections between the line for cost = \$210 and equations for various $s_{\bar{Y}}^2$'s give the $s_{\bar{Y}}^2$, n, and b for all possible experiments costing \$210. What is wanted in our case is the one with the smallest $s_{\bar{Y}}^2$. One can see that no line representing an experiment with $s_{\bar{Y}}^2 < 6.41$ can intersect the cost equation, so that $s_{\bar{Y}}^2 = 6.41$ is the best we can expect. These two lines intersect at $n = 4$ and $b = 7.8$ (≈ 8), which is the best combination of n and b for our requirements.

(b) If we wish to achieve a given variance, as, for example, the value of 18.06 resulting from design (2), what is the least expensive way to do so? We substitute the known and given terms in Expression (10.3) and solve for b:

$$s_{\bar{Y}}^2 = \frac{s^2}{nb} + \frac{s_{B\subset A}^2}{b}$$

$$18.06 = \frac{100}{4b} + \frac{25}{b}$$

$$= \frac{100 + (4 \times 25)}{4b}$$

$$b = \frac{100 + 100}{4 \times 18.06} = 2.8 \approx 3$$

Thus 3 dogs with 4 readings per dog will give the most economic design for a treatment mean square approximating 18.06. When we check this result we find

$$s_{\bar{Y}}^2 = \frac{100}{(3 \times 4)} + \frac{25}{3} = 16.67$$

which is a value less than the aimed-at variance (because we used whole numbers of replicates) but the new cost would be

$$C = (3 \times \$15) + (4 \times 3 \times \$3) = \$81$$

Thus a 6% decrease in cost would be accompanied by an 8% increase in relative efficiency.

This result may also be determined graphically by finding the smallest cost equation that intersects the line of the equation for the desired $s_{\bar{Y}}^2$.

Box 10.6 summarizes the computations of optimal allocation of resources

BOX 10.6

Optimal allocation of resources in two and three level sampling. (Outline of procedure. For explanation see Section 10.4.)

Two levels

In a study of variation in the aphid *Pemphigus populi-transversus*, Sokal (1962) measured forewing length of each of 2 alates for each of 15 galls from each of 23 localities in eastern North America: $n = 2$; $b = 15$. The estimates of the variance components were:

Galls within localities	$s_{B\subset A}^2 = 10.507$
Within galls	$s^2 = 6.813$

In this instance the costs of obtaining another gall or another alate in the same gall are roughly equal. Therefore $c_{R\subset B} = c_{B\subset A}$.

Optimum sample size n per subgroup is computed from Expression (10.7):

$$n = \sqrt{\frac{c_{B\subset A}s^2}{c_{R\subset B}s_{B\subset A}^2}} = \sqrt{\frac{6.813}{10.507}} = \sqrt{0.6484} = 0.805$$

Although $n < 1$ we must use at least one alate per gall in further studies.

BOX 10.6 continued

The number of galls to be measured depends on the desired variance of locality means. This will be

$$s_{\bar{Y}}^2 = \frac{6.813}{1 \times b} + \frac{10.507}{b}$$

Thus the variance of a locality mean based on 10 galls would be

$$\frac{6.813}{1 \times 10} + \frac{10.507}{10} = 0.6813 + 1.0507 = 1.7320$$

Three levels

In Box 10.3 we studied the effect of 3 treatments on glycogen content in rat livers. The experiment was structured into b rats within treatments ($s_{B \subset A}^2 = 36.1$), c preparations within rats ($s_{C \subset B}^2 = 14.2$), and n readings within preparations ($s^2 = 21.1$). Let us assume that a rat costs $c_{B \subset A} = \$5$, a preparation $c_{C \subset B} = \$20$, and a reading $c_{\subset CR} = \$2$. Extending Expression (10.6) for two-level samples, the cost C of a design is

$$\mathrm{C} = bc_{B \subset A} + cbc_{C \subset B} + ncbc_{R \subset C}$$

where b, c, and n are the number of replicates at the B, C, and lowest levels, and $c_{B \subset A}$, $c_{C \subset B}$, and $c_{R \subset C}$ are the costs per replicate at these levels, respectively.

Corresponding to Expression (10.7) we have

$$c = \sqrt{\frac{c_{B \subset A} s_{C \subset B}^2}{c_{C \subset B} s_{B \subset A}^2}} \qquad n = \sqrt{\frac{c_{C \subset B} s^2}{c_{R \subset B} s_{C \subset B}^2}}$$

as the optimal number of replicates c and n in the subgroups and subsubgroups, respectively.

Solving for our case,

$$c = \sqrt{\frac{5 \times 14.2}{20 \times 36.1}} = \sqrt{\frac{71.0}{722}} = \sqrt{0.0983} = 0.3136$$

$$n = \sqrt{\frac{20 \times 21.2}{2 \times 14.2}} = \sqrt{\frac{424.0}{28.4}} = \sqrt{14.926} = 3.864$$

Thus employing 4 readings per preparation and one preparation per rat will give us the most efficient design.

If we can spend \$200 per treatment we substitute in the cost formula above and solve for b:

$$200 = (b \times 5) + (1 \times b \times 20) + (4 \times 1 \times b \times 2) = 5b + 20b + 8b$$

$$b = \frac{200}{33} = 6.06$$

We should use 6 rats per treatment, make one preparation from each rat and carry out 4 readings on each preparation.

If we wish to obtain the most efficient design for a given variance, say 22, we proceed as follows. Substituting in Expression (10.4) and solving for b we get

BOX 10.6 continued

$$22 = \frac{21.2}{4 \times 1 \times b} + \frac{14.2}{1 \times b} + \frac{36.1}{b} = \frac{21.2 + [4(14.2 + 36.1)]}{4b}$$

so that

$$b = \frac{222.4}{4 \times 22} = \frac{222.4}{88} = 2.53$$

We decide to take 3 rats per treatment.

Thus our expected variance of a treatment mean based on 3 rats, one preparation, and 4 readings is

$$s^2_{\bar{Y}} = \frac{21.2}{4 \times 1 \times 3} + \frac{14.2}{1 \times 3} + \frac{36.1}{3} = 1.767 + 4.733 + 12.03 = 18.533$$

It is not exactly 22, since $b = 3$ is fairly far off from the solution to the equation $b = 2.53$. Cost for this design would be

$$C = (3 \times 5) + (1 \times 3 \times 20) + (4 \times 1 \times 3 \times 2) = \$99.$$

Let us compare this with the original design in Box 10.3. In Box 10.3 we employed 2 rats per treatment, 3 preparations per rat, and 2 readings per preparation. By Expression (10.4) this yields an expected MS

$$\frac{21.2}{2 \times 3 \times 2} + \frac{14.2}{3 \times 2} + \frac{36.1}{2} = 1.767 + 2.367 + 18.050 = 22.183.$$

When multiplied by $ncb = 12$ we obtain 266.2, which differs from the actually calculated value of 265.9 because of rounding errors. The relative efficiency of the optimal design computed above is $RE = 22.183/18.533 = 119.7\%$. The cost of the design in Box 10.3 is

$$C = (2 \times 5) + (3 \times 2 \times 20) + (2 \times 3 \times 2 \times 2) = \$154$$

55.6% higher than the optimal design. Clearly 1 preparation per rat and 4 readings per preparation are a more efficient allocation of resources for this study.

for two levels and three levels of sampling. The formulas in the latter case are similar to those in the former and should be self-explanatory.

It should be pointed out that whenever the cost of an experiment is of little consideration, but the design of the most efficient experiment (lowest variance) for a given total number of observations is most important, then the most efficient procedure will always be to take a single observation for each of the subgroups; that is, set $n = 1$ in a two-level anova and both n and $c = 1$ in a three-level anova. But, of course, such a study would yield no information on the magnitude of variation at lower levels.

Exercises 10

10.1 J. A. Weir took blood pH readings on mice of two strains that had been selected for high and low blood pH. The following data on male litter mates are extracted from unpublished records. Only litters with at least four males were considered, and four males were selected at random whenever more than four males were present in a litter. Data are presented on seven litters for each strain. Analyze, show whether the pHH (high) strain has a higher average pH value than the pHL (low) strain, and compute variance components and percent variation among males within litters and litters within strains.

Strain	*Litter code number*	*pH readings*			
pHH	387	7.43	7.38	7.49	7.49
	388	7.39	7.46	7.50	7.55
	389	7.53	7.50	7.63	7.47
	401	7.39	7.39	7.44	7.55
	402	7.48	7.43	7.47	7.44
	404	7.43	7.55	7.44	7.50
	405	7.49	7.49	7.51	7.54
pHL	392	7.40	7.46	7.43	7.42
	408	7.35	7.40	7.46	7.38
	413	7.51	7.39	7.42	7.43
	414	7.46	7.53	7.49	7.45
	415	7.48	7.53	7.52	7.43
	434	7.43	7.40	7.48	7.47
	446	7.53	7.47	7.50	7.53

10.2 The following are unpublished results from a study by R. R. Sokal on geographic variation in the aphid *Pemphigus populi-transversus*. Mean length for antennal segment IV of stem mothers was computed for each of 75 localities in Eastern North America. The means were grouped and coded 1 through 8. The data were subdivided according to drainage system along which the localities occur. A hierarchic classification was set up as follows: The drainages were divided into three coasts: East Coast, Gulf Coast, and Great Lakes. Each coast was subdivided into a number of areas, one of which, the Mississippi area, was further subdivided into ten river systems. Analyze the data to determine whether differences in antennal segment IV length occur among the coasts. At what hierarchic level is there the greatest amount of variation? Interpret your findings. This problem might appear to be a Model I anova throughout. However, for our purposes the areas and river systems are considered random samples from the available equivalent sampling units and the emphasis is on the relative amounts of variation at these hierarchic levels. The data are presented in summary form for ease of computation.

Coast	*Area*	n_{ijk}	$\sum^{n_{ijk}} Y_{ijkl}$
East	New England	3	13
	Hudson	2	12
	Chesapeake Bay	2	11
	North Carolina	1	4
	Florida	1	1

Coast	*Area*	*River system*	n_{ijk}	$\sum^{n_{ijk}} Y_{ijkl}$
Gulf	East		4	14
	Central		4	12
	Mississippi	Red	2	6
		Arkansas	7	29
		Lower Mississippi	10	33
		Illinois	6	32
		Missouri	3	15
		Upper Mississippi	6	34
		Lower Ohio	5	13
		Tennessee-Cumberland	3	9
		Wabash	2	9
		Middle Ohio	7	27
Great Lakes			7	39

$$\sum^{a}\sum^{b_i}\sum^{c_{ij}}\sum^{n_{ijk}} Y_{ijkl}^2 = 1533$$

ANS. $MS_{\text{Areas}\subset\text{Coasts}} = 4.0133$; EMS for this MS is $\sigma^2 + 2.533\sigma^2_{\text{R}\subset\text{A}} + 3.544\sigma^2_{\text{A}\subset\text{C}}$; $s^2_{\text{A}\subset\text{C}}$ is negative.

10.3 Hasel (1938) made a study of sampling practices in timber surveys. The variate studied is board feet of timber in a ponderosa pine forest. He studied 9 sections, each subdivided into 4 quarter sections [each of which can be subdivided into 4 forty-acre sections, or 16 ten-acre sections or 32 five-acre sections or 64 basic plots ($2\frac{1}{2}$-acre section)]. If sampling and distribution of timber were entirely at random, no added variance should exist beyond the basic plots.

Interpret the analysis of variance given below and isolate variance components. If it is standard practice among lumbermen to sample 16 basic plots of a forest, how would you (with the limited information at your disposal, and assuming that cost is not important) recommend that the plots be distributed in order to make efficient comparisons in a new experiment to compare differences between forests? Sampling all plots within the same forty-acre section or only 1 basic plot per section are the two extreme attitudes. Compute their efficiencies relative to each other. To keep the notation uniform, identify variance components as follows: s^2 for basic plots, s_5^2 for fives, s_{10}^2 for tens, s_{40}^2 for forties, σ_Q^2 for quarter sections, and s_S^2 for sections.

Source of variation	*df*	*SS*
Between sections	8	109,693.35
Quarter sections within sections	27	57,937.40
Forties within quarter sections	108	66,569.53
Tens within forties	432	161,109.77
Fives within tens	576	121,940.82
Basic plots within fives	1,152	133,337.74
Total	2,303	650,588.61

After trying several intermediate solutions between the extreme sampling procedures, you will have convinced yourself that sampling one basic plot per section is the most efficient (in terms of obtaining the lowest variance). However, this would involve sampling from 16 different sections of land and would require considerable work and travel. Thus, to assume that cost is unimportant is not realistic in this problem, since work and travel must be equated to cost. Design a sampling procedure that would involve less travel, but would still result in a small variance of the mean (< 50.0).

10.4 How could you improve the design of Weir's experiment (Exercise 10.1) using the same number of mice he did? First assume that costs are not important, then rework the problem, assuming that the extra cost of obtaining an additional litter of mice is about three times the cost of measuring an additional mouse. Were your improvements worthwhile? That is, if the difference between the means of the two strains were the same in the new experiment as it was in the previous one, would you now expect to be able to reject the null hypothesis? ANS. No; one would not expect to reject the H_0 in the new experiments.

10.5 Hanna (1953) studied hair pigment concentration (measured with a spectrophotometer and expressed as optical densities) in 39 pairs of monozygous and 40 pairs of dizygous twins. Two hair samples were taken from each person and hydrolyzed separately. Three readings were taken for each sample. Throughout the study samples were randomized to avoid bias in reading of optical densities.

	Monozygous twins		*Dizygous twins*	
Source of variation	*df*	*MS*	*df*	*MS*
Among pairs	38	2676.1734	39	2509.9174
Between twins within pairs	39	44.2005	40	307.0163
Between samples within twins	78	3.2133	80	2.5451
Among readings within samples (error)	320	0.0662	320	0.0334

Complete the analysis of variance and estimate the added variance components. Compare and interpret the differences in the estimated variance components found for monozygous and dizygous twins. ANS. $s^2_{T \subset P}$ equals 6.8312 for the monozygous twins and 50.7425 for the dizygous twins.

11 TWO-WAY ANALYSIS OF VARIANCE

This chapter covers designs in which the effects of two factors are considered simultaneously. Single factors were discussed in Chapter 9. Although Chapter 10 tested subclasses representing a separate source of variation, these were hierarchic, or nested within the major classification. In this chapter the two sources of variation are of equal rank. Thus, for example, we can group the variates by two separate kinds of treatments—different doses of a drug and different temperatures at which these drugs are tested. Further details about this design are given in Section 11.1. The computation of such an anova for replicated subclasses (more than one variate per subclass or factor combination) is shown in Section 11.2, which also contains a discussion of the meaning of interaction as used in statistics. Significance testing in a two-way anova is the subject of Section 11.3. This is followed by a section (11.4) on two-way anova without replication, or with only a single variate per subclass. The well-known method of paired comparisons is treated in Section 11.5 as a special case of a two-way anova without replication. Section 11.6 deals with complications arising from unequal subclass sizes in a two-way anova, while the last section (11.7) presents a method for analyzing the design of Section 11.4 in which values for some subclasses are missing.

11.1 Two-way anova: design

From the single classification anova of Chapter 9 we progress to the two-way anova of the present chapter by a single logical step. Individual items may be grouped into classes representing the different possible combinations of two treatments or factors. Thus the housefly wing lengths studied in earlier chapters, which yielded samples representing different medium formulations, might also be divided into males and females. We would not only like to know whether medium 1 induced a different wing length than medium 2 but also whether male houseflies differed in wing length from females. Obviously each combination of factors should be represented in a sample of flies. Thus for seven media and two sexes we need at least $7 \times 2 = 14$ samples. Similarly, the experiment testing five sugar treatments on pea sections (Box 9.4) might have been carried out at three different temperatures. This would have resulted in a *two-way analysis of variance* of the effects of sugars as well as of temperatures.

It is the assumption of this method of anova that a given temperature and a given sugar each contribute a certain amount to the growth of a pea section and that these two contributions add their effects without influencing each other. In the next section we shall see how departures from the assumption are measured. A consideration of the expression for decomposing variates in a two-way anova is also deferred to Section 11.2.

The two factors in the present design may represent either Model I or Model II effects or one of each, in which case we talk of a *mixed model.*

Beginners are often confused between a two-level nested anova and a two-way anova. Thus, for example, if we subdivide several strains of houseflies into males and females (a two-way anova), in what way does this differ from the previously cited cases of nested anova in which, for example, each strain of houseflies may be represented by two randomly chosen jars? We represent the design of the two-way anova as follows:

Strains of houseflies

	1	2	3	4	• • •	a
Sex 1 = ♀						
Sex 2 = ♂						

The nested anova is represented by the following arrangement:

Strains of houseflies

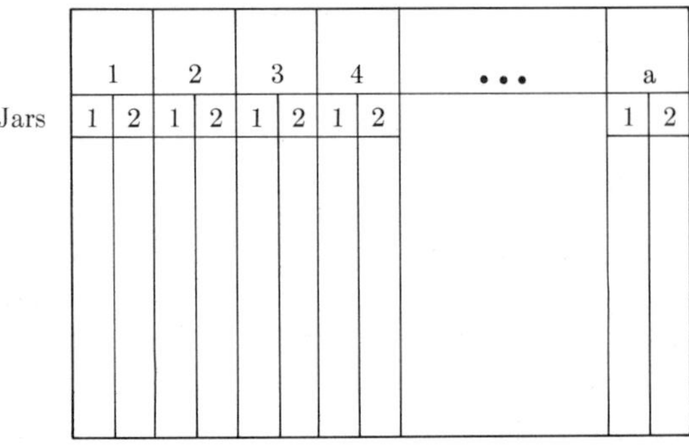

Why can we not rearrange these data into a two-way table with jars at right angles to strains as in the upper table? And why not rearrange sexes nested inside strains as in the lower table? The reason we cannot do this is that the first (two-way) arrangement implies that the two sex classes are *common* to the entire study. Thus maleness is logically the same for all the strains of houseflies and so is femaleness, although their effects may not be the same in all strains for the variable tested. Arranging the sexes nested within each strain would imply that the two sexes per strain were random samples from all possible sexes and that sex 1 in strain 1 is not the same as sex 1 in strain 2, which is, of course, nonsense. Conversely, if we tried to arrange the data from the nested analysis as a two-way anova, we would imply that jars 1 and 2 had the same meaning for all the strains in the experiment. This is not so. Jar 1 in strain 1 has no closer relation to jar 1 in strain 2 than it does to jar 2 in that strain. Jars 1 and 2 are simply arbitrary designations for the two randomly selected jars that represent each strain. By contrast, if all jars 1 contained medium prepared on one day, while all jars labeled 2 were filled with medium prepared on the following day, the "1" and "2" would represent common information for the study that should properly be arranged as a two-way anova. The critical question to be asked is always, "Does the arrangement of the data into a two-way table falsely imply a correspondence across the classes?" If it does, and we recognize that the factor represents only random subdivisions of the classes of another factor (as the jars within strains), we have a nested anova. If there is correspondence across classes, the two-way design is appropriate.

We shall now proceed to illustrate the computation of a two-way anova. You will obtain closer insight into the structure of this design as we explain the computations.

11.2 Two-way anova with replication: computation

We shall learn the computation by means of the simplest possible example—two factors, each divided into two classes. In a study of the inactivation of the effects of Vitamin A by rancid fat it became important to study differences in food consumption when rancid lard was substituted for fresh lard in the diet of rats. Table 11.1 shows the results of an experiment conducted on 12 rats, 6 males and 6 females, three of each sex being fed fresh lard, the other three rancid lard. The following analysis of these data is designed to be as instructive as possible. It is not, however, the most elegant method of carrying out the computational steps; this is illustrated in Box 11.1 and will be considered in due course. Notice some minor changes in symbolism in Table 11.1. We have called the number of rows in the table r and the number of columns c. This will be easy to remember for now, although in Box 11.1 we shall adopt a more general symbolism consistent with that of previous and later chapters.

We commence by computing a single classification anova of the four sub-

TABLE 11.1

Two-way anova with replication. Differences in food consumption when rancid lard was substituted for fresh lard in the diet of rats. Food eaten (in grams) during 73 days by 12 rats aged 30 to 34 days at the start of the experiment. The data are classified in two ways, by fats (fresh *vs* rancid lard) and by sex (male *vs* female).

r sexes ($r = 2$)	*c fats* ($c = 2$) *Fresh*	*Rancid*	Σ
♂	709 679 699	592 538 476	
	2087	1606	3693
♀	657 594 677	508 505 539	
	1928	1552	3480
Σ	4015	3158	7173

Preliminary computations

1. $\sum^{r}\sum^{c}\sum^{n} Y = 7173$
2. $\sum^{r}\sum^{c}\sum^{n} Y^2 = 4{,}365{,}231$
3. $\dfrac{\sum^{r}\sum^{c}(\sum^{n} Y)^2}{n} = \dfrac{(2087^2 + 1606^2 + 1928^2 + 1552^2)}{3} = 4{,}353{,}564.33$
4. $CT = \dfrac{(\text{quantity } \mathbf{1})^2}{abn} = 4{,}287{,}660.75$
5. $SS_{\text{total}} = \text{quantity } \mathbf{2} - \text{quantity } \mathbf{4} = 77{,}570.25$
6. $SS_{\text{subgr}} = \text{quantity } \mathbf{3} - \text{quantity } \mathbf{4} = 65{,}903.58$
7. $SS_{\text{within}} = \text{quantity } \mathbf{5} - \text{quantity } \mathbf{6} = 11{,}666.70$

Preliminary anova

	Source of variation	*df*		*SS*	*MS*
$\bar{Y} - \bar{\bar{Y}}$	Among subgroups	3	$rc - 1$	65,903.58	21,967.85**
$Y - \bar{Y}$	Within subgroups (error)	8	$rc(n - 1)$	11,666.70	1,458.34
$Y - \bar{\bar{Y}}$	Total	11	$rcn - 1$	77,570.25	

The analysis proceeds as follows:

8. $\dfrac{\sum^{r}(\sum^{c}\sum^{n} Y)^2}{cn}$ = row sums squared, summed, and divided by number of items per

row $= \dfrac{(3693^2 + 3480^2)}{6} = 4{,}291{,}441.50$

9. $\dfrac{\sum^{c}(\sum^{r}\sum^{n} Y)^2}{rn}$ = column sums squared, summed, and divided by number of items per

column $= \dfrac{(4015^2 + 3158^2)}{6} = 4{,}348{,}864.83$

TABLE 11.1 *continued*

10. SS_{rows} (SS due to sex) = quantity **8** − quantity **4** = 3780.75
11. SS_{columns} (SS due to fats) = quantity **9** − quantity **4** = 61,204.08
12. $SS_{\text{interaction}}$ (SS due to sex × fats) = quantity **6** − quantity **10** − quantity **11**
= 65,903.58 − 3780.75 − 61,204.08 = 918.75

Completed anova

	Source of variation	*df*		*SS*	*MS*
$\bar{R} - \bar{\bar{Y}}$	Between rows (sexes)	1	$(r - 1)$	3780.75	3780.75 *ns*
$\bar{C} - \bar{\bar{Y}}$	Between columns (fats)	1	$(c - 1)$	61,204.08	61,204.08***
$\bar{Y} - \bar{R} - \bar{C} + \bar{\bar{Y}}$	Interaction (sexes × fats)	1	$(r - 1)(c - 1)$	918.75	918.75 *ns*
$Y - \bar{Y}$	Error	8	$rc(n - 1)$	11,666.70	1458.34
$\bar{Y} - \bar{\bar{Y}}$	Total	11	$rcn - 1$	77,570.25	

Source: Data by Powick (1925).

groups or subclasses, each representing readings on three rats. Such an anova will test whether there is any variation among the four subgroups over and above the variance within the subgroups. For if there were no such added variation it would be unlikely that either sex or freshness of lard would significantly affect food intake. The quantities **1** through **7** are computed in the familiar manner of Box 9.4 although the symbolism changes slightly, since in place of a groups we now have rc subgroups. The preliminary anova shown in Table 11.1 clearly indicates that there is considerable added variation among subgroups, making it likely that we shall find significant effects for at least one of the factors.

The computation is continued by finding the sums of squares for rows and columns of the table. This is done by the general formula stated at the end of Section 9.1. Thus for rows, we square the row sums, sum these squares, and divide them by six, the number of items per row. From this quotient we subtract the correction term, which is the previously computed quantity **4**. The column sum of squares is computed similarly. Since the rows and columns are based on equal sample sizes, we do not have to obtain separate quotients for the square of each row or column sum but carry out a single division after accumulating the squares of the sums.

Let us return for a moment to the preliminary analysis of variance, which divided the total sum of squares into two parts, the sum of squares among the four subgroups and that within the subgroups, the error sum of squares. The new sums of squares pertaining to row and column effects clearly are not part of the error, but must contribute to the differences that comprise the sum of squares among the four subgroups. We therefore subtract row and column SS from the subgroup SS. The latter is 65,903.58. The row

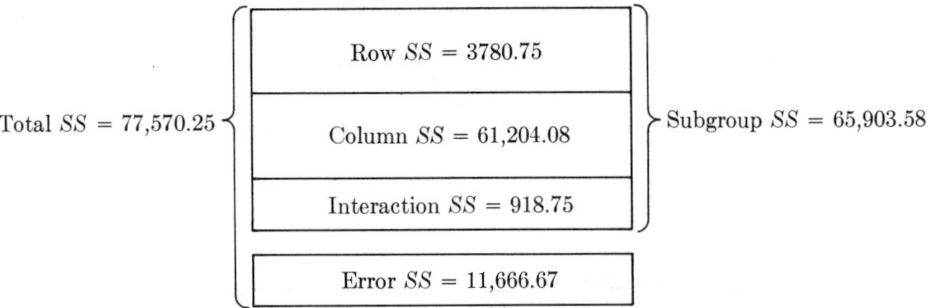

FIGURE 11.1 Diagrammatic representation of the partitioning of the total sums of squares in a two-way orthogonal anova. The areas of the subdivisions are not shown proportional to the magnitudes of the sums of squares.

SS is 3780.75, and the column SS is 61,204.08. Together they add up to 64,984.83, almost but not quite the value of the subgroup sum of squares. The difference represents a third sum of squares, called the *interaction sum of squares,* whose value in this case is 918.75. We shall discuss the meaning of this new sum of squares presently. At the moment let us say only that it is almost always present (but not necessarily significant) and generally need not be independently computed but may be obtained as illustrated above—by the subtraction of the row SS and the column SS from the subgroup SS. This procedure is shown graphically in Figure 11.1, which illustrates the decomposition of the total sum of squares into the subgroup SS and error SS, the former being subdivided into the row SS, column SS, and interaction SS. The relative magnitudes of these sums of squares will differ from experiment to experiment. In Figure 11.1 they are not shown proportional to their actual values in the rat experiment; otherwise the area representing the row SS would have to be about $\frac{1}{16}$th of that allotted to the column SS. Before we can intelligently test for significance in this anova we must understand the meaning of interaction.

We can best explain interaction in a two-way anova by means of an artificial illustration based on the rat data we have just studied. If we interchange the readings for rancid and fresh lard *for the male rats only,* we obtain the data table shown in Table 11.2. Only the sums of the subgroups, rows, and columns are shown. We complete the analysis of variance in the manner presented above and note the results at the foot of Table 11.2. The total, subgroup, and error SS are the same as before (Table 11.1). This should not be surprising, since we are using the same data. All that we have done is to interchange the contents of the upper two cells of the table. When we partition the subgroup SS we do find some differences. We note that the SS between sexes (between rows) is unchanged. Since the change we made was within one row, the total for that row was not altered and consequently the row SS did not change. However, the sums of the columns have been altered appreciably as a result of the interchange of the readings for fresh and rancid

TABLE 11.2

An artificial example to illustrate the meaning of interaction. The readings for rancid and fresh lard of the male rats in Table 11.1 have been interchanged. Only subgroup and marginal totals are given below.

Sexes	*Fats*		Σ
	Fresh	*Rancid*	
♂	1606	2087	3693
♀	1928	1552	3480
Σ	3534	3639	7173

Completed anova

Source of variation	*df*	*SS*	*MS*
Sexes	1	3780.75	3780.75 *ns*
Fats	1	918.75	918.75 *ns*
S × F	1	61,204.05	61,204.05***
Error	8	11,666.70	1458.34
Total	11	77,570.25	

lard in the males. The sum for fresh lard is now very close to that for rancid lard and the difference between the fats, previously quite marked, is now no longer so. By contrast, the interaction *SS*, obtained by subtracting the sums of squares of rows and columns from the subgroup *SS*, is now a large quantity. Remember that the subgroup *SS* is the same in the two examples. In the first example we subtracted sums of squares due to the effects of both fats and sexes, leaving only a tiny residual representing the interaction. In the second example these two *main effects* (fats and sexes) account only for little of the subgroup sum of squares, leaving the interaction sum of squares as a substantial residual. What is the essential difference between these two examples?

In Table 11.3 we have shown the subgroup and marginal means for the original data from Table 11.1 and for the "doctored" data of Table 11.2. The original results are quite clear: males eat slightly more than females and this is true for both fresh and rancid fats. We note further that fresh lard is eaten in considerably greater quantity than rancid lard, and again this is true of both males as well as females. Thus our statements about differences due to sex or to the freshness of the lard can be made independently of each other. However, if we had to interpret the artificial data (lower half of Table 11.3), we would note that although males still eat more fat than females (since row sums have not changed), this difference depends greatly on the nature of the fat. Males eat considerably less fresh lard than the females, but of the rancid fat they eat considerably more. Thus we are no

TABLE 11.3

Comparison of means of the data in Tables 11.1 and 11.2.

	Fats		
Sexes	*Fresh*	*Rancid*	*Mean*
Original data from Table 11.1			
♂	695.67	535.33	615.50
♀	642.67	517.33	580.00
Mean	669.17	526.33	597.75
Artificial data from Table 11.2			
♂	535.33	695.67	615.50
♀	642.67	517.33	580.00
Mean	589.00	606.50	597.75

longer able to make an unequivocal statement about the amount of fat eaten by the two sexes. We have to qualify our statement by the type of lard that is consumed. For fresh fat, $\bar{Y}_{♂} < \bar{Y}_{♀}$, but for rancid fat, $\bar{Y}_{♂} > \bar{Y}_{♀}$. If we examine the effects of fresh versus rancid fats in the artificial example, we notice that slightly more rancid than fresh lard was consumed. However, again we have to qualify this statement by the sex of the consuming rat; the males ate more rancid fat, the females ate more fresh fat.

This dependence of the effect of one factor on the level of another factor is called *interaction*. It is a common and fundamental scientific idea. It indicates that the effects of the two factors are not simply additive but that any given combination of levels of factors such as fresh lard combined with male sex contributes a positive or negative increment to the level of expression of the variable. In common biological terminology a large positive increment of this sort is called *synergism*. When drugs act synergistically, the result of the interaction of the two drugs may be above and beyond the separate effects of each drug. When a combination of levels of two factors inhibit each other's effects, we call it *interference*. Synergism and interference will both tend to magnify the interaction SS.

Testing for interaction is an important procedure in analysis of variance. If the artificial data of Table 11.2 were real, it would be of little value to state that rancid lard was consumed in slightly greater quantities than fresh lard. This statement would cover up the important differences in the data, which are that males eat more rancid fat and females prefer fresh fat.

We are now able to write an expression symbolizing the decomposition of a single variate in a two-way analysis of variance in the manner of Expressions (8.3) and (10.2) for single classification and nested anovas. The expression below assumes that both factors represent fixed treatment effects, Model I. This would seem reasonable, since sex as well as the freshness of the lard are fixed treatments. Variate Y_{ijk} is the kth item in the subgroup representing the ith group of treatment A and the jth group of treatment B. It is decomposed as follows:

$$Y_{ijk} = \mu + \alpha_i + \beta_j + (\alpha\beta)_{ij} + \epsilon_{ijk} \tag{11.1}$$

where μ equals the parametric mean of the population, α_i is the fixed treatment effect for the ith group of treatment A, β_j is the fixed treatment effect of the jth group of treatment B, $(\alpha\beta)_{ij}$ is the interaction effect in the subgroup representing the ith group of factor A and the jth group of factor B, and ϵ_{ijk} is the error term of the kth item in subgroup ij. We make the usual assumption that ϵ_{ijk} is normally distributed with a mean of 0 and a variance of σ^2. If one or both of the factors are Model II, we replace the α_i and β_j in the formula by A_i and/or B_j.

In previous chapters we have seen that each sum of squares represents a sum of squared deviations. What actual deviations does an interaction SS represent? We can see this easily by referring back to the anovas of Table 11.1. The variation among subgroups is represented by $\bar{Y} - \bar{\bar{Y}}$, where $\bar{Y}$ stands for the subgroup mean and $\bar{\bar{Y}}$ for the grand mean. When we subtract the deviations due to rows $\bar{R} - \bar{\bar{Y}}$ and columns $\bar{C} - \bar{\bar{Y}}$ from those of subgroups we obtain

$$\begin{aligned}(\bar{Y} - \bar{\bar{Y}}) - (\bar{R} - \bar{\bar{Y}}) - (\bar{C} - \bar{\bar{Y}}) &= \bar{Y} - \bar{\bar{Y}} - \bar{R} + \bar{\bar{Y}} - \bar{C} + \bar{\bar{Y}} \\ &= \bar{Y} - \bar{R} - \bar{C} + \bar{\bar{Y}}\end{aligned}$$

This somewhat involved expression is the deviation due to interaction. When we evaluate one such expression for each subgroup, square it, sum these squares, and multiply the sum by n, we obtain the interaction SS. This partition of the deviations also holds for their squares. This is so because the sums of the products of the separate terms cancel out.

The actual deviations leading to the sums of squares are shown in Table 11.4. You should calculate some of these deviations to confirm for yourself where they come from. Notice the large deviations due to interaction and their alternating signs, minus in the left upper and right lower quadrants and plus in the right upper and left lower quadrants. A simple method for revealing the nature of the interaction present in the data is to inspect the means of the original data table. We can do this in Table 11.3. The original data, showing no interaction, would yield the following pattern of relative magnitudes:

	Fresh		*Rancid*
♂		>	
♀		>	

The relative magnitudes of the means in the lower table yielding interaction can be summarized as follows:

	Fresh		*Rancid*
♂		$<$	
♀		$>$	

When the pattern of signs expressing relative magnitudes is not uniform, as in the lower table, interaction is indicated. As long as the pattern of means

TABLE 11.4

Deviations from means. Artificial data of Table 11.2. Differences from values obtained in Tables 11.1 and 11.2 are due to rounding errors.

		Fats		
		Fresh	*Rancid*	$\bar{R} - \bar{\bar{Y}}$
Sexes	♂	−62.42 −71.42	+97.92 +71.42	+17.75
	♀	+44.92 +71.42	−80.42 −71.42	−17.75
	$\bar{C} - \bar{\bar{Y}}$	−8.75	+8.75	

Left upper number in each square: $\bar{Y} - \bar{\bar{Y}}$ (= subgroups). *Right lower number in each square:* $\bar{Y} - \bar{R} - \bar{C} + \bar{\bar{Y}}$ (= interaction). $\bar{R} - \bar{\bar{Y}}$ (= rows); $\bar{C} - \bar{\bar{Y}}$ (= columns).

$n \sum^{rc} (\bar{Y} - \bar{\bar{Y}})^2 = 65{,}909.30$ $\qquad n \sum^{rc} (\bar{Y} - \bar{R} - \bar{C} + \bar{\bar{Y}})^2 = 61{,}209.80$

is uniform, as in the upper table, interaction may not be present. However, interaction is often present without change in the *direction* of the differences; only the relative magnitudes may be affected. Thus visual inspection should not take the place of the statistical test.

In summary, when the effect of two treatments applied together cannot be predicted from the average responses of the separate factors, statisticians call this phenomenon interaction and test its significance by means of an interaction mean square. This is a very common scientific relationship. If we say that the effect of density on the fecundity or weight of a beetle depends on its genotype, we imply that a genotype × density interaction is present. If the geographic variation of a parasite depends on the nature of the host species it attacks, we speak of a host × locality interaction. If the effect of temperature on a metabolic process is independent of the effect of oxygen concentration, we say that temperature × oxygen interaction is absent.

Significance testing in a two-way anova will be deferred until the next section. We shall now proceed to a formal presentation of an efficient computational method for carrying out such an anova. Before doing so we should point out that the computational steps **8** and **9** of Table 11.1 could have been shortened by employing the simplified formula for a sum of squares between *two* groups illustrated in Box 9.5. In an analysis with only two rows and two columns the interaction SS can be computed directly as

$$(\text{sum of one diagonal} - \text{sum of other diagonal})^2/rcn$$

which would yield the correct value $[(2087 + 1552) - (1606 + 1928)]^2/12 = 918.75$ for the data of Table 11.1.

The computation of a two-way anova with replication is illustrated in Box 11.1. The example is of oxygen consumption of two species of limpets at three concentrations of seawater. The two species represent one factor and the three concentrations of seawater are the other. Clearly both are fixed treatment effects, hence this is a Model I anova. The outline of the computations, shown step by step in Box 11.1, follows the familiar format of previous analyses. We fill in the analysis of variance table and proceed to the tests of significance described in the next section.

11.3 Two-way anova: significance testing

Before we can test hypotheses about the sources of variation isolated in Box 11.1, we must become familiar with the expected mean squares for this design. In the anova table of Box 11.1 we first show the expected mean squares for Model I, both species differences and seawater concentrations being fixed treatment effects. Incidentally, this model would also be appropriate for the earlier example of fat consumption in rats (sex and freshness of lard are fixed treatments). Note that the within-subgroups or error MS again estimates the parametric variance of the items. The most important fact to remember about a Model I anova is that the mean square at each level of variation carries only the added effect due to that level of treatment; except for the parametric variance of the items, it does not contain any term from a lower line. Thus the expected MS of factor A contains only the parametric variance of the items plus the added term due to factor A, but does not also include interaction effects. In Model I the significance test is therefore simple and straightforward. Any source of variation is tested by the variance ratio of the appropriate mean square over the error MS. Thus for the appropriate tests we employ variance ratios $\mathbf{A}$**/Error,** $\mathbf{B}$**/Error** and ($\mathbf{A} \times \mathbf{B}$)**/Error,** where each boldface term signifies a mean square. Thus $\mathbf{A} = MS_A$, **Error** $= MS_{\text{within}}$.

When we do this in the example of Box 11.1 we find only factor B,

BOX 11.1

Two-way anova with replication.

Oxygen consumption rates of two species of limpets, *Acmaea scabra* and *A. digitalis,* at three concentrations of seawater. The variable measured is $\mu l\ O_2$/mg dry body weight/min at 22°C. There are eight replicates per combination of species and salinity ($n = 8$). This is a Model I anova.

Factor B: seawater concentrations ($b = 3$)	*Factor A: Species* ($a = 2$)				
	Acmaea scabra		*Acmaea digitalis*		Σ
100%	7.16	8.26	6.14	6.14	
	6.78	14.00	3.86	10.00	
	13.60	16.10	10.40	11.60	
	8.93	9.66	5.49	5.80	
	$\Sigma = 84.49$		$\Sigma = 59.43$		143.92
75%	5.20	13.20	4.47	4.95	
	5.20	8.39	9.90	6.49	
	7.18	10.40	5.75	5.44	
	6.37	7.18	11.80	9.90	
	$\Sigma = 63.12$		$\Sigma = 58.70$		121.82
50%	11.11	10.50	9.63	14.50	
	9.74	14.60	6.38	10.20	
	18.80	11.10	13.40	17.70	
	9.74	11.80	14.50	12.30	
	$\Sigma = 97.39$		$\Sigma = 98.61$		196.00
Σ	245.00		216.74		461.74

Preliminary computations

1. Grand total $= \sum^{a}\sum^{b}\sum^{n} Y = 461.74$

2. Sum of the squared observations $= \sum^{a}\sum^{b}\sum^{n} Y^2 = (7.16)^2 + \cdots + (12.30)^2 = 5065.1530$

3. Sum of the squared subgroup (cell) totals, divided by the sample size of the subgroups

$$= \frac{\sum^{a}\sum^{b}\left(\sum^{n} Y\right)^2}{n} = \frac{[(84.49)^2 + \cdots + (98.61)^2]}{8} = 4663.6317$$

4. Sum of the squared column totals divided by the sample size of a column $= \dfrac{\sum^{a}\left(\sum^{b}\sum^{n} Y\right)^2}{bn}$

$$= \frac{[(245.00)^2 + (216.74)^2]}{(3 \times 8)} = 4458.3844$$

5. Sum of the squared row totals divided by the sample size of a row $= \dfrac{\sum^{b}\left(\sum^{a}\sum^{n} Y\right)^2}{an}$

$$= \frac{[(143.92)^2 + (121.82)^2 + (196.00)^2]}{(2 \times 8)} = 4623.0674$$

6. Grand total squared and divided by the total sample size = correction term CT

$$= \frac{\left(\sum^{a}\sum^{b}\sum^{n} Y\right)^2}{abn} = \frac{(\text{quantity } \mathbf{1})^2}{abn} = \frac{(461.74)^2}{(2 \times 3 \times 8)} = 4441.7464$$

7. $SS_{\text{total}} = \sum^{a}\sum^{b}\sum^{n} Y^2 - CT$ = quantity **2** − quantity **6** = 5065.1530 − 4441.7464 = 623.4066

8. $SS_{\text{subgr}} = \dfrac{\sum^{a}\sum^{b}\left(\sum^{n} Y\right)^2}{n} - CT$ = quantity **3** − quantity **6** = 4663.6317 − 4441.7464 = 221.8853

BOX 11.1 continued

9. SS_A (SS of columns) $= \dfrac{\sum^{a}\left(\sum^{b}\sum^{n} Y\right)^2}{bn} - CT =$ quantity **4** − quantity **6** = 4458.3844 − 4441.7464 = 16.6380

10. SS_B (SS of rows) $= \dfrac{\sum^{b}\left(\sum^{a}\sum^{n} Y\right)^2}{an} - CT =$ quantity **5** − quantity **6** = 4623.0674 − 4441.7464 = 181.3210

11. $SS_{A \times B}$ (Interaction SS) $= SS_{\text{subgr}} - SS_A - SS_B =$ quantity **8** − quantity **9** − quantity **10**
= 221.8853 − 16.6380 − 181.3210 = 23.9263

12. SS_{within} (Within subgroups; error SS) $= SS_{\text{total}} - SS_{\text{subgr}} =$ quantity **7** − quantity **8** = 623.4066 − 221.8853 = 401.5213

□□ The machine computation of these quantities is self-evident and is similar to that of earlier anovas. The following relations for some of the above quantities must hold: **2** ≥ **3** ≥ **4** ≥ **6**; **3** ≥ **5** ≥ **6**. □□

Now fill in the anova table.

	Source of variation	*df*	*SS*	*MS*	*Expected MS (Model I)*
$\bar{Y} - \bar{\bar{Y}}$	Subgroups	$ab - 1$	**8**	$\dfrac{\mathbf{8}}{(ab-1)}$	
$\bar{Y}_A - \bar{\bar{Y}}$	A (columns)	$a - 1$	**9**	$\dfrac{\mathbf{10}}{(a-1)}$	$\sigma^2 + \dfrac{nb}{a-1}\sum^{a}\alpha^2$
$\bar{Y}_B - \bar{\bar{Y}}$	B (rows)	$b - 1$	**10**	$\dfrac{\mathbf{9}}{(b-1)}$	$\sigma^2 + \dfrac{na}{b-1}\sum^{b}\beta^2$
$\bar{Y} - \bar{Y}_A - \bar{Y}_B + \bar{\bar{Y}}$	$A \times B$ (interaction)	$(a-1)(b-1)$	**11**	$\dfrac{\mathbf{11}}{(a-1)(b-1)}$	$\sigma^2 + \dfrac{n}{(a-1)(b-1)}\sum^{ab}(\alpha\beta)^2$
$Y - \bar{Y}$	Within subgroups	$ab(n-1)$	**12**	$\dfrac{\mathbf{12}}{ab(n-1)}$	σ^2
$Y - \bar{\bar{Y}}$	Total	$abn - 1$	**7**		

Since the present example is a Model I anova for both factors the expected MS above are correct. Below are the corresponding expressions for other models.

Source of variation	Model II	Mixed Model (A fixed, B random)
A	$\sigma^2 + n\sigma^2_{A\times B} + nb\sigma^2_A$	$\sigma^2 + n\sigma^2_{A\times B} + \frac{nb}{a-1}\sum^{a}\alpha^2$
B	$\sigma^2 + n\sigma^2_{A\times B} + na\sigma^2_B$	$\sigma^2 \qquad + na\sigma^2_B$
$A \times B$	$\sigma^2 + n\sigma^2_{A\times B}$	$\sigma^2 + n\sigma^2_{A\times B}$
Within subgroups	σ^2	σ^2

Anova table

Source of variation	*df*	*SS*	*MS*	F_s
Subgroups	5	221.8853	44.377	
A (columns; species)	1	16.6380	16.638	1.740 *ns*
B (rows; salinities)	2	181.3210	90.660	9.483***
$A \times B$ (interaction)	2	23.9263	11.963	1.251 *ns*
Within subgroups (error)	42	401.5213	9.560	
Total	47	623.4066		

$F_{.05[1,42]} = 4.07 \qquad F_{.05[2,42]} = 3.22 \qquad F_{.001[2,42]} = 8.18$

Since this is a Model I anova, all mean squares are tested over the error MS. For a discussion of significance tests see Section 11.3.

Conclusions.—Oxygen consumption does not differ significantly between the two species of limpets but differs with the salinity. At 50% seawater the O_2 consumption is increased. Salinity appears to affect the two species equally, for there is insufficient evidence of a species $\times$ salinity interaction.

salinity, significant. Neither factor A nor the interaction are significant. We conclude that the differences in oxygen consumption are induced by varying salinities (the two variables appear inversely related), and there does not appear to be sufficient evidence for species differences in oxygen consumption. A tabulation of the relative magnitudes of the means in the manner of the previous section,

	Seawater concentrations				
	100%		*75%*		*50%*
A. scabra		>		<	
A. digitalis		>		<	

shows that the pattern of signs in the two lines is identical. However, this may be misleading, since the mean of *A. scabra* is far higher at 100% seawater than at 75%, but that of *A. digitalis* is only very slightly higher. Although the oxygen consumption curves of the two species when graphed appear far from parallel (see Figure 11.2), this suggestion of a species × salinity interaction cannot be shown to be significant when compared to the within-subgroups variance. Finding a significant difference among salinities does not conclude the analysis. The data suggest that at 75% salinity there is a real reduction in oxygen consumption. Whether this is really so could be tested by the methods of Section 9.7. However, a preferred method of analysis for response curves of this type will be taken up in Sections 14.10 and 14.11.

When we test the anova of fat consumption in rats (Table 11.1) we notice again that only one mean square (between columns, representing fats) is significant. We conclude, therefore, that the apparent difference in fat consumption between males and females in rats is not significant, but that the higher consumption of fresh as compared with rancid lard is highly signifi-

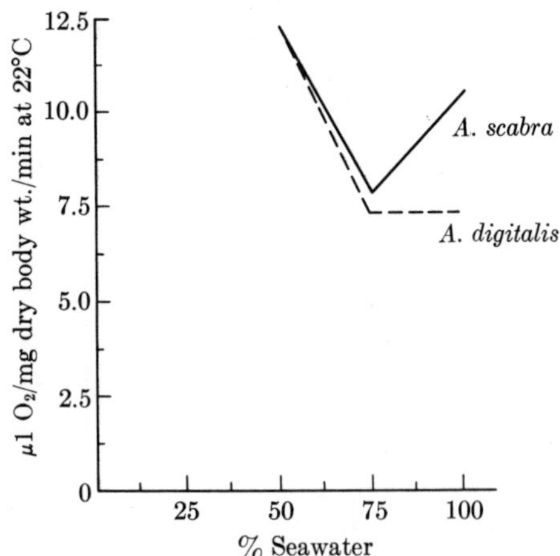

FIGURE 11.2 Oxygen consumption by two species of limpets at three salinities. Data from Box 11.1.

cant. Absence of interaction means that this trend is the same in the two sexes. When we analyze the results of the artificial example in Table 11.2, we find only the interaction *MS* significant. Thus we would conclude that the response to freshness of lard differs in the two sexes. This is brought out by inspection of the data, which show that males prefer rancid fat and females prefer fresh fat.

In the last (artificial) example the mean squares of the two factors (main effects) are not significant in any case. However, many statisticians would not even test them once they found the interaction mean square to be significant, since in such a case an overall statement for each factor would have little meaning. A simple statement of preference for either rancid or fresh lard would be unclear. The presence of interaction makes us qualify our statements: rancid fat is preferred by males, and fresh fat is preferred by females. Similarly, interaction in the example of Box 11.1 would have meant that the pattern of response to changes in salinity differed in the two species. We would consequently have to describe separate, nonparallel response curves for the two species. Occasionally it becomes important to test for overall significance in a Model I anova in spite of the presence of interaction. Take, for example, the situation illustrated by Figure 11.3. It shows the effects of increasing proportions in fresh flour of two types of "conditioned" flour on survival of wild type *Tribolium castaneum* beetles. Conditioned flour is medium that has been heavily contaminated by waste products of the beetles. The two types of flour had been conditioned by two different strains of beetles. The shapes of the response curves are quite different. Undoubtedly there is interaction between the strains conditioning the medium and the concentrations of the conditioned flour as these affect survival of the beetles. Yet the overall effects of increasing the proportion of conditioned flour are quite clear. As the concentration of conditioned flour increases, survival decreases in conditioned flour from both sources. To support this contention we might wish to test the mean square among concentrations (over the error *MS*), regardless of whether the interaction *MS* is significant.

Box 11.1 also lists expected mean squares for a Model II and a mixed model two-way anova. In the Model II note that the two main effects contain the variance component of the interaction as well as their own variance component. Such an arrangement is familiar from nested analysis of variance. When testing significance in a Model II we always test $(\boldsymbol{A} \times \boldsymbol{B})/\mathbf{Error}$. If the interaction is significant we continue testing $\boldsymbol{A}/(\boldsymbol{A} \times \boldsymbol{B})$ and $\boldsymbol{B}/(\boldsymbol{A} \times \boldsymbol{B})$, but when $\boldsymbol{A} \times \boldsymbol{B}$ is not significant we are faced with a decision of whether to pool the interaction *SS* with the error *SS* before testing the mean squares of the main effects. In such a case we proceed according to the rules of Box 10.2. Only one type of mixed model is shown in which factor A is assumed to be fixed and factor B to be random. If the situation is reversed the expected mean squares change accordingly. Here it is the mean square

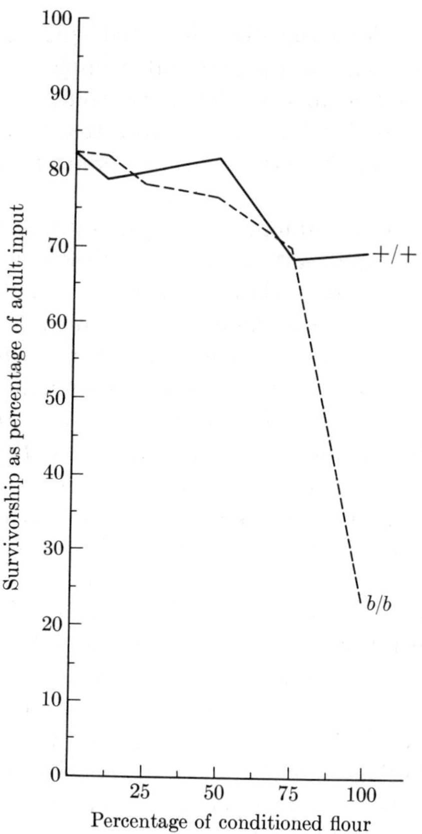

FIGURE 11.3 Adult survivorship as percentage of egg input of wild type *Tribolium castaneum* reared on medium conditioned by either wild type (+/+) or black (*b*/*b*) beetles in varying concentrations. (Extracted from Karten, 1965.)

representing the fixed treatment that carries with it the variance component of the interaction, while the mean square representing the random factor contains only the error variance and its own variance component and does not include the interaction component. We therefore test the MS of the random main effect over the error but test the fixed treatment MS over the interaction if the latter is significant, or if Box 10.2 contraindicates pooling; otherwise we test over the pooled error.

You may wonder how you would be able to write down the formulas for expected mean squares for cases other than those given in the boxes. We shall learn general formulas for these in the next chapter (Section 12.3).

An example of a Model II two-way anova is featured in Table 11.5. The data represent measurements of blood *p*H of mice. Five inbred strains were measured repeatedly. In experiment I, the mice were tested every other day for seven test periods, and in experiment II they were tested at six weekly

TABLE 11.5

A Model II example of a two-way anova with replication. Strain differences and daily differences in blood *p*H for 5 inbred strains of mice. In experiment I samples of 2 mice from each strain were tested every other day until 7 such groups had been tested; in experiment II samples of 5 mice from each strain were tested 6 times at one week intervals. Only the anova tables are shown.

Experiment I ($a = 5$, $b = 7$, $n = 2$).

Source of variation	*df*	*MS*	F_s	*Expected MS*
Mouse strains	4	0.0427	11.24***	$\sigma^2 + 2\sigma^2_{MD} + 14\sigma^2_M$
Days of test	6	0.0148	3.89**	$\sigma^2 + 2\sigma^2_{MD} + 10\sigma^2_D$
Interaction	24	0.0038	1.03 *ns*	$\sigma^2 + 2\sigma^2_{MD}$
Mice treated alike (error)	35	0.0037		σ^2

Experiment II ($a = 5$, $b = 6$, $n = 5$).

Source of variation	*df*	*MS*	F_s	*Expected MS*
Mouse strains	4	0.0920	17.69***	$\sigma^2 + 5\sigma^2_{MD} + 30\sigma^2_M$
Days of test	5	0.0101	1.94 *ns*	$\sigma^2 + 5\sigma^2_{MD} + 25\sigma^2_D$
Interaction	20	0.0052	1.53 *ns*	$\sigma^2 + 5\sigma^2_{MD}$
Mice treated alike (error)	120	0.0034		σ^2

Variance components

	Experiment I		*Experiment II*	
Strains	0.00278	36.4%	0.00289	42.2%
Days	0.00110	14.4%	0.00020	2.9%
Interaction	0.00005	0.7%	0.00036	5.3%
Error	0.0037	48.5%	0.0034	49.7%

Source: Data from Weir (1949).

intervals. The purpose of these experiments was to estimate the proportion of the variation that is genetic. This is an example of a Model II analysis because the experimenter was not concerned with the differences between strains 3 and 4, for example, but was considering the five strains to be a "random" sample of the inbred mouse strains available. The sampling periods are considered to be random as well, since the investigator did not wish to establish that day 1 produced blood *p*H values different from those of day 2, but merely tried to estimate the magnitude of fluctuations in blood *p*H over the period studied. There is no reason to believe in a directed trend of the blood *p*H over the period of the study.

The expected mean squares of the two experiments are shown in Table 11.5, with the actual coefficients of the variance components. The symbolism differs somewhat from that given for Model II in Box 11.1. It is important that you become familiar with some variations in usage. Instead of using A and B as symbols for the two factors, we use M and D to represent mouse strains and days. Such mnemonic notation is frequently used. In such cases investigators often also use a different symbolism for the coefficients; *m* might be used in place of *a* and *d* for *b* (to correspond to *M* and *D*, respectively). Notice also that the source of variation column (copied from the original publication) does not have prepositions such as "among" or "between" and also that the error term is described as "mice treated alike." This is in fact what the error term represents in the present experiment: variation among mice of a given strain within a given day. You will have to become used to differences in naming the sources of variation.

Significance testing in this example proceeds as follows. We first test interaction over the error; this is nonsignificant in both experiments. Following the rules in Box 10.2, assuming no knowledge about the relative magnitudes of the interaction and error variances, we continue to test the main effects over the interaction. Actually, even had we pooled the two sums of squares, the results would not be substantially different. In experiment I, mouse strains as well as days of test have a highly significant variance component. In experiment II, however, only mouse strains have such a component. This is reflected in the magnitudes of the estimated variance components given at the bottom of Table 11.5 as absolute values and as percentages. The error variance is remarkably alike in the two experiments, showing that the experimental techniques were well controlled and gave comparable results in both experiments. The major difference between the experiments is that a substantial percentage (14.4%) of variation is among days in experiment I, but this variance component comprises only 2.9% of the overall variation in experiment II. Clearly there are more fluctuations when readings are taken every two days. A far-fetched explanation might be that there is a rhythmic fluctuation in blood *p*H values with regular cycles every seven days. In such a case samples taken every other day would fluctuate widely, while those taken once a week would be very constant. There is, however, no evidence for such a phenomenon in these mice. Experiments I and experiments II were run at different seasons, and it is quite possible that during the 14 days necessary for experiment I there were considerable day-to-day fluctuations in climate. An inquiry to the investigator revealed that the mice were not maintained in temperature-controlled rooms; hence daily fluctuations during experiment I might have resulted in this large value of the variance. During the 42 days of experiment II the weather might have been considerably more constant. Although these explanations may be incorrect, they illustrate how an anova furnishes insight into the structure of an experiment as well as clues to further questions to be asked and to experiments to be undertaken.

A simple rearrangement of the data of Table 11.5 will provide some insight into the relations between two-way and nested anovas. In the design of experiment II in this table, 5 strains were observed on 6 separate days. In each strain 5 mice had their blood *p*H measured on one day. This design is shown at the head of Table 11.6. We arrange it as a two-way anova be-

TABLE 11.6

Experiment II of Table 11.5 reanalyzed to show relations between a two-way anova and a nested anova.

Correct design *(two-way anova with replication: 5 mice [*] in each cell).*

Days	Strains				
	1	*2*	*3*	*4*	*5*
1	*	*	*	*	*
2	*	*	*	*	*
3	*	*	*	*	*
4	*	*	*	*	*
5	*	*	*	*	*
6	*	*	*	*	*

Anova table

Source of variation	*df*	*SS*	*MS*	F_s
Mouse strains	4	0.3680	0.0920	17.69***
Days of test	5	0.0505	0.0101	1.94 *ns*
Interaction	20	0.1040	0.0052	1.53 *ns*
Error	120	0.4080	0.0034	
Total	149	0.9305		

Incorrect design *(nested anova—days within strains: 5 mice [*] in each cell).*

	Strains				
	1	*2*	*3*	*4*	*5*
Days	1 2 3 4 5 6	1 2 3 4 5 6	1 2 3 4 5 6	1 2 3 4 5 6	1 2 3 4 5 6
Mice	* * * * * *	* * * * * *	* * * * * *	* * * * * *	* * * * * *

Anova table

Source of variation	*df*	*SS*	*MS*	F_s
Mouse strains	4	0.3680	0.0920	14.84***
Days within strains	25	0.1545	0.0062	1.82*
Error	120	0.4080	0.0034	
Total	149	0.9305		

cause the days are common factors throughout the analysis. Day 1 is the same day for strain 1 as it is for strain 3. The anova table for this experiment is reproduced once more in Table 11.6, showing also the numerical values for the sums of squares.

Suppose, however, we had misinterpreted this experiment. We might have thought that the days were nested within the strains as shown in the next design in Table 11.6. Such a design would imply that we randomly chose six days on which to measure the blood pH of each strain and that the six days chosen for the tests on strain 1 were different from the test days for strain 2, and so on. Thus, if we had randomly chosen the days without any thought of making day 1 the same for all the strains, a nested analysis would have been appropriate. Taking the same data and analyzing them (incorrectly, in view of the actual nature of design) as a two-level nested anova, we obtain the anova table shown at the bottom of Table 11.6. Note that the subordinate level (subgroups within groups, which is days within strains in this example) is the summation of two levels from the two-way anova. Degrees of freedom and sums of squares of "days of test" and "interaction" add to the corresponding values for "days within strains." Thus in a hierarchic analysis of variance the subordinate level represents not only differences among the subgroups but also the interaction term. If differences among days exist but the nature of these differences varies from strain to strain, we would have a substantial contribution to interaction. This would not be separately detectable in the nested anova but would contribute to the magnitude of the subgroups-within-groups SS. When we incorrectly consider this example as a nested anova, the days within strains MS is significant at the 5% level, but in the two-way anova neither days of test nor interaction is significant, although the appropriate F_s-ratios are suspiciously large.

11.4 Two-way anova without replication

In many experiments there will be no replication for each combination of factors represented by a cell in the data table. In such cases we cannot easily talk of "subgroups," since each cell contains a single reading only. Frequently it may be too difficult or too expensive to obtain more than one reading per cell, or the measurements may be known to be so repeatable that there is little point in estimating their error. As we shall see below, a two-way anova without replication can be properly applied only with certain assumptions. For some models and tests in anova we must assume that there is no interaction present.

Our illustration for this design is taken from limnology. In Box 11.2 we show temperatures of a lake taken at about the same time on four successive summer afternoons. These temperature measurements were made at ten different depths and were known to be very repeatable. Therefore only single readings were taken at each depth on any one day. What is the appropriate model for this anova? Clearly, the depths are Model I. The four days, however, are not likely to be of specific interest. It is improbable that an investigator would ask whether the water was colder on the 30th of July than on the 31st of July. A more meaningful way of looking at this problem would be to consider these four summer days as random samples

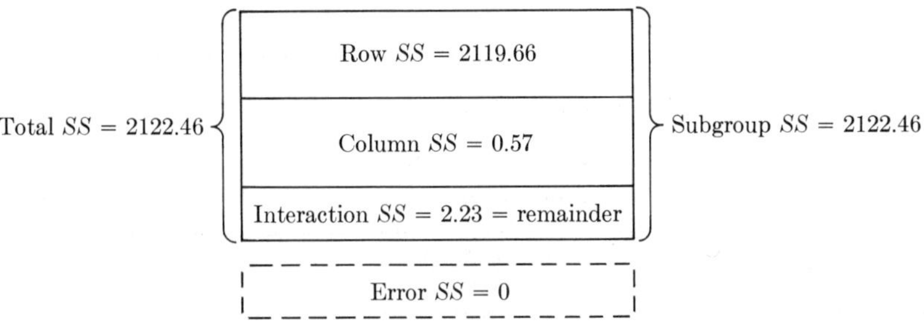

FIGURE 11.4 Diagrammatic representation of the partitioning of the total sums of squares (SS) in a two-way orthogonal anova without replication. The areas of the subdivisions are not shown proportional to the magnitudes of the sums of squares.

that enable us to estimate the day-to-day variability of the temperature stratification in the lake during the stable midsummer period.

The computations are shown in Box 11.2. They are the same as those in Box 11.1 except that the expressions to be evaluated are considerably simpler. Since $n = 1$, much of the summation can be omitted. The subgroup sum of squares in this example is the same as the total sum of squares. If this is not immediately apparent, consult Figure 11.4, which, when compared with Figure 11.1, illustrates that the error sum of squares based on variation within subgroups is missing in this example. Thus, after we subtract the sum of squares for columns (factor A) and for rows (factor B) from the total SS, we are left with only a single sum of squares, which is the equivalent of the previous interaction SS but which is now the only source for an error term in the anova. This SS is known as the *remainder* SS or the *discrepance.*

If you will refer to the expected mean squares for the two-way anova in Box 11.1, you will discover why we made the statement earlier that for some models and tests in a two-way anova without replication we must assume that the interaction is not significant. If interaction is present, only a Model II anova can be entirely tested, while in a mixed model only the fixed level can be tested over the remainder mean square. But in a pure Model I, or for the random factor in a mixed model, it would be improper to test the main effects over the remainder unless we could reliably assume that no added effect due to interaction is present. General inspection of the data in Box 11.2 convinces us that the temperature trends present on any one day are faithfully reproduced on the other days. Thus, interaction is unlikely to be present. If, for example, a severe storm had churned up the lake on one day, changing the temperature relationships at various depths, interaction would have been apparent, and the test of the mean square among days carried out in Box 11.2 would not have been legitimate.

Since we assume no interaction, the row and column mean squares are tested over the error MS. The results are not surprising; casual inspection of the data would have predicted our findings. Added variance is not signifi-

BOX 11.2

Two-way anova without replication.

Temperatures (°C) of Rot Lake on four early afternoons of the summer of 1952 at 10 depths. This is a mixed model anova.

Factor B: Depth in meters ($b = 10$)	*Factor A: Days* ($a = 4$)				
	29 July	*30 July*	*31 July*	*1 August*	Σ
0	23.8	24.0	24.6	24.8	97.2
1	22.6	22.4	22.9	23.2	91.1
2	22.2	22.1	22.1	22.2	88.6
3	21.2	21.8	21.0	21.2	85.2
4	18.4	19.3	19.0	18.8	75.5
5	13.5	14.4	14.2	13.8	55.9
6	9.8	9.9	10.4	9.6	39.7
9	6.0	6.0	6.3	6.3	24.6
12	5.8	5.9	6.0	5.8	23.5
15.5	5.6	5.6	5.5	5.6	22.3
Σ	148.9	151.4	152.0	151.3	603.6

Source: Data from Vollenweider and Frei (1953).

The four sets of readings are treated as replications (blocks) in this analysis. Depth is a fixed treatment effect, days are considered as random effects, hence this is a mixed model anova.

Preliminary computations

1. Grand total $= \sum^{a}\sum^{b} Y = 603.6$

2. Sum of the squared observations $= \sum^{a}\sum^{b} Y^2$

$$= (23.8)^2 + \cdots + (5.6)^2 = 11{,}230.78$$

3. Sum of squared column totals divided by sample size of a column

$$= \frac{\sum^{a}\left(\sum^{b} Y\right)^2}{b} = \frac{[(148.9)^2 + \cdots + (151.3)^2]}{10} = 9108.89$$

4. Sum of squared row totals divided by sample size of a row

$$= \frac{\sum^{b}\left(\sum^{a} Y\right)^2}{a} = \frac{[(97.2)^2 + \cdots + (22.3)^2]}{4} = 11{,}227.98$$

5. Grand total squared and divided by the total sample size = correction term

$$CT = \frac{\left(\sum^{a}\sum^{b} Y\right)^2}{ab} = \frac{(\text{quantity } \mathbf{1})}{ab} = \frac{(603.6)^2}{40} = 9108.32$$

BOX 11.2 continued

6. $SS_{\text{total}} = \sum^{a}\sum^{b} Y^2 - CT$

$= \text{quantity } \mathbf{2} - \text{quantity } \mathbf{5} = 11{,}230.78 - 9108.32 = 2122.46$

7. SS_A (SS of columns) $= \dfrac{\sum^{a}\left(\sum^{b} Y\right)^2}{b} - CT$

$= \text{quantity } \mathbf{3} - \text{quantity } \mathbf{5} = 9108.89 - 9108.32 = 0.57$

8. SS_B (SS of rows) $= \dfrac{\sum^{b}\left(\sum^{a} Y\right)^2}{a} - CT$

$= \text{quantity } \mathbf{4} - \text{quantity } \mathbf{5} = 11{,}227.98 - 9108.32 = 2119.66$

9. SS_{error} (remainder; discrepance) $= SS_{\text{total}} - SS_A - SS_B$

$= \text{quantity } \mathbf{6} - \text{quantity } \mathbf{7} - \text{quantity } \mathbf{8} = 2122.46 - 0.57 - 2119.66$

$= 2.23$

Anova table

Source of variation		*df*	*SS*	*MS*	F_s	*Expected MS*
$\bar{Y}_A - \bar{\bar{Y}}$	A (column; days)	3	0.57	0.190	2.30 *ns*	$\sigma^2 + b\sigma^2_A$
$\bar{Y}_B - \bar{\bar{Y}}$	B (rows; depths)	9	2119.66	235.5	2851.1***	$\sigma^2 + \sigma^2_{AB} + \frac{a}{b-1}\Sigma\beta^2$
$Y - \bar{Y}_A - \bar{Y}_B + \bar{\bar{Y}}$	Error (remainder; discrepance)	27	2.23	0.0826		$\sigma^2 + \sigma^2_{AB}$
$Y - \bar{\bar{Y}}$	Total	39	2122.46			

Conclusions.—Highly significant decreases in temperature occur as depth increases. For testing "days" we must assume interaction between days and depths to be zero. No added variance component among days can be demonstrated.

cant among days, but the differences in temperature due to depths are highly significant, yielding a value of $F_s = 2851.1$. It would be extremely unlikely that chance alone could produce such differences.

A common application of two-way anova without replication is the *repeated testing of the same individuals.* By this we mean that the same group of individuals is tested repeatedly over a period of time. The individuals are one factor (usually considered as random and serving as replication) and the time dimension is the second factor, a fixed treatment effect. For example, we might measure growth of a structure in ten individuals at regular intervals. When testing for the presence of an added variance component (due to the random factor) such a model again assumes that there is no interaction between time and the individuals; that is, the responses of the several individuals are parallel through time. Another use of this design is found in various physiological and psychological experiments in which we test the same group of individuals for the appearance of some response after treatment. Examples include increasing immunity after antigen inoculations, altered responses after conditioning, and measures of learning after a number of trials. Thus we may study the speed with which ten rats, repeatedly tested on the same maze, reach the end point. The fixed treatment effect would be the successive trials to which the rats have been subjected. The second factor, the ten rats, is random, presumably representing a random sample of rats from the laboratory population. When the response variables are two-state attributes rather than being measurable, repeated testing of the same individuals is carried out by the method of Section 16.7.

The method of *randomized complete blocks* is a type of two-way anova (usually without replication) especially common in agricultural research, but also with numerous applications in other biological field and laboratory experiments. When we apply a fertilizer to a small plot of land we usually get a single reading expressed as yield of crop. Similarly, an application of an insecticide to a plot of land is measured as a single mortality among the insects in the plot or a single value for yield of crop. To test differences among five types of fertilizers (or insecticide treatments) we must have at least five plots available. However, the five resulting readings would not yield an error variance useful for testing the significance of differences among treatments. Thus we need replication of the treatments to provide us with an error term. Let us assume for the sake of this discussion that we wish to replicate each treatment four times. We therefore need $5 \times 4 = 20$ plots to test five treatments. An area encompassing 20 plots of land is quite large relative to basic plot size and therefore is likely to be heterogeneous in its soil and in other microclimatic conditions. As a general rule adjacent plots would be more like each other. We therefore would not wish to put all the replicates of treatment A into the same general area; otherwise what may appear to be a significantly different effect of treatment A may actually turn out to be due to the general area in which the plots of A are situated.

The method of randomized complete blocks is designed to overcome this difficulty. In our specific example we lay out four *blocks* within the land available for our experiment, hoping that each of the blocks represents as homogeneous an environment as possible. Each block is then subdivided into five plots and the treatments are allocated at random to these plots. Such an arrangement is shown diagrammatically in Figure 11.5. The five treatments are here called A, B, C, D, and E. The random arrangement of the treatments within each block can be obtained by permutating the five letters. There are $5! = 120$ permutations of the letters A through E. These can be numbered sequentially and by consulting a table of random numbers four permutations could be chosen. The random arrangement of the treatments among

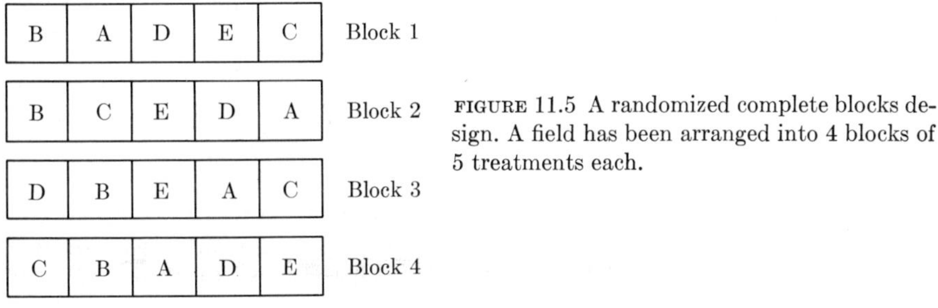

FIGURE 11.5 A randomized complete blocks design. A field has been arranged into 4 blocks of 5 treatments each.

the plots is necessary so that any given treatment would not always maintain the identical relative position in the field.

For analysis the data are entered in a treatments × blocks table, all the treatments of one kind being entered in the appropriate block. If the $b = 4$ blocks are considered as rows (Factor B) and the $a = 5$ treatments as columns (Factor A), an anova table of the example just mentioned would look as follows:

Source of variation	*df*	*Expected MS*
Blocks	$b - 1$	$\sigma^2 + a\sigma^2_B$
Treatments	$a - 1$	$\sigma^2 + \frac{b}{a-1}\sum^{a}\alpha^2$
Error	$(a - 1)(b - 1)$	σ^2

The blocks are considered as random effects. An individual observation in a randomized complete block design can therefore be decomposed as

$$Y_{ij} = \mu + \alpha_i + B_j + \epsilon_{ij} \tag{11.2}$$

where α_i is the treatment effect and B_j is the random block effect. Note that in this expression the interaction term has been omitted because it would serve no useful purpose. When testing a randomized complete block design we simply test the treatment MS over the error. The block MS is rarely tested and we would have to assume that interaction is not present before we could test it. If it is not significantly greater than the error MS, there was less heterogeneity among the blocks than anticipated or the blocks were laid out in such a way that most of the heterogeneity was within blocks instead of among blocks. If we do find added variance among the blocks this does not disturb us, since we did after all expect to find such differences. In fact, if we found that the differences among blocks were small, then in future experiments we might not want to use a randomized blocks design but rather employ a single classification anova.

The randomized complete block design may also be applied to types of experiments other than the agricultural ones for which it was first developed. If an experiment is too big to be undertaken at one time, it may have to be carried out over a period of days (each day representing a block), or it may have to be carried out in different laboratories (each serving as a block).

When treatments are applied to different age classes of animals, these classes can become blocks. Whenever the environment in which the experiment is to be carried out or the research materials are possibly heterogeneous, these might be subdivided into more homogeneous blocks and the experiment carried out and analyzed as randomized complete blocks.

↛ It is important to note that each treatment must be present in every block. If this cannot be done, or when one has too many treatments, different combinations of treatments are applied to a series of blocks. In such cases a more complex analysis must be carried out. Such designs, called *incomplete blocks*, are described in Federer (1955, Chapters XI, XII, and XIII). ↚

An example of a randomized complete block experiment is shown in Table 11.7. This is based on data extracted from a more extensive experiment by Sokal and Karten (1964) in which dry weights of three genotypes of *Tribolium castaneum* beetles were recorded. The experimental work involved measurement of other variables and was very laborious. Therefore the requisite amount of replication could not be carried out at one time, and four series of experiments, several months apart, were undertaken. Within one series of experiments the order of weighing different genotypes was essentially random. We thus have a randomized complete block design with four blocks (series) and three treatments (genotypes).

These data and their analysis are shown in Table 11.7. From the anova table it is clear that there is a significant difference in weight among the genotypes, the heterozygotes being the heaviest, followed by the wild type, with the homozygous mutant weighing least. In Table 11.7 we have not actually tested the differences between pairs of means by multiple comparisons tests. From the analysis shown we only know that the three means jointly are heterogeneous. However, Sokal and Karten tested these differences and found them significant. Interestingly, the series mean square is also significant. Thus environmental factors among the series, minor changes in technique, or other factors were such that differences in weight resulted. By separating out this heterogeneity we were able to perform a much more sensitive test. Had we ignored differences among series and simply obtained four random replicates for each genotype, our error variance would have been that for a completely randomized design:

$$MS_{E(CR)} = \frac{b(a-1)MS_{E(RB)} + (b-1)MS_B}{ab-1} \tag{11.3}$$

In this expression $MS_{E(CR)}$ is the expected error mean square in the completely randomized design, $MS_{E(RB)}$ is the observed error mean square in the randomized block design, and MS_B is the observed mean square among blocks. This formula is derived in Appendix A1.6. In Table 11.7 this expression gives a value for $MS_{E(CR)} = 0.002451$. This is considerably greater than the ob-

TABLE 11.7

Example of a randomized complete block experiment. Mean dry weights (in mg) of 3 genotypes of beetles, *Tribolium castaneum,* reared at a density of 20 beetles per gram of flour. Four series of experiments represent blocks.

b series (*b blocks*; $b = 4$)	*a genotypes* ($a = 3$)			Σ
	$++$	$+b$	bb	
1	0.958	0.986	0.925	2.869
2	0.971	1.051	0.952	2.974
3	0.927	0.891	0.829	2.647
4	0.971	1.010	0.955	2.936
Σ	3.827	3.938	3.661	11.426

$\sum^{a}\sum^{b} Y = 11.426$ $\quad \sum^{a}\sum^{b} Y^2 = 10.914{,}748$ $\quad \sum^{a}(\sum^{b} Y)^2/b = 10.889{,}173$

$\sum^{b}(\sum^{a} Y)^2/a = 10.900{,}847$ $\quad CT = 10.879{,}456$

Anova table

	Source of variation	*df*	*SS*	*MS*	F_s
MS_B	Series	3	0.021,391	0.007,130	10.23**
MS_A	Genotypes	2	0.009,717	0.004,858	6.97*
$MS_{E(RB)}$	Error	6	0.004,184	0.000,697	

Estimate of error variance in a completely randomized design following Expression (11.3)

$$MS_{E(CR)} = \frac{b(a-1)MS_{E(RB)} + (b-1)MS_B}{ab-1}$$

$$= \frac{8 \times 0.000697 + 3 \times 0.007130}{11} = 0.002451$$

Relative efficiency of randomized complete block error variance $MS_{E(RB)}$ with respect to completely randomized error variance $MS_{E(CR)}$

$$RE = \frac{100 \times MS_{E(CR)}}{MS_{E(RB)}} = 351.6\%$$

When the degrees of freedom of the mean squares from which the relative efficiency is calculated are less than 20 (degrees of freedom of $MS_{E(RB)} = 6$ in our case), we use a correction factor for RE (see Cochran and Cox, 1957). Calculate

$$\frac{(df_l + 1)(df_u + 3)}{(df_l + 3)(df_u + 1)}$$

where df_u and df_l are the degrees of freedom of the upper and lower means squares, respectively, and multiply it times the RE. In our case

$$\frac{(6+1)(9+3)}{(6+3)(9+1)} = \frac{(7 \times 12)}{(9 \times 10)} = 0.93333$$

$$RE = 351.6 \times 0.93333 = 328.2\%$$

Source: From more extensive data by Sokal and Karten (1964).

served error in the randomized block design $MS_{E(RB)} = 0.000697$. We can now calculate the relative efficiency of the randomized block error variance with respect to the completely randomized error variance. This is done at the foot of Table 11.7, where we note that randomized blocks in this case are 328.2% as efficient as the completely randomized design. Clearly it was well worthwhile to divide the experiment into series (blocks) and to segregate the variance due to series.

The computer program A3.5 can be used to perform the computations described in Sections 11.2 and 11.4. A general description of the use of this program is given in Section 12.5. However, in most cases the computations are simple enough that the use of a computer is not justified unless the data are extensive or there are many such analyses to be performed.

11.5 Paired comparisons

The randomized complete blocks designs studied in the previous section may differ in the number of rows (blocks) and treatment classes. However, one special case is sufficiently common to merit discussion in a separate section. This is the case of randomized complete blocks in which there are only two treatments or groups ($a = 2$). This case is also known as *paired comparisons* because each observation for one treatment is paired with one for the other treatment. This pair is composed of the same individuals tested twice or of two individuals with common experiences so that we can legitimately arrange the data as a two-way anova.

Let us elaborate on this point. Suppose we test the muscle tone of a group of individuals, subject them to severe physical exercise, and measure their muscle tone once more. Since the same group of individuals will have been tested twice, we can arrange our muscle tone readings in pairs, each pair representing readings on one individual (before and after exercise). Such data are appropriately treated by a two-way anova without replication, which in this case would be a paired comparisons test because there are only two treatment classes. This "before and after treatment" comparison is a very frequent design leading to paired comparisons. Another design simply measures two stages in the development of a group of organisms, time being the treatment intervening between the two stages. The example in Box 11.3 that follows is of this nature. It measures lower face width in a group of girls five years old and in the same group of girls when they are six years old. The paired comparison is for each individual girl, between her face width when she is five years old and her face width at six years.

Paired comparisons often result from dividing an organism or other individual unit so that half receives treatment 1 and the other half treatment 2, which may be the control. Thus if we wish to test the strength of two antigens or allergens we might inject one into each arm of a single individual and measure the diameter of the red area produced. It would not be wise from the point of view of experimental design to test antigen 1 on individual 1 and antigen 2 on individual 2. These individuals may be differentially susceptible to these antigens and we may learn little about the relative potency of the antigens, since this would be confounded by the differential responses of the subjects. A much better design would be to inject antigen 1 into the left arm and antigen 2 into the right arm of a group of n individuals and to analyze the data as a randomized complete block design with n rows (blocks) and 2 columns (treatments). It is probably immaterial whether an antigen is injected into the right or left arm, but if we were designing such an experiment and knew little about the reaction of humans to antigens we might, as a precaution, randomly allocate antigen 1 to the left or right arm for different subjects, antigen 2 being injected into the opposite arm. A similar example is the testing of certain plant viruses by rubbing a certain concentration of the virus over the surface of a leaf and counting the resulting lesions. Since different leaves are differentially susceptible, a conventional way of measuring the strength of the virus is to wipe it over half of the leaf on one side of the midrib, rubbing the other half of the leaf with a control or standard solution.

Another design leading to paired comparisons is when the treatment is given to two individuals sharing a common experience, be this genetic or environmental. Thus a drug or a psychological test might be given to groups of twins or sibs, one of each pair receiving the treatment, the other one not.

Finally, the paired comparisons technique may be used when the two individuals to be compared share a single experimental unit and are thus subjected to common environmental experiences. If we have a set of rat cages, each of which holds two rats, and we are trying to compare the effect of a hormone injection with a control, we might inject one of each pair of rats with the hormone and use his cage mate as a control. This would yield a $2 \times n$ anova in which the n cages would represent the blocks.

One reason for featuring the paired comparisons test in a separate section is that it alone among the two-way anovas without replication has an equivalent, alternative method of analysis, the t-test for paired comparisons, which is the traditional method of solving it. The t-test will be discussed after we have first illustrated treatment of this problem by the method of randomized complete blocks.

The paired comparisons case shown in Box 11.3 analyzes face widths of five- and six-year-old girls, as already mentioned. The question being asked is whether the faces of six-year-old girls are significantly wider than those of five-year-old girls. The data are shown in columns (1) and (2) of Box 11.3 for 15 individual girls. Column (3) features the row sums that are necessary for the analysis of variance. The computations for the two-way anova without replication in Box 11.3 are the same as those already shown for Box 11.2. Remember that since there is no replication the total sum of squares is identical to the subgroup sum of squares. The anova table shows that there is a

BOX 11.3

Paired comparisons (randomized blocks with $a = 2$).

Lower face width (skeletal bigonial diameter in cm) for 15 North American white girls measured when 5 and again when 6 years old.

Individuals	(*1*) *5-year-olds*	(*2*) *6-year-olds*	(*3*) Σ	(*4*) $D = Y_{i1} - Y_{i2}$ (*difference*)
1	7.33	7.53	14.86	0.20
2	7.49	7.70	15.19	.21
3	7.27	7.46	14.73	.19
4	7.93	8.21	16.14	.28
5	7.56	7.81	15.37	.25
6	7.81	8.01	15.82	.20
7	7.46	7.72	15.18	.26
8	6.94	7.13	14.07	.19
9	7.49	7.68	15.17	.19
10	7.44	7.66	15.10	.22
11	7.95	8.11	16.06	.16
12	7.47	7.66	15.13	.19
13	7.04	7.20	14.24	.16
14	7.10	7.25	14.35	.15
15	7.64	7.79	15.43	.15
ΣY	111.92	114.92	226.84	3.00
ΣY^2	836.3300	881.8304	3435.6992	0.6216

Source: From a larger study by Newman and Meredith (1956).

I. This example is first analyzed as a *randomized complete blocks design.* It is a mixed model with the $a = 2$ ages as fixed treatment effects and the b individuals constituting the randomly chosen blocks.

Preliminary computations

1. Grand total $= \sum^a \sum^b Y = 111.92 + 114.92 = 226.84$

2. Sum of the squared observations $\sum^a \sum^b Y^2$

$$= (7.33)^2 + \cdots + (7.79)^2 = 836.3300 + 881.8304 = 1718.1604$$

3. Sum of squared column totals divided by sample size of a column

$$= \frac{\sum^a \left(\sum^b Y\right)^2}{b} = \frac{[(111.92)^2 + (114.92)^2]}{15} = \frac{25732.6928}{15} = 1715.5129$$

4. Sum of squared row totals divided by sample size of a row

$$= \frac{\sum^b \left(\sum^a Y\right)^2}{a} = \frac{[(14.86)^2 + \cdots + (15.43)^2]}{2} = \frac{3435.6992}{2} = 1717.8496$$

BOX 11.3 continued

5. Grand total squared and divided by the total sample size = correction term

$$CT = \frac{\left(\sum^{a}\sum^{b} Y\right)^2}{ab} = \frac{(226.84)^2}{2 \times 15} = 1715.2129$$

6. SS_{total} (= SS_{subgr}, since there is only one observation per cell)

$$= \sum^{a}\sum^{b} Y^2 - CT = \text{quantity } \mathbf{2} - \text{quantity } \mathbf{5}$$
$$= 1718.1604 - 1715.2129 = 2.9475$$

7. SS_A (SS of columns; treatments) $= \dfrac{\sum^{a}\left(\sum^{b} Y\right)^2}{b} - CT$

$$= \text{quantity } \mathbf{3} - \text{quantity } \mathbf{5} = 1715.5129 - 1715.2129 = 0.3000$$

8. SS_B (SS of rows; blocks) $= \dfrac{\sum^{b}\left(\sum^{a} Y\right)^2}{a} - CT$

$$= \text{quantity } \mathbf{4} - \text{quantity } \mathbf{5} = 1717.8496 - 1715.2129 = 2.6367$$

9. SS_{error} (remainder; discrepance; residual SS) $= SS_{\text{total}} - SS_A - SS_B$
$= \text{quantity } \mathbf{6} - \text{quantity } \mathbf{7} - \text{quantity } \mathbf{8} = 2.9475 - 2.6367 - 0.3000$
$= 0.0108$

Anova table

Source of variation	*df*	*SS*	*MS*	F_s	*Expected MS*
Ages (columns factor A)	1	0.3000	0.3000	389.11***	$\sigma^2 + \sigma^2_{AB} + \frac{b}{a-1}\Sigma\alpha^2$
Individuals (rows factor B)	14	2.6367	0.1883	242.02***	$\sigma^2 + a\sigma^2_B$
Remainder	14	0.0108	0.000771		$\sigma^2 + \sigma^2_{AB}$
Total	29	2.9475			

$F_{.001[1,14]} = 17.1 \qquad F_{.001[14,14]} = 5.94$

Conclusions.—The variance ratio for ages is highly significant. We conclude that faces of 6-year-old girls are wider than those of 5-year-olds. If we are willing to assume that the interaction σ^2_{AB} is zero, we may test for an added variance component among individual girls and would find it significant.

II. This same analysis can be carried out using the *t-test for paired comparisons,*

$$t_s = \frac{\bar{D} - (\mu_1 - \mu_2)}{s_{\bar{D}}}$$

where $\bar{D}$ is the mean difference between the paired observations

$(\bar{D} = \sum D/b = 3.00/15 = 0.20)$

BOX 11.3 continued

and $s_{\bar{D}} = s_D/\sqrt{b}$ is the standard error of $\bar{D}$ calculated from the observed differences in column (4).

$$s_D = \sqrt{\frac{\sum D^2 - (\sum D)^2/b}{b-1}} = \sqrt{\frac{0.6216 - (3.00^2/15)}{14}} = \sqrt{0.0216/14}$$

$$= \sqrt{0.001543} = 0.0392810$$

$$s_{\bar{D}} = \frac{s}{\sqrt{b}} = \frac{0.0392810}{\sqrt{15}} = 0.0101423$$

We assume that the true difference between the means of the two groups, $\mu_1 - \mu_2$, equals zero.

$$t_s = \frac{\bar{D} - 0}{s_{\bar{D}}} = \frac{0.20 - 0}{0.0101423} = 19.7194 \qquad \text{with } b - 1 = 14\ df, \text{ and } P \ll 0.001.$$

$$t_s^2 = 388.85$$

which agrees with the previous F_s within an acceptable rounding error.

highly significant difference in face width between the two age groups. If interaction is assumed to be zero, there is a large added variance component among the individual girls, undoubtedly representing genetic as well as environmental differences.

The other method of analyzing paired comparisons designs is the well-known t-*test for paired comparisons.* It is quite simple to apply and is illustrated in the second half of Box 11.3. It tests whether the mean of sample differences between pairs of readings in the two columns is significantly different from a hypothetical mean, which the null hypothesis puts at zero. The standard error over which this is tested is the standard error of the mean difference. The difference column has to be calculated and is shown in column (4) of the data table in Box 11.3. The computations are quite straightforward and the conclusions are the same as for the two-way anova. This is another instance in which we obtain the value of F_s within rounding error when we square the value of t_s.

Although the paired comparisons *t*-test is the traditional method of solving this type of problem, we prefer the two-way anova. Its computation is no more time-consuming, it avoids finding a square root, and it has the advantage of providing a measure of the variance component among the blocks. This is useful knowledge because if there is no significant added variance component among blocks one might simplify the analysis and design of future, similar studies by employing a completely randomized anova.

In the example in Box 11.3 we see again how important it is that we control the blocks (individuals) in this design. We could simply have analyzed these data as a completely randomized anova or as a *t*-test of the difference between two means. If we had done so we would have found that the two age groups are not significantly different. A completely randomized anova would yield an F_s-ratio for ages of 3.172, which is not significantly greater than $F_{.05[1,28]}$. We could have predicted such an outcome. If we estimate the completely randomized error mean square by Expression (11.3) from the anova table in Box 11.3, we obtain a value of $MS_{E(CR)} = 0.091323$, which yields a relative efficiency of

$$RE = \frac{100 \times 0.091323}{0.000771} = 11{,}844.7\%$$

for the paired comparisons (= randomized blocks) design over the completely randomized design, a tremendous improvement. Consequently it does not surprise us that failure to allow for differences among individual girls will lead us to erroneous results about the differences in face width.

11.6 Unequal subclass sizes

All examples of two-way anovas considered so far have had equal sample sizes for each subclass. However, even with the best of intentions it is frequently impossible for an investigator to produce such an evenly balanced design. A given single replicate may be lost or ruined, or the data may vary naturally beyond the control of the investigator. For instance, in an experiment on the effects of treatment of parents on weight of their offspring, the number of offspring per parent are likely to differ among parents and will not be under the control of the investigator.

How can we analyze such data? As we will see, the analysis of two-way anovas with unequal subclass sizes is quite a complicated matter. If at all possible we try to avoid such problems, sometimes even at the expense of losing some information. Thus if in an analysis with four rows and three columns we have at least seven replicates per subclass, but in some subclasses have eight or nine replicates, we might be tempted to reduce the sample size of all subclasses to seven and proceed by the methods learned in the previous sections. Such a procedure is legitimate only if the subclass sample size itself does not affect the variable under study. For example, the number of offspring per mother might affect their weight. Another consideration is that removal of any individuals from subclasses (to equalize sample sizes) must be done at random and each individual within a subclass must have an equal likelihood of being removed in the process of reducing sample size. In some instances, reducing the subclasses to a common sample size would not be advisable because error variance is very great and all available degrees of freedom are needed for the analysis of variance, or possibly because the original data were scarce and expensive to obtain and the relative time and expense of a complete analysis would be insignificant in comparison with the cost of obtaining the data. Thus if the readings are measurements of fossils collected as the result of a long and expensive expedition to a distant land, clearly the investigator would wish to use every single specimen, even though subclass sizes would be unequal and the analysis consequently tedious.

There is one instance of unequal subclass sizes that is relatively simple to analyze. These are two-way anovas with *unequal but proportional subclass sizes* (*numbers*) in which the ratios of the sample sizes of any one pair of rows are the same for all columns (or conversely, the ratios of sample sizes of any pair of columns are the same for all rows). Box 11.4 illustrates a simple example of this sort. It represents an experiment studying the effect of thyroxin

BOX 11.4

Two-way anova with unequal but proportional subclass numbers.

Influence of thyroxin injections on seven-week weight of chicks (in grams). Sample sizes: $n_{11} = n_{12} = 12$; $n_{21} = n_{22} = 8$; total number of items $= \sum^{a}\sum^{b} n_{ij} = 40$. This is a Model I anova.

Sexes (*b = 2*)	*Treatment (a = 2)* Control	Thyroxin injection	Σ
	560	410	
	500	540	
	350	340	
	520	580	
	540	470	
Males	620	550	
	600	480	
	560	440	
	450	600	
	340	450	
	440	420	
	300	550	
	5780	5830	11,610
	530	550	
	580	420	
	520	370	
Females	460	600	
	340	440	
	640	560	
	520	540	
	560	520	
	4150	4000	8,150
Σ	9930	9830	19,760

Source: Data from Radi and Warren (1938).

BOX 11.4 continued

Preliminary computations

1. Grand total $= \sum^{a}\sum^{b}\sum^{n_{ij}} Y = 19{,}760$

2. Sum of the squared observations $= \sum^{a}\sum^{b}\sum^{n_{ij}} Y^2$

$$= (560)^2 + \cdots + (520)^2 = 10{,}061{,}800$$

3. Sum of the squared subgroup (cell) totals each divided by its sample size n_{ij}

$$= \sum^{a}\sum^{b}\frac{\left(\sum^{n_{ij}} Y\right)^2}{n_{ij}} = \frac{(5780)^2}{12} + \frac{(5830)^2}{12} + \frac{(4150)^2}{8}$$

$$+ \frac{(4000)^2}{8} = 9{,}769{,}254.17$$

4. Sum of the squared column totals each divided by its sample size $\sum^{b} n_{ij}$

$$= \sum^{a}\frac{\left(\sum^{b}\sum^{n_{ij}} Y\right)^2}{\sum^{b} n_{ij}} = \frac{(9930)^2}{20} + \frac{(9830)^2}{20}$$

$$= 9{,}761{,}690.00$$

5. Sum of the squared row totals each divided by its sample size $\sum^{a} n_{ij}$

$$= \sum^{b}\frac{\left(\sum^{a}\sum^{n_{ij}} Y\right)^2}{\sum^{a} n_{ij}} = \frac{(11{,}610)^2}{24} + \frac{(8150)^2}{16}$$

$$= 9{,}767{,}743.75$$

6. Grand total squared and divided by total sample size = correction term

$$CT = \frac{\left(\sum^{a}\sum^{b}\sum^{n_{ij}} Y\right)^2}{\sum^{a}\sum^{b} n_{ij}} = \frac{(\text{quantity } \mathbf{1})^2}{\sum^{a}\sum^{b} n_{ij}}$$

$$= \frac{(19{,}760)^2}{40} = 9{,}761{,}440.00$$

7. $SS_{\text{total}} = \sum^{a}\sum^{b}\sum^{n_{ij}} Y^2 - CT = \text{quantity } \mathbf{2} - \text{quantity } \mathbf{6}$

$$= 10{,}061{,}800 - 9{,}761{,}440.00 = 300{,}360.00$$

8. $SS_{\text{subgr}} = \sum^{a}\sum^{b}\frac{\left(\sum^{n_{ij}} Y\right)^2}{n_{ij}} - CT = \text{quantity } \mathbf{3} - \text{quantity } \mathbf{6}$

$$= 9{,}769{,}254.16 - 9{,}761{,}440.00 = 7814.16$$

9. SS_A (SS of columns) $= \sum^{a}\frac{\left(\sum^{b}\sum^{n_{ij}} Y\right)^2}{\sum^{b} n_{ij}} - CT$

$$= \text{quantity } \mathbf{4} - \text{quantity } \mathbf{6} = 9{,}761{,}690.00 - 9{,}761{,}440.00 = 250.00$$

BOX 11.4 continued

10. SS_B (SS or rows) $= \sum^{b} \frac{\left(\sum^{a}\sum^{n_{ij}} Y\right)^2}{\sum^{a} n_{ij}} - CT$

= quantity **5** − quantity **6** = 9,767,743.75 − 9,761,440.00 = 6303.75

11. SS_{AB} (Interaction SS) = $SS_{\text{subgr}} - SS_A - SS_B$ = quantity **8** − quantity **9** − quantity **10** = 7814.16 − 6303.75 − 250.00 = 1260.41

12. SS_{within} (within subgroups; error SS) = $SS_{\text{total}} - SS_{\text{subgr}}$ = quantity **7** − quantity **8** = 300,360.00 − 7814.16 = 292,545.84

Anova table

Source of variation	*df*	*SS*	*MS*
Subgroups	3	7,814.16	2,604.72
B (rows, sex)	1	6,303.75	6,303.75
A (columns, treatments)	1	250.00	250.00
$A \times B$ (interaction)	1	1,260.41	1,260.41
Within subgroups (error)	36	292,545.84	8,126.27
Total	39	300,360.00	

Since the error MS is greater than any other mean square, none of the latter is significant. We therefore conclude that thyroxin injections do not have any effect on the weight of seven-week-old chicks and also that the sexes do not differ in this respect at this stage.

The method for testing significance in such an anova is discussed in Section 11.6.

injections *versus* a control on the weight of seven-week-old chicks. These effects were studied for 12 male chicks and 8 female chicks each. Note that the ratio of number of males over females (the two rows), which is 12/8 = 1.5, is the same for both columns—for the controls as well as the experimentals (thyroxin-injected). We can also demonstrate proportionality of subclass sizes by the converse relationship. The ratio of the two columns (control and thyroxin) is the same (12/12 = 8/8 = 1.0) for both rows (sexes).

The computation of the sums of squares, outlined in Box 11.4, follows the familiar series of steps explained in Box 11.1. There is a minor difference in symbolism because of the inequality of the subclass sample sizes, and quantity **3** is computed differently. Each squared subgroup total is divided by its own sample size because this differs among the subgroups. The completed analysis of variance is shown at the end of the box. It needs no special testing because the error variance in this example is far greater than the treatment mean squares.

↔ However, how should we have tested if the treatment mean squares had been greater than the error variance? The procedure to be followed would be that of Section 11.3, illustrated in Box 11.1. There is only one further complication: the coefficients of the variance components differ for each level of variation. We first encountered this phenomenon in Section 10.3 when we worked out examples of nested analysis with unequal subclass sizes. Whenever a given mean square has to be tested over the error variance, this difference in coefficients is of no importance because it does not enter into the equation. However, as soon as we test over interaction (Model II and fixed treatment effects of a mixed model), we have to allow for the differences in coefficients. The evaluation of these coefficients is somewhat tedious and we shall not present it here. The reader who runs into such a problem should consult Snedecor (1956, Section 12.13) where such an example is worked out in great detail.

In two-way anovas with *disproportional as well as unequal subclass sizes* the computational procedures are considerably more complicated and beyond the scope of the present text. The reasons for the complication is that such analyses are no longer orthogonal—that is, the separate effects are not independent and their sums of squares do not add up to yield the total sums of squares. The reader faced with an anova with disproportional unequal sample sizes is advised to consult chapter 13 in Steel and Torrie (1960), where this subject is treated in some detail. These computations were once discouragingly tedious, but can now be carried out routinely and efficiently by means of a computer. The *BMD* programs, made available by the Health Sciences Computing Facility, University of California, Los Angeles, are a good example of programs for this type of analysis. ↚

11.7 Missing values in a randomized blocks design

Another irregularity that may be encountered (for some of the same reasons as those explained in the previous section) is the problem of missing values in a two-way analysis without replication, usually in a randomized complete blocks design. Since there is no replication, a missing value means that one of the factor combinations is not represented in the analysis. Such an example is shown in Box. 11.5, in which the amount of lactic acid production is measured in frog embryos of six developmental stages. The embryos had been obtained from four different egg clutches, which for the purposes of this experiment could be considered to have been randomly obtained; the developmental stages clearly represent fixed treatments. The anova is therefore a mixed model.

BOX 11.5

Single missing observation in a two-way anova without replication.

Homogenate glycolysis for different developmental stages of frog embryos. Measurement of the variate (lactic acid production) is made at 6 stages: 0, 360, 720, 1200, 1600, and 2000 minutes after first cleavage (temperature constant at 20°C). Total sample size minus missing value $= ab - 1 = 23$. This is a mixed model anova.

b stages $(b = 6)$	*a clutches* $(a = 4)$ *1*	*2*	*3*	*4*	ΣY	$\bar{Y}$
I	21.4	. . .	7.0	9.5	(37.9)	12.6
II	14.3	13.5	5.4	6.6	39.8	10.0
III	23.4	14.1	5.9	7.1	50.5	12.6
IV	29.1	8.2	4.2	3.2	44.7	11.2
V	26.6	13.5	4.9	6.0	51.0	12.8
VI	21.7	5.2	6.6	5.9	39.4	9.8
ΣY	136.5	(54.5)	34.0	38.3	(263.3)	
$\bar{Y}$	22.8	10.9	5.7	6.4		

Source: Data from Cohen (1954).

"Clutches" refers to egg clutches randomly chosen from material available for study. Since stages are fixed treatment effects, this is a mixed model.

Before the analysis can be carried out we must calculate an estimate for the missing observation for stage I, clutch 2:

$$\hat{Y}_{ij} = \frac{b\left(\sum^{a-1} Y\right)_i + a\left(\sum^{b-1} Y\right)_i - \sum^{ab-1} Y}{(a-1)(b-1)}$$

$$\hat{Y}_{1,2} = \frac{6(37.9) + 4(54.5) - 263.3}{(6-1)(4-1)} = \frac{182.1}{15} = 12.1$$

This value is then entered into the above table and the anova is carried out as usual, except that the degrees of freedom for error and total are each decreased by unity and the row *SS* and column *SS* are both adjusted by a special correction before their *MS* is computed.

The anova table *before* the adjustment of *SS* (but using the estimated value of $\hat{Y}_{ij} = 12.1$):

Source of variation	*df*	*SS*
A (Columns; blocks; clutches)	3	1121.57
B (Rows; treatments)	5	36.22
Error	15	184.72
Total	23	1342.51

BOX 11.5 continued

$$\text{Adjusted } SS_A \text{ (blocks)} = SS_A - \frac{\left[\left(\sum^{a-1} Y\right)_j + a\left(\sum^{b-1} Y\right)_i - \sum^{ab-1} Y\right]^2}{a(a-1)(b-1)^2}$$

$$= 1121.57 - \frac{[37.9 + 4(54.5) - 263.3]^2}{4 \times 3 \times (5)^2}$$

$$= 1121.57 - \frac{(255.9 - 263.3)^2}{300} = 1121.57 - 0.18$$

$$= 1121.39$$

$$\text{Adjusted } SS_B \text{ (treatments)} = SS_B - \frac{\left[\left(\sum^{b-1} Y\right)_i + b\left(\sum^{a-1} Y\right)_j - \sum^{ab-1} Y\right]^2}{b(b-1)(a-1)^2}$$

$$= 36.22 - \frac{[54.5 + 6(37.9) - 263.3]^2}{6 \times 5 \times (3)^2}$$

$$= 36.22 - \frac{(281.9 - 263.3)^2}{270}$$

$$= 34.94$$

Completed anova

Source of variation	*df*	*SS*	*MS*	F_s
A (Columns; blocks; clutches)	3	1121.39	373.797	28.34**
B (Rows; treatments)	5	34.94	6.988	
Error	14	184.72	13.19	
Total	22	1342.51		

Conclusions.—Lactic acid production did not differ among developmental stages but was quite heterogeneous among clutches of eggs (added variance component significant if interaction is assumed to be absent).

In this example the missing value is that for stage I, clutch 2. Such a situation may have come about from accidental loss of this replicate or because in this clutch all the individuals had already developed into stage II. The first step in solving such a problem is to find an estimated value $\hat{Y}_{ij}$ for the missing variate. This is done by means of the estimation equation shown in Box 11.5. Evaluating this formula for the present example we obtain an estimated lactic acid production of 12.1 units for stage I, clutch 2. The column and row sums used in this estimation equation are the sums of the row and column in which the missing value occurs. It is for this reason that the summation signs carry the index $a - 1$ and $b - 1$. Once the missing value has been estimated the computations proceed as previously indicated (Box 11.2). The anova table obtained is also shown in Box 11.5. However, a further adjustment to the sums of squares has to be made. This adjustment for both main effects SS is also shown in Box 11.5. The adjustment for SS_B is small and that for SS_A is quite trivial, but in other instances this would not necessarily be so. Note that in the completed anova one degree of freedom has been removed from the error SS and the total SS because of the missing value. Since the missing value was estimated from the marginal totals, it obviously does not represent an independent degree of freedom. Thus, although this is a 6 × 4 table, the total degrees of freedom are only 22.

↠ When two values are missing in a randomized complete blocks design a similar but iterative procedure is used, which is described in Section 8.5 of Steel and Torrie (1960) or Section 11.9 of Snedecor (1956). Other approaches (such as the analysis of covariance) are possible, and you will be led to them through these references. ↞

Exercises 11

11.1 Swanson, Latshaw, and Tague (1921) determined soil pH electrometrically for various soil samples from Kansas. An extract of their data (acid soils) is shown below. Do subsoils differ in pH from surface soils (assume that there is no interaction between localities and depth for pH reading)? ANS. $F_s = 0.894$.

County	*Soil type*	*Surface pH*	*Subsoil pH*
Finney	Richfield silt loam	6.57	8.34
Montgomery	Summit silty clay loam	6.77	6.13
Doniphan	Brown silt loam	6.53	6.32
Jewell	Jewell silt loam	6.71	8.30
Jewell	Colby silt loam	6.72	8.44
Shawnee	Crawford silty clay loam	6.01	6.80
Cherokee	Oswego silty clay loam	4.99	4.42
Greenwood	Summit silty clay loam	5.49	7.90
Montgomery	Cherokee silt loam	5.56	5.20
Montgomery	Oswego silt loam	5.32	5.32
Cherokee	Bates silt loam	5.92	5.21
Cherokee	Cherokee silt loam	6.55	5.66
Cherokee	Neosho silt loam	6.53	5.66

11.2 The following data were extracted from a Canadian record book of purebred dairy cattle. Random samples of 10 mature (five-year-old and older) and 10 two-year-old cows were taken from each of five breeds (honor roll, 305-day class). The average butterfat percentages of these cows were recorded. This gave us a total of 100 butterfat percentages, broken down into five breeds and into two age classes. The 100 butterfat percentages are given below. Analyze and discuss your results. You will note that the tedious part of the calculation has been done for you.

	Breed									
	Ayrshire		*Canadian*		*Guernsey*		*Holstein-Friesian*		*Jersey*	
	Mature	*2-yr*	*Mature*	*2-yr*	*Mature*	*2-yr*	*Mature*	*2-yr*	*Mature*	*2-yr*
	3.74	4.44	3.92	4.29	4.54	5.30	3.40	3.79	4.80	5.75
	4.01	4.37	4.95	5.24	5.18	4.50	3.55	3.66	6.45	5.14
	3.77	4.25	4.47	4.43	5.75	4.59	3.83	3.58	5.18	5.25
	3.78	3.71	4.28	4.00	5.04	5.04	3.95	3.38	4.49	4.76
	4.10	4.08	4.07	4.62	4.64	4.83	4.43	3.71	5.24	5.18
	4.06	3.90	4.10	4.29	4.79	4.55	3.70	3.94	5.70	4.22
	4.27	4.41	4.38	4.85	4.72	4.97	3.30	3.59	5.41	5.98
	3.94	4.11	3.98	4.66	3.88	5.38	3.93	3.55	4.77	4.85
	4.11	4.37	4.46	4.40	5.28	5.39	3.58	3.55	5.18	6.55
	4.25	3.53	5.05	4.33	4.66	5.97	3.54	3.43	5.23	5.72
$\sum Y$	40.03	41.17	43.66	45.11	48.48	50.52	37.21	36.18	52.45	53.40
$\bar{Y}$	4.003	4.117	4.366	4.511	4.848	5.052	3.721	3.618	5.245	5.340

$$\sum^{abn} Y^2 = 2059.6109$$

11.3 King et al. (1964) gave the following results for a study of the amount of cotton (in grams) used for nesting material in both sexes of two subspecies of the deer-mouse *Peromyscus maniculatus.*

		P. m. gracilis	*P. m. bairdii*
♂	$\bar{Y}$	2.9	1.7
	s	1.4	0.9
	n	24	24
♀	$\bar{Y}$	2.6	2.1
	s	1.0	1.0
	n	26	26

Analyze and interpret. Note that you will have to compute the error mean square as a weighted average of the individual variances. ANS. $MS_{\text{sex}} = 0.0624$, $MS_{\text{within}} = 1.1845$.

11.4 Blakeslee (1921) studied length/width ratios of second seedling leaves of two types of Jimson weed called globe (G) and nominal (N). Three seeds of each type were planted in 16 pots. Is there sufficient evidence to conclude that globe and nominal differ in length/width ratio?

Pot identification number	*Types*					
	G			*N*		
16533	1.67	1.53	1.61	2.18	2.23	2.32
16534	1.68	1.70	1.49	2.00	2.12	2.18
16550	1.38	1.76	1.52	2.41	2.11	2.60
16668	1.66	1.48	1.69	1.93	2.00	2.00
16767	1.38	1.61	1.64	2.32	2.23	1.90
16768	1.70	1.71	1.71	2.48	2.11	2.00
16770	1.58	1.59	1.38	2.00	2.18	2.16
16771	1.49	1.52	1.68	1.94	2.13	2.29
16773	1.48	1.44	1.58	1.93	1.95	2.10
16775	1.28	1.45	1.50	1.77	2.03	2.08
16776	1.55	1.45	1.44	2.06	1.85	1.92

(continued)

Pot identification number	*Types*					
	G			*N*		
16777	1.29	1.57	1.44	2.00	1.94	1.80
16780	1.36	1.22	1.41	1.87	1.87	2.26
16781	1.47	1.43	1.61	2.24	2.00	2.23
16787	1.52	1.56	1.56	1.79	2.08	1.89
16789	1.37	1.38	1.40	1.85	2.10	2.00

12 MULTIWAY ANALYSIS OF VARIANCE

We now extend the two-way analysis of variance of the previous chapter to the simultaneous consideration of three or more factors. In Section 12.1 we briefly discuss the nature of the design and the new problems encountered during the analysis. A three-factor (three-way) anova is illustrated in Section 12.2. The next section (12.3) treats multiway anovas with more than three main effects. Section 12.4 enumerates some of the more common experimental designs other than those already learned, which might be employed by the reader of this book. Appropriate instances for application are given and references are cited to a detailed exposition of each design. Finally, in Section 12.5, we discuss computer methods for the analysis of variance.

12.1 The factorial design

In the last chapter we analyzed the effects of two factors on a sample. There is, however, no reason to restrict the design of such an analysis to a consideration of only two factors. Three or more factors may be analyzed simultaneously. Such an analysis is often called *factorial analysis of variance.* Thus, in the example in the next section we shall analyze the effects of temperature, oxygen concentration, and cyanide ion concentration on survival time in minnows. The number of main effects is theoretically unlimited. However, an analysis involving as many as five main effects is a rarity for the following reasons. Even without replication within a subgroup, the number of experimental units necessary becomes very large, and it is frequently impossible or prohibitive in cost to carry out such an experiment. By way of an illustration, imagine an experiment with only three factors. If each of the three factors is divided into four classes or levels, we would have to carry out the experiment on $4 \times 4 \times 4 = 64$ experimental units to represent each combination of factors. This is a substantial number. You will realize that this would not permit any measurement of the basic experimental error and we would have to employ an interaction term as a measurement of experimental error (on the assumption that no added interaction effect is present). In the minimal case of two levels per factor, a five-factor factorial anova would require $2^5 = 32$ experimental units.

Another problem which accompanies a factorial anova with several main effects is the large number of possible interactions. We saw that a two-way anova (two-factor factorial) had only one interaction, $A \times B$. A three-factor factorial has three *first-order interactions*, $A \times B$, $A \times C$, and $B \times C$; it also has a *second-order interaction*, $A \times B \times C$. A four-factor factorial has six first-order interactions, $A \times B$, $A \times C$, $A \times D$, $B \times C$, $B \times D$, and $C \times D$; four second-order interactions, $A \times B \times C$, $A \times B \times D$, $A \times C \times D$, and $B \times C \times D$; and one *third-order interaction*, $A \times B \times C \times D$. These numbers go up rapidly. $C(k, m + 1) = k!/[(m + 1)!(k - m - 1)!]$ mth-order interactions exist in a k-factor factorial, representing the number of combinations of $m + 1$ items out of the k factors. Not only is the computation of these interactions tedious but the testing of their significance and, more importantly, their interpretation becomes exceedingly complex.

The assumptions underlying multiway anova are the same as those for two-way anova. The expected value for a single observation in a (replicated) three-factor case in which all three main effects, A, B, and C, are fixed treatment effects (Model I) is

$$Y_{ijkl} = \mu + \alpha_i + \beta_j + \gamma_k + (\alpha\beta)_{ij} + (\alpha\gamma)_{ik} + (\beta\gamma)_{jk} + (\alpha\beta\gamma)_{ijk} + \epsilon_{ijkl}$$

where μ equals the parametric mean of the population; α_i, β_j, γ_k are the fixed treatment effects for the ith, jth, and kth groups of treatments A, B, and C, respectively; $(\alpha\beta)_{ij}$, $(\alpha\gamma)_{ik}$, and $(\beta\gamma)_{jk}$ are first-order interaction effects in the subgroups represented by the indicated combinations of the ith group of factor A, the jth group of factor B, and the kth group of factor C; $(\alpha\beta\gamma)_{ijk}$ is the second-order interaction effect in the subgroup representing the ith, jth, and kth groups of factors A, B, and C, respectively; and ϵ_{ijkl} is the error term of the lth item in subgroup ijk. This expression is analogous to Expressions (8.3), (10.2), and (11.1). As before, all the main effects could be random (Model II), or only some of them could be, in which case the factorial anova is a mixed model.

We now proceed to analyze a three-way factorial to illustrate the procedures and problems encountered.

12.2 A three-way factorial anova

The example chosen here comes from a study of pollution by factory effluents. The authors measured the effects of different concentrations of cyanide ion on the survival time of minnows. Five concentrations of cyanide were combined with three concentrations of oxygen and three temperatures. The details of the experiment are given in Box 12.1. For each combination of factors ten replicate fishes were examined. However, the data shown here give only the sum of the ten readings and thus do not provide an estimate of within-subgroup variance.

The basic data are tabulated in Box 12.1. Note that since this is a three-factor example, the third factor (oxygen concentration) is arranged within each class or level of the first factor (temperature), because we are limited by the two-dimensionality of the printed page. However, this arrangement does not imply that this is a nested anova; the factors are orthogonal to each other.

Next we arrange the data as three two-way tables as shown in Box 12.1. These tables are produced by summing all levels of the third factor for each two-factor combination. Thus, the value for cyanide level 3 against temperature level 2 is the sum of all the oxygen concentrations for that combination of temperature and cyanide.

The computations shown in Box 12.1 are longer than those of previous anovas but are essentially no more complicated. Some are carried out on the original data table, others on the three two-way tables. Please note that in this example there is no replication for each subgroup; hence there will be no error mean square and, as in the example of Box 11.2, the subgroup SS is the same as the total SS. Thus steps **9** and **18** are not carried out in the present example but would be undertaken in a three-way anova *with* replication.

The layout of the anova table is shown next in Box 12.1. Note that the source of variation is subdivided into three main effects, A, B, and C, three first-order interactions, $A \times B$, $A \times C$, and $B \times C$, and one second-order interaction, $A \times B \times C$. Had there been replication in these data we would also have had a within-subgroups (error) line below the second-order interaction. The experiment with the minnows is clearly Model I. Therefore, the expected mean squares in the table consist simply of the error variance σ^2 and the added effect due to treatment for the particular source of variation. The significance test is very simple. The mean square for any source of variation is tested over the error mean square. Since we do not have a pure measure of error in this example, we must use the second-order interaction mean square on the assumption that the added effect due to the $A \times B \times C$ interaction $\sum^{abc}(\alpha\beta\gamma)^2/(a-1)(b-1)(c-1)$ is zero. The meaning of a second-order interaction will be explained presently. As mentioned in the last chapter, main effects in a Model I anova are rarely tested if the interactions are significant. However, if one wishes for such a test it is simply the ratio of the main effect mean square over the error mean square.

Box 12.1 also shows expected mean squares of a three-factor factorial

BOX 12.1

Three-way anova without replication. (A 5 × 3 × 3 factorial anova.)

Time to intoxication by cyanide in *Phoxinus laevis*, a European minnow, using 5 concentrations of CN ion, 3 oxygen concentrations, and 3 temperatures. The variable is a transformation into logarithms of readings in minutes of survival time, coded and summed for 10 replicate fishes.

Factor A ($a = 3$) *Temperatures in °C*		*Factor B* ($b = 5$) *CN concentrations in mg CN^-/liter*		*Factor C* ($c = 3$) *Oxygen concentrations in mg O_2/liter*	
T_1	5	Cy_1	0.16	O_1	1.5
T_2	15	Cy_2	0.8	O_2	3.0
T_3	25	Cy_3	4.0	O_3	9.0
		Cy_4	20.0		
		Cy_5	100.0		

Source: Data by Wuhrmann and Woker (1953).

This is an example in which all three main effects are fixed treatments; it is therefore a Model I anova.

Factor A	*Factor C*	*Factor B*					
		Cy_1	Cy_2	Cy_3	Cy_4	Cy_5	Σ
T_1	O_1	201	150	131	130	97	709
	O_2	246	164	138	136	102	786
	O_3	271	170	149	127	99	816
		718	484	418	393	298	2311
T_2	O_1	124	104	86	89	60	463
	O_2	158	111	99	91	74	533
	O_3	207	117	81	87	72	564
		489	332	266	267	206	1560

T_3	O_1	79	63	50	51	32	275
	O_2	129	54	51	52	46	332
	O_3	142	93	62	51	52	400
		350	210	163	154	130	1007
Σ		1557	1026	847	814	634	4878

For ease in computation, rearrange the data in three two-way tables as shown below. The value in each cell of a two-way table represents the sum of the items for the particular two-factor combination over all levels of the third factor. Thus, the value 718 in the $A \times B$ table represents the combination $T_1 \times Cy_1$ summed for the three oxygen concentrations O_1, O_2, and O_3 as follows: $201 + 246 + 271 = 718$. The marginal totals are computed as checks.

Two-way tables

$A \times B$ (T × Cy *in this example*)

	Cy_1	Cy_2	Cy_3	Cy_4	Cy_5	Σ
T_1	718	484	418	393	298	2311
T_2	489	332	266	267	206	1560
T_3	350	210	163	154	130	1007
Σ	1557	1026	847	814	634	4878

$A \times C$ (T × O *in this example*)

	O_1	O_2	O_3	Σ
T_1	709	786	816	2311
T_2	463	533	564	1560
T_3	275	332	400	1007
Σ	1447	1651	1780	4878

$B \times C$ (Cy × O *in this example*)

	O_1	O_2	O_3	Σ
Cy_1	404	533	620	1557
Cy_2	317	329	380	1026
Cy_3	267	288	292	847
Cy_4	270	279	265	814
Cy_5	189	222	223	634
Σ	1447	1651	1780	4878

BOX 12.1 continued

Preliminary computations

1. Grand total $= \sum^{a}\sum^{b}\sum^{c} Y = 201 + 246 + \cdots + 46 + 52 = 4878$

2. Sum of the squared observations $= \sum^{a}\sum^{b}\sum^{c} Y^2 = (201)^2 + (246)^2 + \cdots + (46)^2 + (52)^2 = 655{,}278$

The following quantities are most conveniently computed from the two-way tables:

3. Sum of the squared totals for factor A (temperatures) divided by bc, the sample size of each total for A

$$= \frac{\sum^{a}\left(\sum^{b}\sum^{c} Y\right)^2}{bc} = \frac{[(2311)^2 + (1560)^2 + (1007)^2]}{(3 \times 5)} = 585{,}891.33$$

4. Sum of the squared totals for factor B (cyanide concentrations) divided by ac, the sample size of each total for B

$$= \frac{\sum^{b}\left(\sum^{a}\sum^{c} Y\right)^2}{ac} = \frac{[(1557)^2 + (1026)^2 + (847)^2 + (814)^2 + (634)^2]}{(3 \times 3)} = 584{,}320.67$$

5. Sum of the squared totals for factor C (oxygen concentrations) divided by ab, the sample size of each total for C

$$= \frac{\sum^{c}\left(\sum^{a}\sum^{b} Y\right)^2}{ab} = \frac{[(1447)^2 + (1651)^2 + (1780)^2]}{(5 \times 3)} = 532{,}534.00$$

6. Sum of the squared cell totals divided by c, the sample size of each cell, from the $A \times B$ (temperature $\times$ cyanide) two-way table

$$= \frac{\sum^{a}\sum^{b}\left(\sum^{c} Y\right)^2}{c} = \frac{[(718)^2 + (489)^2 + \cdots + (206)^2 + (130)^2]}{3} = 645{,}122.67$$

7. Sum of the squared cell totals divided by b, the sample size of each cell, from the $A \times C$ (temperature $\times$ oxygen) two-way table

$$= \frac{\sum^{a}\sum^{c}\left(\sum^{b} Y\right)^2}{b} = \frac{[(709)^2 + (786)^2 + \cdots + (332)^2 + (400)^2]}{5} = 589{,}747.20$$

8. Sum of the squared cell totals divided by a, the sample size of each cell, from the $B \times C$ (cyanide $\times$ oxygen) two-way table

$$= \frac{\sum^{b}\sum^{c}\left(\sum^{a} Y\right)^2}{a} = \frac{[(404)^2 + (533)^2 + \cdots + (222)^2 + (223)^2]}{3} = 593{,}344.00$$

9. If we had more than a single observation per cell in the original three-way table, we would at this stage compute the sum of the squared cell totals for the three-way table divided by n, the sample size of each cell in that table. In our case $\frac{\sum^{a}\sum^{b}\sum^{c}\left(\sum^{n} Y\right)^2}{n}$, quantity **9**, is identical to quantity **2**.

10. Grand total squared and divided by the total sample size = correction term $CT = \frac{\left(\sum^{a}\sum^{b}\sum^{c} Y\right)^2}{abc}$

$$= \frac{(4878)^2}{(3 \times 5 \times 3)} = 528{,}775.20$$

11. SS_A (SS of temperatures) $= \frac{\sum^{a}\left(\sum^{b}\sum^{c} Y\right)^2}{bc} - CT =$ quantity **3** − quantity **10** = 585,891.33 − 528,775.20 = 57,116.13

12. SS_B (SS of cyanide concentrations) $= \frac{\sum^{b}\left(\sum^{a}\sum^{c} Y\right)^2}{ac} - CT =$ quantity **4** − quantity **10**

$$= 584{,}320.67 - 528{,}775.20 = 55{,}545.47$$

13. SS_C (SS of oxygen concentrations) $= \frac{\sum^{c}\left(\sum^{a}\sum^{b} Y\right)^2}{ab} - CT =$ quantity **5** − quantity **10**

$$= 532{,}534.00 - 528{,}775.20 = 3758.80$$

14. SS_{AB} (temperature $\times$ cyanide interaction SS) $= \frac{\sum^{a}\sum^{b}\left(\sum^{c} Y\right)^2}{c} - \frac{\sum^{a}\left(\sum^{b}\sum^{c} Y\right)^2}{bc} - \frac{\sum^{b}\left(\sum^{a}\sum^{c} Y\right)^2}{ac} + CT$

= quantity **6** − quantity **3** − quantity **4** + quantity **10**
= 645,122.67 − 585,891.33 − 584,320.67 + 528,775.20 = 3685.87

This is equivalent to subtracting SS_A and SS_B [quantities **11** and **12**] from quantity **6**.

BOX 12.1 continued

15. SS_{AC} (temperature × oxygen interaction SS) $= \dfrac{\sum^{a}\sum^{c}\left(\sum^{b} Y\right)^2}{b} - \dfrac{\sum^{a}\left(\sum^{b}\sum^{c} Y\right)^2}{bc} - \dfrac{\sum^{c}\left(\sum^{a}\sum^{b} Y\right)^2}{ab} + CT$

= quantity **7** − quantity **3** − quantity **5** + quantity **10**
= 589,747.20 − 585,891.33 − 532,534.00 + 528,775.20 = 97.07

16. SS_{BC} (cyanide × oxygen interaction SS) $= \dfrac{\sum^{b}\sum^{c}\left(\sum^{a} Y\right)^2}{a} - \dfrac{\sum^{b}\left(\sum^{a}\sum^{c} Y\right)^2}{ac} - \dfrac{\sum^{c}\left(\sum^{a}\sum^{b} Y\right)^2}{ab} + CT$

= quantity **8** − quantity **4** − quantity **5** + quantity **10**
= 593,344.00 − 584,320.67 − 532,534.00 + 528,775.20 = 5264.53

17. SS_{ABC} (temperature × cyanide × oxygen interaction SS)

$$= \sum^{a}\sum^{b}\sum^{c} Y^2 - \frac{\sum^{a}\sum^{b}\left(\sum^{c} Y\right)^2}{c} - \frac{\sum^{a}\sum^{c}\left(\sum^{b} Y\right)^2}{b} - \frac{\sum^{b}\sum^{c}\left(\sum^{a} Y\right)^2}{a} + \frac{\sum^{a}\left(\sum^{b}\sum^{c} Y\right)^2}{bc}$$

$$+ \frac{\sum^{b}\left(\sum^{a}\sum^{c} Y\right)^2}{ac} + \frac{\sum^{c}\left(\sum^{a}\sum^{b} Y\right)^2}{ab} - CT$$

= quantity **2** − quantities **6**, **7**, and **8** + quantities **3**, **4**, and **5** − quantity **10**
= 655,278 − 645,122.67 − 589,747.20 − 593,344.00 + 585,891.33 + 584,320.67 + 532,534.00 − 528.775.20 = 1034.93

If this had been an anova *with* replication, we would have employed quantity **9** $= \dfrac{\sum^{a}\sum^{b}\sum^{c}\left(\sum^{n} Y\right)^2}{n}$ in place of quantity **2** $= \sum^{a}\sum^{b}\sum^{c} Y^2$.

18. SS_{within} (within subgroups; error SS). Since this example is an anova without replication, there is no SS_{within}. However, if there had been replication we could have calculated such an SS as follows:

$$SS_{\text{within}} = \sum^{a}\sum^{b}\sum^{c} Y^2 - \frac{\sum^{a}\sum^{b}\sum^{c}\left(\sum^{n} Y\right)^2}{n} = \text{quantity } \mathbf{2} - \text{quantity } \mathbf{9}.$$

□□ The machine computation of these quantities is self-evident. The following relations for some of the above quantities must

Now fill in the anova table.

Source of variation		*df*	*SS*	*MS*	Expected *MS* (Model I)
Main effects					
$\bar{Y}_A - \bar{\bar{Y}}$	A	$a - 1$	**11**	$\frac{\mathbf{11}}{(a-1)}$	$\sigma^2 + \frac{bc}{a-1}\sum^{a}\alpha^2$
$\bar{Y}_B - \bar{\bar{Y}}$	B	$b - 1$	**12**	$\frac{\mathbf{12}}{(b-1)}$	$\sigma^2 + \frac{ac}{b-1}\sum^{b}\beta^2$
$\bar{Y}_C - \bar{\bar{Y}}$	C	$c - 1$	**13**	$\frac{\mathbf{13}}{(c-1)}$	$\sigma^2 + \frac{ab}{c-1}\sum^{c}\gamma^2$
First-order interactions					
$(\bar{Y}_{AB} - \bar{Y}_A - \bar{Y}_B + \bar{\bar{Y}})$	$A \times B$	$(a-1)(b-1)$	**14**	$\frac{\mathbf{14}}{(a-1)(b-1)}$	$\sigma^2 + \frac{c}{(a-1)(b-1)}\sum^{ab}(\alpha\beta)^2$
$(\bar{Y}_{AC} - \bar{Y}_A - \bar{Y}_C + \bar{\bar{Y}})$	$A \times C$	$(a-1)(c-1)$	**15**	$\frac{\mathbf{15}}{(a-1)(c-1)}$	$\sigma^2 + \frac{b}{(a-1)(c-1)}\sum^{ac}(\alpha\gamma)^2$
$(\bar{Y}_{BC} - \bar{Y}_B - \bar{Y}_C + \bar{\bar{Y}})$	$B \times C$	$(b-1)(c-1)$	**16**	$\frac{\mathbf{16}}{(b-1)(c-1)}$	$\sigma^2 + \frac{a}{(b-1)(c-1)}\sum^{bc}(\beta\gamma)^2$
Second-order interaction $(Y - \bar{Y}_{AB} - \bar{Y}_{AC} - \bar{Y}_{BC} +$ $\bar{Y}_A + \bar{Y}_B + \bar{Y}_C - \bar{\bar{Y}})$	$A \times B \times C$	$(a-1)(b-1)(c-1)$	**17**	$\frac{\mathbf{17}}{(a-1)(b-1)(c-1)}$	$\sigma^2 + \frac{1}{(a-1)(b-1)(c-1)}\sum^{abc}(\alpha\beta\gamma)^2$

If the anova had been replicated, the symbol Y in the second-order interaction term above at left should have been replaced by $\bar{Y}$, the subgroup mean, and all expected mean squares would have their added components multiplied by n. We would also have had an error term as follows:

		df	*SS*	*MS*	Expected *MS*
$Y - \bar{Y}$	Within subgroups	$abc(n-1)$	**18**	$\frac{\mathbf{18}}{abc(n-1)}$	σ^2
$Y - \bar{\bar{Y}}$	Total	$abcn - 1$	**2 – 10**		

The expected mean squares shown above are for a Model I anova. We now show expected mean squares for Model II and mixed models. Being most general, the replicated design is shown. If you have a case without replication, simply eliminate the within subgroups (error) term and set $n = 1$.

Expected mean squares for a three-way factorial anova (other models).

Source of variation	Model II	Mixed Model (A and B fixed, C random)
A	$\sigma^2 + n\sigma^2_{ABC} + nc\sigma^2_{AB} + nb\sigma^2_{AC} + nbc\sigma^2_A$	$\sigma^2 + nb\sigma^2_{AC} + \frac{nbc}{a-1}\sum^{a}\alpha^2$
B	$\sigma^2 + n\sigma^2_{ABC} + nc\sigma^2_{AB} + na\sigma^2_{BC} + nac\sigma^2_B$	$\sigma^2 + na\sigma^2_{BC} + \frac{nac}{b-1}\sum^{b}\beta^2$
C	$\sigma^2 + n\sigma^2_{ABC} + nb\sigma^2_{AC} + na\sigma^2_{BC} + nab\sigma^2_C$	$\sigma^2 + nab\sigma^2_C$
$A \times B$	$\sigma^2 + n\sigma^2_{ABC} + nc\sigma^2_{AB}$	$\sigma^2 + n\sigma^2_{ABC} + \frac{nc}{(a-1)(b-1)}\sum^{ab}(\alpha\beta)^2$
$A \times C$	$\sigma^2 + n\sigma^2_{ABC} + nb\sigma^2_{AC}$	$\sigma^2 + nb\sigma^2_{AC}$
$B \times C$	$\sigma^2 + n\sigma^2_{ABC} + na\sigma^2_{BC}$	$\sigma^2 + na\sigma^2_{BC}$
$A \times B \times C$	$\sigma^2 + n\sigma^2_{ABC}$	$\sigma^2 + n\sigma^2_{ABC}$
Within subgroups	σ^2	σ^2

Anova table

Source of variation		*df*	*SS*	*MS*	F_s
A	Temperature	2	57,116.1	28,558.0	
B	Cyanide	4	55,545.5	13,886.4	
C	Oxygen	2	3,758.8	1,879.4	
$A \times B$	T × Cy	8	3,685.9	460.7	7.121***
$A \times C$	T × O	4	97.1	24.3	< 1 *ns*
$B \times C$	Cy × O	8	5,264.5	658.1	10.172***
$A \times B \times C$	T × Cy × O	16	1,034.9	64.7	
Total		44	126,502.8		

$F_{.001[8,16]} = 6.19$

Since this is a Model I anova, all mean squares are tested over MS_{ABC}, which represents the error variance σ^2, on the assumption that the temperature $\times$ cyanide $\times$ oxygen interaction is zero.

Conclusions.—Temperature $\times$ cyanide interaction and cyanide $\times$ oxygen interaction are significant; that is, the effect of cyanide ion on survival time depends on the oxygen concentration and the temperature of the water. When, as in this instance, interactions involving all main effects are significant, it is usually of no interest to test the main effect mean squares. A statement about the differences due to cyanide will not mean much unless we qualify it by specifying temperature and oxygen concentration. However, if we should wish to assess the gross effects of these three variables, we could test them over the error MS and we would find them highly significant in this example.

assuming Model II and a mixed model in which factors A and B are fixed and factor C is random. In these expected mean squares we have taken the general case in which there is replication within subgroups. For anovas without replication one simply eliminates the last line of the table (the within-subgroups mean square) and sets the coefficient $n = 1$ throughout the table.

Testing significance in Model II and mixed model anovas is more complicated. Let us look at Model II by way of illustration. Inspection of the expected mean squares in Box 12.1 shows that **$(A \times B \times C)$/Error** is the appropriate test for the second-order interaction. If this is present, we test first-order interactions over the second-order interactions. However, there is no exact test for the main effects. There is no mean square in the table which, when subtracted from the mean square of a main effect, will leave only the added variance component due to the main effect. Thus, to test main effects we have to create a synthetic denominator mean square, which can be done in several ways. For example, to test the mean square of A we could construct a denominator MS as the sum of the mean square of $A \times B$ and $A \times C$ minus the mean square of $A \times B \times C$ (in order to subtract the extra values of $\sigma^2 + n\sigma^2_{ABC}$ from the denominator). Significance tests involving such synthetic mean squares are carried out by the approximate method learned in Chapter 10 (see Boxes 10.4 and 10.5). We shall illustrate such a test in the next section.

In a mixed model some mean squares can be tested directly over an appropriate MS, but others will need construction of a synthetic denominator mean square (in the example specified in Box 12.1 all mean squares can be tested over an appropriate error MS).

The actual analysis of the minnow data is shown in an anova table at the end of Box 12.1. Results of the significance tests indicate that there is a highly significant temperature × cyanide interaction, as well as a cyanide × oxygen interaction, but no temperature × oxygen interaction. These relationships can be clearly seen in Figure 12.1, in which the data from the three two-factor tables of Box 12.1 have been graphed. Note that lines depicting the relation between cyanide and temperature and cyanide and oxygen are not parallel, but those depicting the temperature × oxygen relationship are more or less parallel. At the low concentrations of cyanide the effects of oxygen and temperature are much more marked than they are at the higher concentrations. This illustrates the meaning of interaction in these data. We could test the significance of the three main effects, all of which would be highly significant, but in view of the interactions this would not be very meaningful. From Figure 12.1 it is clear that there is a substantial overall effect at least of cyanide and temperature. As cyanide concentration increases, survival time of the fish decreases; as temperature increases, survival time decreases also.

The meaning of a second-order interaction requires some explanation. Interaction $A \times B \times C$ can be interpreted as indicating that interaction

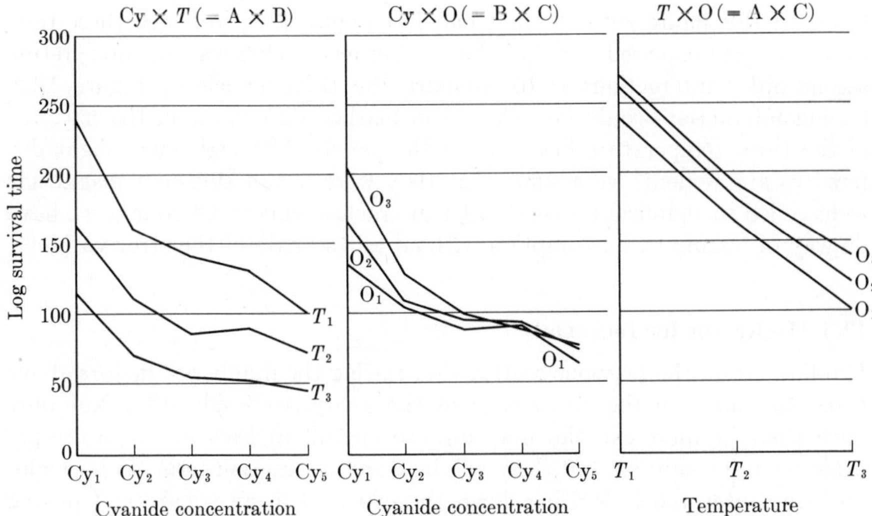

FIGURE 12.1 Graphic demonstration of interaction. Data from Box 12.1.

$A \times B$ differs depending on the level of factor C (or $A \times C$ depending on level of factor B, and so on). We can see this in Figure 12.2 for the minnow data. The three panels of this figure depict the relationship between cyanide concentration and temperature separately for each oxygen concentration.

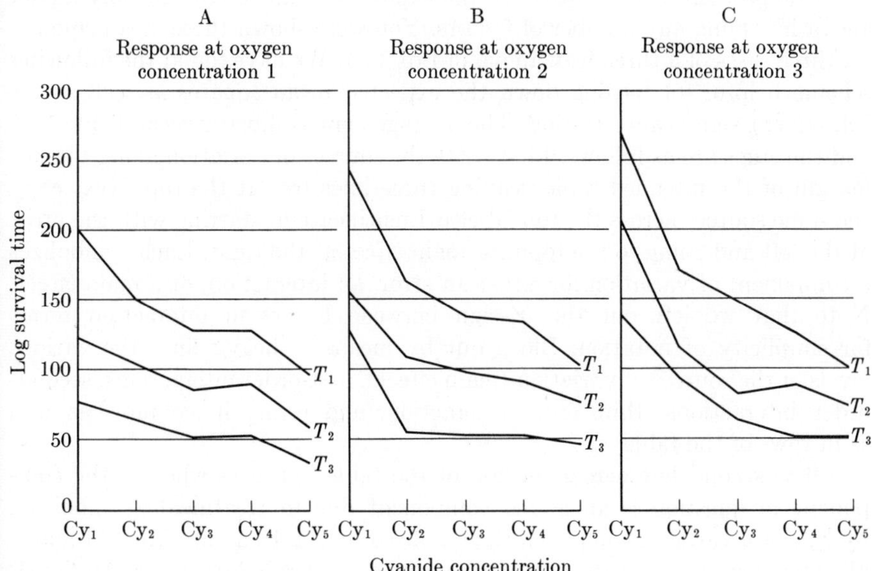

FIGURE 12.2 Graphic demonstration of meaning of second-order interaction. Data from Box 12.1.

Contrast this figure with the first panel of Figure 12.1, in which these relations were summarized for the whole experiment. One way of interpreting second-order interactions is to compare the three panels of Figure 12.2. Strong interaction would be reflected in marked differences in the patterns of the three temperature lines among the panels. Although some slight differences are evident, we assume that they were not of sufficient magnitude to have led to significant second-order interaction effects. Of course, we have no way of testing this assumption without an estimate of the error variance.

12.3 Higher-order factorials

It follows from the previous section that raising the number of factors above three will increase the complexity of the analysis considerably. Not only must there be more experimental units to provide at least one replicate per experimental condition, but there will be many interaction terms to be evaluated and interpreted. Writing down the expected mean square and testing the significance of each source of variation in the anova is more complex than in any example that we have tackled so far. However, if you carefully follow the rules we shall describe, you should not have too much difficulty.

Let us assume that you have a four-factor factorial without replication. We shall label the factors A, B, C, and D. Let us further assume that factors A, B, and C are fixed treatment effects (Model I) and factor D is random. The number of levels for each corresponding main effect will be a, b, c, and d.

One problem is to write down the expected mean squares for any model anova involving any number of factors. You were shown these mean squares for three cases of a three-way anova in Box 12.1. We have found the following scheme helpful for writing down the expected mean squares as well as for subsequent significance testing. The arrangement is illustrated in Table 12.1 and the steps are as follows. First write the sources of variation along the left margin of the intended table, leaving three lines free at the top. Next copy the same sources across the top labeled Identification, starting with the error at the left and going to the topmost main effect at the right. Each symbolizes a component of variation for either an error, an interaction, or a main effect. Note that we left out the $\times$-sign between letters in interaction terms for simplicity of notation. Block out by means of heavy lines the various levels of the sources of variation (main effects, first-order interactions, second-order interactions, third-order interaction, and error) in columns as well as in rows of the table.

The second line across the top of the table indicates whether the component in question is an added component due to treatment (symbolized by $\sum^2$) or a variance component (symbolized by σ^2). In a pure Model I anova all components, except the one representing error, are $\sum^2$. In a pure Model II, all components are σ^2. In a mixed model, any component whose identification does not contain a letter representing a random effect will be an added com-

Expected mean squares in a four-factor factorial anova. Mixed model: A, B, C are fixed treatments; D is random.

Identification[a]	Error	ABCD	BCD	ACD	ABD	ABC	CD	BD	BC	AD	AC	AB	D	C	B	A
Component	σ^2	σ^2	σ^2	σ^2	σ^2	Σ^2	σ^2	σ^2	Σ^2	σ^2	Σ^2	Σ^2	σ^2	Σ^2	Σ^2	Σ^2
Coefficient	1	n	na	nb	nc	nd	nab	nac	nad	nbc	nbd	ncd	$nabc$	$nabd$	$nacd$	$nbcd$
Source of variation																
A	*									*						*
B	*							*							*	
C	*						*							*		
D	*												*			
AB	*				*							*				
AC	*			*							*					
AD	*									*						
BC	*		*						*							
BD	*							*								
CD	*						*									
ABC	*					*										
ABD	*				*											
ACD	*			*												
BCD	*		*													
ABCD	*	*														
Error	*															

[a] Identifying letters are used as subscripts.

ponent due to treatment; all others will be variance components. Thus, in the example of Table 12.1, since the interaction component AC does not contain the random effect D, it is an added component due to treatment and is identified by $\sum^2$. On the other hand, interaction component ACD, which contains D, is a variance component identified by σ^2.

Next we work out the coefficients to be placed in front of each component other than the error variance. These coefficients can be written down by a very simple rule. The maximum possible coefficient is *nabcd*. Omit from this any letters whose corresponding capitals occur in the first line. Thus, for the component ABD, the coefficient is *nc*. Similarly, for the component of B, the coefficient is *nacd*. In an anova without replication, we can set $n = 1$ and ignore it when writing down the coefficients. We are now able to write down any component in its proper form with the correct coefficients. Thus the component of the interaction ACD from Table 12.1 is $nb\sigma^2_{ACD}$, and the component for the main effect C is $nabd \sum\gamma^2/(c - 1)$. Remember that added components due to treatment are labeled by Greek letters and are always divided by the corresponding degrees of freedom.

We now proceed to work out the complete expected mean square for any source of variation by considering the following five rules. (1) *All sources of variation contain the error variance* as shown by an asterisk in that column. (2) *They also contain the component corresponding to their name.* Thus the main effect B will obviously contain the component for B, interaction $A \times B \times C$ that for ABC. The appropriate asterisks form a diagonal across the table and no mean square will have any component to the right of this diagonal.

Next, examine all components located between the error component and the final component in each line. (3) *Consider only those in blocks to the left of the block of the final component.* Thus, if you are working out a first-order interaction, consider only second- and higher-order interactions, and so on. (4) *In the blocks of interest consider only those components that contain within them all the letters identifying the final component.* Thus, for main effect B consider only AB, BC, and BD as well as all higher-order interactions containing B; do not consider AC, AD, or CD. The final rule: (5) *If any of the letters in the interaction terms being considered (other than those identifying the final component in each line) represent a fixed treatment effect, this component is not included in the expected MS; if they all represent random effects, include the component.*

Now let us apply these rules. In a pure Model I anova all letters represent fixed effects. It is obvious therefore by rule (5) that all components except the final component in the line [rule (2)] and the error variance [rule (1)] will be omitted from each expected mean square. This yields the characteristic expected mean squares of a Model I anova containing only the error variance and the component representing a given source of variation (see Box 12.1). By contrast, a pure Model II will contain all possible interaction terms. In a four-factor Model II anova with replication the expected mean square for factor B would therefore read

$$\sigma^2 + n\sigma^2_{ABCD} + na\sigma^2_{BCD} + nc\sigma^2_{ABD} + nd\sigma^2_{ABC} + nac\sigma^2_{BD} + nad\sigma^2_{BC} + ncd\sigma^2_{AB} + nacd\sigma^2_{B}$$

Expected mean squares of mixed model anovas depend on the particular combination of fixed and random factors. We shall go through several mean squares in the mixed model example of Table 12.1 to make certain that the rules are understood. The mean square of A, a fixed treatment effect, will lead to consideration of three first-order interactions containing an A [rules (3) and (4)]. AB and AC include letters representing fixed factors B and C, and hence are omitted [rule (5)], but since D is a random factor the AD term is retained. Any second- or third-order interaction including A also contains at least one letter representing a fixed effect. Thus, ABD contains BD in addition to A; though D is random, B is fixed and the component is excluded [rule (5)]. Therefore, the expected mean square of A (EMS_A) consists of three components as indicated by the asterisks in Table 12.1. $EMS_A = \sigma^2 + nbc\sigma^2_{AD} + nbcd \sum \alpha^2/(a-1)$. The expected mean square for main effect D contains only the error variance and the final component, since all other letters represent fixed effects and their components would be omitted [rule (5)]. Finally, let us work out the expected mean square for the first order interaction $A \times B$. We find AB contained in the second order terms ABC and ABD [rule (4)]. In ABC the letter C represents a fixed effect, hence the term is omitted [rule (5)]. In ABD the D is a random factor and the term is included.

All included terms are represented by asterisks in Table 12.1. When the table is filled in completely it is useful for pointing out how significance tests for given sources of variation should be carried out. From Table 12.1 it is clear that we would test the main effect of C over the $C \times D$ mean square or the interaction $B \times C$ over the $B \times C \times D$ interaction MS. The mixed model in Table 12.1 is a relatively simple case. If two of the main effects had been random, the situation would have been more complicated. One of the problems frequently encountered in Model II or mixed factorials is that there is no appropriate error term against which to test a given mean square. For instance, in the expected mean squares of the Model II anova shown near the end of Box 12.1, there is no exact test for the expected mean square of main effect B. The appropriate significance test must be done by an approximate procedure balancing the variance components as follows: $F_s = MS_B/(MS_{AB} + MS_{BC} - MS_{ABC})$, where the appropriate degrees of freedom for the denominator mean square have to be calculated as in Box 10.4.

Another complication in calculating multifactor factorials is the computation of the higher-order interactions. We saw in Box 12.1 how this is done for a three-way anova. For higher-order interactions similar arrangements into subtables of the original data table are useful. However, the interaction sum of squares can be directly computed from a formula. Box 12.2 gives computational formulas for first- through third-order interactions (in a four-factor factorial), from which the reader should have no difficulty generating yet-higher-order interactions in factorial anovas of any number of factors. Notice the regularity in the summation signs before and within the parentheses, in the denominators, and in the plus and minus signs. Note also that for a given anova the last term of each expression is nothing but the correction term CT.

BOX 12.2

Calculation of higher-order interaction sums of squares.

For a k-factor factorial, these sums of squares are computed as a summation of a series of terms, each consisting of a numerator of the general form $\sum^{a}\sum^{b}\cdots\left(\sum^{k}\sum^{n} Y\right)^2$ and a denominator of the general form $ab\cdots kn$. There will be as many summation signs as there are factors in the factorial, and $\sum^{n}$ and n will be present when the factorial is replicated. The number of $\sum$ signs outside and inside the parentheses in the denominator varies in a regular pattern as shown below, but the total number of $\sum$ signs is the same in any one factorial anova. The coefficients in the denominators correspond to the superscripts of the $\sum$ signs *within* parentheses in the numerator; they are the sample sizes over which the totals being squared have been summed. We shall first write down the formulas for some interaction sums of squares in a replicated four-factor factorial, then point out the regularities in these expressions and the actual ways in which they might be computed:

$$SS_{AB} = \frac{\sum^{a}\sum^{b}\left(\sum^{c}\sum^{d}\sum^{n} Y\right)^2}{cdn} - \frac{\sum^{a}\left(\sum^{b}\sum^{c}\sum^{d}\sum^{n} Y\right)^2}{bcdn} - \frac{\sum^{b}\left(\sum^{a}\sum^{c}\sum^{d}\sum^{n} Y\right)^2}{acdn} + \frac{\left(\sum^{a}\sum^{b}\sum^{c}\sum^{d}\sum^{n} Y\right)^2}{abcdn}$$

$$SS_{ABC} = \frac{\sum^{a}\sum^{b}\sum^{c}\left(\sum^{d}\sum^{n} Y\right)^2}{dn} - \frac{\sum^{a}\sum^{b}\left(\sum^{c}\sum^{d}\sum^{n} Y\right)^2}{cdn} - \frac{\sum^{a}\sum^{c}\left(\sum^{b}\sum^{d}\sum^{n} Y\right)^2}{bdn} - \frac{\sum^{b}\sum^{c}\left(\sum^{a}\sum^{d}\sum^{n} Y\right)^2}{adn} + \frac{\sum^{a}\left(\sum^{b}\sum^{c}\sum^{d}\sum^{n} Y\right)^2}{bcdn} + \frac{\sum^{b}\left(\sum^{a}\sum^{c}\sum^{d}\sum^{n} Y\right)^2}{acdn} + \frac{\sum^{c}\left(\sum^{a}\sum^{b}\sum^{d}\sum^{n} Y\right)^2}{abdn} - \frac{\left(\sum^{a}\sum^{b}\sum^{c}\sum^{d}\sum^{n} Y\right)^2}{abcdn}$$

BOX 12.2 continued

$$SS_{ABCD} = \frac{\sum^a\sum^b\sum^c\sum^d\left(\sum^n Y\right)^2}{n} - \frac{\sum^a\sum^b\sum^c\left(\sum^d\sum^n Y\right)^2}{dn}$$

$$- \frac{\sum^a\sum^b\sum^d\left(\sum^c\sum^n Y\right)^2}{cn} - \frac{\sum^a\sum^c\sum^d\left(\sum^b\sum^n Y\right)^2}{bn}$$

$$- \frac{\sum^b\sum^c\sum^d\left(\sum^a\sum^n Y\right)^2}{an} + \frac{\sum^a\sum^b\left(\sum^c\sum^d\sum^n Y\right)^2}{cdn}$$

$$+ \frac{\sum^a\sum^c\left(\sum^b\sum^d\sum^n Y\right)^2}{bdn} + \frac{\sum^a\sum^d\left(\sum^b\sum^c\sum^n Y\right)^2}{bcn}$$

$$+ \frac{\sum^b\sum^c\left(\sum^a\sum^d\sum^n Y\right)^2}{adn} + \frac{\sum^b\sum^d\left(\sum^a\sum^c\sum^n Y\right)^2}{acn}$$

$$+ \frac{\sum^c\sum^d\left(\sum^a\sum^b\sum^n Y\right)^2}{abn} - \frac{\sum^a\left(\sum^b\sum^c\sum^d\sum^n Y\right)^2}{bcdn}$$

$$- \frac{\sum^b\left(\sum^a\sum^c\sum^d\sum^n Y\right)^2}{acdn} + \frac{\sum^c\left(\sum^a\sum^b\sum^d\sum^n Y\right)^2}{abdn}$$

$$- \frac{\sum^d\left(\sum^a\sum^b\sum^c\sum^n Y\right)^2}{abcn} + \frac{\left(\sum^a\sum^b\sum^c\sum^d\sum^n Y\right)^2}{abcdn}$$

Note that in each term the same summation signs recur in the numerators, but differ in their arrangement. Those outside the parentheses are determined by the nature of the interaction (for the first term) and represent successively smaller subsets from the first term in all combinations. Thus, if the interaction is $B \times C \times D$ the summation outside the parentheses will be $\sum^b\sum^c\sum^d$ for the first term and $\sum^b\sum^c$, $\sum^b\sum^d$, $\sum^c\sum^d$, $\sum^b$, $\sum^c$, and $\sum^d$ for the succeeding terms, with no summation outside the parentheses for the last term. The sign preceding the terms are always in the sequence $+-+-+-\cdots$, with a sign change whenever we come to a new class of subsets of the summation signs. A term such as $\sum^a\left(\sum^b\sum^c\sum^d\sum^n Y\right)^2/bcdn$ simply represents the sum of the squared marginal totals for factor A divided by the sample size of each total. We could easily obtain $\sum^a\sum^b\left(\sum^c\sum^d\sum^n Y\right)^2/cdn$ from a two-way table for factors A and B by squaring the cell sums, summing them, and dividing by their sample size. However, for $\sum^a\sum^b\sum^c\left(\sum^d\sum^n Y\right)^2/dn$ and $\sum^a\sum^b\sum^c\sum^d\left(\sum^n Y\right)^2/n$, we have to lay out more complicated three-way and four-way tables, which can be confusing. When the calculation is done on a computer using a compiler such as FORTRAN IV, which permits at least four-dimensional subscripts, such summing is quite simple without laying out separate tables. Obviously $\left(\sum^a\sum^b\sum^c\sum^d\sum^n Y\right)^2/abcdn$ is the correction term CT.

⇸ In agricultural research, in which many standard designs are repeatedly used, a shorthand notation for indicating the nature of the factorial is frequently used. A three-factor factorial with each factor expressed at two levels is indicated as a $2 \times 2 \times 2$ factorial or 2^3 factorial. A two-way anova with each factor at three levels is a 3^2 factorial. A $2 \times 3 \times 4$ factorial has three factors, one being at two levels, a second at three levels, and the third at four levels. For many standard factorials especially efficient computational methods have been developed, although their importance is lessening with the increasing use of computers. Factorial anovas involving treatments at only two levels can be computed with special elegance. Computations for a 2×2 factorial two-way anova, each factor at two levels with replication are illustrated very completely in Steel and Torrie (1960, Section 11.3). Federer (1955) discusses the general 2^k case in section VII-4 of his book. ⇷

12.4 Other designs

⇸ The analyses of variance studied so far are only an introduction to the field of *experimental design*, which covers a great variety of designs developed in response to the needs of experimental scientists. A detailed consideration of these is beyond the scope of this book and the average biologist would probably not find a study of these designs sufficiently rewarding or their use sufficiently frequent to make it worth his while to become intimately acquainted with them. Therefore, if experiments are planned that do not seem to fit the simple designs so far discussed, you are strongly urged to consult a professional statistician. However, we shall briefly outline some of the more common designs, the reasons for using them, and sources where a discussion in depth and a computational outline can be found.

In Chapter 11 we first became acquainted with the randomized complete block design, in which the rows of a two-way anova represent blocks of an experimental layout. In many instances it becomes necessary to lay out two-dimensional blocks. This is easiest seen in agricultural experiments, where fields may be heterogeneous not only in a north-south direction, but also in an east-west direction. In such a case, to test a single treatment (factor) at k levels the *Latin square design* is especially appropriate. We divide the field into k row blocks and k column blocks as shown in Figure 12.3. Each level

Drier ⟶ Wetter

Low N content ↓	T_1	T_4	T_3	T_2
	T_3	T_2	T_1	T_4
	T_2	T_3	T_4	T_1
High N content	T_4	T_1	T_2	T_3

FIGURE 12.3 An example of four chemical treatments (T_1 through T_4) laid out in a field according to a Latin square design to control known nitrogen content and moisture gradients.

of the treatment T_i is replicated only once in a given row or column. The analysis of a Latin square experiment is shown in sections 8.9 and 8.10 of Steel and Torrie (1960). This design need not be restricted to agricultural research. Suppose we wish to make quantitative chemical analyses of four related substances in a search for chemical differences among them. The analysis of a single sample is a tedious procedure that takes all day and ties up one unit of the analytical equipment such as a Warburg apparatus. Assuming that we have four Warburg setups in our laboratory, we can now run our analyses over four days, arranging the four treatments in the manner of a Latin square; that is, any one substance would be tested only once on any one day and only once in any one Warburg apparatus. The resulting anova table would yield effects representing day-to-day differences as well as differences among apparatuses, but the mean square of greatest interest would be that relating to differences among the four substances.

Another common design is the *split plot design*. This design is for two factors. Each level of one factor (factor A) is assigned to a whole plot within a block, and each plot is then subdivided into as many subplots as there are levels in B, which are then randomly allocated to these subplots. Two different error mean squares are computed: one, expected to be larger, for the plots, and a smaller one for the subplots. This design permits us to obtain more precise information on the factor allocated to the split plots or subplots at the expense of losing information on the factor assigned to the whole plots. For instance, we may be studying the differences between five genetic strains at three temperatures. The three temperatures may be produced in different temperature chambers in one laboratory. The five genetic strains may be in separate culture bottles in the chambers. The reader may ask why such an analysis could not be done as a simple factorial of strains against temperatures. It could indeed, but it would require 15 different temperature chambers, one for each combination of temperatures and strains. Split plot designs are quite often incorrectly analyzed as factorial anovas. They are discussed in many texts. We find the discussion in section 12.11 of Snedecor (1956) quite clearly presented. Steel and Torrie (1960) devote a whole chapter to this subject.

Many other designs will be found in textbooks of experimental design. For example, when in a design for which a randomized block is appropriate, we find it impossible to include all treatments in every block because of the amount of work or space, we are faced with an *incomplete block design*. If only a fraction of combinations of factors (not all of them) are tested in a factorial design, it is called a *fractional factorial*. The computations and significance tests for these analyses are rather complex, and they all require certain assumptions for the analyses to be meaningful. A simple book on experimental design is Pearce (1965). Of the more comprehensive texts on experimental design, we would expect that nonstatistical readers would find Federer (1955) most readable. ↞

12.5 Anova by computer

You should by now be impressed by the multiplicity of designs possible in analysis of variance. For most of the common designs the computations are relatively simple and straightforward. Unless you have to do very many of them, processing on a computer is not worthwhile. For a complicated factorial or an analysis of variance with missing values or disproportionate subclass sizes, the computation gets heavy and the use of a computer would be very welcome. However, such programs are not easily written in a general way applicable to a great variety of cases, and we have refrained from presenting one here.

The main difficulty in writing programs for analysis of variance is to make them general enough. One would not wish to write a separate program for each analysis of variance carried out. It would be much faster to do the computation on a desk calculator than to write a program for the analysis. There is need for a general anova program that would handle most common analyses likely to be encountered. Attempts at writing such programs have not been very successful. Often the instructions to users of an elaborate general anova program are so complex that it takes longer to study the instructions than it does to carry out a single analysis on a desk calculator.

TABLE 12.2

Computation of various anova designs in the form of a factorial anova without replication. For use with computer program in Appendix A3.5.

Design	*Simulated factors*	*Intended anova table in terms of SS and df of factorial anova*	
Single classification anova (equal n)			
a groups	Groups = A	Groups	= A
n items per group	Items = B	Within (error)	= B + (AB)
2-way anova with replication			
a columns	Columns = A	Columns	= A
b rows	Rows = B	Rows	= B
n replicates	Replicates = C	Interaction	= (AB)
		Within subgroups	= C + (AC) + (BC) + (ABC)
Nested anova			
a groups	Groups = A	Groups	= A
b subgroups per group	Subgroups = B	Subgroups within groups	= B + (AB)
n items per subgroup	Items = C	Within subgroups	= C + (AC) + (BC) + (ABC)

Source: Modified from Hartley (1962).

Hartley (1962) has suggested using a factorial anova without replication and with an indefinite number of factors (as many as necessary) to represent many other designs on the computer—that is, although our actual analysis may follow a different design, we present our data laid out as though they were a factorial anova and proceed to analyze them in that manner. This results in replication being represented as a (dummy) factor. The computer analyzes the data as a factorial anova without replication and we use the results of that anova to reconstitute an anova table for the design we had originally intended.

This strategy will become clearer from consideration of Table 12.2, which lists the steps to be taken when using the computer program in Appendix A3.5 for some of the common designs we have studied. Let us look at the single classification anova with equal n shown in the first row of this table. As you will recall from Section 9.3, such an anova has a groups and n items per group. We set up our data as though groups were factor A of the factorial and the replicated items factor B. The data as set up would now have the appearance of a two-way anova without replication, as in Box 11.2. The simulated factors are listed in column 2 of Table 12.2. In the last column we are told how to reconstitute the degrees of freedom and sums of squares of the intended anova table. For our example df and SS among groups are simply the corresponding degrees of freedom and sums of squares for A; the within (error) term for degrees of freedom or sums of squares is the sum of the B term and the AB interaction. The other designs can be analyzed similarly. A little study of the relationships will show you clearly how they were derived and will also give you interesting information about the relationships among the various designs.

This scheme permits the computation of a great variety of balanced designs (with no missing replicates), using a single anova program. For further details, consult the paper by Hartley (1962).

Exercises 12

12.1 Price (1954) investigated the survival of *Bacterium tularense* in lice under three relative humidities (R.H. 100%, 50%, and 0%), four temperatures (37, 29, 20, and 4°C) and on three substrates (killed lice, starved lice, and louse feces). Two replicate experiments were run. Write out the expected mean squares and make the appropriate tests of significance. Interpret. ANS. EMS for MS_T is $\sigma^2 + hs\sigma^2_{ET} + ehs\Sigma\tau^2/(t-1)$.

Analysis of variance

Source of variation	*df*	*MS*
Main effects		
Experiments (E)	1	0.1108
Humidities (H)	2	0.7613
Substrates (S)	2	0.0376
Temperatures (T)	3	3.1334

First-order interactions		
$E \times H$	2	0.1185
$E \times S$	2	0.1266
$E \times T$	3	0.0060
$H \times S$	4	0.1985
$H \times T$	6	0.0666
$S \times T$	6	0.0654
Second-order interactions		
$E \times H \times S$	4	0.0130
$E \times H \times T$	6	0.0601
$E \times S \times T$	6	0.0206
$H \times S \times T$	12	0.0475
Third-order interaction		
$E \times H \times S \times T$	12	0.0517

12.2 Analyze the data of Table 11.1 as if the experiment had been in the form of a three-factor anova without replication; that is, pretend that the three rats in each subgroup represent a dummy factor. In the resulting anova table show which *SS* and *df* must be pooled in order to obtain the correct results shown at the end of Table 11.1.

12.3 Analyze the data given in Box 10.1 as if the experiment followed the form of a three-factor anova without replication. To do this you must introduce the dummy factors "females" (with four levels) and "measurements" (with two levels). In the resulting anova table indicate which *SS* and *df* must be pooled in order to obtain the correct anova. ANS. *SS* among females within cages = *SS* females + $SS_{F \times C}$.

12.4 Write out the expected mean squares for a five-factor experiment in which factors *A*, *B*, and *C* are random and factors *D* and *E* are fixed treatment effects.

12.5 Skinner and Allison (1923) studied the effect on cotton of date of planting and of the addition of borax to fertilizer. For each treatment combination a single measurement was made of the weight of the green plants in pounds. Analyze and interpret. For the analysis, assume that the second-order interaction has negligible effects. Was the addition of borax beneficial?

		Method of application		
		I	*II*	*III*
Date of planting	*Amount of borax (pounds)*	*In drill; seed planted 1 week later*	*In drill; seed planted immediately*	*Broadcast; seed planted immediately*
	0	61	60	73
June 2	5	56	59	80
	10	58	61	55
	0	67	72	72
June 9	5	69	62	79
	10	67	58	68
	0	62	69	68
June 18	5	62	70	71
	10	57	72	64
	0	40	45	54
July 7	5	37	49	51
	10	29	33	41
	0	26	28	35
July 15	5	26	24	36
	10	21	23	33
	0	10	13	17
Aug. 3	5	10	11	14
	10	8	12	14

13 ASSUMPTIONS OF ANALYSIS OF VARIANCE

We have blithely proceeded to study design after design in analysis of variance without bothering to discuss whether the analysis and especially its test of significance are appropriate to a given set of data. We shall now examine the underlying assumptions of the analysis of variance, methods for testing whether these assumptions are valid, the consequences for an anova if the assumptions are violated, and steps to be taken if the assumptions cannot be met. By taking up the fundamental assumptions of anova last in our consideration of the subject, we may seem to be building the superstructure before the foundation. Rigorous mathematical treatment clearly demands statement of the model prior to a treatment of the subject. Yet, we have found that the generally nonmathematical audiences we have been addressing will learn better if they first understand the structure and purpose of analysis of variance and are able to carry out the computations before completing their survey by a study of the assumptions. However, we should stress that before carrying out any anova on an actual research problem you should reassure yourself that the assumptions listed in this chapter seem reasonable and, if they are not, you should carry out one of several possible alternative steps to remedy the situation.

The first section (13.1) refreshes your memory on a fundamental assumption, random sampling. In each of the next four sections we explain a special assumption of analysis of variance, describe a procedure for testing it, briefly discuss the consequences if the assumption does not hold, and (if it doesn't) give instructions on how to proceed. Section 13.2 treats the assumption of independence, 13.3 homogeneity of variances, 13.4 normality, and 13.5 additivity.

In many cases departure from the assumptions of analysis of variance can be rectified by transformation of the original data by use of a new scale.

The rationale behind this is presented in a brief general introduction to transformations (13.6). The next three sections cover the most common transformations. Section 13.7 explains the logarithmic transformation, 13.8 the square root transformation, and 13.9 the arcsine or angular transformation.

When transformations are unable to make the data conform to the assumptions of analysis of variance, other techniques of analysis, analogous to the intended anova, must be employed. These are the nonparametric techniques briefly mentioned in Chapter 3. As stated there, these techniques are sometimes used by preference even when the parametric method (anova in this case) can be legitimately employed. Rapidity of computation and a preference for the generally simple assumptions of the nonparametric analyses cause many research workers to turn to them. However, when the assumptions of anova are met, these methods are less efficient than anova. Section 13.10 examines several nonparametric methods in lieu of single classification anova, and Section 13.11 features nonparametric methods in lieu of two-way anova.

13.1 A fundamental assumption

All anovas require that sampling of individuals be at random. Thus, in a study of the effects of three doses of a drug (plus a control) on five rats each, the five rats allocated to each treatment must be selected at random. If the five rats employed as controls are either the youngest or the smallest or the heaviest rats, while those allocated to some other treatment are selected in some other way, it is clear that the results are not apt to yield an unbiased estimate of the true treatment effects. Nonrandomness of sample selection may well be reflected in lack of independence of the items (see Section 13.2) or in heterogeneity of variances (Section 13.3) or nonnormal distribution (Section 13.4). Adequate safeguards to ensure random sampling during the design of an experiment, or when sampling from natural populations, are essential.

13.2 Independence

An assumption stated in each explicit expression for the expected value of a variate [for example, Expression (8.3) was $Y_{ij} = \mu + \alpha_i + \epsilon_{ij}$] is that the error term ϵ_{ij} is a random normal variable. In addition, for completeness we should also add the statement that it is assumed that the ϵ's are independently and identically (see the next section) distributed.

Thus, if you arrange the variates within any one group in some logical order independent of their magnitude (such as the order in which the measurements were obtained), we would expect the ϵ_{ij}'s to succeed each other in a random sequence. Consequently, we assume a long sequence of large positive values followed by an equally long sequence of negative values to be quite unlikely. We would also not expect positive and negative values to alternate with regularity.

How could departures from independence arise? An obvious example would be an experiment in which the experimental units were plots of ground laid out in a field. In such a case it is often found that adjacent plots of ground give rather similar yields. It would thus be important not to group all the plots containing the same treatment into an adjacent series of plots but rather to randomize the allocation of treatments among the experimental plots. The physical process of randomly allocating the treatments to the experimental plots ensures that the ϵ's will be independent.

Lack of independence of the ϵ's can result from correlation in time rather than space. In an experiment we might measure the effect of a treatment by recording weights of ten individuals. Our balance may suffer from a maladjustment that results in giving successive underestimates, compensated for by several overestimates. Conversely, compensation by the operator of the balance may result in regularly alternating over- and underestimates of the true weight. Here again randomization may overcome the problem of nonindependence of errors. For example, we may determine the sequence in which individuals of the various groups are weighed according to some random procedure.

If lack of independence of errors is suspected, a test of it can usually be carried out by means of a runs test (see Section 17.2).

There is no simple adjustment or transformation to overcome the lack of independence of errors. The basic design of the experiment or the way in which it was performed must be changed. We have seen how a randomized blocks design often overcomes lack of independence of error by randomizing the effects of differences in soils or cages. Similarly, in the experiment with the biased balance we could obtain independence of errors by redesigning the experiment, using different times of weighing as blocks. Of course, if a source of error is suspected or known, attempts can be made to remove it; knowing that the balance is biased, we may have it fixed. If the ϵ's are not independent, the validity of the usual F-test of significance can be seriously impaired.

13.3 Homogeneity of variances

In Section 9.4 and Box 9.6, in which we described the t-test for the difference between two means, you were told that the statistical test was valid only if we could assume that the variances of the two samples were equal. Although we have not stressed it so far, this assumption that the ϵ_{ij}'s have identical variances also underlies the equivalent anova test for two samples—and in fact any type of anova. *Equality of variances* in a group of samples is an important precondition for several statistical tests. Synonyms for this condition are *homogeneity of variances* or *homoscedasticity*, a jawbreaker that makes students in any biometry class sit up and take notice. The term is coined from

BOX 13.1

Tests of homogeneity of variances.

Lengths of third molar of eight species of the condylarth *Hyopsodus*.

Samples ($a = 8$)	(1) $df = n_i - 1$	(2) s^2_i	(3) *Coded* s^2_i	(4) *log coded* s^2_i
1	17	0.0707	7.07	0.84942
2	12	0.1447	14.47	1.16047
3	16	0.0237	2.37	0.37475
4	15	0.0836	8.36	0.92221
5	7	0.2189	21.89	1.34025
6	10	0.1770	17.70	1.24797
7	9	0.0791	7.91	0.89818
8	9	0.2331	23.31	1.36754
	95			

Source: Data from Olson and Miller (1958).

Bartlett's test of homogeneity of variances

1. Look up the common logarithms of the $a = 8$ variances. Before finding the logarithm it is convenient to use a multiplicative code to make the variances all greater than unity and thus avoid negative logarithms. In the present case, the variances were multiplied by 100. This coding has no effect upon the test.
2. Sum the degrees of freedom

$$\sum^{a} (n_i - 1) = 17 + 12 + \cdots + 9 = 95$$

3. Compute a weighted average variance

$$s^2 = \frac{\sum^{a} (n_i - 1)s_i^2}{\sum^{a} (n_i - 1)} = \frac{[17(7.07) + 12(14.47) + \cdots + 9(23.31)]}{95}$$
$$= 11.25$$

and look up its logarithm,

$$\log 11.25 = 1.05115$$

4. Compute

$$\sum^{a} (n_i - 1) \log s_i^2 = 17(0.84942) + 12(1.16047) + \cdots + 9(1.36754)$$
$$= 90.44786$$

5. $X^2 = 2.3026 \left[\sum^{a} (n_i - 1) \log s^2 - \sum^{a} (n_i - 1) \log s_i^2\right]$

$$= 2.3026[(\text{quantity } \mathbf{2} \times \text{quantity } \mathbf{3}) - \text{quantity } \mathbf{4}]$$
$$= 2.3026[95(1.05115) - 90.44786] = 2.3026(9.41139) = 21.671$$

The constant 2.3026 transforms the common logarithms to natural ones, which are needed for the formula.

BOX 13.1 continued

After X^2 is divided by a correction factor, quantity **6** below, it is distributed as $\chi^2_{[a-1]}$ when H_0 is true. Since the correction factor is usually slightly greater than unity, there is no need to apply it unless the X^2 lies slightly above the borderline of significance.

Since $\chi^2_{.01[7]} = 18.475$, our value of X^2 is clearly significant and the correction factor is unlikely to change this decision. We compute it simply to illustrate the formula.

6. Correction factor $C = 1 + \frac{1}{3(a-1)}\left[\sum^a \frac{1}{(n_i - 1)} - \frac{1}{\sum^a (n_i - 1)}\right]$

$$= 1 + \frac{1}{3(8-1)}\left[\frac{1}{17} + \frac{1}{12} + \cdots + \frac{1}{9} - \frac{1}{95}\right]$$

$$= 1 + \frac{1}{21}[0.72588] = 1.03457$$

□□ Evaluate term in square brackets first by accumulating sum and differences of quotients as shown in Appendix A2.14. □□

$$\text{Adjusted } X^2 = \frac{X^2}{C} = \frac{21.671}{1.03457} = 20.947 \qquad \text{with } a - 1 = 7\ df.$$

Since $\chi^2_{.01[7]} = 18.475$, we conclude that the variances of the eight samples are significantly heterogeneous.

The F_{max}-test

Find the greatest variance, $s^2_{max} = 0.2331$, and the smallest, $s^2_{min} = 0.0237$. Then compute the maximum variance ratio $s^2_{max}/s^2_{min} = 0.2331/0.0237 = 9.84$. Table **T** (cumulative probability distribution of $F_{max\,\alpha[a,n-1]}$) assumes that the df of all variances are equal, but we can make an approximate test using the lesser of the degrees of freedom of the two variances needed in computing the variance ratio ($n_8 - 1 = 9$). $F_{max\,.05[8,9]} = 8.95$ so that the observed variance ratio would be considered significant at $P < 0.05$. The variances of the eight samples are heterogeneous.

Greek roots meaning equal scatter; the converse condition (inequality of variances among samples) is called *heteroscedasticity*. Because we assume that each sample variance is an estimate of the same parametric error variance, the assumption of homogeneity of variances makes intuitive sense.

We have already seen how to test whether two samples are homoscedastic prior to a t-test of the differences between two means or a two-sample analysis of variance: we use an F-test for the hypotheses $H_0: \sigma_1^2 = \sigma_2^2$ and $H_1: \sigma_1^2 \neq \sigma_2^2$, as illustrated in Section 8.3 and Box 8.1. When there are more than two groups, *Bartlett's test for homogeneity of variances* is frequently suggested. This test is illustrated in Box 13.1 on lengths of molars for samples of eight species

BOX 13.2

Approximate test of equality of means when the variances are heterogeneous.

Length of third molar of eight species of the condylarth *Hyopsodus*. The data were given in part in Box 13.1.

(1)	*(2)*	*(3)*	*(4)*	*(5)*	*(6)*	*(7)*
Samples $(a = 8)$	n_i	$\bar{Y}_i$	s^2_i	$w_i = n_i/s^2_i$	$w_i\bar{Y}_i$	$1 - (w_i/\sum^a wi)$
1	18	3.88	0.0707	254.60	987.8480	0.83263
2	13	4.61	0.1447	89.84	414.1624	0.94094
3	17	4.79	0.0237	717.30	3435.8670	0.52845
4	16	4.73	0.0836	191.39	905.2747	0.87418
5	8	4.92	0.2189	36.55	179.8260	0.97597
6	11	4.96	0.1770	62.15	308.2640	0.95914
7	10	5.20	0.0791	126.42	657.3840	0.91689
8	10	6.58	0.2331	42.90	282.2820	0.97180
				1521.15	7170.9081	

Computations

1. Make a table giving n, $\bar{Y}$, and s^2 for each sample as shown in columns (2), (3), and (4) above.

2. In column (5) compute weights for each sample, $w_i = n_i/s_i^2$. Thus, for the first sample, $18/0.0707 = 254.60$.

3. In column (6) compute weighted means $w_i\bar{Y}_i$ for each sample. Thus, for the first sample, $254.60 \times 3.88 = 987.8480$.

4. The sum of the weights, $\sum^a w_i = 1521.15$, may be accumulated simultaneously with step **3**.

5. Sum column (6) to obtain $\sum^a w_i\bar{Y}_i = 7170.9081$.

6. The weighted grand mean, $\bar{\bar{Y}}_w = \dfrac{\sum^a w_i\bar{Y}_i}{\sum^a w_i}$

$$= \frac{\text{quantity 5}}{\text{quantity 4}} = \frac{7170.9081}{1521.15} = 4.7141.$$

7. Compute $\sum^a w_i\bar{Y}_i^2$ as $\sum^a [\bar{Y}_i(w_i\bar{Y}_i)]$

$$= 3.88(987.8469) + 4.61(414.1624) + \cdots + 6.58(282.2820)$$

$$= 34{,}171.4384.$$

BOX 13.2 continued

8. Compute the weighted correction term, $CT_w = \dfrac{\left(\sum^a w_i\bar{Y}_i\right)^2}{\sum^a w_i}$

$$= \frac{[\text{quantity } \mathbf{5}]^2}{\text{quantity } \mathbf{4}} = \frac{(7170.9081)^2}{1521.15} = 33{,}804.6366.$$

9. The weighted $SS_{\text{groups}} = \sum^a w_i\bar{Y}_i^2 - CT_w$

$$= \text{quantity } \mathbf{7} - \text{quantity } \mathbf{8} = 34{,}171.4369 - 33{,}804.6366 = 366.8003.$$

10. In column (7) compute $1 - \dfrac{w_i}{\sum^a w_i}$ for each sample. For sample 1, this is

$$1 - \frac{254.60}{1521.15} = 0.83263.$$

□□ If the desk calculator has a constant divisor feature, it should be used for this calculation. If not, compute the reciprocal of $\sum^a w_i$ and use it as a constant multiplier. □□

11. Square the quantities in column (7), divide each by $(n_i - 1)$, and accumulate. (This can all be done in a single sequence of operations on most desk calculators.)

$$\sum^a \frac{\left[1 - \left(w_i/\sum^a w_i\right)\right]^2}{n_i - 1} = \frac{0.83263^2}{17} + \cdots + \frac{0.97180^2}{9} = 0.60937.$$

12. $$F'_s = \frac{\text{quantity } \mathbf{9}/(a-1)}{1 + \left[\dfrac{2(a-2)}{a^2-1} \times \text{quantity } \mathbf{11}\right]}$$

$$= \frac{366.8003/7}{1 + \left[\dfrac{2 \times 6}{64 - 1} \times (0.60937)\right]} = \frac{52.40004}{1.11606} = 46.950$$

13. The degrees of freedom are:

$$\nu_1 = a - 1 = 7$$

$$\nu_2 = \frac{a^2 - 1}{3 \times \text{quantity } \mathbf{11}} = \frac{63}{3(0.60937)} = 34.5$$ where the 3 in the expression is a constant.

F'_s is distributed approximately as $F_{[\nu_1,\nu_2]}$. Since F'_s is far greater than $F_{.001[7,30]} = 4.82$, we can reject the null hypothesis that the samples were drawn from populations with equal means. Note that we say "populations," not "*a* population" as we usually do. Since the variances are unequal, it is evident that the samples could not come from any one statistical population.

BOX 13.3

Test of equality of the means of two samples whose variances are assumed to be unequal.

Comparison of chemical composition of the urine of apes. Milligrams of glutamic acid per milligram of creatinine.

	n	$\bar{Y}$	$s_{\bar{Y}}$
Chimpanzees	37	0.115	0.017
Gorillas	6	0.511	0.144

Source: Data by Gartler, Firschein, and Dobzhansky (1956).

We first reconstruct the variances of the two samples: $s^2 = s_{\bar{Y}}^2 \times n$

Chimpanzees $\quad s^2 = (0.017)^2 \times 37 = 0.010693$

Gorillas $\quad s^2 = (0.144)^2 \times 6 = 0.124416$

By the method of Box 8.1, $F_s = 0.124416/0.010693 = 11.64$. Since $F_{.001[5,30]} = 5.53$, it is clearly not necessary to interpolate for the exact degrees of freedom. It is most improbable ($P \ll 0.001$) that the two variances were sampled from the same population. We therefore cannot use the t-tests of Section 9.4 and Box 9.6 and use the following approximate test,

$$t'_s = \frac{(\bar{Y}_1 - \bar{Y}_2) - (\mu_1 - \mu_2)}{\sqrt{\dfrac{s_1^2}{n_1} + \dfrac{s_2^2}{n_2}}}$$

where t'_s is expected to be distributed approximately as t' when the null hypothesis is true and the other quantities have their conventional meaning.

The critical value of t'_α for type I error α is computed as

$$t'_\alpha = \frac{t_{\alpha[\nu_1]}\dfrac{s_1^2}{n_1} + t_{\alpha[\nu_2]}\dfrac{s_2^2}{n_2}}{\dfrac{s_1^2}{n_1} + \dfrac{s_2^2}{n_2}} = \frac{t_{\alpha[\nu_1]}s_{\bar{Y}_1}^2 + t_{\alpha[\nu_2]}s_{\bar{Y}_2}^2}{s_{\bar{Y}_1}^2 + s_{\bar{Y}_2}^2}$$

We compute

$$t'_s = \frac{(0.115 - 0.511)}{\sqrt{(0.017)^2 + (0.144)^2}} = \frac{-0.396}{\sqrt{0.000289 + 0.020736}}$$

$$= \frac{-0.396}{\sqrt{0.021025}} = \frac{0.396}{0.1450} = -2.73$$

Since $t_{.05[36]} = 2.028$ and $t_{.05[5]} = 2.571$, $s_{\bar{Y}_1}^2 = 0.000289$, and $s_{\bar{Y}_2}^2 = 0.020736$,

$$t'_{.05} = \frac{2.028(0.000289) + 2.571(0.020736)}{0.021025} = 2.564$$

Since $|t'_s| > t'_{.05}$, we conclude that the two means are significantly different. The gorillas excrete more glutamic acid in their urine than the chimpanzees.

Note that the formula for the critical value t'_α is a weighted average whose value must lie between $t_{\alpha[\nu_1]}$ and $t_{\alpha[\nu_2]}$. Therefore, if, as in our case, t'_s is greater

BOX 13.3 continued

than the t_α for the smaller number of degrees of freedom, we need not calculate t'_α but can accept t'_s to be significant. Conversely, if t'_s is smaller than t_α for the larger number of degrees of freedom, we can declare it not significant without further test. Only when t'_s lies between $t_{\alpha[\nu_1]}$ and $t_{\alpha[\nu_2]}$ need we calculate t'_α.

When the sizes of the two samples are equal, $n_1 = n_2 = n$, we can simplify the formula for t'_s to

$$t'_s = \frac{(\bar{Y}_1 - \bar{Y}_2) - (\mu_1 - \mu_2)}{\sqrt{\frac{1}{n}\,(s_1^2 + s_2^2)}}$$

which equals Expression (9.3) encountered in Section 9.4 where the variances were assumed to be equal. However, for the homoscedastic case $df = 2(n - 1)$. In this case with unequal variances the appropriate $df = n - 1$. We apply the formula to the survival time of the female and male roaches of Box 8.1. Although we found these variances not to be significantly different, we might have reason to suspect that the variances of males and females were in fact different. Since the test of the present box is always more conservative than that of Box 9.6, in which we assume homoscedasticity, we might feel safer to assume $\sigma_1^2 \neq \sigma_2^2$. We repeat the original data:

Females ($n_1 = 10$)	$\bar{Y}_1 = 8.5$ days	$s_1^2 = 3.6$
Males ($n_2 = 10$)	$\bar{Y}_2 = 4.8$ days	$s_2^2 = 0.9$

$$t'_s = \frac{(8.5 - 4.8)}{\sqrt{\frac{1}{10}(3.6 + 0.9)}} = \frac{3.7}{\sqrt{0.45}} = \frac{3.7}{0.6708} = 5.52$$

In view of the critical value $t'_{.001[9]} = 4.781$, females clearly survived significantly longer than males.

of fossil mammals. The computation is explained in the box. The final statistic of significance is X^2, which is compared with a critical value of chi square. We note that the samples are highly heteroscedastic and that an ordinary analysis of variance is therefore not appropriate in this case. Regrettably, Bartlett's test is unduly sensitive to departures from normality in the data (see Section 13.4) and a significant X^2 may therefore indicate nonnormality rather than heteroscedasticity. For this reason many statisticians no longer recommend this test.

There is also a "quick and dirty" method which, while not as efficient, will be preferred by many because of its simplicity. This is the F_{max}-test. This test relies on the tabled cumulative probability distribution of a statistic which is the variance ratio of the largest to the smallest of several sample variances. This distribution is shown in Table **T**. The F_{max}-test shown near the end of Box 13.1 corroborates the findings of the more exact test by Bartlett. The variances are significantly heterogeneous.

What may be the reasons for such heterogeneity? In this case, we suspect that some of the populations are inherently more variable than others. Some species are relatively uniform for one character, while others are quite variable for the same character. In an anova representing the results of an experiment, it may well be that one sample was obtained under less standardized conditions than the others and hence has a greater variance. There are also many cases in which the heterogeneity of variances is a function of an improper choice of measurement scale. With some measurement scales, variances vary as functions of means. Thus differences among means bring about heterogeneous variances. For example, in variables following the Poisson distribution the variance is in fact equal to the mean, and populations with greater means will therefore have greater variances. Such departures from the assumption of homoscedasticity can often be easily corrected by a suitable transformation, as discussed later in this chapter.

A very rapid first inspection for heteroscedasticity checks for correlation between the means and variances or between the means and the ranges of the samples. If the variances increase with the means (as in a Poisson distribution), the ratios $s^2/\bar{Y}$ or $s/\bar{Y} = CV$ will be approximately constant for the samples. If mean and variance are independent, these ratios will vary widely.

The consequences of moderate heterogeneity of variances are not too serious for the overall test of significance, but single degree of freedom comparisons may be far from accurate.

What can we do if our data are inherently heteroscedastic? We may carry out an approximate test of equality of means on the assumption of heterogeneity of variances as shown in Box 13.2 (based on Snedecor, 1956). The method differs from an ordinary single classification anova in that the means are weighted according to the reciprocal of the variance of the sample from which they were taken, and a special error MS must be used to take the weighting into account. We find that the differences among the means are highly significant (as would have been found if the ordinary anova had been performed).

When there are only two means to be tested (Section 9.4), we frequently test the hypothesis of equality of means by a t-test. Such a test assumes the equality of the two sample variances. When this assumption is not valid, we can perform an approximate t-test, which is illustrated in Box 13.3. This test calculates an approximate t-value, t'_s, for which the critical value is calculated as a weighted average of the critical values of t based on the corresponding degrees of freedom of the two samples. The computation is quite simple and is illustrated in the box. A related test based on fiducial probability is the Behrens-Fisher test, which can be found in the explanatory comments for Table VI in Fisher and Yates (1963).

13.4 Normality

We have assumed that the error terms, ϵ_{ij}, of the variates in each sample will be independent, that the variances of the error terms of the several samples will be equal, and, finally, that the error terms are normally distributed. Tests for the last assumption are familiar to you. Unless each sample is very large, it would be pointless to fit a normal distribution to it by the computations of Section 6.5. If there is serious question about the normality of the data, a graphic test as illustrated in Section 6.7 might be applied to each sample separately.

The consequences of nonnormality of error are not too serious. Only very skewed distribution would have a marked effect on the significance level of the F-test or on the efficiency of the design. The best way to correct for lack of normality is to carry out a transformation that will make the data normally distributed, as explained in later sections of this chapter. If no simple transformation is satisfactory, a nonparametric test as carried out in Sections 13.10 and 13.11 should be substituted for the analysis of variance.

13.5 Additivity

In two-way or higher-order anova without replication it is necessary to assume that interaction is not present if one is to make tests of the main effects in a Model I anova. This assumption of no interaction in a two-way anova is sometimes also referred to as the assumption of additivity of the main effects. By this we mean that any single observed variate can be decomposed into additive components representing the treatment effects of a particular row and column as well as a random term special to it. If interaction is actually present, then the F-test will be very inefficient and possibly misleading if the effect of the interaction is very large. A check of this assumption requires either more than a single observation per cell (so that an error mean square can be computed) or an independent estimate of the error mean square from previous *comparable* experiments.

Interactions can be due to a variety of causes. Most frequently it means that a given treatment combination, such as level 2 of factor A when combined with level 3 of factor B, makes a variable deviate from the expected value. Such a deviation is regarded as an inherent property of the natural system under study, as in examples of synergism or interference. Similar effects occur when a given replicate is quite aberrant, as may happen if an exceptional plot is included in an agricultural experiment, if a diseased individual is included in a physiological experiment, or if by mistake an individual from a different species is included in a biometric study. Finally, an interaction term will result if the effect of the two factors A and B on the response variable Y are multiplicative rather than additive. An example will make this clear.

In Table 13.1 we show the additive and multiplicative treatment effects in a hypothetical two-way anova. Let us assume that the expected population

TABLE 13.1

Illustration of additive and multiplicative effects.

Factor B	Factor A			
	$\alpha_1 = 1$	$\alpha_2 = 2$	$\alpha_3 = 3$	
	2	3	4	Additive effects
$\beta_1 = 1$	1	2	3	Multiplicative effects
	0	0.30	0.48	Log of multiplicative effects
	6	7	8	Additive effects
$\beta_2 = 5$	5	10	15	Multiplicative effects
	0.70	1.00	1.18	Log of multiplicative effects

mean μ is zero. Then the mean of the sample subjected to treatment 1 of factor A and treatment 1 of factor B should be 2 by the conventional additive model. This is so because each factor at level 1 contributes unity to the mean. Similarly, the expected subgroup mean subjected to level 3 for factor A and level 2 for factor B is 8, the respective contributions to the mean being 3 and 5. However, if the process is multiplicative rather than additive, as occurs in a variety of physicochemical and biological phenomena, the expected values would be quite different. For treatment A_1B_1, the expected value equals 1, which is the product of 1 and 1. For treatment A_3B_2, the expected value is 15, the product of 3 and 5. If we were to analyze multiplicative data of this sort by a conventional anova, we would find that the interaction sum of squares would be greatly augmented because of the nonadditivity of the treatment effects. In this case, there is a simple remedy. By transforming the variable into logarithms (Table 13.1), we are able to restore the additivity of the data. The third item in each cell gives the logarithm of the expected value, assuming multiplicative relations. Notice that the increments are strictly additive again ($SS_{A \times B} = 0$). As a matter of fact, on a logarithmic scale we could simply write $\alpha_1 = 0$, $\alpha_2 = 0.30$, $\alpha_3 = 0.48$, $\beta_1 = 0$, $\beta_2 = 0.70$. Here is a good illustration of how transformation of scale, discussed in detail in later sections of this chapter, helps us meet the assumptions of analysis of variance.

If we wish to ascertain whether the interaction found in a given set of data could be explained in terms of multiplicative main effects, then we can perform a test due to Tukey (1949), which is illustrated in Box 13.4. This test is also useful when testing for nonadditivity in a two-way Model I anova without replication in experiments where it is reasonable to assume that interaction, if present at all, could only be due to multiplicative main effects. This test partitions the interaction sum of squares into one degree of freedom due to multiplicative effects of the main effects and a residual sum of squares

BOX 13.4

Tukey's test for nonadditivity.

Oxygen consumption of two species of limpets (*Acmaea*) at three salinities. Since only the subgroup cell totals are needed for computation, the individual observations are not given here. Data from Box 11.1. Each subgroup cell total is based on $n = 8$ observations.

	(*1*)	(*2*)	(*3*)	(*4*)
	Factor A: Species ($a = 2$)			
Factor B: Seawater concentration ($b = 3$)	*A. scabra*	*A. digitalis*	Sum/$a = n\bar{Y}_B$	$n\bar{Y}_B - n\bar{\bar{Y}}$
100%	84.49	59.43	71.96	− 5.00
75%	63.12	58.70	60.91	−16.05
50%	97.39	98.61	98.00	21.04
Sum/$b = n\bar{Y}_A$	81.67	72.25	$76.96 = n\bar{\bar{Y}}$	
$n\bar{Y}_A - n\bar{\bar{Y}}$	4.71	−4.71		
Q_i	613.5596	835.4694		

Computations

1. Form a two-way table of subgroup (cell) totals. Compute row totals divided by a, column totals divided by b, and the grand total divided by ab. These yield values of $n\bar{Y}_B$, of $n\bar{Y}_A$, and a value of $n\bar{\bar{Y}}$, respectively. Also evaluate deviations of the $n\bar{Y}_B$'s and $n\bar{Y}_A$'s from $n\bar{\bar{Y}}$.

2. For each column compute

$$Q_i = \sum^{b}\left(\sum^{n} Y_{ij}\right)(n\bar{Y}_B - \bar{\bar{Y}})$$

$$Q_1 = 84.49(-5.00) + 63.12(-16.05) + 97.39(21.04) = 613.5596$$

$$Q_2 = 59.43(-5.00) + 58.70(-16.05) + 98.61(21.04) = 835.4694$$

3. Compute

$$Q = \sum^{a} Q_i(n\bar{Y}_A - \bar{\bar{Y}}) = 613.5596(4.71) + 835.4694(-4.71)$$
$$= -1045.1952$$

4. Compute

$$K = \left[\sum^{a} (n\bar{Y}_A - \bar{\bar{Y}})^2\right] \times \left[\sum^{b} (n\bar{Y}_B - \bar{\bar{Y}})^2\right]$$
$$= [(4.71)^2 + (-4.71)^2] \times [(-5.00)^2 + (-16.05)^2 + 21.04^2]$$
$$= (44.3682) \times (725.2841) = 32{,}179.5500$$

5. SS for nonadditivity $SS_{\text{nonadd}} = Q^2/Kn$

$$= (-1045.1952)^2/(32{,}179.5500 \times 8) = 4.2435$$

This SS with one degree of freedom is a part of the interaction SS (calculated in Box 11.1).

BOX 13.4 continued

Source of variation	*df*	*SS*	*MS*
$A \times B$	2	23.9263	
Nonadditivity	1	4.2435	4.2435
Residual	1	19.6828	19.6828

$F_s = MS_{\text{nonadd}}/MS_{\text{resid}}$ is less than one and hence clearly not significant. We have no evidence of nonadditivity (multiplicative effects of the factors A and B) in these data.

to represent the other possible interactions or to serve as error in case the anova has no replication. The computations as outlined in Box 13.1 are straightforward. We find that in the example shown (oxygen consumption of two species of limpets), which did not have any significant interaction, the single degree of freedom SS for nonadditivity due to multiplicative effects is also not significant. If the test for nonadditivity had been significant and the residual SS nonsignificant, then an analysis of log oxygen consumption would be preferred since it would yield a simpler analysis.

13.6 Transformations

If the evidence indicates that the assumptions for an analysis of variance or a t-test cannot be maintained, two courses of action are open to us. We may carry out a different test not requiring the rejected assumptions, such as the distribution-free tests in lieu of anova, discussed at the end of this chapter. A second approach would be to transform the variable to be analyzed in such a manner that the resulting transformed variates meet the assumptions of the analysis. Let us look at a simple example of what transformation will do. A single variate of the simplest kind of anova (completely randomized, single classification, Model I) decomposes as follows: $Y_{ij} = \mu + \alpha_i + \epsilon_{ij}$. In this model the components are additive with the error term ϵ_{ij} normally distributed. However, we might encounter a situation in which the components were multiplicative in effect so that $Y_{ij} = \mu\alpha_i\epsilon_{ij}$, the product of these three terms. In such a case the assumptions of normality and of homoscedasticity would break down. The general parametric mean μ is constant in any one anova, but the treatment effect α_i differs from group to group. Clearly, the scatter among the variates Y_{ij} would double in a group in which α_i is twice as great as in another. Assume that $\mu = 1$, the smallest $\epsilon_{ij} = 1$, and the greatest 3; then if $\alpha_i = 1$, the range of the Y's will be $3 - 1 = 2$. However, when $\alpha_i = 4$, the corresponding range will be four times as wide from $4 \times 1 = 4$ to $4 \times 3 = 12$, a range of 8. Such data will be heteroscedastic. We can correct this situation simply by transforming our model into logarithms. We would therefore obtain $\log Y_{ij} = \log \mu + \log \alpha_i + \log \epsilon_{ij}$, which is additive and homoscedastic. The entire analysis of variance would then be carried out on the transformed variates.

At this point many of you will feel more or less uncomfortable about what we have done. Transformation seems too much like "data grinding." When you learn that often a statistical test may be made significant after transformation of a set of data, though it would not have been so without such a transformation, you may feel even more suspicious. What is the justification for transforming the data? It takes some getting used to the idea, but there is really no scientific necessity to employ the common linear or arithmetic scale to which we are accustomed. Teaching of the "new math" in elementary schools has done much to dispel the naive notion that the decimal system of numbers is the only "natural" one. It takes extensive experience in science and in the handling of statistical data to appreciate the fact that the linear scale, so familiar to all of us from our earliest experience, occupies a similar position with relation to other scales of measurement as does the decimal system of numbers with respect to the binary or octal numbering systems and others. If a system is multiplicative on a linear scale, it may make much more sense to think of it as an additive system on a logarithmic scale. The square root of a variable is another frequent transformation. The square root of the surface area of an organism is often a more appropriate measure of the fundamental biological variable subjected to physiological and evolutionary forces than is the area. This is reflected in the normal distribution of the square root of the variable as compared to the skewed distribution of areas. In many cases experience has taught us to express experimental variables not in linear scale but as logarithms, square roots, reciprocals, or angles. Thus, *p*H values are logarithms and dilution series in microbiological titrations are expressed as reciprocals. As soon as you are ready to accept the idea that the scale of measurement is arbitrary, you simply have to look at the distributions of transformed variates to decide which transformation most closely satisfies the assumptions of the analysis of variance before carrying out an anova.

A fortunate fact about transformations is that very often several departures from the assumptions of anova are simultaneously cured by the same transformation to a new scale. Thus frequently, simply by making the data homoscedastic, we also make them approach normality and insure additivity of the treatment effects.

When a transformation is applied, tests of significance are performed on the transformed data, but estimates of means are usually given in the familiar untransformed scale. Since the transformations discussed in this chapter are nonlinear, confidence limits computed in the transformed scale and changed back to the original scale would be asymmetrical. Stating the standard error in the original scale would therefore be misleading. In reporting results of research with variables that require transformation, furnish means in the untransformed scale followed by their (asymmetrical) confidence limits rather than by their standard errors.

An easy way to find out whether a given transformation will yield a distribution satisfying the assumptions of anova is to plot the cumulative distributions of the several samples on probability paper. By changing the scale of the second coordinate axis from linear to logarithmic, square root or any other one, we can see whether a previously curved line, indicating skewness, straightens out to indicate normality (you may wish to refresh your memory on these graphic techniques studied in Section 6.7). We can look up class marks on transformed scales or, easier still, use a variety of available probability graph papers whose second axis is in logarithmic, angular, or other scale. Thus we not only test whether the data become more normal through transformation, but we can also get an estimate of the standard deviation under transformation as measured by the slope of the estimated line. The assumption of homoscedasticity implies that the slopes for the several samples should be the same. If the slopes are very heterogeneous, homoscedasticity has not been achieved.

Three transformations will be discussed below: the logarithmic transformation (Section 13.7), the square root transformation (Section 13.8), and the angular or arcsine transformation (Section 13.9). Two other transformations will be taken up in connection with regression analysis in Section 14.12.

13.7 The logarithmic transformation

The most common transformation applied is conversion of all variates into logarithms, usually common logarithms. Whenever the mean is positively correlated with the variance (greater means are accompanied by greater variances), the logarithmic transformation is quite likely to remedy the situation and make the variance independent of the mean. Frequency distributions skewed to the right are often made more symmetrical by transformation to logarithmic scale. We saw in the previous section and in Table 13.1, that logarithmic transformation is also called for when effects are multiplicative.

In Table 13.2 we illustrate the effect of a logarithmic transformation. These data are lengths of juvenile Silver Salmon, checked in a downstream trap during four different two-week periods. Note that the variances increase as the means do. This could also have been foreseen from the increase in range of each sample with increase in mean if one compares samples based on approximately the same number of fish, such as the first and third samples. As soon as the variates are transformed to logarithms, we note that the variance of a sample is independent of its mean. We could now proceed and carry out an analysis of variance of these data if we wished to. To report these data, you should transform the means back into linear scale by looking up their antilogarithms. Remember that a statement of the standard error

TABLE 13.2

An application of the logarithmic transformation. Length of the juvenile Silver Salmon *Oncorhynchus kisutch* checked in a downstream trap during four different two-week periods (pooled over period 1933 to 1942).

(1) Length in mm Y	*(2)* log of length $\log Y$	*(3)* Nov. 12–25 f	*(4)* Jan. 7–20 f	*(5)* Apr. 1–14 f	*(6)* Apr. 29–May 12 f
60	1.778	1	2	—	—
65	1.813	4	—	—	—
70	1.845	33	1	—	—
75	1.875	94	4	—	1
80	1.903	125	2	1	—
85	1.929	98	12	5	5
90	1.954	44	5	6	20
95	1.978	18	5	5	105
100	2.000	3	—	16	328
105	2.021	1	—	25	757
110	2.041	—	—	45	1371
115	2.061	—	—	57	1562
120	2.079	—	—	87	1372
125	2.097	—	—	65	781
130	2.114	—	—	20	388
135	2.130	—	—	5	159
140	2.146	—	—	5	51
145	2.161	—	—	5	18
150	2.176	—	—	1	6
155	2.190	—	—	—	4
160	2.204	—	—	1	1
165	2.217	—	—	—	1
n		421	31	349	6930
Untransformed variable					
$\bar{Y}$		80.962	83.710	116.977	115.551
s^2		45.92	83.280	124.672	82.541
log transformation					
$\overline{\log Y}$		1.907	1.920	2.066	2.061
$s^2_{\log Y}$		0.001312	0.00261	0.001868	0.002062
Back-transformed means					
antilog $(\overline{\log Y})$		80.72	83.18	116.4	115.1

Source: Data from Shapovalov and Taft (1954).

is of no value in these data. Rather, you first compute confidence limits in logarithmic scale, retransform the confidence limits to linear scale, and report the reliability of your sample estimate as such. Thus for the first sample ($n = 421$) we would report the mean and 95% confidence limits as follows:

$$\bar{Y} = \text{antilog}\ (\overline{\log Y}) = \text{antilog}\ (1.907) = 80.72$$

$$L_1 = \text{antilog}\ [\overline{\log Y} - t_{.05[420]}\sqrt{s^2_{\log Y}/n}] = \text{antilog}\,[1.907 - 1.966\ (0.00177)]$$
$$= \text{antilog}\ (1.9035) = 80.08$$

$$L_2 = \text{antilog}\ [1.907 + 1.966\ (0.00177)] = \text{antilog}\ (1.9105) = 81.38$$

Note that these limits are asymmetrical around the mean: $L_2 - \bar{Y} = 81.38 - 80.72 = 0.66$, $\bar{Y} - L_1 = 80.72 - 80.08 = 0.64$. These limits differ from confidence limits computed from the original measurements, which would be as follows:

$$\bar{Y} = 80.962$$

$$L_1 = \bar{Y} - t_{.05[420]}\sqrt{s^2/n} = 80.962 - 1.966\ (0.3303) = 80.313$$

$$L_2 = 80.962 + 1.966\ (0.3303) = 81.611$$

When the variates to be transformed include zeros, a technical problem arises, since the logarithm of zero is negative infinity. In such cases the transformation $\log\ (Y + 1)$ will avoid the problem. When values include numbers between zero and one, it may be desirable to code the variates by multiplying by 10, 100, or some higher power of 10 to avoid negative characteristics in the logarithms. Although logarithms to any base are acceptable, common logarithms are simplest to use. For logarithmic transformation a two- or three-place mantissa is generally adequate when working by desk calculator.

Logarithmic transformations are frequently needed in the analysis of variables related to the growth of organisms, such as the example in Table 13.2.

13.8 The square root transformation

When the data are counts, as of insects on a leaf or blood cells in a hemacytometer, we frequently find the square root transformation of value. You will remember that such distributions are likely to be Poisson rather than normally distributed and that in a Poisson distribution the variance is the same as the mean. Therefore, the mean and variance cannot be independent but will vary identically. Transforming the variates to square roots will generally make the variances independent of the means. When the counts include zero values, it has been found desirable to code all variates by adding 0.5. The transformation then is $\sqrt{Y + \frac{1}{2}}$.

Table 13.3 shows an application of the square root transformation. Again the sample with the greater mean has a significantly greater variance prior to transformation. After transformation the variances are not significantly different. For reporting of means the transformed means are squared again and confidence limits are reported in lieu of standard errors.

TABLE 13.3

An application of the square root transformation. The data represent the number of adult *Drosophila* emerging from single pair cultures for two different medium formulations (medium A contained DDT).

(1) Number of flies emerging Y	(2) Square root of number of flies $\sqrt{Y}$	(3) Medium A f	(4) Medium B f
0	0.00	1	—
1	1.00	5	—
2	1.41	6	—
3	1.73	—	—
4	2.00	3	—
5	2.24	—	—
6	2.45	—	—
7	2.65	—	2
8	2.83	—	1
9	3.00	—	2
10	3.16	—	3
11	3.32	—	1
12	3.46	—	1
13	3.61	—	1
14	3.74	—	1
15	3.87	—	1
16	4.00	—	2
		15	15

	Medium A	Medium B
Untransformed variable		
$\bar{Y}$	1.933	11.133
s^2	1.495	9.410
Square root transformation		
$\overline{\sqrt{Y}}$	1.297	3.307
$s^2_{\sqrt{Y}}$	0.2630	0.2089

Tests of equality of variances

$$F_s = \frac{s^2_2}{s^2_1} = \frac{9.410}{1.495} = 6.294* \qquad F_{.025[14,14]} = 2.98 \qquad F_s = \frac{s^2_{\sqrt{Y_1}}}{s^2_{\sqrt{Y_2}}} = \frac{0.2630}{0.2089} = 1.259\ ns$$

Back-transformed (squared) means

	Medium A	Medium B
$(\overline{\sqrt{Y}})^2$	1.682	10.936
95% confidence limits		
$L_1 = \overline{\sqrt{Y}} - t_{.05}s_{\overline{\sqrt{Y}}}$	$1.297 - 2.145\sqrt{\frac{0.2630}{15}}$ $= 1.013$	$3.307 - 2.145\sqrt{\frac{0.2089}{15}}$ $= 3.054$
$L_2 = \overline{\sqrt{Y}} + t_{.05}s_{\overline{\sqrt{Y}}}$	1.581	3.560
Back-transformed (squared) confidence limits		
L_1^2	1.026	9.327
L_2^2	2.500	12.674

Source: Data from unpublished study by R. R. Sokal.

13.9 The arcsine transformation

This transformation, also known as the *angular transformation*, is especially appropriate to percentages and proportions. You may remember from Section 5.2 that the standard deviation of a binomial distribution is $\sigma = \sqrt{pq/k}$. Since $\mu = p$, $q = 1 - p$, and k is constant for any one problem, it is clear that in a binomial distribution the variance would be a function of the mean. The arcsine transformation prevents this.

The transformation finds $\theta = \arcsin \sqrt{p}$, where p is a proportion. The transformation has been tabulated in Table **K**. The term arcsin is synonymous with inverse sine or $\sin^{-1}$, which stands for the angle whose sine is the given quantity. Thus, if we look up arcsin 0.431 we find 41.03°, the angle whose sine is 0.431. These relationships are stated in conventional mathematical notation as: $\sin 41.03° = 0.431$; $\arcsin 0.431 = 41.03°$. The arcsine transformation stretches out both tails of a distribution of percentages or proportions and compresses the middle. An example of its application is shown in Table 13.4, where percent fertilities from a sample of 100 vials of 10 eggs each

TABLE 13.4

An application of the use of the arcsine transformation. Percent fertility of eggs of the CP strain of *Drosophila melanogaster* raised in 100 vials of 10 eggs each.

(1)	*(2)*	*(3)*	*(4)*[a]
% fertility 100p	*arcsin* $\sqrt{p}$	*f*	*Cumulative frequencies F*
0	0	1	1
10	18.44	3	4
20	26.56	8	12
30	33.21	10	22
40	39.23	6	28
50	45.00	15	43
60	50.77	14	57
70	56.79	12	69
80	63.44	13	82
90	71.56	9	91
100	90.00	9	100
		100	

Source: Data from Sokal (1966); see Figure 13.1.
[a] Needed for preparation of Figure 13.1.

are shown for a strain of *Drosophila*. Figure 13.1 shows quite clearly that the distribution is not normal (it is platykurtic). When the percentages are transformed to angles in column (2) of Table 13.4, the distribution closely approximates the normal. The expected variance of such a distribution of arcsines is

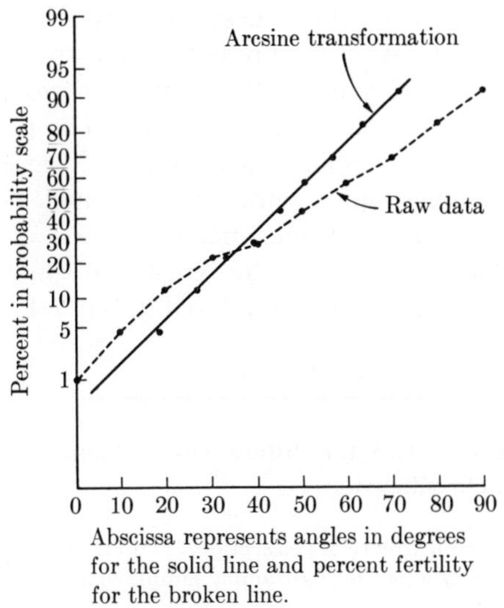

FIGURE 13.1 Data from Table 13.4 plotted on arithmetic probability paper to show the effect of the arcsine transformation.

$\sigma_\theta^2 = 180^2/4\pi^2 n = 820.8/n$, where θ represents the angles in which the arcsines are expressed. When θ is measured in radians rather than degrees, then $\sigma_\theta^2 = \frac{1}{4}n$. In reporting means we need to convert them back to proportions or percentages. This can be done by means of Table **L,** which gives arcsin $\sqrt{p}$ as arguments and the proportions as functions; that is, it is a table of $p = [\sin(\theta)]^2$. When the percentages in the original data fall between 30 and 70%, it is generally not necessary to apply the arcsine transformation.

13.10 Nonparametric methods in lieu of single classification anova

If none of the above transformations manage to make our data meet the assumptions of analysis of variance, we may resort to an analagous nonparametric method. These techniques are also called *distribution-free methods,* since they are not dependent on a given distribution (such as the normal in the case of anova), but usually will work for a wide range of different distributions. They are called nonparametric methods because their null hypothesis is not concerned with specific parameters (such as the mean in analysis of variance) but only with the distribution of the variates. In recent years, nonparametric analysis of variance has become quite popular because it is simple to compute and permits freedom from worry about the assumptions of an anova. Yet we should point out that in cases where these assumptions hold entirely or even approximately, the analysis of variance is generally the more efficient statistical test for detecting departures from the null hypothesis.

The easiest way to get a feel for these nonparametric tests is to become acquainted with one. The tests of the present section are all based on the idea of *ranking* the variates in an example after pooling all groups and considering them as a single sample for purposes of ranking. The present section considers analogues of single classification analysis of variance—that is, the general case with a samples and n_i variates per sample. For such a case the *Kruskal-Wallis test* is generally recommended.

Box 13.5 gives an illustration of the Kruskal-Wallis test applied to the rate of growth of pea sections first encountered in Box 9.4. This is an example

BOX 13.5

Kruskal-Wallis test. A test for differences of location in ranked data grouped by single classification.

Effect of different sugars on growth of pea sections. Data from Box 9.4: $a = 5$ groups; n_i = number of items in group i (in this example sample sizes are equal, $n_i = n = 10$).

Computations

1. Rank all observations from smallest to largest, pooled together into a single group. In case of ties, compute average ranks. For example, the 4 variates $Y = 59$ represent ranks 18, 19, 20, and 21. Their average rank therefore equals $(18 + 19 + 20 + 21)/4 = 19.5$.

Average Rank	*Rank*	*Y*	*Average Rank*	*Rank*	*Y*	*Average Rank*	*Rank*	*Y*
2	1	56	19.5	18	59		35	64
	2	56		19	59	37.5	36	65
	3	56		20	59		37	65
7	4	57		21	59		38	65
	5	57	23.5	22	60		39	65
	6	57		23	60		40	66
	7	57		24	60	42.5	41	67
	8	57		25	60		42	67
	9	57	27.5	26	61		43	67
	10	57		27	61		44	67
14	11	58		28	61		45	68
	12	58		29	61		46	70
	13	58	31.5	30	62		47	71
	14	58		31	62	48.5	48	75
	15	58		32	62		49	75
	16	58		33	62		50	76
	17	58		34	63			

2. Replace each observation in the original data table by its rank or average rank.

BOX 13.5 continued

		Treatments							
Control		*2% Glucose added*		*2% Fructose added*		*1% Glucose + 1% Fructose added*		*2% Sucrose added*	
Y	*Rank*	Y	*Rank*	Y	*Rank*	Y	*Rank*	Y	*Rank*
75	48.5	57	7	58	14	58	14	62	31.5
67	42.5	58	14	61	27.5	59	19.5	66	40
70	46	60	23.5	56	2	58	14	65	37.5
75	48.5	59	19.5	58	14	61	27.5	63	34
65	37.5	62	31.5	57	7	57	7	64	35
71	47	60	23.5	56	2	56	2	62	31.5
67	42.5	60	23.5	61	27.5	58	14	65	37.5
67	42.5	57	7	60	23.5	57	7	65	37.5
76	50	59	19.5	57	7	57	7	62	31.5
68	45	61	27.5	58	14	59	19.5	67	42.5
$\overset{n_i}{(\Sigma R)_i}$	450		196.5		138.5		131.5		358.5

3. Sum the ranks separately for each group $\left[= \left(\sum^{n_i} R\right)_i\right]$;

for example, $\left(\sum^{n_1} R\right)_1 = 48.5 + 42.5 + \cdots + 50 + 45 = 450$

4. Compute Expression (13.1). The numbers 12 and 3 are constants.

$$H = \left[\frac{12}{\left(\sum^{a} n_i\right)\left(\sum^{a} n_i + 1\right)} \sum^{a} \frac{\left(\sum^{n_i} R\right)_i^2}{n_i}\right] - 3\left(\sum^{a} n_i + 1\right)$$

$$= \frac{12}{50(50+1)}\left[\frac{450.0^2}{10} + \frac{196.5^2}{10} + \cdots + \frac{358.5^2}{10}\right] - 3(50+1)$$

$$= \frac{12}{50(51)}(40{,}610.9) - 3(51) = 191.11 - 153 = 38.11$$

5. Since there were ties, this H value must be corrected by dividing it by

$$D = 1 - \frac{\sum^{m} T_j}{\left(\sum^{a} n_i - 1\right)\sum^{a} n_i\left(\sum^{a} n_i + 1\right)}$$

where T_j is a function of the t_j, the number of variates tied in the jth group of ties. (This t has no relation to Student's t.) The function is $T_j = t_j^3 - t_j$, computed easiest as $(t_j - 1)t_j(t_j + 1)$. Since in most cases the tied group will range from $t = 2$ to $t = 10$ ties, we give a small table of T over this range; the summation of T_j is over the m different ties.

t_j	2	3	4	5	6	7	8	9	10
T_j	6	24	60	120	210	336	504	720	990

BOX 13.5 continued

For example, for the first tied group in the table of ranks $t_j = 3$, since there are 3 variates of equal magnitude.

The t_j's for the present example are shown below together with the corresponding T_j's:

t_j	3	7	7	4	4	4	4	4	4	2
T_j	24	336	336	60	60	60	60	60	60	6

$$\sum^{m} T_j = 24 + 336 + \cdots + 6 = 1062$$

$$D = 1 - \frac{1062}{(50 - 1)50(50 + 1)} = 1 - \frac{1062}{49(50)51} = 1 - \frac{1062}{124{,}950}$$

$$= 1 - 0.00850 = 0.99151$$

$$\text{Adjusted } H = \frac{H}{D} = \frac{38.11}{0.99151} = 38.44$$

If the null hypothesis (that the a groups do not differ in "location") is true, H is distributed approximately as $\chi^2_{[a-1]}$. Since H is greater than $\chi^2_{.01[4]} = 13.277$, we may reject the null hypothesis and conclude that different sugars affect rate of growth of pea sections differentially. Box 13.7 gives a method for a posteriori testing of differences among treatments.

---

of single classification with equal sample sizes, but the computational formulas shown in Box 13.5 are general and applicable to varying sample sizes as well. We first rank all variates from the smallest to the largest, ignoring the division into groups. During such a procedure ties are a frequent problem. You will note several ties for almost every value on the measurement scale of the variable. For these ties we calculate the average of the ranks occupied by the tied values, as shown in Box 13.5. We next reconstitute the original data table but replace each original variate by its rank or average rank as the case may be; this is shown in Box 13.5. Next we evaluate the following expression:

$$H = \left[\frac{12}{\sum^{a} n_i \left(\sum^{a} n_i + 1\right)} \sum^{a} \frac{\left(\sum^{n_i} R\right)_i^2}{n_i}\right] - 3\left(\sum^{a} n_i + 1\right) \tag{13.1}$$

This formula seems formidable, but most of the quantities are familiar. $\sum^{a} n_i$ is simply the sum of the sample sizes of the entire analysis. The numbers 12 and 3 are constants in every problem and $\left(\sum^{n_i} R\right)_i$ is the sum of the ranks for the ith group. The statistic H as shown by Expression (13.1) is appropriate for data without ties, but must be divided by a correction factor D when ties are present. This correction factor is computed from a formula shown in Box 13.5. H is distributed approximately as $\chi^2_{[a-1]}$ for large samples when the null hypothesis is true. For tests at $\alpha = 0.10$ or $\alpha = 0.05$, the χ^2 approximation is very good even with n as small as 5. However, for $\alpha = 0.01$, the test is conservative for small values of n; it rejects less than 1% of the tests if the hypothesis is true (personal communication from K. R. Gabriel, based on an unpublished Monte Carlo study). When $a = 3$ and the sample sizes n_i are each less than five, the statistic H is not distributed as chi square. An exact distribution has been tabled and can be found in Kruskal and Wallis (1952) or as Table O in Siegel (1956).

If the populations are not different from each other, we would expect their rank sums to be approximately the same (allowing for differences in sample size where such exist). We note in the example of Box 13.5 that we are led to reject the null hypothesis, which is that the true "location" of the several populations is the same. The conclusion reached by the Kruskal-Wallis test is the same as by the regular anova in Box 9.4. The null hypothesis is decisively rejected.

When the test is only between two samples (such a design would give rise to a t-test or anova with two classes), we employ either of two nonparametric tests, which yield the same statistic and give the same results. The tests are called the *Mann-Whitney* U-*test* or the *Wilcoxon two-sample test.* The null hypothesis is that the two samples come from populations having the same distribution. In Box 13.6 we show how to apply both tests. The data are morphological measurements on two samples of chigger nymphs. The Mann-Whitney U-test as illustrated in Box 13.6 is a semigraphical test and is quite simple to apply. It will be especially convenient when the data are already graphed and there are not too many items in each sample.

Please note that the methods of Box 13.6 do not really require that each individual observation represent a precise measurement. So long as you can order the observations you are able to perform these tests. Thus, for example, suppose you placed some meat out in the open and studied the arrival times of individuals of two species of blowflies. You could record exactly the time of arrival of each individual fly, starting from a point zero in time when the meat was set out. On the other hand, you might simply rank arrival times of the two species, noting that individual 1 of species B came first, 2 individuals from species A next, then 3 individuals of B, followed by the simultaneous arrival of one of each of the two species (a tie), and so forth. While such ranked or ordered data could not be analyzed by the parametric methods studied earlier, the techniques of Box 13.6 are entirely applicable.

The method of calculating the sample statistic U_s for the Mann-Whitney and the Wilcoxon tests is straightforward, as shown in Box 13.6. The critical values for $U_{\alpha[n_1,n_2]}$ are shown in Table **CC**, which is adequate for cases in which the larger sample size $n_1 \leq 20$. The probabilities in Table **CC** assume a one-tailed test. For a two-tailed test you should double the value of the probability shown in that table. When $n_1 > 20$, compute the expression shown near the bottom of Box 13.6. Since this expression is distributed as a normal deviate, consult the table of t (Table **Q**), for $t_{\alpha[\infty]}$ using one- or two-tailed probabilities, depending on your hypothesis. A further complication arises from tied observations, which require the more elaborate formula shown at the bottom of Box 13.6. However, it takes a substantial number of ties to affect the outcome of the test appreciably. Corrections for ties increase the t_s-value slightly; hence the uncorrected formula is more conservative.

BOX 13.6

Mann-Whitney *U*-test and Wilcoxon two-sample test for two samples, ranked observations, not paired.

Two samples of nymphs of the chigger *Trombicula lipovskyi*. Variate measured is length of cheliceral base stated as micrometer units.

(1)	*(2)*	*(3)*	*(4)*
Sample A		*Sample* B	
Y	*Rank* (R)	*Y*	*Rank* (R)
104	2	100	1
109	7	105	3
112	9	107	4.5
114	10	107	4.5
116	11.5	108	6
118	13.5	111	8
118	13.5	116	11.5
119	15	120	16
121	17.5	121	17.5
123	19.5	123	19.5
125	21		
126	22.5		$91.5 = \sum^{n_1} R$
126	22.5		
128	25		
128	25		
128	25		
	$259.5 = \sum^{n_2} R$		

Source: Data by D. A. Crossley.

Designate sample size of larger sample as n_1 and that of the smaller sample as n_2. In this case, $n_1 = 16$, $n_2 = 10$. If the two samples are of equal size, it does not matter which is designated as 1.

There are two equivalent procedures for carrying out a test of equality of "location" of two samples.

Mann-Whitney U-test

1. List the observations from the smallest to the largest in such a way that the two samples may be easily compared. A convenient method is to make a graph as shown below.

BOX 13.6 continued

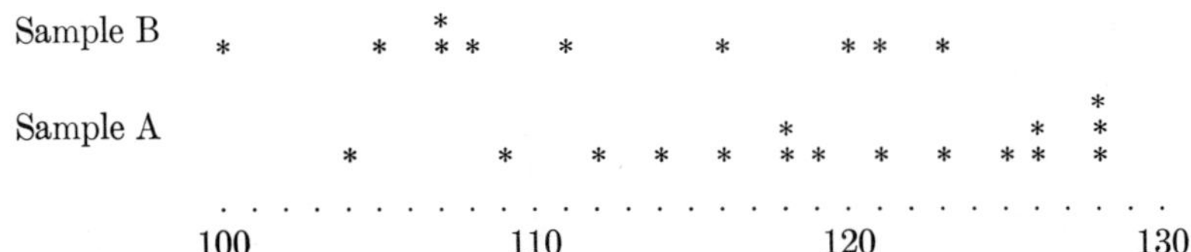

2. For each observation in one sample (it is convenient to use the smaller sample), count the number of observations in the other sample which are lower in value (to the left). Count $\frac{1}{2}$ for each tied observation. For example, there are zero observations in Sample A less than the first observation in Sample B; 1 observation less than the second, third, fourth, and fifth observations in Sample B; 2 observations in A less than the sixth in B; 4 observations in A less than the seventh in B, but one is equal (tied) with it, so we count $4\frac{1}{2}$. Continuing in a similar manner, we obtain counts of 8, $8\frac{1}{2}$, and $9\frac{1}{2}$. The sum of these counts $C = 36\frac{1}{2}$. The Mann-Whitney statistic U_s is the greater of the two quantities C and $(n_1n_2 - C)$, in this case $36\frac{1}{2}$ and $[(16 \times 10) - 36\frac{1}{2}] = 123\frac{1}{2}$.
3. We compare the value of $U_s = 123\frac{1}{2}$ with the critical values for $U_{\alpha[n_1,n_2]}$ in Table **CC**. The null hypothesis is rejected if the observed value is too large. Since $U_{.025[16,10]} = 118$ and $U_{.01[16,10]} = 124$, the two samples are significantly different at $0.05 > P > 0.02$ (we double the probabilities because this is a two-tailed test).

The Wilcoxon two-sample test

1. Rank all of the observations together from low to high, as shown in columns (2) and (4) above. Give average ranks in case of ties.
2. Sum the ranks of the smaller sample (sample size n_2).
3. Compute the Wilcoxon statistic as

$$C = n_1n_2 + \frac{n_2(n_2+1)}{2} - \sum^{n_2} R = 16(10) + \frac{10(10+1)}{2} - 91.5$$

$$= 160 + 55 - 91.5 = 123.5$$

where n_1 is the size of the larger sample and n_2 is the size of the smaller. This statistic must be compared with $n_1n_2 - C$ and the greater of the two quantities chosen as a test statistic U_s. In this case, $160 - 123.5 = 36.5$, so we use 123.5, which is identical to the value obtained in the Mann-Whitney test and is significant when compared to the critical value in Table **CC**.

In cases where $n_1 > 20$, calculate the following quantity

$$t_s = \frac{\left(U_s - \frac{n_1n_2}{2}\right)}{\sqrt{\frac{n_1n_2(n_1+n_2+1)}{12}}}$$

which is approximately normally distributed. The denominator 12 is a constant. Look up the significance of t_s in Table **Q** against critical values of $t_{\alpha[\infty]}$ for a

BOX 13.6 continued

one-tailed test or two-tailed test as required by the hypothesis. When tied values occur, the above formula is modified as follows:

$$t_s = \frac{\left(U_s - \frac{n_1 n_2}{2}\right)}{\sqrt{\left(\frac{n_1 n_2}{(n_1 + n_2)(n_1 + n_2 - 1)}\right)\left(\frac{(n_1 + n_2)^3 - (n_1 + n_2) - \sum^{m} T_j}{12}\right)}}$$

where $\sum^{m} T_j$ has the same meaning as in Box 13.5.

It is desirable to obtain an intuitive understanding of the rationale behind these tests. In the Mann-Whitney test we can conceive of two extreme situations: in one case the two samples overlap and coincide entirely, in the other they are quite separate. In the latter case, if we take the sample with the lower-valued variates there will be no points of the contrasting sample to the left of it; that is, we can go through every observation in the lower-valued sample without having any items of the higher-valued one to the left of it. Conversely, all the points of the lower-valued sample would be to the left of every point of the higher-valued one had we started out with the latter. Our total count would therefore be the total count of one sample multiplied by every observation in the second sample, which yields $n_1 n_2$. Thus, since we are told to take the greater of the two values, the sum of the counts C or $n_1 n_2 - C$, our result in this case would be $n_1 n_2$. On the other hand, if the two samples coincide completely, then for each point in one sample we would have those points below it plus a half point for the tied value representing that observation in the second sample which is at exactly the same level as the observation under consideration. A little experimentation will show this value to be $[n(n - 1)/2] + (n/2) = n^2/2$. Clearly the range of possible U-values must be between this and $n_1 n_2$ and the critical value must be somewhere within this range.

In computing the Wilcoxon statistic the magnitude of the sum of the ranks determines the outcome of the test. If this sum is unusually large or unusually small, it must mean that the sample under consideration is near the upper or lower extreme of the overall distribution of ranks of the two samples considered together.

Our conclusion as a result of the tests in Box 13.6 is that the two samples differ significantly in the distribution of cheliceral base length. It is obvious that the chiggers in sample A have longer cheliceral bases than those of sample B.

Finally, let us caution you against a possible source of confusion when comparing our writeup of the Wilcoxon two-sample test with accounts found in other sources. The nature of Table **CC** is such that we had to choose the greater of the two possible values for U_s. Often the corresponding table gives

the lower bounds for the critical value (for example Table K in Siegel, 1956) and therefore requires that the lower of the two values for U_s be tested against the critical value in the table. In such a case, for two samples to be significantly different the sample statistic must be *less* than the critical value. We prefer the formulation given here because it retains the familiar form of a significance test in which the sample statistic is greater than a tabled critical value. The two approaches to the significance test are, however, mathematically equivalent.

Now that we have learned a technique for testing two groups, we can return to the case of several groups and show how we could carry out multiple comparisons between pairs of treatments based on a nonparametric test of significance. This is the nonparametric analogue to the a posteriori tests we studied in Section 9.7. The approach we have taken here is the simultaneous test procedure of Dwass (1960) as further developed by K. R. Gabriel (unpublished). The technique, described in Box 13.7, finds the Wilcoxon-Mann-Whitney U-statistic for pairwise comparisons for each of the treatments and compares these with a critical value found by a formula shown in Box 13.7. The computation of the statistic, U_s, is quite simple. We again take the greater of the two quantities C or $(n^2 - C)$. In Box 13.6 in place of n^2 we used n_1n_2, but the *STP* method of Box 13.7 applies only to equal sample sizes; hence n_1n_2 becomes n^2. The actual computations are very simple and lead to the conclusion that glucose, fructose, and the mixed sugar are not different from each other, but are different from sucrose, which in turn is different from the control. This is the same conclusion that was reached in Box 9.9, but of course this nonparametric method need not always give the same conclusion as a corresponding anova.

Two additional points should be noted. We proceeded to test all pairwise comparisons in this example because of certain statistical properties of the *STP* in the nonparametric case. When significant pairs are members of larger sets, these larger sets must themselves be significant and, conversely, significant larger sets cannot be broken up into pairs all of which are nonsignificant. That is, there must be at least one significant pair causing the significance of the larger set. Therefore, by knowing which pairs are nonsignificant we can construct the maximally nonsignificant sets as we have done in this example. The second point is that the Kruskal-Wallis test of Box 13.5 is not based on the same test criterion as the nonparametric *STP*. Therefore, although this is unlikely, it is possible that a nonsignificant Kruskal-Wallis analysis among several groups might yield at least some significant pairs when investigated by the nonparametric *STP* of Box 13.7.

13.11 Nonparametric methods in lieu of two-way anova

Nonparametric tests exist for randomized blocks, the most common design for two-way analysis of a variance. The tests are for designs which, if they were to be done by anova, would be Model I for one factor, while the other one represents the blocks in the experiment. A Model II design, estimating the variance components, would clearly not be appropriate in a test based on ranks, since we are only concerned with location of the various samples in the overall ranking.

The test illustrated in Box 13.8 is *Friedman's method for randomized blocks*. The example is the lake temperatures first encountered in Box 11.2, where they were used to illustrate the computation of a randomized blocks

BOX 13.7

Nonparametric multiple comparisons by *STP* (an a posteriori test for equal sample sizes).

This method is based on U, the Wilcoxon-Mann-Whitney statistic, and requires equal sample sizes.

The technique is applied to the data on pea sections in Box 13.5. As in the ordinary Mann-Whitney test, it is simplest to prepare a graph of the data as shown below (see also Box 13.6).

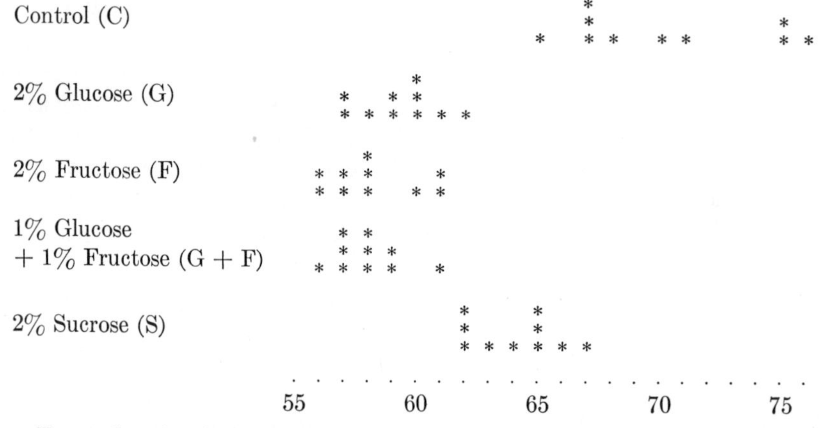

For each pair of samples, compute U_s as follows. For each observation in one sample, count the number of observations in the other samples which are lower in value (to the left). Count a $\frac{1}{2}$ for each tied observation. For example, Control (C) versus Glucose (G): 10 observations in G are below the first observation in C, 10 observations in G are less than the second observation in C, and so on. The sum of the counts is $C(\text{C, G}) = 10 + 10 + \cdots + 10 = 100$. Obviously, the same sum will be found for $C(\text{C, F})$ and $C(\text{C, G + F})$.

$$C(\text{C, S}) = 6\tfrac{1}{2} + 9\tfrac{1}{2} + 9\tfrac{1}{2} + 9\tfrac{1}{2} + 10 + 10 + 10 + 10 + 10 + 10 = 95$$

$$C(\text{G, F}) = 3 + 3 + 5\tfrac{1}{2} + 7 + 7 + 7\tfrac{1}{2} + 7\tfrac{1}{2} + 7\tfrac{1}{2} + 9 + 10 = 67$$

$$C(\text{G, G + F}) = 2\tfrac{1}{2} + 2\tfrac{1}{2} + 5\tfrac{1}{2} + 8 + 8 + 9 + 9 + 9 + 9\tfrac{1}{2} + 10 = 73$$

$$C(\text{G, S}) = 0 + 0 + 0 + 0 + 0 + 0 + 0 + 0 + 0 + 1\tfrac{1}{2} = 1\tfrac{1}{2}$$

$$C(\text{F, G + F}) = \tfrac{1}{2} + \tfrac{1}{2} + 2\tfrac{1}{2} + 2\tfrac{1}{2} + 5\tfrac{1}{2} + 5\tfrac{1}{2} + 5\tfrac{1}{2} + 9 + 9\tfrac{1}{2} + 9\tfrac{1}{2} = 50\tfrac{1}{2}$$

BOX 13.7 continued

C(F, S): All points of F are below those of S, so the count must be the minimum possible; namely, zero. This is also true of C(G + F, S).

The greater of the two quantities C or $(n^2 - C)$ is now entered as U_s into the table below (n is the sample size for each group).

Summary of U_s values

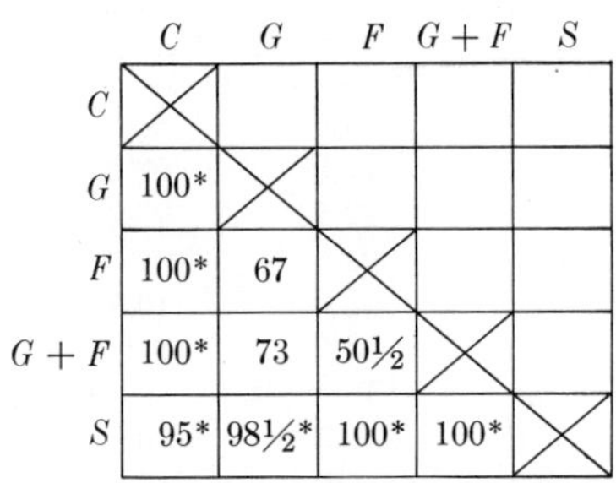

	C	G	F	$G + F$	S
C					
G	100*				
F	100*	67			
$G + F$	100*	73	50½		
S	95*	98½*	100*	100*	

*Significant at P = 0.05 by test described below.

The critical value for U_s is:

$$U_{\alpha[a,n]} = \frac{n^2}{2} + Q_{\alpha[a,\infty]}n\sqrt{\frac{(2n + 1)}{24}}$$

where a is the number of samples, n is the number of observations within each sample, and $Q_{\alpha[a,\infty]}$ is the α significance level for the studentized range for a groups and ∞ df (see Table **U**).

In our case,

$$U_{.05[5,10]} = \frac{10^2}{2} + (3.858)10\sqrt{\frac{[2(10) + 1]}{24}} = 50 + 38.58\sqrt{\frac{21}{24}}$$

$$= 50 + 38.58\sqrt{0.875} = 50 + 36.09 = 86.09$$

This quantity can always be rounded back to the nearest $\frac{1}{2}$. Therefore $U_{.05[5,10]} = 86$. All U_s-values greater than this are significant and have been marked with an asterisk in the above table. All sets of groups containing a significant U_s between at least one pair are significantly heterogeneous. Thus, we find that the set (G, F, G + F) is not significantly heterogeneous. Since C and S were heterogeneous with any other treatment, any set containing either C or S (or both) would be significantly heterogeneous. The data therefore break up into the following partition:

(G, F, G + F) (S) (C)

This is the same conclusion that was reached in Box 9.9 and Section 9.7, although such complete agreement cannot always be expected.

BOX 13.8

Friedman's method for randomized blocks.

Temperatures (°C) of Rot Lake on four early afternoons of the summer of 1952 at 10 depths. Data from Box 11.2.

Depth in meters ($a = 10$)	*Days* ($b = 4$)			
	29 July	*30 July*	*31 July*	*1 August*
0	23.8	24.0	24.6	24.8
1	22.6	22.4	22.9	23.2
2	22.2	22.1	22.1	22.2
3	21.2	21.8	21.0	21.2
4	18.4	19.3	19.0	18.8
5	13.5	14.4	14.2	13.8
6	9.8	9.9	10.4	9.6
9	6.0	6.0	6.3	6.3
12	5.8	5.9	6.0	5.8
15.5	5.6	5.6	5.5	5.6

Procedure

The four sets of readings are treated as blocks in this analysis.

1. Assign ranks to items within each of the $b = 4$ blocks (columns in this case) separately, resulting in the table that follows. If there are ties, treat them in the conventional manner (by average ranks). There are $a = 10$ treatments (depths; rows in this case).

Depth in meters	*Days*				
	29 July	*30 July*	*31 July*	*1 August*	$\sum^{b} R_{ij}$
0	10	10	10	10	40
1	9	9	9	9	36
2	8	8	8	8	32
3	7	7	7	7	28
4	6	6	6	6	24
5	5	5	5	5	20
6	4	4	4	4	16
9	3	3	3	3	12
12	2	2	2	2	8
15.5	1	1	1	1	4

2. Sum the ranks for each of the $a = 10$ groups (rows in the present case) $= \sum^{b} R_{ij}$. This yields values such as 40, 36, 32, ...

BOX 13.8 continued

3. Compute

$$X^2 = \left[\frac{12}{ab(a+1)} \sum^{a} \left(\sum^{b} R_{ij}\right)^2\right] - 3b(a+1)$$

$$= \frac{12}{10(4)(10+1)} (40^2 + 36^2 + \cdots + 8^2 + 4^2) - 3(4)(10+1)$$

$$= \frac{12}{440}(6160) - 132 = 35.983$$

This value is to be compared with $\chi^2_{[a-1]}$. Since $\chi^2_{.005[9]} = 23.589$, our observed value is significant at $P < 0.005$, as was to be expected in view of the consistency of the data.

anova without replication. You will recall that the four dates are considered as blocks and the depths as the treatment effects. In Friedman's method the variates are ranked *within each block*, as shown in Box 13.8. Note that because of the regularity of the thermal stratification of the lake, the ranking for the four days is entirely uniform. We sum the ranks for each of the treatment groups and compute a statistic X^2, which is distributed as $\chi^2_{[a-1]}$, where a is the number of treatments (depths in this example). Clearly this value will be maximal in our example because the ranks are ordered uniformly in each of the blocks. In an example in which the variates would not have been as uniformly ranked in each block, the marginal sums (row sums in this case) would not have been as high or as low as in this instance but would have tended more toward intermediate values. Since these sums are being squared, the large values exert a disproportionate influence upon the sample statistic X^2. We conclude that the effect of depth on water temperature is highly significant, which agrees with our earlier finding by analysis of variance.

A special case of randomized blocks is the paired comparisons design, discussed in Section 11.5 and illustrated in Box 11.3. It can be carried out by Friedman's test, but a more widely used method is that of *Wilcoxon's signed-ranks test*, illustrated in Box 13.9. The example to which it is applied has not yet been encountered. It records mean litter size in two strains of guinea pigs kept in large colonies during the years 1916 through 1924. Each of these values is the average of a large number of litters. Note the parallelism in the changes in the variable in the two strains. During 1917 and 1918 (war years for the U.S.), a shortage of caretakers and of food resulted in a decrease in the number of offspring per litter. As soon as better conditions returned, the mean litter size increased again. Notice that a subsequent drop in 1922 is again mirrored in both lines, suggesting that these fluctuations are environmentally caused. It is therefore quite appropriate that the data be treated as randomized blocks, with years as blocks and the strain differences as the fixed

BOX 13.9

Wilcoxon's signed-ranks test for two groups, arranged as paired observations.

Mean litter size of two strains of guinea pigs, compared over $n = 9$ years.

	(1)	*(2)*	*(3)*	*(4)*
Year	*Strain B*	*Strain 13*	D	*Rank* (R)
1916	2.68	2.36	+0.32	+9
1917	2.60	2.41	+0.19	+8
1918	2.43	2.39	+0.04	+2
1919	2.90	2.85	+0.05	+3
1920	2.94	2.82	+0.12	+7
1921	2.70	2.73	−0.03	−1
1922	2.68	2.58	+0.10	+6
1923	2.98	2.89	+0.09	+5
1924	2.85	2.78	+0.07	+4
	Absolute sum of negative ranks			1
	Sum of positive ranks			44

Source: Data by S. Wright.

Procedure

1. Compute the differences between the n pairs of observations. These are entered in column (3), labeled D.
2. Rank these differences from the smallest to the largest *without regard to sign.*
3. Assign to the ranks the original signs of the differences.
4. Sum the positive and negative ranks separately. The sum that is smaller in absolute value, T_s, is compared with the values in Table **DD** for $n = 9$.

Since $T_s = 1$, which is equal to or less than the entry for one-tailed $\alpha = 0.005$ in the table, our observed difference is significant at the 1% level. Litter size in strain B is significantly different from that of strain 13.

For large samples ($n > 50$) compute

$$t_s = \frac{T_s - \dfrac{n(n+1)}{4}}{\sqrt{\dfrac{n(n+\frac{1}{2})(n+1)}{12}}}$$

where T_s is as defined in step **4** above. Compare the computed value with $t_{\alpha[\infty]}$ in Table **Q**.

treatments to be tested. Column (3) in Box 13.9 lists the differences on which a conventional paired comparisons t-test could be performed. For Wilcoxon's test these differences are ranked *without regard to sign* so that the smallest absolute difference is ranked 1, and the largest absolute difference (of the nine differences) is ranked 9. Tied ranks are computed as averages as usual. After the ranks have been computed the original sign of each difference is assigned to the corresponding rank. The sum of the positive or of the negative ranks, whichever one is smaller in absolute value, is then computed (it is labeled T_s) and is compared with the critical value T in Table **DD** for the corresponding sample size. In view of the significance of the rank sum, it is clear that strain B has a litter size different from that of strain 13. This is a very simple test to carry out, but it is, of course, not as efficient as the corresponding parametric t-test, which should be preferred if the necessary assumptions hold. Note that one needs minimally six differences in order to carry out Wilcoxon's signed-ranks test. In six paired comparisons, all differences must be of like sign for the test to be significant at the 5% level.

For a large sample an approximation to the normal curve is available, which is given in Box 13.9. Note that the absolute magnitudes of these differences play a role only so far as they affect the ranks of the differences.

A still simpler test is the *sign test*, which counts the number of positive and negative signs among the differences (omitting all differences of zero). We then test the hypothesis that the n plus and minus signs are sampled from a population in which the two kinds of signs are present in equal proportions, as might be expected if there were no true difference between the two paired samples. Such sampling should follow the binomial distribution, and the test of the hypothesis that the parametric frequency of the plus signs is $\hat{p} = 0.5$ can be made in a number of ways. Let us learn these by applying the sign test to the guinea pig data of Box 13.9. There are nine differences, of which eight are positive and one is negative. We could follow the methods of Section 5.2 (illustrated in Table 5.3) in which we calculate the expected probability of sampling one minus sign in a sample of nine on the assumption of $\hat{p} = \hat{q} = 0.5$. The probability of such an occurrence and all "worse" outcomes equals 0.0195. Since we have no a priori notions that one strain should have a greater litter size than the other, this is a two-tailed test and we double the probability to 0.0390. Clearly, this is an improbable outcome and we reject the null hypothesis that $\hat{p} = \hat{q} = 0.5$.

Since the computation of the exact probabilities may be quite tedious if no table of cumulative binomial probabilities is at hand, a second approach is to make use of Table **W**, which furnishes confidence limits for $\hat{p}$ for various sample sizes and sampling outcomes. Looking up sample size 9 and $Y = 1$ (number showing the property), we find the 95% confidence limits to be 0.0028 and 0.4751 by interpolation, thus excluding the value $\hat{p} = \hat{q} = 0.5$ postulated by the null hypothesis. At least at the 5% significance level we can conclude that it is unlikely that the number of plus and minus signs is equal. The confidence limits imply a two-tailed distribution; if we intend a one-tailed test, we can infer a 0.025 significance level from the 95% confidence limits and a 0.005 level from the 99% limits. Obviously, such a one-tailed test would be carried out only if the results were in the direction of the alternative hypothesis. Thus, if the alternative hypothesis were that strain 13 in Box 13.9 had greater litter size than strain B, we would not have bothered testing this example at all, since the observed proportion of years showing this relation was less than half. For larger samples, we can use the normal approximation to the binomial distribution as follows: $t_s = (Y - \mu)/\sigma_Y = (Y - kp)/\sqrt{kpq}$, substituting the mean and standard deviation of the binomial distribution learned in Section 5.2. In our case, we let n stand for k and assume that $p = q = 0.5$. Therefore, $t_s = (Y - \frac{1}{2}n)/\sqrt{\frac{1}{4}n} = (Y - \frac{1}{2}n)/\frac{1}{2}\sqrt{n}$. The value of t_s is then compared with $t_{\alpha[\infty]}$ in Table **Q**, using one tail or two tails of the distribution as warranted. When the sample size $n \geq 12$, this is a satisfactory approximation.

A third approach would be to test the departure from expectation $\hat{p} = \hat{q} = 0.5$ by one of the methods of Chapter 16.

We can employ the sign test also for slightly more complicated hypotheses. If the two columns of data in a paired-comparisons case are labeled Y_1 and Y_2 respectively, we can test whether the response Y_1 is greater than Y_2 by K units. Simply evaluate the difference $D = Y_1 - (Y_2 + K)$ and carry out a sign test on the signs of the D-values. A second hypothesis might be that the response Y_1 is greater than Y_2 by $100p$ %. In such a case the sign test should be performed on the differences $D = Y_1 - (1 + p)Y_2$.

Exercises 13

13.1 In a study of flower color in Butterflyweed (*Asclepias tuberosa*) Woodson (1964) obtained the following results:

Geographic region	$\bar{Y}$	n	s
C1	29.3	226	4.59
SW2	15.8	94	10.15
SW3	6.3	23	1.22

The variable recorded was a color score (ranging from 1 for pure yellow to 40 for deep orange red) obtained by matching flower petals to sample colors in Maerz and Paul's *Dictionary of Color*. Analyze and interpret (note that the standard deviations appear unequal). ANS. $F_s = 1666.391$, $df = 2$ and 133.

13.2 Allee and Bowen (1932) studied survival time of goldfish (in minutes) when placed in colloidal silver suspensions. Experiment No. 9 involved 5 replications, and experiment No. 10 involved 10 replicates. Do the results of the two experiments differ? Addition of urea, NaCl, and Na_2S to a third series of suspensions apparently prolonged the life of the fish.

Colloidal silver		
Experiment No. 9	*Experiment No. 10*	*Urea and salts added*
210	150	330
180	180	300
240	210	300

Colloidal silver		*Urea and*
Experiment No. 9	*Experiment No. 10*	*salts added*
210	240	420
210	240	360
	120	270
	180	360
	240	360
	120	300
	150	120

Analyze and interpret. Check assumptions of normality and equality of variances. Compare anova results with those obtained using the Kruskal-Wallis test. ANS. $H = 12.0047$.

13.3 Number of bacteria in 1 cc of milk from three cows counted at three periods (data from Park, Williams, and Krumwiede, 1924).

	At time of milking	*After 24 hours*	*After 48 hours*
Cow No. 1	12,000	14,000	57,000
2	13,000	20,000	65,000
3	21,500	31,000	106,000

(a) Calculate means and variances for the three periods and examine the relation between these two statistics. Transform the variates to logarithms and compare means and variances based on the transformed data. Discuss.
(b) Carry out an anova on transformed and untransformed data. Discuss your results.
(c) Analyze the data by Friedman's method for randomized blocks.

13.4 Test for a difference in surface and subsoil pH in the data of Exercise 11.1, using Wilcoxon's signed-ranks test. ANS. $\Sigma R = 38$; $P > 0.10$.

13.5 In a study of a sea cucumber (*Stichopus* sp.) measurements were made of the lengths of "buttons" (in micrometer units) in samples of tissue from three body regions for two specimens (unpublished data by David Olsen).

Specimen	*Dorsal*				*Body regions* *Lateral*		*Ventral*			
No. 2	115	91	102	89	92	90	96	90	72	93
	88	84	80	73	90	91	98	114	89	81
	100	98	77	100	90	91	80	90	91	124
	96	87	91	111	93	82	93	87	111	92
	88	91	85	94	85	91	96	128	91	108
No. 9	81	108	92	88	113	112	91	104	96	107
	75	90	90	110	92	90	117	105	85	98
	90	96	100	91	96	105	104	91	90	80
	95	119	100	97	85	91	91	85	82	83
	99	88	91	95	88	92	81	80	89	90

Are the variances homogeneous? Are the distributions normal? Make histograms for the data in each cell of the two-way table and also perform graphic analyses. Analyze by a two-way anova. Why can you not show the effect of body regions to be significant? Is there anything disturbing about these data?

14 REGRESSION

Our studies so far have dealt with only one variable at a time. However, we frequently measure two or more variables on each individual and we consequently would like to be able to express more precisely the nature of the relationships between these variables. This brings us to the subjects of *regression* and *correlation*. In regression we estimate the relationship of one variable with another by expressing the one in terms of a linear (or a more complex) function of the other. In correlation analysis, which is sometimes confused with regression, we estimate the degree to which two variables vary together. Chapter 15 deals with correlation, and we shall postpone our effort to clarify the relation and distinction between regression and correlation until then. The variables involved in regression and correlation are continuous or, if meristic, they are treated as though they were continuous. If the variables are qualitative (that is, attributes), then the methods of regression and correlation cannot be used.

In Section 14.1, we review the notion of mathematical functions and introduce the new terminology required for regression analysis. This is followed in Section 14.2 by a discussion of the appropriate statistical models for regression analysis. The basic computations in simple linear regression are shown in Section 14.3. Machine computation for the case of one dependent variate for each independent variate is outlined in Section 14.4, and tests of significance and computation of confidence intervals for regression problems are discussed in Section 14.5. The case with several dependent variates for each independent variate is treated in Section 14.6.

Section 14.7 serves as a summary of regression and discusses the various uses of regression analysis in biology. In this section we also show how to estimate the relative efficiency of experimental designs in which more than one variable is measured. The following section (14.8) treats a procedure known as inverse prediction, in which the most probable value of the independent variate is estimated from an observed dependent variate. In our next section (14.9) we test differences among two or more regression coefficients.

In Section 14.10 we relate regression to the analysis of variance, showing how a single degree of freedom for linear contrast can be partitioned from a mean square among groups and demonstrating that these contrasts are really regression computations. Linear regression can be treated as such a contrast and the residual degrees of freedom can be used to test for departures from linearity in the regressional relationship. The method of orthogonal polynomials (Section 14.11) is a general approach to testing for curvilinearity in regression and to fitting curves to data. Section 14.12 shows how transformation of scale can straighten out curvilinear relationships for ease of analysis.

The methods described above assume that the independent variable is measured without error. When both variables are measured with error, the so-called Model II case, techniques for estimating regression equations are somewhat different and are discussed in Section 14.13. Finally, Section 14.14 briefly mentions several advanced topics in regression such as multiple and partial regression, discriminant functions, and multivariate analysis, with signposts to sources where these topics can be pursued in depth.

14.1 Introduction to regression

Much scientific thought concerns the relations between pairs of variables hypothesized to be in a cause-and-effect relationship. We shall not be concerned here with the philosophical requirements for establishing whether the relationship between two variables is really one of cause and effect. We shall be content with establishing the form and significance of *functional relationships* between two variables, leaving the demonstration of cause-and-effect relationships to the established procedures of the scientific method. A *function* is a mathematical relationship enabling us to predict what values of a variable Y correspond to given values of a variable X. Such a relationship, generally written as $Y = f(X)$, is familiar to all of us from our general scientific and cultural experience. Nevertheless, we shall briefly review functions as an appropriate introduction to the subject of regression.

The simplest type of regression follows the equation $Y = X$, illustrated in Figure 14.1. This shows the relation between number of growth rings on a tree as a function of age in years (assuming we can estimate fractions of growth rings corresponding to fractions of years). The graph is self-explanatory, but we might examine the properties of the function $Y = X$. It is obvious that whatever the value of X (age in years), the value of Y (number of growth rings) will be of corresponding magnitude. Thus, if the tree is 10 years old, it will have 10 growth rings; a 15-year-old tree will have 15

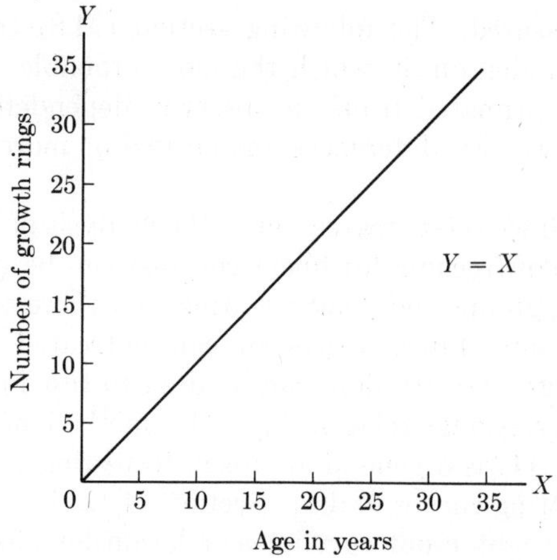

FIGURE 14.1 Number of growth rings on a tree as a function of age in years.

growth rings. When $X = 0$, Y will also equal zero, and the line describing the function goes through the origin of the coordinate system illustrated in Figure 14.1. This means that a tree zero years old has no growth rings, a fact reasonable in terms of our knowledge of the biology of trees. Clearly, with the relationship $Y = X$ we can predict accurately the number of growth rings, given the age of a tree. We call the variable Y the *dependent variable,* while X is called the *independent variable.* The magnitude of Y depends on the magnitude of X and can therefore be predicted from the independent variable, which presumably is free to vary. Remember that although a cause would always be considered an independent variable and an effect a dependent variable, a functional relationship observed in nature may not actually be a cause-and-effect relationship. It is conventional in statistics to label the dependent variable Y and the independent variable X.

Figure 14.2 is another functional relationship indicated by the equation $Y = bX$. Here the independent variable X is multiplied by a coefficient b, a slope factor. In the example chosen, which relates the height of a plant in centimeters to its age in days, we find that the slope factor $b = \frac{1}{7} = 0.143$. This means that for an increase of 7 units of X, there will be an increase of one unit of Y. Thus, a 14-day-old plant will be 2.0 cm tall. For any fixed scale, the slope of the function line will depend on the magnitude of b, the slope factor. As b increases, the slope of the line becomes steeper. We note

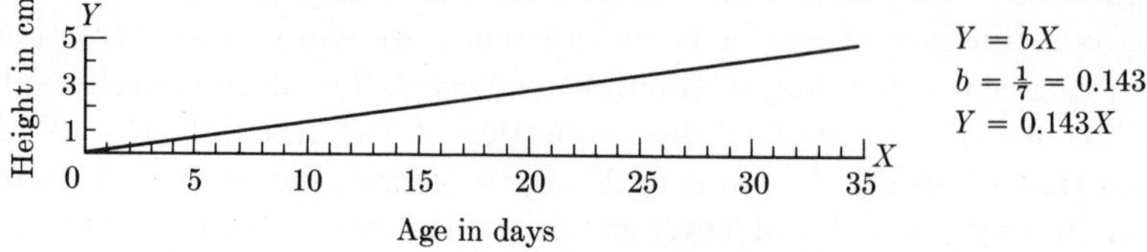

FIGURE 14.2 Height of a plant in centimeters as a function of age in days.

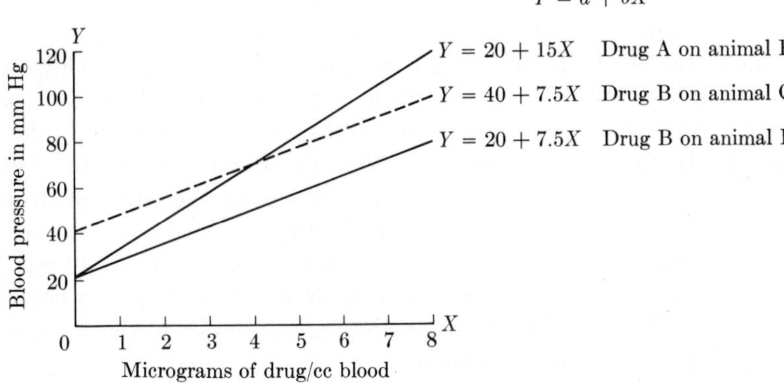

FIGURE 14.3 Blood pressure of an animal in mm Hg as a function of drug concentration in μg per cc of blood.

again that when $X = 0$, the dependent variable also equals zero. This is entirely reasonable; a plant that is zero days old has no height.

In biology, such a relationship can clearly be appropriate over only a limited range of values of X. Negative values of X are meaningless and it is unlikely that the plant will continue to grow at a uniform rate. Quite probably the slope of the functional relationship will flatten out as the plant gets older and approaches its adult size. But for a limited portion of the range of variable X (age in days) the linear relationship $Y = bX$ may be an adequate description of the functional dependence of Y on X.

Finally, we turn to a situation such as that depicted in Figure 14.3, which illustrates the effect of two drugs on blood pressure in two species of animals. The relationships depicted in this graph can be expressed by the formula $Y = a + bX$. The highest line is of the relationship $Y = 20 + 15X$, which represents the effect of drug A on animal P. The quantity of drug is measured in micrograms, the blood pressure in millimeters mercury. From the equation given, it is easy to calculate the blood pressure to be expected in the animal after a given amount of drug has been administered. Thus, after 4 μg of the drug have been given, the blood pressure would be $Y = 20 + (15)(4) = 80$ mm Hg. Note that by this formula, when the independent variable equals zero, the dependent variable does not also equal zero, but equals a. In Figure 14.3, when $X = 0$, the function just studied will yield a blood pressure of 20 mm, which is the normal blood pressure of animal P in the absence of the drug. It would not be biologically reasonable to assume that the animal had no blood pressure in the absence of the drug. The magnitude 20 mm can be obtained by solving the equation after substituting zero for X or, what amounts to the same thing, examining the graph for the intersection of the function line with the Y-axis. This point yields the magnitude of a, which is therefore called the Y-*intercept*.

The two other functions in Figure 14.3 show the effects of varying both

a, the Y-intercept, and b, the slope. In the lowest line, $Y = 20 + 7.5X$, the Y-intercept remains the same, but the slope has been halved. We visualize this as the effect of a different drug B on the same organism P. Obviously, when no drug is administered, the blood pressure should be at the same Y-intercept, since the identical organism is being studied. However, different drugs are likely to exert different hypertensive effects, as reflected by the different slopes. The third relationship also describes the effect of drug B, which is assumed to remain the same, but the experiment is carried out on a different species Q, whose normal blood pressure is assumed to be 40 mm Hg. Thus, the equation for the effect of drug B on species Q is written as $Y = 40 + 7.5X$. We notice that this line is exactly parallel to the one studied previously since it has the same slope.

You have probably noticed that the equation represented in Figure 14.3, $Y = a + bX$, is the most general linear equation and contains within it the two simpler versions from Figures 14.1 and 14.2. In $Y = bX$ we simply assume that $a = 0$; in $Y = X$, $a = 0$ and $b = 1$.

Those of you familiar with analytical geometry will have recognized the slope factor b as the *slope* of the function $Y = a + bX$, generally symbolized by m. If you know calculus, you will, of course, recognize b as the *derivative* of that same function ($dY/dX = b$). In biometry, b is called the *regression coefficient* and the function is called a *regression equation*. The origin of the word "regression" dates back to certain biometric studies made by Galton in which the literal meaning of regression as "returning to" or "going back to" was appropriate; however, the accepted meaning in statistics now relates to the dependence of the means of variable Y on the independent variable X by some mathematical equation. When we wish to stress that the regression coefficient is of variable Y on variable X, we write $b_{Y \cdot X}$. If we would ever wish to regress X on Y (if this were legitimate), the proper symbol for the coefficient is $b_{X \cdot Y}$.

14.2 Models in regression

In any real example observations would not lie perfectly along a regression line because of random error in measuring and because of the effects of unpredictable environmental factors. Thus in regression a functional relationship does not mean that given an X the value of Y must be $a + bX$, but rather that the mean (or expected value) of Y is $a + bX$.

Significance tests in regression are based on the following two models. The more common of these, *Model I regression*, is especially suitable in experimental situations. It is based on four assumptions.

1. The independent variable X is measured without error. We therefore say that the X's are "fixed." We mean by this that only Y, the dependent variable, is a random variable; X does not vary at random, but is under the

control of the investigator. Thus in the example of Figure 14.3 we varied dose of drug at will and studied the response of the random variable blood pressure. We can manipulate X in the same way that we were able to manipulate the treatment effect in a Model I anova. As a matter of fact, as you shall see later, there is a very intimate relationship between Model I anova and Model I regression.

2. The expected value for the variable Y for any given value X is described by the linear function $\mu_Y = \alpha + \beta X$. This is the same relation we have encountered before, but we use Greek letters for a and b, since we are describing a parametric relationship. Another way of stating this assumption is that the parametric means μ_Y of the values of Y are a function of X and lie on a straight line described by this equation.

3. For any given value of X the Y's are independently and normally distributed. This can be represented by the equation $Y = \alpha + \beta X + \epsilon$, where ϵ is assumed to be a normally distributed error term with a mean of zero. Figure 14.4 illustrates this concept with a regression line similar to the ones in Figure 14.3. A given experiment can be repeated several times. Thus, for instance, we could administer 2, 4, 6, 8, and 10 μg of the drug to each of 20 individuals of an animal species and obtain a frequency distribution of blood pressure responses Y to the independent variates $X = 2, 4, 6, 8$, and 10 μg. In view of the inherent variability of biological material, it is obvious that the responses to each dosage would not be the same in every individual; you would obtain a frequency distribution of values of Y (blood pressure) around the expected value. Assumption 3 states that these sample values would be independently and normally distributed. This is indicated by the normal curves, which are superimposed about several points in the regression line in Figure 14.4. A few are shown to give you an idea of the scatter about the regression line. In actuality there is, of course, a continuous scatter, as though these separate normal distributions were stacked right next to each other, there being, after all, an infinity of intermediate values of X between any two dosages. In those rare cases in which the independent variable is discontinuous, the distributions of Y would be physically separate from each other and would occur only along those points of the abscissa corresponding to independent variates. An example of such a case would be weight of

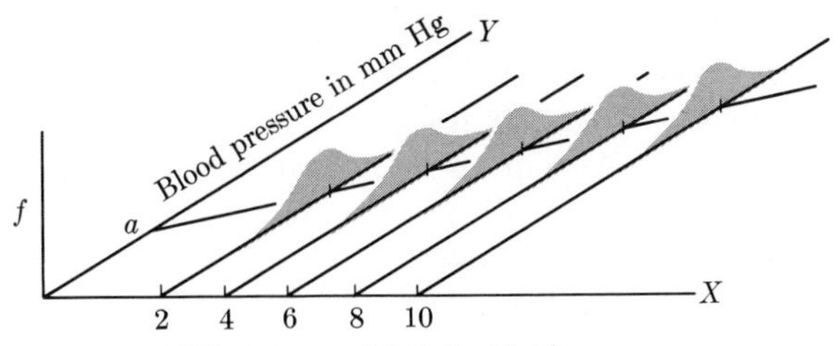

FIGURE 14.4 Blood pressure of an animal in mm Hg as a function of drug concentration in μg per cc of blood. Repeated sampling for a given drug concentration.

offspring (Y) as a function of number of offspring (X) in litters of mice. There may be three or four offspring per litter, but there would be no intermediate value of X representing 3.25 mice per litter.

Not every experiment will have more than one replicate of Y for each value of X. In fact, the basic computations we shall learn in the next section are for only one value of Y per X, this being the more common case. However, you should realize that even in such instances the basic assumption of Model I regression is that the single variate of Y corresponding to the given value of X is a sample from a population of independently and normally distributed variates.

4. The final assumption is a familiar one. We assume that these samples along the regression line are homoscedastic; that is, they have a common variance, σ^2, which is the variance of the ϵ's in the above expression. Thus we assume that the variance around the regression line is constant and independent of the magnitude of X or Y.

In *Model II regression* the independent variable is also measured with error. We do not consider X to be fixed and at the control of the investigator. Suppose you sample a population of female flies and measure a number of ovarioles and total weight of each individual. The distributions of these variables will probably differ. You might be interested in studying a number of ovarioles as a function of weight. In this case the weight, which you assume to be the independent variable, is not fixed and certainly not the "cause" of ovarian development. Although, as will be discussed in the next chapter, cases of this sort are much better analyzed by the methods of correlation analysis, we frequently wish to describe the functional relationship between such variables. To do so, we need to resort to the special techniques of Model II regression discussed in Section 14.13.

We shall now discuss the rationale behind the computational steps in Model I regression.

14.3 The basic computations

To learn the basic computations necessary to carry out a Model I linear regression, we shall choose an example with only one Y-value per independent variate X, as this is computationally simpler. The extension to replicated values of Y per single value of X is shown in Section 14.6. We should stress that the computations of the present section illustrated in Table 14.1 are shown only for pedagogical reasons to facilitate an understanding of the meaning of regression. Just as in the case of the standard deviation, there are simple computational formulas, which will be presented in Section 14.4 after the principles of regression have been mastered.

The data on which we shall learn regression come from a study of water loss in *Tribolium confusum*, the confused flour beetle. Nine batches of 25 beetles were weighed (individual beetles could not be weighed with available

TABLE 14.1

Basic computations in regression. Weight loss (in mg) of nine batches of 25 *Tribolium* beetles after six days of starvation at nine different humidities.

	(1)	(2)	(3)	(4)	(5)	(6)	(7)	(8)	(9)	(10)	(11)	(12)
	Percent relative humidity	*Weight loss in mg*	$x =$	$y =$					$d_{Y \cdot X} =$		$\hat{y} =$	
	X	Y	$(X - \bar{X})$	$(Y - \bar{Y})$	x^2	xy	y^2	$\hat{Y}$	$Y - \hat{Y}$	$d^2_{Y \cdot X}$	$\hat{Y} - \bar{Y}$	$\hat{y}^2$
	0	8.98	−50.39	2.958	2539.1521	−149.0536	8.7498	8.7038	0.2762	0.0763	2.6818	7.1921
	12	8.14	−38.39	2.118	1473.7921	− 81.3100	4.4859	8.0652	0.0748	0.0056	2.0432	4.1747
	29.5	6.67	−20.89	0.648	436.3921	− 13.5367	0.4199	7.1338	−0.4638	0.2151	1.1118	1.2361
	43	6.08	− 7.39	0.058	54.6121	− 0.4286	0.0034	6.4153	−0.3353	0.1124	0.3933	0.1547
	53	5.90	2.61	−0.122	6.8121	− 0.3184	0.0149	5.8831	0.0169	0.0003	−0.1389	0.0193
	62.5	5.83	12.11	−0.192	146.6521	− 2.3251	0.0369	5.3776	0.4524	0.2047	−0.6444	0.4153
	75.5	4.68	25.11	−1.342	630.5121	− 33.6976	1.8010	4.6857	−0.0057	0.0000	−1.3363	1.7857
	85	4.20	34.61	−1.822	1197.8521	− 63.0594	3.3197	4.1801	0.0199	0.0004	−1.8419	3.3926
	93	3.72	42.61	−2.302	1815.6121	− 98.0882	5.2992	3.7543	−0.0343	0.0012	−2.2677	5.1425
Sum	453.5	54.20	− 0.01	0.002	8301.3889	−441.8176	24.1307	54.1989	0.0011	0.6160	0.0009	23.5130
Mean	50.39	6.022						6.022				
Sum/$(n - 1)$					1037.6736	− 55.2272	3.0163			0.0880[a]		

Source: Nelson (1964).
[a] Sum divided by $n - 2$.

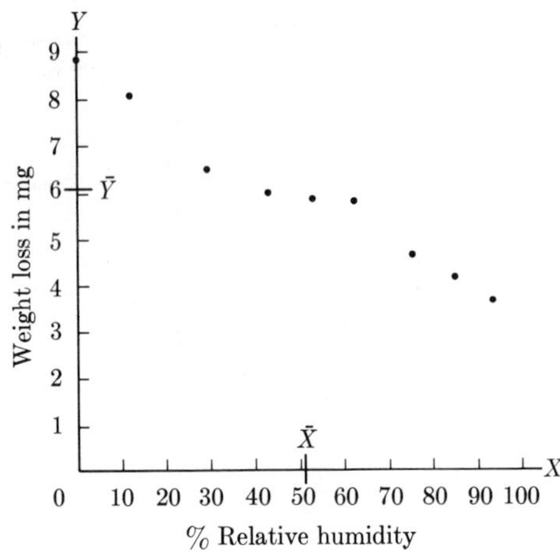

FIGURE 14.5 Weight loss (in mg) of nine batches of 25 *Tribolium* beetles after six days of starvation at nine different relative humidities. Data from Table 14.1 after Nelson (1964).

equipment), kept at different relative humidities, and weighed again after six days of starvation. Weight loss in milligrams was computed for each batch. This is clearly a Model I regression in which the weight loss is the dependent variable Y and the relative humidity is the independent variable X, a fixed treatment effect under the control of the experimenter. The purpose of the analysis is to establish whether the relationship between relative humidity and weight loss can be adequately described by a linear regression of the general form $Y = a + bX$. The original data are shown in columns (1) and (2) of Table 14.1. They are plotted in Figure 14.5, from which it appears that a negative relationship exists between weight loss and humidity: as the humidity increases, the weight loss decreases. The means of weight loss and relative humidity, $\bar{Y}$ and $\bar{X}$, respectively, are marked along the coordinate axes. The average humidity is 50.39%, and the average weight loss is 6.022 mg. How can we fit a regression line to these data, permitting us to estimate a value of Y for a given value of X? Unless the actual observations lie exactly on a straight line, we will need a criterion for determining the best possible placing of the regression line. Statisticians have generally followed the principle of least squares, which we first encountered in Chapter 4 when learning about the arithmetic mean and the variance. If we were to draw a horizontal line through $\bar{X}$, $\bar{Y}$ (that is, a line parallel to the X-axis at the level of $\bar{Y}$), then deviations to that line drawn parallel to the Y-axis would represent the deviations from the mean for these observations with respect to variable Y (see Figure 14.6). We learned in Chapter 4 that the sum of these observations, $\sum(Y - \bar{Y}) = \sum y = 0$, and the sum of squares of these deviations, $\sum(Y - \bar{Y})^2 = \sum y^2$, is less than that from any other horizontal line. Another way of saying this is that the arithmetic mean of Y represents the least squares horizontal line. Any horizontal line drawn through the data at a point other than $\bar{Y}$ would yield a sum of deviations other than zero and a sum of deviations squared greater than $\sum y^2$. Therefore a mathematically correct but impractical method for finding the mean of Y would be to draw a series of horizontal lines across a graph, calculate the sum of squares of deviations from it, and choose that line yielding the smallest sum of squares.

In linear regression, we still draw a straight line through our observations, but it is no longer necessarily horizontal. A sloped regression line will indicate for each value of the independent variable X_i an estimated value of the dependent variable. We should distinguish the estimated value of Y_i, which we shall hereafter designate as $\hat{Y}_i$ (read: Y-hat or Y-caret), and the observed values, conventionally designated as Y_i. The regression equation therefore should read

$$\hat{Y} = a + bX \tag{14.1}$$

which indicates that for given values of X, this equation calculates estimated values $\hat{Y}$ (as distinct from the observed values Y in any actual case). The deviation of an observation Y_i from the regression line is $(Y_i - \hat{Y}_i)$ and is generally symbolized as $d_{Y \cdot X}$. These deviations can still be drawn parallel to the Y-axis, but since the regression line is sloped, they meet it at an angle (see Figure 14.7). The sum of these deviations is again zero ($\sum d_{Y \cdot X} = 0$), and the sum of their squares yields a quantity $\sum(Y - \hat{Y})^2 = \sum d^2_{Y \cdot X}$ analogous to the sum of squares $\sum y^2$. For reasons that will become clear later, $\sum d^2_{Y \cdot X}$ is called the unexplained sum of squares. The least squares *linear regression line* through a set of points is defined as that straight line which results in $\sum d^2_{Y \cdot X}$ being at a minimum. Geometrically, the basic idea is that one would prefer using a line that is in some sense close to as many points as possible. For purposes of regression analysis, it is most useful to define closeness in terms of the vertical distances from the points to a line and to use the line making the sum of these squared deviations as small as possible. A convenient consequence of this criterion is that the line must pass through the point $\bar{X}$, $\bar{Y}$. Again, it would be feasible but impractical to calculate the correct regression

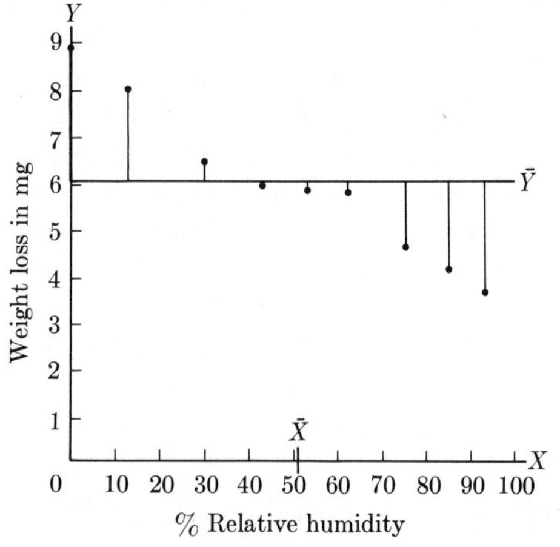

FIGURE 14.6 Deviations from the mean (of Y) for the data of Figure 14.5.

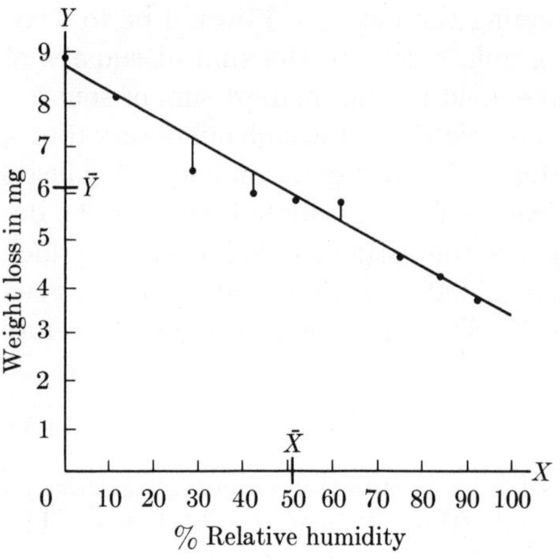

FIGURE 14.7 Deviations from the regression line for the data of Figure 14.5.

slope by pivoting a ruler around the point $\bar{X}$, $\bar{Y}$, and calculating the unexplained sum of squares, $\sum d^2_{Y \cdot X}$, for each of the innumerable possible positions. Whichever position gave the minimal value of $\sum d^2_{Y \cdot X}$ would be the least squares regression line.

To derive the formula for the slope of a line based on a minimal value of $\sum d^2_{Y \cdot X}$, we would have to resort to calculus. We shall, therefore, not present the derivation of the regression coefficient here, but shall try to learn its formula through an intuitive approach.

Let us examine the values of Y for values of X other than $\bar{X}$. For example, we note that when $X = 12$, $Y = 8.14$. Thus, for a deviation $x = (X - \bar{X}) = (12 - 50.39) = -38.39$ units of X, we find a corresponding deviation $y = (Y - \bar{Y}) = (8.14 - 6.022) = +2.118$ units of Y. Now suppose the deviations y were always the same in sign and magnitude as the deviations x. This means that for a deviation of $x = 5$, we would obtain a corresponding deviation of $y = 5$. This would be the case in a regression equation of the type $Y = X$ or $Y = a + X$ (when $b = 1$). If the deviations were equal in magnitude but unequal in sign, the appropriate equations would be $Y = -X$ or $Y = a - X$, with $b = -1$. How could we construct a statistic with a numerical value of positive or negative unity from our deviations x and y? Suppose we multiply our deviations together to give xy. Since we wish to derive our statistic from all the data available in a problem, we would sum over all available deviations from the mean, giving $\sum xy$. But since we are postulating that $y = x$ for every variate, it follows that $\sum xy$ would equal $\sum x^2$. To make the sum of squares yield a coefficient with the desired value of unity, we must divide it by another quantity. The obvious choice is the sum of squares of X, yielding $\sum x^2/\sum x^2 = 1$. Therefore, when $y = x$, $\sum xy/\sum x^2 = 1$. Similarly, when $y = -x$, $\sum xy = -\sum x^2$ and $\sum xy/\sum x^2 = -1$. Now if for each deviation x we get a greater deviation y, then obviously the regression equation will be $Y = bX$ or $Y = a + bX$, where $b > 1$. Since

$y > x$, $\sum xy$ is necessarily $> \sum x^2$ and $\sum xy/\sum x^2 > 1$. Similarly, when $y < x$, $\sum xy < \sum x^2$ and $\sum xy/\sum x^2 < 1$. This provides us with a formula for b which at least seems to have the desirable properties, although we have in no way proven that it is the slope of the line yielding the least squares fit to the data. Let us calculate $b = \sum xy/\sum x^2$ for our weight loss data.

We first compute the deviations from the respective means of X and Y as shown in columns (3) and (4) of Table 14.1. The sums of these deviations, $\sum x$ and $\sum y$, are slightly different from their expected value of zero because of rounding errors. The squares of these deviations yield sums of squares and variances in columns (5) and (7). In column (6) we have computed the products xy, which in this example are all negative because the deviations are of unlike sign. The sum of these products $\sum^{n} xy$ is a new quantity, called the *sum of products*. This is a poor, but well-established term, referring to $\sum xy$, the sum of the products of the deviations rather than $\sum XY$, the sum of the products of the variates. You will recall that $\sum y^2$ is called the sum of squares, while $\sum Y^2$ is the sum of the squared variates. The sum of products is analogous to the sum of squares. When divided by the degrees of freedom, it yields the *covariance* by analogy with the variance resulting from a similar division of the sum of squares. You may recall first having encountered covariances in Section 8.4. Note that the sum of products can be negative as well as positive. If it is negative, this indicates a negative slope of the regression line: as X increases, Y decreases. In this respect it differs from a sum of squares, which can only be positive. From Table 14.1 we find that $\sum xy = -441.8176$, $\sum x^2 = 8301.3889$, and $b = \sum xy/\sum x^2 = -0.05322$. Thus, for a one-unit increase in X, there is a decrease of 0.05322 units of Y. Relating it to our actual example, we can say that for a 1% increase in relative humidity, there is a reduction of 0.05322 mg in weight loss.

How can we complete the equation $Y = a + bX$? We have stated that the regression line will go through the point $\bar{X}$, $\bar{Y}$. Therefore, when X is at its mean, we estimate that Y should also be at its mean. Whether this is actually so in our data we cannot tell, since we do not have an X variate exactly at the mean. Even if we did, it would not be very likely that the observed value of Y would be exactly at the mean. Remember, after all, that the Y-values of our observations are only a sample from a population centering around μ_Y. At $\bar{X} = 50.39$, $\hat{\bar{Y}} = 6.022$; that is, we use $\bar{Y}$, the observed mean of Y, as an extimate $\hat{\bar{Y}}$ of the mean. We can substitute these means into Expression (14.1):

$$\hat{Y} = a + bX$$

$$\bar{Y} = a + b\bar{X}$$

$$a = \bar{Y} - b\bar{X}$$

$$a = 6.022 - (-0.05322)(50.39)$$

$$= 8.7038$$

Therefore $\hat{Y} = 8.7038 - 0.05322X$.

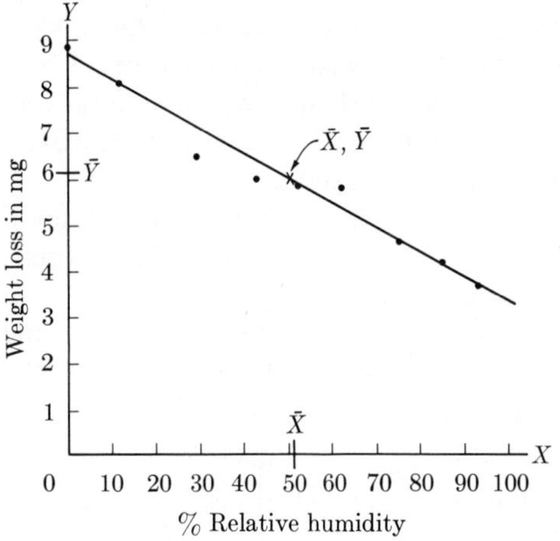

FIGURE 14.8 Linear regression fitted to data of Figure 14.5.

This is the equation that relates weight loss to relative humidity. Note that when X is zero (humidity zero), the estimated weight loss is greatest. It is then equal to $a = 8.7038$ mg. But as X increases to a maximum of 100, the weight loss would decrease to 3.3818 mg.

We can use the regression formula to draw the regression line: simply estimate $\hat{Y}$ at two convenient points of X such as $X = 0$ and $X = 100$ and draw a straight line between them. This line has been added to the observed data and is shown in Figure 14.8. Note that it goes through the point $\overline{X}$, $\overline{Y}$. In fact, for drawing the regression line, we frequently use the intersection of the two means and one other point.

Now bear with us through some very elementary algebra. Since

$$a = \overline{Y} - b\overline{X}$$

we can write Expression (14.1), $\hat{Y} = a + bX$, as

$$\begin{aligned}\hat{Y} &= (\overline{Y} - b\overline{X}) + bX \\ &= \overline{Y} + b(X - \overline{X})\end{aligned}$$

Therefore

$$\hat{Y} = \overline{Y} + bx \tag{14.2}$$

Also,

$$\hat{Y} - \overline{Y} = bx$$

$$\hat{y} = bx \tag{14.3}$$

where $\hat{y}$ is defined as the deviation $\hat{Y} - \overline{Y}$. Next, using Expressions (14.1) or (14.2), we estimate $\hat{Y}$ for every one of our given values of X. The estimated values $\hat{Y}$ are shown in column (8) of Table 14.1. Compare them with the observed values of Y in column (2). Overall agreement between the two columns of values is good. Note that except for rounding errors, $\sum \hat{Y} = \sum Y$ and hence $\overline{\hat{Y}} = \overline{Y}$. However, our actual Y-values usually are different from the estimated values $\hat{Y}$. This is due to individual variation around regression.

The regression line is a better base from which to compute deviations than the arithmetic average $\bar{Y}$, since the value of X has been taken into account in constructing it.

When we compute deviations of each observed Y-value from its estimated value $(Y - \hat{Y}) = d_{Y \cdot X}$ and list these in column (9), we notice that these deviations exhibit one of the properties of deviations from a mean: they sum to zero except for rounding errors. Thus $\sum d_{Y \cdot X} = 0$, just as $\sum y = 0$. Next, we compute in column (10) the squares of these deviations of observed values of Y from values estimated by regression and sum them to give a new sum of squares, $\sum d^2_{Y \cdot X} = 0.6160$. When we compare $\sum (Y - \bar{Y})^2 = \sum y^2 = 24.1307$ with $\sum (Y - \hat{Y})^2 = \sum d^2_{Y \cdot X} = 0.6160$, we note that the new sum of squares is much less than the previous old one. What has caused this reduction? The regression line is a set of averages (one for each value of X) as compared with the single arithmetic mean of Y. Allowing for different magnitudes of X has eliminated most of the variance of Y from the sample. Remaining is the *unexplained sum of squares* $\sum d^2_{Y \cdot X}$, which expresses that portion of the total SS of Y which is not accounted for by differences in X. It is unexplained with respect to X. The difference between the total SS, $\sum y^2$, and the unexplained SS, $\sum d^2_{y \cdot x}$, is not surprisingly called the *explained sum of squares*, $\sum \hat{y}^2$, and is based on the deviations $\hat{y} = \hat{Y} - \bar{Y}$. The computation of this deviation and its square is shown in columns (11) and (12). Note that $\sum \hat{y}$ approximates zero and that $\sum \hat{y}^2 = 23.5130$. Add the unexplained $SS = 0.6160$ to this and you obtain $\sum y^2 = \sum \hat{y}^2 + \sum d^2_{y \cdot x} = 24.1290$, which is equal (except for rounding errors) to the independently calculated value of 24.1307 in column (7). We shall return to the meaning of the unexplained and explained sums of squares in later sections.

We now proceed to demonstrate more efficient machine methods for computation of a regression equation using the same example with single values of Y for each value of X.

14.4 Machine computation: single Y for each value of X

The regression coefficient $\sum xy / \sum x^2$ can be rewritten as

$$b_{Y \cdot X} = \frac{\sum^n (X - \bar{X})(Y - \bar{Y})}{\sum^n (X - \bar{X})^2} \qquad (14.4)$$

The denominator of this expression is the sum of squares of X. Its computational formula as first encountered in Section 4.9 is $\sum x^2 = \sum X^2 - (\sum X)^2/n$. We shall now learn an analogous formula for the numerator of Expression (14.4), which is the sum of products developed in the last section. The customary formula is

$$\sum^n xy = \sum^n XY - \frac{(\sum^n X)(\sum^n Y)}{n} \qquad (14.5)$$

The quantity $\sum XY$ is simply the accumulated product of the two variables. Expression (14.5) is derived in Appendix A1.8. The actual machine computations for a regression equation (single value of Y per value of X) are illustrated in Box 14.1, employing the weight loss data of Table 14.1.

□□ To compute a regression equation, we need six quantities. These are n, $\sum X$, $\sum X^2$, $\sum Y$, $\sum Y^2$, and $\sum XY$. Of these, n is usually known in advance of the computation. Unless the variates are given to more than three-figure accuracy, it is possible to obtain the other five quantities simultaneously on most desk calculators. For each pair of (X, Y) observations: (a) enter X at the extreme left of the keyboard and Y at the extreme right. For example, in a keyboard with 10 columns the following ten digits would be entered for the second pair of variates in Box 14.1: (12.0 | 00008.14). We are using the second pair in this illustrative example because the first pair contains $X_1 = 0$, which, though calculated in the same way, would not demonstrate the computation as clearly as a nonzero variate, since X^2 and $2XY$ are also zero. (b) Square this quantity. If your calculator carries 21 places in the accumulating (long) dials, these should now contain the following 21 digits (00144.00 | 0195.360 | 066.2596). The 144.00 is X_2^2, the 66.2596 is Y_2^2, and 195.360 is $2X_2Y_2$. If your short counter dials provide eleven places, they should contain (012.0 | 00008.14). After this process has been repeated for each pair of observations and the results accumulated, the final results will be: $\sum X^2$, $2\sum XY$, $\sum Y^2$ in the accumulating dials (31152.75 | 4578.520 | 350.5350) and $\sum X$, $\sum Y$ in the counter dials (453.5 | 00054.20). To obtain $\sum XY$, compute $\frac{1}{2}$ (4578.520) = 2289.260. As can be seen, care must be taken that one result does not overflow into another one. In the present example, we were able to keep the factors separate, but if our sample size had been much larger, we would have had to copy down intermediate sums, clear the dials, and then continue until the results again threatened to overlap. To check the results add the values of X and Y to obtain $\sum X$ and $\sum Y$ once more. If these agree with your previous values, we can assume that $\sum X^2$, $\sum Y^2$, and $2\sum XY$ are correct (compensating errors would not be detected).

Whenever the number of variates and digits to be processed is greatly in excess of the capacity of the calculating machine, the five quantities cannot be computed simultaneously, and the computation should be segmented as follows: (1) n, $\sum X$, $\sum X^2$ by the method appropriate for your calculator as discussed in Section 4.9; (2) n, $\sum Y$, $\sum Y^2$ by the same method; and (3) $\sum XY$ by a method analogous to the 1, 5-zeros method. Put (01 | 0000012.0) into the multiplier and (00000008.14) into the multiplicand (keyboard). The product is (8.14 | 00097.680) in the long dials and (1 | 0000012.0) in the short dials. After all nine pairs have been accumulated, the long dials would read (54.20 | 02289.260) and the short ones (9 | 0000453.5) to represent $\sum Y$, $\sum XY$, n, and $\sum X$, respectively. Since $\sum X$ and $\sum Y$ are known already from steps (1) and (2), they serve as checks on the computations. This technique should

BOX 14.1

Machine computation of regression statistics. Single value of Y for each value of X.

Data from Table 14.1.

Weight loss in mg (Y)	8.98	8.14	6.67	6.08	5.90	5.83	4.68	4.20	3.72
Percent relative humidity (X)	0	12.0	29.5	43.0	53.0	62.5	75.5	85.0	93.0

Basic computations

1. Compute sample size, sums, sums of the squared observations, and the sum of the XY's.

$$n = 9 \qquad \sum X = 453.5 \qquad \sum Y = 54.20$$

$$\sum X^2 = 31{,}152.75 \qquad \sum Y^2 = 350.5350 \qquad \sum XY = 2289.260$$

□□ The method of obtaining these sums on a calculating machine is described in the text (Section 14.4). □□

2. The means, sums of squares, and sum of products are

$$\bar{X} = 50.389 \qquad \bar{Y} = 6.022$$

$$\sum x^2 = 8301.3889 \qquad \sum y^2 = 24.1306$$

$$\sum xy = \sum XY - \frac{(\sum X)(\sum Y)}{n}$$

$$= 2289.260 - \frac{(453.5)(54.20)}{9} = -441.8178$$

3. The regression coefficient is

$$b_{Y \cdot X} = \frac{\sum xy}{\sum x^2} = \frac{-441.8178}{8301.3889} = -0.05322$$

4. The Y-intercept is

$$a = \bar{Y} - b_{Y \cdot X}\bar{X} = 6.022 - (-0.05322)(50.389) = 8.7037$$

5. The explained sum of squares is

$$\sum \hat{y}^2 = \frac{(\sum xy)^2}{\sum x^2} = \frac{(-441.8178)^2}{8301.3889} = 23.5145$$

6. The unexplained sum of squares is

$$\sum d^2_{Y \cdot X} = \sum y^2 - \sum \hat{y}^2 = 24.1306 - 23.5145 = 0.6161$$

also be employed when the calculator is not equipped with an automatic squaring feature.

As in the case of the sum of squares, the machine method for computation of the sum of products is efficient for relatively small samples. When $n \geq 150$, it usually pays to set up a (two-way) frequency distribution and compute the sum of products from it. Since such cases are more frequent in correlation analysis, we postpone our discussion of this procedure until Section 15.3. □□

Box 14.1 also shows how to compute the explained sum of squares $\sum \hat{y}^2 = \sum(\hat{y} - \bar{Y})^2$ and the unexplained sum of squares $\sum d^2_{Y \cdot X} = \sum(Y - \hat{Y})^2$. That

$$\sum d^2_{Y \cdot X} = \sum y^2 - \frac{(\sum xy)^2}{\sum x^2} \tag{14.6}$$

is demonstrated in Appendix A1.9. From this demonstration it also becomes obvious that the explained sum of squares is

$$\sum \hat{y}^2 = \sum b^2 x^2 = b^2 \sum x^2 = \frac{(\sum xy)^2}{(\sum x^2)^2} \sum x^2$$

$$\sum \hat{y}^2 = \frac{(\sum xy)^2}{\sum x^2} \tag{14.7}$$

14.5 Tests of significance in regression

We have so far interpreted regression as a method for providing an estimate, $\hat{Y}_1$, given a value of X_1. Another interpretation is as a method for explaining some of the variation of the dependent variable Y in terms of the variation of the independent variable X. A sample of Y values exhibits variance s^2_Y, computed by summing and squaring deviations $y = Y - \bar{Y}$. In Figure 14.9 we can see that the deviation y can be decomposed into two parts, $\hat{y}$ and $d_{Y \cdot X}$. It is also clear from Figure 14.9 that the deviation $\hat{y} = \hat{Y} - \bar{Y}$ represents the deviation of the estimated value $\hat{Y}$ from the mean of Y. The height of $\hat{y}$ is clearly a function of x. We have already seen that

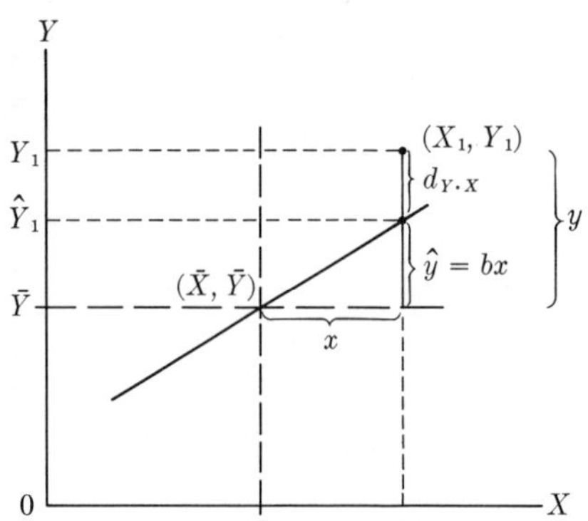

FIGURE 14.9 Schematic diagram to show relations involved in partitioning the sum of squares of the dependent variable.

$\hat{y} = bx$ [Expression (14.3)]. Those of you familiar with analytical geometry will, of course, recognize this as the point-slope form of the equation. If b, the slope of the regression line, were steeper, $\hat{y}$ would be relatively longer for a given value of x. The remaining portion of the deviation y is $d_{Y \cdot X}$. It represents the residual variation of the variable Y after the explained variation has been subtracted. We can see that $y = \hat{y} + d_{Y \cdot X}$ by writing out these deviations explicitly as $Y - \bar{Y} = (\hat{Y} - \bar{Y}) + (Y - \hat{Y}) = Y - \bar{Y}$.

For each of these deviations we can compute a corresponding sum of squares. Appendix A1.9 gives the computational formula for the unexplained sum of squares,

$$\sum d_{Y \cdot X}^2 = \sum y^2 - \frac{(\sum xy)^2}{\sum x^2}.$$

Transposed, this yields

$$\sum y^2 = \frac{(\sum xy)^2}{\sum x^2} + \sum d_{Y \cdot X}^2.$$

Of course, $\sum y^2$ corresponds to y, $\sum d_{Y \cdot X}^2$ to $d_{Y \cdot X}$, and

$$\sum \hat{y}^2 = \frac{(\sum xy)^2}{\sum x^2}$$

to $\hat{y}$ (as shown in the previous section). Thus we are able to partition the sum of squares of the dependent variable in regression in an analogous way to the partition of the total SS in analysis of variance. You may wonder how the additive relation of the deviations can be matched by an additive relation of their squares without the presence of any cross products. Some simple algebra in Appendix A1.10 will show that the cross products cancel out. The magnitude of the unexplained deviation $d_{Y \cdot X}$ is independent of the magnitude of the explained deviation $\hat{y}$ just as in anova the magnitude of the deviation of an item from the sample mean is independent of the magnitude of the deviation of the sample mean from the grand mean. This relationship between regression and analysis of variance can be carried further. We can undertake an analysis of variance of the partitioned sums of squares as follows:

	Source of variation	*df*	*SS*	*MS*
$\hat{Y} - \bar{Y}$	Explained (estimated Y from mean of Y)	1	$\sum \hat{y}^2 = \frac{(\sum xy)^2}{\sum x^2}$	$s_{\hat{Y}}^2$
$Y - \hat{Y}$	Unexplained, error (observed Y from estimated Y)	$n - 2$	$\sum d_{Y \cdot X}^2 = \sum y^2 - \sum \hat{y}^2$	$s_{Y \cdot X}^2$
$Y - \bar{Y}$	Total (observed Y from mean of Y)	$n - 1$	$\sum y^2 = \sum Y^2 - \frac{(\sum Y)^2}{n}$	s_Y^2

The *explained mean square*, or *mean square due to linear regression*, measures the amount of variation in Y accounted for by variation of X. It is tested over the *unexplained mean square*, which measures the residual variation and is used as an error MS. The mean square due to linear regression, $s_{\hat{Y}}^2$, is based on one degree of freedom, and consequently $n - 2$ df remain for the

BOX 14.2

Standard errors of regression statistics and their degrees of freedom.

For explanation of this box, see Sections 14.5 and 14.6: ν identifies degrees of freedom; a = number of values of X when there are n_i Y-values for each X; n = sample size when there is a single Y-value for each value of X.

Statistic	s	*More than one Y-value for each value of X*		*Single Y-value for each value of X*	
$b_{Y\cdot X}$ (regression coefficient)	s_b	$\sqrt{\dfrac{s^2_{Y\cdot X}}{\Sigma x^2}}$	$\nu = a - 2$	$\sqrt{\dfrac{s^2_{Y\cdot X}}{\Sigma x^2}}$	$\nu = n - 2$
$\bar{Y}_i$ (sample mean)	$s_{\bar{Y}}$	At any value X_i: $\sqrt{\dfrac{MS_{\text{within}}}{n_i}}$	$\nu = \Sigma n_i - a$	At $\bar{X}$: $\sqrt{\dfrac{s^2_{Y\cdot X}}{n}}$	$\nu = n - 2$
$\hat{Y}_i$ (estimated Y for a given value X_i)	$s_{\hat{Y}}$	$\sqrt{s^2_{Y\cdot X}\left[\dfrac{1}{\Sigma n_i} + \dfrac{(X_i - \bar{X})^2}{\Sigma x^2}\right]}$	$\nu = a - 2$ ($\nu = \Sigma n_i - 2$ if pooled $s^2_{Y\cdot X}$ is employed)	$\sqrt{s^2_{Y\cdot X}\left[\dfrac{1}{n} + \dfrac{(X_i - \bar{X})^2}{\Sigma x^2}\right]}$	$\nu = n - 2$
Predicted Y_i for a given value X_i	$\hat{s}_Y$	$\sqrt{s^2_{Y\cdot X}\left[\dfrac{1}{\Sigma n_i} + \dfrac{(X_i - \bar{X})^2}{\Sigma x^2}\right] + MS_{\text{within}}}$	$\nu = \Sigma n_i - 2$	$\sqrt{s^2_{Y\cdot X}\left[1 + \dfrac{1}{n} + \dfrac{(X_i - \bar{X})^2}{\Sigma x^2}\right]}$	$\nu = n - 2$
Predicted mean $\bar{Y}_i$ of k items for a given value X_i	$\hat{s}_{\bar{Y}}$	$\sqrt{s^2_{Y\cdot X}\left[\dfrac{1}{\Sigma n_i} + \dfrac{(X_i - \bar{X})^2}{\Sigma x^2}\right] + \dfrac{MS_{\text{within}}}{k}}$	$\nu = \Sigma n_i - 2$	$\sqrt{s^2_{Y\cdot X}\left[\dfrac{1}{k} + \dfrac{1}{n} + \dfrac{(X_i - \bar{X})^2}{\Sigma x^2}\right]}$	$\nu = n - 2$

Note: If $F_s = s^2_{Y\cdot X}/MS_{\text{within}}$ is not significant, you may wish to pool the two mean squares as follows:

$$\text{pooled variance } s^2_{Y\cdot X} = \frac{SS_{\text{within}} + \sum d^2_{Y\cdot X}}{\sum n_i - 2}$$

error MS since the total sum of squares possesses $n - 1$ degrees of freedom. The test is of the null hypothesis $H_0{:}\beta = 0$. When we carry out such an anova on the weight loss data of Box 14.1, we obtain the following results:

Source of variation	*df*	*SS*	*MS*	F_s
Explained—due to linear regression	1	23.5145	23.5145	267.18***
Unexplained—error around regression line	7	0.6161	0.08801	
Total	8	24.1306		

The significance test is $F_s = s_{\hat{Y}}^2/s_{Y\cdot X}^2$. It is clear from the observed value of F_s that a large and significant portion of the variance of Y has been explained by regression on X.

We now proceed to the standard errors for various regression statistics, their employment in tests of hypotheses, and the computation of confidence limits. Box 14.2 lists these standard errors in two columns. The left column is for the case of more than one Y-value for each value of X and will be taken up in Section 14.6. The right column is for the case with a single Y-value for each value of X. The first row of the table considers the *standard error of the regression coefficient*, which is simply the square root of the quotient of the unexplained variance divided by the sum of squares of X. Note that the unexplained variance $s_{Y\cdot X}^2$ is a fundamental quantity that is a part of all standard errors in regression. The standard error of the regression coefficient permits us to test various hypotheses and to set confidence limits to our sample estimate of b. The computation of s_b is illustrated in step **1** of Box 14.3, using the weight loss example of Box 14.1.

The significance test illustrated in step **2** tests the *"significance" of the regression coefficient;* that is, it tests the null hypothesis that the sample value of b comes from a population with a parametric value $\beta = 0$ for the regression coefficient. This is a t-test, the appropriate degrees of freedom being $n - 2 = 7$. If we cannot reject the null hypothesis, there is no evidence that the regression is significantly deviant from zero in either the positive or negative direction. Our conclusions for the weight loss data are that a highly significant negative regression is present. We saw earlier (Section 9.4) that $t^2 = F$. When we square $t_s = 16.345$ from Box 14.3, we obtain 267.16, which (within rounding error) equals the value of F_s found in the anova earlier in this section. The significance test in step **2** of Box 14.3 could, of course, also have been used to test whether b is significantly different from a parametric value β other than zero.

Setting confidence limits to the regression coefficient presents no new features. The computation is shown in step **3** of Box 14.3. In view of the small magnitude of s_b, the confidence interval is quite narrow. The confidence limits are shown in Figure 14.10 as dotted lines representing the 95% bounds of the slope. Note that the regression line as well as its confidence limits pass through the means for X and Y. Variation in b therefore rotates the regression line about the point $\bar{X}$, $\bar{Y}$.

BOX 14.3

Significance tests and computation of confidence limits of regression statistics. Single value of Y for each value of X.

Based on standard errors and degrees of freedom of Box 14.2; using example of Box 14.1.

$n = 9 \qquad \bar{X} = 50.389 \qquad \bar{Y} = 6.022$

$b_{Y \cdot X} = -0.05322 \qquad \sum x^2 = 8301.3889$

$$s^2_{Y \cdot X} = \frac{\sum d^2_{Y \cdot X}}{(n-2)} = \frac{0.6161}{7} = 0.08801$$

1. Standard error of the regression coefficient

$$s_b = \sqrt{\frac{s^2_{Y \cdot X}}{\sum x^2}} = \sqrt{\frac{0.08801}{8301.3889}} = \sqrt{0.000{,}010{,}602} = 0.003{,}2561$$

2. Testing significance of the regression coefficient

$$t_s = \frac{(b-0)}{s_b} = \frac{-0.05322}{0.003{,}2561} = -16.345$$

$t_{.001[7]} = 5.408 \qquad P < 0.001$

3. 95% confidence limits for regression coefficient

$$t_{.05[7]}s_b = 2.365(0.003{,}2561) = 0.00770$$

$$L_1 = b - t_{.05[7]}s_b = -0.05322 - 0.00770 = -0.06092$$

$$L_2 = b + t_{.05[7]}s_b = -0.05322 + 0.00770 = -0.04552$$

4. Standard error of the sampled mean $\bar{Y}$ (at $\bar{X}$)

$$s_{\bar{Y}} = \sqrt{\frac{s^2_{Y \cdot X}}{n}} = \sqrt{\frac{0.08801}{9}} = 0.098{,}8883$$

5. 95% confidence limits for the mean μ_Y corresponding to $\bar{X}(\bar{Y} = 6.022)$

$$t_{.05[7]}s_{\bar{Y}} = 2.365(0.098{,}8883) = 0.233871$$

$$L_1 = \bar{Y} - t_{.05[7]}s_{\bar{Y}} = 6.022 - 0.2339 = 5.7881$$

$$L_2 = \bar{Y} + t_{.05[7]}s_{\bar{Y}} = 6.022 + 0.2339 = 6.2559$$

6. Standard error of $\hat{Y}$, an estimated Y for a given value of X_i

$$s_{\hat{Y}} = \sqrt{s^2_{Y \cdot X}\left[\frac{1}{n} + \frac{(X_i - \bar{X})^2}{\sum x^2}\right]}$$

for example, for $X_i = 100\%$ relative humidity,

$$s_{\hat{Y}} = \sqrt{0.08801\left[\frac{1}{9} + \frac{(100 - 50.389)^2}{8301.3889}\right]}$$

$$= \sqrt{0.08801[0.40760]} = \sqrt{0.035{,}873} = 0.18940$$

7. 95% confidence limits for μ_{Y_i} corresponding to the estimate $\hat{Y}_i = 3.3817$ at $X_i = 100\%$ relative humidity

BOX 14.3 continued

$t_{.05[7]}s_{\hat{Y}} = 2.365(0.18940) = 0.44793$

$L_1 = \hat{Y}_i - t_{.05[7]}s_{\hat{Y}} = 3.3817 - 0.4479 = 2.9338$

$L_2 = \hat{Y}_i + t_{.05[7]}s_{\hat{Y}} = 3.3817 + 0.4479 = 3.8296$

8. Standard error of a predicted mean $\bar{Y}_i$ to be obtained in a new experiment run at $X_i = 100\%$ relative humidity. Our best prediction for this mean would be $\bar{Y}_i = \hat{Y}_i = 3.3817$. If the new experiment were based on a sample size of $k = 5$, the standard error of the predicted mean would be

$$\hat{s}_{\bar{Y}} = \sqrt{s^2_{Y \cdot X}\left[\frac{1}{k} + \frac{1}{n} + \frac{(X_i - \bar{X})^2}{\sum x^2}\right]}$$

$$= \sqrt{0.08801\left[\frac{1}{5} + \frac{1}{9} + \frac{(100 - 50.389)^2}{8301.3889}\right]}$$

$$= \sqrt{0.08801[0.60760]} = \sqrt{0.05347} = 0.23124$$

9. 95% prediction limits for a sample mean of 5 weight losses at 100% relative humidity (using the standard error computed above)

$t_{.05[7]}\hat{s}_{\bar{Y}} = 2.365(0.23124) = 0.5469$

$L_1 = \hat{Y}_i - t_{.05[7]}\hat{s}_{\bar{Y}} = 3.3817 - 0.5469 = 2.8348$

$L_2 = \hat{Y}_i + t_{.05[7]}\hat{s}_{\bar{Y}} = 3.3817 + 0.5469 = 3.9286$

Next, we calculate a *standard error to the observed sample mean* $\bar{Y}$. You will recall from Section 7.1 that $s^2_{\bar{Y}} = s^2_Y/n$. However, now that we have regressed Y on X, we are able to account for some of the variation of Y in terms of the variation of X. The variance of Y around the point $\bar{X}, \bar{Y}$ on the regression line is less than s^2_Y; it is $s^2_{Y \cdot X}$. At $\bar{X}$ we may therefore compute confidence limits of $\bar{Y}$, using as a standard error of the mean $s_{\bar{Y}} = \sqrt{s^2_{Y \cdot X}/n}$ with $n - 2$ degrees of freedom. This standard error is computed in step 4 of Box 14.3, and 95% confidence limits for the sampled mean $\bar{Y}$ at $\bar{X}$ are

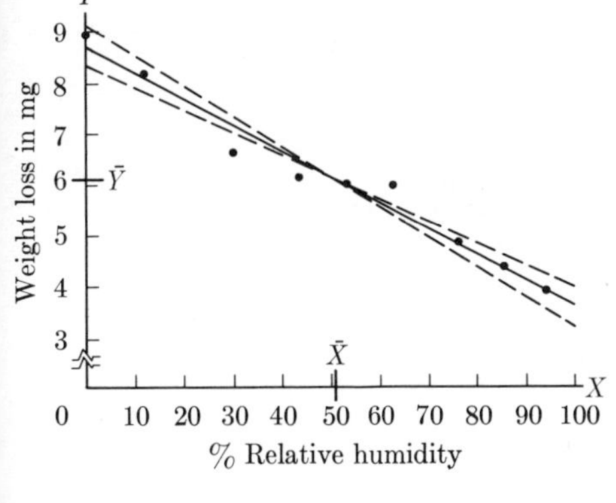

FIGURE 14.10 95% confidence limits to regression line of Figure 14.8.

calculated in step **5**. These limits (5.7881 – 6.2559) are considerably narrower than the confidence limits for the mean based on the conventional standard error $s_{\bar{Y}}$, which would be from 4.687 to 7.357. Clearly, differences in relative humidity explain much of the variation in weight loss.

The standard error for $\bar{Y}$ is only a special case of the *standard error for any estimated value* $\hat{Y}$ *along the regression line.* A new factor now enters the error variance, whose magnitude is in part a function of the distance of a given value X_i from its mean $\bar{X}$. Thus, the farther away X_i is from its mean, the greater will be the error of estimate. This factor is seen in the third row of Box 14.2 as the deviation $X_i - \bar{X}$, squared and divided by the sum of squares of X. The standard error for an estimate $\hat{Y}_i$ for a relative humidity $X_i = 100\%$ is given in step **6** of Box 14.3. The 95% confidence limits for $\mu_{\hat{Y}_i}$, the parametric value corresponding to the estimate $\hat{Y}_i$ are shown in step **7** of that box. Note that the width of the confidence interval is 3.8296 – 2.9338 = 0.8958, considerably wider than the confidence interval at $\bar{X}$ calculated in step **5**, which was 6.2559 – 5.7881 = 0.4678. This illustrates the point that confidence limits are wider away from the mean than at the mean. If we calculate a series of confidence limits for different values of X_i, we obtain a biconcave confidence belt as shown in Figure 14.11. The farther we get away from the mean, the less reliable are our estimates of Y because of the uncertainty about the true slope, β, of the regression line.

Furthermore, the linear regressions that we fit are often only rough approximations to the more complicated functional relationships between biological variables. Very often there is an approximately linear relation along a certain range of the independent variable, beyond which range the slope changes rapidly. For example, heartbeat of a poikilothermic animal will be directly proportional to temperature over a range of tolerable temperatures, but beneath and above this range the heartbeat will eventually decrease as the animal freezes or suffers heat prostration. Hence common sense indicates that one should be very cautious about extrapolating from a

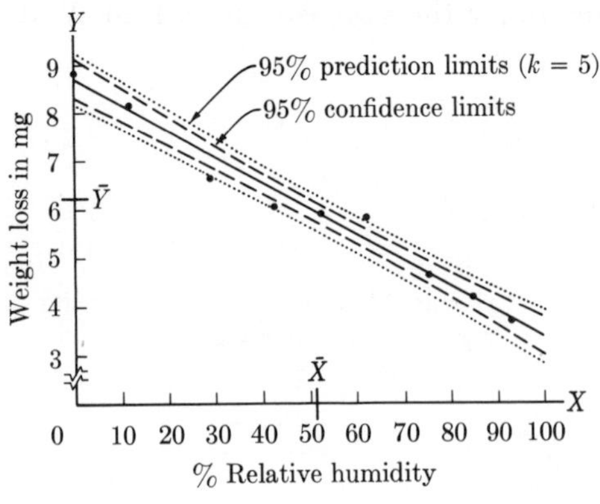

FIGURE 14.11 95% confidence limits to regression estimates for data of Figure 14.8.

regression equation if one has any doubts about the linearity of the relationship.

The confidence limits for α, the parametric value of a, are a special case of that for $\mu_{\hat{Y}_i}$ at $X_i = 0$.

In certain situations, we may require that a regression line pass through the origin. This would be proper for cases in which the parametric Y-intercept α is really zero, as in growth curves where the size of a structure is effectively zero at age zero, or in biochemical work where none of a substance is produced by time zero. Belief that regression through the origin is the proper way to proceed can be supported by observing that the confidence limits for α include zero. The necessary computations for fitting a regression through the origin are shown in several textbooks (for example, Steel and Torrie, 1960, section 9.13).

The next two standard errors have somewhat different applications. They relate to *predictions made on the basis of the regression equation.* Once you have established a functional relationship between two variables, as between weight loss and relative humidity in this example, you can use this equation to predict the outcome of future experiments. If, for example, you wish to run another batch of *Tribolium* beetles at 100% relative humidity, your best estimate for the weight loss would be $\hat{Y} = 3.3817$ mg, based on the regression equation of Box 14.1. You should expect a standard error greater than that of the estimate $\hat{Y}$ because there is now added the error variance of weight losses within the new sample at $X_i = 100\%$ relative humidity. The formulation in the fourth row of Box 14.2 is most frequently given in statistics texts, but it is of limited usefulness because it describes the expected variance of the sample items—that is, your prediction of the variance of future individual weight loss readings. More useful is the formulation in the last row of the box that describes the *standard error of the predicted sample mean based on k items.* Thus, if you were to repeat the experiment on weight loss and run five batches of beetles at 100% relative humidity, you could expect the standard error of your sample mean of the five batches to be 0.23124 as calculated in step **8** of Box 14.3. The prediction limits for the sample mean are shown in step **9**. Note that they are wider than the confidence limits of the corresponding estimated value $\hat{Y}$ for the relative humidity of $X_i = 100\%$. The biconcave prediction belts are graphed in Figure 14.11 and are wider than the confidence belts of the estimated Y-values. As the size of the intended sample increased, the standard error would diminish and the prediction limits would approach the confidence limits shown in step **7**. When the sample size decreases to unity, one computes the prediction limits from the standard error in the fourth row in Box 14.2. These prediction limits yield bounds for the sample items.

In the next section we shall show how to compute regression when there is more than one Y-value per value of X. We shall also go through the computation of standard errors for such cases based on the formulas in the left column of Box 14.2.

14.6 More than one value of Y for each value of X

We now take up the case of Model I regression as originally defined in Section 14.2 and illustrated by Figure 14.4. For each value of the treatment X we sample Y repeatedly, obtaining a sample distribution of Y-values at each of the chosen points of X. We have selected an experiment from the laboratory of one of us (Sokal) in which *Tribolium* beetles were reared from eggs to adulthood at four different densities. The percentage survival to adulthood was calculated for varying numbers of replicates at these densities. Following Section 13.9, these percentages were given arcsine transformations, which are listed in Box 14.4. These transformed values are more likely to be normally distributed than percentages. The arrangement of these data is very much like that of a single classification Model I anova. There are four different densities and several replicated survival values at each density. We now would like to determine whether there are differences in survival among the four groups, and also whether we can establish a regression of survival on density.

A first approach, therefore, is to carry out an analysis of variance, using the methods of Section 9.2 and Box 9.1. Our aim in doing this is illustrated diagrammatically in Figure 14.12. If the analysis of variance were not significant, this would indicate that the means are not significantly different from each other as shown in Figure 14.12, A, and it would be unlikely that a regression line fitted to these data would have a slope significantly different from zero. Occasionally, when the means increase or decrease slightly as X increases, they may not be different enough for the mean square among groups to be significant by anova, yet a significant regression can be found. These are usually cases on the borderline of statistical significance. When we find a marked regression of the means on X, as shown in Figure 14.12, B, we usually will find a significant difference among the means by an anova. However, we cannot turn this argument around and say that a significant difference among means as shown by an anova necessarily indicates that a significant linear regression can be fitted to these data. In Figure 14.12, C, the means follow a **U**-shaped function (a parabola). Though the means would likely be significantly different from each other, clearly a straight line fitted to these data would be a horizontal line halfway between the upper and the lower points. In such a set of data, *linear* regression can explain only little of the variation of the dependent variable. However, a curvilinear regression (see Section 14.11) would fit these data and remove most of the variance of Y. A similar case is shown in Figure 14.12, D, in which the means describe a periodically changing phenomenon, rising and falling alternatingly. Again the regression line for these data has slope zero. A curvilinear (cyclical)

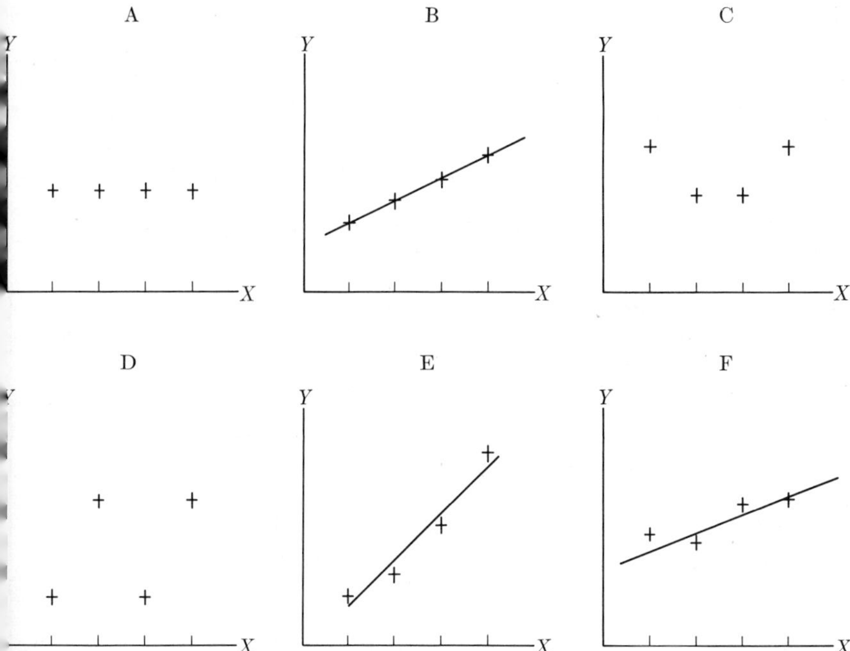

FIGURE 14.12 Differences among means and linear regression. General trends only are indicated by these figures. Significance of any of these would depend on the outcomes of appropriate tests. (For further explanation see text.)

regression could also be fitted to such data, but our main purpose in showing this example is to indicate that there could be heterogeneity among the means of Y apparently unrelated to the magnitude of X. Remember that in real examples you will rarely ever get a regression as clear-cut as the linear case in 14.12, B, or the curvilinear one in 14.12, C, nor will you necessarily get heterogeneity of the type shown in 14.12, D, in which any straight line fitted to the data would be horizontal and halfway between the upper and lower points. You are more likely to get data in which linear regression can be demonstrated, but which will not fit a straight line well. The residual deviations of the means around linear regression might be removed by changing from linear to curvilinear regression (as is suggested by the pattern of points in Figure 14.12, E) or they remain as inexplicable residual heterogeneity around the regression line, as indicated in Figure 14.12, F.

We carry out the computations following the by now familiar outline for analysis of variance and obtain the anova table shown in Box 14.4. The three degrees of freedom among the four groups yield a mean square that is highly significant when tested over the within-groups mean square. It therefore is clearly worthwhile to continue the analysis to see whether a significant regression of survival on density exists. The additional steps for the regression analysis follow in Box 14.4. We compute the sum of squares of X, the sum of products of X and Y, the explained sum of squares of Y, and the unexplained sum of squares of Y. The formulas will look unfamiliar because of the complication of the several Y's per value of X. The computations for

BOX 14.4

Computation of regression with more than one value of Y per value of X. General case with unequal sample sizes.

The variates Y are arcsine transformations of the percentage survival of the beetle *Tribolium castaneum* at 4 densities (X = number of eggs per gram of flour medium).

	Density = X ($a = 4$)			
	5/g	*20*/g	*50*/g	*100*/g
	61.68	68.21	58.69	53.13
	58.37	66.72	58.37	49.89
Survival; in degrees	69.30	63.44	58.37	49.82
	61.68	60.84		
	69.30			
$\sum^{n_i} Y$	320.33	259.21	175.43	152.84
n_i	5	4	3	3
$\bar{Y}_i$	64.07	64.80	58.48	50.95
$\sum^{a} n_i = 15$				

Source: Data by Sokal (1967).

Anova computations

The steps are the same as those described in Box 9.1.

1. $\sum^{a}\sum^{n_i} Y = 907.81$

2. $\sum^{a}\sum^{n_i} Y^2 = 55{,}503.6547$

3. $\sum^{a} \frac{\left(\sum^{n_i} Y\right)^2}{n_i} = 55{,}364.9680$

4. $CT_Y = \frac{\left(\sum^{a}\sum^{n_i} Y\right)^2}{\sum^{a} n_i} = 54{,}941.2664$

5. SS_{total} = quantity **2** − quantity **4**
$= 55{,}503.6547 - 54{,}941.2664 = 562.3883$

6. SS_{groups} = quantity **3** − quantity **4**
$= 55{,}364.9680 - 54{,}941.2664 = 423.7016$

BOX 14.4 continued

7. SS_{within} = quantity **5** − quantity **6** = 562.3883 − 423.7016 = 138.6867

Anova table

Source of variation	*df*	*SS*	*MS*	F_s
$\bar{Y} - \bar{\bar{Y}}$ Among groups	3	423.7016	141.2339	11.20**
$Y - \bar{Y}$ Within groups	11	138.6867	12.6079	
$Y - \bar{\bar{Y}}$ Total	14	562.3883		

The groups differ significantly.

We proceed to test whether the differences among the survival values can be accounted for by linear regression on density. If no significant differences among groups had been found, it would be possible, although unlikely, for linear regression to be significant. If $SS_{\text{groups}} < (MS_{\text{within}} \times F_{[1, \sum^a n_i - a]})$, it is impossible for regression to be significant.

Computations for regression analysis

8. Sum of X weighted by sample size $= \sum^a n_i X$

$$= 5(5) + 4(20) + 3(50) + 3(100) = 555$$

9. Sum of X^2 weighted by sample size $= \sum^a n_i X^2$

$$= 5(5)^2 + 4(20)^2 + 3(50)^2 + 3(100)^2 = 39{,}225$$

10. Sum of products of X and $\bar{Y}$ weighted by sample size $= \sum^a n_i X \bar{Y}$

$$= \sum^a X\left(\sum^{n_i} Y_i\right) = 5(320.33) + \cdots + 100(152.84) = 30{,}841.35$$

11. Correction term for $X = CT_X = \dfrac{\left(\sum^a n_i X\right)^2}{\sum^a n_i}$

$$= \frac{(\text{quantity } \mathbf{8})^2}{\sum^a n_i} = \frac{(555)^2}{15} = 20{,}535.00$$

12. Sum of squares of $X = \sum x^2 = \sum^a n_i X^2 - CT_X$

$$= \text{quantity } \mathbf{9} - \text{quantity } \mathbf{11} = 39{,}225 - 20{,}535 = 18{,}690$$

13. Sum of products $= \sum xy$

$$= \sum^n X\left(\sum^{n_i} Y_i\right) - \frac{\left(\sum^a n_i X\right)\left(\sum^a \sum^{n_i} Y\right)}{\sum^a n_i}$$

$$= \text{quantity } \mathbf{10} - \frac{\text{quantity } \mathbf{8} \times \text{quantity } \mathbf{1}}{\sum^a n_i}$$

$$= 30{,}841.35 - \frac{(555)(907.81)}{15} = -2747.62$$

BOX 14.4 continued

14. Explained sum of squares $= \sum \hat{y}^2 = \dfrac{(\sum xy)^2}{\sum x^2}$

$$= \frac{(\text{quantity } \mathbf{13})^2}{\text{quantity } \mathbf{12}} = \frac{(-2747.62)^2}{18{,}690} = 403.9281$$

15. Unexplained sum of squares $= \sum d^2_{Y \cdot X} = SS_{\text{groups}} - \sum \hat{y}^2$
$= \text{quantity } \mathbf{6} - \text{quantity } \mathbf{14} = 423.7016 - 403.9281 = 19.7735$

Completed anova table with regression

Source of variation	*df*	*SS*	*MS*	F_s
$\bar{Y} - \bar{\bar{Y}}$ Among densities (groups)	3	423.7016	141.2339	11.20**
$\hat{Y} - \bar{\bar{Y}}$ Linear regression	1	403.9281	403.9281	40.86*
$\bar{Y} - \hat{Y}$ Deviations from regression	2	19.7735	9.8868	<1 *ns*
$Y - \bar{Y}$ Within groups	11	138.6867	12.6079	
$Y - \bar{\bar{Y}}$ Total	14	562.3883		

In addition to the familiar mean squares, MS_{groups} and MS_{within}, we now have the mean square due to linear regression, $MS_{\hat{Y}}$, and the mean square for deviations from regression, $MS_{Y \cdot X}$ $(= s^2_{Y \cdot X})$. To test if the deviations from linear regression are significant, compare the ratio $F_s = MS_{Y \cdot X}/MS_{\text{within}}$ with $F_{\alpha[a-2, \sum^a n_i - a]}$. Since we find $F_s < 1$, we accept the null hypothesis that the deviations from linear regression are zero. Using the rules of Box 10.2 we do not pool MS_{within} with $s^2_{Y \cdot X}$.

To test for the presence of linear regression, we therefore tested $MS_{\hat{Y}}$ over the mean square of deviations from regression $s^2_{Y \cdot X}$ and, since $F_s = 403.9281/9.8868 = 40.86$ is greater than $F_{.05[1,2]} = 18.5$, we clearly reject the null hypothesis that there is no regression, or that $\beta = 0$. If we had pooled, we would have compared the ratio $F_s = MS_{\hat{Y}}/s^2_{Y \cdot X} = 403.9281/12.1892 = 33.14$ with $F_{.001[1,13]} = 17.8$ and would have reached the same conclusion.

16. Regression coefficient (slope of regression line) $= b_{Y \cdot X} = \dfrac{\sum xy}{\sum x^2}$

$$= \frac{\text{quantity } \mathbf{13}}{\text{quantity } \mathbf{12}} = \frac{-2747.62}{18{,}690} = -0.14701$$

17. Y-intercept $= a = \bar{\bar{Y}} - b_{Y \cdot X}\bar{X}$

$$= \frac{\text{quantity } \mathbf{1}}{\sum^a n_i} - \frac{\text{quantity } \mathbf{16} \times \text{quantity } \mathbf{8}}{\sum^a n_i}$$

$$= \frac{907.81}{15} - \frac{(-0.14701)555}{15} = 60.5207 + 5.4394 = 65.9601$$

Hence, the regression equation is $\hat{Y} = 65.9601 - 0.14701X$

the sum of squares of X involve the multiplication of X by the number of items in the study. Thus, though there may appear to be only four densities, there are, in fact, as many densities (although of only four magnitudes) as there are values of Y in the study. Having completed the computations, we again present the results in the form of an anova table, as shown in Box 14.4. Note that the major quantities in this table are the same as in a single classification anova, but in addition we now have a sum of squares representing linear regression, which is always based on one degree of freedom. This sum of squares is subtracted from the SS among groups, leaving a residual sum of squares (of two degrees of freedom in this case) representing the deviations from linear regression.

We should become quite certain what these sources of variation represent. The SS due to linear regression represents that portion of the SS among groups that is explained by linear regression on X. The SS due to deviations from regression represents the residual variation or scatter around the regression line as illustrated by the various examples in Figure 14.12. The SS within groups is a measure of the variation of the items around each group mean.

We first test whether the mean square for deviations from regression ($MS_{Y \cdot X} = s^2_{Y \cdot X}$) is significant by computing the variance ratio of $MS_{Y \cdot X}$ over the within-groups MS. In our case, the deviations from regression are clearly not significant, since the mean square for deviations is less than that within groups. This is a situation analogous to testing mean squares for subgroups in a hierarchic anova and, in view of the nonsignificance of the mean square for deviations, we might consider pooling it with the error MS. We follow the rules in Box 10.2. Since we have no prior knowledge of the variance ratio, and $F_s > F_{.75}$, we do not pool. We now test the mean square for regression, $MS_{\hat{Y}}$, over the mean square for deviations from regression and find it to be highly significant. Linear regression on density has clearly removed a significant portion of the variation of survival values. When one pools, one tests the regression mean square over the pooled mean square representing the SS of deviations from regression and within-groups SS. Had we done so in this case, we would have obtained the same result.

Significance of the mean square for deviations from regression could mean either that Y is a curvilinear function of X or that there is a large amount of random heterogeneity around the regression line (as already discussed in connection with Figure 14.12; actually a mixture of both conditions may prevail). If the $MS_{Y \cdot X}$ is significant and curvilinearity is suspected, a curvilinear regression (Section 14.11) may be fitted. If heterogeneity is responsible for a significant $MS_{Y \cdot X}$, we use it as the denominator mean square against which to test the significance of linear regression.

We complete the computation of the regression coefficient and regression equation as shown at the end of Box 14.4. Our conclusions are that as density increases, survival decreases and that this relationship can be expressed by a significant linear regression of the form $\hat{Y} = 65.9601 - 0.14701X$, where

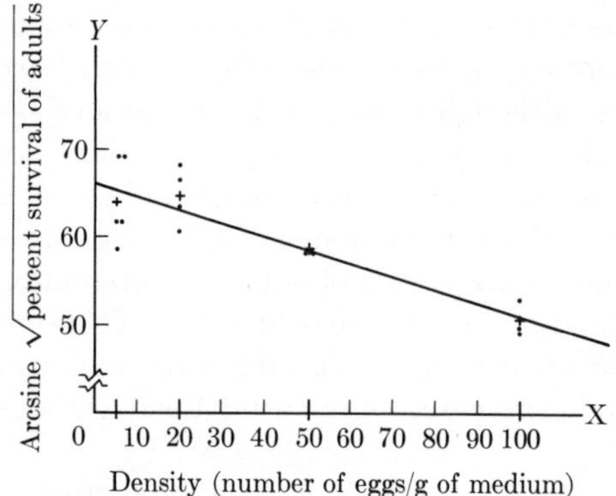

FIGURE 14.13 Linear regression fitted to data of Box 14.4. Sample means are identified by + signs.

X is density per gram and $\hat{Y}$ is the arcsine transformation of percentage survival. This relation is graphed in Figure 14.13.

We now proceed to compute standard errors and confidence limits for the example we have just discussed. The pertinent standard errors are found in the left column of Box 14.2. Since we decided not to pool the mean square for deviations from regression with the within-groups mean square, we use the former and its 2 degrees of freedom and the MS_{within} with its 11 *df*, with the appropriate degrees of freedom as indicated in the box. In other cases, if we did pool, we would use the pooled mean square of deviations around regression throughout when $s^2_{Y \cdot X}$ is called for in Box 14.2. The computation of the *standard error of the regression coefficient and of the 95% confidence limits of* β are shown in steps **1** and **2** of Box 14.5. We calculate the *standard error of the sample mean for any one value of* X based on the MS_{within}. Thus in step **3** we compute the standard error for the mean survival of 64.07 at density $5/g$ (expressed in degrees; remember this is an arcsine transformation)

BOX 14.5

Computation of standard errors and confidence limits of regression statistics. More than one value of Y for each value of X.

Data from Box 14.4.

$$b_{Y \cdot X} = -0.14701 \qquad s^2_{Y \cdot X} = 9.8868$$

$$\sum x^2 = 18{,}690 \qquad MS_{\text{within}} = 12.6079$$

In all of the computations below, the MS for deviations from regression $s^2_{Y \cdot X} = 9.8868$ with 2 *df* is used. The formulas for the standard errors are from the left half of Box 14.2.

1. Standard error of the regression coefficient

$$s_b = \sqrt{\frac{s^2_{Y \cdot X}}{\sum x^2}} = \sqrt{\frac{9.8868}{18{,}690}} = \sqrt{0.0005290} = 0.02300$$

BOX 14.5 continued

2. 95% confidence limits for regression coefficient β

$$t_{.05[2]}s_b = 4.303(0.02300) = 0.09897$$

$$L_1 = b - t_{.05[2]}s_b = -0.14701 - 0.09897 = -0.24598$$

$$L_2 = b + t_{.05[2]}s_b = -0.14701 + 0.09897 = -0.04804$$

3. Standard error of a sample mean at any value X_i

$$s_{\bar{Y}} = \sqrt{\frac{MS_{\text{within}}}{n_i}}$$

For the first sample at density $X_1 = 5/g$, we find $n_1 = 5$. The standard error would be

$$s_{\bar{Y}} = \sqrt{\frac{12.6079}{5}} = \sqrt{2.52158} = 1.58795$$

4. The 95% confidence interval for the mean of the first sample ($X_1 = 5/g$; $\bar{Y}_1 = 64.07$)

$$t_{.05[11]}s_{\bar{Y}} = 2.201(1.58795) = 3.4951$$

$$L_1 = \bar{Y}_1 - t_{.05[11]}s_{\bar{Y}} = 64.07 - 3.4951 = 60.5749$$

$$L_2 = \bar{Y}_1 + t_{.05[11]}s_{\bar{Y}} = 64.07 + 3.4951 = 67.5651$$

5. Standard error of $\hat{Y}_i$, an estimated mean for a given value of X_i. As an example we shall compute $\hat{Y}_i$ for $X_i = 59/g$.

Since $\hat{Y} = a + bx$, for $X_i = 59$, $\hat{Y}_i = 65.9601 - 0.14701(59) = 57.2865$

$$s_{\hat{Y}} = \sqrt{s^2_{Y\cdot X}\left[\frac{1}{\sum n_i} + \frac{(X_i - \bar{X})^2}{\sum x^2}\right]}$$

$\bar{X}$ in this formula refers to

$$\frac{\sum^a n_i X}{\sum^a n_i} = \frac{555}{15} = 37.0 \qquad \text{(values from Box 14.4).}$$

For a density of $X_i = 59/g$

$$s_{\hat{Y}} = \sqrt{9.8868\left[\frac{1}{15} + \frac{(59-37)^2}{18{,}690}\right]} = \sqrt{9.8868[0.09256]}$$

$$= \sqrt{0.91512} = 0.95662$$

6. 95% confidence interval for μ_{Y_i} corresponding to the estimate $\hat{Y}_i = 57.2865$ at $X_i = 59/g$

$$t_{.05[2]}s_{\hat{Y}} = 4.303(0.95662) = 4.1163$$

$$L_1 = \hat{Y}_i - t_{.05[2]}s_{\hat{Y}} = 57.2865 - 4.1163 = 53.1702$$

$$L_2 = \hat{Y}_i + t_{.05[2]}s_{\hat{Y}} = 57.2865 + 4.1163 = 61.4028$$

95% of such intervals should contain the true mean μ_{Y_i} for $X_i = 59$. These limits are plotted in Figure 14.14 for values of X_i from 0 to 100.

BOX 14.5 continued

7. Standard error of a predicted mean $\bar{Y}_i$ to be obtained in a new experiment run at a density of $X_i = 59/g$. Our best prediction for this mean would be $\bar{Y}_i = \hat{Y} = 57.2865$. If the new experiment were based on a sample of size $k = 10$, the standard error of the predicted mean would be

$$\hat{s}_{\bar{Y}} = \sqrt{s^2_{Y \cdot X}\left[\frac{1}{\sum n_i} + \frac{(X_i - \bar{X})^2}{\sum x^2}\right] + \frac{MS_{\text{within}}}{k}}$$

$$= \sqrt{9.8868\left[\frac{1}{15} + \frac{(59 - 37)^2}{18{,}690}\right] + \frac{12{,}6079}{10}}$$

$$= \sqrt{0.91514 + 1.26079} = \sqrt{2.17593} = 1.47510$$

8. 95% prediction limits for the mean of a sample of 10 survival readings at density $59/g$ (using the standard error computed above)

$$t_{.05[13]}\hat{s}_{\bar{Y}} = 2.160(1.4751) = 3.1862$$

$$L_1 = \hat{Y}_i - t_{.05[13]}\hat{s}_{\bar{Y}} = 57.2865 - 3.1862 = 54.1003$$

$$L_2 = \hat{Y}_i + t_{.05[13]}\hat{s}_{\bar{Y}} = 57.2865 + 3.1862 = 60.4727$$

Although the standard error of these prediction limits is greater than that of the estimated $\hat{Y}_i$ in step **5**, these limits (also shown in Figure 14.14) are narrower than those found in step **6** because of the greater number of degrees of freedom and consequent decrease in magnitude of t.

as 1.58795, and the 95% confidence limits computed in step **4** are 60.5749–67.5651.

Next we estimate the mean survival $\hat{Y}_i$ for another value of the density, say $X_i = 59/g$. Note that this value of X_i was not one of the four employed in the original experiment. So long as density X is a continuous variable, we can estimate Y for any biologically reasonable value of X that does not involve undue extrapolation beyond the range of observed X-values. The estimate of $\hat{Y}_i$ for $X_i = 59/g$ can be obtained from the regression equation and yields a survival value of 57.2865°. In step **5** we calculate the *standard error of the estimated mean*, followed by the *confidence limits for the estimated mean* (53.1702–61.4028) in step **6**. We next predict the mean that we would obtain in another experiment. At density $59/g$ our best estimate for such a survival value is, of course, again the regression estimate of 57.2865°. However, the standard error is as shown in the last line of Box 14.2. Let us assume that the mean to be obtained in our next experiment will be based on ten replicates. In steps **7** and **8** we obtain the *standard error for the predicted mean* ($\hat{s}_{\bar{Y}} = 1.4751$) and *prediction limits* of 54.1003 and 60.4727, which in this unusual case are narrower than the confidence limits for the estimated mean obtained previously because of the greater number of degrees of freedom of the standard error and consequent decrease of t. Confidence and prediction bands around the regression line are shown in Figure 14.14.

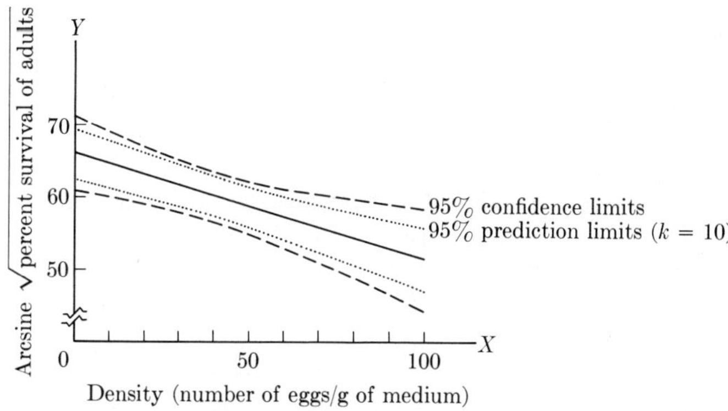

FIGURE 14.14 Confidence limits to regression estimates for data of Figure 14.13. Note the unusual circumstance in which the confidence limits are outside the prediction limits. This is so because in this problem the latter are based on substantially more degrees of freedom.

Finally, in Box 14.6 we have an example of equal sample sizes of Y-values for each value of X. As it happens, these data also involve the effect of density—this time on weight of brown trout grown at different densities. From each density, 25 trout have been sampled to yield the data presented here. The preliminary computations are familiar from single classification anova with equal sample sizes (Box 9.4) and need not be reviewed. The analysis of variance indicates a significant difference in body length among the four densities. We therefore proceed to test for linear regression, going through computational steps **8** to **15** in a manner similar to that of Box 14.4. The formulas are somewhat simplified because of the equality of sample sizes. Note that the sum of products is again negative, yielding yet another negative regression. The reader may begin to believe that all sums of products and regression slopes are negative, but this is only an accident of choice of the examples considered so far. In later examples in this chapter and in the homework exercises, positive regression coefficients will be encountered. The completed anova table is shown in Box 14.6. Deviations from regression are not significant, but, following the rules of Box 10.2, the mean square is not pooled with the error mean square. We therefore test the regression MS over the deviations MS and find that the variance ratio, though substantial, is not significant. Although inspection of the means (Figure 14.15) appears to reveal a clear relation between density and size of the fish, we cannot consider this as significant. If the establishment of such a relationship is important to us, it is clear that we need to obtain several more densities (X-values) to yield more degrees of freedom against which to test the regression. In view of the lack of significance of the findings, we need not proceed to calculate a regression equation, although the computation is shown in Box 14.6 to complete the computational outline. Significance tests for the various regression statistics could be carried out by using the formulas in Box 14.2, simplifying whenever possible to take advantage of the equality of sample sizes.

BOX 14.6

Computation of regression with more than one value of *Y* per value of *X*. Equal sample sizes.

Effect of crowding on weight of *Salmo trutta*, the brown trout. (X = number of fish per container; Y = weight in grams; n = 25 observations per group; a = 4 groups.) Original data not given here to save space.

	Density $= X$				
	175	*350*	*525*	*700*	*Sum*
$\sum^{n} Y$	87.806	92.755	73.170	63.490	317.221
$\sum^{n} Y^2$	355.91689	362.99238	233.21971	175.51487	1127.64385

Source: From more extensive data of J. Calaprice.

Preliminary computations

The computations are like those in Box 14.4 except that certain steps may be simplified because sample sizes are equal.

1. $\sum^{a}\sum^{n} Y = 317.221$

2. $\sum^{a}\sum^{n} Y^2 = 1127.64385$

3. $\frac{1}{n}\sum^{a}\left(\sum^{n} Y\right)^2 = \frac{87.806^2 + \cdots + 63.490^2}{25}$

$$= \frac{25{,}698.21266}{25} = 1027.92851$$

4. $CT_Y = \left(\frac{\sum^{a}\sum^{n} Y}{an}\right)^2 = \frac{\text{quantity } \mathbf{1}}{4(25)} = 1006.29163$

5. SS_{total} = quantity **2** − quantity **4** = 121.35222

6. SS_{groups} = quantity **3** − quantity **4** = 21.63688

7. SS_{within} = quantity **5** − quantity **6** = 99.71534

Complete the preliminary anova table in the usual manner.

Anova table

Source of variation	*df*	*SS*	*MS*	F_s
Among densities	3	21.63688	7.21229	6.944**
Within densities	96	99.71534	1.03870	
Total	99	121.35222		

BOX 14.6 continued

There is clearly evidence for differences among the densities. The following computations test for the existence of a linear relationship. The numbered quantities correspond to those identically numbered in Box 14.4.

8. $n\sum^{a} X = 25(175 + 350 + \cdots + 700) = 25(1{,}750) = 43{,}750$

9. $n\sum^{a} X^2 = 25(175^2 + 350^2 + \cdots + 700^2) = 25(918{,}750) = 22{,}968{,}750$

10. $\sum^{a} X\left(\sum^{n} Y\right) = 175(87.806) + \cdots + 700(63.490) = 130{,}687.550$

11. $CT_X = \dfrac{\left(n\sum^{a} X\right)^2}{an} = \dfrac{(\text{quantity } \mathbf{8})^2}{an} = \dfrac{(43{,}750)^2}{100} = 19{,}140{,}625$

12. $\sum x^2 = n\sum^{a} X^2 - CT_X = \text{quantity } \mathbf{9} - \text{quantity } \mathbf{11}$

$= 22{,}968{,}750 - 19{,}140{,}625 = 3{,}828{,}125$

13. $\sum xy = \sum^{a} X\left(\sum^{n} Y\right) - \dfrac{\left(n\sum^{a} X\right)\left(\sum^{a}\sum^{n} Y\right)}{an}$

$= \text{quantity } \mathbf{10} - \dfrac{\text{quantity } \mathbf{8} \times \text{quantity } \mathbf{1}}{an}$

$= 130{,}687.550 - \dfrac{(43{,}750)(317.221)}{100}$

$= 130{,}687.550 - 138{,}784.1875 = -8096.6375$

14. $\sum \hat{y}^2 = \dfrac{(\sum xy)^2}{\sum x^2} = \dfrac{(\text{quantity } \mathbf{13})^2}{\text{quantity } \mathbf{12}}$

$= \dfrac{(-8096.6375)^2}{3{,}828{,}125} = 17.12471$

15. $\sum d^2_{Y\cdot X} = SS_{\text{groups}} - \sum \hat{y}^2 = \text{quantity } \mathbf{6} - \text{quantity } \mathbf{14}$

$= 21.63689 - 17.12471 = 4.51218$

Completed anova table with regression

Source of variation	*df*	*SS*	*MS*	F_s
$\bar{Y} - \bar{\bar{Y}}$ Among densities	3	21.63688	7.21229	6.944**
$\hat{Y} - \bar{\bar{Y}}$ Linear regression	1	17.12471	17.12471	7.590 *ns*
$\bar{Y} - \hat{Y}$ Deviations from regression	2	4.51218	2.25609	2.172 *ns*
$Y - \bar{Y}$ Within groups	96	99.71534	1.03870	
$\bar{Y} - \bar{\bar{Y}}$ Total	99	121.35222		

Testing MS of deviations from regression, we find it to be not significant. The rules of Box 10.2 indicate that we should *not* pool. Therefore, we test regression MS over deviations MS, which is not significant. Thus, although the means are

BOX 14.6 continued

not all equal, we have no evidence that a linear regression exists or that the deviations from regression are significant.

Although we would not ordinarily proceed to computations of a regression line when the results are not significant, we do so here to complete the computational outline for this box.

16. Regression coefficient $= b_{y \cdot x} = \dfrac{\sum xy}{\sum x^2} = \dfrac{\text{quantity } \mathbf{13}}{\text{quantity } \mathbf{12}}$

$$= \frac{-8096.6375}{3{,}828{,}125} = -0.00211504$$

17. Y-intercept $= a = \bar{\bar{Y}} - b\bar{X}$

$$= \frac{\text{quantity } \mathbf{1}}{an} - \text{quantity } \mathbf{16} \left(\frac{\text{quantity } \mathbf{8}}{an}\right)$$

$$= \frac{317.221}{100} - \frac{(-0.00211504)(43{,}750)}{100}$$

$$= 3.17221 + 0.00211504(437.5) = 4.09754$$

The regression line $\hat{Y} = 4.0975 - 0.0021150X$ is fitted to the data points in Figure 14.15.

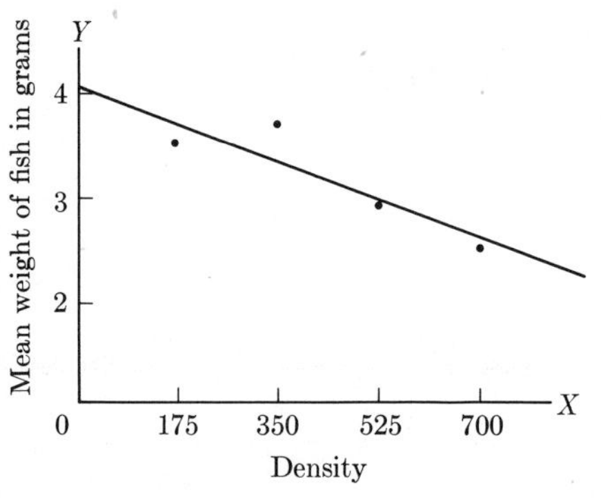

FIGURE 14.15 A regression line fitted to the data of Box 14.6. Although regression on density is not significant in this instance, the line has been fitted nevertheless for illustrative purposes.

14.7 The uses of regression

We have been so busy learning the mechanics of regression analysis that we have not had time to give much thought to the various applications of regression. We shall take up five more or less distinct applications in this section. All are discussed in terms of Model I regression. When applying these techniques you must be careful that any given case is in fact one in which the independent variable is measured without error. From a logical and philosophical point of view, some of the applications below would also fit a Model II case. However, the mathematical assumptions underlying the methodology would not be satisfied.

First, we might mention the *study of causation*. If we wish to know whether variation in a variable Y is caused by changes in another variable X, we manipulate X in an experiment and see whether we can obtain a significant regression of Y on X. The idea of causation is a complex, philosophical one that we shall not go into here. You have undoubtedly been cautioned from your earliest scientific experience not to confuse concomitant variation with causation. Variables may vary together, yet this covariation may be accidental or both may be functions of a common cause affecting them. The latter cases are usually Model II regression with both variables varying freely. When we manipulate one variable and find that such manipulations affect a second variable, we generally are satisfied that the variation of the independent variable X is the cause of the variation of the dependent variable Y (not the cause of the variable!). However, even here it is best to be cautious. When we find that heartbeat rate in a cold-blooded animal is a function of ambient temperature, we may conclude that temperature is one of the causes of differences in heartbeat rate. There may well be other factors affecting rate of heartbeat. A possible mistake is to invert the cause-and-effect relationship. It is unlikely that anyone would suppose that heartbeat rate affects the temperature of the general environment, but we might be mistaken about the cause-and-effect relationships between two chemical substances in the blood, for instance. If these two are related by a feedback relationship, then administration of chemical A, the precursor, might increase production of chemical B, the product. However, injection of chemical B might inhibit production of the precursor. If we did not know which was the precursor and which the product, we might be confused about cause-and-effect relationships here. Despite these cautions, regression analysis is a commonly used device for screening out causal relationships. While a significant regression of Y on X does not prove that changes in X are the cause of variations of Y, the converse statement is true. When we find no significant regression of Y on X, we can in all but the most complex cases infer quite safely (allowing for the possibility of type II error) that deviations of X do not affect Y.

The *description of scientific laws* and *prediction* are a second general area of application of regression analysis. Science aims at mathematical description of relations between variables in nature, and Model I regression analysis permits us to estimate functional relationships between variables, one of which is subject to error. These functional relationships do not always have clearly interpretable biological meaning. Thus, in many cases it may be difficult to assign a biological interpretation to the statistics a and b, or their corresponding parameters α and β. When we do so, we speak of a *structural mathematical model*, one whose component parts have clear scientific meaning. However, mathematical curves that are not structural models are also of value in science. Most regression lines are *empirically fitted curves*, in which the functions simply represent the best mathematical fit (by a criterion such as least squares) to an observed set of data. The constants necessary to fit these curves do not possess any clear inherent meaning. Biologists generally do not look with favor upon such empirically fitted curves. Yet when they adequately and precisely describe the relations between natural phenomena, they are of real value as temporary devices until enough insight into the phenomena is obtained to postulate a hypothesis from which a structural model can be constructed. Such models are preferred for a number of reasons, although they may actually not be markedly better predictors.

Examples from the history of science are not hard to find. Good empirical formulas for the motions of the heavenly bodies were available in the time of Kepler and Copernicus; the Newtonian ideas on gravitation altered the nature of the formulas but improved the prediction only relatively mildly. Subsequent improvements in the structural formulas have continued to lessen the error of prediction. A biological analogue might be the use for many years of heart-muscle extract as a simulated antigen in the Wassermann test for syphilis. There seems no logical or biological reason why such an extract should provide an antigen for a spirochaete-induced disease. Yet, the method proved to be reasonably reliable. It is with the same sort of philosophy that we often use regression equations to describe scientific laws, knowing full well that the structure of the equations does not represent the workings of nature. Regression is useful in establishing such prediction equations, preferably simple, but if necessary more complex, as we progress beyond simple linear regression into curvilinear and multiple regression. These equations will not only predict values of Y for given values of X, but will also estimate the error in these predictions.

Comparison of dependent variates is another application of regression. As soon as it is established that a given variable is a function of another one, as in Box 14.6 in which we found weight of trout to be a function of density, one is bound to ask to what degree any observed weight difference between two trout or two samples of trout is a function of the density at which they have been raised. It would be unfair to compare a trout raised at very high density (and expected to be relatively small) with one raised under optimal conditions of low density. This is the same point of view that makes us disinclined to compare the mathematical knowledge of a fifth-grader with that of a college student. Since we could undoubtedly obtain a regression of mathematical knowledge on years of schooling in mathematics, we should be comparing how far a given individual deviates from his expected value based on such a regression. Thus, relative to his classmates and age group, the fifth-grader may be far better than is the college student relative to his peer group. This suggests that we calculate *adjusted* Y*-values* that allow for the magnitude of the independent variable X. A conventional way of calculating such adjusted Y-values is as the overall mean of the population $\bar{Y}$ plus the deviation $d_{Y \cdot X} = (Y_i - \hat{Y}_i)$ of the given Y-value Y_i from the regression line. As you know, such deviations may be either positive or negative; therefore adjusted Y-values can be above or below the mean. We shall formally define an adjusted Y-value as

$$Y_{\text{adj}} = \bar{Y} + d_{Y \cdot X} = Y - bx \tag{14.8}$$

Let us apply this formula to the data of Table 14.1. Comparing the weight loss in the *Tribolium* sample raised at 29.5% relative humidity ($Y = 6.67$ mg) and that at 62.5% relative humidity ($Y = 5.83$ mg), we are led to conclude that weight loss in the more humid environment was less by 0.84 mg. The estimated values $\hat{Y}$ at these two relative humidities bear this out. According to our regression line, we expect 7.1338 mg weight loss at 29.5% R.H., while at 62.5% R.H. we expect a loss of 5.3776 mg. Thus, the expected difference should be greater (1.76 mg). To calculate adjusted Y-values for these two relative humidities, we need the following quantities:

X_i	29.5	62.5	% R.H.
Y_i	6.67	5.83	mg
$\hat{Y}_i$	7.1338	5.3776	mg
$d_{Y \cdot X}$	−0.4638	0.4524	mg
$Y_{\text{adj}} = \bar{Y} + d_{Y \cdot X}$	5.558	6.474	mg

Remembering that $\bar{Y} = 6.022$, we obtain the adjusted Y-values as shown above. These lead to quite different interpretations than do the ordinary Y-values. We now find a higher adjusted weight loss at the higher humidity. What does this mean? It indicates that the actual sample used in the experiment at the higher humidity happened to deviate strongly above the regression line. It lost more weight than was expected. On the other hand, the sample at 29.5% relative humidity had a lower weight loss than expected. Thus, although overall weight loss was greater at the lower humidity, adjusted weight loss allowing for the regression relationship is quite the reverse for these two samples. Comparatively speaking, the sample of beetles tested at 62.5% R.H. lost more weight than the sample at the lower humidity.

In this manner, adjusted means are useful for making comparisons among dependent variates after allowing for differences among independent variates that affect them. As the standard error of an adjusted mean, we use the standard error of estimated Y for a given value of X_i, shown in the third row of Box 14.2. Sometimes we do not even bother to calculate adjusted Y. We can simply state $d_{Y \cdot X}$ and compare the unexplained deviations among the variates. A large and positive unexplained deviation means that this individual or sample lies considerably above expectation by regression; a negative value of $d_{Y \cdot X}$ indicates the converse. Since the deviations $d_{Y \cdot X}$ are in absolute measurement units, it may be desirable to divide them by $s_{Y \cdot X}$ to obtain standardized scores. These are quite frequently used in educational statistical research to provide an estimate in standard deviation units of an individual's relative standing with respect to his group, allowing for an independent variable such as age or amount of schooling.

Differences among adjusted Y-values are not necessarily scientifically meaningful. They may merely represent random error around the regression line—the resultant of unknown and probably unknowable factors differentiating the responses of the individuals concerned. However, such differences may also reflect biologically meaningful distinctions among the individuals leading to recognition of new and important causal factors. Statistical analysis cannot by itself distinguish between these alternatives. Further experimentation and analysis are required to reveal these new insights.

Statistical control is an application of regression that is not widely known among biologists and represents a scientific philosophy that is not well established in biology outside agricultural circles. Biologists frequently categorize work as either descriptive or experimental, with the implication that only the latter can be analytical. However, statistical approaches applied to descriptive work can, in a number of instances, take the place of experimental techniques quite adequately—occasionally they are even to be preferred. These approaches are attempts to substitute statistical manipulation of a concomitant variable for control of the variable by experimental means. An example will clarify this technique.

Let us assume that we are studying the effects of various diets on blood pressure in rats. We find that the variability of blood pressure in our rat population is considerable, even before we introduce differences in diet. Further study reveals that the variability is largely due to differences in age among the rats of the experimental population. This can be demonstrated by a significant linear regression of blood pressure on age. To reduce the variability of blood pressure in the population, we should keep the age of the rats constant. The reaction of most biologists at this point will be to repeat the experiment using rats of only one age group; this is a valid, commonsense approach, which is part of the experimental method. An alternative approach is superior in some cases, when it is not practical or too costly to hold the variable constant. We might continue to use rats of variable ages and simply record the age of each rat as well as its blood pressure. Then we regress blood pressure on age and use an adjusted mean as the basic blood pressure reading for each individual. We can now evaluate the effect of differences in diet on these adjusted means. Or we can analyze the effects of diet on unexplained deviations, $d_{Y \cdot X}$, after the experimental blood pressures have been regressed on age (which amounts to the same thing).

What are the advantages of such an approach? Often it will be impossible to secure adequate numbers of individuals all of the same age. By using regression we are able to utilize all the individuals in the population. The use of statistical control assumes that it is relatively easy to record the independent variable X and, of course, that this variable can be measured without error, which would be generally true of such a variable as age of a laboratory animal. Statistical control may also be preferable because we obtain information over a wider range of both Y and X and also because we obtain added knowledge about the relations between these two variables, which would not be so if we restricted ourselves to a single age group. Furthermore, in situations where the assumed additivity of treatment effects does not hold, this can only be learned over a range of age groups. Suppose that the effect of diet on blood pressure is noticeable only in older rats and not in younger ones. Clearly we would not discover this unless we did use rats of different age groups.

Regression permits us to predict what the variance of an organism would be under statistical or experimental control. Instead of s_Y^2, the previous variance of Y, we can now use the unexplained residual variance $s_{Y \cdot X}^2$. By regressing Y on X we increase the information per individual by $100 s_Y^2 / s_{Y \cdot X}^2 - 100\%$. This is very much like the relative efficiency in design discussed in Section 10.4. In calculating relative efficiency in regression, it is customary to take the ratio of total variance over unexplained variance.

Let us work out an actual example with the trout weights of Box 14.6. Since these data are structured into variance among densities and within densities, we have to be careful which variance to use here. The variance among densities was reduced by regression on density, and we should compare it with the unexplained mean square around regression, called in Box 14.6 "the mean square of deviations from regression." The relative efficiency is $RE = (MS_{\text{groups}} \times 100)/MS_{Y \cdot X} = (7.21229 \times 100)/2.25609 = 319.7\%$. You may recall that regression in the trout example was not significant, so we cannot be certain of the significance of the increase in efficiency here. If it were significant, we could improve the efficiency of our design by $319.7 - 100 = 219.7\%$ by recording the density at which the trout were raised, and by regressing weight on density. If information on density is easily available, this procedure would be well worthwhile. A similar increase in efficiency will result from using only trout reared at the previous mean density.

Finally, *substitution of variables* is a special application that may occasionally be useful. Suppose we are interested in a response variable that is very difficult or expensive to measure. Let us assume that we wish to measure blood pressure in mice, which might be quite complex with the equipment at our disposal. Let us assume, furthermore, that after much effort we have accurately measured the blood pressures of 25 mice of known, but differing ages. We now regress blood pressure on age and find that a substantial portion of the variance of blood pressure was in fact a function of age. Instead of continuing to measure blood pressure, we could simply record the ages of the mice and predict blood pressure from them. Obviously, this would not be as efficient as measuring the blood pressures directly. But if our initial experiment has been properly carried out, we might be able to predict to a satisfactory degree of accuracy what the blood pressures of mice should be, given their age distribution. We might, therefore, have to use a greater number of mice to predict blood pressure from age than if we actually measured the blood pressure, but it may be much more economical to record the ages of larger samples of mice than to measure the blood pressures of smaller samples. Whenever the dependent variable is difficult and costly to measure, you might consider the substitution of variables approach.

14.8 Estimation of X from Y

Occasionally we have a problem in which we know the value of Y for an individual and wish to estimate the corresponding value of X for it. You might think that this problem is quite simple and requires us only to reverse the regression equation, writing an equation of the type $\hat{X} = a' + b_{X \cdot Y}Y$, where $a \neq a'$ and $b_{X \cdot Y} = \sum xy / \sum y^2$. However, such an approach would be quite improper because our initial assumptions were that X is measured without error and Y is the dependent, random, and normally distributed variable. Unless this situation were reversed, it would not be legitimate to regress X on Y. The appropriate procedure is quite simple, although the computations (outlined in Box 14.7) are somewhat tedious.

Since $\hat{Y} = a + b_{Y \cdot X}X_i$, we can estimate X_i by rearranging this equation

---

BOX 14.7

Estimation of X from Y.

Data from Table 14.1 and Box 14.1.

The problem.—Given a weight loss reading of $Y_i = 7$ mg for a sample of 25 *Tribolium* beetles, what can we infer about the relative humidity under which they had been kept (assuming that the experimental setup was identical to that previously used)?

Since $\hat{Y}_i = a + b_{Y \cdot X}X_i$

$$\hat{X}_i = \frac{(\mathrm{Y} - a)}{b_{Y \cdot X}} = \frac{7.0 - 8.7037}{-0.05322} = 32.0124$$

The 95% confidence limits of this estimate are computed as follows:

We define a quantity D as

$$D = b_{Y \cdot X}^2 - t_{.05[n-2]}^2 s_b^2 = (-0.05322)^2 - (2.365)^2(0.0032561)^2 = 0.002773$$

where $n = 9$, the number of samples (of beetles) in the previous study (see Table 14.1); and another quantity H as

$$H = \frac{t_{.05[n-2]}}{D}\sqrt{s_{Y \cdot X}^2\left[D\left(1 + \frac{1}{n}\right) + \frac{(Y_i - \bar{Y})^2}{\sum x^2}\right]}$$

$$= \frac{2.365}{0.002773}\sqrt{0.08801\left[(0.002773)\left(1 + \frac{1}{9}\right) + \frac{(7.0 - 6.022)^2}{8301.3889}\right]}$$

$$= 852.8669\sqrt{0.08801[0.0030811 + 0.0001152]}$$

$$= 852.8669\sqrt{0.00028131} = 852.8669(0.016772) = 14.30428$$

BOX 14.7 continued

The 95% confidence limits are

$$L_1 = \bar{X} + \frac{b_{Y \cdot X}(Y_i - \bar{Y})}{D} - H$$

$$= 50.389 + \frac{(-0.05322)(7 - 6.022)}{0.002773} - 14.30428$$

$$= 50.389 - 18.76998 - 14.30428 = 31.619 - 14.30428 = 17.315$$

$$L_2 = \bar{X} + \frac{b_{Y \cdot X}(Y_i - \bar{Y})}{D} + H = 31.619 + 14.30428 = 45.923$$

Note that the limits are symmetrical about $\bar{X} + [b_{Y \cdot X}(Y_i - \bar{Y})/D]$, not about $X_i = (Y_i - a)/b_{Y \cdot X}$. If we wish to estimate X_i for a $\bar{Y}_i$ based on a sample of size n in a regression analysis with more than one Y-value per value of X (as in Box 14.4, for example), H becomes

$$H = \frac{t_{\alpha[\nu]}}{D} \sqrt{s^2_{Y \cdot X} \left[D \left(\frac{1}{n} + \frac{1}{a} \right) + \frac{(\bar{Y}_i - \bar{Y})^2}{\sum x^2} \right]}$$

where α is the significance level chosen, $s^2_{Y \cdot X}$ is the MS of deviations around regression (or a pooled MS), a is the number of groups in the anova, and ν is the number of degrees of freedom for $s^2_{Y \cdot X}$.

to yield $\hat{X}_i = (Y_i - a)/b_{Y \cdot X}$. In Box 14.7, we assume that we have a weight loss of 7 mg in a sample of 25 *Tribolium* beetles. We wish to estimate the relative humidity $\hat{X}_i$ at which these beetles had been kept, assuming that the experimental setup was the same as before. Following the formula just given, we estimate the relative humidity at 32.0124%. It is not appropriate to assign standard errors to such an estimate, but there is a method for providing it with confidence limits. Since computational formulas for these limits are rather unwieldy, we divide the computation into three steps. We first compute quantity D as defined in Box 14.7, then proceed to a second quantity H. The limits are relatively simple functions of D and H, as shown in the box. However, please note that these confidence limits are unusual in one respect—they are *not* symmetrical around the estimate $\hat{X}_i$, but are symmetrical around another value, $\bar{X} + [b_{Y \cdot X}(Y_i - \bar{Y})/D]$, which is close to $\hat{X}_i$ but not identical to it. By the way, all values of t in Box 14.7 are shown as $t_{.05}$ to provide 95% confidence limits. For other $100(1 - \alpha)\%$ confidence limits, you should use t_α. Note the wide confidence interval: 17.3–45.9% relative humidity after rounding (the estimate $\hat{X}_i$ should also be rounded to 32.0% R.H.).

Box 14.7 also provides a formula that is appropriate when the estimate of X is based on an initial study of replicated Y-values per single value of X and you have more than one value of Y from which to estimate $\hat{X}_i$.

This method of estimating X from Y, also called *inverse prediction*, is frequently applied in the statistical analysis of dosage-mortality problems in bioassay. Such a study involves a regression of cumulative mortalities of organisms on dosage of a substance. Thus dosage X_1 will cause a mortality of Y_1 percent; dosage $X_2 > X_1$, a mortality of $Y_2 > Y_1$ percent; $X_3 > X_2$, a mortality of $Y_3 > Y_2$ percent, and so forth. As one reaches a certain dosage, mortality is 100%; the entire sample of organisms is killed. Frequently, such data are transformed—the mortalities to a so-called probit scale (see Section 14.12) and the dosages to logarithms to make the regression of mortality on dosage linear. A common measure of the potency of the substance (or of the tolerance of the organisms) is the dosage required to kill 50% or 95% of the organisms. Such a point is called an LD_{50} or LD_{95} of the organisms, the 50% or 95% lethal dose. This is clearly an inverse prediction problem. Given a value of Y, namely 50% or 95% mortality, estimate a corresponding value of X, the dosage. Confidence limits to the estimate are set in the manner indicated above.

14.9 Comparison of regression lines

Often an investigator obtains two or more regression lines from similar data and wishes to know whether the functional relationships described by the regression equations are the same. For example, he may have established a regression of blood pressure on age in a sample of animals and may now wish to compare this regression equation with that in three other samples, each of which has been subjected to a different diet or drug. The basic design of such a test is that of the analysis of variance. There will be a samples representing the treatment groups and the control. There is one major new aspect, however. In previous analyses we encountered only one variable, Y. In this example Y would be the blood pressure. However, in addition, for each reading of Y we also have a reading of X, the age of the animal. Thus two separate analyses of variance are possible, one for each variable, and also a joint one analyzing the covariance between X and Y. Such an analysis in its complete form is called the *analysis of covariance* (abbreviated *ancova,* to match anova).

An analysis of covariance provides answers to several questions. First of all, it furnishes the regression coefficients of Y on X for each of the samples and tests whether the slopes of the several regression lines could have come from populations with the same slope. It tests whether the means of the dependent variable are significantly different among groups and whether this difference is due to differences in the independent variable among the groups. An ancova fits a common regression line to the group means of Y and tests whether there is significant heterogeneity among these means around this regression line. Finally, it compares this regression to a pooled regression of all the Y items on the X items. To elaborate on just the last of these ideas, it is not necessarily true that means will have the same regression slopes as the items on which they are based. An example from physical anthropology will make this clear. Within any one race, there is a positive (Model II) regression of the dimensions of most bones on each other. A large-headed person will have a large jaw. On the other hand, if we take means of various races, the racial differences in skeletal structure come to the fore and there may be no regression at all, or even negative regression, between the means of two structures over all human races. Thus, increasing cheekbones in the Mongoloid races do not necessarily imply increase in jaw size. Whenever two or more variables are studied in a sampling design appropriate to an analysis of variance, an analysis of covariance is possible and sometimes very informative. The complete ancova is beyond the scope of the present text, and we would refer the reader to Snedecor (1956, chapter 13), where a very readable account is presented. ↤+

In this chapter we shall only take up the first step in an analysis of covariance: testing two or more regression lines for homogeneity of slope. Why would we be interested in testing differences between regression slopes? We might find that different toxicants yield different dosage-mortality curves or that different drugs yield different relationships between dosage and response (see, for example, Figure 14.3). In the example presented for this section, genetically differing cultures yielded different responses to increasing density, an important fact in understanding the effect of natural selection in these cultures. The regression slope of one variable on another is as fundamental a statistic of a sample as is the mean or the standard deviation, and in comparing samples it may be as important to compare regression coefficients as it is to compare these other statistics.

We have deliberately chosen a rather involved example to permit you to review some of the things you have learned about regression in previous sections. The steps are explained extensively in Box 14.8, and only a few comments need be furnished here. The example is taken from a study of the effects of increasing densities of eggs (independent variable X) on survival to adulthood (dependent variable Y) in *Tribolium* beetles. This experiment was performed in cultures at five gene frequencies of b^+, the wild type allele of the *black* locus. Varying numbers of replicates, somewhat below 100, were used for each gene frequency. Some of the replicates were run at the same densities; therefore the methods of Box 14.4 for repeated Y-values for the same value of X had to be employed for the separate regression analyses at each gene frequency. The preliminary computations leading to the basic regression analyses are shown in the early portion of Box 14.8. From these, we learn that only gene frequency 0 has a significant added component in survival values due to differences among densities (groups), but a significant regression of survival on density could be demonstrated for the 0, 0.25, and 0.50 gene frequencies. Heterogeneity among the residual means around the regression line was observed at gene frequency 0, which showed marked departure from linearity in the response of survival to density. The remaining statistics

BOX 14.8

Tests for equality of slopes of several regression lines.

The effect of varying density on survival of beetles to adulthood. Angular transformations of proportion surviving to adulthood (Y) in *Tribolium castaneum* regressed on density of egg input (X). Beetles reared at five gene frequencies in 40 g of flour. These data had more than one replicate (Y-value) per density X and were initially analyzed in the manner of Box 14.4, separately for each gene frequency.

Preliminary computations

The voluminous original data are not shown here, but the basic statistics needed for further computations are given in the following table.

Basic statistics

	Gene frequencies of b^+				
	0.00	*0.25*	*0.50*	*0.75*	*1.00*
$\sum^{a}\sum^{n_i} Y$	2727.920	4849.780	4928.730	5030.490	3635.200
$\sum^{a}\sum^{n_i} Y^2$	116,000.4508	254,386.4652	271,743.4087	283,463.7025	208,178.3130
$\sum^{a}\frac{(\sum^{n_i} Y)^2}{n_i}$	114,411.4130	251,930.4305	268,492.1059	281,965.6767	206,987.9839
$\sum^{a} n_i$	68	94	91	90	64
a (no. of groups)	25	30	27	26	20
$\sum^{a} n_i X$	1684.000	2508.000	2215.000	2066.000	1171.000
$\sum^{a} n_i X^2$	48,310	75,262	60,367	53,384	25,497
$\sum^{a} X(\sum^{n_i} Y)$	62,996.700	127,344.100	118,406.550	115,036.440	66,441.970

Source: Data from Sokal and Sonleitner (1968).

The basic analyses of regression separately for each of the five gene frequencies of b^+ are supplied in the following table. The computations are explained in Box 14.4. Asterisks, representing significance values as conventionally employed, are appended to sums of squares. To conserve space, neither mean squares nor degrees of freedom are shown. Significance of the SS_{groups} (of Y) indicates that survival proportions differ among the various densities. If the explained SS is significant, then the regression of survival on density is significant for that gene frequency. Significance of the unexplained SS (tested over SS_{error}) indicates residual heterogeneity of means around the regression line, which could be due to added effects that influence various densities differentially. In this case, the heterogeneity of $\sum d^2_{Y \cdot X}$ at gene frequency 0 is due to nonlinearity of survival response to density.

Regression statistics (*computed by the methods of Box* 14.4)

	Gene frequencies of b^+				
	0.00	*0.25*	*0.50*	*0.75*	*1.00*
Σx^2	6606.235	8346.426	6452.440	5957.822	4071.359
Σxy	−4559.436	−2052.158	−1561.988	−441.253	−70.830
SS_{groups}	4976.891***	1713.770	1542.882	789.792	508.624
$\Sigma \hat{y}^2$	3146.793***	504.570***	378.122***	32.680	1.232
$\Sigma d^2_{Y \cdot X}$	1830.098*	1209.200	1164.760	757.112	507.392
SS_{within}	1589.038	2456.035	3251.303	1498.026	1190.329
$b_{Y \cdot X}$	−0.6902	−0.2459	−0.2421	−0.0741	−0.0174
$\bar{\bar{Y}}$	40.116	51.593	54.162	55.894	56.800
$\bar{\bar{X}}$	24.765	26.681	24.341	22.956	18.297
a (Y-intercept)	57.209	58.154	60.055	57.595	57.118

Test of equality amonq k regression coefficients

Compute the machine formula for the sum of squares among b's with $k - 1$ degrees of freedom. From Expression (14.10a),

$$SS_{\text{among } b\text{'s}} = \sum^{k} \left(\sum \hat{y}^2 \right) - \sum \hat{y}^2_{\text{pooled}}$$

The following required quantities are computed [see Expression (14.10)]:

BOX 14.8 continued

1. $\sum^{k}\left(\sum \hat{y}^2\right) = 3146.793 + \cdots + 1.232 = 4063.397$

2. $\sum^{k}\sum xy = (-4559.436) + \cdots + (-70.830) = -8685.665$

3. $\sum^{k}\sum x^2 = 6606.235 + \cdots + 4071.359 = 31{,}434.282$

4. $\sum \hat{y}^2_{\text{pooled}} = \dfrac{\left(\sum^{k}\sum xy\right)^2}{\sum^{k}\sum x^2} = \dfrac{(\text{quantity } \mathbf{2})^2}{\text{quantity } \mathbf{3}} = \dfrac{(-8685.665)^2}{31{,}434.282} = 2399.95227$

5. $\bar{b} = \dfrac{\sum^{k}\sum xy}{\sum^{k}\sum x^2} = \dfrac{\text{quantity } \mathbf{2}}{\text{quantity } \mathbf{3}} = \dfrac{-8685.665}{31{,}434.282} = -0.27631$

6. $SS_{\text{among b's}} = \sum^{k}\left(\sum \hat{y}^2\right) - \sum \hat{y}^2_{\text{pooled}} = \text{quantity } \mathbf{1} - \text{quantity } \mathbf{4} = 4063.397 - 2399.9523 = 1663.4447$

This sum of squares has $k - 1 = 5 - 1 = 4$ *df*.

7. $MS_{\text{among } b\text{'s}} = \dfrac{\text{quantity } \mathbf{6}}{(k-1)} = \dfrac{1663.4447}{4} = 415.8612$

8. $\bar{s}^2_{Y\cdot X}$, a weighted average $s^2_{Y\cdot X}$ for all groups $= \dfrac{\sum^{k}\sum d^2_{Y\cdot X}}{\left(\sum^{k} a - 2k\right)}$ (where a refers to the number of groups)

$$= \frac{(1830.098 + \cdots + 507.392)}{(128 - 10)} = \frac{5468.562}{118} = 46.3437$$

The degrees of freedom pertaining to this mean square are $\sum^{k} a - 2k = 118$. We can now display our results in the form of an anova table.

Source of variation	*df*	*SS*	*MS*	F_s
Among b's (variation among regressions)	4	1663.4447	415.8612	8.973**
Weighted average of deviations from regression (average varia-				

9. $F_s = \dfrac{MS_{\text{among } b\text{'s}}}{\bar{s}^2_{Y\cdot X}} = \dfrac{\text{quantity } \mathbf{7}}{\text{quantity } \mathbf{8}} = \dfrac{415.8612}{46.3437} = 8.973 \gg F_{.01[4,118]}$, since $F_{.01[4,120]} = 3.48$

We conclude that the five groups were not sampled from populations of equal slopes. The beetles differ in their survival response to density depending on their gene frequencies.

If the sample regression coefficients had been homogeneous we could have used $\bar{b}$, the average regression coefficient computed in step **5**, as an estimate of the common slope. Since we have evidence that the regression coefficients are not all equal, we are interested in a detailed analysis of the differences among the b's. As discussed in Section 9.6, there are two approaches that might be taken, a priori and a posteriori tests. The experimenter might wish to test certain hypotheses specified before the data were examined. For example, in the present case it is of interest to test whether the magnitude of the slopes is, perhaps, a function of the initial gene frequency in the population (this is tested below). Or one may wish to test for the difference in slopes of a particular pair of regression lines that the investigator has reason to believe may differ. The other approach tests all possible combinations of regression lines in order to ascertain which differ from which. An STP procedure (see Section 9.7) is given in Box 14.9.

Test to determine whether the differences among the regression coefficients can be explained by a linear regression of slope on the criterion variable of each class (in this case initial gene frequency).

$SS_{\hat{b}}$, the sum of squares explained by a linear regression of b on gene frequency (let Z stand for gene frequency), is

$$\sum \hat{b}^2 = \frac{\left[\sum^k \sum x^2(b - \bar{b})(Z - \bar{Z})\right]^2}{\sum^k \sum x^2(Z - \bar{Z})^2} = \frac{\left[\sum (b - \bar{b})z\right]^2}{\sum z^2}$$

where $\bar{b}$ is as defined above and $\bar{Z} = \sum^k Z \sum x^2 / \sum^k \sum x^2$.

The machine formulas are

$$\sum (b - \bar{b})z = \sum^k \left(Z \sum xy\right) - \frac{\left[\sum^k \left(Z \sum x^2\right)\right]\left[\sum^k \sum xy\right]}{\sum^k \sum x^2}$$

$$\sum z^2 = \sum^k \left(Z^2 \sum x^2\right) - \frac{\left[\sum^k \left(Z \sum x^2\right)\right]^2}{\sum^k \sum x^2}$$

BOX 14.8 continued

The necessary quantities for the following computational steps are furnished in the above table of regression statistics (remember that gene frequencies are labeled Z).

$$\sum^{k}\left(Z\sum xy\right) = 0.00(-4559.436) + \cdots + 1.00(-70.830) = -1695.80325$$

$$\sum^{k}\left(Z\sum x^2\right) = 0.00(6606.235) + \cdots + 1.00(4071.359) = 13{,}852.55200$$

$$\sum (b-\bar{b})z = \sum^{k}(Z\sum xy) - \frac{\left[\sum^{k}(Z\sum x^2)\right]\times \text{quantity } \mathbf{2}}{\text{quantity } \mathbf{3}} = -1695.80325 - \frac{(13{,}852.55200)(-8685.665)}{31{,}434.282} = 2131.82119$$

$$\sum^{k}\left(Z^2\sum x^2\right) = (0.00)^2(6606.235) + \cdots + (1.00)^2(4071.359) = 9557.39550$$

$$\sum z^2 = \sum^{k}\left(Z^2\sum x^2\right) - \frac{\left[\sum^{k}\left(Z\sum x^2\right)\right]^2}{\text{quantity } \mathbf{3}} = 9557.39550 - \frac{(13{,}852.55200)^2}{31{,}434.282} = 3452.81207$$

Therefore, $\sum \hat{b}^2 = SS_{\hat{b}} = \dfrac{[\sum (b-\bar{b})z]^2}{\sum z^2} = \dfrac{(2131.82119)^2}{3452.81207} = 1316.22037$

This SS has 1 degree of freedom.

The SS for difference among b's (quantity **6**) is 1663.4447 with 4 degrees of freedom. If we subtract from it $SS_{\hat{b}}$, we are left with $SS_{b\cdot Z} = 347.22433$, which is a SS with 3 degrees of freedom representing deviations from linear regression.

$$MS_{b\cdot Z} = \frac{347.22433}{3} = 115.741$$

$$F_s = \frac{MS_{b\cdot Z}}{\text{quantity } \mathbf{8}} = \frac{115.741}{46.3437} = 2.497$$

The critical F-value is $F_{\alpha[3,118]}$. We look up $F_{.05[3,120]} = 2.68$, showing that the deviations from linear regressions are not significant. Finally, testing the significance of $MS_{\hat{b}}$, we do not pool $MS_{b.Z}$ and $\bar{s}^2_{Y\cdot X}$ (following the rules of Box 10.2) and test

$$F_s = \frac{MS_{\hat{b}}}{MS_{b \cdot z}} = \frac{1316.22037}{115.741} = 11.37$$

The critical F-value is $F_{.05[1,3]} = 10.1$. $MS_{\hat{b}}$ is clearly significant. The regression slope is itself a linear function of the gene frequency.

F-test for difference between two regression coefficients

For $k = 2$ regression coefficients, we can simplify the test for differences among b's to

$$F_s = \frac{(b_1 - b_2)^2}{\dfrac{\sum x_1^2 + \sum x_2^2}{(\sum x_1^2)(\sum x_2^2)} \bar{s}^2_{Y \cdot X}}$$

where $\bar{s}^2_{Y \cdot X}$ is the weighted average $s^2_{Y \cdot X}$ from step **8** in the previous computation. Since there are only two groups we can write its formula as

$$\frac{\sum y_1^2 - \dfrac{(\sum xy)_1^2}{\sum x_1^2} + \sum y_2^2 - \dfrac{(\sum xy)_2^2}{\sum x_2^2}}{n_1 + n_2 - 4}$$

Compare F_s with $F_{\alpha[1, n_1+n_2-4]}$. Since there is a single degree of freedom in the numerator, $t_s = \sqrt{F_s}$.

necessary for the test of equality among the regression coefficients are shown in the second table of Box 14.8.

To test for equality among k regression coefficients, we need a sum of squares among regression coefficients, obtained by the formula

$$SS_{\text{among } b\text{'s}} = \sum^{k} [\sum x^2(b - \bar{b})^2] \tag{14.9}$$

where $\bar{b}$ is the pooled or common slope and is equal to

$$\bar{b} = \frac{\sum^{k} (b\sum x^2)}{\sum^{k} (\sum x^2)} = \frac{\sum^{k} (\sum xy)}{\sum^{k} (\sum x^2)}$$

The machine formula (presented without proof) for $SS_{\text{among } b\text{'s}}$ is

$$\sum^{k} \frac{(\sum xy)^2}{\sum x^2} - \frac{\left(\sum^{k} \sum xy\right)^2}{\sum^{k} \sum x^2} \tag{14.10}$$

$$= \sum^{k} (\sum \hat{y}^2) - \sum \hat{y}^2_{\text{pooled}} \tag{14.10a}$$

Following the outline in Box 14.8 we are led to the conclusion that the five groups were not sampled from populations with equal slopes. The layout in Box 14.8 can, of course, be adapted to simpler types of problems in which each sample for which a regression coefficient has been computed has only one Y per value of X. Such adaptations should be possible from the box without difficulty.

We now run a special a priori test *that is of interest in this particular example only*, although similar cases may occur in other problems. Since we know that the regressions are different, we might wish to test whether the regression slopes are themselves a function of the differences in the criterion variable defining each class. In this case this is gene frequency. This is done by calculating a linear regression of each value of b on the gene frequency of its sample. When we do that, we find a significant linear regression. This shows that as gene frequency increases, the slope decreases. By subtracting the sum of squares for linear regression of b on gene frequency from the sum of squares for differences among b's, we obtain a heterogeneity SS around the new regression line (of b on gene frequency), which when tested is shown to lack significance. A similar example might be regression of growth (Y) on time (X) at different temperatures. Growth rates ($b_{Y \cdot X}$) might differ and be linear functions of the temperatures.

Finally, at the end of Box 14.8 we show a simple test for the significance of the *difference between two regression coefficients*. This can be done either by a t-test or an F-test. The test shown there is an F-test.

In Box 14.9 we show an *a posteriori test of all k regression coefficients* to examine their differences for significance. In this special example, there is very little purpose in doing so because the a priori tests were adequate to interpret the data. However, we show how this could be done in cases where

BOX 14.9

A posteriori tests for differences among a set of regression coefficients by the Simultaneous Test Procedure.

Data from Box 14.8.

For a set of k regression coefficients, b, to be just significant at the α level of significance, the $SS_{\text{among } b\text{'s}}$ must be equal to or greater than

$$SS_{\text{CRIT}} = (k - 1)\bar{s}^2_{Y\cdot X}F_{\alpha[k-1, \sum^k a - 2k]}$$

where $\bar{s}^2_{Y\cdot X}$ is the weighted $s^2_{Y\cdot X}$ for all groups (quantity **8**) from Box 14.9 with $df = \sum^k a - 2k$. We may test all combinations and sets of b's and by comparing the sums of squares for the various contrasts to this critical SS, we can make decisions about the significance of each comparison. The probability of any Type I error at all (experiment-wise error rate) is α.

For the example in Box 14.8 the critical SS, using $F_{.01[4,118]} = 3.483$, is

$$SS_{\text{CRIT}} = (5 - 1) \times (46.3437) \times (3.483) = 645.660$$

We now proceed to a systematic testing of all sets of b's that could be significant.

A. *Sets of 5 b's*

This is, of course, the total set which is significant, since the $SS_{\text{among } b\text{'s}}$ computed in Box 14.8 is $1663.4447 > SS_{\text{CRIT}}$.

B. *Sets of 4 b's*

The second table in Box 14.8 gives the statistics $\sum \hat{y}^2$, $\sum xy$, $\sum x^2$ needed for the tests.

1. Set b_1, b_2, b_3, and b_4

$$\sum^4 (\sum \hat{y}^2) = 3146.793 + \cdots + 32.680 = 4062.165$$

$$\sum^4 \sum xy = (-4559.436) + \cdots + (-441.253) = 8614.835$$

$$\sum^4 \sum x^2 = 6606.235 + \cdots + 5987.822 = 27{,}362.923$$

$$SS = 4062.165 - [(8614.835)^2/27{,}362.923]$$
$$= 1349.904 > SS_{\text{CRIT}}$$

There are significant differences among b_1, b_2, b_3, and b_4.

2. Test of last four b's (b_2, b_3, b_4, and b_5)

$$\sum^4 \sum \hat{y}^2 = 504.570 + \cdots + 1.232 = 916.604$$

$$\sum^4 \sum xy = (-2052.158) + \cdots + (-70.830) = -4126.229$$

$$\sum^4 \sum x^2 = 8346.426 + \cdots + 4071.359 = 24{,}828.047$$

$$SS = 916.604 - [(-4126.229)^2/24{,}828.047]$$
$$= 230.857 < SS_{\text{CRIT}}$$

This is not significant. Therefore further tests need not be made on subsets of b_2, b_3, b_4, and b_5.

BOX 14.9 continued

C. *Between b_1 and other b's*

The only tests left to be made are between b_1 and the various combinations of the other b's. For the following tests the computation is identical, so only the sums of squares are given.

1. b_1, b_2, and b_3

$$SS = 4029.483 - [(-8173.582)^2/21{,}405.101] = 908.384 > SS_{\text{CRIT}}$$

2. b_1, b_2

$$SS = 3651.363 - [(-6611.594)^2/14{,}952.661] = 727.925 > SS_{\text{CRIT}}$$

Though further tests could be made (1, 3; 1, 4; 1, 2, 4; 1, 3, 4), it is clear at this point that b_2, b_3, b_4, and b_5 are not significantly different from one another and b_1 is significantly different from all the others.

no a priori tests might suggest themselves, such as regressions induced in different samples by qualitatively different treatments (no criterion variable), and where it would still be important to screen the response curves for parallelisms and differences. The method employed is the simultaneous test procedure (*STP*), and the computations are quite simple and completely laid out in Box 14.9.

14.10 Linear comparisons in anova

In previous sections we have already noted the close relationship between regression and analysis of variance. This will become even more obvious in the present section. Our discussion here relates to a priori tests in analysis of variance, a line of analysis introduced in Section 9.6. You will recall that we were able to subdivide the sum of squares and the degrees of freedom among groups into separate sums of squares, each based on a single degree of freedom and furnishing certain desired information about the overall differences in the analysis of variance. The sum of squares among groups (treatments) could be decomposed into a set of orthogonal single degree of freedom comparisons—meaning that their *SS* and *df* were independent of each other and additive (added up to the *SS* and *df* among groups). Other individual degree of freedom comparisons were nonorthogonal, one comparison cutting across the other, so that their sums of squares did not add up to the among groups *SS*. An important idea due to R. A. Fisher is that all degrees of freedom and sums of squares in an anova can eventually be partitioned into single degrees of freedom with corresponding sums of squares that can be identified with certain contrasts. Not all of these contrasts will be scientifically meaningful, and it is generally not practicable to carry out so complete a decomposition of the sums of squares, yet the general idea has considerable attraction and, as we shall see in this section and the next, we are able in many cases to probe rather deeply into the nature of the treatment and interaction sums of squares by means of individual degrees of freedom comparisons.

We shall now provide a more general method, related to regression, for performing such individual degree of freedom comparisons, and at the same time lay the groundwork for the study of orthogonal polynomials taken up in the next section. If you are by now vague about the subject matter of Section 9.6, we urge you to reread it before proceeding.

In Section 14.6 on replicated Y's per single value of X, we studied a single classification anova in which survival of *Tribolium* beetles was regressed on density. The density classes were the groups of the anova, but at the same time these classes had numerical values of their own. Thus one density was $5/g$, another $20/g$, and so forth. It was therefore quite natural to consider them as an independent variable X and to regress the survival values Y on density X. However, we can also carry out a regression of response variable Y in an anova where the groups as such are not expressly quantified. We do this by employing dummy variables to symbolize the groups and then regressing the response means $\bar{Y}_i$ on these dummy variables. This sounds like a futile exercise, but the establishment of a significant regression of Y on a carefully chosen dummy variable will yield important analytical insights. An example will clarify this point.

Table 14.2 lists the treatment sums of the data from Box 9.4, which

TABLE 14.2

Means and treatment sums from data of Box 9.4. Length of pea sections grown in tissue culture. Sample size: $n = 10$. Coefficients c_{ij} for two linear comparisons given below means.

	(1)	*(2)*	*(3)*	*(4)*	*(5)*
	Control	*2% glucose*	*2% fructose*	*1% glucose + 1% fructose*	*2% sucrose*
$\bar{Y}$	70.1	59.3	58.2	58.0	64.1
$\sum^{n} Y$	701	593	582	580	641
Comparisons					
Control *vs* treatments (c_{i1})	+4	−1	−1	−1	−1
Mixed *vs* pure sugars (c_{i2})	0	−1	−1	+3	−1

are lengths of pea sections grown in tissue culture with different sugar solutions. This is the example for which we provided an a priori comparison among means in Section 9.6. Each treatment is based on the same sample size, $n = 10$. Underneath the sums of Table 14.2, there are two rows of single digits, the dummy variables, all *coefficients of linear comparisons.* Symbolism for these coefficients is even less standardized than that of most other statistics. We follow two recent texts in labeling them c_{ij}. The subscript i refers to the group $(1, \ldots, a)$ in the anova, while j identifies the particular linear comparison $(1, \ldots, a - 1)$.

How are these coefficients chosen? For purposes of convenience, they are generally expressed as integers though they may be fractional. For any one comparison, the coefficients must sum to zero; $\sum_{i=1}^{a} c_{ij} = 0$. The purpose of the first row of coefficients is to test whether the control is different from the mean of the four treatment groups. Note that the four treatment groups are given a minus sign, and the control group is given a plus sign. This is done to indicate the contrast. To obtain the mean of the four treatment groups we have to compute $\frac{1}{4}(\bar{Y}_2 + \bar{Y}_3 + \bar{Y}_4 + \bar{Y}_5)$. Thus, each treatment mean is weighted by a coefficient of $\frac{1}{4}$ as contrasted with the coefficient 1 for the control. To avoid fractions we multiply the coefficients by 4.

An easy way of working out the coefficient to be assigned to each sum in a linear comparison is to assign it the integer representing *the number of samples in the contrasted set.* Thus in this example there is a control set and a sugar set. The control set has one sample, the sugar set has four samples. Therefore assign a coefficient 4 to the control set and a coefficient 1 to each member of the sugar set. The signs of the two sets must be unlike, but computationally it makes no difference whether the plus is assigned to the 1's or to the 4. To reinforce what we have just learned, before explaining how to use these dummy variables, let us work out the second row of coefficients in Table 14.2. We now wish to contrast mixed versus pure sugars, that is, to test the mean of the three pure sugars against the single mixed sugar, leaving out the control group altogether. Since the control group does not enter in this comparison, we assign it a value $c_{1,2} = 0$. The comparison is between $\frac{1}{3}(\bar{Y}_2 + \bar{Y}_3 + \bar{Y}_5)$ versus $\bar{Y}_4$. By the rule of thumb given above, the single mixed sugar is assigned a coefficient 3 and the pure sugars each a coefficient 1, since the comparison is between a set of three pure sugars and a set of one mixed sugar. Again, we assign negative signs to the 1's and a plus sign to the 3, but we could just as well switch the signs around.

The significance tests for these differences, which in Section 9.6 were carried out as conventional analyses of variance, are now carried out as regressions on these dummy variables. Corresponding to the conventional formula for the explained sum of squares $(\sum xy)^2/\sum x^2$, we can write an explained SS for regression of $\bar{Y}$ on the dummy variable, $(\sum^{a} c\bar{y})^2/\sum^{a} c^2$, omitting the subscripts from c_{ij} for simplicity's sake. Note that the dummy variable c is already written in lower case; it is a deviate as well as a variate, since its sum and mean are zero. The computational formula for an explained SS is

$$(\sum xy)^2/\sum x^2 = \left[\sum XY - \frac{1}{n}\sum X \sum Y\right]^2 \Big/ \left[\sum X^2 - \frac{1}{n}(\sum X)^2\right]$$

If we substitute $\bar{Y}$ for Y and c for X in this expression and ignore all terms that would involve $\sum^{a} c$, we can simplify the explained SS for regression of Y on the dummy variable as follows:

$$\left(\sum^{a} c\bar{Y}\right)^2 \Big/ \sum^{a} c^2 = \left(\sum^{a} c \sum^{n} Y\right)^2 \Big/ n^2 \sum^{a} c^2$$

The final step is to multiply the sum of squares times n because we have to make it equivalent to a sum of squares of *groups*, not of *means*, in order to make it compatible with the rest of the anova of which it is a part. Therefore the final formula for the explained SS based on a linear comparison is

$$\left(\sum^{a} c \sum^{n} Y\right)^2 \Big/ n \sum^{a} c^2 \tag{14.11}$$

Now we apply Expression (14.11) to the two comparisons we have planned in Table 14.2. We compute

$$[(4 \times 701) + (-1 \times 593) + (-1 \times 582) + (-1 \times 580)$$
$$+ (-1 \times 641)]^2 \div 10[4^2 + (-1)^2 + (-1)^2 + (-1)^2 + (-1)^2]$$
$$= [408]^2/200 = 837.32 \quad \text{and}$$
$$[(0 \times 701) + (-1 \times 593) + (-1 \times 582) + (3 \times 580) + (-1 \times 641)]^2$$
$$\div 10[0^2 + (-1)^2 + (-1)^2 + 3^2 + (-1)^2] = [76]^2/120 = 48.13$$

If you check back, you will find that these sums of squares for the difference between treatments and the control and between the pure and mixed sugars are the same as were found in Section 9.6. They are incorporated into the anova as was shown there and once more prove to be significant.

Why all this bother when we were able to carry out such an analysis by the more conventional method of obtaining sums of squares learned in the analysis of variance chapter? Although it took some time to explain how to evaluate a sum of squares from a linear comparison of treatments as shown above, the actual setup of the data and the computation are generally somewhat simpler than the corresponding techniques of Box 9.8, once the idea has been mastered. A major advantage is that the new technique permits a comprehensive view of all comparisons through the table of coefficients, which often helps to avoid inconsistencies. Furthermore, a table of coefficients permits an easy test for orthogonality of comparisons. You will recall (from Section 9.6) the distinction between orthogonal and nonorthogonal comparisons. The former add up to the treatment sum of squares; the latter do not. On the basis of their table of coefficients two comparisons are orthogonal if and only if the sum of the sample sizes times the products of their coefficients equals zero, or expressed as a formula, $\sum_{i=1}^{a} n_i c_{ij} c_{ik} = 0$. The subscripts j and k refer to two separate comparisons and i to the various groups. Let us test the two comparisons in Table 14.2, which we have just completed. Since n_i is constant ($= 10$) for all groups, this factor can be ignored.

$$(4 \times 0) + (-1 \times -1) + (-1 \times -1) + (-1 \times 3) + (-1 \times -1) = 0$$

It is obvious that these two comparisons are orthogonal, as already stated in Section 9.6. Finally, understanding the use of these coefficients will make our task of learning about curvilinear regression in the next section far easier.

Now, we shall go over some completely worked-out cases of individual degree of freedom comparisons illustrated in Box 14.10. Let us emphasize again that these are all a priori comparisons. The first example comes from Box 9.8 and is per diem fecundity in three selected lines of drosophila. The first test is of selected versus nonselected lines. The coefficients are 1 for lines R.S. as well as S.S. because there is one line in the contrasted set. However, line N.S. has a coefficient of 2, since there are two lines, R.S. and S.S., in its contrasted set. We arbitrarily assign a plus sign to one of the contrasted sets and minus to the other. In testing resistant versus susceptible lines, we set R.S. $= +1$ and S.S. $= -1$ and put a coefficient of zero for the nonselected line, which is not affected by this test. The computations are straightforward and outlined in Box 14.10. The results are the same as in Box 9.8. We find a significant difference in fecundity due to selection, but no difference in fecundity between the two selected lines.

The next analysis in Box 14.10 is a more extensive study in which the number of abdominal bristles in eight strains of houseflies differing in *DDT*-resistance is analyzed. The strains are arranged in four groups differing in their degree of resistance to the toxicant. This study was undertaken in the hope that a morphological correlate such as bristle number could be found for *DDT*-resistance, which could be used (by inverse prediction) to indicate the resistance status of a housefly population. This is a nested analysis of variance in which differences among jars within strains were tested as well as differences among strains. However, we see from the overall analysis of variance that only differences among strains were highly significant; the number of bristles was not affected by environmental variations in the culture medium among different jars. Following the rules in Box 10.2 we do not pool the two mean squares but test the linear comparisons over the mean square of jars within strains. We now wish to partition the *SS* among strains based on seven degrees of freedom into separate single degree of freedom comparisons to learn more about the differences existing in this study. For example, the first comparison suggested by the design of the study is resistant versus nonresistant strains—that is, groups 1 and 2 versus groups 3 and 4. The coefficients for this comparison are 5 for members of groups 1 and 2 because there are five strains comprising the contrasted set, groups 3 and 4; and, conversely, 3 for members of groups 3 and 4, because there are three strains in groups 1 and 2. Similar considerations govern the choice of coefficients for the other comparisons. The computations at the right side of the table are no different than before and are not discussed in detail.

We find that resistant flies differ from nonresistant ones in bristle number, that the very resistant group 1 differs from the slightly resistant group 2, and that the latter group is heterogeneous, since strain *LDD* is significantly different from strain *RKS*. The difference between group 3 and group 4 is not significant. The investigator had no logical basis for designing contrasts within group 3, consisting of three strains responsible for two degrees of freedom among them. He therefore calculated the sum of squares pertaining to the two degrees of freedom among the three strains in the conventional, previously learned manner, as shown in a line beneath the main table in Box 14.10. The mean square among the three strains of group 3 is not significant. Finally, we find no difference between the two strains of group 4. We may conclude that selection for resistance has affected abdominal bristle number variously in different strains, apparently raising it most in the highly resistant strain, while among the nonresistant strains there are no differences for this variable. In the last line of the table of housefly comparisons, there is yet another contrast that clearly cannot be orthogonal to the previous ones, since these already used up the available 7 degrees of freedom. This is the contrast between groups 2 and 3. Having already divided the data into groups 1 and 2 versus groups 3 and 4, a contrast between groups 2 and 3 can no longer be orthogonal. We can easily convince ourselves of this fact by accumulating the product of the coefficients of the first row with those of the last row: $\sum c_{i1}c_{i6} = (5 \times 0) + (5 \times 3) + (5 \times 3) + (-3 \times -2) + (-3 \times -2) + (-3 \times -2) + (-3 \times 0) + (-3 \times 0) = 48$. They do not sum to zero, and hence are not orthogonal. However, we may still be interested in testing such a contrast. If so, the computations are carried out as before and we find that the mean square is significant at the 5% level. Thus, there is a suggestion of a significant difference between groups 2 and 3. Although it is generally desirable to work only with orthogonal comparisons and there is, furthermore, some question about the validity of the significance tests when nonorthogonal comparisons are permitted, the final arbiter in these matters should still be the investigator, who should know best which tests are of interest to him.

The last example in Box 14.10 shows the minor complications that arise when sample sizes are unequal. The data are length of larval period in three strains of *Drosophila melanogaster*, one selected for short larval period (*SL*), one for long larval period (*LL*), and a control strain (*CS*). The sample sizes and sums are given in the table. To carry out the computations for such a case, it is simplest first to write down the "unweighted" coefficients such as would be written were the sample sizes the same. This gives us a general idea of the contrasts we would like to compute. Note that the first contrast is of strain *SL* versus strain *LL*, to test the overall effect of selection. If the strains selected for short and long larval period were not even different, selection clearly would have been ineffective. The second comparison tests the symmetry of the results of selection. If selection for short larval period had been as effective as selection for long larval period, then the control (unselected)

BOX 14.10

A priori individual degree of freedom comparisons of means in analysis of variance.

Per diem fecundity in *Drosophila melanogaster*. Data from anova of Box 9.8 (equal n): $a = 3$; $n = 25$.

Comparisons	$\sum^{n} Y$	*Resistant* (R.S.) *Line* 631.4	*Susceptible* (S.S.) *Line* 590.7	*Nonselected* (N.S.) *Line* 834.3	*(1)* $\sum^{a} c \sum^{n} Y$	*(2)* $\sum c^2$	*(3)* $n\sum c^2$	*(4)* $\sum \hat{y}^2 = (1)^2/(3)$
Selected *vs* nonselected	c_{i1}	+1	+1	−2	−446.5	6	150	1329.08
Resistant *vs* susceptible	c_{i2}	+1	−1	0	+40.7	2	50	33.13

Anova table

Source of variation	*df*	*SS*	*MS*
Strains	2	1362.21	681.10***
Selected *vs* nonselected	1	1329.08	1329.08***
Resistant *vs* susceptible	1	33.13	33.13 *ns*
Error	72	5659.02	78.60

Computational steps

See text (Section 14.10) for method of assignment of correct coefficients of linear comparison.

1. $\sum^{a} c \sum^{n} Y = (1 \times 631.4) + (1 \times 590.7) + (-2 \times 834.3) = -446.5$ and $(1 \times 631.4) + (-1 \times 590.7) = 40.7$

2. $\sum c^2 = (1)^2 + (1)^2 + (-2)^2 = 6$ and $(1)^2 + (-1)^2 = 2$

3. $n \sum c^2 = 25(6) = 150$ and $25(2) = 50$

4. $\sum \hat{y}^2 = (\text{quantity } \mathbf{1})^2/\text{quantity } \mathbf{3} = (-446.5)^2/150 = 1329.08$ and $(40.7)^2/50 = 33.13$

Partitioning of degrees of freedom in a larger example (degrees of freedom for strains not completely subdivided).

Abdominal bristle number in housefly strains. Strains are subdivided on the basis of resistance to DDT: $a = 8$ strains; $n = 24$ flies per strain.

		Strains											
		Group 1 Strongly resistant	*Group 2 Slightly resistant*		*Group 3 Normal*			*Group 4 Susceptible*		*(1)*	*(2)*	*(3)*	*(4)*
Meaningful comparisons		OL	LDD	RKS	RH	LC	BS	NKS	NH	$\sum^a c \sum^n Y$	$\sum c^2$	$n\sum c^2$	$\sum \hat{y}^2 = (1)^2/(3)$
	$\sum^n Y$	822	651	790	705	659	648	732	661				
Resistant *vs* nonresistant strains	c_{i1}	+5	+5	+5	−3	−3	−3	−3	−3	1100	120	2880	420.14
Group 1 *vs* group 2	c_{i2}	+2	−1	−1	0	0	0	0	0	203	6	144	286.17
LLD *vs* RKS (within group 2)	c_{i3}	0	+1	−1	0	0	0	0	0	−139	2	48	402.52
Group 3 *vs* group 4	c_{i4}	0	0	0	+2	+2	+2	−3	−3	−155	30	720	33.37
Among strains of group 3	(*SS* for 2 *df*; compute in conventional manner as shown below)												
NKS *vs* NH (within group 4)	c_{i5}	0	0	0	0	0	0	+1	−1	71	2	48	105.02
Group 2 *vs* group 3	c_{i6}	0	+3	+3	−2	−2	−2	0	0	299	30	720	124.17*

Among strains of group 3 $$SS = \frac{(705)^2 + (659)^2 + (648)^2}{24} - \frac{(705 + 659 + 648)^2}{72} = 76.19 \text{ (}MS\text{ significant at } P < 0.05\text{)}$$

Source: Data by Sokal and Hunter (1955).

BOX 14.10 continued

Anova table

Source of variation	*df*	*SS*	*MS*
Strains	7	1323.42	189.06***
Resistant *vs* nonresistant	1	420.14	420.14***
Group 1 *vs* group 2	1	286.17	286.17***
LLD *vs* RKS (within group 2)	1	402.52	402.52***
Group 3 *vs* group 4	1	33.37	33.37 *ns*
Among strains of group 3	2	76.19	38.10 *ns*
NKS *vs* NH (within group 4)	1	105.02	105.02 *ns*
Jars within strains	16	357.25	22.33 *ns*
Within jars	168	4663.25	27.76

Strains and all linear contrasts are tested over the mean square for jars within strains. Although this MS is not significant, following the rules of Box 10.2 we decided not to pool it with the mean square within jars.

Individual degree of freedom comparisons in an anova with unequal n

Length of larval period (in hours) in lines of *Drosophila melanogaster* selected for short (SL) and long (LL) larval period compared with the control strain (CS).

Comparisons		SL	CS	LL	*(1)* $\sum^{a} n_i c_i \sum^{n_i} Y$	*(2)* $\sum^{a} n_i c_i^2$	*(3)* $\sum \hat{y}^2 = \frac{(1)^2}{(2)}$
	n_i	80	69	33			
	$\sum^{n_i} Y$	8078	7291	3640			
Effect of selection (shortest *vs* longest; linear)	Unweighted c_{i1}	(−1	0	+1)			
	Weighted $n_i c_{i1}$	−33	0	+80	24,626	298,320	2032.85
Symmetry of selection (ends *vs* middle; quadratic)	Unweighted c_{i2}	(−1	+2	−1)			
	Weighted $n_i c_{i2}$	−69	+113	−69	15,341	1,419,054	165.85

Source: Data from Hunter (1959).

See text for assignment of correct coefficients of linear comparison. Computation as before except that step **3** is not necessary, since quantity **2**, $\sum^{a} n_i c^2_i$, is already weighted by n_i.

Anova table

Source of variation	*df*	*SS*	*MS*
Among lines	2	2198.70	1099.35***
Effect of selection	1	2032.85	2032.85***
Symmetry of selection	1	165.85	165.85*
Error	179	7055.25	39.41

strain should be exactly in the middle between the two. This can be tested by comparing the control strain with the average of the short and long strains. Having established the contrasts, we now replace the unweighted coefficients by others weighted according to sample sizes. The weighted coefficient written down for each member of one set is the *total sample size* of the contrasted set. Thus, for the first comparison, the weighted coefficient for SL is the sample size of the contrasted LL; conversely, for the LL it is the sample size of the contrasted SL. In the second comparison, the weighted coefficient for both SL and LL is 69, the sample size of the contrasted CS, and the weighted coefficient for CS is the sum of the sample sizes of the contrasted lines, SL and LL, $80 + 33 = 113$. The signs of these coefficients follow those of the unweighted coefficients. The rest of the computations are as before, except that step **3**, which consisted of multiplying n times the SS of the coefficients, is no longer necessary because the coefficients are already weighted by sample size. We learn from the anova table that there is a clear difference in length of larval period between the strains selected in opposite directions for this variable, and also that their distance from the nonselected control strain is not symmetrical.

We are now ready to consider orthogonal polynomials.

14.11 Orthogonal polynomials

In Figure 14.12 we saw that a linear regression is not always sufficient to account for differences among sample means. One often finds significant deviations around linear regression. You learned to test the significance of such deviations in Section 14.6. Figures 14.12,E and 14.12,F illustrate the difference in types of deviations around regression. The former appears to be a curvilinear trend; the latter seems to be added heterogeneity without any apparent regularity, which might arise from a situation such as the following. Suppose we measure the weight of four samples of mice, each sample representing a different age group. We would expect them to be of different weights. Yet, if we also fed the four samples different rations, we could expect the sample means to deviate significantly around the regression on age because of the differences in amount or quality of food eaten.

On the other hand, the relation between weight and age might not be linear; but could best be described by fitting a *curvilinear regression*, which usually is expressed as a polynomial function of the following general form:

$$\hat{Y} = a + bX + cX^2 + dX^3 + \cdots \tag{14.12}$$

Such an expression uses increasing powers of X, the independent variable, and a different regression coefficient preceding each power of X. Several points should be made regarding such a curvilinear regression equation. If it only involves terms of X and X^2, it will yield a parabola with one inflection point. As increasing powers of X are used, the curve becomes more and more complex and will be able to fit a given set of data increasingly well. However, with each added power of X, the mean square over which the regression MS must be tested loses another degree of freedom. Remember that the test for linear regression employs a mean square for deviations from regression based on $a - 2$ degrees of freedom. When there are only $a = 5$ means to be regressed, the highest order polynomial we can fit is a cubic, since a cubic polynomial would leave us with an MS for deviations with a single degree of freedom $(5 - 2 - 1 - 1 = 1)$. Even when there are numerous groups, fifth and even higher powers are fitted only occasionally. For most biological work, terms of X no higher than cubic are used.

Generally, such polynomial regressions are empirical fits. That is, in most cases, we cannot read structural meaning into the terms of X^2 and X^3. We simply use them to obtain a better fitting regression line to a set of points for any of the purposes mentioned in Section 14.7. Since our aim is to find the best fitting line, we first fit a linear regression to a set of data, then see if we can remove a significant portion from the residual sum of squares by adding a quadratic term to the regression equation. We do the same for the cubic and possibly for higher terms. Curve fitting is often, therefore, a stepwise procedure, with a significance test for each increase in powers of X to find out whether a significant improvement in fit has been made. Of course, if we had some a priori basis for expecting, say, a quadratic relationship, then we could proceed directly to test this. In carrying out such tests in a stepwise manner, we are handicapped by the fact that as we increase the number of terms in the polynomial the coefficients of the powers of X also change. To be precise in symbolism, we must write the first three polynomial regressions as follows:

$$\hat{Y} = a_0 + b_0X$$

$$\hat{Y} = a_1 + b_1X + c_1X^2$$

$$\hat{Y} = a_2 + b_2X + c_2X^2 + d_2X^3$$

where in most cases $a_0 \neq a_1 \neq a_2$, $b_0 \neq b_1 \neq b_2$, and $c_1 \neq c_2$. This makes the mechanics of fitting Y to successively higher terms of X very cumbersome, because each time the entire regression has to be recomputed from the beginning. You cannot simply add the next highest power of X.

The computation of curvilinear regression is quite tedious, and persons embarking upon the fitting of a higher-order polynomial to a set of data should be quite clear about their purposes. As part of the computation, one needs to solve a series of simultaneous equations, which for cubic or higher equations is generally done by a mathematical procedure known as matrix inversion and nowadays is carried out almost entirely on computers. For a second-degree polynomial (one involving only X and X^2), desk calculator methods are found in many texts. These techniques are extensions of multiple regression analysis, so some knowledge of that topic would be desirable before carrying out such computations. Steel and Torrie (1960) work out the second-degree case by formula in sections 16.4 and 16.5 of their book, and their account is probably most suitable for a reader without a background in multiple regression. Computation of third- and higher-degree regression equations is usually practical only on digital computers or by the method of orthogonal polynomials to be discussed below. Most computation centers have programs specifically for this purpose. But even if they do not, one can use a multiple regression analysis or a stepwise regression analysis program (both standard library programs) using X, X^2, X^3, and so on, as variables for input to the program. ↮

Since the fitting of increasing powers of X involves so much recomputation, statisticians have developed methods to circumvent this inconvenience. In lieu of Expression (14.12), they write another polynomial regression equation,

$$\hat{Y} = A + B\xi_1 + C\xi_2 + D\xi_3 + \cdots \tag{14.13}$$

where the capital letters are again constant coefficients analogous to regression coefficients, and the ξ_j's are coefficients of special *orthogonal polynomials* of X; that is, the ξ_j's are complicated functions of the powers of X. These coefficients are independent of each other; by contrast, the powers of X in Expression (14.12) are clearly correlated. Thus $\sum \xi_j = 0$ and $\sum \xi_j \xi_k = 0$. Since they are orthogonal, it is possible to add successive terms to the equation until no significant improvement in fit is obtained. Using the simple procedures presented in this book, orthogonal polynomials can be used only for data in which the values of the independent variable X are evenly spaced with an equal number of replicates.

There are separate sets of these orthogonal polynomials for each sample size n and for increasing powers of X. In Table **H** we list orthogonal polynomials that permit the fitting of maximally third-degree regression equations to sample sizes up to $n = 12$. Coefficients for sample sizes up to 75 are found in Fisher and Yates (1963, Table XXIII). Pearson and Hartley (1956, Table 47) furnish coefficients for fitting sixth-degree equations up to sample size 52. Since the polynomial coefficients are frequently fractional, they are multiplied by the coefficients λ_j furnished underneath each column of polynomials, as is the sum of squares of the polynomials, which it is convenient to have available for computation. Remember that since the sum and hence the mean of each column of polynomial coefficients is zero, the sum of their squares $\sum \xi_j^2$ is also their sum of squares. The coefficients that have been multiplied by λ_j are generally symbolized ξ'_j. The coefficients listed in Table **H** are not the only sets of orthogonal polynomials, but they are those commonly employed for curvilinear regression.

These orthogonal polynomials are used as dummy variables on which Y

is regressed in the manner learned in the last section. They are in fact one type of coefficients of linear comparisons, whose values for different powers of X are mutually orthogonal for any given sample size n. The unweighted coefficients given in the example on length of larval period in drosophila in Box 14.10 are actually orthogonal polynomial coefficients. By successively regressing the Y-values on linear, quadratic, cubic, and higher coefficients, we learn whether the data show significant curvilinear regression and to what degree.

Table 14.3 illustrates such a computation for the trout data from Box 14.6, which are set up in the manner of the linear comparisons of Box 14.10. As coefficients of linear comparison, we use the coefficients of the orthogonal polynomials given in Table **H** for sample size 4. The computation of the explained sum of squares due to the linear, quadratic, and cubic components is as shown in Box 14.10 and is not further discussed here. The completed anova is shown at the bottom of Table 14.3. It is obvious that there is a significant difference among densities, and there is a highly significant mean square for linear regression. When we tested this mean square before (in Box 14.6) we found it to be nonsignificant. How can we explain this discrepancy? If we test for successively higher orders of polynomials we use the residual MS as a denominator mean square. This is what was done in Box

TABLE 14.3

Testing for curvilinear regression by orthogonal polynomials. This method applies to equally spaced X-values with equal sample sizes at each X. (Trout data from Box 14.6; length of trout regressed on density; $a = 4$, $n = 25$). The coefficients for the orthogonal polynomials are found in Table **H** under sample size 4, since there are four means being regressed. These coefficients are used in linear comparisons of the kind carried out in Box 14.10.

Regression		*Density X*				*(1)*	*(2)*	*(3)*	*(4)*
		175	350	525	700				
	$\sum^{n} Y$	87.806	92.755	73.170	63.490	$\sum^{a} c \sum^{n} Y$	Σc^2	$n\Sigma c^2$	$\Sigma \hat{y}^2 = (1)^2/(3)$
Linear	c_{iL}	−3	−1	1	3	−92.533	20	500	17.12471
Quadratic	c_{iQ}	1	−1	−1	1	−14.629	4	100	2.14008
Cubic	c_{iC}	−1	3	−3	1	34.439	20	500	2.37209

Completed anova

Source of variation	*df*	*SS*	*MS*	F_s
Among densities	3	21.63688	7.212	6.941**
Linear regression	1	17.12471	17.124	16.481***
Quadratic regression	1	2.14008	2.140	2.060 *ns*
Cubic regression	1	2.37209	2.372	2.283 *ns*
Within densities	96	99.71535	1.039	

14.6. Since the residual MS is large in this example, we find linear regression not significant. If, on the other hand, we decide to test a priori for several individual degree of freedom comparisons, we assume by implication that there is no random heterogeneity around linear regression, and that the departures from linearity are due to the curvilinear functional relationships. We then test over the error MS. The model changes and the interpretation changes accordingly. Decomposing the SS for deviations around linear regression into quadratic and cubic terms showed neither to be significant, which is not surprising in view of the lack of significance of the MS for deviations. Thus a linear regression seems adequate to describe the relation between length of trout and density at which it has been reared.

↛ In this book we restrict our discussion of orthogonal polynomials in relation to curvilinear regression to testing for degree of curvilinearity. It is possible, however, to obtain a curvilinear regression equation expressed in terms of orthogonal polynomials as given by Expression (14.13). Once an orthogonal polynomial regression equation has been computed, it can be retransformed into the conventional regression coefficients and powers of X. The actual computation of the regression equation is somewhat tedious, however, even on a desk calculator this is not a prohibitively time-consuming task, and readers will find the account at the end of Section 15.6 of Snedecor (1956) quite helpful. ↚

We can also use orthogonal polynomials to decompose an interaction sum of squares into single degree of freedom contrasts. An example of this is shown in Box 14.11, which is largely self-explanatory. It is a typical two-way

BOX 14.11

Partitioning of interaction in a two-way anova by means of orthogonal polynomials.

Lengths of pea sections in millimeters after 20 hours of growth in tissue culture. They had been subjected to two treatments: (1) presence or absence of auxin and (2) kept in sucrose for various periods of time. Ten sections were run per treatment combination.

Table of subgroup sums for treatment combinations based on 10 items.

	Hours in sucrose						
Auxin	*0*	*2*	*4*	*6*	*8*	*10*	Σ
Absent	589	617	652	671	687	707	3923
Present	735	698	674	647	642	650	4046
Σ	1324	1315	1326	1318	1329	1357	

Source: Unpublished data by W. Purves.

BOX 14.11 continued

The means are plotted in Figure 14.16. A two-way anova yielded the following results.

Anova table

Source of variation	*df*	*SS*	*MS*	F_s
Time in sucrose	5	56.5417	11.3083	<1 *ns*
Auxin	1	126.0750	126.0750	4.120*
Time × Auxin interaction	5	1584.4750	316.8950	10.355**
Within subgroups	108	3304.9917	30.6018	

In view of the significant interaction, it is not especially interesting to partition the sums of squares for the effect of time in sucrose, but this is done below because the sums of squares explained by linear, quadratic, and cubic regression are needed for subsequent computations.

Computation in the manner of Box 14.10: $n = 10$ observations per time period; $a = 6$ time periods; $b = 2$ auxin treatments.

	X = hours in sucrose						*(1)*	*(2)*	*(3)*	*(4)*
Comparisons	*0*	*2*	*4*	*6*	*8*	*10*	$\sum^{a} c \sum\sum^{bn} Y$	$\sum c^2$	$bn\sum c^2$	$\sum \hat{y}^2 = \frac{(1)^2}{(3)}$
$\sum\sum^{bn} Y$	1324	1315	1326	1318	1329	1357				
Linear (c_{iL})	−5	−3	−1	1	3	5	199	70	1400	28.2864
Quadratic (c_{iQ})	5	−1	−4	−4	−1	5	185	84	1680	20.3720
Cubic (c_{iC})	−5	7	4	−4	−7	5	99	180	3600	2.7225
										51.3809 = Σ

The coefficients of linear comparisons are the orthogonal polynomial coefficients from Table **H** for sample size 6.

SS for deviations from cubic regression

$$= SS \text{ (time in sucrose)} - SS \text{ (linear + quadratic + cubic)}$$
$$= 56.5417 - 51.3809 = 5.1608$$

None of the regression terms is significant.

Since the sums of squares unexplained after regression was quite small, higher polynomials were not fitted.

There is only one degree of freedom for the effects of auxin. Therefore no further partitions are possible.

The interaction *SS* has 5 *df*, however, so that it is possible to subdivide it further to see whether some particular aspect of the interaction is of more importance than others.

There are many logical ways to partition the interaction, depending upon the nature of the factors. One way would be to study the differences in the linear,

BOX 14.11 continued

quadratic, cubic, and possibly higher regressions in the two treatments. To do this, perform the analysis carried out above on the column totals separately for each row treatment as follows.

No auxin

	X = *hours in sucrose*						(1)	(2)	(3)	(4)
Comparisons	*0*	*2*	*4*	*6*	*8*	*10*	$\sum^{a} c \sum^{n} Y$	Σc^2	$n\Sigma c^2$	$\Sigma \hat{y}^2 = \frac{(1)^2}{(3)}$
$\sum^{n} Y$	589	617	652	671	687	707				
Linear (c_{iL})	−5	−3	−1	1	3	5	819	70	700	958.2300
Quadratic (c_{iQ})	5	−1	−4	−4	−1	5	−116	84	840	16.0190
Cubic (c_{iC})	−5	7	4	−4	−7	5	24	180	1800	0.3200

With auxin

	X = *hours in sucrose*						(1)	(2)	(3)	(4)
Comparisons	*0*	*2*	*4*	*6*	*8*	*10*	$\sum^{a} c \sum^{n} Y$	Σc^2	$n\Sigma c^2$	$\Sigma \hat{y}^2 = \frac{(1)^2}{(3)}$
$\sum^{n} Y$	735	698	674	647	642	650				
Linear (c_{iL})	−5	−3	−1	1	3	5	−620	70	700	549.1429
Quadratic (c_{iQ})	5	−1	−4	−4	−1	5	301	84	840	107.8583
Cubic (c_{iC})	−5	7	4	−4	−7	5	75	180	1800	3.1250

To test if the two treatments differ in the slope of their regression lines, add together the sums of squares explained by the separate regressions for each treatment row and subtract the sums of squares explained by the common regression line (computed from the column totals). Computed in this manner, each SS has $b - 1$ degrees of freedom.

$$SS_{\text{time}_L \times \text{auxin}} = 958.2300 + 549.1429 - 28.2864 = 1479.0865$$

$$F_s = 1479.0865/30.6018 = 48.333^{**}$$

The slope of the regression of length on hours in sucrose (linear regression) differs, depending on the presence or absence of auxin. To test if the shapes of the two lines differ also, we perform similar operations on the sums of squares explained by higher-order regression terms (in this case quadratic and cubic terms).

$$SS_{\text{time}_Q \times \text{auxin}} = 16.0190 + 107.8583 - 20.3720 = 103.5053$$

$$F_s = 103.5053/30.6018 = 3.382 \quad (0.1 > P > 0.05)$$

$$SS_{\text{time}_C \times \text{auxin}} = 0.3200 + 3.1250 - 2.7225 = 0.7225$$

$$F_s < 1 \; ns$$

Since the SS explained by the remaining terms is so small (1584.4756 −

BOX 14.11 continued

1479.0864 − 103.5053 − 0.7225 = 1.1614), it is not worthwhile to compute the individual sums of squares for higher-order components of the time × auxin interaction ($F_s < 1$). If any of the above SS were significant, we could conclude that the lines differed in shape.

Complete anova table

Source of variation	*df*	*SS*	*MS*	F_s
Subgroups	11	1767.0920		
Time in sucrose	5	56.5417	11.3083	<1 *ns*
Time_L	1	28.2864	28.2864	<1 *ns*
Time_Q	1	20.3720	20.3720	<1 *ns*
Time_C	1	2.7225	2.7225	<1 *ns*
Residual	2	5.1608	2.5804	<1 *ns*
Auxin	1	126.0750	126.0750	4.120*
Time × Auxin	5	1584.4750	316.8950	10.355**
Time_L × Auxin	1	1479.0865	1479.0865	48.333**
Time_Q × Auxin	1	103.5053	103.5053	3.382 ($0.1 > P > 0.05$)
Time_C × Auxin	1	0.7225	0.7225	<1 *ns*
Residual	2	1.1614	0.5807	<1 *ns*
Within subgroups	108	3304.9917	30.6018	
Total	119			

anova with replication (as studied in Chapter 11) with a significant interaction between the two factors (presence or absence of auxin and number of hours in sucrose) affecting growth of pea sections. There is no overall linear regression of length of pea sections on number of hours in sucrose, but when we partition the interaction and test for the difference in linear regression on hours in sucrose between those samples run with added auxin and those. without auxin, we find a highly significant difference of the linear component. This becomes clear when we regard Figure 14.16, in which it is obvious that number of hours in sucrose increased the length of pea sections without auxin but decreased the length of those with auxin. When the data are considered together, as they would be in an overall analysis of variance, hours in sucrose do not seem to affect the length of the pea sections, which is an entirely misleading finding. Thus the careful study of interactions is frequently repaid with considerable insight into the mechanism of a given scientific process.

Remember that the method we have presented using tables of orthogonal polynomials can only be applied with equally spaced values of X and equally replicated samples of Y. Unequal sample sizes and unequally spaced values of X require more complicated procedures if orthogonal polynomials are to be fitted.

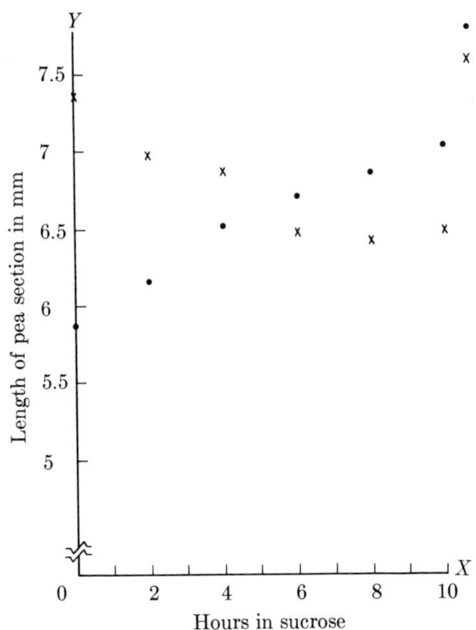

FIGURE 14.16 Mean lengths of pea sections grown with or without auxin and treated with sugar for varying periods. Data from Box 14.11.

So far, our discussion of curvilinear regression has dealt with methods for detecting it and we have given signposts to methods for fitting curvilinear regression. However, one common way of circumventing the tedious computations of curvilinear regression is to transform the date to linearity. This is the subject of the next section.

14.12 Transformations in regression

In transforming either or both variables in regression, we aim at simplifying a curvilinear relationship to a linear one. As a by-product of such a procedure the proportion of the variance of the dependent variable explained by the independent variable is generally increased and the distribution of the deviations of points around the regression line tends to become normal and homoscedastic. Rather than fit a complicated curvilinear regression to points plotted on an arithmetic scale, it is far more expedient to compute a simple linear regression for variates plotted on a transformed scale. A general test of whether transformation will improve linear regression is to graph the points to be fitted on ordinary graph paper as well as on other graph paper in a scale suspected to improve the relationship. If the function straightens out and the systematic deviation of points around a visually fitted line is reduced, the transformation is worthwhile.

We shall briefly discuss a few of the transformations commonly applied in regression analysis. Square root and arcsine transformations (Sections 13.8 and 13.9) are not mentioned below, but they are also effective in regression cases involving data suited to such transformations (as in Box 14.8).

The *logarithmic transformation* is the most frequently used. Anyone doing statistical work is therefore well advised to keep a supply of semilog paper handy. Most frequently we transform the dependent variable Y. This transformation is indicated when percentage changes in the dependent variable vary directly with changes in the independent variable. Such a relationship is indicated by the equation $\hat{Y} = ae^{bX}$, where a and b are constants and e is the base of the natural logarithm. After the transformation, we obtain $\log \hat{Y} = \log a + b(\log e)X$. In this expression $\log e$ is a constant which when multiplied times b yields a new constant factor b', which is equivalent to a regression coefficient. Similarly, $\log a$ is a new Y-intercept, a'. We can then simply regress $\log Y$ on X to obtain the function $\log \hat{Y} = a' + b'X$ and obtain all our prediction equations and confidence intervals in this form. Figure 14.17 shows an example of transforming the dependent variate to logarithmic form, which results in considerable straightening of the response curve.

A logarithmic transformation of the independent variable in regression is effective when proportional changes in the independent variable produce

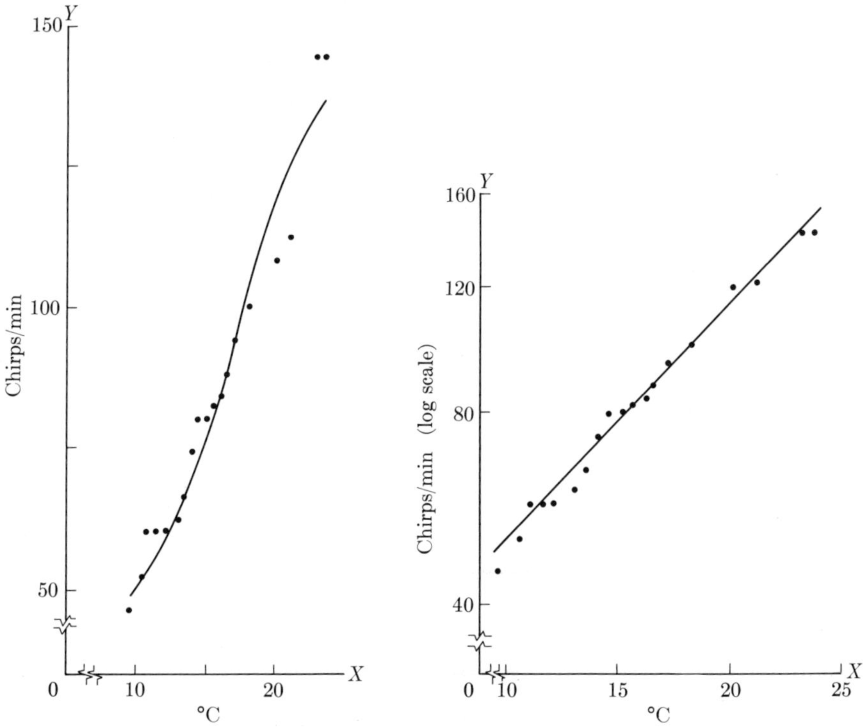

FIGURE 14.17 Logarithmic transformation of a dependent variable in regression. Chirp-rate as a function of temperature in males of the tree cricket *Oecanthus fultoni*. Each point represents the mean chirp-rate/min for all group observations at a given temperature in °C. Original data in left panel, Y plotted on logarithmic scale in right panel. (Data from Block, 1966.)

linear responses in the dependent variable. An example might be the decline in weight of an organism as density increases, where the successive increases in density need to be in a constant ratio in order to effect equal decreases in weight. This belongs to a well-known class of biological phenomena, another example of which is the Weber-Fechner law in physiology and psychology, which states that a stimulus has to be increased by a constant proportion in order to produce a constant increment in response. Figure 14.18 illustrates how logarithmic transformation of the independent variable results in the straightening of the regression line. For computations one would transform X into logarithms.

Logarithmic transformation for both variables is applicable in situations in which the true relationship can be described by the formula $\hat{Y} = aX^b$. This is a curve of the general shape shown in the left panel of Figure 14.19, which becomes straight when both variables are transformed to the logarithmic scale. The regression equation is rewritten as $\log \hat{Y} = \log a + b \log X$ and the computation is done in the conventional manner. This equation is the well-known *allometric growth curve* applicable in many organisms where the ratio between increments in structures of different size remains roughly constant, yielding a relatively great increase of one variable with respect to the other on a linear scale. Examples are the greatly disproportionate growth of various organs in some organisms, such as the sizes of antlers of deer or horns of stag beetles, with respect to their general body sizes. We should point out that the symbolism employed by us here is the converse of the conventional symbolism for allometric growth, but we felt it important to retain the symbol a for Y-intercept and b for the regression coefficient. For comparison of structures subject to allometric growth, it is best to compare adjusted and transformed means after the allometric organ has been regressed on some measure of general body size. The analysis of allometric growth

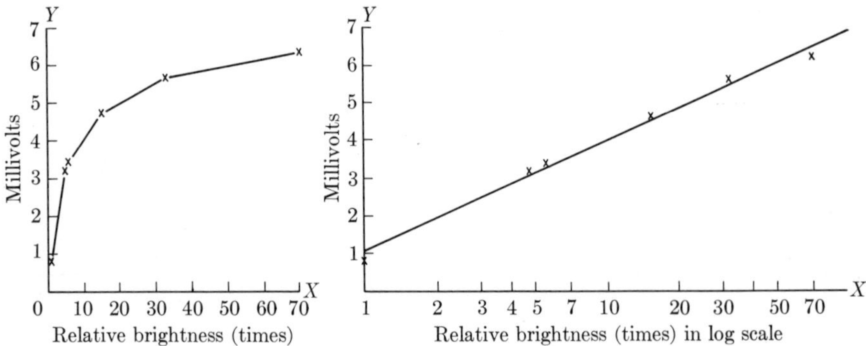

FIGURE 14.18 Logarithmic transformation of the independent variable in regression. This illustrates size of electrical response to illumination in cephalopod eye. Ordinate millivolts; abscissa relative brightness of illumination. A proportional increase in X (relative brightness) produces a linear electrical response Y. (Data in Fröhlich, 1921.)

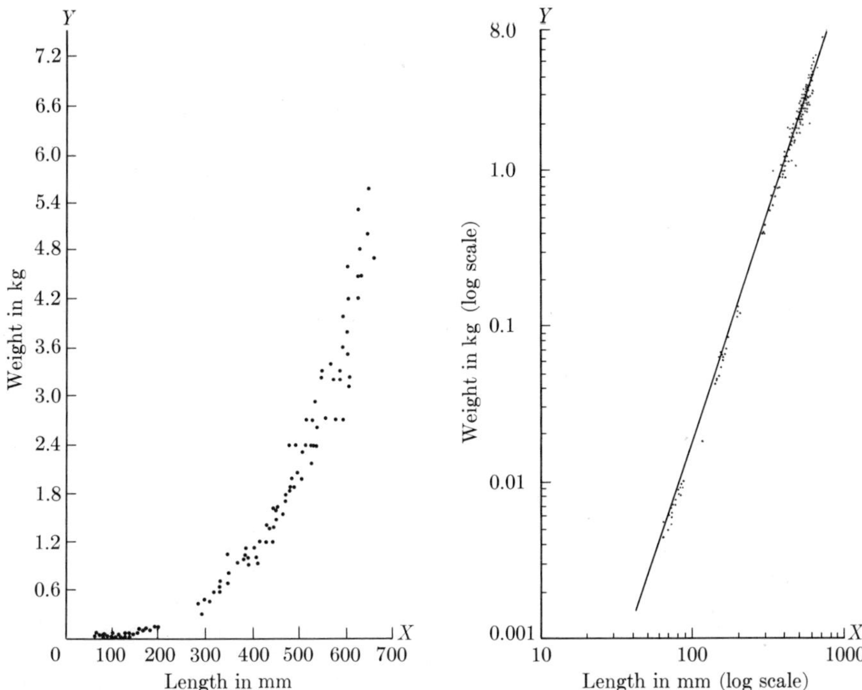

FIGURE 14.19 Logarithmic transformation of both variables in regression. Weight versus length in the cabezon *Scorpaenichthys marmoratus*. (Data from O'Connell, 1953.)

relationships is a subject of considerable depth that we cannot pursue here. The interested reader is referred to an elementary account in Simpson, Roe, and Lewontin (1960) and to a more advanced review of the field by Teissier (1960). An example of a double logarithmic transformation carried out on log-log graph paper is shown in Figure 14.19. In all logarithmic transformations rounding to two- or three-place values for desk calculator operations is advisable.

Many rate phenomena (a given performance per unit of time or per unit of population) such as wing beats per second or number of eggs laid per female, will yield hyperbolic curves when plotted in original measurement scale. Thus, they form curves described by the general mathematical equations $bXY = 1$ or $(a + bX)Y = 1$. From these we can derive $1/Y = bX$ or $1/Y = a + bX$. By transforming the dependent variable into its reciprocal, we can frequently obtain straight-line regressions. For an illustration of the *reciprocal transformation* see Figure 14.20.

Finally, some cumulative curves can be straightened by the *probit transformation*. Refresh your memory on the cumulative normal curve shown in Figure 6.5. Remember that by changing the ordinate of the cumulative normal into probability scale we were able to straighten out this curve. We

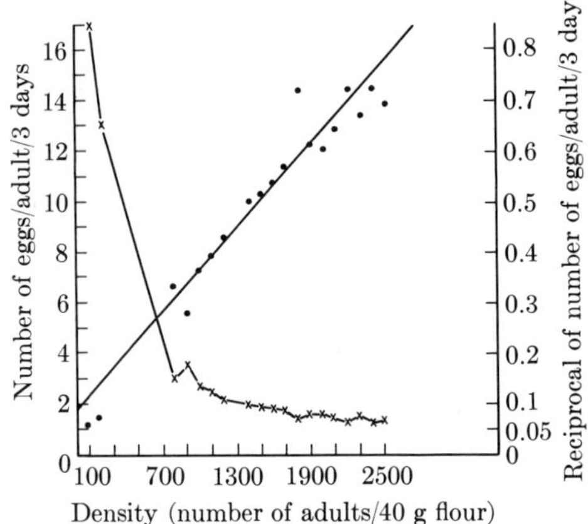

FIGURE 14.20 The reciprocal transformation applied to the dependent variable. Fecundity in *Tribolium* beetles expressed as number of eggs per adult per 3 days graphed as a function of density. Transforming the dependent variable into reciprocal scale changes the relationship from a hyperbolic to a linear one. Crosses are plots in original scale, circles in reciprocal scale. (Data by R. R. Sokal.)

do the same thing here except that we graduate the probability scale in standard deviation units. Thus, the 50% point becomes 0 standard deviations, the 84.13% point becomes +1 standard deviation, and the 2.27% point becomes −2 standard deviations. Such standard deviations, corresponding to a cumulative percentage are called *normal equivalent deviates* (*N.E.D.*). If we use ordinary graph paper and mark the ordinate in *N.E.D.* units, we would obtain a straight line when plotting the cumulative normal curve against it. *Probits* are simply normal equivalent deviates coded by the addition of 5.0, which will avoid negative values for most deviates. Thus, the probit value 5.0 corresponds to a cumulative frequency of 50%, probit value 6.0 corresponds to a cumulative frequency of 84.13%, and probit value 3.0 corresponds to a cumulative frequency of 2.27%. Probit tables, giving the probit equivalents of cumulative percentages, are available in Fisher and Yates (1963) and Pearson and Hartley (1956).

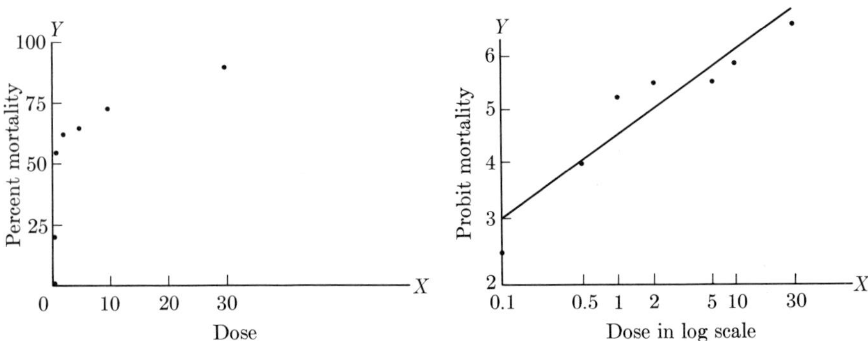

FIGURE 14.21 Dosage mortality data illustrating an application of the probit transformation. Data are mean mortalities for two replicates. Twenty *Drosophila melanogaster* per replicate were subjected to seven doses of an "unknown" insecticide in a class experiment. The point at dose 0.1 which yielded 0% mortality has been assigned a probit value of 2.5 in lieu of $-\infty$, which cannot be plotted.

Figure 14.21 shows an example of mortality percentages for increasing doses of an insecticide. These represent differing points of a cumulative frequency distribution. With increasing dosages an ever greater proportion of the sample dies until at a high enough dose the entire sample is killed. It is often found that if the doses of toxicants are transformed into logarithms, the tolerances of many organisms to these poisons are approximately normally distributed. These transformed doses are often called *dosages*. Increasing dosages lead to a cumulative normal distribution of mortalities, often called *dosage-mortality curves*. These curves are the subject matter of an entire field of biometric analysis, *bioassay*, to which we can only refer in passing here. The most common technique in this field is *probit analysis*, which until relatively recently was a very tedious computational procedure. Dissatisfaction with the tedium led to graphic approximations on so-called *probit paper*, which is probability graph paper in which the abscissa has been transformed into logarithmic scale. A regression line is fitted by eye to dosage-mortality data graphed on probit paper (see Figure 14.21). From the eye-fit line the 50% kill is estimated by the process of inverse prediction discussed in Section 14.8.

→ Recent availability of digital computers has made the full computation even simpler than graphic approximations. A good program is available in the BMD package (Dixon, 1965). The reader who is interested in probit analysis will find a good discussion in chapter 18 of Goulden (1952) or in the latter half of chapter 4 in Goldstein (1964). The reader who wishes to pursue the subject in greater depth should consult Finney (1952). ↔

14.13 Model II regression

We now turn to the somewhat complex subject of regression in which both the dependent as well as the independent variable are subject to error. First, let us see why this should lead to complications. We return to the conventional regression equation of Expression (14.1), which by now should be thoroughly familiar: $\hat{Y} = a + bX$. Our computations are based on the observed Y-values, Y_i, which should be written without the caret on top. What we are really after, however, is a relation between the "true" values of Y and X from which observational error has been removed. Such a relation can be written as

$$\eta = \alpha + \beta\xi \qquad (14.14)$$

where η (eta) is the true value of Y and ξ (xi) is the true value of X (the symbol ξ is now used in a different meaning than in orthogonal polynomials), and α and β are the parametric values of a and b, respectively. In the Model I case we further assume that the independent variable X is measured without error. Therefore $\xi = X$ and $\sigma^2_\xi = \sigma^2_X$. Please note that for simplicity's sake we have dropped the i-subscript from the notation, although to be meaningful each of these relations refers to a single variate or pair of variates. We assume that the dependent variable Y is measured with error. Therefore, we can write

$$Y = \eta + \epsilon \tag{14.15}$$

where ϵ is the error of measurement of Y. We obtain our estimate of β from b calculated as $\sum xy/\sum x^2$ or (dividing numerator and denominator of this expression by $n - 1$) from s_{XY}/s_X^2, where s_{XY} is the covariance of X and Y. Above we stated that s_X^2 equals σ_ξ^2, and we hope that s_{XY} will estimate $\sigma_{\xi\eta}$ so that s_{XY}/s_X^2 will estimate $\beta = \sigma_{\xi\eta}/\sigma_\xi^2$.

In the case of Model I regression s_{XY} will be our estimate of $\sigma_{\xi\eta}$ since ϵ, the error portion of Y, is independent of ξ and will therefore not contribute to the covariance beween X and Y. Since s_X^2 equals σ_ξ^2, the estimate of β will be unbiased.

In Model II regression both X and Y are subject to error and therefore

$$Y = \eta + \epsilon$$

and

$$X = \xi + \delta$$

where δ is the error term for X. The covariance of X and Y should still be an unbiased estimate of $\sigma_{\xi\eta}$, since

$$\begin{aligned}\sum XY &= \sum (\xi + \delta)(\eta + \epsilon) \\ &= \sum\xi\eta + \sum\xi\epsilon + \sum\eta\delta + \sum\delta\epsilon\end{aligned}$$

and all but the first of these products is expected to be zero on the basis of Model II regression (since the error parts are independent of ξ and η and of each other). However, s_X^2 will now estimate $\sigma_X^2 = \sigma_\xi^2 + \sigma_\delta^2$ and the computed regression coefficient b is expected to be lower in absolute value than the true slope of the functional relationship. Furthermore, it can be shown that deviations $(Y_i - \hat{Y}_i)$ are now no longer independent of the magnitude of $\hat{Y}_i$. This invalidates the conventional tests of significance, which assume homogeneity of error.

There is one special case of apparent Model II regression that fortunately permits us to apply Model I methods for tests of significance. This is the so-called *Berkson case,* for which independent variables are measured with error but are controlled by the experimenter. This occurs frequently in experimental work. The reader may have been uneasy earlier about the application of Model I regression to cases where different doses of a hormone or different densities were applied to organisms. How can we be sure that such independent variables are measured without error? Not only will we have some error in administering the dose of hormone or reading the density, but surely the effective dose or density (the dose or density directly interacting with the organism) will not necessarily be that intended by us. Thus, there is again an error δ that attaches itself to each of our intended independent variates X. We can now write $X = \xi + \delta$, where X is the intended or nominal value of the independent variate, ξ is the actual or effective value, and δ is the error term making the difference between the intended and actual variates. In this model, X and δ are not expected to be correlated, since there is no reason to suppose, according to our model, that the magnitude of the intended variate and its error of application should be correlated. We may, therefore, use the ordinary Model I regression procedures for estimation and tests of significance. A readable discussion of these problems can be found in Mandel (1964, section 12.5).

There are several methods for obtaining solutions to the regression equation in a Model II case, depending upon one's knowledge of the respective error variances σ_ϵ^2 and σ_δ^2 or their ratio. Thus, for example, if the ratio of the errors of measurement for Y and X is known—that is, if we can estimate $\lambda = \sigma_\epsilon^2/\sigma_\delta^2$—a method is provided in section 12.5 of Mandel (1964). This is a situation not too likely to arise in the biological sciences, and we do not pursue it here. A relatively simple approach, in which no knowledge of these variances is assumed, is *Bartlett's three-group method*, described in Box 14.12 below. This method does not yield a conventional least squares regression line, and consequently special methods must be used for significance testing.

BOX 14.12

Bartlett's three-group method for Model II regression.

Weights (X) of unspawned female cabezon (a California fish, *Scorpaenichthys marmoratus*) and the number of eggs they subsequently produced (Y). Sample size, $n = 11$.

	Weight (*to nearest 100 grams*) X	*Eggs* (*in thousands*) Y
	14	61
	17	37
	24	65
	25	69
	27	54
	33	93
	34	87
	37	89
	40	100
	41	90
	42	97
Sum	334	842
Mean	30.4	76.5

Source: Data from O'Connell (1953).

BOX 14.12 continued

Both variables are subject to error. It is desired to predict number of eggs produced as a function of weight before spawning.

Computational steps

1. Arrange data by magnitude of X.

2. Divide the data pairs into thirds. $k = n/3 = 11/3 = 3.66 \approx 4$. If, as in this case, n is not evenly divisible by 3, the first and the last third should have the same sample size k. Find the mean $\bar{X}_1$ and $\bar{Y}_1$ for the first third, $\bar{X}_3$ and $\bar{Y}_3$ for the last third, and $\bar{X}$ and $\bar{Y}$ for the total data.

First third			***Last third***		
	X_1	Y_1		X_3	Y_3
	14	61		37	89
	17	37		40	100
	24	65		41	90
	25	69		42	97
Sum	80	232	Sum	160	376
Mean	20.0	58.0	Mean	40.0	94.0

3. The slope of the regression line is

$$b' = \frac{\bar{Y}_3 - \bar{Y}_1}{\bar{X}_3 - \bar{X}_1} = \frac{94.0 - 58.0}{40.0 - 20.0} = \frac{36.0}{20.0} = 1.80$$

4. The Y-intercept is

$$a' = \bar{Y} - b'\bar{X} = 76.5 - (1.80)30.4 = 21.78$$

Hence, the prediction equation is

$$\hat{Y} = a' + b'X = 21.78 + 1.80X$$

This line is shown in Figure 14.22, fitted to the eleven data points. Had the data been inappropriately analyzed as a Model I regression, the equation would have been rather similar:

$$\hat{Y} = 19.652 + 1.870X$$

5. For confidence limits of β', the following statistics must be computed:

i. $\sum^{3}\sum^{k_i} X^2 = 14^2 + 17^2 + \cdots + 41^2 + 42^2 = 11{,}074$

ii. $\sum^{3}\sum^{k_i} Y^2 = 61^2 + 37^2 + \cdots + 90^2 + 97^2 = 68{,}640$

iii. $\sum^{3}\sum^{k_i} XY = 14 \times 61 + 17 \times 37 + \cdots + 41 \times 90 + 42 \times 97$
$= 27{,}310$

BOX 14.12 continued

$$\textbf{iv.}\ \sum^{3}\frac{\left(\sum^{k_i} X\right)^2}{k_i} = \frac{80^2}{4} + \frac{94^2}{3} + \frac{160^2}{4} = 10{,}945.33$$

$$\textbf{v.}\ \sum^{3}\frac{\left(\sum^{k_i} Y\right)^2}{k_i} = \frac{232^2}{4} + \frac{234^2}{3} + \frac{376^2}{4} = 67{,}052.00$$

$$\textbf{vi.}\ \sum^{3}\frac{\left(\sum^{k_i} X\right)\left(\sum^{k_i} Y\right)}{k_i} = \frac{80 \times 232}{4} + \frac{94 \times 234}{3} + \frac{160 \times 376}{4}$$

$$= 27{,}012.00$$

Then

$$\sum y^2 = \sum^{3}\sum^{k_i} Y^2 - \sum^{3}\frac{\left(\sum^{k_i} Y\right)^2}{k_i} = \text{quantity } \textbf{ii} - \text{quantity } \textbf{v}$$

$$= 68{,}640.00 - 67{,}052.00 = 1{,}588.00$$

$$\sum x^2 = \sum^{3}\sum^{k_i} X^2 - \sum^{3}\frac{\left(\sum^{k_i} X\right)^2}{k_i} = \text{quantity } \textbf{i} - \text{quantity } \textbf{iv}$$

$$= 11{,}074.00 - 10{,}945.33 = 128.67$$

$$\sum xy = \sum^{3}\sum^{k_i} XY - \sum^{3}\frac{\left(\sum^{k_i} X\right)\left(\sum^{k_i} Y\right)}{k_i} = \text{quantity } \textbf{iii} - \text{quantity } \textbf{vi}$$

$$= 27{,}310.00 - 27{,}012.00$$

$$= 298.00$$

6. The confidence limits for β' are

$$\frac{(b'C - \sum xy) \pm \sqrt{C[b'^2 \sum x^2 + \sum y^2 - 2b' \sum xy] - \sum x^2 \sum y^2 - (\sum xy)^2}}{C - \sum x^2}$$

where C is $k(\bar{X}_3 - \bar{X}_1)^2(n-3)/2\, t^2_{\alpha[n-3]}$. The $+$ term yields the upper limit L_2, and the $-$ term, the lower limit L_1.

In the present example, if we wish 95% confidence limits to β',

$$t_{.05[11-3]} = 2.26 \qquad k = 4 \qquad C = \frac{4(20)^2(11-3)}{2(2.26)^2} = 1253.035$$

Therefore,

$$L_1 = \left[\frac{1}{(1253.035 - 128.67)}\right]\Big\{[1.80(1253.035) - 298.00]$$

$$- \sqrt{1253.035[1.80^2(128.67) + 1588.00 - 2(1.80)298.0] - [128.67(1588.00) - (298.0)^2]}\Big\}$$

$$= \frac{1957.4630 - \sqrt{874{,}810.436}}{1124.365} = \frac{1957.4630 - 935.3130}{1124.365} = 0.9091$$

$$L_2 = \frac{1957.4630 + 935.3130}{1124.365} = 2.5728$$

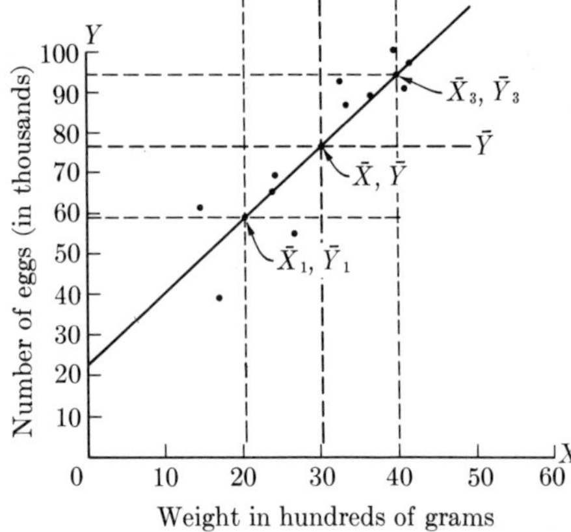

FIGURE 14.22 Egg production and weight of eleven cabezon (from Box 14.12) with equation $\hat{Y} = 21.78 + 1.80X$ fitted by Model II regression. The dashed lines indicate means of X and Y for the first and last third in the array of the data and for the total data.

The data are from a sample of eleven fish whose weights before spawning and subsequent egg production were recorded. These data are shown in Box 14.12 already arrayed by magnitude of the X variate, which is the first step of Bartlett's procedure. We then divide the array of pairs into three equal groups, if possible, or at least into two equally sized end groups, as was done in Box 14.12. The remaining computations are quite simple and are shown in the box. They yield a value of b', which is an estimate of the slope of the regression. The Y-intercept is computed in the conventional manner except that we call it a' to distinguish it from a in Model I linear regression. The regression line is fitted to the variables in Figure 14.22. Methods for setting confidence limits to the regression coefficient are also shown in Box 14.12. These are somewhat tedious to compute, but contain no new complexities. A new symbol is introduced: k corresponds to $n/3$; when the three groups have unequal sample sizes, as in this case, we designate it as k_i.

14.14 Advanced topics in regression

As you may have gathered from the length of this chapter, regression analysis is an important subject in statistics and an invaluable tool in biometric work. Numerous other techniques could have been described in this chapter, but in view of the introductory nature of this text, we had to draw the line somewhere. In this section we discuss briefly three other important techniques to give you some idea of their applications and refer you to suitable references where you may learn more about these methods if you find that your work suggests their application.

A corollary of the uses of regression for the study of causation and prediction (as described in Section 14.7) is its application to reduce the unexplained error variance of a variable. Prior to regression on X, all of σ_Y^2, the variance of Y, is unexplained. After regression the unexplained variance, $\sigma_{Y \cdot X}^2$, is smaller. We might visualize regressing the unexplained deviations

$d_{Y \cdot X}$ on a third variable Z in order to remove more of the variance of Y. We would then obtain a new unexplained variance of Y, $\sigma^2_{Y \cdot XZ}$, in which that portion of the fluctuations of Y determined by X and by Z has been removed. The new unexplained deviations $d_{Y \cdot XZ}$ would be those that neither X nor Z can explain. You can see that this process could go on indefinitely. In fact, this process of successively regressing a dependent variable Y on a series of independent variables, X, Z, W, . . . , or more conventionally, X_1, X_2, X_3, . . . , is analogous to the fundamental process of science. Scientists examine a phenomenon, study its variation, and successively reduce its unexplained variation as more and more of its causes are understood. We might state tongue-in-cheek that the aim of science is to reduce unexplained variances $\sigma^2_{Y \cdot X_i \ldots}$ to zero.

Regressing a variable Y on a series of independent variables could be done by successively regressing deviations in the manner indicated. However, generally, if we suspect several variables of being functionally related to Y, we try to regress Y on all of these simultaneously. This leads to a *multiple regression equation*, which has the form

$$Y = a + b_{Y1.}X_1 + b_{Y2.}X_2 + b_{Y3.}X_3 + \cdots$$

in which the X_1, X_2, X_3, . . . refer to separate, independent variables, which, however, need not be uncorrelated with each other. A regression coefficient such as $b_{Y1.}$ denotes the regression coefficient of Y on variable X_1 that one would expect to obtain if the other variables in this regression equation, X_2, X_3, . . . had been held constant experimentally. It is called a *partial regression coefficient*. In this way we are able to predict a value of Y as a function of several independent variables X_i and generally make a more successful prediction of Y. Figure 14.23 illustrates multiple regression with two inde-

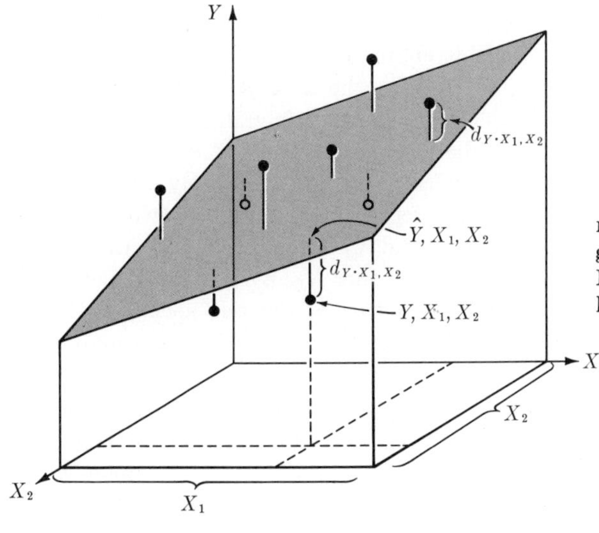

FIGURE 14.23 Multiple regression of Y on X_1 and X_2. Note that the regression line has become a plane.

pendent variables, X_1 and X_2. Note that the regression line of simple linear regression has now become a plane—a tilted board, as it were, from which deviations are calculated as distances dropped onto the plane parallel to the Y-axis. This could not easily be represented graphically for more than two independent variables, but the principle remains the same for any number of independent variables. Until a few years ago, the computation of multiple regression equations was a tedious procedure on which no one embarked without compelling reasons. The availability of computer programs to handle the heavy computation involved—especially the solution of simultaneous equations—has made the method generally accessible to the extent that some caution has to be expressed against its uncritical application. Readers wishing to become familiar with multiple regression will find a short account in chapter 14 of Steel and Torrie (1960) and a more detailed treatment in Volume II of Li (1964). An introduction to computer applications is given by Cooley and Lohnes (1962). Among the numerous multiple regression programs available at computation centers there is one that has become especially popular. *Stepwise multiple regression analysis* regresses in an exploratory manner a variable Y on variables $X_1, X_2, X_3, \ldots$, taking various combinations of these to obtain a minimum of unexplained residual variance in terms of the smallest number of independent variables by dropping any potential independent variables that do not remove a significant independent portion of the variation. Most of the major computer manufacturers advertise such a program for their larger computers.

The technique of *discriminant functions*, although known since R. A. Fisher's work in the thirties, has only recently (due to the availability of digital computers) been much applied in various biological fields, especially in systematics. The basic problem it solves is simple to explain. Suppose we have two samples representing different populations, such as different sexes, or possibly different species. We have measured one character for them and find that though their means for this character are not identical, their distributions overlap considerably, so that on the basis of this character one could not, with any degree of accuracy, identify an unknown specimen as belonging to one or the other of the two populations. A second character might also differentiate them somewhat, but not absolutely. An example of this is shown in Figure 14.24. This represents samples of two species of limpets plotted against two variables X_1 and X_2 commonly used to distinguish them. Regarded from the point of view of either X_1 or X_2, it would be impossible to assign an unknown specimen to either of the two species with any reasonable degree of certainty. The histograms along the coordinate axes make this abundantly clear.

Discriminant function analysis computes a new variable Z, which is a linear function of both variables X_1 and X_2. This function is of the type $Z = \lambda_1 X_1 + \lambda_2 X_2$ and in general $Z = \lambda_1 X_1 + \lambda_2 X_2 + \lambda_3 X_3 + \cdots$, which is the equation of a line cutting across the intermixed cluster of points representing

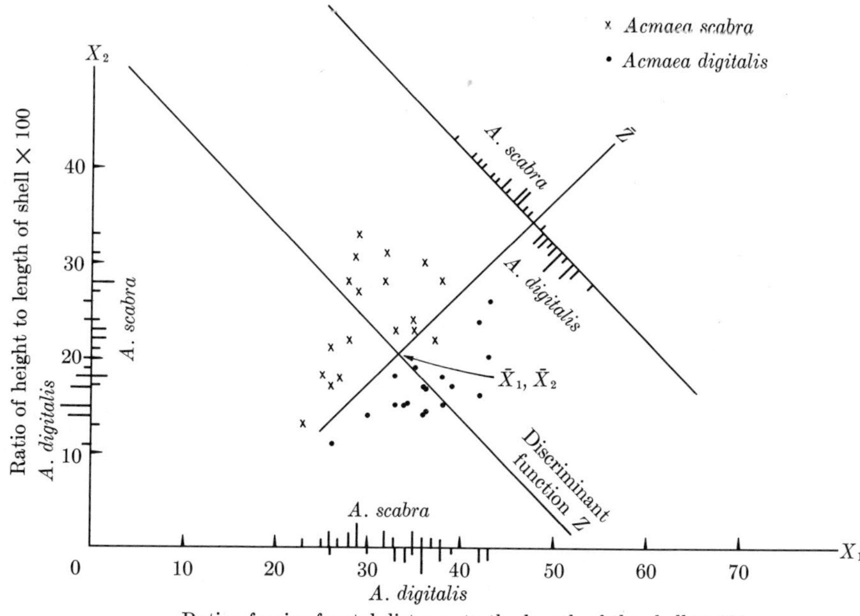

FIGURE 14.24 Illustration of a discriminant function. Histograms for *Acmaea digitalis* below abscissa and to left of ordinate, those for *Acmaea scabra* above abscissa and to right of ordinate. (For explanation, see text.)

the two species. This function is constructed in such a way that as many as possible of the members of one population have high values for Z and as many as possible of the members of the other have low values, so that Z serves as a much better discriminant of the two populations than does variable X_1 or X_2 taken singly. This can be seen in Figure 14.24 by the line drawn parallel to the discriminant function on which a histogram of the Z-values (discriminant function scores) has been graphed. Note that except for a single individual of *Acmae scabra*, which is included with *A. digitalis*, all other members of *A. scabra* are in a group by themselves. Clearly, the discriminant function has served admirably to separate the two groups. If you are faced with a new, unknown specimen, you would measure X_1 and X_2 for it, from these calculate a value of Z employing the previously calculated coefficients λ_1 and λ_2, and with an estimable degree of accuracy allocate the new individual to the correct group. As you can see, this is a very useful device whenever we need to identify unknown specimens and assign them to previously recognized groups. This would be true in taxonomic identification where previous taxa have already been established. Where the previous groups have not been established, discriminant function is of no assistance, and other techniques, collectively called numerical taxonomy, have to be employed (see Section 15.8). Another example might be a case where by means of an intricate technique it might be possible to determine the sex of an immature form. Yet, by measuring several easily obtained external characteristics, we might by means of a discriminant function be able to allocate an immature to one sex or the other. Good write-ups of discriminant functions are found in Goulden (1952), Morrison (1967), and Seal (1965). Cooley and Lohnes (1962), Dixon (1964), and most computer manufacturers furnish such a computer program.

Discriminant functions is a topic in the general area of *multivariate analysis*. This is the branch of statistics dealing with the simultaneous variation of two or more variables. As computer facilities have become widely available, these methods are ever more widely applied, and it will not be many years before various types of multivariate analysis will be among the arsenal of most research biologists. At the moment, the degree of statistical sophistication required for work of this sort is still considered relatively high, and we do not feel that these techniques merit inclusion in an introductory text. One technique that might be singled out by name is *multivariate analysis of variance*, or *manova*, for short, which carries out the equivalent of an anova in cases where a number of variables have been measured for several samples. Three books providing introductions to multivariate analysis in biology are Morrison (1967), Rao (1952), and Seal (1964). Useful summaries with enclosed FORTRAN programs are also found in Cooley and Lohnes (1962). ↮

Exercises 14

14.1 The following temperatures (Y) were recorded in a rabbit at various times (X) after being inoculated with rinderpest virus (data from Carter and Mitchell, 1958).

Time after injection (hrs)	*Temperature (°F)*
24	102.8
32	104.5
48	106.5
56	107.0
72	103.9
80	103.2
96	103.1

Graph the data. Clearly the last three data points represent a different phenomenon from the first four pairs. *For the first four points:* (a) Calculate b. (b) Calculate the regression equation and draw in the regression line. (c) Test the hypothesis that $\beta = 0$ and set 95% confidence limits. (d) Set 95% confidence limits to your estimate of the rabbit's temperature 50 hours after the injection. ANS. $b. = 0.1300$, $\hat{Y}_{50} = 106.5$.

14.2 The following table is extracted from data by Sokoloff (1955). Adult weights of female *Drosophila persimilis* reared at 24°C are affected by their density as larvae. Carry out an anova among densities. Then calculate the regression of weight on density and partition the sums of squares among groups into that explained and unexplained by linear regression. Graph the data with the regression line fitted to the means. Interpret your results.

Larval density	*Mean weight of adults (in mg)*	*s of weights (not $s_{\bar{Y}}$)*	n
1	1.356	0.180	9
3	1.356	0.133	34
5	1.284	0.130	50
6	1.252	0.105	63
10	0.989	0.130	83
20	0.664	0.141	144
40	0.475	0.083	24

14.3 Using the complete data given in Exercise 14.1, calculate the regression equation and compare it with the one obtained earlier. Discuss the effect of the inclusion of the last three points in the analysis.

14.4 Davis (1955) reported the following results in a study of the amount of energy metabolized by the English sparrow, *Passer domesticus*, under various constant temperature conditions and a ten-hour photoperiod.

Temperature (°C)	*Calories* $\bar{Y}$	n	s
0	24.9	6	1.77
4	23.4	4	1.99
10	24.2	4	2.07
18	18.7	5	1.43
26	15.2	7	1.52
34	13.7	7	2.70

Analyze and interpret.

14.5 The following results were obtained in a study of oxygen consumption (microliters/mg dry weight/hr) in *Heliothis zea* by Phillips and Newsom (1966) under controlled temperatures and photoperiods.

Temperature (°C)	*Photoperiod (hrs)*	
	10	14
18	0.51	1.61
21	0.53	1.64
24	0.89	1.73

Compute regression for each photoperiod separately and test for homogeneity of slopes. Confirm your computation of $s^2_{Y \cdot X}$ for the ten-hour photoperiod by computing the sums of squares explained by quadratic regression (use orthogonal polynomials). ANS. $s^2_{Y \cdot X} = 0.019267$ and 0.00060 for 10- and 14-hour photoperiods.

14.6 Length of developmental period (in days) of the potato leafhopper, *Empoasca fabae*, from egg to adult at various constant temperatures (Kouskolekas and Decker, 1966). The original data were weighted means, but for purposes of this analysis we shall consider them as ordinary means.

Temperature °*F*	*Mean length of developmental period in days* $\bar{Y}$	*Number reared* n
59.8	58.1	10
67.6	27.3	20
70.0	26.8	10
70.4	26.3	9
74.0	19.1	43
75.3	19.0	4
78.0	16.5	20
80.4	15.9	21
81.4	14.8	16
83.2	14.2	14
88.4	14.4	27
91.4	14.6	7
92.5	15.3	4

Analyze and interpret. Compute deviations from the regression line $(\bar{Y}_i - \hat{\bar{Y}}_i)$ and plot against temperature.

14.7 Webber (1955) studied the relation between pupal weight (in mg) and the number of ovarioles in females of the blowfly *Lucilia cuprina*.

Pupal weight	*Number of ovarioles*
6.30	62
7.91	76
8.20	81
8.25	80
8.65	83
9.48	105
16.08	142
18.15	152
21.70	184
23.12	186
21.88	182
26.20	230
28.82	193
32.00	260

Clearly both variables are subject to error but the count of the number of ovarioles is probably fairly accurate and can be used as the independent variable in a Model I regression. However, the functional relationship desired would be a prediction equation for the number of ovarioles, given the pupal weight. This is so because it is technically far easier to weigh a pupa than to dissect an adult and count the number of ovarioles. Estimate the number of ovarioles expected in a female pupa weighing 25 mg and set 95% confidence limits to your estimate. ANS. The 95% confidence limits are 173.2 and 236.0.

14.8 Given the lower face width (skeletal bigonial diameter in centimeters) for the 15 North American white females at age 5, it is desired to be able to predict their lower face width at age 6. Use the data given in Box 11.3 for a Model II regression analysis.

14.9 Test whether the regression slopes of temperature on depth for the Rot Lake data given in Box 11.2 are equal over the four days.

14.10 The experiment cited in Exercise 14.4 was repeated using a 15-hour photoperiod, and the following results were obtained.

Temperature (°C)	*Calories* $\bar{Y}$	n	s
0	24.3	6	1.93
10	25.1	7	1.98
18	22.2	8	3.67
26	13.8	10	4.01
34	16.4	6	2.92

Test for the equality of slopes of the regression lines for the 10-hour and 15-hour photoperiod. ANS. $F_s = 0.003$.

15 CORRELATION

In this chapter we continue our discussion of bivariate statistics. The chapter on regression dealt with the functional relation of one variable upon the other; the present chapter treats the measurement of the amount of association between two variables. This general topic is called correlation analysis.

It is not always obvious which type of analysis—regression or correlation—one should employ in a given problem. There has been considerable confusion in the minds of investigators and also in the literature on this topic. We shall try to make the distinction between these two approaches clear at the outset in Section 15.1. In the following section (15.2) you will be introduced to the product-moment correlation coefficient, the common correlation coefficient of the literature. We shall derive a formula for this coefficient and give you something of its theoretical background. The close mathematical relationship between regression and correlation analysis and the algebraic relations between the t-test for paired comparisons and correlation will also be examined in this section. There is a fair amount of algebra to be absorbed but none of it is difficult and its mastery will improve your overall understanding of several subjects in this and previous chapters.

The actual computation of a product-moment correlation coefficient for small as well as large samples is the subject of Section 15.3, and various tests of significance involving correlation coefficients are treated in the next section (15.4). Having learned something about correlation coefficients, it is time to discuss their applications in Section 15.5.

The next section (15.6) contains some special computations with bivariate scattergrams, yielding the so-called principal axes of the ellipse of scattered points as well as the construction of concentric ellipses based on an observed sample of points. In Section 15.7 several nonparametric methods for association are offered for use in cases in which the necessary assumptions for tests involving correlation coefficients do not hold, or where quick but less than fully efficient tests are preferred for reasons of speed in computation or for convenience. Finally, Section 15.8 features signposts to advanced topics in correlation, together with brief discussions of the meaning and usefulness of each. These include the topics of multiple and partial correlation, correlation matrices, factor analysis, cluster analysis, and numerical taxonomy.

15.1 Correlation and regression

There has been much confusion on the subject matter of correlation and regression. Quite frequently correlation problems are treated as regression in the scientific literature, and the converse is equally true. There are several reasons for this confusion. First of all, the mathematical relations between the two methods of analysis are quite close and mathematically one can easily move from one to the other. Hence, the temptation to do so is great. Second, earlier texts had not made the distinction between the two approaches sufficiently clear, and this liability has still not been entirely overcome. A recently published text synonymizes the two, a step that we feel can only compound the confusion. Finally, while the approach chosen by an investigator may be correct in terms of his intentions, the data available for analysis may be such as to make one or the other of the techniques inappropriate.

Let us examine these points at some length. The many and close mathematical relations between regression and correlation will be detailed in the following section. It suffices for now to state that for any given problem the majority of the computational steps are the same whether one carries out a regression or a correlation analysis. You will recall that the fundamental quantity required for regression analysis is the sum of products. This is the very same quantity that serves as the base for the computation of the correlation coefficient. There are some simple mathematical relations between regression coefficients and their corresponding correlation coefficients. Thus the temptation exists to compute a correlation coefficient corresponding to a given regression coefficient. Yet, as we shall see below, this would be wrong unless our intention at the outset had been to study association and the data were appropriate for such a computation.

An analogy from analysis of variance comes to mind. You will recall the fundamental distinctions in purpose of Model I and Model II anova. Yet the computations for the two models were essentially the same, in some cases up to and including the initial significance tests. Only subsequent to the establishment of overall significance did the procedural paths separate. In Model I we estimated means, subdivided the differences among treatments into separate comparisons, and established confidence limits. In Model II we estimated variance components. Although it would be perfectly possible to carry out the computation for estimating a variance component in a significant Model I anova, the quantity obtained would not be an added variance component but a fixed sum of squares whose magnitude is determined by treatment effects. Calling it a variance component would serve no useful purpose. Conversely, testing the significance between the means of random samples in a Model II anova by individual degree of freedom comparisons might yield some significant pairs. Yet the establishment of such differences would be meaningless in a Model II in view of the random sampling of the groups.

In a similar manner we can easily compute a correlation coefficient from data that were properly analyzed by Model I regression. Such a coefficient is meaningless as an estimate of any population correlation coefficient. Conversely, we can evaluate a regression coefficient of one variable on another in data that have been properly computed as correlations. Not only would construction of such a functional dependence for these variables not meet our intentions, but we have already seen that a conventional regression coefficient computed from data in which both variables are measured with error furnishes biased estimates of the functional relation.

Let us then look at the intentions or purposes behind the two types of analyses. In regression we intend to describe the dependence of a variable Y on an independent variable X. As we have seen, we employ regression equations to lend support to hypotheses regarding the possible causation of changes in Y by changes in X; for purposes of prediction, of Y in terms of X; and for purposes of explaining some of the variation of Y by X, by using the latter variable as a statistical control. Studies of the effects of temperature on heartbeat rate, nitrogen content of soil on growth rate in a plant, age of an animal on blood pressure, or dose of an insecticide on mortality of the insect population are all typical examples of regression for the purposes named above.

In correlation, by contrast, we are concerned largely whether two variables are interdependent or *covary*—that is, vary together. We do not express one as a function of the other. There is no distinction between independent and dependent variables. It may well be that of the pair of variables whose correlation is studied, one is the cause of the other, but we neither know nor assume this. A more typical (but not essential) assumption is that the two variables are both effects of a common cause. What we wish to estimate is the degree to which these variables vary together. Thus we might be interested in the correlation between arm length and leg length in a population of mammals or between body weight and egg production in female blowflies or between days to maturity and number of seeds in a weed. Reasons why we would wish to demonstrate and measure association between pairs of variables need not concern us yet. We shall take this up in Section 15.5. It suffices for now to state that when we wish to establish the degree of association between pairs of variables in a population sample, correlation analysis is the proper approach.

Even if we attempt the correct method in line with our purposes we may run afoul of the nature of the data. Thus we may wish to establish cholesterol content of blood as a function of weight, and to do so we may take a random sample of men of the same age group, obtain each individual's cholesterol content and weight, and regress the former on the latter. However, both these variables will have been measured with error. Individual variates of the supposedly independent variable X were not deliberately chosen or controlled by the experimenter. The underlying assumptions of Model I regression do not hold, and fitting a Model I regression to the data is not legitimate, although you will have no difficulty finding instances of such improper practices in the published research literature. If it is really an equation describing the dependence of Y on X that we are after, we should carry out a Model II regression. However, if it is the degree of association between the variables (interdependence) that is of interest, then we should carry out a correlation analysis, for which these data are suitable. The converse difficulty is trying to obtain a correlation coefficient from data that are properly computed as a regression—that is, when X is fixed. An example would be heartbeats of a poikilotherm as a function of temperature, where several temperatures have been applied in an experiment. Such a correlation coefficient is easily obtained mathematically but would simply be a numerical value, not an estimate of a parametric measure of correlation. There is an interpretation that can be given to the square of the correlation coefficient that has some relevance to a regression problem. However, it is not in any way an estimate of a parametric correlation.

This discussion is summarized in Table 15.1, which shows the relations between correlation and regression. The two columns of the table indicate the two conditions of the pair of variables: in one case one random and measured

TABLE 15.1

The relations between correlation and regression. This table indicates the correct computation for any combination of purposes and variables, as shown.

Purpose of investigator	*Nature of the two variables*	
	Y random, X fixed	*Y_1, Y_2 both random*
Establish and estimate dependence of one variable upon another	Model I regression.	Model II regression. (Model I generally inappropriate. For an exception, the Berkson case, see Section 14.13.)
Establish and estimate association (interdependence) between two variables	Meaningless for this case. If desired, an estimate of the proportion of the variation of Y explained by X can be obtained as the square of the correlation coefficient between X and Y.	Correlation coefficient. (Significance tests entirely appropriate only if Y_1, Y_2 are distributed as bivariate normal variables.)

with error, the other variable fixed; in the other case, both variables random. We have departed from the usual convention of labeling the pair of variables Y and X or X_1, X_2 for both correlation and regression analyses. In regression we continue the use of Y for the dependent variable and X for the independent variable, but in correlation both of the variables are in fact random variables, which we have throughout the text designated as Y. We therefore refer to the two variables as Y_1 and Y_2. The rows of the table indicate the intention of the investigator in carrying out his analysis and the four quadrants of the table indicate the appropriate procedures for a given combination of intention of investigator and nature of the pair of variables.

15.2 The product-moment correlation coefficient

There are numerous correlation coefficients in statistics. The most common of these is called the *product-moment correlation coefficient*, which in its current formulation is due to Karl Pearson. We shall derive its formula through an intuitive approach similar to the one we used to introduce the regression coefficient (Section 14.3).

You have seen that the sum of products is a measure of covariation and it is, therefore, likely that this will be the basic quantity from which to obtain a formula for the correlation coefficient. We shall label the variables whose correlation is to be estimated as Y_1 and Y_2. Their sum of products will, therefore, be $\sum y_1 y_2$ and their covariance $[1/(n-1)]\sum y_1 y_2 = s_{12}$. The latter quantity is analogous to a variance, that is, a sum of squares divided by its degrees of freedom.

A standard deviation is expressed in original measurement units such as inches, grams, or cubic centimeters. Similarly, a regression coefficient is expressed as so many units of Y per unit of X, such as 5.2 grams/day. However, a measure of association should be independent of the original scale of measurement so that we can compare the degree of association in one pair of variables with that in another. One way to accomplish this is to divide the covariance by the standard deviations of variables Y_1 and Y_2. This results in dividing each deviation y_1 and y_2 by its proper standard deviation and making it into a standardized deviate. The expression now becomes the sum of the products of standardized deviates divided by $n - 1$

$$r_{Y_1Y_2} = \frac{\sum y_1 y_2}{(n-1)s_{Y_1}s_{Y_2}} \tag{15.1}$$

This is the formula for the product-moment correlation coefficient $r_{Y_1Y_2}$ between variables Y_1 and Y_2. We shall simplify the symbolism to

$$r_{12} = \frac{\sum y_1 y_2}{(n-1)s_1 s_2} = \frac{s_{12}}{s_1 s_2} \tag{15.2}$$

This notation is adequate unless you have two-digit variable numbers, in which case commas are necessary. Thus, $r_{12,14}$ is the correlation between variables Y_{12} and Y_{14}. Expression (15.2) can be rewritten in another common form. Since

$$s\sqrt{n-1} = \sqrt{s^2(n-1)} = \sqrt{\frac{\sum y^2}{n-1}(n-1)} = \sqrt{\sum y^2}$$

Expression (15.2) can be rewritten as

$$r_{12} = \frac{\sum y_1 y_2}{\sqrt{\sum y_1^2 \sum y_2^2}} \tag{15.3}$$

which is preferable for computation because it does not require you to compute two separate standard deviations, each requiring the finding of a square root, but employs the two sums of squares (obtained earlier than standard deviations) and requires the extraction of only a single square root. To state Expression (15.2) more generally for variables Y_j and Y_k, we can write it as

$$r_{jk} = \frac{\sum y_j y_k}{(n-1)s_j s_k} \tag{15.4}$$

The correlation coefficient r_{jk} can range from $+1$ for perfect association to -1 for perfect negative association. This is intuitively obvious when we consider the correlation of a variable Y_j with itself. Expression (15.4) would then yield $r_{jj} = \sum y_j y_j / \sqrt{\sum y_j^2 \sum y_j^2} = \sum y_j^2 / \sum y_j^2 = 1$, which yields a perfect correlation of $+1$. If deviations in one variable were paired with opposite but equal deviations representing another variable, this would yield a correlation of -1 because the sum of products in the numerator would be negative. Proof that the correlation coefficient is bounded by $+1$ and -1 will be given shortly.

We can compute r_{jk} for any set of paired values of Y_j and Y_k, whatever their underlying distribution and even if these are data that properly should be analyzed by regression. However, in such cases, the correlation coefficient is simply a mathematical index, not a sample statistic estimating an unknown parameter, and would be of little interest. If, however, the variates follow a specified distribution, the *bivariate normal distribution*, the correlation coefficient r_{jk} will estimate a parameter of that distribution symbolized by ρ_{jk}. Actually, r_{jk} is slightly biased and underestimates ρ_{jk} in small samples. A correction for bias is suggested as

$$r_{jk}^* = r_{jk}\left\{1 + \frac{1 - r_{jk}^2}{2(n-4)}\right\}$$

by Kendall and Stuart (1961), who discuss this problem in their section 26.17. In the above expression r_{jk}^* is the unbiased estimate of ρ_{jk}, and r_{jk} and n have their previous meanings. For most practical applications this bias can be ignored. The probability density function for the bivariate normal distribution has a formidable-looking formula, which we shall not reproduce here.

Let us approach the distribution empirically. Suppose you have sampled a hundred items and measured two variables on each item, obtaining two samples of 100 variates in this manner. If you plot these 100 items on a graph in which the variables Y_1 and Y_2 are the coordinates, you will obtain a scattergram of points as in Figure 15.3, A, below. Let us assume that both variables,

Y_1 and Y_2, are normally distributed, and also that they are quite independent of each other, so that the fact that one individual happens to be greater than the mean in character Y_1 has no effect whatsoever on its value for variable Y_2. Thus this same individual may be greater or less than the mean for variable Y_2. If there is absolutely no relation between Y_1 and Y_2 and if the two variables are standardized to make their scales comparable, you would find that the outline of the scattergram is roughly circular. Of course, for a sample of 100 items, the circle would be only imperfectly outlined; but the larger the sample, the more clearly could one discern a circle with the central area around the intersection $\bar{Y}_1$, $\bar{Y}_2$ heavily darkened because of the aggregation there of many points. If you keep sampling, you will have to super-impose new points upon previous points, and if you visualize these points in a physical sense, such as grains of sand, a mound peaked in a bell-shaped fashion would gradually accumulate. This is a three-dimensional realization of a normal distribution shown in perspective in Figure 15.1. Regarded from either coordinate axis, the mound would present a two-dimensional appearance, and its outline would be that of a normal distribution curve, the two perspectives giving the distributions of Y_1 and Y_2, respectively.

If we assume that the two variables Y_1 and Y_2 are not independent but are positively correlated to some degree, then if a given individual has a large value of Y_1, it is more likely than not to have a large value of Y_2 as well. Similarly, a small value of Y_1 will likely be associated with a small value of Y_2. Were you to sample items from such a population, the resulting scattergram (shown in Figure 15.3, D, below) would become elongated in the form of an ellipse. This is so because those parts of the circle that formerly included individuals high for one variable and low for the other (and vice versa), are now scarcely represented. Continued sampling (with the sand grain model) yields a three-dimensional elliptic mound shown in Figure 15.2. If correlation is perfect, all the data would fall along a single regression line (the identical line would describe the regression of Y_1 on Y_2 and of Y_2 on Y_1), and if we let them pile up in a physical model, they would result in a flat, essentially two-dimensional normal curve lying on this regression line.

The circular or elliptical shape of the outline of the scattergram and of the resulting mound is clearly a function of the degree of correlation between the two variables, and this is the parameter ρ_{jk} of the bivariate normal distribution. By analogy with Expression (15.2), the parameter ρ_{jk} can be defined as

$$\rho_{jk} = \sigma_{jk}/\sigma_j\sigma_k \tag{15.5}$$

where σ_{jk} is the parametric covariance of variables Y_j and Y_k, and σ_j and σ_k are the parametric standard deviations of variables Y_j and Y_k, as before. When two variables are distributed according to the bivariate normal, a sample correlation coefficient r_{jk} estimates the parametric correlation coefficient ρ_{jk}. We can make some statements about the sampling distribution of ρ_{jk} and set confidence limits to it.

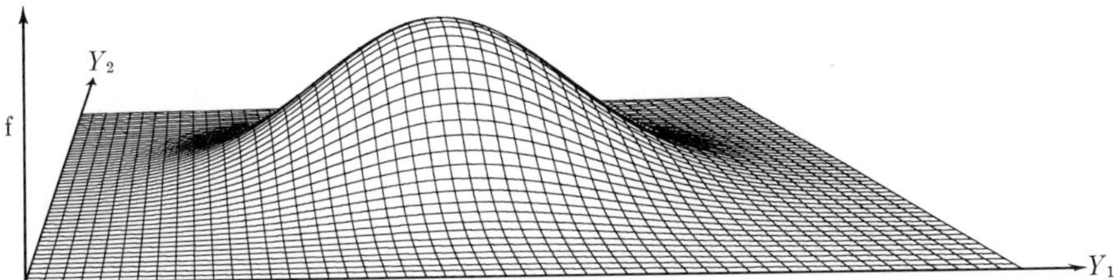

FIGURE 15.1 Bivariate normal frequency distribution. The parametric correlation ρ between variables Y_1 and Y_2 equals zero. The frequency distribution may be visualized as a bell-shaped mound.

Regrettably, the elliptical shape of scattergrams of correlated variables is not usually very clear unless either very large samples have been taken or the parametric correlation ρ_{jk} is very high. To illustrate this point, we show in Figure 15.3 several graphs illustrating scattergrams resulting from samples of 100 items from bivariate normal populations with differing values of ρ_{jk}. Note that in the first graph (Figure 15.3, A), with $\rho_{jk} = 0$, the circular distribution is only very vaguely outlined. A far greater sample is required to demonstrate the circular shape of the distribution more clearly. No substantial difference is noted in Figure 15.3, B, based on $\rho_{jk} = 0.3$. Knowing that this depicts a positive correlation, one can visualize a positive slope in the scattergram; but without prior knowledge this would be difficult to detect visually. The next graph (Figure 15.3, C, based on $\rho_{jk} = 0.5$) is somewhat clearer, but still does not exhibit an unequivocal trend. In general, correlation cannot be inferred from inspection of scattergrams based on samples from populations with ρ_{jk} between -0.5 and $+0.5$ unless the sample is very numerous. This point is illustrated in the last graph (Figure 15.3, G), also sampled from a population with $\rho_{jk} = 0.5$ but based on a sample of 500. Here, the positive slope and elliptical outline of the scattergram are quite evident.

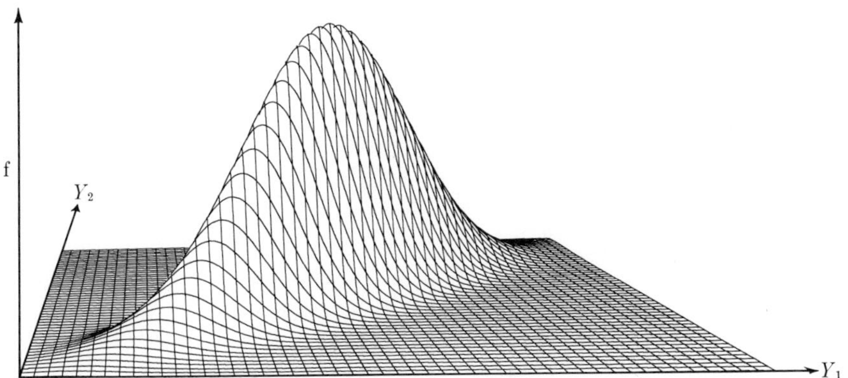

FIGURE 15.2 Bivariate normal frequency distribution. The parametric correlation ρ between variables Y_1 and Y_2 equals 0.9. The bell-shaped mound of Figure 15.1 has become elongated.

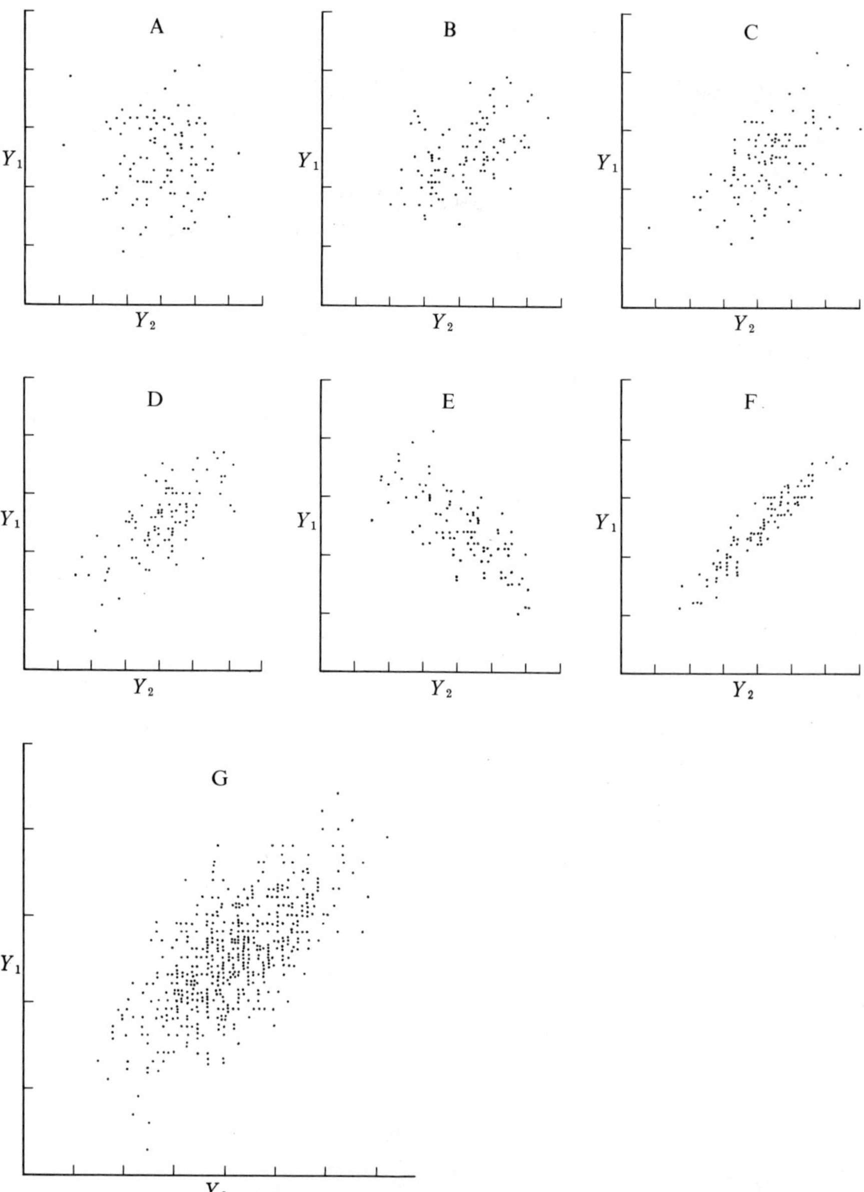

FIGURE 15.3 Random samples from bivariate normal distributions with varying values of the parametric correlation coefficient ρ. Sample sizes $n = 100$ in all graphs except 15.3, G, which has $n = 500$. A. $\rho = 0$. B. $\rho = 0.3$. C. $\rho = 0.5$. D. $\rho = 0.7$. E. $\rho = -0.7$. F. $\rho = 0.9$. G. $\rho = 0.5$.

Figure 15.3, D, based on $\rho_{jk} = 0.7$ and $n = 100$, shows the trend more clearly. Note that the next graph (Figure 15.3, E), based on the same magnitude of ρ_{jk} but representing negative correlation, also shows the trend but is more strung out than Figure 15.3, D. The difference in shape of the ellipse has no relation to the negative nature of the correlation; it is simply a function of sampling error, and the comparison of these two figures should give you some idea of the variability to be expected on random sampling from a bivariate normal distribution. Finally, Figure 15.3, F, representing a correlation of $\rho_{jk} = 0.9$, shows a tight association between the variables and a reasonable approximation to an ellipse of points. (Incidentally, these figures were generated by a pseudorandom number generator on a computer.)

Now let us return to the expression for the sample correlation coefficient shown in Expression (15.3). Squaring this expression results in

$$r_{12}^2 = \frac{(\sum y_1 y_2)^2}{\sum y_1^2 \sum y_2^2}$$

$$= \frac{(\sum y_1 y_2)^2}{\sum y_1^2} \cdot \frac{1}{\sum y_2^2}$$

Look at the left term of the last expression. It is the square of the sum of products of variables Y_1 and Y_2, divided by the sum of squares of Y_1. If this were a regression problem, this would be the formula for the explained sum of squares of variable Y_2 on variable Y_1, $\sum \hat{y}_2^2$. In the symbolism of the regression chapter, it would be $\sum \hat{y}^2 = (\sum xy)^2/\sum x^2$. Thus, we can write

$$r_{12}^2 = \frac{\sum \hat{y}_2^2}{\sum y_2^2} \tag{15.6}$$

The square of the correlation coefficient, therefore, is the ratio of the explained sum of squares of variable Y_2 divided by the total sum of squares of variable Y_2 or, equivalently,

$$r_{12}^2 = \sum \hat{y}_1^2 / \sum y_1^2 \tag{15.6a}$$

which could be derived as easily. Remember that since we are not really regressing one variable on the other, it is just as legitimate to have Y_1 explained by Y_2 as the other way around. This ratio is a proportion between zero and one. This becomes obvious after a little contemplation of the meaning of this formula. The explained sum of squares of any variable must be smaller than its total sum of squares or, maximally, if all the variation of a variable has been explained, it can be as great as the total sum of squares, but certainly no greater. Minimally, it will be zero if none of the variable can be explained by the other variable with which the covariance has been computed. Thus, we obtain an important measure of the proportion of the variation of one variable determined by the variation of the other. This quantity, the square of the correlation coefficient, r_{12}^2, is called the *coefficient of determination*. It ranges from zero to 1 and must be positive regardless of whether the correlation coefficient is negative or positive. Incidentally, here is proof that the correlation coefficient cannot vary beyond -1 and $+1$. Since its square is the coefficient of determination, and we have just shown that the bounds of the latter are zero to 1, it is obvious that the bounds of its square root will be ± 1.

The coefficient of determination is useful also when one is considering the relative importance of correlations of different magnitudes. As can be seen by a re-examination of Figure 15.3, the rate at which the scatter diagrams go from a distribution with a circular outline to that of an ellipse seems to be more directly proportional to r^2 than to r itself. Thus, in Figure 15.3, B, with $\rho^2 = 0.09$, it is difficult to detect the correlation visually. However, by the time we reach Figure 15.3, D, with $\rho^2 = 0.49$, the presence of correlation is very apparent.

The coefficient of determination is a quantity that may be useful in regression analysis also. You will recall that in a regression we used anova to partition the total sum of squares into explained and unexplained sums of squares. Once such an analysis of variance has been carried out, one can obtain the ratio of the explained sums of squares over the total SS as a measure of the proportion of the total variation that has been explained by the regression. However, as already discussed in Section 15.1, it would not be meaningful to take the square root of such a coefficient of determination and consider it as an estimate of the parametric correlation of these variables.

We can invert Expression (15.6) and use it to compute the explained and unexplained sum of squares in cases where we know the coefficients of correlation or determination. Thus, given a sample in which you know that the correlation between two variables is r_{12}, and given the total sum of squares of variable Y_1 as $\sum y_1^2$, you can compute the explained SS as $\sum \hat{y}_1^2 = r_{12}^2 \sum y_1^2$ and the unexplained SS as $\sum d_{1\cdot 2}^2 = (1 - r_{12}^2)\sum y_1^2$ and subject these to an analysis of variance. The quantity $1 - r_{12}^2$ is sometimes known as the *coefficient of nondetermination*, expressing the proportion of variance of a variable that has not been explained by another variable. The square root of this coefficient, $\sqrt{1 - r_{12}^2}$, is known as the *coefficient of alienation*, measuring the lack of association between variables Y_1 and Y_2.

We shall now take up some mathematical relations between the coefficients of correlation and regression. At the risk of being repetitious, we should stress again that though we can easily convert one coefficient into the other, this does not mean that the two types of coefficients can be used interchangeably on the same sort of data. One important relationship between the correlation coefficient and the regression coefficient can be derived as follows from Expression (15.3):

$$r_{12} = \frac{\sum y_1 y_2}{\sqrt{\sum y_1^2 \sum y_2^2}} = \frac{\sum y_1 y_2}{\sqrt{\sum y_1^2}} \cdot \frac{1}{\sqrt{\sum y_2^2}}$$

Multiplying numerator and denominator of this expression by $\sqrt{\sum y_1^2}$, we obtain

$$r_{12} = \frac{\sum y_1 y_2}{\sqrt{\sum y_1^2}\sqrt{\sum y_1^2}} \cdot \frac{\sqrt{\sum y_1^2}}{\sqrt{\sum y_2^2}} = \frac{\sum y_1 y_2}{\sum y_1^2} \cdot \frac{\sqrt{\sum y_1^2}}{\sqrt{\sum y_2^2}}$$

Dividing numerator and denominator of the right term of this expression by $\sqrt{n-1}$, we obtain

$$r_{12} = \frac{\sum y_1 y_2}{\sum y_1^2} \cdot \frac{\sqrt{\dfrac{\sum y_1^2}{n-1}}}{\sqrt{\dfrac{\sum y_2^2}{n-1}}} = b_{2\cdot 1}\frac{s_1}{s_2} \tag{15.7}$$

Similarly, we could demonstrate that

$$r_{12} = b_{1\cdot 2}\frac{s_2}{s_1} \tag{15.7a}$$

and hence

$$b_{2\cdot 1} = r_{12}\frac{s_2}{s_1} \qquad b_{1\cdot 2} = r_{12}\frac{s_1}{s_2} \tag{15.7b}$$

In these expressions $b_{2\cdot 1}$ is the regression coefficient for variable Y_2 on Y_1. We see, therefore, that the correlation coefficient is the regression slope multiplied by the ratio of the standard deviations of the variables. The correlation coefficient may thus be regarded as a standardized regression coefficient. If the two standard deviations are identical, both regression coefficients and the correlation coefficient will be identical in value.

A second relation between the two coefficients is the following: multiplying together the $b_{2\cdot 1}$ and $b_{1\cdot 2}$ terms of Expression (15.7b), we obtain

$$b_{2\cdot 1}b_{1\cdot 2} = r_{12}\frac{s_2}{s_1}r_{12}\frac{s_1}{s_2} = r_{12}^2$$

and therefore,

$$r_{12} = \pm\sqrt{b_{2\cdot 1}b_{1\cdot 2}} \tag{15.8}$$

Thus, if we know the two regression coefficients, we can easily compute the correlation coefficient as the geometric mean (see Section 4.2) of the two regression coefficients. The sign of the correlation coefficient will be that of the regression coefficients and is, of course, determined by the sign of the sum of products $\sum y_1 y_2$.

If the two regression lines are drawn into the same graph, correlation may be regarded as a measure of the angle between the two regression lines. If this angle is a right angle, there can be no correlation in the data. We obtain a circular scattergram. If this angle is very small, then correlation is very high. In the case of perfect correlation, the two regression lines coincide and the angle between them becomes zero.

Having engaged in more than the usual amount of algebra for this book, we hope the reader is in a frame of mind to tolerate a little more. Now that we know about the coefficient of correlation, some of the earlier work on paired comparisons (see Section 11.5) can be put into proper perspective. In Appendix A1.11 we show for the corresponding parametric expressions that the variance of a sum of two variables is

$$s^2_{(Y_1+Y_2)} = s_1^2 + s_2^2 + 2r_{12}s_1s_2 \tag{15.9}$$

where s_1 and s_2 are standard deviations of Y_1 and Y_2, respectively, and r_{12} is the correlation coefficient between these variables. Similarly, for a difference between two variables, we obtain

$$s^2_{(Y_1-Y_2)} = s_1^2 + s_2^2 - 2r_{12}s_1s_2 \tag{15.9a}$$

What Expression (15.9) indicates is that if we make a new composite variable that is the sum of two other variables, the variance of this new variable will be the sum of the variances of the variables of which it is composed plus an added term, which is a function of the standard deviations of these two variables and of the correlation between them. It is shown in Appendix A1.11 that this added term is twice the covariance of Y_1 and Y_2. When the two variables being summed are uncorrelated, this added covariance term will also be zero, and the variance of the sum will simply be the sum of the variances of the two variables. This is the reason why, in an anova or in a t-test of the difference between the two means, we had to assume the independence of the two variables to permit us to add their variances. Otherwise we would have had to allow for a covariance term. By contrast, in the paired comparisons technique we expect correlation between the variables, since each pair shares a common experience. This point will become clearer if we look at the data in Table 15.2. These are the guinea pig data we analyzed by a nonparametric technique in Box 13.9. We first calculate the correlation coefficient between variables Y_1 and Y_2 and find it to be quite high, 0.882. If these data had been (improperly) analyzed by a t-test for difference of means (Section 9.4), the result, as shown in Table 15.2, would not have been significant. This is so because the standard error for the t-test was too large, being based on the sum of the variances of the two variables without subtracting a term due to the correlation. The paired comparisons test (Section 11.5 and Box 11.3) automatically subtracts such a term, resulting in a smaller standard error and consequently in a larger value of t_s, since the numerator of the ratio remains the same. Thus, whenever correlation between two variables is positive, the variance of their differences will be considerably smaller than the sum of their variances; this is the reason why the paired comparisons test has to be used in place of the t-test for difference of means. These considerations are equally true for the corresponding analyses of variance, single classification, and two-way anova.

At the foot of Table 15.2, we show how one could obtain an estimate of the correlation coefficient from the variances of the variables and of their differences. Although this is not a recommended method for obtaining the

TABLE 15.2

Relations between the correlation coefficient and the paired comparisons test. Litter size in guinea pigs. (Data from Box 13.9.)

Year	*Strain B* Y_1	*Strain 13* Y_2	$D = Y_1 - Y_2$
1916	2.68	2.36	+0.32
1917	2.60	2.41	+0.19
1918	2.43	2.39	+0.04
1919	2.90	2.85	+0.05
1920	2.94	2.82	+0.12
1921	2.70	2.73	−0.03
1922	2.66	2.58	+0.08
1923	2.98	2.89	+0.09
1924	2.85	2.78	+0.07
ΣY	24.74	23.81	+0.93
$\bar{Y}$	2.749	2.646	+0.103

1. Computation of correlation coefficient between Y_1 and Y_2

$$n = 9$$

$$\sum y_1^2 = 0.2619 \qquad \sum y_2^2 = 0.3638$$

$$s_1^2 = 0.03274 \qquad s_2^2 = 0.04548$$

$$\sum y_1 y_2 = 0.2723 \qquad \sqrt{\sum y_1^2 \sum y_2^2} = 0.3087$$

$$r_{12} = 0.882$$

2. t-test for difference of means (in the manner of Box 9.6)

$$t_s = \frac{(\bar{Y}_1 - \bar{Y}_2) - (\mu_1 - \mu_2)}{\sqrt{\frac{1}{n}(s_1^2 + s_2^2)}} = \frac{0.103}{0.093226} = 1.105\ ns$$

3. t-test for paired comparisons (in the manner of Box 11.3)

$$s_D^2 = 0.01015 \qquad s_{\bar{D}} = 0.03364$$

$$t_s = \frac{\bar{D} - 0}{s_{\bar{D}}} = \frac{0.103}{0.03365} = 3.06 > t_{.02[8]}$$

The difference in outcome of the two tests shows presence of correlation. We can calculate r_{12} in the usual way (step **1**) or as follows:

$$s_D^2 = s^2_{(Y_1 - Y_2)} = s_1^2 + s_2^2 - 2r_{12}s_1s_2 \quad \text{[from Expression (15.9a)]}$$

$$2r_{12}s_1s_2 = s_1^2 + s_2^2 - s_D^2$$

$$r_{12} = \frac{s_1^2 + s_2^2 - s_D^2}{2s_1s_2}$$

$$= \frac{0.03274 + 0.04548 - 0.01015}{2(0.1809)(0.2132)}$$

$$= \frac{0.06807}{0.077136}$$

$$= 0.882$$

correlation coefficient, it does provide further insight into the relationships that we have just discussed.

Finally, you might ask what effects negative correlation would have on the considerations of the past three paragraphs. Such a situation might arise with a compensatory phenomenon. Thus, if a pair of animals share a cage and the total amount of food available to the two is constant, the more one animal eats, the less the other animal will get. If the pair of animals can be differentiated, as by sex, then over a series of cages there will be a negative correlation between the amount of food eaten by males and females. In such a case the standard error of the paired comparisons test will be greater than that of the difference of means test because the covariance term is now added to the sum of the two variances (the negative sign before the covariance term and the negative sign of the covariance itself together yield a plus sign). Therefore, with negative correlations a paired comparisons test is less likely to result in rejection of the null hypothesis than is the test for difference of means. This is another instance where the two-way analysis of variance would be more useful than a t-test. In this case, an overall test for difference of means of the variables (treatments) has little meaning. Also differences due to rows, the classification factor grouping them into pairs, will not mean much in view of the negative correlation. The important aspect of the analysis is the negative interaction between rows and columns. However, to establish the significance of such an interaction, one needs to have replication of each subgroup. Otherwise, the only way to demonstrate such a relationship is through the actual computation of a correlation coefficient between the variables.

We now proceed to learn efficient methods of computing the product-moment correlation coefficient.

15.3 Computation of the product-moment correlation coefficient

The computation of a product-moment correlation coefficient is quite simple. The basic quantities needed are the same six required for computation of the regression coefficient (Section 14.4). Box 15.1 illustrates how the coefficient should be computed for small and medium-sized samples. The example is based on a sample of twelve crabs in which gill weight Y_1 and body weight Y_2 have been recorded. We wish to know whether there is a correlation between the weight of the gill and that of the body, the latter representing a measure of overall size. The existence of a positive correlation might lead you to conclude that a bigger-bodied crab with its resulting greater amount of metabolism would require larger gills in order to provide the necessary oxygen. The computations are illustrated in Box 15.1. The correlation coefficient of 0.87 agrees with the clear slope and narrow elliptical outline of the scattergram for these data in Figure 15.4.

This example has an added interesting feature. It is a *part-whole correla-*

BOX 15.1

Computation of the product-moment correlation coefficient for small and medium-sized samples.

Relationships between gill weight and body weight in the crab *Pachygrapsus crassipes*. $n = 12$.

(1) Y_1 *Gill ueight in milligrams*	(2) Y_2 *Body weight in grams*
159	14.40
179	15.20
100	11.30
45	2.50
384	22.70
230	14.90
100	1.41
320	15.81
80	4.19
220	15.39
320	17.25
210	9.52

Source: Unpublished data by L. Miller.

Computation

1. $\sum Y_1 = 159 + \cdots + 210 = 2347$

2. $\sum Y_1^2 = 159^2 + \cdots + 210^2 = 583{,}403$

3. $\sum Y_2 = 14.40 + \cdots + 9.52 = 144.57$

4. $\sum Y_2^2 = (14.40)^2 + \cdots + (9.52)^2 = 2204.1853$

5. $\sum Y_1Y_2 = 14.40(159) + \cdots + 9.52(210) = 34{,}837.10$

□□ The machine computation of these quantities is discussed in Section 14.4. □□

6. Sum of squares of $Y_1 = \sum y_1^2 = \sum Y_1^2 - \dfrac{(\sum Y_1)^2}{n}$

$$= \text{quantity } \mathbf{2} - \frac{(\text{quantity } \mathbf{1})^2}{n} = 583{,}403 - \frac{(2347)^2}{12}$$

$$= 124{,}368.9167$$

7. Sum of squares of $Y_2 = \sum y_2^2 = \sum Y_2^2 - \dfrac{(\sum Y_2)^2}{n}$

$$= \text{quantity } \mathbf{4} - \frac{(\text{quantity } \mathbf{3})^2}{n} = 2204.1853 - \frac{(144.57)^2}{12}$$

$$= 462.4782$$

BOX 15.1 continued

8. Sum of products $= \sum y_1y_2 = \sum Y_1Y_2 - \dfrac{(\sum Y_1)(\sum Y_2)}{n}$

$$= \text{quantity } \mathbf{5} - \frac{\text{quantity } \mathbf{1} \times \text{quantity } \mathbf{3}}{n}$$

$$= 34{,}837.10 - \frac{(2347)(144.57)}{12} = 6561.6175$$

9. Product-moment correlation coefficient [by Expression (15.3)] =

$$r_{12} = \frac{\sum y_1y_2}{\sqrt{\sum y_1^2 \sum y_2^2}} = \frac{\text{quantity } \mathbf{8}}{\sqrt{\text{quantity } \mathbf{6} \times \text{quantity } \mathbf{7}}}$$

$$= \frac{6561.6175}{\sqrt{(124{,}368.9167)(462.4782)}} = \frac{6561.6175}{\sqrt{57{,}517{,}912.7314}}$$

$$= \frac{6561.6175}{7584.0565} = 0.8652 \approx 0.87$$

tion. Although gill weight is only a small fraction of body weight, it is a part of it, and the correlation is, therefore, between two variables, one of which is a part of another. We can rewrite these variables as follows: Y_1 is gill weight as before, but Y_2, the body weight, is made up of Y_0, body weight *minus* gill weight, and Y_1, gill weight. Since $Y_2 = Y_0 + Y_1$, Y_1 and Y_2 should be positively correlated to some degree because Y_1 is part of Y_2 and will clearly be positively correlated with itself. The correlation between Y_1 and Y_2 depends

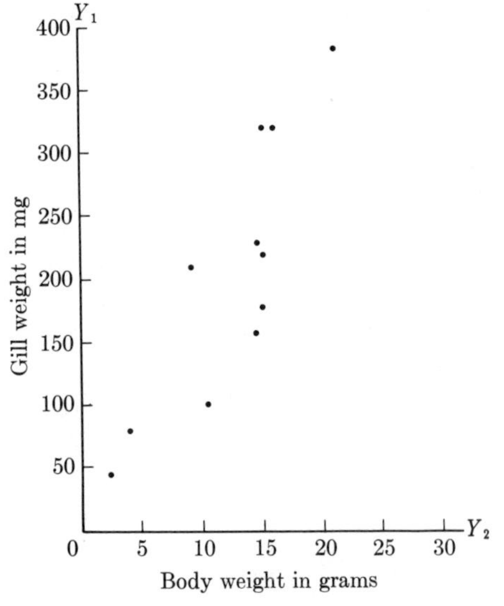

FIGURE 15.4 Scatter diagram for crab data of Box 15.1.

on the degree of correlation between Y_0 and Y_1, and can be predicted from the formula

$$r_{12} = \frac{s_1 + r_{01}s_0}{\sqrt{s_0^2 + 2r_{01}s_0s_1 + s_1^2}} \tag{15.10}$$

Thus, the correlation of a part with a whole is a function of s_1 and s_0, the standard deviations of the part and of the whole minus that part, as well as of r_{01}, the correlation between these variables. When the correlation $r_{01} = 0$, the formula for a part-whole correlation simplifies to

$$r_{12} = \frac{1}{\sqrt{1 + (s_0^2/s_1^2)}} \tag{15.10a}$$

and the correlation is then simply a function of the ratio of the variances. If the variances were equal in addition to no correlation between the part and its complement, then the formula for the part-whole correlation simplifies to $r_{12} = 1/\sqrt{1+1} = 1/\sqrt{2} = 0.707$, which shows you that when two parts are equally variable, the correlation of one with their sum will be quite high. In the past, such correlations were called spurious. They are not really that, but rather are logical consequences of the way in which the variables were defined. As long as this is kept in mind, there is nothing necessarily wrong with calculating such correlations, except that one should not be surprised to find a sizable correlation of a part with a whole. When the part has a much smaller variance than the whole, Expression (15.10) shows that the expected value of the part-whole correlation is not likely to be great unless r_{01} is substantial.

When sample sizes are large, the method of Box 15.1 becomes relatively tedious. In such cases one sets up the data as *two-way frequency distributions*. The considerations here are very much the same as those we first encountered in Section 2.5, when we advised you about grouping variates into simple frequency distributions. If it would take longer to set the data up in a two-way frequency distribution than it would to compute them directly on a desk calculator with an automatic squaring device, then clearly the method to be discussed is inefficient. In our experience, at sample sizes between 150 and 200, straight desk calculator computation as per Box 15.1 becomes inefficient and a two-way frequency distribution analyzed as in Box 15.2 will yield an answer more rapidly. Of course, if we wish to inspect the two-way scattergram of items for these two variables, we must set them up in a two-way frequency distribution in any case.

The data in Box 15.2 are wing vein lengths from the right and left forewings of 500 worker bees. Obviously, we would expect a correlation between these variables. Larger bees are expected to have larger wings on both sides of the body, and wing vein length will reflect this. The degree to which the right and left sides are correlated is a measure of developmental homeostasis.

BOX 15.2

Computation of the product-moment correlation coefficient for large samples, using a two-way frequency distribution.

Lengths of the median cross vein in the right (Y_1) and left (Y_2) forewings of 500 worker bees. The variates are coded ($mm \times 50$). Data by Phillips (1929).

To form a two-way frequency distribution, you must establish a system of class intervals for both variables. These classes are listed at the margins of a two-way table such as the one given below, and you tally the number of observations which fall into each cell of the table, representing a class of Y_1- and Y_2-values.

Several computational schemes are available for obtaining the basic six quantities for regression or correlation from a two-way frequency distribution. We believe that the procedure adopted here is computationally the simplest, but it does require a previously prepared data sheet with products of the coded class marks entered in the cells in small print. Such sheets can be prepared in quantity by mimeographing or other duplication. If you need to compute only one such two-way frequency distribution, with no immediate prospect of further work of this kind, you might use a simple data sheet and carefully copy the product numbers from the cells. We believe this is still faster than the conventional approach, which is briefly described after step **5** below.

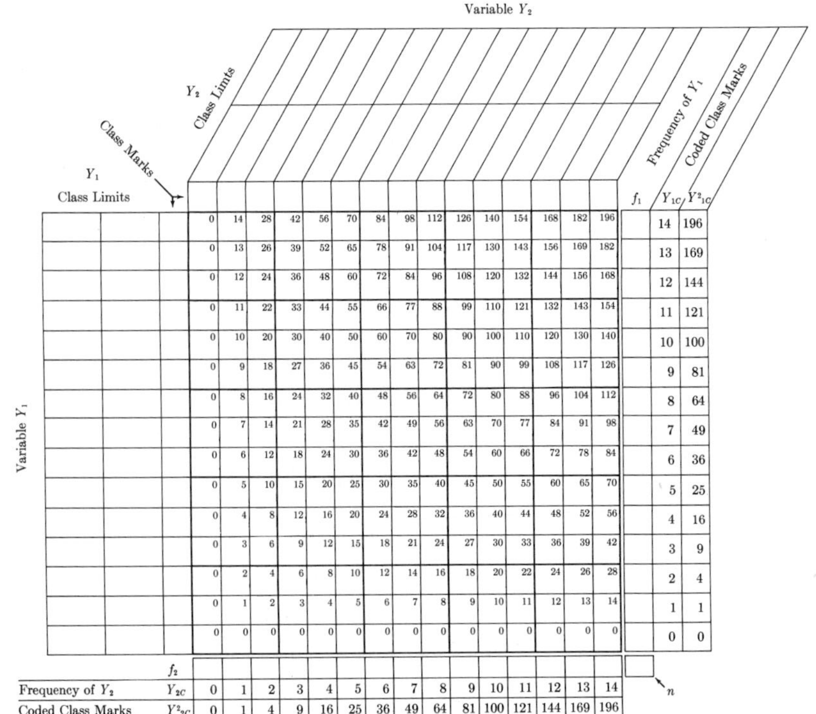

Y_1 Class Limits	Class Marks																	f_1	Y_{1C}	Y^2_{1C}
			0	14	28	42	56	70	84	98	112	126	140	154	168	182	196		14	196
			0	13	26	39	52	65	78	91	104	117	130	143	156	169	182		13	169
			0	12	24	36	48	60	72	84	96	108	120	132	144	156	168		12	144
			0	11	22	33	44	55	66	77	88	99	110	121	132	143	154		11	121
			0	10	20	30	40	50	60	70	80	90	100	110	120	130	140		10	100
			0	9	18	27	36	45	54	63	72	81	90	99	108	117	126		9	81
			0	8	16	24	32	40	48	56	64	72	80	88	96	104	112		8	64
			0	7	14	21	28	35	42	49	56	63	70	77	84	91	98		7	49
			0	6	12	18	24	30	36	42	48	54	60	66	72	78	84		6	36
			0	5	10	15	20	25	30	35	40	45	50	55	60	65	70		5	25
			0	4	8	12	16	20	24	28	32	36	40	44	48	52	56		4	16
			0	3	6	9	12	15	18	21	24	27	30	33	36	39	42		3	9
			0	2	4	6	8	10	12	14	16	18	20	22	24	26	28		2	4
			0	1	2	3	4	5	6	7	8	9	10	11	12	13	14		1	1
			0	0	0	0	0	0	0	0	0	0	0	0	0	0	0		0	0
		f_2																		
Frequency of Y_2		Y_{2C}	0	1	2	3	4	5	6	7	8	9	10	11	12	13	14			
Coded Class Marks		Y^2_{2C}	0	1	4	9	16	25	36	49	64	81	100	121	144	169	196			

BOX 15.2 continued

Class Marks

Length of vein m, left forewing (Y_2)

Length of vein m, right forewing (Y_1)

Y_1 ↓ \ Y_2 →	74	75	76	77	78	79	80	81	82	83	84	85	86	87		f_1	Y_{1C}	Y^2_{1C}
	0	14	28	42	56	70	84	98	112	126	140	154	168	182	196		14	196
	0	13	26	39	52	65	78	91	104	117	130	143	156	169	182		13	169
86	0	12	24	36	48	60	72	84	96	108	120	132	144	156 1	168	1	12	144
85	0	11	22	33	44	55	66	77	88 1	99 1	110 2	121 2	132 3	143	154	9	11	121
84	0	10	20	30	40	50	60	70	80 3	90 7	100 2	110 1	120 1	130	140	14	10	100
83	0	9	18	27	36	45	54 1	63 4	72 6	81 16	90 3	99 1	108	117	126	31	9	81
82	0	8	16	24	32	40 3	48 13	56 19	64 12	72 5	80 2	88	96	104	112	54	8	64
81	0	7	14	21	28	35 8	42 18	49 22	56 19	63 3	70 1	77	84	91	98	71	7	49
80	0	6	12	18 2	24 6	30 23	36 46	42 23	48 1	54 1	60	66	72	78	84	102	6	36
79	0	5	10	15 10	20 17	25 34	30 25	35 8	40	45	50	55	60	65	70	94	5	25
78	0	4	8 2	12 14	16 19	20 12	24 11	28 1	32	36	40	44	48	52	56	59	4	16
77	0	3 1	6 4	9 19	12 14	15 6	18 1	21	24	27	30	33	36	39	42	45	3	9
76	0 1	2 5	4 2	6 4	8 3	10 1	12	14	16	18	20	22	24	26	28	16	2	4
75	0	1	2 1	3 1	4	5 1	6	7	8	9	10	11	12	13	14	3	1	1
74	0	0 1	0	0	0	0	0	0	0	0	0	0	0	0	0	1	0	0
f_2	1	7	9	50	59	88	115	77	42	33	10	4	4	1		500 ← n		
Y_{2C}	0	1	2	3	4	5	6	7	8	9	10	11	12	13	14			
Y^2_{2C}	0	1	4	9	16	25	36	49	64	81	100	121	144	169	196			

The form as shown in this box provides cells for as many as 14 classes for each of the two variables. This should be adequate in most instances, but if finer subdivision into more classes seems desirable for very numerous data, the prepared table can be expanded to any size and you will have no difficulty in working out the product coefficients within the cells, which are explained below. We first show the entire empty form, and then those parts of the form essential for computation with the data on wing vein length filled in. The actual tally marks are not shown; only their numerical values are given.

By summing the rows of the table of frequencies that result, you obtain a frequency distribution for variable Y_1, and by summing the columns you obtain the frequencies for the Y_2 variable.

Computation

As in ordinary frequency distributions, it is usually convenient to code the class marks. The column Y_{1C} gives the coded class marks for Y_1, and the row

BOX 15.2 continued

Y_{2C} gives the coded class marks for Y_2. The squares of Y_{1C} and Y_{2C} are also given for convenience. In our sample, $n = \sum f_1 = \sum f_2 = \sum f_{12} = 500$. The following quantities may now be calculated:

1. $\sum f_1 Y_{1C} = 1(0) + 3(1) + \cdots + 9(11) + 1(12) = 2947$

2. $\sum f_1 Y_{1C}^2 = 1(0) + 3(1) + \cdots + 9(121) + 1(144) = 19{,}517$

3. $\sum f_2 Y_{2C} = 1(0) + 7(1) + \cdots + 4(12) + 1(13) = 2918$

4. $\sum f_2 Y_{2C}^2 = 1(0) + 7(1) + \cdots + 4(144) + 1(169) = 19{,}140$

The numbers in small print in each cell of the table are $Y_{1C}Y_{2C}$. These are used for the computation of $\sum f_{12} Y_{1C} Y_{2C}$ (= each cell frequency times the $Y_{1C}Y_{2C}$ values for each cell, summed over all cells that contain nonzero frequencies).

5. $\sum f_{12} Y_{1C} Y_{2C} = 1(156) + 1(88) + 1(99) + 2(110) + \cdots + 1(5) + 1(0)$

$= 18{,}981$

If you do not use the prepared sheet with the built-in products of $Y_{1C}Y_{2C}$, you can work out these products individually on a calculating machine as you accumulate them for all cells which contain frequencies $f_{12} > 0$. This can easily be done on calculators with transfer devices that permit the computation of a product $a \times b \times c$. By proceeding systematically, by row or by column (possibly locking the Y_C-value of the given row or column in the memory dials if the calculator has such), you can go through this computation rapidly.

6. Sum of squares of $Y_1 = \sum y_1^2 = \sum f_1 Y_1^2 - \dfrac{(\sum f_1 Y_1)^2}{n}$

$$= \text{quantity } \mathbf{2} - \frac{(\text{quantity } \mathbf{1})^2}{n} = 19{,}517 - \frac{(2947)^2}{500}$$

$$= 2147.382$$

7. Sum of squares of $Y_2 = \sum y_2^2 = \sum f_2 Y_2^2 - \dfrac{(\sum f_2 Y_2)^2}{n}$

$$= \text{quantity } \mathbf{4} - \frac{(\text{quantity } \mathbf{3})^2}{n} = 19{,}140 - \frac{(2918)^2}{500}$$

$$= 2110.552$$

8. Sum of products of Y_1 and $Y_2 = \sum y_1 y_2$

$$= \sum f_{12} Y_{1C} Y_{2C} - \frac{(\sum f_1 Y_{1C})(\sum f_2 Y_{2C})}{n}$$

$$= \text{quantity } \mathbf{5} - \frac{\text{quantity } \mathbf{1} \times \text{quantity } \mathbf{3}}{n}$$

$$= 18{,}981 - \frac{(2947)(2918)}{500} = 1782.308$$

9. Correlation coefficient $= r_{12} = \dfrac{\sum y_1 y_2}{\sqrt{\sum y_1^2 \sum y_2^2}}$

BOX 15.2 continued

$$= \frac{\text{quantity } \mathbf{8}}{\sqrt{\text{quantity } \mathbf{6} \times \text{quantity } \mathbf{7}}}$$

$$= \frac{1782.308}{\sqrt{(2147.382)(2110.552)}} = \frac{1782.308}{\sqrt{4{,}532{,}161.374864}}$$

$$= \frac{1782.308}{2128.887} = 0.8372 \approx 0.837$$

Although these particular data are quite old, the measurement of the degree of symmetry of organisms by correlation of bilaterally paired organs has received considerable attention in recent years. The computational steps for obtaining the six basic quantities from the two-way frequency distribution are illustrated in Box 15.2. Once the basic quantities have been obtained, the rest of the computation of the correlation coefficient is the same as in Box 15.1 and need not be specially taken up here. We find the correlation to be quite high; yet remember that a correlation of 0.8372 yields a coefficient of determination of only 0.7009. Thus, only 70% of the variance of wing vein length on one side is accounted for by variation of the wing vein length in the opposite side.

The preparation of such a frequency distribution by tallying is practical up to 500 or maximally 1000 observations. When we get beyond that figure, mechanical aids should be resorted to. By punching the data on cards, they could be easily sorted and a two-way frequency distribution prepared with the help of the counter on the sorting machine. If more than one correlation coefficient has to be computed, it is usually worthwhile for the computation to be done on a computer. Numerous correlation programs exist, and one is furnished in Appendix A3.13. The advantage of the computer becomes especially noticeable when more than two variables have been recorded for each item. For j variables the outcome of the study will be a $j \times j$ *correlation matrix*—that is, a table of correlation coefficients of variable 1 against variable 2, variable 1 against variable 3, and so on until variable 1 is correlated with variable j. This would be followed by the correlation of 2 against 3, 2 against 4, up to 2 against j, the final value in the correlation matrix being the correlation of variable $j - 1$ versus variable j. The computation of such correlation matrices in the precomputer days was a very difficult and tedious procedure, often taking several months for sizable studies. The ease with which these can now be obtained has given impetus to considerable research employing multivariate analysis, some of which is discussed in Sections 14.14 and 15.8.

15.4 Significance tests in correlation

The most common significance test is whether a sample correlation coefficient could have come from a population with a parametric correlation coefficient of zero. The null hypothesis is, therefore, $H_0: \rho = 0$. This implies that the two variables are uncorrelated. If the sample comes from a bivariate normal distribution and $\rho = 0$, the standard error of the correlation coefficient is $s_r = \sqrt{(1 - r^2)/(n - 2)}$. The hypothesis is tested as a t-test with $n - 2$ degrees of freedom, $t_s = (r - 0)/\sqrt{(1 - r^2)/(n - 2)} = r\sqrt{(n - 2)/(1 - r^2)}$. We should emphasize that this standard error applies only when $\rho = 0$, so that it cannot be applied to testing a hypothesis that ρ is a specific value other than zero. The t-test for the significance of r is mathematically equivalent to the t-test for the significance of b, in either case measuring the strength of the association between the two variables being tested. This is somewhat analogous to the situation in Model I and Model II single classification anova where the same F-test establishes the significance regardless of the model.

Significance tests following this formula have been carried out systematically and are tabulated in Table **Y**, which permits the direct inspection of a sample correlation coefficient for significance without further computation. Box 15.3 illustrates tests of the hypothesis $H_0: \rho = 0$, using Table **Y** as well as the t-test discussed at first.

BOX 15.3

Tests of significance and confidence limits for correlation coefficients.

Test of the null hypothesis H_0: $\rho = 0$ *versus* H_1: $\rho \neq 0$

The simplest procedure is to consult Table **Y**, where the critical values of r are tabulated for $df = n - 2$ from 1 to 1000. If the absolute value of the observed r is greater than the tabulated value in the column for two variables, we reject the null hypothesis.

Examples.—In Box 15.1 we found the correlation between body weight and gill weight to be 0.8652, based on a sample of $n = 12$. For 10 degrees of freedom the critical values are 0.576 at the 5% level and 0.708 at the 1% level of significance. Since the observed correlation is greater than both of these, we can reject the null hypothesis, H_0: $\rho = 0$, at $P < 0.01$.

In Box 15.2 we obtained the correlation $r = 0.8372$, based upon $n = 500$ observations. This is far greater than the critical values (0.088 and 0.115) given in Table **Y**. We again reject the null hypothesis that $\rho = 0$ at $P \ll 0.01$.

Table **Y** is based upon the following test, which may be carried out when the table is not available or when an exact test is needed at significance levels or at degrees of freedom other than those furnished in the table. The null hypothesis is tested by means of the t-distribution (with $n - 2$ df) by using the standard error of r. When $\rho = 0$,

$$s_r = \sqrt{(1 - r^2)/(n - 2)}$$

Therefore,

$$t_s = \frac{(r - 0)}{\sqrt{(1 - r^2)/(n - 2)}} = r\sqrt{(n - 2)/(1 - r^2)}$$

For the data of Box 15.1, this would be

BOX 15.3 continued

$$t_s = 0.8652\sqrt{(12 - 2)/(1 - 0.8652^2)} = 0.8652\sqrt{10/0.25143}$$

$$= 0.8652\sqrt{39.7725} = 0.8652\ (6.3065) = 54.564 > t_{.01[10]}$$

For a one-tailed test the 0.10 and 0.02 values of t should be used for 5% and 1% significance tests, respectively. Such tests would apply if the alternative hypothesis were H_1: $\rho > 0$ or H_1: $\rho < 0$, rather than H_1: $\rho \neq 0$.

When n is greater than 50, we can also make use of the z-transformation described in the text. For the data of Box 15.2, since $\sigma_z = 1/\sqrt{n - 3}$, we test

$$t_s = \frac{z - 0}{1/\sqrt{n - 3}} = z\sqrt{n - 3}$$

Since z is normally distributed and we are using a parametric standard deviation, we compare t_s with $t_{\alpha[\infty]}$ or employ Table **P**, areas of the normal curve. In Table **M** we find that for $r = 0.837$, $z = 1.2111$ and

$$t_s = 1.2111\sqrt{497} = 26.997$$

This value, when looked up in the table of areas of a normal curve (Table **P**), yields a very small probability ($< 10^{-6}$).

When $n < 50$ (or in critical cases when $n < 100$), more exact procedures using z are given in the last section of this box.

Test of the null hypothesis H_0: $\rho = \rho_1$, where $\rho_1 \neq 0$

To test this hypothesis we cannot use Table **Y** or the t-test given above, but must make use of the z-transformation.

Suppose we wish to test the null hypothesis H_0: $\rho = +0.50$ versus H_1: $\rho \neq +0.50$ for the data of Box 15.2. We would use the following expression:

$$t_s = \frac{z - \zeta}{1/\sqrt{n - 3}} = (z - \zeta)\sqrt{n - 3}$$

where z and ζ are the z-transformations of r and ρ, respectively. Again we compare t_s with $t_{\alpha[\infty]}$ or look it up in Table **P**. From Table **M** we find

For $r = 0.837$ $\qquad z = 1.2111$

For $\rho = 0.500$ $\qquad \zeta = 0.5493$

Therefore

$$t_s = (1.2111 - 0.5493)(\sqrt{497}) = 14.7538$$

The probability of obtaining such a value of t_s by random sampling is $P < 10^{-6}$ (see Table **P**). It is most unlikely that the parametric correlation between right and left wing veins is 0.5.

Confidence limits

If $n > 50$, we can set confidence limits to r using the z-transformation. We first convert the sample r to z, set confidence limits to this z, and then transform these limits back to the r-scale. We shall find 95% confidence limits for the data of Box 15.2.

For $r = 0.837$, $z = 1.2111$; $\alpha = 0.05$

BOX 15.3 continued

$$L_1 = z - t_{\alpha[\infty]}s_z = z - \frac{t_{.05[\infty]}}{\sqrt{n-3}} = 1.2111 - \frac{1.960}{22.2953}$$

$$= 1.2111 - 0.0879 = 1.1232$$

$$L_2 = z + \frac{t_{.05[\infty]}}{\sqrt{n-3}} = 1.2111 + 0.0879 = 1.2990$$

Using Table **N**, we retransform these z-values to the r-scale:

$L_1 \approx 0.808$ and $L_2 \approx 0.862$

are the 95% confidence limits around $r = 0.837$.

The use of the z-transformation when sample sizes are less than 50

For such smaller-sized samples, the calculation of exact probabilities are difficult. The following modified z-transformation has been suggested by Hotelling (1953) for use in small samples:

$$z^* = z - \frac{3z + r}{4n}$$

$$\sigma_{z^*} = \frac{1}{\sqrt{n-1}}$$

The distribution z^* is closer to a normal distribution than z. It is not known for how small a sample size this transformation is adequate, but most likely it should not be used for $n < 10$. As an example, z^* and σ_{z^*} are computed for the data of Box 15.1.

$r = 0.8652$ $n = 12$ $z = 1.3137$

$$z^* = z - \frac{(3z + r)}{4n} = 1.3137 - \frac{3(1.3137) + 0.8652}{4(12)}$$

$$= 1.3137 - 0.1001 = 1.2136$$

$$\sigma_{z^*} = \frac{1}{\sqrt{n-1}} = \frac{1}{\sqrt{11}} = \frac{1}{3.31662} = 0.30151$$

The differences between z and z^* and between their standard errors are slight and will be important only for critical tests near the borderline of significance. For example, to test the null hypothesis H_0: $\rho = 0$ versus H_1: $\rho \neq 0$,

$$t_s = (z^* - \zeta^*)\sqrt{n-1}$$

where ζ^* is the z^*-transformation applied to the ρ of the null hypothesis

$$\zeta^* = \frac{\zeta - (3\zeta + \rho)}{4n}$$

which equals zero since ζ and ρ are both zero. We again compare t_s with $t_{\alpha[\infty]}$ or look it up in Table **P**. Applying this formula to the gill-weight–body-weight correlation of Box 15.1, we obtain

$$t_s = (1.2136)\sqrt{11} = 1.2136(3.31662) = 4.0251 \qquad P < 0.001$$

If we had used the ordinary z-transformation, our result would have been

$$t_s = (z - \zeta)\sqrt{n-3} = 1.3137\sqrt{9} = 3.9411$$

leading to the same conclusion as before.

When $\rho \neq 0$, the distribution of sample values of r is markedly asymmetrical, and, although a standard error has been found for r in such cases, it should not be applied unless the sample is very large ($n > 500$), a most infrequent case of little interest. To overcome this difficulty, we transform r to a function z, developed by Fisher. The formula for z is

$$z = \frac{1}{2} \ln \left(\frac{1 + r}{1 - r} \right) \tag{15.11}$$

You may recognize this as $z = \tanh^{-1} r$, the formula for the inverse hyperbolic tangent of r. This function has been tabulated in Table **M**, where values of z corresponding to absolute values of r are given. Inspection of Expression (15.11) will show that when $r = 0$, z will also equal zero, since $\frac{1}{2} \ln 1$ equals zero. However, as r approaches 1, $(1 + r)/(1 - r)$ approaches infinity; consequently, z approaches infinity. Therefore, substantial differences between r and z occur at the higher values for r. Thus, when r is 0.115, $z = 0.1155$. For $r = -0.531$, we obtain $z = -0.5915$; $r = 0.972$ yields $z = 2.1273$. Note by how much z exceeds r in this last pair of values. For convenience in subsequent computations, we also furnish the inverse table (Table **N**) in which corresponding values of r can be looked up, given a value of z. Thus, $z = 0.70$ corresponds to $r = 0.6044$, and a value of $z = -2.76$ corresponds to $r = -0.9920$.

The advantage of the z-transformation is that while correlation coefficients are distributed in skewed fashion for values of $\rho \neq 0$, the values of z are approximately normally distributed for any value of its parameter, which we call ζ (zeta), following the usual convention. The expected variance of z is

$$\sigma_z^2 = \frac{1}{n - 3} \tag{15.12}$$

This is an approximation adequate for sample sizes $n \geq 50$ and a tolerable approximation even when $n \geq 25$. An interesting aspect of the variance of z evident from Expression (15.12) is that it is independent of the magnitude of r, but is simply a function of sample size n. Where the significance tests or confidence limits are critical and a better approximation is desired, Hotelling (1953) recommends another transformation z^*, which is computed as

$$z^* = z - \frac{3z + r}{4n} \tag{15.13}$$

Its variance is given as

$$\sigma_{z^*}^2 = \frac{1}{n - 1} \tag{15.14}$$

This is actually another approximation for a more involved expression but should be satisfactory for sample sizes $n \geq 10$.

As shown in Box 15.3, for sample sizes greater than 50 we can also use the z-transformation to test the significance of a sample r employing the hypothesis $H_0: \rho = 0$. In the second section of Box 15.3 we show the test of a null hypothesis that $\rho \neq 0$. We may have a hypothesis that the true correlation between two variables is a given value ρ different from zero. Such hypotheses about the expected correlation between two variables are frequent in genetic work, and we may wish to test observed data against such a hypothesis. Although there is no a priori reason to assume that the true correlation between right and left sides of the bee wing vein lengths is 0.5, we show the test of such a hypothesis to illustrate the method. Corresponding to $\rho = 0.5$, there is ζ, the parametric value of z. It is the z-transformation of ρ. We note that the probability that the sample r of 0.837 could have been sampled from a population with $\rho = 0.5$ is vanishingly small.

Next, in Box 15.3 we see how to set confidence limits to a sample correlation coefficient r. This is done by means of the z-transformation; it will result in asymmetrical confidence limits when these are retransformed to r-scale, as when setting confidence limits with variables subjected to square root or logarithmic transformations.

Occasionally one has measured two variables in several samples and obtained correlation coefficients for each sample. It may be of interest to know whether these correlation coefficients can be considered as samples from a population exhibiting a common correlation among the variables. One way of stating the null hypothesis is to say that the k sampled correlation coefficients are homogeneous and estimate a common parametric value of ρ. Such problems arise not infrequently in systematics, where taxonomic differentiation results in differences in correlation between body structures. Thus, not only do characters vary among taxa, or geographically within a taxon, but

BOX 15.4

Test of homogeneity among two or more correlation coefficients.

Correlation coefficients between length of wing and width of band on wing in females of ten populations of the butterfly *Heliconius charitonius* based on varying sample sizes.

(1)	(2)	(3)	(4)	(5)	(6)
n_i	$n_i - 3$	r_i	z_i	Weighted $z_i = (n_i - 3)z_i$	Weighted $z^2_i = (n_i - 3)z^2_i$
100	97	0.29	0.2986	28.9642	8.6487
46	43	0.70	0.8673	37.2939	32.3450
28	25	0.58	0.6625	16.5625	10.9727
74	71	0 56	0 6328	44.9288	28.4309
33	30	0.55	0.6184	18.5520	11.4726
27	24	0.67	0.8107	19.4568	15.7736
52	49	0.65	0.7753	37.9897	29.4534
26	23	0.61	0.7089	16.3047	11.5584
20	17	0.64	0.7582	12.8894	9.7727
17	14	0.56	0.6328	8.8592	5.6061
423	393			241.8012	164.0341

Source: Data from Brown and Comstock (1952).

BOX 15.4 continued

We wish to test whether the $k = 10$ sample r's could have been taken from the same population; if so, we shall combine them into an estimate of ρ.

Computational steps

1. Average z:

$$\bar{z} = \frac{\sum^{k} \text{weighted } z_i}{\sum (n_i - 3)} = \frac{\text{sum of column (5)}}{\text{sum of column (2)}} = \frac{241.8012}{393} = 0.61527$$

2. To test that all r's are from the same population:

Calculate

$$\sum^{k} (n_i - 3)z_i^2 = \text{sum of column (6)} = 164.0341$$

Compute correction term

$$\bar{z} \sum^{k} (n_i - 3)z_i = \text{quantity } \mathbf{1} \times \text{sum of column (5)}$$
$$= (0.61527) \times (241.8012) = 148.7730$$

Compute

$$X^2 = \sum^{k} (n_i - 3)z_i^2 - \bar{z} \sum^{k} (n_i - 3)z_i = 164.0341 - 148.7730 = 15.2611$$

X^2 is to be compared with values in the χ^2-table for $k - 1$ df:

$\chi^2_{.10[9]} = 14.7$ $\qquad$ $\chi^2_{.05[9]} = 16.9$

We do not quite have sufficient evidence to reject the null hypothesis of no heterogeneity among the correlation coefficients.

3. Estimate of common ρ:

From $\bar{z} = 0.61527$ $\qquad$ $\bar{r} = 0.5477 \approx 0.55$

The special case of two correlation coefficients

For only two correlation coefficients we may test $H_0\colon \rho_1 = \rho_2$ versus $H_1\colon \rho_1 \neq \rho_2$ as follows:

$$t_s = \frac{z_1 - z_2}{\sqrt{\dfrac{1}{n_1 - 3} + \dfrac{1}{n_2 - 3}}}$$

Since $z_1 - z_2$ is normally distributed and we are using a parametric standard deviation, we compare t_s with $t_{\alpha[\infty]}$ or employ Table **P**, areas of the normal curve.

For example, the correlation between body weight and wing length in *Drosophila pseudoobscura* was found by Sokoloff (1966) to be 0.552 in a sample of $n_1 = 39$ at the Grand Canyon and 0.665 in a sample of $n_2 = 20$ at Flagstaff, Arizona.

Grand Canyon: $\quad z_1 = 0.6213$ $\qquad$ Flagstaff: $\quad z_2 = 0.8017$

$$t_s = \frac{0.6213 - 0.8017}{\sqrt{\frac{1}{36} + \frac{1}{17}}} = \frac{-0.1804}{\sqrt{0.086601}} = \frac{-0.1804}{0.29428} = -0.6130$$

BOX 15.4 continued

By linear interpolation in Table **P**, we find the probability of a t_s being between ± 0.6130 to be about $2(0.22941) = 0.45882$, so we clearly have no evidence on which to reject the null hypothesis.

Procedures to follow for small sample sizes are suggested in the text (Section 15.4).

the correlation between pairs of characters also changes in different populations. The example we have chosen in Box 15.4 to illustrate the test of homogeneity among correlation coefficients is of this kind. It shows correlation between wing length and width of a band on the wing in females of ten populations of a species of butterfly. These ten populations were taken from different islands and localities in the Caribbean area. Note that the sample correlation coefficients differ considerably from a low of 0.29 to a high of 0.70. The computations are quite simple and in effect consist of calculating a weighted sum of squares of the z-values corresponding to the correlation coefficients. The sum of squares so calculated is called X^2 and is distributed as χ^2 with $k - 1$ degrees of freedom.

It may not be immediately obvious why the weighted sum of squares of z should be distributed as χ^2. It represents the quantity

$$X^2 = \sum^{k} \left(\frac{z - \bar{z}}{1/\sqrt{n_i - 3}} \right)^2 = \sum^{k} (n_i - 3)(z - \bar{z})^2$$

The denominator in the parentheses is the parametric standard deviation of z, σ_z. If we had equal sample sizes, so that the standard deviations were all equal, we could rewrite this expression as $(1/\sigma_z^2) \sum^{k} (z - \bar{z})^2$. Looked at in this light, the expression becomes analogous to Expression (7.7), $(1/\sigma^2) \sum (Y_i - \bar{Y})^2$, which, as you may recall, followed the χ^2-distribution with $n - 1$ degrees of freedom. Thus, if the z's represent a random sample from a single normally distributed population, the X^2 obtained above should follow the χ^2-distribution with $k - 1$ degrees of freedom.

In this specific example, we find that X^2 is smaller than the 5% critical value of chi square. We therefore do not reject the null hypothesis. The correlation coefficients in the ten populations should, therefore, be considered homogeneous (even though the $r = 0.29$ appears quite deviant from the rest). This result is rather typical; the r-values have to be very different or based on larger sample sizes before this test can detect differences. If the sample correlations are adjudged homogeneous we can use them to estimate an overall average to obtain a common estimate of ρ. This is done by looking up the r-value corresponding to $\bar{z}$ in Table **N**. The estimated common correlation is 0.5477. Note that this average is weighted by the reciprocals of the variances, which are a function of sample size. If some sample sizes are low and an especially precise estimate of ρ is desired, we must make allowance for a bias introduced in our sample estimate of ζ. An unbiased estimate, z_u, is given by

$z_u = z - [\rho/2(n-1)]$. You will note that to obtain such an estimate we need to know ρ, which is usually impossible; if we did know ρ, then we would, of course, not be interested in z_u. We therefore obtain a more approximate estimate of ζ from the sample r by the following formula: $z_u = z - [r/(2n-5)]$. These biases are of the same general kind as the bias in estimating the population variance as $\sum y^2/n$, first encountered in Section 4.6. In practice we would correct the estimate $\bar{z}$ by subtracting from it

$$\frac{\bar{r}\sum^{k}\left(\frac{n_i-3}{2n_i-5}\right)}{\sum^{k}(n_i-3)}$$

For the example in Box 15.4 this correction amounts to 0.01112, which results in a new estimate of $\bar{z} = 0.61527 - 0.01112 = 0.60415$. The new estimate of $\bar{r}$ is, therefore, 0.5400, which is only slightly different from the original $\bar{r}$ of 0.5477. Although workers are cautioned in the literature against applying the homogeneity test (and the two-sample test discussed below) to data with small sample sizes, no clear instructions can be found on what constitutes "small sample sizes" or on how to proceed in such cases. One approach might be to use the z^*-transformation of Hotelling, substituting z^* for z and $n_i - 1$ in place of $n_i - 3$ in the various formulas.

A test for the significance of the difference between two sample correlation coefficients is a special case of the test for homogeneity among r's and can be carried out in this manner. A simpler procedure is shown in the second half of Box 15.4. A standard error for the difference is computed and tested against a table of areas of the normal curve. In the example the correlation between body weight and wing length in two *Drosophila* populations was tested, and the difference in correlation coefficients between the two populations was found not significant. The formula given is an acceptable approximation when the smaller of the two samples is greater than 25. It is frequently used with even smaller sample size, as shown in our example in Box 15.4.

15.5 Applications of correlation

The purpose of correlation analysis is to measure the intensity of association observed between any pair of variables and to test whether it is greater than could be expected by chance alone. Once established, such an association is likely to lead to reasoning about causal relationships between the variables. Students of statistics are told at an early stage not to confuse significant correlation with causation. We are also warned about so-called nonsense correlations, a well-known case being the positive correlation between the number of Baptist ministers and the per capita liquor consumption in cities with populations of over 10,000 in the United States. Individual cases of correlation must be carefully analyzed before inferences are drawn from them. It is useful to distinguish correlations in which one variable is the entire, or more likely, the partial cause of another, from others in which the two correlated variables have a common cause and from more complicated situations involving both direct influence and common causes (see Figure 15.5).

Almost all observed correlations can be described in terms of one of the structural designs of this figure or a more complicated structure composed of these elements. There are probably few situations in nature in which two variables are totally correlated, either because one is the cause of the other (Figure 15.5, A) or both arise from a common cause (Figure 15.5, B). Exceptions are some trivial cases such as age of tree (Y_2) and number of growth rings (Y_1, Figure 15.5, A). The model described by Figure 15.5, C is more frequent. The correlation between age (Y_2) and weight (Y_1) in an animal is likely to be of this type. Factors other than age (Y_3, Y_4) will affect the animal, so the correlation between these two variables is unlikely to be perfect. Weight Y_1) and size (Y_2) of an animal usually have age (Y_4) as a common factor, leading to a model as shown in Figure 15.5, D. Most correlations found in nature are likely to be of the type shown in Figure 15.5, E in which two or more common causes (Y_4 and Y_5) lead to correlation of variables Y_1 and Y_2. Correlation between morphological parts in organisms is surely of this type, with many morphogenetic forces exerting an influence on the correlated variables. The next degree of complexity is shown in Figure 15.5, F, where the correlation between Y_1 and Y_2 is due both to the direct effect of one of the variables (Y_2) as well as to a common cause (Y_4). Adult weight and length of developmental period are correlated in many insects. There is a direct effect of length of developmental period on adult weight: the longer the immature insect feeds, the heavier the adult will be. However, the density at which the insects were

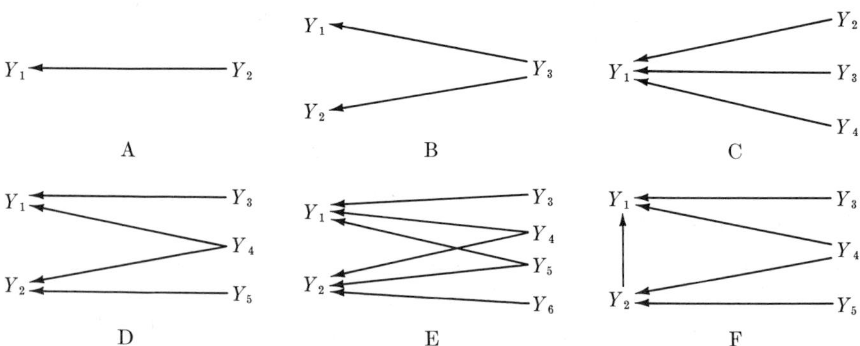

FIGURE 15.5 Different structural explanations for the observed correlation r_{12} between Y_1 and Y_2 (assuming only linear relations between variables). A. Y_2 is the entire cause of the variation of Y_1. In such a case r_{12} must equal to 1 and r_{12}^2 must also equal 1. B. The common cause Y_3 totally determines variables Y_1 and Y_2. Again, r_{12} must be 1. C. In this case, Y_2 is one of several causes of Y_1, and r_{12} will be less than 1. D. The correlation between variables Y_1 and Y_2 is due to a common cause Y_4. Since other causes, Y_3 and Y_5, also determine Y_1 and Y_2, respectively, the correlation between these variables will not be perfect. E. The correlation between variables Y_1 and Y_2 is due to two common causes Y_4 and Y_5. F. The correlation between variables Y_1 and Y_2 is due to the direct effect of Y_2 on Y_1 as well as to a common cause Y_4 affecting both variables.

reared will affect both larval period and weight. It will reduce the weight because of reduction in food, but generally increase the developmental period. Thus, the same common cause will increase one but lower the other of the two correlated variables, thus decreasing the correlation caused by the direct effect of density on weight. You can see that models of this sort can rapidly become quite complex.

The establishment of a significant correlation does not tell us which of these structural models is appropriate. Further analysis is needed to discriminate between the various models.

The traditional distinction of real versus nonsense or illusory correlation is of little use. In the supposedly legitimate correlations, the relations were of one of the types shown in Figure 15.5 and causal connections were known or at least believed to be clearly understood. In so-called illusory correlations, no reasonable connection between the variables can be found; or if one is demonstrated, it is of no real interest or may be shown to be an artifact of the sampling procedure. Thus, the correlation between Baptist ministers and liquor consumption is simply a consequence of city size. The larger the city, the more Baptist ministers it will contain on the average and the greater will be the liquor consumption. This is really an example of Figure 15.5, D, the common cause being city size. The correlation is of little interest to anyone studying either the distribution of Baptist ministers or the consumption of alcohol. Some correlations have time as the common factor, and processes that change with time are frequently likely to be correlated, not because of any functional biological reasons but simply because the change with time in the two variables under consideration happens to be in the same direction. Thus, size of an insect population building up through the summer may be correlated with the height of some weeds, but this may simply be a function of the passage of time. There may be no ecological relation between the plant and the insects.

Perhaps the only correlations properly called nonsense or illusory are those assumed by popular belief or scientific intuition which, when tested by proper statistical methodology using adequate sample sizes, are found to be not significant. Thus, if we can show that there is no significant correlation between amount of saturated fats eaten and the degree of atherosclerosis, we can consider this to be an illusory correlation. Remember also that when testing significance of correlations at conventional levels of significance, you must allow for type I error, which will lead to a certain percentage of correlations being judged significant when in fact the parametric value of $\rho = 0$.

Correlation coefficients have a history of extensive use and application dating back to the English biometric school at the turn of the century. Recent years have seen somewhat less application of this technique as increasing segments of biological research have become experimental. In experiments in which one factor is varied and the response of another variable to the deliberate variation of the first is examined, the method of regression is more appropriate, as has already been discussed. However, large areas of biology and of other sciences remain, where the experimental method is not suitable because variables cannot be brought under control of the investigator. There are many areas of ecology, systematics, evolution, and other fields in which experimental methods are difficult to apply. As yet the weather cannot be controlled nor can historical evolutionary factors be altered. Nevertheless, we need an understanding of the scientific mechanisms underlying these phenomena as much as of those in biochemistry or experimental embryology. In such cases, correlation analysis serves as a first descriptive technique estimating the degrees of association among the variables involved. Other techniques such as factor analysis, briefly discussed in Section 15.8, lead to further analytical insights into the mechanisms of such phenomena even without application of the experimental method.

15.6 Principal axes and confidence regions

We have used the correlation coefficient as a measure of intensity of association between two variables in a bivariate scattergram of the type shown in Figure 15.3. Occasionally it is of interest to characterize the trend of the scattered points by a line. It is also desirable to have a measure of the reliability of the two-way scattergram. Simple confidence limits are not adequate, since points on the scattergram can vary in two dimensions. We therefore talk of *confidence regions* around the sample mean $(\bar{Y}_1, \bar{Y}_2)$ and around individual observations in the scattergram. In the past, research workers have often used regression lines to characterize two-way scattergrams. However, in correlation analysis, in which both variables are random, regression lines are not appropriate. Moreover, it would be difficult to decide which of the pair of variables to regress on the other one. Sometimes both regression lines are shown cutting through the scattergram, and the viewer can visualize a trend line intermediate between them. The method of principal axes described below provides a single axis to represent the trend expressed by the scattergram and is preferable from a number of other considerations, several of which will be mentioned below.

We saw in Section 15.2 that the bivariate normal distribution can be represented by means of concentric ellipses describing the topography of a bell-shaped mound. In the univariate work of Chapter 6 we assumed that observed samples had been taken from a normal distribution and, consequently, calculated expected frequencies (areas under the normal curve). Analogously, we shall calculate concentric ellipses in these data representing various volumes under the bell-shaped surface of the bivariate normal distribution. In effect, the concentric ellipses are the contour lines of the three-dimensional model of Figure 15.2. In the univariate case, we set confidence limits to the mean and also calculated expected frequencies for normal distributions whose mean and standard deviation were identical to the sample

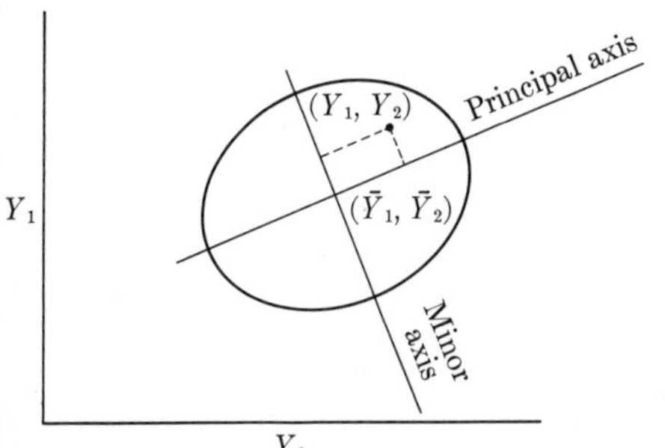

FIGURE 15.6 Equal frequency ellipse in which principal and minor axes and deviations of an observation (Y_1, Y_2) from these axes are shown. Note that the deviations are measured in a direction perpendicular to the principal and minor axes rather than perpendicular to the abscissa, as was done in regression analysis.

statistics. To do this, we needed two parameters: a fixed point, the mean μ for which our best estimate is $\bar{Y}$, and a measure of variation, the standard deviation σ which we estimate as s. The corresponding information necessary in the bivariate case is the sample mean $(\bar{Y}_1, \bar{Y}_2)$, the standard deviations s_1 and s_2, and the covariance s_{12}. You will recall from analytical geometry that an ellipse can be described by two principal axes, the major axis and the minor axis at right angles to each other, the major axis being the longest possible axis of the ellipse. Once the above parameters are estimated, we can draw concentric ellipses around the mean.

Our first task, therefore, is to find the slope and equation of the major axis from a sample. A criterion of fitting a regression line through bivariate data was that it must pass through $(\bar{Y}_1, \bar{Y}_2)$ and that the SS of the deviations of the observed Y-values from regression was to be minimal. These deviations are parallel to the Y-axis (see Figure 14.7). A corresponding criterion for locating the principal axis is to require that the SS of deviations, which represent perpendiculars to the principal axis, be at a minimum. These perpendiculars, illustrated in Figure 15.6, take into consideration the deviation of a given point with respect to both $\bar{Y}_1$ and $\bar{Y}_2$, by contrast with the regression case in which the deviation was measured only with respect to Y but not to X.

The equation of the principal axis defined in this way is $Y_1 = \bar{Y}_1 + b_1(Y_2 - \bar{Y}_2)$. The expression involves the means of the two variables and b_1, the slope of the principal axis. This slope is computed from $b_1 = s_{12}/(\lambda_1 - s_1^2)$, where s_{12} is the covariance $\sum y_1 y_2/(n - 1)$, s_1^2 is the variance of Y_1, and λ_1 is a new quantity defined as follows in terms of variances and covariance of Y_1 and Y_2:

$$\lambda_1 = \tfrac{1}{2}[s_1^2 + s_2^2 + \sqrt{(s_1^2 + s_2^2)^2 - 4(s_1^2 s_2^2 - s_{12}^2)}]$$

The slope, b_2, of the minor axis, which is at right angles to the slope of the major axis, can be found easily as $b_2 = -1/b_1$. The details of the computation are illustrated in Box 15.5.

To describe the shape of an ellipse, we need also to know the ratio of the lengths of its major and minor axes (this is a constant for all the concentric ellipses). This can be computed as the ratio λ_1/λ_2, where $\lambda_2 = s_1^2 + s_2^2 - \lambda_1$.

BOX 15.5

Principal axes and confidence regions for a bivariate scattergram.

Data from Box 15.1.

1. Compute s_1^2, s_2^2, s_{12}, $\bar{Y}_1$ and $\bar{Y}_2$. Using the sums of squares and cross products given in Box 15.1, we find:

$$s_1^2 = \frac{\sum y_1^2}{n-1} = \frac{124{,}368.9167}{11} = 11{,}306.265$$

$$s_2^2 = \frac{\sum y_2^2}{n-1} = \frac{462.4782}{11} = 42.043$$

$$s_{12} = \frac{\sum y_1 y_2}{n-1} = \frac{6561.6175}{11} = 596.511$$

$$\bar{Y}_1 = \frac{\sum Y_1}{n} = \frac{2347}{12} = 195.583$$

$$\bar{Y}_2 = \frac{\sum Y_2}{n} = \frac{144.57}{12} = 12.048$$

2. The eigenvalues (characteristic roots, latent roots) are found as follows (note that s_{12}^2 is the square of the covariance s_{12}):

Let $D = \sqrt{(s_1^2 + s_2^2)^2 - 4(s_1^2 s_2^2 - s_{12}^2)}$

Then $\lambda_1 = \dfrac{s_1^2 + s_2^2 + D}{2}$

and

$$\lambda_2 = \frac{s_1^2 + s_2^2 - D}{2}$$

or

$$\lambda_2 = s_1^2 + s_2^2 - \lambda_1$$

$$\begin{aligned} D &= \sqrt{(11{,}306.265 + 42.043)^2 - 4[11{,}306.265(42.043) - 596.511^2]} \\ &= \sqrt{11{,}348.308^2 - 4(119{,}523.926)} = \sqrt{128{,}305{,}998.759} \\ &= 11{,}327.224 \end{aligned}$$

$$\lambda_1 = \frac{11{,}348.308 + 11{,}327.224}{2} = 11{,}337.766$$

$$\lambda_2 = \frac{11{,}348.308 - 11{,}327.224}{2} = 10.542$$

3. The slope of the principal axis is

$$b_1 = \frac{s_{12}}{\lambda_1 - s_1^2} = \frac{596.511}{11{,}337.766 - 11{,}306.265} = \frac{596.511}{31.501} = 18.936$$

The equation of this axis is

$$\begin{aligned} Y_1 &= \bar{Y}_1 + b_1(Y_2 - \bar{Y}_2) = 195.583 + 18.936(Y_2 - 12.048) \\ &= -32.558 + 18.936Y_2 \end{aligned}$$

and it is plotted in Figure 15.7.

BOX 15.5 continued

4. The slope of the minor axis (at right angles to the principal axis) is

$$b_2 = \frac{-1}{b_1} = \frac{-1}{18.936} = -0.0528$$

The equation of the minor axis is consequently

$$\begin{aligned} Y_1 &= \bar{Y}_1 + b_2(Y_2 - \bar{Y}_2) = 195.583 - 0.0528(Y_2 - 12.048) \\ &= 196.219 - 0.0528Y_2 \end{aligned}$$

5. Computation of 95% confidence limits for β_1, the slope of the principal axis:

$$\text{Let } H = \frac{\chi^2_{\alpha[1]}}{[(\lambda_1/\lambda_2) + (\lambda_2/\lambda_1) - 2]\,n}$$

$$= \frac{3.841}{[(11{,}337.766/10.542) + (10.542/11{,}337.766) - 2]12}$$

$$= \frac{3.841}{(1075.485 + 0.00093 - 2)12} = \frac{3.841}{12{,}881.831} = 0.0002982$$

and

$$A = \sqrt{\frac{H}{(1 - H)}} = \sqrt{\frac{0.0002982}{0.9997018}} = \sqrt{0.0002983} = 0.01727$$

The 95% confidence limits to the slope β_1 are

$$L_1 = \frac{b_1 - A}{1 + b_1 A} = \frac{18.936 - 0.01727}{1 + 18.936(0.01727)} = \frac{18.91873}{1.327} = 14.257$$

$$L_2 = \frac{b_1 + A}{1 - b_1 A} = \frac{18.936 + 0.01727}{1 - 0.327} = \frac{18.95327}{0.673} = 28.162$$

6. Confidence ellipses for the bivariate mean (μ_1, μ_2):
The confidence region describes an ellipse around the point $(\bar{Y}_1, \bar{Y}_2)$. The equation for a $100(1 - \alpha)\%$ confidence region is

$$P\{s_1^2(Y_2 - \mu_2)^2 - 2s_{12}(Y_1 - \mu_1)(Y_2 - \mu_2) + s_2^2(Y_1 - \mu_1)^2 \leq C_\alpha\} = 1 - \alpha$$

where μ_1 is the true mean for Y_1, μ_2 is the true mean for Y_2, and

$$C_\alpha = \frac{\lambda_1\lambda_2(n - 1)2}{n(n - 2)} F_{\alpha[2,n-2]}$$

For a 95% confidence region applied to our example, $C_{.05}$ would be computed as follows:

$$F_{.05[2,10]} = 4.10$$

$$C_{.05} = \frac{(11{,}337.766)(10.542)(12 - 1)2}{12(12 - 2)} 4.10$$

$$= \frac{119{,}522.729(22)}{120} 4.10 = 21{,}912.500(4.10) = 89{,}841.250$$

The simplest way to make use of the confidence region is to graph its boundary. Several shortcuts are possible by computing the coordinates of particular points for which most of the above equation reduces to zero.

Given below are the coordinates of points a to h on Figure 15.7.

BOX 15.5 continued

a. $Y_2 = \bar{Y}_2 = 12.048$

$$Y_1 = \bar{Y}_1 + \sqrt{\frac{C_\alpha}{s_2^2}} = 195.583 + \sqrt{\frac{89{,}841.250}{42.043}}$$

$$= 195.583 + 46.227 = 241.810$$

b. $Y_2 = \bar{Y}_2 = 12.048$

$$Y_1 = \bar{Y}_1 - \sqrt{\frac{C_\alpha}{s_2^2}} = 195.583 - 46.227 = 149.356$$

c. $Y_1 = \bar{Y}_1 = 195.583$

$$Y_2 = \bar{Y}_2 + \sqrt{\frac{C_\alpha}{s_1^2}} = 12.048 + \sqrt{\frac{89{,}841.250}{11{,}306.265}}$$

$$= 12.048 + 2.819 = 14.867$$

d. $Y_1 = \bar{Y}_1 = 195.583$

$$Y_2 = \bar{Y}_2 - \sqrt{\frac{C_\alpha}{s_1^2}} = 12.048 + 2.819 = 9.229$$

e. $$Y_2 = \bar{Y}_2 + \sqrt{\frac{C_\alpha}{\lambda_2(1 + b_1^2)}} = 12.048 + \sqrt{\frac{89{,}841.250}{10.542(1 + 18.936^2)}}$$

$$= 12.048 + 4.868 = 16.916$$

$$Y_1 = \bar{Y}_1 + b_1(Y_2 - \bar{Y}_2) = 195.583 + 18.936(4.686)$$
$$= 195.583 + 88.734 = 284.317$$

f. $$Y_2 = \bar{Y}_2 - \sqrt{\frac{C_\alpha}{\lambda_2(1 + b_1^2)}} = 12.048 - 4.868 = 7.18$$

$$Y_1 = \bar{Y}_1 - b_1(Y_2 - \bar{Y}_2) = 195.583 - 88.734 = 106.849$$

g. $$Y_2 = \bar{Y}_2 + \sqrt{\frac{C_\alpha}{\lambda_1(1 + b_2^2)}} = 12.048 + \sqrt{\frac{89{,}841.250}{11{,}337.766[1 + (-0.05280^2)]}}$$

$$= 12.048 + 2.811 = 14.859$$

$$Y_1 = \bar{Y}_1 + b_2(Y_2 - \bar{Y}_2) = 195.583 - 0.05280(2.811) = 195.435$$

h. $$Y_2 = \bar{Y}_2 - \sqrt{\frac{C_\alpha}{\lambda_1(1 + b_2^2)}} = 12.048 - 2.811 = 9.237$$

$$Y_1 = \bar{Y}_1 - b_2(Y_2 - \bar{Y}_2) = 195.583 + 0.148 = 195.731$$

When these points are plotted and the axes drawn, it will become apparent that the two axes will not appear to be at right angles unless Y_1 and Y_2 are drawn to the same scale. This was not practical for Figure 15.7, since the units of measurement were very different.

It is sometimes of interest to draw "equal frequency ellipses" to describe the covariation found in a sample. It is the bivariate analogue of the $100(1 - \alpha)\%$ expected range in the dice-grams. One can construct an ellipse such that about

BOX 15.5 continued

100(1 − α)% of the observations in the given sample are expected to be enclosed. To do this, set

$$C_\alpha = \frac{\lambda_1\lambda_2(n-1)2}{(n-2)} F_{\alpha[2,n-2]}$$

and use the procedures described above for confidence regions.

The quantities λ_1 and λ_2 are important quantities in mathematics and statistics, known under a great variety of names. Among other terms, they are called eigenvalues, latent roots, or characteristic roots. In the type of problem we are discussing here, the eigenvalues are quantities analogous to variances, and λ_1 and λ_2 measure variability along the major and minor axes, respectively. In data that are very highly correlated and, consequently, are represented by a very narrow, elongated ellipse, most of the variance can be accounted for by the major axis, and therefore the value of λ_1 would be very great with respect to the magnitude of λ_2. If the quantities λ_1 and λ_2 were equal, the major and minor axes would be of equal length, the data would be represented by a circle, and hence there would be no correlation between the variables. In Figure 15.7 the major and minor axes do not appear to be orthogonal (at right angles) because Y_1 and Y_2 are not drawn to the same scale. In Figure 15.6, where both axes are of identical scale, the orthogonality is evident.

Step 5 of Box 15.5 illustrates the computation of approximate confidence

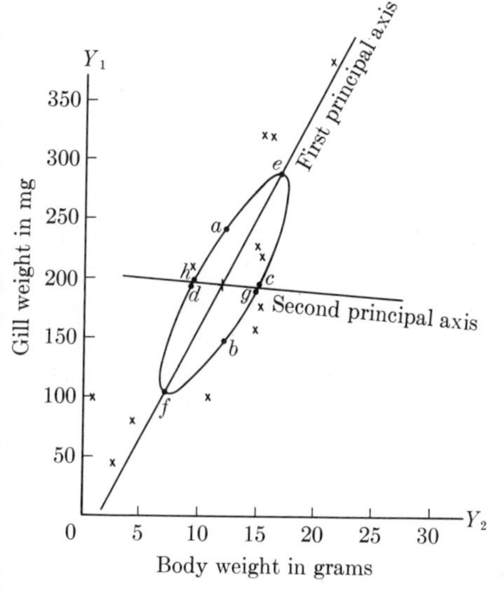

FIGURE 15.7 95% confidence region for the bivariate mean (μ_1, μ_2), using the crab data of Box 15.1. Axes do not appear at right angles since the ordinate and abscissa are not drawn to the same scale. (See Section 15.6 for an explanation.)

limits to the slope of the principal axis. It is exact only for large n. With small n and low r, A may be imaginary and the limits may not exist.

In step 6 of Box 15.5 is shown a method for finding confidence ellipses for the bivariate mean (μ_1, μ_2). Since two variables are involved, we cannot state univariate confidence intervals as was our previous practice, but need to identify a confidence region describing an ellipse around the point $(\bar{Y}_1, \bar{Y}_2)$. The shape of the ellipse is a function of the correlation between the variables, and the size (area) of the ellipse is a function of the confidence coefficient $1 - \alpha$. The equation describing the ellipse is shown in Box 15.5. We expect $100\ (1 - \alpha)\%$ of such ellipses to contain the true mean (μ_1, μ_2). Drawing an ellipse is a tedious procedure as one has to plot a sufficient number of points to be able to draw a curve connecting them. We substitute our sample estimates of $\bar{Y}_1$ for μ_1 and $\bar{Y}_2$ for μ_2, evaluate C_α as shown in the box, and can then solve for various sample values of Y_1, given possible values of Y_2. These equations would be difficult to work out; therefore we try to plot only certain points in which most of the equation can be especially simplified. We have labeled such points with lower case letters in the accompanying Figure 15.7. At points a and b, Y_2 equals $\bar{Y}_2$, at c and d the ellipse is cut at $Y_1 = \bar{Y}_1$, at e and f the ellipse is cut by the major axis, and at g and h it is cut by the minor axis. These points are evaluated for the example of Box 15.5 and plotted in Figure 15.7.

This is an equal-frequency ellipse of means; that is, the probability of obtaining a sample mean equal to or further from the mean $\bar{Y}_1$, $\bar{Y}_2$ is the same, α, for all points on the ellipse. If we desire an equal-frequency ellipse for observations analogous to the expected 95% range in a Dice-gram, we use a different formula for C_α, which has the same relation to the previous C_α as the variance of items has to the variance of means (see Box 15.5 for the formula). All of these computations are performed by the computer program given in Appendix A3.13.

The technique of principal axes is important in multivariate analysis, where instead of ellipses we encounter clouds of observations describing hyperellipsoids in a multidimensional space. To simplify the description of these clouds of points, we calculate principal axes through the hyperellipsoids. An important property which we have not so far emphasized is that the eigenvalues, which represent the variance along the principal axes, are always maximized, so that as we take successive principal axes representing the major axis, the second major axis, and so forth of the hyperellipsoid, we are successively removing the greatest, second greatest, and successively smaller sources of variation. This is the technique of principal component analysis, a branch of factor analysis discussed in Section 15.8.

15.7 Nonparametric tests for association

Occasionally data are known not to be bivariate normally distributed, yet we wish to test for the significance of association between the two variables. One method of analyzing such data is by ranking the variates and calculating a coefficient of rank correlation. This approach belongs to the general family of nonparametric methods we encountered in Chapter 13, where we learned methods for analyses of ranked variates paralleling anova. In other cases especially suited to ranking methods, we cannot measure the variable on an absolute scale, but only on an ordinal scale. This is typical of data in which we estimate relative performance, as in assigning positions in a class. We can say that A is the best student, B the second best student, C and D are both equal to each other and next best, and so on. Two instructors may independently rank a group of students and we can then test whether these two sets of rankings are correlated, as they should be if the judgments of the instructors are based on objective evidence. Of greater biological interest are the following examples. We might wish to correlate order of emergence in a sample of insects with a ranking in size, or order of germination in a sample of plants with rank order of flowering. A geneticist might predict the rank order of performance of a series of n genotypes he synthesizes and would wish to show the correlation of his prediction with the rank orders of the realized performance of these genotypes. A taxonomist might wish to array n organisms from those most like form X to those least like it. Will a similar array prepared by a second taxonomist be significantly correlated with the first—that is, are the taxonomic judgments of the two observers correlated?

There are several rank correlation coefficients. We present in Box 15.6 *Kendall's coefficient of rank correlation,* generally symbolized by τ (tau), although it is a sample statistic, not a parameter. The formula for Kendall's

BOX 15.6

Kendall coefficient of rank correlation, τ.

Computation of a rank correlation coefficient between the total length (Y_1) of 15 aphid stem mothers and the mean thorax length (Y_2) of their parthenogenetic offspring (based on measurement of four alates, or winged forms): $n = 15$ pairs of observations.

(1) Stem mother	*(2)* Y_1	*(3)* R_1	*(4)* Y_2	*(5)* R_2	*(1)* Stem mother	*(2)* Y_1	*(3)* R_1	*(4)* Y_2	*(5)* R_2
1	8.7	8	5.95	9	8	6.5	2	4.18	1
2	8.5	6	5.65	4	9	6.6	3	6.15	13
3	9.4	9	6.00	10	10	10.6	12	5.93	8
4	10.0	10	5.70	6.5	11	10.2	11	5.70	6.5
5	6.3	1	4.70	2	12	7.2	4	5.68	5
6	7.8	5	5.53	3	13	8.6	7	6.13	12
7	11.9	15	6.40	15	14	11.1	13	6.30	14
					15	11.6	14	6.03	11

Source: Data from a more extensive study by R. R. Sokal.

BOX 15.6 continued

Computational steps

1. Rank variables Y_1 and Y_2 separately and then replace the original variates with the ranks (assign tied ranks if necessary). These ranks are listed in columns (*3*) and (*5*) above. There was one tie in variable Y_2, so the 4th and 11th variates were assigned an average rank of 6.5.

2. Write down the n ranks of one of the two variables in order, paired with the rank values assigned for the other variable (as shown below). If only one variable has ties, order the pairs by the variable without ties (as in the present example). If both variables have ties, it does not matter which of the variables is ordered. There are three equivalent methods for computing the rank correlation coefficient.

3a. The conventional method is to obtain a sum of the counts C_i, as follows. Examine the first value in the column of ranks paired with the ordered column. In our case, this is rank 2. Count all ranks subsequent to it which are higher than the rank being considered. Thus, in this case, count all ranks greater than 2. There are fourteen ranks following the 2 and all of them except rank 1 are greater than 2. Therefore, we count a score of $C_1 = 13$. Now we look at the next rank (rank 1) and find that all thirteen subsequent ranks are greater than it; therefore, C_2 is also equal to 13. However, C_3 is only equal to 2 since only ranks 14 and 15 are higher than rank 13. Continue in this manner, taking each rank of the variable in turn and count the number of higher ranks subsequent to it. This can usually be done in one's head, but we show it explicitly below so that the method will be entirely clear. In the case of ties, count a $\frac{1}{2}$. Thus, for C_{10}, there are $4\frac{1}{2}$ ranks greater than the first rank of 6.5.

R_1	R_2	*Subsequent ranks greater than pivotal rank* R_2	*Counts* C_i
1	2	13, 5, 3, 4, 12, 9, 10, 6.5, 6.5, 8, 14, 11, 15	13
2	1	13, 5, 3, 4, 12, 9, 10, 6.5, 6.5, 8, 14, 11, 15	13
3	13	14, 15	2
4	5	12, 9, 10, 6.5, 6.5, 8, 14, 11, 15	9
5	3	4, 12, 9, 10, 6.5, 6.5, 8, 14, 11, 15	10
6	4	12, 9, 10, 6.5, 6.5, 8, 14, 11, 15	9
7	12	14, 15	2
8	9	10, 14, 11, 15	4
9	10	14, 11, 15	3
10	6.5	(6.5), 8, 14, 11, 15	$4\frac{1}{2}$
11	6.5	8, 14, 11, 15	4
12	8	14, 11, 15	3
13	14	15	1
14	11	15	1
15	15		0
			$78\frac{1}{2} = \sum^{n} C_i$

We then need the following quantity:

BOX 15.6 continued

$$N = 4\sum^{n} C_i - n(n-1) = 4(78\tfrac{1}{2}) - 15(14) = 314 - 210 = 104$$

3b. An equivalent procedure, which is simpler when there are no ties, is to compute a sum of the following counts k_i, each being equal to the number of ranks *of any magnitude* to the right of (or below) rank i. Thus, there are $k_1 = 13$ ranks after rank 1. Rank 1 should then be crossed out. Count k_2 equal to the number of ranks which follow rank 2 (ignoring rank 1, which was crossed out). Thus, k_2 equals 13 also. Rank 2 is then crossed out. Count $k_3 = 10$, since ten ranks follow rank 3, and so forth:

$$\sum^{n} k_i = 13 + 13 + 10 + 9 + 9 + 5 + 4 + 3 + 1 + 2 + 2 + 1 + 0$$
$$= 79$$

The quantity N is now defined as

$$N = 4\sum^{n} k_i - \sum^{m} T_2 - n(n-1)$$

The new quantity $\sum^{m} T$ is a correction term needed to correct for the presence of ties in the R_2 ranks (it is zero if there are no ties). A T-value equal to $t(t-1)$ is computed for each group of t tied variates and summed over m such groups. In our case, there is $m = 1$ group of $t = 2$ tied variates; therefore

$$\sum^{m} T_2 = T = 2(1) = 2,$$
$$N = 4(79) - 2 - 15(14) = 316 - 2 - 210 = 104$$

Note that $\sum^{n} C_i = \sum^{n} k_i - \dfrac{\sum^{m} T_2}{4}$.

3c. The following semigraphical procedure is often very convenient. Draw lines connecting identical ranks. In the case of ties, draw the lines from the tied ranks in such a way that they do not intersect. Count the number of intersections formed by the lines ($X = 26$ in the present case).

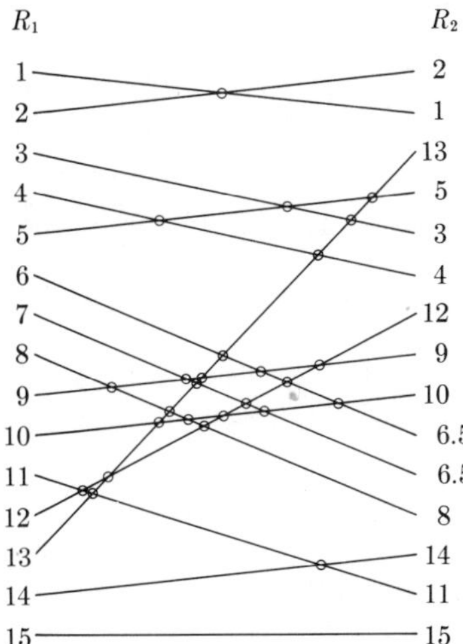

BOX 15.6 continued

The quantity N previously found by counting ranks can now be computed as follows:

$$N = n(n-1) - 4X - \sum^{m} T_2$$

where $\sum^{m} T_2$ is the correction term for ties defined above. As we have seen, in the present example $\sum^{m} T_2 = 2$. Therefore

$$N = 15(14) - 4(26) - (2) = 210 - 4(26) - 2 = 210 - 104 - 2 = 104$$

If there is a negative correlation between the ranks, it will simplify the counting of intersections if one of the variables is ranked in reverse order. The sign of N will then have to be reversed.

4. The Kendall coefficient of rank correlation, τ, can be found as follows:

$$\tau = \frac{N}{\sqrt{\left[n(n-1) - \sum^{m} T_1\right]\left[n(n-1) - \sum^{m} T_2\right]}}$$

where $\sum^{m} T_1$ and $\sum^{m} T_2$ are the sums of correction terms for ties in the ranks of variable Y_1 and Y_2, respectively, defined as in step **3b**. In our example $\sum^{m} T_1 = 0$, since there were no ties in the ranks R_1, $\sum^{m} T_2 = 2$ because of one group of $t = 2$ tied ranks R_2. $T = t(t-1) = 2(2-1) = 2$. Had there been more groups of ties, we would have summed the T's.

$$\tau = \frac{104}{\sqrt{[15(14)][15(14) - 2]}} = \frac{104}{\sqrt{210(208)}} = \frac{104}{\sqrt{43{,}680}}$$

$$= \frac{104}{208.9976} = 0.4976$$

If there are no ties, the equation can be simplified to

$$\tau = \frac{N}{n(n-1)}$$

5. To test significance for sample sizes >10, we can make use of a normal approximation to test the null hypothesis that the true value of $\tau = 0$:

$$t_s = \frac{\tau}{\sqrt{2(2n+5)/9n(n-1)}} = \frac{0.4976}{\sqrt{2[2(15)+5]/9(15)(14)}}$$

$$= \frac{0.4976}{\sqrt{70/1890}} = \frac{0.4976}{0.19253} = 2.59 \qquad \text{compared with } t_{\alpha[\infty]}$$

When this value is looked up in Table **P** (areas of the normal curve), we find the probability of such a t_s arising by chance to be 0.0096 (both tails).

When n is ≤ 10, the approximation given above is not adequate and the special table given below must be used. The table gives the 5 and 1% (two-tailed) critical values for N (the numerator of τ) for $n = 5$ to 10. These are exact only if there are no ties. If there are ties, a special table must be consulted (see Burr, 1960).

BOX 15.6 continued

Critical value of N such that $P(N_s > N_\alpha) \leq \alpha$ where N_s and N_α refer to sample estimates and critical values of N, respectively. This table is based on Table Q in Siegel (1956).

	N_α	
n	$\alpha = 0.05$	$\alpha = 0.01$
5	20	—
6	26	30
7	30	38
8	36	44
9	40	52
10	46	58

coefficient of rank correlation is $\tau = N/n(n - 1)$, where n is the conventional sample size and N is a count of ranks which can be obtained in a variety of ways (we show three simple approaches in Box 15.6). If a second variable Y_2 is perfectly correlated with the first variable Y_1, then the variates Y_2 should be in the same order as the Y_1-variates. However, if the correlation is less than perfect, the order of the variates Y_2 will not entirely correspond to that of Y_1. The quantity N measures how well the second variable corresponds to the order of the first. It has a maximal value of $n(n - 1)$ and a minimal value of $-n(n - 1)$. The following small example will make this clear. Suppose we have a sample of five individuals which have been arrayed by rank of variable Y_1:

Y_1 1 2 3 4 5

Y_2 1 3 2 5 4

Note that the ranking by variable Y_2 is not totally concordant with that by Y_1. One of the techniques in Box 15.6 (step 3a) is to count the number of higher ranks following any given rank, sum this quantity for all ranks, multiply the sum $\sum^n C_i$ by four, and subtract from it a correction factor $n(n - 1)$ to obtain a statistic N. For variable Y_1 we find $\sum^n C_i = 4 + 3 + 2 + 1 + 0 = 10$, then compute $N = 4 \sum^n C_i - n(n - 1) = 40 - 5(4) = 20$, to obtain the maximum possible score $N = n(n - 1) = 20$. Obviously, Y_1 being ordered is always perfectly concordant with itself. However, for Y_2 we only obtain $\sum^n C_i = 4 + 2 + 2 + 0 + 0 = 8$, and $N = 4(8) - 5(4) = 12$. Since the maximum score of N is $n(n - 1) = 20$ and the observed score 12, an obvious coefficient suggests itself as $N/n(n - 1) = [4 \sum^n C_i - n(n - 1)]/n(n - 1) = 12/20 = 0.6$. Kendall's τ is the converse of another coefficient called the *coefficient of disarray* $= -N/n(n-1)$, which measures the degree to which two sets of rankings are in disarray, or not ordered in the same way. Frequently, ties present minor computational complications that are dealt with in Box 15.6. The correlation in that box is between total body size of aphid stem mothers and mean thorax length of their offspring. In this case, there was no special need to turn to rank correlation except that there is some evidence that these data are bimodal and not normally distributed. The significance of τ for sample sizes greater than 10 can easily be tested by a standard error shown in Box 15.6. Where sample size is less than 10, look up critical values of N at the end of Box 15.6.

Another rank correlation coefficient is that by Spearman. It is computed for data arranged in a similar manner. There is no simple mathematical relation between the two coefficients. Spearman's coefficient, r_s, can be computed directly from the differences between the ranks R_1 and R_2 of paired variables 1 and 2 as follows:

$$r_s = \frac{1 - \sum^{n}(R_1 - R_2)^2}{n(n^2 - 1)}$$

BOX 15.7

Olmstead and Tukey's corner test for association.

A graphic "quick and dirty" method for determining the presence but not the magnitude of correlation. Data from Box 15.6: $n = 15$.

1. Prepare scattergram of the data to be tested. Scattergram in Figure 15.8 shows total length of stem mother along the ordinate Y_1 and mean thorax length of alates along the abscissa Y_2.

2. Draw the medians of Y_1 and Y_2 into the scattergram. When n is an odd number, the median lines will run across the median item of each array. When n is an even number, the median lines will pass between the pair of central variates of each array. Label the upper right and lower left quadrants $+$ and the upper left and lower right quadrants $-$, respectively.

3. Apply a ruler to the left side of the scattergram. Note the leftmost point and the sign of the quadrant in which it is located. Now slowly move the ruler to the right. Count 1 for each point, including the first, preceded by the sign of the quadrant in which it is located. Stop counting as soon as your next point crosses the median perpendicular to your ruler (the Y_1 median in this case). In other words, if your first point is $+1$, stop counting when you reach a negative point or vice versa. Enter the count ("quadrant count") of your points preceded by the appropriate sign at the left side of a square table.

Repeat this procedure starting at the bottom of the scattergram. Enter the quadrant count at the bottom of the table. Continue with the right side and

BOX 15.7 continued

top of the scattergram and enter quadrant counts at the right and top sides of the table, respectively.

When we follow this procedure on our scattergram, we obtain +5 for the left quadrant, +2 for the bottom, and +2 for the right. Obtaining the top quadrant count involves some complications.

Since the sample size is odd in this example, two points will lie on the medians (unless the same point is the median observation for both variables). The fourth point from the top lies on the Y_2 median at $Y_2 = 5.93$. In such a case, we are told to ignore this point (and also the other point on the Y_1 median at $Y_1 = 8.8$) and replace them by a single point at $Y_1 = 8.8$ and $Y_2 = 5.93$. This new point is marked by a triangle in the figure. The top quadrant count clearly becomes +4.

The broken lines in Figure 15.8 have been drawn to indicate the limits across which points change sign.

4. Sum the four quadrant counts, take the absolute value of this "quadrant sum" and read the probability from Table **FF**. In our case the absolute value of the quadrant sum $|S| = 13$, which is significant at 0.02. It is quite improbable that the association observed is due to sampling error.

Note: Another complication, not encountered in our example, occurs when on moving the ruler inward we are faced with a tie—two or more points on the same level and carrying opposite signs. In such a case, count the number of points in the tied group favorable to inclusion in the quadrant count and divide by (1 + the number of points unfavorable to inclusion in the quadrant count). In the case of two tied points, this will give $\frac{1}{2}$.

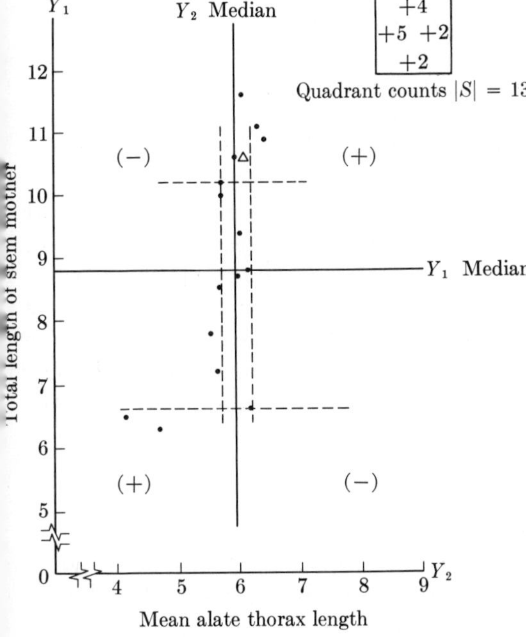

FIGURE 15.8 Diagram illustrating procedure for Olmstead and Tukey's corner test for association. (See Box 15.7 for explanation.) Scattergram of aphid morphological data from Box 15.6.

If the two variables Y_1 and Y_2 are actually independent, then the numerical value of Spearman's coefficient of rank correlation is highly correlated with that of Kendall's coefficient ($\rho_{r_s\tau} = 1$ for $n = 2$, it decreases to 0.98 for $n = 5$, and then increases to 1.0 again as $n \to \infty$). If the true correlation is not zero, then one would expect the two coefficients to be sensitive to different kinds of departures from independence. It is not possible, at present, to give any suggestions as to when each coefficient would best be used. When n is ≤ 10 special tables are needed for significance testing. For $n > 10$, one can test r_s as an ordinary product moment correlation coefficient with relatively little error.

It is sometimes desirable to establish the significance of the association between two variables without computation by inspection of a scattergram. *Olmstead and Tukey's corner test for association* is designed for this purpose and permits a significance test for plotted points, regardless of whether the exact numerical values are known to the tester. The method is described in Box 15.7 and Figure 15.8.

15.8 Advanced topics in correlation

In Section 14.14, we mentioned the subject of multiple and partial regression, where a dependent variable Y is regressed on several independent variables $X_1, X_2, X_3, \ldots$. We can extend the concept of correlation in a similar manner. The correlation coefficient between Y_1 and Y_2 is a measure of covariation between these two variables. A *multiple correlation coefficient* measures the joint covariation of a variable Y_1 with several other variables Y_2, Y_3, and Y_4. The square of a multiple correlation coefficient is the *coefficient of multiple determination.* The greater the number of variables associated with Y_1, the greater will be the multiple correlation coefficient, since additional correlated variables will increase the proportion of explained variance of Y_1 (unless the added variables are uncorrelated with Y, in which case the multiple correlation coefficient remains the same).

A *partial correlation coefficient* measures the correlation between any pair of variables when other, specified variables, have been kept constant. Thus, the correlation $r_{13\cdot245}$ is the correlation between variables Y_1 and Y_3 when variables Y_2, Y_4, and Y_5 are kept constant. Keeping these other variables constant can be done either experimentally or statistically and should give equivalent results. Again, we encounter the parallel approaches of experimental control and statistical control. It might not be immediately obvious why correlation should differ depending on whether other variables are free to vary or kept constant. If you refer back to Figure 15.5, this will become obvious. In Figure 15.5, D the correlation between Y_1 and Y_2 depends entirely on the common cause, Y_4. If we calculate the partial correlation $r_{12\cdot4}$, we should expect this to be zero because there is no reason why Y_1 and Y_2 should be correlated when Y_4 does not vary. Keeping variable Y_4 constant in Figure

15.5, E would enhance the relative correlating influcnce of common cause Y_5 on Y_1 and Y_2, because a larger proportion of the variation of Y_1 and Y_2 would now be accounted for by the common element Y_5. If both Y_4 and Y_5 are kept constant, the partial correlation $r_{12 \cdot 45}$ would again be zero. Partial correlations are simple to compute when only three variables are involved. With more than three variables, the computations become quite tedious and complex. They are carried out nowadays almost entirely on computers. Methods for calculating partial and multiple correlations can be found in sections 14.7 and 14.15 of Steel and Torrie (1960).

Figure 15.9 gives a graphic illustration of the meaning of partial correlations. The latter are fairly useful in a general descriptive way in trying to understand the interrelationships among a set of variables. Thus, we may wish to know the correlation between length of arm and length of leg with general size of the organism being kept constant. It is obvious that arm and leg length will be highly correlated because of general size. Thus, a tall individual will have long arms and long legs, while a short individual will have short extremities. However, even if we were to select individuals all of the same general size or if we statistically control general size, we would expect some residual correlation between arm length and leg length. This is probable in vertebrates because they are determined embryologically as limb buds, and homologous embryological mechanisms are responsible for the differentiation and determination of these buds. Therefore there will be some correlation

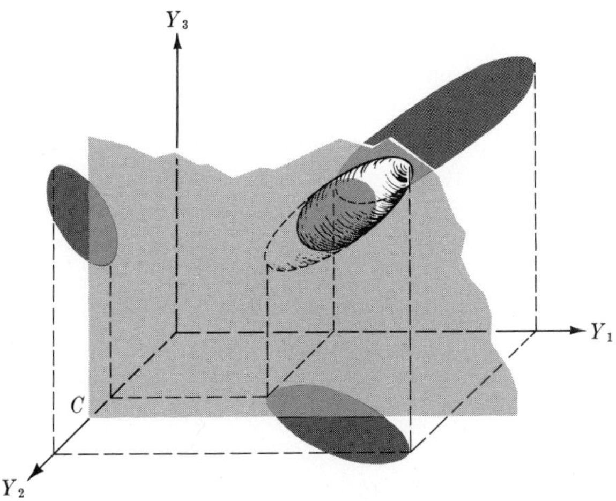

FIGURE 15.9 A geometric interpretation of partial correlation. A three-dimensional equal-frequency ellipsoid is shown in the center of the figure with its two-dimensional projections (shadows) shown on each wall. Cutting through the figure at the point $Y_2 = C$ (a constant) is a plane. The trace of the ellipsoid on this plane is an equal-frequency ellipse (the central shaded ellipse in the figure). This represents the covariation in variables Y_1 and Y_3 when variable Y_2 is held constant at C. The magnitude of the partial correlation between Y_1 and Y_3 is reflected in the shape of this ellipse. The more elongate this ellipse is, the higher will be the partial correlation.

between these even in the absence of the common cause, general size. If a significant correlation between two variables is changed to a nonsignificant partial correlation when a third variable is kept constant, this suggests but does not prove that the variable held constant is the common cause behind the correlation of the other two. The method of factor analysis discussed below is generally more suitable for the purpose of unraveling such causal nexuses among a set of variables.

Extensive studies in multiple and partial correlation usually start from a *correlation matrix*, which is a symmetrical table of the correlation coefficients of each variable in a set of variables with every other one. Such a matrix serves as a point of departure for numerous statistical techniques. Given the standard deviations of each variable, we may use it to obtain covariances. Or we may proceed from such a matrix to a study in multivariate analysis such as discriminant functions.

From a matrix of correlations emerge patterns of structure. Various methods of grouping the variables according to the magnitudes and interrelationships among their correlation coefficients have been developed. These methods are generally known as *cluster analysis*. Early attempts in this direction date back to the turn of the century, but there have recently been extensive developments in this field. A comprehensive recent review paper is Ball (1965).

While cluster analysis groups those variables that are highly correlated and similar to each other and excludes from clusters those variables that are unlike, *factor analysis* aims to express covariation in terms of k underlying factors that explain a large part of the variance and covariance of the original variables. The number of factors is much less than that of the number of variables in the study. There are several different methods for extracting factors from correlation matrices. The techniques of factor analysis are quite complex and time-consuming. However, the advent of computers has made the wide employment of factoı analytical procedures feasible. A suitable introductory text is Fruchter (1954), but Harman (1960) is a more comprehensive and up-to-date account of the subject, including a discussion of computer techniques. Seal (1964) discusses the application of factor analytical procedures in biology, emphasizing one school of factor analysis, that of principal components. In biology factor analysis aims at (1) resolving complex relationships into the interaction of fewer and simpler factors, be these physiological and environmental variables underlying behavioral correlations or morphometric trends underlying morphological correlations, (2) isolation and identification of causal factors behind biological correlations. A discussion of the application of factor analysis to work in biological systematics is given by Sokal (1965).

The analysis of correlation matrices representing similarities among taxonomic units has, in the last few years, developed into an extensive field of research called *numerical taxonomy*. This is a quantitative method for classification defined as the "numerical evaluation of the affinity or similarity between taxonomic units and the ordering of these units into taxa on the basis of their affinities." The procedure in numerical taxonomy consists of computing similarity matrices, frequently composed of product-moment correlation coefficients, and clustering these matrices into taxonomic systems based on these similarities. The computational procedures all require so much time as to make any but the smallest studies necessarily computer-oriented. This is a rapidly developing field. Interested readers should consult Sokal and Sneath (1963), and the current literature in the journals *Systematic Zoology*, *Taxon*, and *Evolution*. The statistical background required for most of the work in numerical taxonomy is such that readers of this book should have no difficulty in following the procedures. ↚

Exercises 15

15.1 Graph the following data in the form of a bivariate scatter diagram. Compute the correlation coefficient and set 95% confidence intervals to ρ. The data were collected for a study of geographic variation in the aphid *Pemphigus populi-transversus*. The values in the table represent locality means based on equal sample sizes for 23 localities in eastern North America. The variables, extracted from Sokal and Thomas (1965), are expressed in millimeters. Y_1 = tibia length, Y_2 = tarsus length. The correlation coefficient will estimate correlation of these two variables over localities.

Locality code number	Y_1	Y_2
1	0.631	0.140
2	.644	.139
3	.612	.140
4	.632	.141
5	.675	.155
6	.653	.148
7	.655	.146
8	.615	.136
9	.712	.159
10	.626	.140
11	.597	.133
12	.625	.144
13	.657	.147
14	.586	.134
15	.574	.134
16	.551	.127
17	.556	.130
18	.665	.147
19	.585	.138
20	.629	.150
21	.671	.148
22	.703	.151
23	.662	.142

ANS. $r = 0.910$.

15.2 This two-way frequency distribution was given by Band and Ives (1963) in a study of viability of different combinations of chromosomes in drosophila. Each variate represents means of paired cultures run at two temperatures.

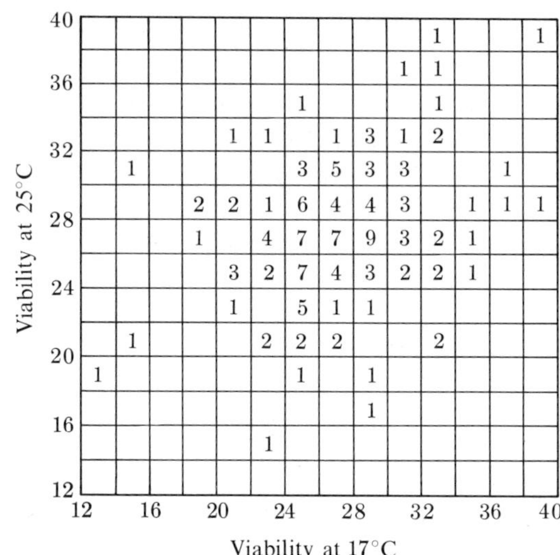

Note that the figures given in the margin are class limits, not class marks. Compute the correlation coefficient. Test the null hypothesis $\rho = 0$ versus the alternative hypothesis $\rho > 0$.

15.3 The following data were extracted from a larger study by Brower (1959) on speciation in a group of swallowtail butterflies. Morphological measurements are in millimeters coded $\times$ 8.

Species	*Specimen number*	Y_1 *Length of 8th tergite*	Y_2 *Length of superuncus*
Papilio multicaudatus	1	24.0	14.0
	2	21.0	15.0
	3	20.0	17.5
	4	21.5	16.5
	5	21.5	16.0
	6	25.5	16.0
	7	25.5	17.5
	8	28.5	16.5
	9	23.5	15.0
	10	22.0	15.5
	11	22.5	17.5
	12	20.5	19.0
	13	21.0	13.5
	14	19.5	19.0
	15	26.0	18.0
	16	23.0	17.0
	17	21.0	18.0
	18	21.0	17.0
	19	20.5	16.0
	20	22.5	15.5

Species	Specimen number	Y_1 Length of 8th tergite	Y_2 Length of superuncus
Papilio rutulus	21	20.0	11.5
	22	21.5	11.0
	23	18.5	10.0
	24	20.0	11.0
	25	19.0	11.0
	26	20.5	11.0
	27	19.5	11.0
	28	19.0	10.5
	29	21.5	11.0
	30	20.0	11.5
	31	21.5	10.0
	32	20.5	12.0
	33	20.0	10.5
	34	21.5	12.5
	35	17.5	12.0
	36	21.0	12.5
	37	21.0	11.5
	38	21.0	12.0
	39	19.5	10.5
	40	19.0	11.0
	41	18.0	11.5
	42	21.5	10.5
	43	23.0	11.0
	44	22.5	11.5
	45	19.0	13.0
	47	22.5	14.0
	48	21.0	12.5
Papilio glaucus	49	17.5	9.0
	50	19.0	9.0
	51	18.0	10.0
	52	19.0	8.0
	53	19.0	8.5
	54	21.0	8.5
	55	19.0	9.0
	56	21.0	9.5
	57	19.0	9.5
	58	17.5	9.0
	64	23.5	10.5
	65	16.0	9.5
	66	19.5	10.0
	67	17.0	7.0
	68	16.0	8.0
	69	18.0	9.0
	70	18.5	8.5
	71	19.0	8.5
	72	16.5	8.5
	73	18.5	9.5
	74	17.5	9.5

(continued)

Species	Specimen number	Y_1 Length of 8th tergite	Y_2 Length of superuncus
Papilio glaucus (continued)	75	19.5	8.0
	76	18.5	8.0
	77	19.0	9.0
	78	20.5	9.0
	79	17.0	9.5
	80	18.5	9.0
	81	16.5	9.0
	82	17.5	9.0
	83	19.0	9.0
	84	18.5	7.5
	85	18.5	8.5
	86	17.5	9.0
	87	18.5	9.5
	88	16.5	11.0
	89	23.0	9.0
	90	22.0	8.0
	91	18.0	10.0
	92	21.5	9.0
	93	18.5	9.0
	94	20.5	8.0
	95	21.0	10.0
	96	21.0	8.0
	97	22.5	9.5
	98	18.0	10.0
Papilio eurymedon	99	17.5	11.5
	100	17.5	11.5
	101	17.0	11.0
	102	18.5	11.5
	103	19.5	10.0
	104	16.0	11.5
	105	17.0	10.5
	106	17.5	10.5
	107	19.0	10.5
	108	19.0	12.0
	109	19.0	11.5
	110	21.0	11.0
	111	19.5	10.5
	112	19.5	10.5
	113	19.5	9.5
	114	19.0	10.5
	115	19.5	12.0
	116	21.5	11.5
	117	16.5	11.0
	118	21.5	13.5

Compute the correlation coefficient separately for each species. Test for homogeneity of the four correlation coefficients. ANS. For *Papilio rutus* $r = 0.1958$.

15.4 Graph the data given in Exercise 15.3 for *Papilio glaucus* in the form of a scatter diagram. (a) Compute the principal and minor axes. (b) Draw the 95% equal frequency ellipse for the scatter diagram. (c) Draw the 95% confidence ellipse for the bivariate mean (μ_1, μ_2).

15.5 Test for the presence of association between tibia length and tarsus length in the data of Exercise 15.1 using (a) the Olmstead and Tukey Corner test of association, (b) Kendall's coefficient of rank correlation. ANS. $|S| = 15$.

15.6 The following four correlation coefficients are extracted from a paper by F. H. Clark (1941). They represent correlations between skull width and skull length in four samples of the deermouse, *Peromyscus maniculatus sonoriensis*, from different localities. Can we consider the four sample r's to have come from a single population with parametric correlation ρ? If so, what is your estimate of ρ?

	n	r
Miller Canyon, Ariz.	71	0.68
Winslow, Ariz.	54	0.54
San Felipe, Calif.	83	0.56
Victorville, Calif.	57	0.43

15.7 The following table of data is from an unpublished morphometric study of the cottonwood, *Populus deltoides*, by T. J. Crovello. One hundred leaves from one tree were measured when fresh and after drying. The variables shown are fresh leaf width (Y_1) and dry leaf width (Y_2), both in millimeters. Group the data in a two-way frequency distribution and calculate r.

Y_1	Y_2	Y_1	Y_2	Y_1	Y_2
90	88	94	87	86	80
88	87	83	81	98	90
55	52	90	88	100	98
100	95	90	84	70	65
86	83	91	90	100	96
90	88	88	86	105	98
82	77	98	94	95	88
78	75	98	94	95	90
115	109	89	85	93	89
100	95	90	86	95	97
110	105	105	102	95	90
84	78	95	90	104	99
76	71	100	93	81	78
100	97	104	99	85	83
110	105	84	78	108	103
95	90	85	81	112	106
99	98	89	92	96	90
104	100	118	110	98	93
92	92	104	97	83	78
80	82	105	100	103	98
110	106	113	108	107	104
105	97	106	100	72	67
101	98	87	83	105	100
95	91	100	96	80	75
80	76	108	103	99	95
103	97	101	97	108	102

(continued)

Y_1	Y_2	Y_1	Y_2	Y_1	Y_2
108	103	92	90	86	84
113	110	84	85	94	92
90	85	89	85	106	104
97	93	120	111	98	104
107	106	96	86	100	96
112	115	97	94	95	90
101	98	86	83	108	102
95	91				

ANS. For the ungrouped data $r = 0.974$.

16 ANALYSIS OF FREQUENCIES

Almost all our work so far has dealt with estimation of parameters and tests of hypotheses in continuous variables. The present chapter treats an important class of cases, tests of hypotheses about frequencies. Biological variables may be distributed into two or more classes, depending on some criterion such as arbitrary class limits in a continuous variable or a set of mutually exclusive attributes. An example of the former would be a frequency distribution of birth weights (a continuous variable arbitrarily divided into a number of contiguous classes); one of the latter would be a qualitative frequency distribution such as the frequency of individuals of ten different species obtained from a soil sample. For any such distribution we may hypothesize that it was sampled from a population in which the frequencies of the various classes represent certain parametric proportions of the total frequency. We need a test of goodness of fit for our observed frequency distribution to the expected frequency distribution representing our hypothesis. You may recall that we first realized the need for such a test in Chapters 5 and 6, where we calculated expected binomial, Poisson, and normal frequency distributions but were unable to decide whether an observed sample distribution departed significantly from the theoretical one.

In Section 16.1 we introduce the idea of goodness of fit, discuss the types of significance tests that are appropriate, the basic rationale behind such tests, and develop general computational formulas for these tests.

Section 16.2 illustrates the actual computations for goodness of fit when the data are arranged by a single criterion of classification, as in a one-way quantitative or qualitative frequency distribution. This applies to cases expected to follow one of the well-known frequency distributions such as the binomial, Poisson, or normal distribution, as well as to expected distributions following some other law suggested by the scientific subject matter under investigation, such as, for example, tests of goodness of fit of observed genetic ratios against expected Mendelian frequencies. The goodness of fit tests discussed in this section are the chi-square, G, and Kolmogorov-Smirnov tests.

An interesting property of the G-test statistic is its additivity, so that the results of several G-tests can be summed to yield meaningful results. Similarly, overall G-tests can be partitioned into separate G-tests representing individual degrees of freedom in a manner analogous to that practiced in anova. Such procedures are treated in Section 16.3 dealing with pooled, total, and interaction G. The X^2-statistic resulting from chi-square tests is also approximately additive so that similar analyses can be made using it.

We proceed to significance tests of frequencies in two-way classifications—called tests of independence. Section 16.4 discusses various two-way classifications including the common tests of 2×2 tables in which each of two criteria of classification divides the frequencies into two classes, yielding a four-cell table. Various procedures for analyzing such cases, including the conventional chi-square test, the G-test, and Fisher's exact test, are presented. Also included is an STP approach to partitioning the differences among the overall departures from expectation. The following section (16.5) is a discussion of procedures for testing frequency distributions classified on the basis of more than two criteria. Section 16.6 presents methods for testing the difference between two percentages and for estimating sample sizes when planning experiments for such a test. Section 16.7 deals with goodness of fit tests in two-way classifications in which successive columns of the two-way table represent repeated measurements of the same individuals rather than different individuals, as in the examples taken up in Section 16.4. This design is analogous to the randomized block design of Section 11.4.

Through the early sections of this chapter we furnish two different ways of carrying out goodness of fit tests—by chi-square tests and by the G-statistic. The former is the traditional way of analyzing such cases, but as is explained at various places throughout the text, G has general theoretical advantages over X^2, as well as being computationally simpler for tests of independence. It may be confusing to the reader to have two alternative tests presented for most types of problems and our inclination would be to drop the chi-square tests entirely and teach G only. However, we are not prepared at this stage to carry out so marked a break with tradition; hence we have retained the chi-square tests in all cases where they are not markedly inferior to the corresponding G-tests. To the newcomer to statistics, however, we would recommend that he familiarize himself principally with the G-tests.

16.1 Tests for goodness of fit: Introduction

The basic idea of a goodness of fit test will be quite simple to convey, in view of your by now extensive experience with statistical hypothesis testing.

Let us assume that a geneticist has carried out a crossing experiment between two F_1 hybrids and obtains an F_2 progeny of 90 offspring, 80 of which appear to be wild type and 10 are mutants. The geneticist assumes dominance and expects a 3:1 ratio of the phenotypes. When we calculate the actual ratios, however, we observe that the data are in a ratio $80/10 = 8:1$. Expected values for p and q are $\hat{p} = 0.75$ and $\hat{q} = 0.25$ for the wild type and mutant, respectively. Note that we use the caret (generally called "hat" in statistics) to indicate hypothetical or expected values of the binomial proportions. However, the observed proportions of these two classes are $p = 0.89$ and $q = 0.11$, respectively. Yet another way of noting the contrast between observation and expectation is to state it in frequencies: the observed frequencies are 80 and 10 for the two phenotypes. Expected frequencies should be $\hat{f}_1 = \hat{p}n = 0.75(90) = 67.5$ and $\hat{f}_2 = \hat{q}n = 0.25(90) = 22.5$, respectively, where n refers to the sample size of offspring from the cross. Note that when we sum the expected frequencies they yield $67.5 + 22.5 = n = 90$, as they should.

The obvious question that comes to mind is whether the deviation from the 3:1 hypothesis observed in our sample is of such a magnitude as to be improbable. In other words, do the observed data differ enough from the expected values to cause us to reject the null hypothesis? For the case just considered, you already know two methods for coming to a decision about the null hypothesis. Clearly, this is a binomial distribution in which p is the probability of being a wild type and q is the probability of being a mutant. It is possible to work out the probability of obtaining an outcome of 80 wild type and 10 mutants as well as all "worse" cases for $\hat{p} = 0.75$ and $\hat{q} = 0.25$, and a sample of $n = 90$ offspring. We use the conventional binomial expression here $(\hat{p} + \hat{q})^n$ except that p and q are hypothesized, and we replace the symbol k by n, which we adopted in Chapter 5 as the appropriate symbol for the sum of all the frequencies in a frequency distribution. In this example, we have only one sample, so what would ordinarily be labeled k in the binomial is, at the same time, n. Such a problem was illustrated in Table 5.3 and Section 5.2, and we can compute the cumulative probability of the tail of the binomial distribution by the machine method of Box 5.1. When this is done, we obtain a probability of 0.00085 for all outcomes as deviant or more deviant from the hypothesis. Note that this is a one-tailed test, the alternative hypothesis being that there are, in fact, more wild type offspring than the Mendelian hypothesis would postulate. Assuming $\hat{p} = 0.75$ and $\hat{q} = 0.25$, the observed sample is, consequently, a very unusual outcome, and we conclude that there is a significant deviation from expectation.

A less time-consuming approach based on the same principle is to look up confidence limits for the binomial proportions as was done for the sign test in Section 13.11. Interpolation in Table **W** shows that for a sample of $n = 90$, an observed percentage of 89% would yield approximate 99% confidence limits of 78 and 96 for the true percentage of wild type individuals. Clearly, the hypothesized value of $\hat{p} = 0.75$ is beyond the 99% confidence bounds.

Now, let us develop a third approach by a goodness of fit test. Table 16.1 illustrates how we might proceed. In the first column are given the observed

TABLE 16.1

Developing the chi-square test for goodness of fit. Observed and expected frequencies from the outcome of a genetic cross, assuming a 3:1 ratio of phenotypes among the offspring.

	(1)	*(2)*	*(3)*	*(4)*	*(5)*
Phenotypes	*Observed frequencies* f	*Expected frequencies* $\hat{f}$	*Deviations from expectation* $f - \hat{f}$	*Deviations squared* $(f - \hat{f})^2$	$\frac{(f - \hat{f})^2}{\hat{f}}$
Wild type	80	$\hat{p}n = 67.5$	12.5	156.25	2.315
Mutant	10	$\hat{q}n = 22.5$	−12.5	156.25	6.944
Sum	90	90.0	0		$X^2 = 9.259$

frequencies f representing the outcome of the experiment. Column (2) shows the expected frequencies $\hat{f}$ based on the particular hypothesis being tested. In this case, the hypothesis is a 3:1 ratio and we have already calculated the expected frequencies under these conditions as $\hat{f}_1 = \hat{p}n = 0.75(90) = 67.5$ and $\hat{f}_2 = \hat{q}n = 0.25(90) = 22.5$.

How can we develop a statistic for testing to what degree the observed frequencies in column (1) differ from the expected frequencies in column (2)? The following test statistic is easily understood and its structure makes intuitive sense. We first measure $f - \hat{f}$, the deviation of observed from expected frequencies. Note that the sum of these deviations equals zero, for reasons very similar to those causing the sum of deviations from a mean to add to zero. Hence in an example with two classes, the deviations are always equal and opposite in sign (proof of this property is developed in Appendix A1.12). Following our previous approach of making all deviations positive by squaring them, we square $(f - \hat{f})$ in column (4) to yield a measure of the magnitude of the deviation from expectation. This quantity must be expressed as a proportion of the expected frequency. After all, if the expected frequency were 13.0, a deviation of 12.5 would be an extremely large one, comprising almost 100% of $\hat{f}$, but such a deviation would only represent 10% of an expected frequency of 125.0. Thus, we obtain column (5) as the quotient of the quantity in column (4) divided by that in column (2). Note that the magnitude of the quotient is greater for the second line, in which the $\hat{f}$ is smaller. Our next step in developing our test statistic is to sum these quotients, which is done at the foot of column (5), yielding a value of 9.259.

What shall we call this new statistic? We have some nomenclatural problems here. Many of you will have recognized this as the so-called *chi-square test*, regularly taught to beginning genetics classes. The name of the test is too well established to make a change seem practical, but, in fact, this quantity just computed, the sum of column (5), could not possibly be a χ^2. The latter is a continuous and theoretical frequency distribution, while our quantity 9.259 is a sample statistic based on discrete frequencies. The latter point is easily seen if you visualize other possible outcomes. For instance, we could have had as few as zero mutants matched by 90 wild type individuals (assuming that the total number of offspring $n = 90$ remains constant), or we could have had 1, 2, 3, or more mutants, in each case balanced by the correct number of wild type offspring to yield a total of 90. Observed frequencies change in unit increments, and since the expected frequencies remain constant, it is clear that deviations, their squares, and the quotients are not continuous variables but can assume only certain values.

The reason why this test has been called the chi-square test and why many persons call the statistic obtained as the sum of column (5) a chi-square, is that the sampling distribution of this sum approximates that of a chi-square distribution with one degree of freedom. We have previously encountered cases of this sort. The tests of homogeneity of variances (Section 13.4) and of homogeneity of correlation coefficients (Section 15.4) employed a sample statistic distributed approximately as a chi-square. However, since the sample statistic is not a chi-square, we have followed the increasingly prevalent convention of labeling the sample statistic X^2 rather than χ^2. We shall follow the same practice here, although in one sense this is misleading. The three tests that we have considered, the homogeneity tests for variances and correlation coefficients, and the goodness of fit test shown in Table 16.1, have only one thing in common—their sample statistic is approximately distributed as chi-square. For this reason, we give it an English letter resembling χ^2. However, structurally the three tests are quite different and to be strictly correct, we should really provide each of these with a separate symbol to emphasize that the three sample statistics are not the same. Since we have followed a policy of conservatism on symbolism, we retain the X^2, but the reader should be clear that the sample statistic X^2 in this chapter is for the test of goodness of fit and is not the same as the X^2 for homogeneity of either variances or correlations.

Why should the distribution of X^2 as computed in Table 16.1 approximate that of χ^2? Rigorous proof requires a knowledge of mathematical statistics beyond the scope of this book, but we shall provide two simple demonstrations that this is so—one through an inspection of the formula for X^2 and the other empirically by means of a sampling experiment. Let us express the computational steps of Table 16.1 by means of a formula:

$$X^2 = \frac{(f_1 - \hat{f}_1)^2}{\hat{f}_1} + \frac{(f_2 - \hat{f}_2)^2}{\hat{f}_2}$$

$$= \sum^{2} \frac{(f_i - \hat{f}_i)^2}{\hat{f}_i} \tag{16.1}$$

Expression (16.1) is the summation of the two quotients of deviations squared over expected frequencies. By some very simple algebra shown in Appendix A1.13, we can change Expression (16.1) into

$$X^2 = \frac{(f_1 - \hat{p}n)^2}{\hat{p}\hat{q}n} \tag{16.2}$$

or

$$X^2 = \frac{(f_2 - \hat{q}n)^2}{\hat{p}\hat{q}n} \tag{16.2a}$$

The numerator of either of these expressions is the square of the deviation between observed and expected frequencies in one class; the denominator is the product of the expected probabilities $\hat{p}$ and $\hat{q}$ multiplied by sample size n. You may recall Expression (7.6), $\frac{1}{\sigma^2}\sum(Y_i - \mu)^2$, which describes the χ^2-distribution. Its numerator is a sum of squares, and its denominator is a parametric variance. Look at Expression (16.2) again. Its numerator is a deviation from expectation squared, which is a sum of squares with one degree of freedom. The denominator is the variance of the binomial distribution, a parametric variance. It therefore appears quite reasonable from Expression (16.2) that X^2 as computed in the above problem is distributed as chi-square with one degree of freedom.

Next, let us study the distribution of X^2 by means of a sampling experiment. Suppose our expectation of a 3:1 ratio is, in fact, correct. What would be the distribution of the sample statistics X^2 if we took a series of samples of 90 items from a population in proportions 3:1 and tested them against the 3:1 hypothesis? If the sample came out exactly in a 3:1 ratio (mathematically impossible for 90 items), then our value of X^2 would be zero. The greater the departure from expectation, the greater the value of X^2. Figure 16.1 (left histogram) shows a distribution of X^2's for 1000 samples of 90 items from a population specified as $\hat{p} = 0.75$ and $\hat{q} = 0.25$. On the distribution of X^2 we have superimposed the probability density function of the χ^2-distribution for one degree of freedom, broken up into corresponding classes for ease of comparison. There is general agreement between the two distributions. It would appear that the X^2-values are distributed approximately as χ^2 with one degree of freedom.

The value of $X^2 = 9.259$ from Table 16.1, when compared with the critical value of χ^2 (Table **R**), is highly significant ($P < 0.005$). The chi-square test is always one-tailed. Since the deviations are squared, negative and

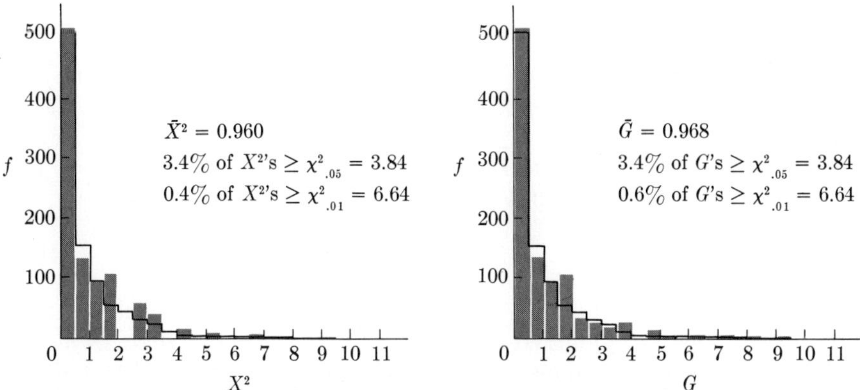

FIGURE 16.1 Results of a sampling experiment in which 1000 samples of size $n = 90$ were drawn from a binomial population with $\hat{p} = 0.75$. The histograms show the distribution of the resulting values of X^2 (left) and G (right) computed for each sample on the hypothesis that $\hat{p} = 0.75$ and $\hat{q} = 0.25$. The expected $\chi^2_{[1]}$-distribution is superimposed for comparison.

positive deviations both result in positive values of X^2. Clearly, we reject the 3:1 hypothesis and conclude that the proportion of wild type is greater than 0.75. The geneticist must, consequently, look for a mechanism explaining this departure from expectation.

The test for goodness of fit can be applied to a distribution with more than two classes. Table 16.2 shows the outcome of a dihybrid cross in tomato genetics in which the expected ratio of phenotypes is 9:3:3:1. The table is set up as before with expected frequencies calculated as $\hat{f}_i = \hat{p}_i n$, where the

TABLE 16.2

The chi-square test for goodness of fit for more than two classes. Observed and expected frequencies in a dihybrid cross between tall, potato-leaf tomatoes and dwarf, cut-leaf tomatoes.

Phenotypes	(1) *Observed frequencies* f	(2) *Expected frequencies* $\hat{f}$	(3) *Deviations from expectation* $f - \hat{f}$	(4) *Deviations squared* $(f - \hat{f})^2$	(5) $\frac{(f - \hat{f})^2}{\hat{f}}$
Tall, cut-leaf	926	906.2	19.8	392.04	0.433
Tall, potato-leaf	288	302.1	−14.1	198.81	0.658
Dwarf, cut-leaf	293	302.1	− 9.1	82.81	0.274
Dwarf, potato-leaf	104	100.7	3.3	10.89	0.108
Sum	1611	1611.1	− 0.1		$X^2 = 1.473$

Source: Data from MacArthur (1931).

values of $\hat{p}_i$ are the expected probabilities for the $a = 4$ classes. Clearly, $\sum^{a}\hat{p}_i = 1$, since the probabilities of all possible outcomes sum to one. The values of $\hat{p}_i$ for this example are $\hat{p}_1 = \frac{9}{16}$, $\hat{p}_2 = \frac{3}{16}$, $\hat{p}_3 = \frac{3}{16}$, $\hat{p}_4 = \frac{1}{16}$. Again, we calculate deviations from expectation, square them, and divide them by expected frequencies. The sum of these quotients yields a value of $X^2 = 1.473$. The operation can be described by the formula

$$X^2 = \sum^{a}\frac{(f_i - \hat{f}_i)^2}{\hat{f}_i} \tag{16.3}$$

which is a generalization of Expression (16.1). What can we conclude about our test for goodness of fit? If we assume for the moment that X^2 in this case is also distributed approximately as χ^2, we need to know how many degrees of freedom there are in this example to enable us to compare it with the appropriate χ^2-distribution.

In the previous example (Table 16.1) we had one degree of freedom. This is evident when we consider the frequencies in the two classes of the table and their sum, $80 + 10 = 90$. In such an example, the total frequency is fixed. Therefore, if we were to vary the frequency of any one class, the other class would have to compensate for changes in the first class to retain a correct total. If the first class had a frequency of 75, the second class would have to contain 15 items to make the total 90. Thus, only one class is able to vary freely, the other class being constrained by the constant sum. Here, the meaning of *one degree of freedom* becomes quite clear. One of the classes is free to vary; the other is not. In the second case with four classes, any *three* of them can vary freely; but the fourth class must constitute the difference between the total sum and the sum of the first three. Thus, in a four-class case we have three degrees of freedom and, in general, when we have a classes, we have $a - 1$ degrees of freedom.

In Table **R** (chi-square) under three degrees of freedom, we find between 90% and 50% of $\chi^2_{[3]}$-values > 1.473. We therefore have no reason for rejecting the null hypothesis. So far as we can tell, the dihybrid cross of Table 16.2 follows the 9:3:3:1 ratio.

In Figure 16.2 we have shown the result of another sampling experiment. One thousand samples of size $n = 165$ were drawn from a population with a 9:3:3:1 ratio. (An $n = 1611$, as in the actual example of Table 16.2, was not used, since it would have required an excessive amount of computer time.) The 1000 X^2-values are shown in the form of a histogram with the probability density function of the $\chi^2_{[3]}$-distribution superimposed for ease of comparison. You will notice that the fit appears better than in Figure 16.1. This is because a larger sample size was used ($n = 165$ versus 90 before).

It is more difficult to specify the alternative hypothesis than it is to specify the null hypothesis. In some genetics experiments, there are only a limited number of possible alternative hypotheses. Thus, if the 3:1 ratio is rejected, a 1:1 ratio or some more complicated ratio might be applicable in

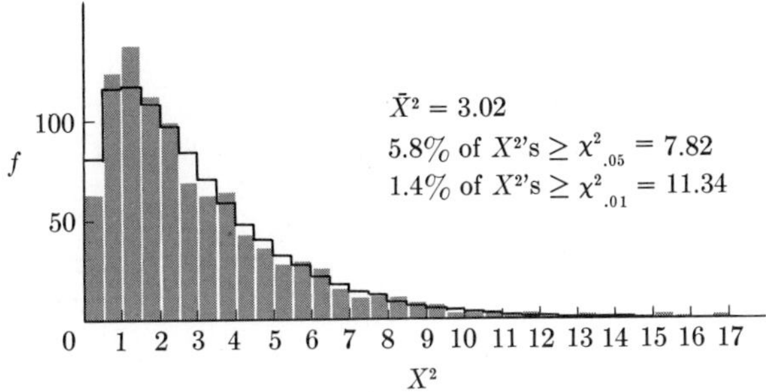

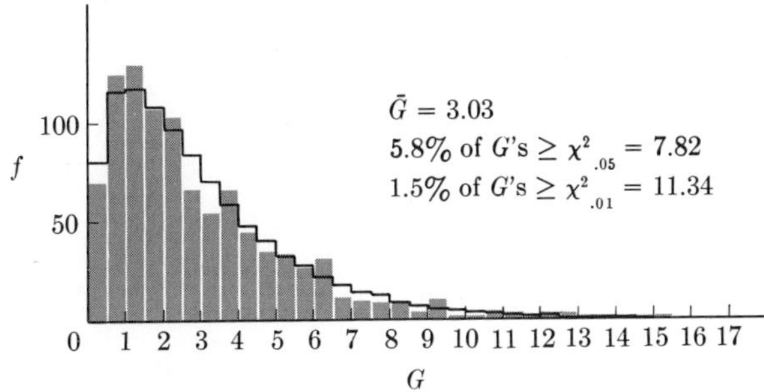

FIGURE 16.2 Results of a sampling experiment in which 1000 samples of size $n = 165$ were drawn from a population divided into four classes in the ratio 9:3:3:1. The histograms show the distribution of the resulting values of X^2 (above) and G (below) computed for each sample on the hypothesis that the parametric ratio is 9:3:3:1. The expected $\chi^2_{[3]}$-distribution is superimposed for comparison.

a given case. However, in other examples, there may be a variety of complex alternative hypotheses that are difficult to specify. In such cases, the main consideration is the acceptance or rejection of the null hypothesis. If we accept a 3:1 ratio, we still do not know absolutely that 3:1 is the true ratio. All we know is that, given our sample, we cannot show that it is not the true ratio.

In some goodness of fit tests, we subtract more than one degree of freedom from the number of classes, a. These are instances where the parameters for the null hypothesis have been extracted from the sample data themselves, in contrast with the null hypotheses encountered so far (in Tables 16.1 and 16.2). In these two cases, the hypothesis to be tested was generated on the basis of the investigator's general knowledge of the specific problem and of Mendelian genetics. The values of $\hat{p} = 0.75$ and $\hat{q} = 0.25$ were dictated by the 3:1 hypothesis and were not estimated from the sampled data. Similarly, the 9:3:3:1 hypothesis was also based on general genetic theory. For this reason, the expected frequencies are based on an *extrinsic hypothesis*, a hypothesis external to the data. By contrast, consider the expected frequencies of birth weights under the assumption of normality (Box 6.1). You will recall that to compute these frequencies, you needed values for μ and σ, which you estimated from the sample mean $\bar{Y}$ and the sample standard deviation s of the birth weights. Therefore the two parameters of the computed normal distribution, the mean and the standard deviation, came from the sampled observations themselves. The expected normal frequencies represent an *intrinsic hypothesis*. In such a case, to obtain the correct number of degrees of freedom for the chi-square test of goodness of fit (and for the G-test described below) we would subtract from a, the number of classes into which the data had been grouped, not only one degree of freedom for n, the sum of the frequencies, but also two further degrees of freedom—one for the estimate of the mean and the other for the estimate of the standard deviation. Thus, in such a case, a sample statistic X^2 would be compared with chi-square for $a - 3$ degrees of freedom. Exact rules on how many degrees of freedom to subtract and computations for such tests are given in the next section.

Before we proceed to the analysis of specific designs in succeeding sections, we must develop a general formula for computing the sample statistic X^2 for testing the goodness of fit. Expression (16.3)

$$X^2 = \sum^{a} \frac{(f_i - \hat{f}_i)^2}{\hat{f}_i}$$

is applicable to any chi-square test of goodness of fit, although its use is tedious and requires the computation of deviations, their squaring, and their division by expected frequencies over all of the classes. This formula, analogous to the theoretical formula for a sum of squares, is not practical for computation on a desk calculator. We therefore require a generally applicable computational formula for X^2. This can be easily derived from Expression (16.3), as shown in Appendix A1.14. It is

$$\sum^{a} \frac{f_i^2}{\hat{f}_i} - n \tag{16.4}$$

which can be easily obtained as the sum of the quotients of the squares of the observed frequencies divided by their expected frequencies. From this sum of quotients is subtracted n, the sum of all the frequencies.

There are other techniques that can be used to test the agreement of observed frequencies to those expected on the basis of an hypothesis. The following method has only very recently come into use, but has properties which will often make it preferable to the standard chi-square test.

Let us consider the example of Table 16.1. Using Expression (5.1) for the expected relative frequencies in a binomial distribution, we can compute two quantities which are of interest to us here:

$$C(90, 80)\left(\frac{80}{90}\right)^{80}\left(\frac{10}{90}\right)^{10} = 0.1327$$

$$C(90, 80)\left(\frac{3}{4}\right)^{80}\left(\frac{1}{4}\right)^{10} = 0.0005514$$

The first quantity is the probability of observing the sampled results (80 wild type and 10 mutants) on the hypothesis that $\hat{p} = p$—that is, that the population parameter equals the observed sample proportion. The second is the probability of observing the sampled results assuming that $\hat{p} = \frac{3}{4}$, as per the Mendelian null hypothesis. Note that these expressions yield the probabilities for the observed outcomes only, *not for observed and all worse outcomes*. Thus, $P = 0.0005514$ is less than the earlier computed $P = 0.00085$, which is the probability of 10 *and fewer* mutants, assuming $\hat{p} = \frac{3}{4}$, $\hat{q} = \frac{1}{4}$.

The first probability (0.1327) is greater than the second (0.0005514), since the hypothesis is based on the observed data. If the observed proportion p is in fact equal to the proportion $\hat{p}$ postulated under the null hypothesis, then the two computed probabilities will be equal and their ratio, L, will equal 1.0. The greater the difference between p and $\hat{p}$ (the expected proportion under the null hypothesis), the higher the ratio will be (the probability based on p is divided by the probability based on $\hat{p}$ or defined by the null hypothesis). This indicates that the ratio of these two probabilities or *likelihoods* can be used as a statistic to measure the degree of agreement between sampled and expected frequencies. A test based on such a ratio is called a *likelihood ratio test*. In our case, $L = 0.1327/0.0005514 = 240.66$. The theoretical distribution of this ratio is, in general, complex and poorly known. However, it has been shown that the distribution of

$$G = 2 \ln L = 2 (\log e) \log L \tag{16.5}$$

can be approximated by the χ^2-distribution when sample sizes are large (for a definition of "large" in this case, see below). The appropriate degrees of freedom for a given test are the same as for the chi-square tests discussed above. In our case,

$$G = 2 \ln L = 2(\log e) \log L = 2(2.302585)(2.38148) = 2(5.48356) = 10.967$$

If we compare this observed value with a χ^2-distribution with one degree of freedom, we find that the result is significant ($P < 0.005$), as was found before for the chi-square test. In general, G will be numerically rather similar to X^2. Notation for the *log likelihood ratio test* is as little standardized as that for the chi-square test. The symbol $2I$ is sometimes used for G. The letter I is used because G can be interpreted as twice the amount of information present in the sample for distinguishing between the $\hat{p}$ of the null hypothesis and that of the alternative hypothesis. The term "information" is used here in the sense defined in the field of information theory. The interested reader who has a knowledge of mathematical statistics may find the book by Kullback (1959) very interesting in this regard. However, the employment of the G-test ($= 2I$-test) does not require any special mathematical sophistication and, as we shall see, this method has some considerable advantages over the conventional chi-square tests.

A disadvantage of G compared with X^2 is that a table of logarithms must be consulted repeatedly, whereas the chi-square test requires only simple arithmetic. But with appropriate tables and with efficient computational formulas, the G-tests often require much less computational effort than is required for the X^2-tests (this is especially so for the tests described in Sections 16.4 and 16.5).

For developing a computational formula, Expression (16.5) for the G-statistic can be rewritten in various ways, depending upon the particular application. In Appendix A1.15 it is shown that for a general goodness of fit test it can be written as follows:

$$G = 2\sum^{a} f_i \ln\left(\frac{f_i}{\hat{f}_i}\right) \tag{16.6}$$

This can be expressed as follows to simplify computation

$$\begin{aligned} G &= 2[\sum^{a} f_i \ln f_i - \sum^{a} f_i \ln \hat{f}_i] \\ &= 2[\sum^{a} f_i \ln f_i - (2.30259)\sum^{a} f_i \log \hat{f}_i] \end{aligned} \tag{16.6a}$$

which is convenient to use with Table **G**. Table **G** gives $f \ln f$ for integral values of f between 0 and 10,000. Thus, the quantities $f \ln f$ or $n \ln n$ can be looked up directly and accumulated on an adding machine. Since there is a chance of miscopying information from the tables, it is most convenient to use a printing calculator so that one need only check the printed record against the table in order to verify the results. If errors are found it is simple to correct the final results without having to repeat all of the computations. Another frequently used computational formula for G is

$$G = 2[\sum^{a} f_i \ln f_i - \sum^{a} f_i \ln \hat{p}_i - n \ln n] \tag{16.7}$$

also derived in Appendix A1.15.

In Figures 16.1 and 16.2 we also show the results of computing the G-statistic for the sampling experiments described above. In Figure 16.1 note that discontinuity is more apparent in X^2 than in G. Also, G appears to follow the χ^2-distribution a bit more closely. In Figure 16.2 (which is based on a larger sample size) both distributions follow the χ^2-distribution fairly closely.

16.2 Single classification goodness of fit tests

Before discussing in detail the computational steps involved in tests of goodness of fit of single classification frequency distributions some remarks on the choice of a test statistic are in order. The traditional method for such a test is the chi-square test for goodness of fit. The new approach by the G-test can be recommended on theoretical grounds and has also been shown to be more efficient for variously structured data by empirical studies. The major advantage of the G-test is in more complicated designs where it is simpler to carry out than the chi-square test and often permits certain types of detailed analyses which are impossible using X^2.

It would therefore seem that the G-test should be uniformly employed for goodness-of-fit tests, and if the computation is done by computer we would recommend its use exclusively. The computer program for goodness-of-fit tests (Program A3.3) furnished by us does in fact use G. There is only one reservation regarding the use of G. In the simplest design, single classification tests for goodness of fit, which are the subject of this section, the computation on a desk calculator is somewhat more tedious for G than for X^2. Thus while G would be the preferred test-statistic, people will tend to go the way of least resistance and will more likely use X^2. We therefore furnish directions for the chi-square test first, followed by an outline of the procedure for the G-test. Even if you are inclined to be satisfied with the chi-square test, you should still go in detail over the account of how to do the G-test, because this will be necessary for an understanding of the more complex designs and tests, many of which we carry out with G alone.

For continuous sample distributions whose goodness of fit is to be tested there is a preferred method—the Kolmogorov-Smirnov test. It is by far the simplest and is described at the end of this section.

Chi-square tests of goodness of fit for single classification frequency distributions are given in Box 16.1, which treats only those cases in which the

BOX 16.1

Tests for goodness of fit. Single classification, expected frequencies based on hypothesis extrinsic to the sampled data.

Chi-square test

1. Frequencies divided into $a \geq 2$ classes.

In a genetic experiment involving a cross between two varieties of the bean *Phaseolus vulgaris*, Smith (1939) obtained the following results:

Phenotypes ($a = 8$)	*Observed frequencies* f	*Expected frequencies* $\hat{f}$
Purple/buff	63	67.8
Purple/testaceous	31	22.6
Red/buff	28	22.6
Red/testaceous	12	7.5
Purple	39	45.2
Oxblood red	16	15.1
Buff	40	45.2
Testaceous	12	15.1
Total	241	241.1

BOX 16.1 continued

The expected frequencies $\hat{f}_i$ were computed on the basis of the expected ratio of 18:6:6:2:12:4:12:4. Compute $\hat{f}_i$ as $\hat{p}_i n$. Thus, $\hat{f}_1 = \hat{p}_1 n = (\frac{18}{64}) \times 241 = 67.8$. Expression (16.4), the formula for machine calculation of X^2, is

$$X^2 = \sum^{a} \frac{f_i^2}{\hat{f}_i} - n$$

In this case,

$$X^2 = \frac{(63)^2}{67.8} + \frac{(31)^2}{22.6} + \cdots + \frac{(12)^2}{15.1} - 241 = 250.4909 - 241 = 9.4909$$

□□ Accumulate quotients $f_i^2/\hat{f}_i$ in the machine and subtract n from the sum of the quotients. □□

This value is to be compared to a χ^2-distribution (Table **R**) with $a - 1$ df ($df = 8 - 1 = 7$ in the present example). Since the above $X^2 < \chi^2_{.05[7]} = 14.067$, the data may be considered consistent with Smith's hypothesis.

2. Special case of frequencies divided into $a = 2$ classes.
Employ

Expression (16.2) $$X^2 = \frac{(f_1 - \hat{p}n)^2}{\hat{p}\hat{q}n}$$

or

Expression (16.2a) $$X^2 = \frac{(f_2 - \hat{q}n)^2}{\hat{p}\hat{q}n}$$

When the expected frequencies are in simple ratios, such as 3:1 or 1:1, employ

Expression (16.8) $$X^2 = \frac{(f_1 - rf_2)^2}{rn} \qquad \text{where } r = \frac{\hat{p}}{\hat{q}}$$

In an F_2 cross in drosophila the following 176 progeny were obtained, of which 130 were wild type flies and 46 ebony mutants. Assuming that the mutant is an autosomal recessive, one would expect a ratio of 3 wild type flies to each mutant fly. To test whether the observed results are consistent with this 3:1 hypothesis, we apply Expression (16.2).

Flies	f	*Hypothesis*	$\hat{f}$
Wild type	$f_1 = 130$	$\hat{p} = 0.75$	$\hat{p}n = 132.0$
Ebony mutant	$f_2 = 46$	$\hat{q} = 0.25$	$\hat{q}n = 44.0$
	$n = 176$		

By Expression (16.2)

$$X^2 = \frac{(130 - 132.0)^2}{0.75 \times 0.25 \times 176} = \frac{4.0}{33} = 0.1212$$

Since the ratio $r = \hat{p}/\hat{q} = 3$ is integral, we might have applied Expression (16.8):

BOX 16.1 continued

$$X^2 = \frac{[130 - 3(46)]^2}{3(176)} = \frac{[-8]^2}{528} = 0.1212 \qquad \text{as before.}$$

Since $X^2 < \chi^2_{.05[1]} = 3.84$ (Table **R**), we clearly do not have sufficient evidence to reject our null hypothesis.

Another equivalent equation, which is sometimes convenient, makes use of the normal approximation to the binomial distribution:

$$t_s = \frac{(p - \hat{p})}{\sqrt{\hat{p}\hat{q}/n}} \qquad X^2 = t_s^2 = \frac{n(p - \hat{p})^2}{\hat{p}\hat{q}}$$

where p is the proportion observed in one of the classes, $\hat{p}$ is the expected proportion for the same class $\hat{q} = 1 - \hat{p}$, and $n = f_1 + f_2$, as before. In the example given above we would have

$$X^2 = \frac{176(0.7386364 - 0.7500000)^2}{0.25 \times 0.75} = \frac{0.022{,}727}{0.1875} = 0.1212 \qquad \text{as before.}$$

It could also have been looked up in the table of areas of the normal curve (Table **P**) or as $t_s = \sqrt{0.1212} = 0.3481$, clearly a nonsignificant value when compared with $t_{\alpha[\infty]}$.

Adjustment for small sample sizes in cases with $a = 2$ classes:

If	*Compute*
$n > 200$	X^2 by Expressions (16.1), (16.2), (16.8), or by the normal approximation to the binomial as shown immediately above.
$200 \geq n > 25$	X^2_{adj} (Yates' correction) by Expressions (16.1Y), (16.2Y), or (16.8Y).
$n < 25$	Exact probabilities of the binomial as shown in Table 5.3, Section 5.2 (employ machine method of Box 5.1).

G-test

1. Frequencies divided into $a \geq 2$ classes

Employ Expressions (16.6a) when $\hat{f}$'s have already been calculated or (16.7) when $\hat{p}$'s are given. Since we have presented the $\hat{f}$'s in the bean example given above we shall use Expression (16.6a):

$$G = 2\left[\sum^a f_i \ln f_i - 2.30259 \sum^a f_i \log \hat{f}_i\right]$$

Using Table **G** we find

$$\sum^a f_i \ln f_i = 63 \ln 63 + \cdots + 12 \ln 12 = 261.017 + \cdots + 29.819$$
$$= 855.206$$

Using Table **C** we can compute

$$\sum^a f_i \log \hat{f}_i = 63 \log 67.8 + \cdots + 12 \log 15.1$$
$$= 63(1.83123) + \cdots + 12(1.17898) = 369.5282$$

BOX 16.1 continued

G can then be computed as

$$G = 2\left[\sum^{a} f_i \ln f_i - 2.30259 \sum^{a} f_i \log \hat{f}_i\right]$$
$$= 2[855.206 - 2.30259(369.5282)] = 2[4.334] = 8.668$$

Since our observed $G = 8.676 < \chi^2_{.05[7]} = 14.067$, we may consider the data consistent with the null hypothesis (as was concluded above by the chi-square test).

2. Special case of frequencies divided into $a = 2$ classes

For the G-test there is no special shortcut equation for the case in which $a = 2$. But, as in the X^2-test, an additional complication is the fact that Yates' correction should usually be applied when n is less than about 200. The computations are carried out as described in part **1** above, but using adjusted f's instead of the original f's if Yates' correction is to be applied (see text). We now apply the G-test to the above drosophila data.

Flies	f	*Adjusted* f	$\hat{f}$
Wild type	130	130.5	132.0
Ebony mutant	46	45.5	44.0
	176	176.0	176.0

Using Table **G***, and adjusted frequencies f_i,

$$\sum^{a} f_i \ln f_i = 130.5 \ln 130.5 + 45.5 \ln 45.5 = 635.714 + 173.706$$
$$= 809.420$$

Using Table **C**,

$$\sum^{a} f_i \log \hat{f}_i = 130.5 \log 132.0 + 45.5 \log 44.0$$
$$= 130.5(2.12057) + 45.5(1.64345) = 351.5114$$
$$G_{\text{adj}} = 2[809.420 - 2.30259(351.5114)] = 0.0668$$

which is obviously not significant (as also found previously).

expected frequencies are based on a hypothesis extrinsic to the data—there are a classes and the expected proportions in each class are assumed on the basis of outside knowledge and are not functions of parameters estimated from the sample. We start with the general case for any number of classes, where the number of classes is symbolized by a, to emphasize the analogy with analysis of variance. The data are the results of a complicated genetic cross expected to result in an 18:6:6:2:12:4:12:4 ratio. A total of 241 progeny was obtained and allocated to the eight phenotypic classes. The expected frequencies can be simply computed by multiplying the total sample size n times the expected probabilities of occurrence. We note that the overall fit is quite good. This is an example of a test for goodness of fit with an extrinsic hypothesis because the genetic ratio tested is based on considerations prior and external to the sample observations tested in the example. The machine computation is quite simple as indicated in the box. Since Expression (16.4) yields X^2 (the sum of squared quantities), the answer must be positive. If you obtain a negative result on the machine, you can be certain of a computational error.

We learned in the previous section that the value of X^2 is to be compared with a critical value of χ^2 for $a - 1$ degrees of freedom. When we compare our result with the χ^2-distribution, we find that the value of X^2 obtained from our sample is not significant. We do not have sufficient evidence to reject the null hypothesis and are led to conclude that the sample is consistent with the specified genetic ratio. However, especially in examples such as the one just analyzed, remember that we have not specified an alternative hypothesis; there are a variety of alternative hypotheses that also could not be excluded if we were to carry out a significance test for them. We have not really proven that the data are distributed as specified, and there are a variety of other genetic ratio hypotheses that could also be plausible.

No general directions are found in the literature regarding how small a sample may be and still be suitable for the chi-square test of goodness of fit. However, caution should be exercised with sample sizes of n less than 50 in interpreting the results. Also, as a general rule, no expected frequency should be less than 5, and whenever classes with expected frequencies of less than 5 occur in a problem, expected and observed frequencies for those classes are generally pooled with an adjacent class in order to obtain a joint class with an expected frequency $\hat{f} > 5$ (in Box 16.2 below such a case is illustrated).

We saw in the last section that there is an especially simple formula for the chi-square test, Expression (16.2), which can be applied to data in which there are only two classes. If, as in the example shown in Box 16.1 (part 2), the expected frequencies are expressed as a ratio $r:1$, an even simpler formula may be applied:

$$X^2 = (f_1 - rf_2)^2/rn \tag{16.8}$$

where r is the ratio $\hat{p}/\hat{q}$. Such a formula is especially useful in genetics where simple ratios such as 3:1 or 1:1 are common.

The example in part 2 of Box 16.1 is a monohybrid cross with an expected ratio of 3 wild type to 1 mutant. Expressions (16.2) or (16.8) test the deviations from expectation; both yield identical results. The expected frequencies can be shown to be very little different from the observed frequencies, and it is no surprise, therefore, that the resulting value of X^2 is 0.1212, which is far less than the critical value of χ^2 at one degree of freedom. Inspection of the chi-square table reveals that roughly 75% of all samples, from a population with the expected ratio, would show greater deviations than the sample at hand. Although calculation of expected frequencies is not necessary by Expression (16.8), knowledge of the size and direction of deviations from expectation aids the formulation of new hypotheses and experiments.

Box 16.1 shows yet another method of analyzing the example just discussed. It could be done as a test of the deviation of the observed proportion from the expected proportion using the normal approximation to the binomial distribution. It is assumed that the quantity $t_s = (p - \hat{p})/\sqrt{\hat{p}\hat{q}/n}$ is normally distributed. This value can be regarded as a normal deviate ($t_{[\infty]}$), but we can also square it yielding $X^2 = (t_s)^2 = n(p - \hat{p})^2/\hat{p}\hat{q}$. By simple algebra it can be shown to equal Expression (16.2).

In tests of goodness of fit involving only two classes, the value of X^2 and G as computed from Expressions (16.1), (16.2), (16.4), (16.5), (16.6), (16.6a), (16.7), or (16.8) will exhibit a bias that can be modified by applying *Yates' correction for continuity*, making the value of X^2 or G approximate the χ^2-distribution more closely. This correction consists of adding or subtracting 0.5 from the observed frequencies in such a way as to minimize the value of X^2 and G. In the case of X^2 this can simply be done by subtracting 0.5 from the absolute values of the deviation $f - \hat{f}$, yielding the following adjusted versions of these formulas (the letter Y follows the expression numbers to indicate Yates' correction).

$$X^2_{\text{adj}} = \frac{\sum^{2}(|f_i - \hat{f}_i| - \frac{1}{2})^2}{\hat{f}_i} \tag{16.1Y}$$

$$X^2_{\text{adj}} = \frac{(|f_1 - \hat{p}_1 n| - \frac{1}{2})^2}{\hat{p}\hat{q}n} \tag{16.2Y}$$

$$X^2_{\text{adj}} = \frac{[|f_1 - rf_2| - \frac{1}{2}(r + 1)]^2}{rn} \tag{16.8Y}$$

The employment of these formulas is suggested as a routine by some, while others suggest that they are necessary only for sample sizes $n < 200$. Actually, Yates' correction makes only a small difference in X^2 or G even when sample sizes are below 200. When we apply formula (16.2Y) to the genetic cross of Table 16.1 in Section 16.1, based on a sample size of $n = 90$, we obtain $X^2_{\text{adj}} = 8.533$, where X^2_{adj} is the value of X^2 with Yates' correction applied. Note that though the X^2_{adj}-value is lower than the unadjusted $X^2 = 9.259$, the significance level does not change, remaining at $P < 0.005$.

We might state the following guide lines for testing cases with two classes. For sample sizes $n > 200$, use regular formulas for X^2 or G. For sample sizes 25 to 200, use Yates' correction as shown in Expressions (16.1Y), (16.2Y), and (16.8Y), or by applying the G-test to adjusted frequencies, and for sample sizes 25 or less work out the exact probabilities as shown in Table 5.3, Section 5.2 (employ machine method of Box 5.1).

The G*-test.* The problems dealt with in Box 16.1 by means of the chi-square test are next analyzed by means of the G-test in the second half of that box. We employ Expression (16.6a) to obtain $G = 8.668$. The computations are simple and straightforward although slightly more tedious than those for X^2. Again our test statistic is less than the critical value of χ^2.

We next learn that for the G-test there is no special simplified equation for the two-cell case ($a = 2$). One employs the general equations [Expressions (16.6a) or (16.7)]. To apply Yates' correction for small sample sizes to the G-test one simply adjusts the f_i, changing them to reduce the difference between them and the corresponding expected frequencies by one half. One then employs Expression (16.6a) or (16.7) as before to obtain G_{adj}. Values of $(f + \frac{1}{2}) \ln (f + \frac{1}{2})$ necessary for the computation of G_{adj} can be found in Table **G***.

Let us compare the results of all these methods for a test for goodness of fit in a small sample. We employ the example of Table 5.3, illustrating a litter of 17 offspring of which 14 were females and 3 were males. We wish to test these data against a hypothetical sex ratio of 1:1. By the rules just enumerated, we should use the exact probability, which was worked out in Table 5.3 as 0.006,363,42, being the probability of a deviation as great or greater in one direction from the 1:1 hypothesis. Since we are interested in a two-tailed test, we double this probability to yield $P \approx 0.0127$. Ordinary X^2 is [by Expression (16.8)]

$$X^2 = \frac{(14 - 3)^2}{17} = \frac{121}{17} = 7.118; P = 0.005$$

and the adjusted X^2 yields [by Expression (16.8Y)]

$$X^2_{adj} = \frac{[14 - 3 - \frac{1}{2}(2)]^2}{17} = \frac{100}{17} = 4.882; P = 0.025$$

These probabilities were looked up in Table **P**, areas of the normal curve, by entering the table with argument $\sqrt{X^2}$ (since $\sqrt{\chi^2_{\alpha[1]}} = t_{\alpha[\infty]}$). In this instance, there is no question by any of the tests that the results deviate significantly from the 1:1 hypothesis. However, it is interesting that in this particular case the ordinary X^2 approximates most closely the exact probability value, while the adjusted X^2 was farther off (on the average, however, the adjusted X^2 will yield more correct probabilities).

Applying the G-test to the example of Table 5.3, we obtain the following results [by Expression (16.7)]:

$$\begin{aligned} G &= 2[14 \ln 14 + 3 \ln 3 - 14 \ln \tfrac{1}{2} - 3 \ln \tfrac{1}{2} - 17 \ln 17] \\ &= 2[3.861] = 7.723; P = 0.0054 \end{aligned}$$

$$\begin{aligned} G_{adj} &= 2[13\tfrac{1}{2} \ln 13\tfrac{1}{2} + 3\tfrac{1}{2} \ln 3\tfrac{1}{2} - 13\tfrac{1}{2} \ln \tfrac{1}{2} - 3\tfrac{1}{2} \ln \tfrac{1}{2} - 17 \ln 17] \\ &= 2[3.1395] = 6.279; P = 0.012 \end{aligned}$$

Thus, the probability resulting from the adjusted G-test is virtually exact in this case.

Figure 16.3 shows the result of another sampling experiment carried out

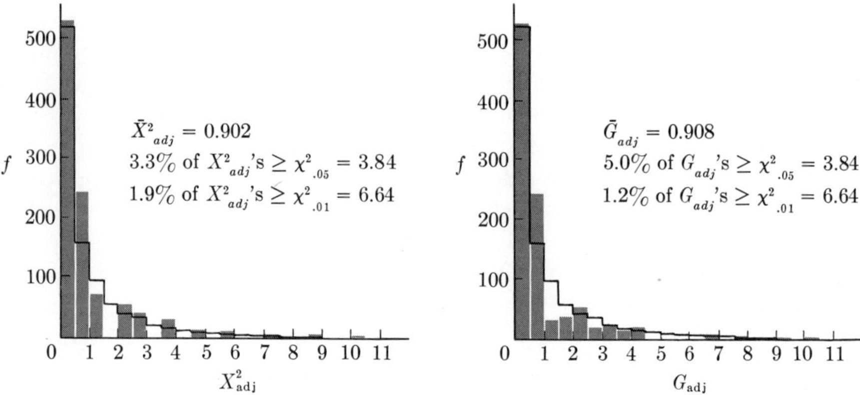

FIGURE 16.3 Results of a sampling experiment in which 1000 samples of size $n = 90$ were drawn from a binomial population with $\hat{p} = 0.75$. Yates' correction was applied when computing X^2 and G.

The histograms show the distribution of the resulting values of X^2_{adj} (left) and G_{adj} (right) computed for each sample on the hypothesis that $\hat{p} = 0.75$ and $\hat{q} = 0.25$. The expected $\chi^2_{[1]}$-distribution is superimposed for comparison.

as described above for Figure 16.1, but with Yates' correction applied. The right half of the figure also shows the results of applying Yates' correction to the G-test.

At the right of each figure is given the observed percentages corresponding to the 5% and 1% tails of the χ^2-distribution. The percentages for the G_{adj} sampling experiment agree very well with the expected 5% and 1% values.

We next apply the tests for goodness of fit to frequencies arrayed in a single classification where the expected frequencies are based on a hypothesis intrinsic to the sampled data. Such a computation is illustrated in Box 16.2, where we have taken the sex ratio data in sibships of twelve, first introduced in Table 5.4, Section 5.2. Only the chi-square method is illustrated, but the G-test could have been employed as well, using Expression (16.6a). As you will recall, the expected frequencies in these data are based on the binomial distribution, with the required parametric proportion of males $\hat{p}_{♂}$ estimated from the observed frequencies of the sample, where $p_{♂} = 0.519{,}215$. The general computation of this case is as previously discussed in Box 16.1. We employ the simple computational formula [Expression (16.4)] for efficient computation. We have deliberately omitted a column of actual deviations $f - \hat{f}$ from the table in Box 16.2 so as to discourage the computation of X^2 by the long method [Expression (16.1)]. Only the signs of the deviations are shown to indicate the pattern of departures from expectation.

Two aspects make this computation different from that in step **1** of Box 16.1. As a general rule, we avoid expected frequencies less than 5. Therefore, the classes of $\hat{f}_i$ at both tails of the distribution are too small. We lump

BOX 16.2

Chi-square test for goodness of fit. Single classification, expected frequencies based on hypothesis intrinsic to the sampled data.

Sex ratio in 6115 sibships of twelve in Saxony. The fourth column gives the expected frequencies computed in Table 5.4, assuming a binomial distribution.

(1)	*(2)*	*(3)*		*(4)*		*(5)*
♂♂	♀♀	f		$\hat{f}$		*Deviation from expectation*
12	0	7	52	2.3	28.4	+
11	1	45		26.1		
10	2	181		132.8		+
9	3	478		410.0		+
8	4	829		854.3		−
7	5	1112		1265.5		−
6	6	1343		1367.2		−
5	7	1033		1085.1		−
4	8	670		628.1		+
3	9	286		258.5		+
2	10	104		71.8		+
1	11	24	27	12.1	13.0	+
0	12	3		0.9		
		6115 = n		6114.7		

Since expected frequencies $\hat{f}_i < 5$ should be avoided, we lump the classes at both tails with the adjacent classes to create classes of adequate size. Corresponding classes of observed frequencies f_i should be lumped to match. The number of classes after lumping is $a = 11$.

Compute X^2 by Expression (16.4):

$$X^2 = \sum^{a} \frac{f_i^2}{\hat{f}_i} - n = \frac{(52)^2}{28.4} + \frac{(181)^2}{132.8} + \cdots + \frac{(27)^2}{13.0} - 6115$$

$$= 6221.2199 - 6115 = 106.2199$$

Since there are $a = 11$ classes, the degrees of freedom are maximally $a - 1 = 10$. However, because the mean $\hat{p}_{♂}$ was estimated from the $p_{♂}$ of the sample a further degree of freedom is removed, and the sample X^2 is compared to a χ^2-distribution with $a - 2 = 11 - 2 = 9$ degrees of freedom:

$$X^2 = 106.2199 > \chi^2_{.005[9]} = 23.589$$

The null hypothesis—that the sample data follow a binomial distribution—is therefore rejected decisively.

Typically, the following degrees of freedom will pertain to chi-square tests for goodness of fit with expected frequencies based on a hypothesis *intrinsic* to the sample data (a is the number of classes after lumping, if any):

BOX 16.2 continued

Distribution	*Parameters estimated from sample*	*df*
Binomial	$\hat{p}$	$a - 2$
Normal	μ, σ	$a - 3$
Poisson	μ	$a - 2$

When the parameters for such distributions are estimated from hypotheses *extrinsic* to the sampled data, the degrees of freedom are uniformly $a - 1$.

These tests could also be carried out by the G-test, employing Expressions (16.6a) or (16.7) and the same critical value of χ^2 as for the chi-square test.

them by adding their frequencies to those in contiguous classes as shown in Box 16.2. Clearly, the observed frequencies must be lumped to match. The number of classes a is the number *after* lumping has taken place. In our case, $a = 11$.

The other new feature is the number of degrees of freedom considered for the significance test. We always subtract one degree of freedom for the fixed sum (in this case $n = 6115$). However, we subtract an additional degree of freedom for every parameter of the expected frequency distribution estimated from the sampled distribution. In this case, we estimated $\hat{p}_{♂}$ from the sample and, therefore, a second degree of freedom is subtracted from a, making the final number of degrees of freedom $a - 2 = 11 - 2 = 9$. Comparing the sample value of $X^2 = 106.2199$ with the critical value of χ^2 at nine degrees of freedom, we find it highly significant ($P < 0.005$ assuming that the null hypothesis is correct). We therefore reject this hypothesis and conclude that the sex ratios are not binomially distributed. As is evident from the pattern of deviations, there appears an excess of sibships in which one sex or the other predominates. Had we applied the G-test to these data the critical value would have been the same ($\chi^2_{\alpha[9]}$).

These methods (X^2 and G) for testing the goodness of fit of a set of data to an expected frequency distribution can be applied not only to the binomial, but to the normal, Poisson, and other distributions as well. For a normal distribution we customarily estimate two parameters μ and σ from the sampled data. Hence the appropriate degrees of freedom are $a - 3$. In the Poisson, only one parameter, μ, must be estimated; the appropriate degrees of freedom are $a - 2$. Remember that the critical issue is whether the expected frequencies are based on an extrinsic or intrinsic hypothesis, not on whether the test is against a normal versus some other distribution. It is conceivable that an investigator has a hypothesis about the mean and standard deviation of a distribution and makes his expected normal frequencies conform to this hypothesis. In such a case, the two extra degrees of freedom are not subtracted, and the degrees of freedom would only be $a - 1$, since estimates of the parameters were not obtained from the sample.

A nonparametric test that is applicable to continuous frequency distributions and in many cases has greater power than the chi-square test for goodness of fit is the *Kolmogorov-Smirnov test*. This test is quite simple to carry out. It is based on the absolute differences between observed and expected cumulative frequency distributions. These differences are expressed as differences between relative cumulative frequencies, and we can look up tabled critical values and decide whether the maximum difference between the observed and expected cumulative frequency distribution is significant. An example will make this clear. We turn to the femur lengths of aphid stem mothers from Box 6.4. In Table 16.3 we have again arrayed the 25 measure-

TABLE 16.3

Cumulative observed (F) and expected frequencies ($\hat{F}$) of femur length of 25 aphid stem mothers. (Data from Box 2.1.) $\bar{Y} = 4.004$; $s = 0.3656$.

(1)	(2)	(3)	(4)	(5)
Y	$(Y - \bar{Y})/s$	$\hat{F}$	F	$d = \lvert F - \hat{F}\rvert$
3.3	−1.93	0.0268	0.04	0.0132
3.5	−1.38	0.0838	0.08	0.0038
3.6	−1.10	0.1335	0.12	0.0135
3.6	−1.10	0.1335	0.16	0.0265
3.6	−1.10	0.1335	0.20	0.0665
3.6	−1.10	0.1335	0.24	0.1065
3.8	−0.56	0.2877	0.28	0.0077
3.8	−0.56	0.2877	0.32	0.0323
3.8	−0.56	0.2877	0.36	0.0723
3.8	−0.56	0.2877	0.40	0.1123
3.9	−0.28	0.3897	0.44	0.0503
3.9	−0.28	0.3897	0.48	0.0903
3.9	−0.28	0.3897	0.52	0.1303 ← d_{max}
4.1	0.26	0.6026	0.56	0.0426
4.1	0.26	0.6026	0.60	0.0026
4.2	0.54	0.7054	0.64	0.0654
4.3	0.81	0.7910	0.68	0.1110
4.3	0.81	0.7910	0.72	0.0710
4.3	0.81	0.7910	0.76	0.0310
4.3	0.81	0.7910	0.80	0.0090
4.4	1.08	0.8599	0.84	0.0199
4.4	1.08	0.8599	0.88	0.0201
4.4	1.08	0.8599	0.92	0.0601
4.5	1.36	0.9131	0.96	0.0469
4.7	1.90	0.9713	1.00	0.0287

ments in order of magnitude and in the next column have shown the standardized deviation of each of these readings from its mean. We should remind

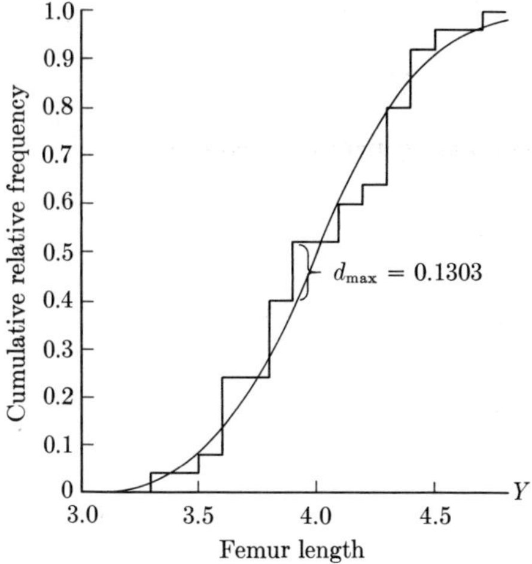

FIGURE 16.4 Cumulative observed frequency distribution of femur lengths of 25 aphid stem mothers (stepped line). The cumulative expected distribution of these data, assuming normality, is superimposed (sigmoid line). Data from Box 6.4 and Table 16.3. The largest difference between the two distributions, d_{max}, is shown on the graph.

you that the mean of these readings is 4.004 and the standard deviation 0.3656. In the third column we record the cumulative relative expected frequency of each of these items. We use the symbols $\hat{F}$ for cumulative expected frequencies and F for cumulative observed frequencies, either relative or absolute (it will be obvious in any given example which of the two it is). The cumulative expected frequencies in column (3) of Table 16.3 are based on the areas of the normal curve corresponding to the deviations in column (2). Column (4) lists the cumulative relative observed frequency of the variates. Since we have a sample of 25 items, each successive item increases the cumulative frequency by 0.04. If we were to graph these cumulative distribution functions, we would obtain the two curves of Figure 16.4. Notice that the curve representing the observed frequencies is a stepped function, because there are only a finite number of observations in a sample. The deviation between the observed and expected cumulative frequencies differs for different values of the abscissa (femur length). The maximum absolute difference, $d_{max} = 0.1303$, is at the third value of $Y = 3.9$. We designate d_{max} as the sample statistic D, look it up in Table **EE**, the table of critical values of the Kolmogorov-Smirnov function, and find that the 5% critical value for $n = 25$ is 0.26404. We have little evidence, therefore, for rejecting the null hypothesis that the femur lengths of the aphid stem mothers are normally distributed. Table **EE** furnishes critical values of D for several values of α and for n up to 100. A formula for computing it for values of $n > 100$ is given in Box 16.3. Although the test is properly applied only to continuous functions as just carried out, in a large sample it is convenient to group the data in a frequency distribution and apply the test to the cumulative frequencies of these classes.

An example of the Kolmogorov-Smirnov test applied to a large sample (the birth weights of Chinese children from Box 6.1) is shown in Box 16.3.

BOX 16.3

Kolmogorov-Smirnov test for goodness of fit. Single classification, hypothesis extrinsic or intrinsic to sampled data.

In Box 6.1 (birth weights of Chinese children) the observed frequencies were qualitatively compared to the frequencies expected on the basis of a normal distribution with $\mu = \bar{Y}$ and $\sigma = s$. The following procedure tests the significance of this agreement.

1. Form cumulative frequency distributions for the observed frequencies and for the expected frequencies (shown in Boxes 6.1 and 6.3). These should both be expressed either as absolute or as relative cumulative frequencies. Using absolute frequencies we obtain the following:

(1)	*(2)*	*(3)*	*(4)*
Y (*oz*)	$\hat{F}$	$\hat{F}$	$d = \lvert F - \hat{F} \rvert$
59.5	2	2.8	0.8
67.5	8	22.7	14.7
75.5	47	118.3	71.3
83.5	432	468.5	36.5
91.5	1320	1368.6	48.6
99.5	3049	3021.2	27.8
107.5	5289	5184.9	104.1 ← $d_{\max}$
115.5	7296	7204.7	91.3
123.5	8529	8532.6	3.6
131.5	9170	9180.0	10.0
139.5	9371	9400.5	29.5
147.5	9445	9454.5	9.5
155.5	9459	9464.0	5.0
163.5	9464	9464.9	0.9
171.5	9465	9464.9	0.1

2. Compute d, the absolute value of the difference between the cumulative frequencies [these values are given in column (4) above] and locate the largest difference, $d_{\max}$. In our case, $d_{\max} = 104.1$.

For absolute frequencies, the Kolmogorov-Smirnov test statistic is

$$D = \frac{d_{\max}}{n} = \frac{104.1}{9465} = 0.0110$$

Had we used relative frequencies, we would not have divided by n. In such a case, the test statistic is $D = d_{\max}$.

For small samples ($n \leq 100$) we can compare the value of D with the entries in Table **EE**. We reject the null hypothesis if the observed D is greater than the critical value of D_α.

For larger samples the critical values can be computed as

$$D_\alpha = \sqrt{\frac{-\ln\left(\frac{1}{2}\alpha\right)}{2n}}$$

BOX 16.3 continued

For $\alpha = 0.05$, this is $1.358/\sqrt{n}$, and for $\alpha = 0.01$ it is $1.628/\sqrt{n}$. In the present case

$$D_{.05} = \frac{1.358}{\sqrt{9465}} = \frac{1.358}{97.2882} = 0.01396$$

Since our observed D is less than this critical value we do not have sufficient evidence to warrant rejecting the null hypothesis.

The Kolmogorov-Smirnov test is properly applied to continuous frequency distributions only.

The computation is straightforward, using a frequency distribution instead of individual observations. In this instance, differences between absolute cumulative frequencies were found, which must be divided by n before they can be compared with the critical values in Table **EE.** In practice it is not necessary to calculate all differences and find the largest one. By looking up the critical value first and keeping it in mind, any difference greater than the critical value will already make the test significant.

Although the Kolmogorov-Smirnov test assumes a continuously distributed variable, it is often used in the case of a discrete one because it has been shown that in such cases the type I error is no greater than that assumed by using the Kolmogorov-Smirnov tables. An especially advantageous use of the Kolmogorov-Smirnov test for goodness of fit is with small sample sizes. In such cases it is not necessary (actually inadvisable) to lump classes.

Another useful property of the Kolmogorov-Smirnov statistic is for setting confidence limits to an entire cumulative frequency distribution. By computing $F_i \pm D_\alpha$ for all values or classes of Y_i, we obtain $100(1 - \alpha)\%$ confidence limits. The 95% confidence limits of the cumulative distribution function of the aphid stem mother data are shown in Figure 16.5. Thus, we

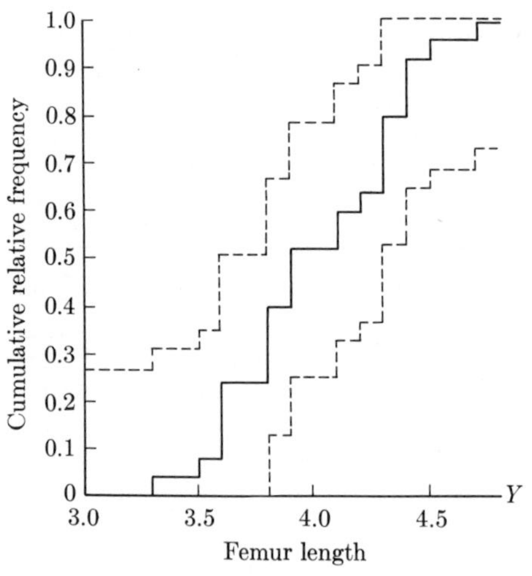

FIGURE 16.5 Confidence interval (95%) for the observed cumulative frequency distribution of Figure 16.4.

would expect 95% of such sets of confidence limits to completely enclose the true cumulative frequency distribution.

The programs in Appendix A3.1, A3.2, and A3.3 perform the computations described in this section. Programs A3.1 and A3.2 are designed for goodness of fit tests on continuous variables (g_1, g_2, and D, the Kolmogorov-Smirnov statistic, are computed). Program A3.3 is designed for frequency type data (the G-test is used to test agreement to Poisson, binomial, or arbitrary expected distributions, or to simple ratios).

16.3 Replicated tests of goodness of fit

Frequently experiments leading to tests of goodness of fit are replicated. For example, the data in Box 16.4 are from a genetic experiment on houseflies in which eight replicate crosses were carried out, with the offspring expected in a ratio of 9 wild type to 7 pale-eyed flies. Although one could simply pool the results from the eight progenies and test the hypothesis of a 9:7 ratio on the grand totals, doing so loses information. The various tests that can be performed on such data are illustrated in Boxes 16.4 and 16.5, using the G-test. It should be pointed out that corresponding chi-square tests can be performed, but these have the disadvantage that the X^2-values are not com-

BOX 16.4

Replicated goodness of fit tests (G-statistic).

Numbers of wild type (N^+) and pale-eyed (ge) flies in $b = 8$ progenies of the housefly, *Musca domestica*. The expected ratio of the $a = 2$ phenotypic classes N^+ and ge is 9:7.

Progeny	N^+	ge	Σ
1	83	47	130
2	77	43	120
3	110	96	206
4	92	58	150
5	51	31	82
6	48	61	109
7	70	42	112
8	85	66	151
Σ	616	444	$1060 = n$

Source: Unpublished data by R. L. Sullivan.

The order of computations may differ somewhat, depending on whether all replicates are obtained essentially simultaneously or over a period of time (see text for further discussion of this issue). The resulting final analysis will be the same in any case. In this example the progenies were obtained essentially simultaneously, and we therefore proceed immediately to an overall analysis of the data, reserving tests of goodness of fit of individual progenies until step **5**.

BOX 16.4 continued

Computation

1. Set up a table as shown below, which is essentially an $f \ln f$ transformation of the original data table. To obtain this, make use of Table **G**. The two columns inside the table (N^+ and ge) are simply $f \ln f$ of the original frequencies. The values in the inner margins are the sums of these functions. The values in the outer margins are $f \ln f$ of the marginal totals of the original table. Clearly, the transformations of the sums of these variables will not be the same as the sum of the transformations.

Progeny	N^+	*ge*	*Sum*	$f \ln f$ *of row sums of original table*
1	366.764	180.957	547.721	632.780
2	334.473	161.732	496.205	574.499
3	517.053	438.177	955.230	1097.542
4	416.005	235.506	651.511	751.595
5	200.523	106.454	306.977	361.351
6	185.818	250.763	436.581	511.357
7	297.395	156.982	454.377	528.472
8	377.625	276.517	654.142	757.609
Sum	2695.656	1807.088	4502.744	$\Sigma = 5215.205$
$f \ln f$ of columns sums of original table	3956.720	2706.546	$\Sigma = 6663.266$	$7383.986 = n \ln n$

The following quantities are necessary for the intended computations:

a. The sum of the transforms of all frequencies = 4502.744

b. The sum of the transforms of the column sums of the original table = 6663.266

c. The sum of the transforms of the row sums of the original table = 5215.205

d. The transform of n, the total number of items in the study = 7383.986

e. For each class compute $\ln \hat{p}_i$:

$$\ln \hat{p}_1 = 2.30259 \log \hat{p}_1 = 2.30259 \log \left(\tfrac{9}{16}\right)$$

$$= 2.30259(9.75012 - 10) = 2.30259(-0.24988) = -0.57537$$

$$\ln \hat{p}_2 = 2.30259 \log \hat{p}_2 = 2.30259 \log \left(\tfrac{7}{16}\right)$$

$$= 2.30259(9.64098 - 10) = 2.30259(-0.35902) = -0.82668$$

There are two main tests of interest, shown in steps **2** and **3**.

2. Are the ratios of the progenies (replicates) homogeneous? Compute G_H (G for heterogeneity).

BOX 16.4 continued

$$G_H = 2[\text{quantity } \mathbf{a} - \text{quantity } \mathbf{b} - \text{quantity } \mathbf{c} + \text{quantity } \mathbf{d}]$$
$$= 2[4502.744 - 6663.266 - 5215.205 + 7383.986]$$
$$= 2[8.259] = 16.518$$

The value is compared to a χ^2-distribution with $(a - 1)(b - 1) = (2 - 1)(8 - 1) = 7$ degrees of freedom. Since $\chi^2_{.05[7]} = 14.067$ there is evidence that the ratios among the progenies are not all homogeneous.

3. A second test is for goodness of fit to the 9:7 hypothesis for the pooled data, or all eight replicates pooled together. These frequencies are $\sum^{b} f_1 = 616$ and $\sum^{b} f_2 = 444$. The test is by Expression (16.7) from Section 16.2. This expression and the corresponding quantity symbols of the present box are listed in parallel rows below, where f_i refers to the pooled observed frequencies rather than those of individual progenies:

$$G = 2[\sum^{a} f_i \ln f_i - \overbrace{\sum^{a} f_i \ln \hat{p}_i} - n \ln n]$$

$$G = 2[\text{quantity } \mathbf{b} - \sum^{a} f_1 \times \text{quantity } \mathbf{e}_1 - \sum^{a} f_2 \times \text{quantity } \mathbf{e}_2 - \text{quantity } \mathbf{d}]$$
$$= 2[6663.266 - 616(-0.57537) - 444(-0.82668) - 7383.986]$$
$$= 2[0.7538] = 1.5076$$

This value is clearly not significant when compared to a χ^2-distribution with $a - 1 -$ (number of parameters estimated from the sample) $= 2 - 1 - 0 = 1$ degree of freedom.

4. Adding the G for heterogeneity and the pooled G, we obtain a total G,

$$G_H + G_P = G_T = 16.518 + 1.508 = 18.026$$

with $b(a - 1) = 8(2 - 1) = 8$ degrees of freedom. This can be used to test whether the data as a whole fit the expected ratio. Since $\chi^2_{.05[8]} = 15.507$, we conclude that they do not. Since G_H was significant and G_P was small, we know that the reason that G_T is significant is due to the significant heterogeneity. There are two analyses that might next be done, depending upon the interests of the investigator. One of these is described below and the other, an a posteriori test of differences among the progenies using an *STP* procedure, is described in Box 16.5.

5. Partitioning the total G into contributions due to individual samples. For each sample compute G. We employ Expression (16.7) here rather than (16.6a) as in Box 16.1 because we already have computed values for $\ln \hat{p}_i$ (which are constant in this problem), and Expression (16.6a) would require $\ln \hat{f}_i$, which differs for every progeny. Note that all quantities needed have been computed or looked up in a table previously.

$$G = 2[\sum^{a} f_i \ln f_i - \sum^{a} f_i \ln \hat{p}_i - n_i \ln n_i]$$

For progeny 1 this would be

BOX 16.4 continued

$$G = 2[83 \ln 83 + 47 \ln 47 - 83 \ln (\tfrac{9}{16}) - 47 \ln (\tfrac{7}{16}) - 130 \ln 130]$$
$$= 2[366.764 + 180.957 - 83(-0.57537) - 47(-0.82668) - 632.780]$$
$$= 2[1.551] = 3.102$$

This has $a - 1 -$ (number of parameters estimated from sample) $= 2 - 1 - 0 = 1$ degrees of freedom. It is not significant.

We can continue testing each progeny in this fashion. In doing so we find only progeny 6 significant, as shown below:

$$G = 2[48 \ln 48 + 61 \ln 61 - 48(-0.57537) - 61(-0.82668) - 109 \ln 109]$$
$$= 6.538^*$$

Since $p_1 = \frac{48}{109} = 0.440$, we see that the discrepancy was in the direction of having too few wild type flies ($\hat{p}_1 = \frac{9}{16} = 0.5625$).

As a computational check we add the eight individual G's and note that their sum equals the total G, $G_T = 18.026$.

6. We can summarize our results in the following tables ($* = 0.01 < P \leq 0.05$).

Tests	*df*	*G*
Pooled	1	1.508
Heterogeneity	7	16.518*
Total	8	18.026*

Progeny	*df*	*G*
1	1	3.102
2	1	3.113
3	1	0.680
4	1	1.595
5	1	1.194
6	1	6.538*
7	1	1.803
8	1	0.001
Total	8	18.026

Note: If individual values of G had been computed first, the heterogeneity G could be obtained simply by subtracting pooled G from the total G (the sum of the individual G's) and need not be independently computed as in step **2** above.

pletely additive so that the anova-like table given in step **6** of Box 16.4 would only be approximately correct. For this reason we have only given the computations for the G-test. Such a test employing X^2 is called an *interaction chi-square test* or a *heterogeneity chi-square test,* and by analogy we might call the corresponding G-test an *interaction* or *heterogeneity G-test.*

The order in which the computations are to be carried out depends in part on the design of the experiment. If the replicates are obtained in sequence with an appreciable time interval between each, the first logical step is likely to be a goodness of fit test of each individual replicate. Thus, if the progenies were obtained a week apart, the investigator would usually wish to carry out a significance test for each of the progenies as he obtained it, reserving the overall analysis for the end of the experiment. On the other hand, if the results are obtained essentially simultaneously, the overall tests of goodness of fit may be carried out first, with individual replicates analyzed if warranted by the data and the interest of the investigator. In the example of Box 16.4, the data were collected essentially simultaneously, and we leave the analysis of the separate progenies until later in the box (step **5**). We should emphasize that the final outcome of the analysis is identical regardless of which way the problem is approached.

First we test whether the outcomes of all the replicates (progenies in this case) are homogeneous. By that we mean whether the sampled proportions (of wild type) in the eight progenies could have been sampled from a single population. If the proportions are not uniform among replicates, they are said to be heterogeneous. Five hypothetical heterogeneous replicates might have the following relations between the proportions p_1 and p_2 of wild type and pale-eyed flies, respectively:

$$p_1 > p_2$$
$$p_1 > p_2$$
$$p_1 = p_2$$
$$p_1 < p_2$$
$$p_1 > p_2$$

We see immediately why such heterogeneity could also be called *interaction*, by analogy with analysis of variance.

The actual test is shown in Box 16.4. The computation of G_H (heterogeneity G) is by a formula entirely analogous to the interaction formula in a two-way orthogonal anova: $\bar{Y} - \bar{R} - \bar{C} + \bar{\bar{Y}}$ (see Section 11.2).

$$\begin{aligned} G_H = 2[&\text{sum of } f \ln f \text{ of frequencies in each cell of the table} \\ &- \text{sum of } f \ln f \text{ of column sums of frequencies} \\ &- \text{sum of } f \ln f \text{ of row sums of frequencies} \\ &+ f \ln f \text{ of total number of items in the table}] \end{aligned}$$

We find that we must reject this null hypothesis for the data of Box 16.4. One or more of the progenies differ in outcome from the others. If we had simply pooled all the data and carried out an overall test we could not, of course, have detected this heterogeneity.

The next test carried out is on the pooled frequencies, which means the eight progenies treated as though they were a single large progeny. The test employed is exactly the same as that for a single classification shown in Section 16.2 using Expression (16.7). We find that the pooled data do not deviate significantly from expectation. This might surprise you in view of the significant heterogeneity of the data, but it happens quite frequently when replicates are heterogeneous in such a way that they tend to compensate for their respective deviations from expectation.

We can illustrate this point by an artificial, deliberately exaggerated example. Suppose we test sex ratios in six bottles of drosophila, three of which had only male offspring and three only female offspring. If we pooled these data (assuming that the number of male and female offspring were approximately the same) we would be unable to reject the 50:50 sex ratio hypothesis for these drosophila, although there clearly is no such sex ratio in these flies. Heterogeneity in this example would, of course, be highly significant.

The scientific relations expressed by these tests for heterogeneity and of pooled data are of a fundamental nature and must be clearly understood. To deepen and confirm your understanding of them, we furnish further hypothetical examples on sex ratios in drosophila in which the differences are more subtle. Instead of using the G-test illustrated in Box 16.4, we shall employ the sample statistic X^2 to show that the so-called chi-square tests can also serve for the analysis of replicated tests for goodness of fit. Four examples are given in Table 16.4.

In example A we have four progenies, each with a slight excess of females, but each not significant by individual chi-square tests, which are easiest carried out by application of Expression (16.8). [Since $r = \hat{p}/\hat{q} = 1$, this formula reduces to $X^2 = (f_1 - f_2)^2/n$ in this case.] Although each individual X^2 is suspiciously high, none of them is significant. However, when we add these values of X^2 together to obtain a total X^2, we note that the test is now significant at the 5% level. When considered together, suspicious results for individual progenies yield an improbable total result. We conclude from the total X^2 that it is unlikely to obtain four sample progenies each of which is as far from expectation as these. This finding is reinforced when we examine the pooled X^2. Since all progenies deviate in the same direction, the pooled frequencies show this deviation clearly and, being based on a larger sample size, yield a highly significant value of X^2. To obtain the heterogeneity X^2, we subtract pooled from total X^2. The resulting heterogeneity X^2 of 0.072 with 3 degrees of freedom is clearly not significant. This tells us that the deviations from expectation of the progenies were in the same direction and not significantly different from each other.

Example B has the same total frequencies but in progenies 2 and 4 the frequencies of males and females have been reversed. The individual values of X^2 are necessarily the same as before, suspicious but not significant. The total X^2, being the sum of the individual ones must also be the same as before, indicating that these progenies do not conform to expectation. The important difference between examples A and B lies in the pooled and heterogeneity X^2. Since two of the progenies in B favored females and the other two favored males, the pooled frequencies of males and females are now almost identical. The pooled X^2 is therefore negligible, 0.002, a suspiciously good fit, unlikely to occur in actual, experimental data. Subtracting such a low pooled X^2 from the total X^2 leaves a large heterogeneity X^2, indicating that the sus-

TABLE 16.4

Some hypothetical replicated tests of goodness of fit to illustrate meaning of total, pooled, and heterogeneity G or X^2. Computation of G is illustrated in Box 16.4, of X^2 in this table, using formulas from Box 16.1. The null hypothesis being tested is a 50:50 sex ratio.

	Progenies	♀	♂	*n*		*df*	X^2
A	1	59	41	100		1	3.240 *ns*
	2	58	42	100		1	2.560 *ns*
	3	57	42	99		1	2.273 *ns*
	4	58	40	98		1	3.306 *ns*
					Total	4	11.379 $P < 0.05$
	Σ	232	165	397	Pooled	1	11.307 $P < 0.005$
					Heterogeneity	3	0.072 *ns*
B	1	59	41	100		1	3.240 *ns*
	2	42	58	100		1	2.560 *ns*
	3	57	42	99		1	2.273 *ns*
	4	40	58	98		1	3.306 *ns*
					Total	4	11.379 $P < 0.05$
	Σ	198	199	397	Pooled	1	0.002 *ns*
					Heterogeneity	3	11.377 $P < 0.01$
C	1	59	41	100		1	3.240 *ns*
	2	58	42	100		1	2.560 *ns*
	3	72	26	98		1	21.592 $P < 0.005$
	4	73	26	99		1	22.313 $P < 0.005$
					Total	4	49.705 $P < 0.005$
	Σ	262	135	397	Pooled	1	40.627 $P < 0.005$
					Heterogeneity	3	9.078 $P < 0.05$
D	1	52	48	100		1	0.160 *ns*
	2	36	64	100		1	7.840 $P < 0.02$
	3	52	47	99		1	0.252 *ns*
	4	52	46	98		1	0.367 *ns*
					Total	4	8.619 *ns*
	Σ	192	205	397	Pooled	1	0.426 *ns*
					Heterogeneity	3	8.193 $P < 0.05$

picious departures from expectation were not in a uniform direction, but in different directions, as we know them to be.

Example C illustrates a case in which all three tests are significant. All four progenies have an excess of females. The first two have only a small (nonsignificant) excess, and the last two deviate highly significantly from expectation. The total X^2 is consequently highly significant. The pooled X^2,

BOX 16.5

A posteriori tests by *STP* of the homogeneity of sets of replicates tested for goodness of fit.

In situations such as that in Box 16.4, where there is significant heterogeneity among replicates, the following technique can be used to test whether all of the replicates are different from one another or whether there are sets of homogeneous replicates with possibly only a few aberrant ones. We shall use the housefly progenies of Box 16.4 to illustrate the technique.

For the heterogeneity test to be just significant at the α level, G_H must equal to $\chi^2_{\alpha[(a-1)(b-1)]}$. This value is used as the constant critical value for G in the tests that follow. In our case, $a = 2$; $b = 8$; therefore $\chi^2_{.05[7]} = 14.067$ can be used as a critical value if a 5% experiment-wise error rate is accepted.

The test procedure consists of a heterogeneity test for all sets of replicates (progenies in this example) taken 2, 3, 4, . . . at a time. If there are many replicates a computer program is useful. However, usually not all of these tests need be carried out, as will be demonstrated below.

1. If not already calculated for the tests described in Box 16.4, the $f \ln f$ values corresponding to all of the observed frequencies and their sums should be looked up in Table **G**. All necessary quantities are contained in the second table of Box 16.4, but for convenience we list them again:

Progeny	*(1)* *Sum of $f \ln f$ for cell frequencies* $\sum^{a}(f_i \ln f_i)$	*(2)* *$f \ln f$ of row sums of original table* $\left(\sum^{a} f_i\right) \ln \left(\sum^{a} f_i\right)$	*(3)* p_i
1	547.721	632.780	0.638
2	496.205	574.499	0.642
3	955.230	1097.542	0.534
4	651.511	751.595	0.613
5	306.977	361.351	0.622
6	436.581	511.357	0.440
7	454.377	528.472	0.625
8	654.142	757.609	0.563

Column (1) lists the sum of the $f \ln f$ values for the cells of each replicate (progeny). Thus, for the first progeny $547.721 = 366.764 + 180.957$. Column (2) lists the transform for the total number in each replicate. Thus 632.780 is the transform of the 130 flies in progeny 1. We list p_i, the observed proportion of wild type flies in column (3) to suggest to us which contrasts are of interest. The expected proportion of wild type, $\hat{p}$, is $9/16 = 0.5625$ for all progenies. In place of p_i, one could also use $f_i - \hat{f}_i$ to suggest tests, but that would involve more calculation.

2. Since progenies 1, 2, 4, 5, and 7 have rather similar p_i's, we shall test this set first.

BOX 16.5 continued

a. Compute the column totals of the frequencies in Box 16.4 for this set of progenies. These are 373 for N^+ and 221 for *ge* and their sum is 594. The corresponding $f \ln f$ values in Table **G** are 2208.749, 1192.994, and 3793.806.

b. Compute G_H analogously to step **2** in Box 16.4.

$$G_H = 2[\text{sum of transforms of observed frequencies in this set}$$
$$- \text{sum of transforms of column sums of frequencies}$$
$$- \text{sum of transforms of row sums of frequencies}$$
$$+ \text{transform of total number of items in the set}]$$
$$= 2[(547.721 + 496.205 + 651.511 + 306.977 + 454.377)$$
$$- (2208.749 + 1192.994)$$
$$- (632.780 + 574.499 + 751.595 + 361.351 + 528.472)$$
$$+ 3793.806]$$
$$= 2[2456.791 - 3401.743 - 2848.697 + 3793.806]$$
$$= 2[0.157] = 0.314$$

which is far less than our critical $G_{.05} = 14.067$; therefore this set of replicates is considered homogeneous.

We now try adding progeny 8 to our set (this progeny has the next lower value of p_i). In computing a new G_H we can make use of some of the intermediate computations given above. The new column totals are $373 + 85 = 458$ and $221 + 66 = 287$, and their sum is 745 (corresponding $f \ln f$ transforms are 2806.106, 1624.271, and 4926.971).

$$G_H = 2[(2456.791 + 654.142) - (2806.106 + 1624.271)$$
$$- (2848.697 + 757.609) + 4926.971]$$
$$= 2[3110.933 - 4430.377 - 3606.306 + 4926.971] = 2[1.221] = 2.442$$

Hence this set is also homogeneous.

Progeny 3 has the next lower value of p_i. Adding it to the set and recomputing G_H as before,

$$G_H = 2[(3110.933 + 955.230) - (568 \ln 568 + 383 \ln 383)$$
$$- (3606.305 + 1097.542) + 951 \ln 951]$$
$$= 2[3.390] = 6.780$$

Again, the set is homogeneous. It is also a maximally nonsignificant set, since the test of the complete set yields a significant G_H of 16.518 as shown in Box 16.4. Therefore, adding progeny 6 to set {1, 2, 3, 4, 5, 7, 8} would yield a significantly heterogeneous set.

We could now try to find a maximally nonsignificant set involving progeny 6. Starting with progenies 6 and 3, we obtain

$$G_H = 2[(436.581 + 955.230) - (158 \ln 158 + 157 \ln 157)$$
$$- (511.357 + 1097.542) + 315 \ln 315]$$
$$= 2[1391.811 - 1593.721 - 1608.899 + 1812.060] = 2[1.251] = 2.502$$

BOX 16.5 continued

This value of G_H is not significant. Continuing in this fashion, we find that the set {3, 4, 5, 6, 7, and 8} is also maximally nonsignificant, since its $G_H = 11.804$ would be increased to 14.450 if we added progeny 1.

Thus, we may consider progenies {1, 2, 3, 4, 5, 7, and 8} and {3, 4, 5, 6, 7, and 8} as being homogeneous sets. Arraying the progenies by proportion of wild type (high to low), we can summarize their relations as follows:

Progeny 2 1 7 5 4 8 3 6

(First underline spans progenies 2 through 3; second underline spans progenies 7 through 6.)

It is now of interest to the investigator to discover why progenies 1 and 2 (high proportion wild type) were different from progeny 6 (low proportion wild type).

in view of the consistent trend in favor of females, is also highly significant and the heterogeneity X^2 is significant as well, showing that the trend, although in all cases favoring females, is not uniform in magnitude.

Finally, in example D we show that the total X^2 is an overall measure of the departure from expectation of the several progenies. While progeny 2 has a significantly larger number of males, 1, 3, and 4 have slight, nonsignificant excesses in favor of females. The total X^2, though suspiciously high, is not significant; the fit of the other progenies was good enough that the departure from expectation by progeny 2 was not sufficient to affect our overall judgment based on the total X^2. The frequencies in the progenies tended to compensate for each other making the pooled X^2 not significant, but on subtraction the heterogeneity X^2 is significant, indicating correctly that the data are heterogeneous, although neither total nor pooled X^2 had been significant.

These very same computations could be carried out with the G-statistic as shown in Box 16.4, and would lead to similar results. One difference between the computations of G and X^2 is that in the latter the separate computation of heterogeneity X^2 in a manner similar to that shown for G in step (2) of Box 16.4 is quite complex and therefore rarely done. Instead, as we have seen, heterogeneity X^2 is obtained by subtraction of pooled from total X^2. However, unlike that obtained with the G-statistic, this yields only an approximate value of heterogeneity X^2.

When G or X^2-values are to be summed or partitioned, Yates' correction is not used since the adjusted G and X^2-values are not additive.

Returning to the example in Box 16.4, we now add heterogeneity G to pooled G to obtain total G. We find that the data as a whole do not fit the expected ratio. Occasionally we might obtain a significant G_T, but nonsignificant though appreciable values of G_H and G_P. In such a case we are led to conclude that the overall fit of the data to the hypothesis is poor, but we are unable to pinpoint the exact nature of the departure from expectation.

When we compute G separately for each of the individual replicates of the study we find only progeny 6 departing significantly from expectation, by having fewer wild type and more mutant flies. The analysis can be summarized (step **6**) by a table similar to an anova table.

Since the replicates are not all homogeneous, it is of interest to know the reason for this heterogeneity. Precisely which set of samples is homogeneous and which samples are significantly different from the rest to cause the heterogeneity? It may sometimes be possible to determine this by inspection, but the methods of Box 16.5 give a routine procedure for carrying out an a posteriori test to locate the sources of the heterogeneity. The computations in Box 16.5 involve an *STP* using the *G*-statistic. In carrying out the computations we find that the significance of the heterogeneity appears to be due to differences between progenies 1 and 2 on one hand and 6 on the other. It is then of interest for the investigator to try to determine why these progenies gave different results. For example, he might investigate whether the medium in which they were reared could have been different or if there were any biological differences among the parents of these progenies.

Although the replicated goodness of fit tests in Boxes 16.4 and 16.5 tested a simple 9:7 ratio, these techniques can also be used for more complicated situations such as single classification data with more than two classes, including replicated goodness of fit tests to binomial or Poisson distributions.

The computations discussed in this section can be carried out using the program in Appendix A3.3 for single classification and replicated goodness of fit tests, and the program in Appendix A3.15 for the a posteriori test described in Box 16.5.

16.4 Tests of independence: two-way tables

The notion of statistical or probabilistic independence was first introduced in Section 5.1, where it was shown that if two events were independent, the probability of their occurring together could be computed as the product of their separate probabilities. Thus, if among the progeny of a certain genetic cross the probability of a kernel of corn being red is $\frac{1}{2}$ and the probability of the kernel being dented is $\frac{1}{3}$, the probability of obtaining a dented and red kernel would be $\frac{1}{2} \times \frac{1}{3} = \frac{1}{6}$, if the joint occurrences of these two characteristics are statistically independent.

The appropriate statistical test for this genetic problem would be to test the frequencies for goodness of fit to the expected ratios of 2 (red, not dented):2 (not red, not dented):1 (red, dented):1 (not red, dented). This would be a simultaneous test of two null hypotheses: that the expected proportions are $\frac{1}{2}$ and $\frac{1}{3}$ for red and dented, respectively, and that these two properties are independent. The first null hypothesis tests the Mendelian model in general. The second tests whether these characters assort independently—that is, whether they are determined by genes located in different linkage groups. If the second hypothesis must be rejected, this is taken as evidence that the characters are linked—that is, located on the same chromosome.

There are numerous instances in biology in which the second hypothesis concerning the independence of two properties is of great interest and the first hypothesis regarding the true proportion of one or both properties is of little interest. In fact, often no hypothesis regarding the parametric values $\hat{p}_i$ can be formulated by the investigator. We shall cite several examples of such situations, which lead to the test of independence to be learned in this section. We employ this test whenever we wish to test whether two different properties, each occurring in two states, are dependent on each other. For instance, specimens of a certain moth may occur in two color phases—light and dark. Fifty specimens of each phase may be exposed in the open, subject to predation by birds. The number of surviving moths is counted after a fixed interval of time. The proportion predated may differ in the two color phases. The two properties in this example are color and survival. We can divide our sample into four classes: light-colored survivors, light-colored prey, dark survivors, and dark prey. If the probability of being preyed upon is independent of the color of the moth, the expected frequencies of these four classes can be simply computed as independent products of the proportion of each color (in our experiment $\frac{1}{2}$) and the overall proportion preyed upon in the entire sample. Should the statistical test of independence explained below show that the two properties are not independent, we are led to conclude that one of the color phases is more susceptible to predation than the other. This is an important biological phenomenon; the exact proportions of the two properties are of little interest here. The proportion of the color phases is arbitrary, and the proportion survival is of interest only insofar as it differs for the two phases.

A second example might relate to a sampling experiment carried out by a plant ecologist. He obtains a random sample of 100 individuals of a fairly rare species of tree distributed over an area of 400 square miles. For each tree he notes whether it is rooted in a serpentine soil, or not, and whether the leaves are pubescent, or smooth. Thus the sample of $n = 100$ trees can be divided into four groups: serpentine-pubescent, serpentine-smooth, nonserpentine-pubescent, and nonserpentine-smooth. If the probability of a tree being pubescent or not is independent of its location, our null hypothesis of the independence of these properties will be upheld. If, on the other hand, the proportion of pubescence differs for the two types of soils, our statistical test will most probably result in rejection of the null hypothesis of independence. Again, the expected frequencies will simply be products of the independent proportions of the two properties—serpentine versus nonserpentine, and pubescent versus smooth. In this instance the proportions may themselves be of interest to the investigator.

The example we shall work out in detail is from immunology. A sample of 111 mice was divided into two groups, 57 that received a standard dose of pathogenic bacteria followed by an antiserum and a control group of 54 that received the bacteria, but no antiserum. After sufficient time had elapsed for an incubation period and for the disease to run its course, 38 dead mice and 73 survivors were counted. Of those that died, 13 had received bacteria *and* antiserum while 25 had received bacteria only. A question of interest is whether the antiserum had in any way protected the mice so that there were proportionally more survivors in that group. Here again the proportions of these properties are of no more interest than in the first example (predation on moths).

Such data are conveniently displayed in the form of a *two-way table* as shown below. Two-way and multiway tables (more than two criteria) are often known as *contingency tables.* This type of two-way table, in which each of the two criteria is divided into two classes, is known as a 2×2 *table.*

	Dead	*Alive*	Σ
Bacteria and antiserum	13	44	57
Bacteria only	25	29	54
Σ	38	73	111

Thus 13 mice received bacteria and antiserum but died, as seen in the table. The marginal totals give the number of mice exhibiting any one property: 57 mice received bacteria and antiserum; 73 mice survived the experiment. Altogether 111 mice were involved in the experiment and constitute the total sample.

In discussing such a table it is convenient to label the cells of the table and the row and column sums as follows:

a	b	$a + b$
c	d	$c + d$
$a + c$	$b + d$	n

From a two-way table one can systematically compute the expected frequencies (based on the null hypothesis of independence) and compare them with the observed frequencies. For example, the expected frequency for cell d (bacteria, alive) would be

$$\hat{f}_{\text{bact,alv}} = n\hat{p}_{\text{bact,alv}} = n\hat{p}_{\text{bact}} \times \hat{p}_{\text{alv}} = n\left(\frac{c+d}{n}\right)\left(\frac{b+d}{n}\right)$$

$$= (c + d)(b + d)/n$$

which in our case would be $(54)(73)/111 = 35.5$, a higher value than the observed frequency of 29. We can proceed similarly to *compute the expected frequencies for each cell in the table by multiplying a row total times a column total, and dividing the product by the grand total.* The expected frequencies can be conveniently displayed in the form of a two-way table:

	Dead	*Alive*	Σ
Bacteria and antiserum	19.5	37.5	57.0
Bacteria only	18.5	35.5	54.0
Σ	38.0	73.0	111.0

You will note that the row and column sums of this table are identical to those in the table of observed frequencies, which should not surprise you since the expected frequencies were computed on the basis of these row and column totals. It should therefore be clear that a test of independence will not test whether any property occurs at a given proportion but can only test whether or not the two properties are manifested independently.

Before we turn to the actual tests of significance, we need to consider different models related to the design of the experiment or sampling procedure. These models relate to the question of whether the marginal totals in the two-way table are fixed by the investigator or are free to vary and reflect population parameters. A reconsideration of the examples taken up so far will make this point clear. In the example of the trees growing in two types of soils and exhibiting two types of leaves, only the total sample size n was fixed; 100 trees had been sampled. The actual proportion of trees growing in serpentine soils was an outcome of the sampling experiment and was not under the control of the investigator. Similarly, the proportion of leaves that were pubescent was not fixed by the investigator but was an outcome of the experiment. For a test of the null hypothesis one should compute the probability of obtaining the frequencies observed in the 2×2 table or worse departures from independence out of all possible 2×2 tables with the same total sample size n. We shall call this design the Model I for a two-way table.

There is another type of Model I in which the same individuals are subjected to two successive tests. Here again only the total sample size is fixed, but in such an experiment the hypotheses to be tested are different. This type of Model I is discussed in Section 16.7.

A second design, Model II, would have the marginal totals for one of the two criteria fixed. An example would be the immunology experiment that we have treated in detail just above. The number of mice given bacteria plus antiserum and that given bacteria only was fixed by the experimenter at 57 and 54 animals, respectively. The proportion that survived the experiment was free to vary, depending on the effect of the treatment. In testing the null hypothesis of independence one should compute the probability of getting the observed results or worse departures from independence out of all possible two-way tables with the same marginal totals for the two treatment classes.

The moth example cited at the beginning of this section is another example of a Model II test for independence. The investigator released 50 specimens of each color phase, and hence the marginal totals for color phase were fixed by the experiment. The proportion predated depended on the natural forces involved and is free to vary.

A third model for the design of a 2×2 table is one in which both marginal totals are fixed by the experiment. We have so far not yet encountered an example of this. Assume that we wish to investigate the preference of weevil larvae for two types of beans differing in their seed coats. The investigator places a fixed number, say 100 beans, divided equally between types

A and B, into a container. He then adds 70 first instar larvae to the jar. Under such density conditions, no more than one larva will attack any one bean. After enough time has elapsed to permit each larva to enter a bean, a count is made of how many beans of each type have been attacked by a weevil larva. Again, we have four frequency classes of beans: A—attacked, B—attacked, A—not attacked, and B—not attacked. The number of beans of both types and the number of beans that were attacked were both fixed by the experimenter. The number attacked was equal to the number of larvae released in the jar. When testing the null hypothesis of independence in this Model III, we should compute the probability of obtaining the observed frequencies and all worse departures from independence out of all possible 2×2 tables with the same marginal totals for both criteria—bean type and presence or absence of attacking larvae.

The actual tests of independence that have been proposed for these 2×2 tables are of three kinds: the chi-square test, the G-test, and an exact probability test known as Fisher's exact test. Conventional formulations of the last two tests were originally designed for specific models; the G-test [Expression (16.12)] was based on Model I, and Fisher's exact test, based on Expression (16.13) and illustrated in Box 16.7, was intended for Model III. The chi-square test, which has been the most frequently used test in the past, was not specifically designed for any one model. In practice, the designs represented by the three models have not been clearly distinguished in textbooks of statistics, and research workers have been applying each of the tests to all three models. The criteria for choice between the tests have not been the model of the design, but rather the sample size of the test and the computational effort involved. Fortunately, the different techniques seem to provide rather similar results even when applied to the inappropriate model.

The chi-square or G-tests test the goodness of fit of the observed cell frequencies to their expectations. For the chi-square test we could employ Expressions (16.3) or (16.5), but there is a much simplified formula for X^2 in 2×2 tables, which will be presented in the next paragraph. A new equation for G, which works for any two-way table, will be presented later on in this section.

The shortcut equation for calculating X^2 in 2×2 contingency tables,

$$X^2 = \frac{(ad - bc)^2 n}{(a + b)(c + d)(a + c)(b + d)} \tag{16.9}$$

can be derived from Expression (16.1). In the immunology experiment this expression would yield

$$X^2 = \frac{[(13 \times 29) - (44 \times 25)]^2 \times 111}{57 \times 54 \times 38 \times 73} = \frac{58{,}022{,}919}{8{,}538{,}372} = 6.796$$

This value is more correct than $X^2 = 6.767$, which would be computed from Expression (16.3) using expected frequencies rounded to one decimal place.

Let us state without explanation that the observed X^2 should be compared with χ^2 for one degree of freedom. We shall examine the reasons for this at the end of this section. The probability of finding a fit as bad, or worse, to these data is $0.005 < P < 0.01$. We conclude, therefore, that mortality in these mice is not independent of the presence of antiserum. We note that the percentage mortality among those animals given bacteria *and* antiserum is $(13)(100)/57 = 22.8\%$, considerably lower than the mortality of $(25)(100)/54 = 46.3\%$ among the mice to whom only bacteria had been administered. Clearly, the antiserum has been effective in reducing mortality.

With small sample sizes it is frequently recommended to use *Yates' correction*, which, as before, means that we should adjust the observed frequencies by adding or subtracting 0.5 in such a way as to reduce each $|f_i - \hat{f}_i|$. In a 2×2 test of independence, this means that we would add 0.5 to cells a and d and subtract 0.5 from cells b and c when the quantity $ad - bc$ is negative. Should this quantity be positive we would subtract 0.5 from cells a and d and add it to cells b and c. This adjustment can also be automatically carried out by using the following expression:

$$X^2_{\text{adj}} = \frac{\left(|ad - bc| - \frac{n}{2}\right)^2 n}{(a + b)(c + d)(a + c)(b + d)} \tag{16.9Y}$$

Recent work by Grizzle (1967) has shown that the application of Yates' correction to chi-square tests with Models I and II almost always results in an unduly conservative test (the type I error is much lower than desired). It would therefore seem that Yates' correction is unnecessary even with quite low sample sizes, such as $n = 20$.

The equations for the standard G-test of independence were designed for Model I and are based on a multinomial distribution. A *multinomial distribution* is a discrete probability distribution in which an attribute has more than two classes. The binomial distribution with two classes of attributes is a special case of it. We do not discuss the multinomial distribution in detail in this text. The interested reader is referred to section 5.12 in Ostle (1963). The probability of observing the results of a 2×2 table assuming a multinomial distribution is the following:

$$\frac{n!}{a!\,b!\,c!\,d!}\left(\frac{a}{n}\right)^a\left(\frac{b}{n}\right)^b\left(\frac{c}{n}\right)^c\left(\frac{d}{n}\right)^d \tag{16.10}$$

If we assume that the row and column classifications are independent, this expression changes to

$$\left[\frac{n!}{a!\,b!\,c!\,d!}\left(\frac{(a+b)(a+c)}{n^2}\right)^a\left(\frac{(a+b)(b+d)}{n^2}\right)^b\right] \times \left[\left(\frac{(a+c)(c+d)}{n^2}\right)^c\left(\frac{(b+d)(c+d)}{n^2}\right)^d\right] \tag{16.11}$$

If we compute the natural logarithm of the ratio of these two probabilities or likelihoods, the result is

$$\ln\left[\frac{a^a\, b^b\, c^c\, d^d\, n^n}{(a+b)^{a+b}(a+c)^{a+c}(b+d)^{b+d}(c+d)^{c+d}}\right]$$
$$= a \ln a + b \ln b + c \ln c + d \ln d + n \ln n - (a+b) \ln (a+b) - (a+c) \ln (a+c) - (b+d) \ln (b+d) - (c+d) \ln (c+d) \tag{16.12}$$

Twice Expression (16.12) is G, which is approximately distributed as χ^2 with one degree of freedom.

There is no shortcut equation for Yates' correction applied to the G-test (see Box 16.6), but one must simply adjust the frequencies a, b, c, and d as described above and look up the new frequencies, all ending in $\frac{1}{2}$, in Table **G**, especially designed for such numbers.

In Box 16.6 we illustrate the G-test (using Yates' correction) applied to the sampling experiment in plant ecology, dealing with trees rooted in two different soils and possessing two types of leaves. The result of the analysis shows clearly that we cannot reject the null hypothesis of independence between cell type and leaf type. The presence of pubescent leaves is

BOX 16.6

2 × 2 test of independence (using the G-statistic).

A plant ecologist samples 100 trees of a rare species from a 400-square-mile area. He records for each tree whether it is rooted in serpentine soils, or not, and whether its leaves are pubescent or smooth. This represents a Model I design (only n is fixed).

Soil	*Pubescent*	*Smooth*	*Totals*
Serpentine	12	22	34
Not Serpentine	16	50	66
Totals	28	72	$100 = n$

The conventional algebraic representation of this table is as follows:

			Σ
	a	b	$a+b$
	c	d	$c+d$
Σ	$a+c$	$b+d$	$a+b+c+d=n$

If $ad - bc$ is positive, as it is in our example, since $(12 \times 50) - (16 \times 22) = 248$, subtract $\frac{1}{2}$ from a and d and add $\frac{1}{2}$ to b and c. If $ad - bc$ is negative, add $\frac{1}{2}$ to a and d and subtract $\frac{1}{2}$ from b and c. This is Yates' correction and may be ignored when $n > 200$. The new 2 × 2 table looks as follows.

BOX 16.6 continued

Soil	*Pubescent*	*Smooth*	*Totals*
Serpentine	$11\frac{1}{2}$	$22\frac{1}{2}$	34
Not serpentine	$16\frac{1}{2}$	$49\frac{1}{2}$	66
Totals	28	72	100

Compute the following quantities using Tables **G** and **G***.

1. $\sum f \ln f$ for the cell frequencies

$$= 11\tfrac{1}{2} \ln 11\tfrac{1}{2} + 22\tfrac{1}{2} \ln 22\tfrac{1}{2} + 16\tfrac{1}{2} \ln 16\tfrac{1}{2} + 49\tfrac{1}{2} \ln 49\tfrac{1}{2}$$

$$= 28.087 + 70.054 + 46.255 + 193.148 = 337.544$$

2. $\sum f \ln f$ for the row and column totals

$$= 34 \ln 34 + 66 \ln 66 + 28 \ln 28 + 72 \ln 72$$

$$= 119.896 + 276.517 + 93.302 + 307.920 = 797.635$$

3. Look up $n \ln n = 100 \ln 100 = 460.517$

4. Using Expression (16.12)

$$G_{\text{adj}} = 2[\text{quantity } \mathbf{1} - \text{quantity } \mathbf{2} + \text{quantity } \mathbf{3}]$$

$$= 2[337.544 - 797.635 + 460.517] = 2[0.426] = 0.852$$

Compare G_{adj} with critical value of χ^2 for one degree of freedom. Since our observed G_{adj} is much less than $\chi^2_{.05[1]} = 3.841$, we accept the null hypothesis that the leaf type is independent of the type of soil in which the tree is rooted.

independent of whether the tree is rooted in serpentine soils or not.

Figure 16.6 shows the results of a sampling experiment from a population in which two independent factors, A and B, each had a probability of occurrence $\hat{p}_A = \hat{p}_B = \frac{2}{3}$. Total sample size was fixed ($n = 20$), but the marginal totals were free to vary (Model I). The distribution of resulting values of X^2 and G are shown and fitted to the expected χ^2-distributions. Figure 16.7 shows the result of applying Yates' correction to X^2 and G computed from a second set of samples. Since the sample size was small, none of the results fit the expected χ^2-distributions very well (although G_{adj} fits rather well in the right tail where the fit is most important). With larger sample sizes the G and chi-square tests are more similar to each other and closer to the χ^2-distribution.

□□□ Although the formula of the chi-square test appears to be simpler than that for G, in actual fact the computation of the latter is frequently less cumbersome when tables of $f \ln f$ are at hand. Computing X^2 requires the running product of the four marginal totals, which is awkward without a

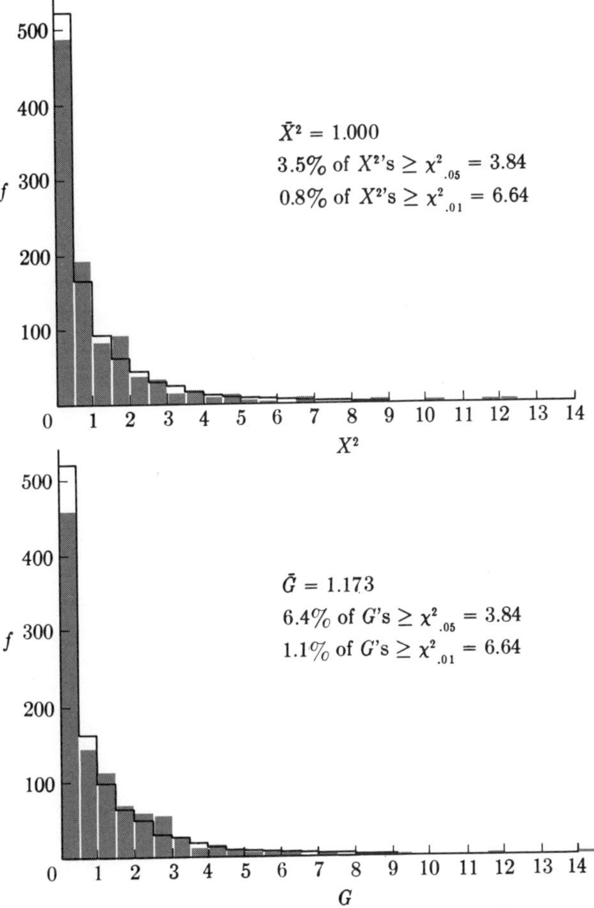

FIGURE 16.6 Results of a sampling experiment in which 1000 samples of size $n = 20$ were drawn from a population in which the presence or absence of two factors A and B is independent. Values of both $\hat{p}_A$ and $\hat{p}_B$, which stand for the presence of A and B, respectively, were $\frac{2}{3}$. The histograms show the distribution of the resulting values of X^2 (above) and G (below) computed for each sample on the hypothesis that factors A and B are independent. The expected $\chi^2_{[1]}$-distribution is superimposed for comparison.

transfer key for chain multiplication. Frequently the resulting product exceeds the capacity of the machine. □□□

The method that is especially designed for a 2 × 2 table of Model III is *Fisher's exact test.* This method computes the probability of obtaining the observed results and all worse cases. These probabilities are computed assuming that the row and column classifications are independent (the null hypothesis) and that the row and column totals are fixed. Fisher's exact test answers the following question. Given two-way tables with the same fixed marginal

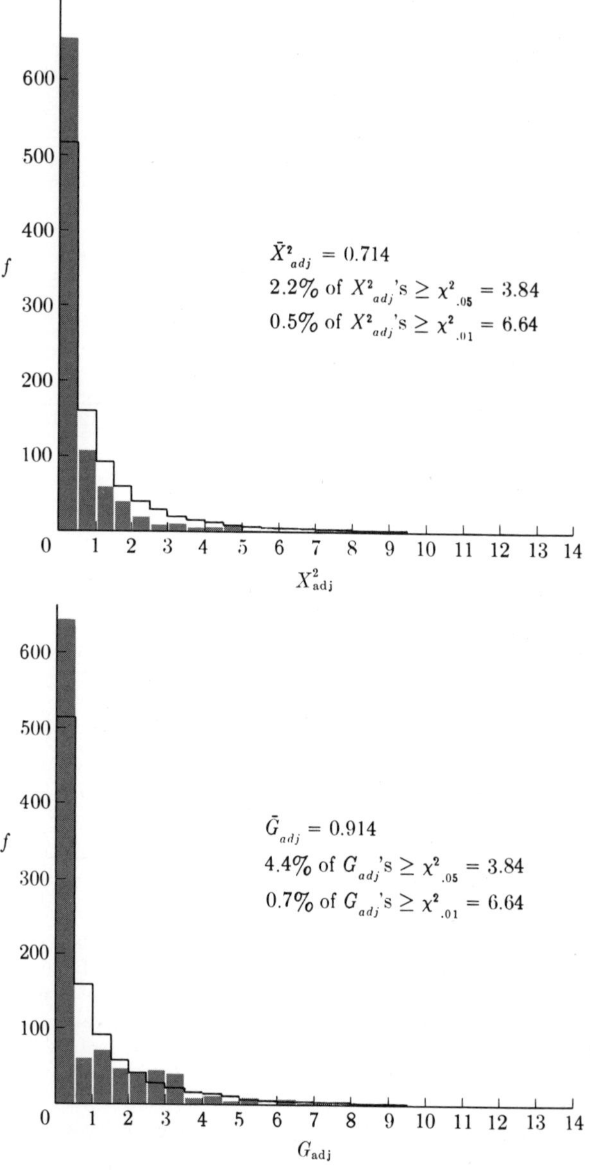

FIGURE 16.7 Results of a sampling experiment in which 1000 samples of size $n = 20$ were drawn from a population in which the presence or absence of two factors A and B is independent. Values of both $\hat{p}_A$ and $\hat{p}_B$, which stand for the presence of A and B, respectively, were $\frac{2}{3}$. Yates' correction was applied when computing tests of goodness of fit. The histograms show the distribution of the resulting values of X^2_{adj} (above) and G_{adj} (below) computed for each sample on the hypothesis that factors A and B are independent. The expected $\chi^2_{[1]}$-distribution is superimposed for comparison.

totals as the observed one, what is the chance of obtaining the observed cell frequencies a, b, c, and d and all cell frequencies that represent a greater deviation from expectation? The total number of ways in which a two-way table with fixed marginal totals can be obtained is

$$C(n, a+b) \times C(n, a+c) = \frac{n!}{(a+b)!\,(c+d)!} \times \frac{n!}{(a+c)!\,(b+d)!}$$

This is simply the product of the number of ways of taking $(a + b)$ items from n, multiplied by the number of combinations of $(a + c)$ items taken from n. From the coefficients of the multinomial distribution it can be shown that there are $n!/a!\,b!\,c!\,d!$ ways of obtaining the cell frequencies which we observed. Therefore the probability of obtaining a 2×2 table with cell frequencies a, b, c, and d is computed as the ratio of the two quantities given above. This can be simplified to

$$P = \frac{(a+b)!\,(c+d)!\,(a+c)!\,(b+d)!}{a!\,b!\,c!\,d!\,n!} \tag{16.13}$$

We must then compute the probability of all "worse" cases. The simplest way of doing this is to list systematically all the worse cases, compute the probability of each and sum these probabilities. Since one usually is interested in a two-tailed test, one should consider all worse cases in both tails, but as a computational convenience one usually computes the probability for the tail that contains the observed result and simply doubles this probability. The computation can be simplified somewhat by noting the fact that the numerator (products of factorials of the marginal totals) and $n!$ in the denominator remain constant for all worse cases. Only the frequencies a, b, c, and d change in each table.

A worked example of this method in Box 16.7 records the results of an experiment with acacia ants. All but 28 trees of two species of acacia were cleared from an area in Central America. There were 15 trees of species A

BOX 16.7

Fisher's exact test for independence in a 2 × 2 table.

Invasion rate of acacia trees of two species by ant colonies. For details on this experiment see text.

Acacia species	*Not invaded*	*Invaded by ant colony*	Σ
A	2	13	15
B	10	3	13
Σ	12	16	28

BOX 16.7 continued

The conventional algebraic representation of this table is as follows:

			Σ
	a	b	$a + b$
	c	d	$c + d$
Σ	$a + c$	$b + d$	$a + b + c + d = n$

1. Using Table **D** (common logarithms of factorials), compute the following sum of transforms of the marginal totals and subtract from it the transform of the total frequency:

$$\begin{aligned} \log (a + b)! &= \log 15! = 12.1165 \\ \log (c + d)! &= \log 13! = 9.7943 \\ \log (b + d)! &= \log 12! = 8.6803 \\ \log (a + c)! &= \log 16! = 13.3206 \\ -\log n! &= -\log 28! = -29.4841 \\ & \overline{\qquad 14.4276} \end{aligned}$$

2. In a similar manner compute the sum of the transforms of the observed cell frequencies:

$$\begin{aligned} \log a! &= \log \ 2! = 0.3010 \\ \log b! &= \log 13! = 9.7943 \\ \log c! &= \log 10! = 6.5598 \\ \log d! &= \log \ 3! = 0.7782 \\ & \overline{\qquad 17.4333} \end{aligned}$$

3. Subtract quantity **2** from quantity **1** and look up its antilogarithm. Since the result will be a probability (which must be ≤ 1), we can add 10 to the characteristic of quantity **1** in order to avoid negative logarithms.

$$\begin{array}{r} 24.4276 - 10 \\ -\ 17.4333 \qquad \\ \hline 6.9943 - 10 \end{array}$$

$$\text{antilog } (6.9943 - 10) = 0.000{,}9870$$

Assuming independence of the invasion rate and the proportion of the two species of acacia trees as well as fixed marginal totals, this is the probability of obtaining the results of the experiment by chance. We next compute the probability of all worse cases.

4. If $ad - bc$ is negative (as in the present example), decrease cells a and d by one and add unity to cells b and c. In this new 2×2 table the cell frequencies depart even more from expected frequencies (on the assumption of independence) than do the experimental results.

BOX 16.7 continued

Acacia species	*Not invaded*	*Invaded by ant colony*	Σ
A	1	14	15
B	11	2	13
Σ	12	16	28

Note that the marginal totals have remained the same. This means that quantity **1** will remain unchanged and we need only compute a new value corresponding to quantity **2**.

If $ad - bc$ had been positive, we would have decreased cells b and c by unity, and increased cells a and d by the same amount.

$$\begin{array}{rl} \log 1! = & 0.0000 \\ \log 14! = & 10.9404 \\ \log 11! = & 7.6012 \\ \log 2! = & 0.3010 \\ \hline & 18.4826 \end{array} \qquad \begin{array}{r} 24.4276 - 10 \\ -\ 18.8426 \\ \hline 5.5850 - 10 \end{array}$$

$$\text{antilog } (5.5850 - 10) = 0.000{,}038{,}46 \approx 0.000{,}038$$

5. Step **4** is repeated until one of the cell frequencies reaches zero. In the present example this happens on the very next step.

Acacia species	*Not invaded*	*Invaded by ant colony*	Σ
A	0	15	15
B	12	1	13
Σ	12	16	28

$$\begin{array}{rl} \log 0! = & 0.0000 \\ \log 15! = & 12.1165 \\ \log 12! = & 8.6803 \\ \log 1! = & 0.0000 \\ \hline & 20.7968 \end{array} \qquad \begin{array}{r} 24.4276 - 10 \\ -\ 20.7968 \\ \hline 3.6308 - 10 \end{array}$$

$$\text{antilog } (3.6308 - 10) = 0.000{,}000{,}427{,}4 \approx 0.000{,}000$$

6. Assuming independence of the invasion rate and the proportions of the two species of acacia trees and also fixed marginal totals, the total probability of obtaining the results of our experiment, as well as all less likely outcomes, is obtained by summing the probabilities computed in steps **3, 4,** and **5** and multiplying this sum by 2 (for a two-tailed test).

$$P = 2(0.000{,}987 + 0.000{,}038 + 0.000{,}000) = 2(0.001{,}025) = 0.002{,}050 \approx 0.002$$

Rather than double the probability to allow for the fact that we have not computed the probabilities of extreme outcomes in the other tail of the

BOX 16.7 continued

distribution, we could compare the exact probability (0.001025) with a significance level of $\alpha/2$, as discussed in Section 7.8 at the end of the discussion of the sex ratio example.

Based on these results, we reject the null hypothesis that invasion rate is independent of the host species (at the 0.2% level of significance).

and 13 of species B. These trees had been freed from ants by insecticide. Next 16 separate colonies of ant species X were brought in that had been obtained from cut-down trees of species A in an area nearby. The colonies were situated roughly equidistant to the 28 trees and were permitted to invade them. The number of trees of each species which was reinfested by ants was recorded. This is clearly a Model III problem, since the total number of trees of the two species, and the number of infested trees are both fixed, the latter because there is a fixed number of ant colonies. (The number of infested trees is fixed because there are only 16 ant colonies; hence 12 trees must remain free of ants. No more than one colony will invade a tree.) It would appear that the invasion rate is much higher for the acacia species A, in which these ants had lived before, there being $(13)(100)/15 = 86.7\%$ invasions of species A, contrasted with $(3)(100)/13 = 23.1\%$ invasions of species B. We conclude in Box 16.7 that the probability of obtaining as bad a fit to expectation as shown in these data (or worse) is approximately 0.002. We therefore reject the null hypothesis of independence and conclude that the difference in invasion rate is related to the host species, acacia species A being more attractive than species B.

The computations are somewhat long, depending on the size of the sample. When the lowest cell frequency in the table is greater than 10, the method is quite impractical without a computer. When no marginal total is greater than 15 (and consequently the total sample size is no greater than 30), we do not have to carry out the computations but can avail ourselves of published tables in which we can look up the probabilities directly. Table 38 of the *Biometrika Tables* (Pearson and Hartley, 1958) serves this purpose; it is copied in slightly simpler form as table I of Siegel (1956). Instructions for the use of these tables can be found in the quoted sources.

When no computer is accessible the chi-square or G-tests can be used to yield approximate probabilities when sample sizes are so large as to make desk calculator operation impractical. When n is smaller, as in Box 16.7, the approximations are not so good. In this example G_{adj} and X^2_{adj} yield probabilities of 0.0010 and 0.0013, respectively, while Fisher's exact test yields a P-value of 0.0020.

Tests of independence need not be restricted to 2×2 tables. In the two-way cases considered in this section, we are only concerned with two properties, but each of these properties may be divided into any number of classes. Thus organisms may occur in four color classes and be sampled from three localities, yielding a 4×3 test of independence. Such a test would examine whether the color proportions exhibited by the marginal totals are independent of the localities at which the individuals have been sampled. Such tests are often called *$R \times C$ tests of independence*, R and C standing for the number of rows and columns in the frequency table. Another case, examined in detail in Box 16.8, concerns bright red color patterns found in samples of a species of tiger beetle on four occasions during the spring and summer. It was of interest to know whether the percentage of bright red

BOX 16.8

R × C test of independence using the *G*-test.

Frequencies of color patterns of a species of tiger beetle (*Cicindela fulgida*) found at various seasons

Season	*Color pattern* ($a = 2$)		*Totals*	*% Bright red*
($b = 4$)	*Bright red*	*Not bright red*		
Early spring	29	11	40	72.5
Late spring	273	191	464	58.8
Early summer	8	31	39	20.5
Late summer	64	64	128	50.0
Totals	374	297	$671 = n$	55.7

Source: Unpublished data from H. L. Willis.

Compute the following sums, using Table **G** for $f \ln f$.

1. Sum of transforms of the frequencies in the body of the contingency table

$$= \sum^{b}\sum^{a} f_{ij} \ln f_{ij} = 29 \ln 29 + 11 \ln 11 + \cdots + 64 \ln 64$$
$$= 97.652 + 26.377 + \cdots + 266.169 = 3314.027$$

2. Sum of transforms of the row totals $= \sum^{b}\left(\sum^{a} f_{ij}\right) \ln \left(\sum^{a} f_{ij}\right)$

$$= 40 \ln 40 + \cdots + 128 \ln 128 = 147.555 + \cdots + 621.060$$
$$= 3760.400$$

3. Sum of the transforms of the column totals $= \sum^{a}\left(\sum^{b} f_{ij}\right) \ln \left(\sum^{b} f_{ij}\right)$

$$= 374 \ln 374 + 297 \ln 297 = 2215.672 + 1691.038 = 3906.710$$

4. Transform of the grand total $= n \ln n = 671 \ln 671 = 4367.384$

5. $G = 2[\text{quantity } \mathbf{1} - \text{quantity } \mathbf{2} - \text{quantity } \mathbf{3} + \text{quantity } \mathbf{4}]$

$$= 2[3314.027 - 3760.400 - 3906.710 + 4367.384 = 2[14.301] = 28.602$$

BOX 16.8 continued

This value is to be compared with a χ^2-distribution with $(a - 1)(b - 1)$ degrees of freedom, where a is the number of columns and b the number of rows in the table. In our case, $df = (2 - 1)(4 - 1) = 3$.

Since $\chi^2_{.005[3]} = 12.838$, our G-value is significant at $P < 0.005$, and we must reject our null hypothesis that frequency of color pattern is independent of season.

individuals changed significantly over the time of observation. We test whether the proportion of beetles colored bright red (55.7% for the entire study) is independent of the time of collection.

Whenever there are more than two rows or columns, the computation of a test of independence by the ordinary chi-square test becomes rather tedious. It would consist of calculating expected frequencies by the formula already learned (for any given cell the expected frequency is the product of the row total times the column total divided by the grand sum of frequencies for the example). The value of X^2 can best be computed by using Expression (16.5). Judged by the amount of computational labor, the G-test is almost always to be preferred. Furthermore, it is usually the more exact test.

As shown in Box 16.8, the following is a simple general rule for computation of the G-test of independence:

$$G = 2[(\sum f \ln f \text{ for the cell frequencies}) - (\sum f \ln f \text{ for the row and column totals}) + n \ln n]$$

The transformations can be looked up in Table **G.** In the formulas in Box 16.8 we employ a double subscript to refer to entries in a two-way table, as in the structurally similar case of two-way anova. The quantity f_{ij} in Box 16.8 refers to the observed frequency in row i and column j of the table.

The results in Box 16.8 show clearly that the frequency of bright red color patterns in these tiger beetles is dependent on the season. We note a decrease of bright red beetles in late spring and early summer, followed by an increase again in late summer. To find out whether the differences between successive points in time are significant, we could subject these data to an *STP* analysis similar to that in Box 16.5. In such an analysis we would test the independence of selected subsets of the data.

The three models defined above apply to the $R \times C$ test of independence as well. Thus the example in Box 16.8 is clearly a Model I, since the investigator did not limit the number of beetles to be collected at any one season, and obviously had no control over the number of beetles in the two color classes. No methods seem to have been especially devised for Models II and III, and such cases are analyzed in the manner described above.

The *degrees of freedom for chi-square and* G*-tests of independence* are always the same and can be computed using the rules given earlier (Section

16.2). There are k cells in the table but we must subtract one degree of freedom for each independent parameter we have estimated from the data. We must, of course, subtract one degree of freedom for the observed total sample size, n. We have also estimated $a - 1$ row probabilities and $b - 1$ column probabilities, where a and b are the number of rows and columns in the table, respectively. Thus, there are $k - (a - 1) - (b - 1) - 1 = k - a - b + 1$ degrees of freedom for the test. But since $k = a \times b$, this expression becomes $(a \times b) - a - b + 1 = (a - 1) \times (b - 1)$, the conventional expression for the degrees of freedom in a two-way test of independence. Thus, the degrees of freedom in the example of Box 16.8, a 4×2 case, was $(4 - 1) \times (2 - 1) = 3$. In all 2×2 cases there is clearly only $(2 - 1) \times (2 - 1) = 1$ degree of freedom.

Another name for test of independence is *test of association*. If two properties are not independent of each other they are *associated*. Thus, in the example testing relative frequency of two plant species on north and south slopes we can speak of an association between species and exposure. In the immunology experiment there is a negative association between presence of antiserum and mortality. *Association* is thus similar to correlation, but it is a more general term, applying to attributes as well as continuous variables.

The computations described in this section can be performed using the computer programs given in the appendix. The program in Appendix A3.14 performs Fisher's exact test and the program in Appendix A3.15 can be used to perform an $R \times C$ test for independence using the G-test criterion.

16.5 Tests of independence: multiway tables

Just as in anova, where we proceeded from two-way classifications to three- and four-way tables, so in the analysis of frequencies we can turn from two-way tests of independence to those involving classification of a sample of observations. In a three-way test of independence $\hat{f}_{ijk}$ equals the product of the totals for row i, column j, and depth k divided by the square of the total sample size, n. By analogy with Expression (16.6), G can be computed as $2 \sum^{abc} f_{ijk} \ln (f_{ijk}/\hat{f}_{ijk})$. The computational equations given in Box 16.9 can be derived by simple algebra. The degrees of freedom can be worked out as before as

$$df = abc - (a - 1) - (b - 1) - (c - 1) - 1 = abc - a - b - c + 2$$

since $(a - 1)$ parameters were estimated for the a columns, $(b - 1)$ for the b rows, $(c - 1)$ for the c depths, and 1 for the grand total, n.

The example in Box 16.9 is an analysis of an experiment in which adult drosophila were classified according to three factors: sex, whether they were healthy or poisoned, and the location in which they had pupated. These data are from an experiment in which drosophila larvae were reared in medium containing DDT. It was of interest to learn whether the two sexes had different tolerances to the poison and also whether the site where the fly had pupated (in the medium, at the margin of the medium, on top of the medium, or away from the medium on the wall of the vial) affected its chance for subsequent survival.

As the analysis is much more complicated by the chi-square test, we have presented only the G-test. The three factors are found not to be independent in the example in Box 16.9, and the box also describes how the overall G-value can be partitioned into components, each testing a separate

BOX 16.9

Tests of independence in a three-way table and a priori tests of partitions.

Emerged drosophila are classified according to pupation site, P (in the medium, IM; at the margin of the medium, AM; on the wall of the vial, OW; and on top of the medium, OM); sex, S; and mortality, M (healthy versus poisoned).

Pupation Site ($a = 4$) P	*Sex* ($b = 2$) S	*Mortality, M* ($c = 2$) *Healthy*	*Poisoned*	*Totals*
IM	♀	55	6	61
	♂	34	17	51
		89	23	112
AM	♀	23	1	24
	♂	15	5	20
		38	6	44
OW	♀	7	4	11
	♂	3	5	8
		10	9	19
OM	♀	8	3	11
	♂	5	3	8
		13	6	19
	Totals	150	44	194

Source: Data by R. R. Sokal.

Sex	*Healthy*	*Poisoned*	*Totals*
♀	93	14	107
♂	57	30	87
Totals	150	44	194

BOX 16.9 continued

The lower table is the $M \times S$ two-way table needed for step **2.** The other 2 two-way tables ($P \times S$ and $P \times M$) are not shown. Their totals can be obtained from the three-way table above.

Numerous analyses can be performed on data such as these to answer specific questions. A question that might be asked first is, "Are the effects of pupation site, sex, and mortality independent of each other?"

Computational steps

1. Arrange the data into a three-way table. For convenience, this should be done in such a way that the totals for the 3 two-way tables can be recorded as well as the totals for the 3 "one-way" tables and the "zero-way" frequency or grand total.

2. For each of these tables, compute $\sum f \ln f$ by summing the $f \ln f$ transforms (Table **G**) of each cell in the table.

a. $P \times S \times M = \sum f \ln f = 55 \ln 55 + 34 \ln 34 + 6 \ln 6 + \cdots + 3 \ln 3$

$= 220.403 + 119.896 + 10.751 + \cdots + 3.296$

$= 581.783$

Cells of two-way tables. (Quantities **b, c,** and **d** are needed only for steps **4, 6,** and **7.** They are shown here for ease of reference.)

b. $P \times S = \sum f \ln f = 61 \ln 61 + 51 \ln 51 + 24 \ln 24 + \cdots + 8 \ln 8$

$= 250.763 + 200.523 + 76.273 + \cdots + 16.636$

$= 673.500$

c. $P \times M = \sum f \ln f = 89 \ln 89 + 23 \ln 23 + 38 \ln 38 + \cdots + 6 \ln 6$

$= 399.489 + 72.116 + 138.228 + \cdots + 10.751$

$= 707.480$

d. $M \times S = \sum f \ln f = 93 \ln 93 + 14 \ln 14 + 57 \ln 57 + 30 \ln 30$

$= 421.532 + 36.947 + 230.454 + 102.036$

$= 790.969$

Marginal totals ("one-way tables")

e. $P = \sum f \ln f = 112 \ln 112 + \cdots + 19 \ln 19 = 528.472 + \cdots + 55.944$

$= 806.864$

f. $M = \sum f \ln f = 150 \ln 150 + 44 \ln 44 = 751.595 + 166.504$

$= 918.099$

g. $S = \sum f \ln f = 107 \ln 107 + 87 \ln 87 = 499.993 + 388.534$

$= 888.527$

h. Grand total $= \sum f \ln f = 194 \ln 194 = 1021.964$

BOX 16.9 continued

3. Compute the following sum:

$$G = 2[\text{quantity } \mathbf{a} - \text{quantities } \mathbf{e}, \mathbf{f}, \mathbf{g} + 2(\text{quantity } \mathbf{h})]$$

$$= 2[581.783 - 2613.490 + 2(1021.964)] = 2(12.2210) = 24.442$$

This value is to be compared to a χ^2-distribution with $abc - (a - 1) - (b - 1) - (c - 1) - 1 = abc - a - b - c + 2 = (4)(2)(2) - 4 - 2 - 2 + 2 = 10$ *df*.

Since our G-value is greater than $\chi^2_{.01[10]} = 23.209$, we reject the hypothesis that the three criteria are jointly independent. It is, therefore, of interest to study the data further in order to understand the nature of this lack of independence.

4. Test of the hypothesis that mortality is independent of pupation site. Lack of independence among the three factors jointly does not necessarily imply that any two lack independence. Computing G from the mortality $\times$ pupation site two-way table, we find

$$G = 2[\text{quantity } \mathbf{c} - \text{quantities } \mathbf{e}, \mathbf{f} + \text{quantity } \mathbf{h}]$$

$$= 2[707.480 - 1724.964 + 1021.964] = 2[4.480] = 8.960$$

$$df = ac - (a - 1) - (c - 1) - 1 = ac - a - c + 1$$

$$= (4)(2) - 4 - 2 + 1 = 3$$

Since G is significant at $0.05 > P > 0.01$, mortality is apparently not independent of pupation site.

5. Because the above test was significant, it is of interest to partition the last G to learn more precisely the nature of the lack of independence. The four pupation sites can be grouped into two classes, those where the pupa is in intimate contact with the medium (including sites IM and AM) and those where the pupa is largely free of medium (sites OW and OM). The three tests below would seem reasonable. Each G-value is to be compared with a χ^2-distribution with a single degree of freedom, since these are basically tests of 2×2 tables.

a. Is mortality independent of those pupation sites in intimate contact with the medium? To test this, form the table, and compute the G-statistic as in Box 16.8.

Pupa	IM	AM	Σ
Healthy	89	38	127
Poisoned	23	6	29
Σ	112	44	156

$$G = 2[89 \ln 89 + \cdots + 6 \ln 6 - (127 \ln 127 + \cdots + 44 \ln 44) + 156 \ln 156]$$

$$= 2[620.584 - 1407.840 + 787.778] = 2[0.522] = 1.044 \; ns$$

b. Is mortality independent of those pupation sites free of the medium?

BOX 16.9 continued

Pupa	OW	OM	Σ
Healthy	10	13	23
Poisoned	9	6	15
Σ	19	19	38

$G = 2[0.499] = 0.998$ *ns*

c. Is mortality independent of whether pupation site is in the medium or outside of it?

Pupa	IM + AM	OW + OM	Σ
Healthy	127	23	150
Poisoned	29	15	44
Σ	156	38	194

$G = 2[3.460] = 6.920^{**}$

Since the third G is greater than $\chi^2_{.01[1]} = 6.635$, we reject its null hypothesis. Clearly, mortality is higher among those flies that pupated free of the medium.

Because these are orthogonal comparisons, the three G's sum to yield (within rounding error) the value of G obtained in step **4**, representing a more inclusive test $(1.044 + 0.998 + 6.920 = 8.962)$.

6. We may carry out a similar analysis for the relationship between pupation site and sex, but the small G-value obtained by the overall test below is clearly not worth partitioning.

$$G = 2[\text{quantity } \mathbf{b} - \text{quantities } \mathbf{e},\ \mathbf{g} + \text{quantity } \mathbf{h}]$$

$$= 2[673.500 - 1695.391 + 1021.964] = 2[0.073] = 0.146$$

$$df = ab - (a - 1) - (b - 1) - 1 = ab - a - b + 1$$

$$= (4)(2) - 4 - 2 + 1 = 3.$$

Since $\chi^2_{.05[3]} = 7.815$, this G is clearly not significant.

7. Is mortality independent of sex? Compute G for the sex by mortality two-way table (one degree of freedom):

$$G = 2[\text{quantity } \mathbf{d} - \text{quantities } \mathbf{f},\ \mathbf{g} + \text{quantity } \mathbf{h}]$$

$$= 2[790.969 - 918.099 - 888.527 + 1021.964] = 2[6.307]$$

$$= 12.614^{***}$$

Since this is greater than $\chi^2_{.05[1]} = 3.841$, we may reject this hypothesis. In examining the data (see the sex-by-mortality table at the beginning of this box), we note that the mortality appears to be much higher in the males than in the females.

BOX 16.9 continued

8. For completeness, we may calculate one final component, the $P \times M \times S$ interaction.

$$G = 2[\text{quantity } \mathbf{a} - \text{quantities } \mathbf{b}, \mathbf{c}, \mathbf{d} + \text{quantities } \mathbf{e}, \mathbf{f}, \mathbf{g} - \text{quantity } \mathbf{h}]$$
$$= 2[581.783 - 2171.949 + 2613.490 - 1021.964] = 2[1.360]$$
$$= 2.720 \; ns$$

It should be pointed out, however, that this quantity, which has $(a - 1)(b - 1)(c - 1)$ degrees of freedom (3 in our case), need not always be positive. This is because the three tests of independence given in steps **4**, **6**, and **7** are not orthogonal (even though each will generally be of interest). If the interaction G is negative, an alternative system of partitions should be performed. These alternative methods are more complex, as discussed on pp. 171–172 of Kullback (1959).

9. Finally, we may summarize our complete analysis in the following table.

Hypothesis tested	*df*	*G*
$P \times S$ independence	3	0.146
$P \times M$ independence	3	8.962*
IM *vs* AM	1	1.044
OW *vs* OM	1	0.998
IM + AM *vs* OW + OM	1	6.920**
$M \times S$ independence	1	12.614***
$P \times M \times S$ interaction	3	2.720
$P \times M \times S$ independence	10	24.442**

aspect of the joint dependence of the three factors. Sex and pupation site are found to be independent, implying that the two sexes did not differ in their choice of locations for pupation. The test of independence between pupation site and mortality is significant, indicating that survival of the flies depended on their choice of a site for pupation, or possibly that healthy and poisoned larvae pupate at different sites. Since three degrees of freedom are available for this test (and G is large enough, 8.962, to yield potentially significant single degree of freedom components), we may further subdivide this test statistic into separate a priori contrasts. The four pupation sites fall into two classes—those where the pupa is in intimate contact with the medium (sites IM and AM) and those where the pupa is largely free of the medium (sites OW and OM). Independence tests of mortality versus the two sites in each of these classes, illustrated in steps **5a** and **5b** of Box 16.9, are not significant. This means that *within* each class of pupation sites there is no differential mortality due to differences in specific sites. However, a test of mortality against the two classes of pupation sites is highly significant, indicating that pupae free of medium had different (higher) mortalities than those located in intimate contact with the medium. The test of independence between mortality and sex is also significant, death rates in the two sexes being rather different.

The final component for the analysis is the three-factor interaction between pupation site, mortality, and sex. The three-factor interaction here has essentially the same meaning as in three-way analysis of variance: significant three-factor interaction means that the degree of association (measured by G for a two-way test of independence) between one pair of factors differs over the several levels or classes of a third factor. The overall test of independence means that the probability of any single outcome can be predicted on the basis of the marginal frequencies. Situations in which the three factors are independent cannot therefore show interaction, but those in which the hypothesis of independence has been rejected may or may not exhibit three-factor interaction. Unlike the analysis of variance interaction, three-way interaction in the G-test can sometimes be negative. When this happens, a different set of partitions of the data may be of greater interest. Such an analysis is more complex (see Kullback, 1959, for a technical account of this infrequent problem).

16.6 Testing equality of two percentages

In the 2×2 tests of independence of Section 16.4, one way of looking for suspected lack of independence was to examine the percentage occurrence of one of the properties in the two classes based on the other property. Thus we compared the percentage of tree species A on north and south slopes, or we studied the percentage mortality with or without antiserum. This way of looking at a test of independence suggests another test of such data. Can we develop a test for the significance of differences between two percentages? Several such tests are found in the statistical literature. We have chosen a method based on the arcsine transformation because it permits us to develop a method for finding the sample size required for a test of equality of percentages. The test statistic is

$$t_s = \frac{\arcsin \sqrt{p_1} - \arcsin \sqrt{p_2}}{\sqrt{820.8\left(\dfrac{1}{n_1} + \dfrac{1}{n_2}\right)}} \qquad (16.14)$$

where p_1 and p_2 are the proportions of the attribute in the two samples, n_1 and n_2 are respective sample sizes, and 820.8 is a constant representing the parametric variance of a distribution of arcsine transformations of proportions or percentages. The application is straightforward and is illustrated in Box 16.10.

When we compare the results indicated by this test with the results given by the previous tests (chi-square, G, and Fisher's exact test), we find that the probabilities it yields are very similar to those given by the chi-square

BOX 16.10

Testing equality of two percentages.

Number of winged and wingless individuals in a species of water bug (*Mesovelia* sp.) sampled from two localities.

Locality	*Winged*	*Wingless*	Σ	p *(of winglessness)*
A	86	29	115	0.252
B	103	9	112	0.080

Source: Data by J. Galbreath.

From Table **K**,

$\arcsin \sqrt{0.252} = 30.13 \qquad \arcsin \sqrt{0.080} = 16.43$

By Expression (16.14),

$$t_s = \frac{\arcsin \sqrt{p_1} - \arcsin \sqrt{p_2}}{\sqrt{820.8\left(\frac{1}{n_1} + \frac{1}{n_2}\right)}} = \frac{30.13 - 16.43}{\sqrt{820.8\left(\frac{1}{115} + \frac{1}{112}\right)}}$$

$$= \frac{13.70}{\sqrt{820.8(0.01762)}} = \frac{13.70}{3.803} = 3.60$$

We compare t_s with a normal deviate which we look up in the table of areas of the normal curve (Table **P**), doubling the probability since we wish to make a two-tailed test (we could also look up $t_{\alpha[\infty]}$ in Table **Q**). We obtain a value of $P = 0.0003$ and conclude that the sample from locality A has a significantly higher proportion of winglessness than the sample from locality B.

When sample sizes are very small, the probabilities obtained by this method depart appreciably from those obtained by more refined techniques.

test (when Yates' correction is *not* used). Therefore, for critical work with smaller sample sizes we would recommend the use of Fisher's exact test or the G-test with Yates' correction.

↛ When we have more than two percentages or proportions, the homogeneity of which we wish to test, one approach is by the method of Brandt and Snedecor illustrated in section 9.9 of Snedecor (1956). ↚

It may be necessary to find the sample size required for a test of equality of percentages. The method described in Box 16.11 is based on the same principles given in Box 9.11 and discussed in Section 9.8 but, since percentages are not expected to be normally distributed, the arcsine transformation is used for computing the quantity δ^2. The constant 820.8 is used as an estimate of the true error variance with infinite degrees of freedom, being the

BOX 16.11

Sample size required to detect a given "true" difference between two percentages.

The rationale behind this method is discussed in Section 16.6. Simply look up the correct coefficient in the table below for α, the significance level desired, and P, the desired probability that the planned experiment will be successful, and divide by $\delta^2 = (\arcsin \sqrt{p_1} - \arcsin \sqrt{p_2})^2$. The quotient is the estimate of the sample size.

P	α			
	.1	.05	.01	.001
.5	4,442.2	6,306.4	8,883.7	10,891.5
.8	10,150.2	12,884.8	16,474.3	19,171.6
.9	14,059.3	17,249.8	21,368.5	24,426.2
.99	25,890.0	30,161.4	35,536.7	39,450.1

For example, what sample size is required in an experiment if one wishes to be 80% certain of detecting a true difference between the two proportions $p_1 = 0.65$ and $p_2 = 0.55$ at the 5% level of significance?

For $\alpha = 0.05$ and $P = 0.80$, the constant in the above table equals 12,884.8.

$$\arcsin \sqrt{p_1} = \arcsin \sqrt{0.65} = 53.73 \qquad \arcsin \sqrt{p_2} = \arcsin \sqrt{0.55} = 47.87$$

$$\delta^2 = (53.73 - 47.87)^2 = 5.86^2 = 34.3396$$

$$n = \frac{12{,}884.8}{34.3396} = 375.2$$

Thus, about 375 observations are needed for each sample.

If one wishes to determine the sample size needed when one of the percentages is a theoretical rather than a sample value (as in genetic research), divide the coefficient from the above table by $2\delta^2$.

For example, suppose one wishes to plan an experiment in which the expected ratio was 3:1. Thus, $p_1 = 0.75$. What sample size should be used so that one could be 80% sure of detecting a possible competing ratio (such as a 1:1 ratio, $p_2 = 0.50$) at the 5% level of significance?

$$\arcsin \sqrt{0.75} = 60.00 \qquad \arcsin \sqrt{0.50} = 45.00$$

The numerator will be the same as before, 12,884.8.

$$2\delta^2 = 2(60.00 - 45.00)^2 = 450$$

$$n = \frac{12{,}884.8}{450} = 28.6$$

Therefore, about 29 observations are required.

These approximations seem to hold fairly well down to a predicted n of 20. Below that they seem consistently to underestimate the required sample size by unity. Therefore, whenever the procedure described above yields $n < 20$, the estimated sample size should be increased by one.

expected variance of the angular transforms. Therefore, an explicit expression for the sample size is possible and one need not iterate to obtain a solution as was required in Section 9.8. The formula used is

$$n \geq 2\left(\frac{\sigma}{\delta}\right)^2 \{t_{\alpha[\nu]} + t_{2(1-P)[\nu]}\}^2 \tag{16.15}$$

Since σ^2 is the constant 820.8 and because degrees of freedom, ν, is infinity, $2\sigma^2\{t_{\alpha[\nu]} + t_{2(1-P)[\nu]}\}^2$ will always be a constant for any given value of α and P. We have provided in Box 16.11 a small table of coefficients which should suffice for most ordinary situations. These coefficients are to be divided by $(\arcsin\sqrt{p_1} - \arcsin\sqrt{p_2})^2$ for any given problem to yield n, the sample size required. Box 16.11 also illustrates how to estimate the required sample size for a test of the difference between two percentages when one of the percentages is a theoretical rather than a sample value. This would apply in a single classification goodness of fit test with only two classes.

Finding sample sizes required for one-way classifications with more than two classes, such as are discussed in Section 16.2, is difficult because of the great variety of possible deviations from expectation.

16.7 Randomized blocks for frequency data

The tests of independence of Section 16.4 can be regarded as tests of the effect of treatments on the percentage of some attribute. In that section, each treatment was applied to a number of independently and randomly selected individuals that were different for the separate treatments. Thus, in the immunology example, two samples of mice were taken and subjected to different treatments—one bacteria alone, the other bacteria and antiserum. These mice had presumably been randomly selected from the pool of such animals available to the investigator. In the example dealing with the species of trees found on the north and south slopes of a mountain, the samples were obviously separate, consisting of different groups of trees on the two slopes.

Sometimes it is neither possible nor desirable to collect data in that manner. For example, consider an experiment to compare the phototactic response of a species of ticks under various experimental conditions. One could take a sample of ticks, divide it into several groups, and test each group under a separate set of conditions. But what if it is known that the behavior of the organisms is rather variable and depends upon such factors as age or condition of the animal? In such a case it might happen that by chance more older individuals were placed into one group than the other or that some other sampling accident has taken place. Thus treatment effects might be confounded with heterogeneity of the experimental subjects. There are three possible solutions to such a problem. If the factors causing heterogeneity (such as age or prior condition) are known, we can try to keep these constant. This may require a large population from which to sample. Thus, in order to obtain animals of uniform age we may have to select from a large base population of animals of various ages, which may be a difficult or expensive undertaking.

If the factors causing heterogeneity of response are unknown, two other approaches may be tried. One is simply to take large enough samples to overcome the heterogeneity. For obvious reasons, this is often not feasible nor desirable.

The alternative procedure, the subject matter of the present section, uses the same individuals for both treatments. In this design, even if there were some heterogeneity in the inherent responses of the organisms, one can still look for and test for a significant *change* in their behavior due to the treatment. This same experimental design was discussed previously in Section 11.4 (randomized blocks), the only difference being that the earlier section dealt with continuous variables that can be analyzed by the analysis of variance, whereas here we are considering attributes for which we can only record the frequency or percentage of individuals possessing a certain characteristic. As in its previous application in Section 11.4, the method here is based on the correlation between treatments over blocks (individuals in the above example) and assumes the absence of treatment × block interaction. Because frequently the same sample of individuals is exposed to two or more treatments, this method is also known as *repeated testing of the same individuals*. Another appropriate name for this technique is *testing of correlated proportions*.

A second analytical situation in which the present design is indicated is when there is some reason to believe that a prior treatment will affect the responses in a subsequent treatment. In such cases clearly the same sample has to be used because the effects of the subsequent treatment have to be compared with the effects of the prior treatment in the same individuals. An example in point is conditioning or learning in animals. Given a certain stimulus, individuals in a group of animals may or may not respond. If the same or a different stimulus is given at a later time, the presence or absence of a response in these animals may be conditioned by their previous response and in this way the occurrence of learning or conditioning may be established.

Our first encounter with repeated testing of the same individuals will be the case of two treatments (only one repetition of the test). In the continuous case this is the paired-comparisons design (Section 11.5). This is a special

BOX 16.12

Randomized blocks for frequency data. (Repeated testing of the same individuals or testing of correlated proportions.)

Individuals tested twice. McNemar test for the significance of changes

The response of the rabbit tick (*Haemaphysalis leporis-palustris*) to light was measured by placing individual ticks in arenas 1 or 2 inches in diameter and

BOX 16.12 continued

recording whether the tick left the arena on the side toward (+) or away from (−) the light. Each tick was scored first in the 1-inch arena and then in the 2-inch one. It was of interest to test whether the response of an individual to light would change depending on the distance a tick had to crawl before its response was recorded.

The results can be summarized as follows:

2-inch arena	*1-inch arena* −	+	Σ
−	8	5	13
+	9	8	17
Σ	17	13	30

Source: Data by V. E. Nelson.

Note that the pattern of response appeared to reverse itself. In the 1-inch arena 13 out of 30 ticks turned toward the light, whereas 17 of the 30 did so in the 2-inch arena.

We label the cells in the above table in the conventional manner:

	−	+	Σ
−	a	b	$a + b$
+	c	d	$c + d$
Σ	$a + c$	$b + d$	n

We wish to test whether the change in proportion between the two trials is significant. The null hypothesis is that the frequencies of the two types of "changers" are equal—that is, that the number of ticks that followed a + response with a − equaled those following a − response with a +. If this condition is satisfied, the proportions in the two trials must necessarily be the same. Concisely stated, the null hypothesis is H_0: $(b - c) = 0$. The appropriate test statistic is computed by Expression (16.16):

$$X^2 = \frac{(b - c)^2}{(b + c)}$$

When $b + c$ is less than 200, Yates' correction should be applied [Expression (16.16Y)]

$$X^2_{\text{adj}} = \frac{(|b - c| - 1)^2}{(b + c)}$$

In our example these quantities are

$$X^2 = \frac{(5 - 9)^2}{(5 + 9)} = \frac{4^2}{14} = 1.1429$$

$$X^2_{\text{adj}} = \frac{(|5 - 9| - 1)^2}{(5 + 9)} = \frac{3^2}{14} = 0.6429$$

BOX 16.12 continued

These X^2's approximate the distribution of χ^2 with one degree of freedom. Clearly, there is little evidence of a significant change in behavioral response.

Individuals tested three or more times. Cochran's Q-test

The data are scores expressing the condition of 24 plants (*Acacia cornigera*) observed in three different months. All of these plants were from an experimental plot in which the growing shoots had been deprived of the ant *Pseudomyrmex ferruginea,* which lives within the stems. A positive rating represents a plant that has all or most of its uppermost shoot apices still intact and in which most of the foliage does not show the feeding damage of phytophagous insects. A negative rating represents a plant that has all of its growing shoot apices eaten off and in which over one-half of the foliage shows severe damage. In the complete study (not shown here) there was also a control plot.

1. The data are arranged as a two-way table, using 0's for negative ratings and 1's for positive ratings. The $b = 24$ rows (blocks) of this table correspond to the individual plants scored and the $a = 3$ columns (treatments) correspond to the repeated scoring of these plants in different months.

Individual plants (blocks)	*Month in which scored (treatments)*			Σ
	March	*June*	*August*	
1	0	1	0	1
2	0	0	0	0
3	0	1	0	1
4	1	0	0	1
5	1	1	0	2
6	1	0	0	1
7	0	0	0	0
8	1	1	0	2
9	1	1	0	2
10	1	0	0	1
11	1	0	0	1
12	1	1	0	2
13	0	1	1	2
14	0	1	1	2
15	0	0	0	0
16	0	1	0	1
17	0	1	0	1
18	1	0	0	1
19	1	0	0	1
20	1	0	0	1
21	1	0	1	2
22	1	1	1	3
23	1	0	0	1
24	1	1	0	2
Σ	15	12	4	31

Source: Data by D. H. Janzen.

BOX 16.12 continued

2. Compute the row and column sums and record them along the margins of the table.

3. Grand total of the frequencies in the table $= \sum^a \sum^b Y = 31$

4. Sum of the row sums squared $= \sum^b \left(\sum^a Y\right)^2$

$$= 1^2 + 0^2 + 1^2 + \cdots + 2^2 = 53$$

5. Sum of the column sums squared $= \sum^a \left(\sum^b Y\right)^2 = 15^2 + 12^2 + 4^2 = 385$

6. Test statistic [Expression (16.17)] $= Q$

$$= \frac{(a-1)\left[a \sum^a \left(\sum^b Y\right)^2 - \left(\sum^a \sum^b Y\right)^2\right]}{a \sum^a \sum^b Y - \sum^b \left(\sum^a Y\right)^2}$$

$$= \frac{(a-1)[a \times \text{quantity } \mathbf{5} - (\text{quantity } \mathbf{3})^2]}{(a \times \text{quantity } \mathbf{3}) - \text{quantity } \mathbf{4}}$$

$$= \frac{(3-1)[3 \times 385 - 31^2]}{(3 \times 31) - 53} = \frac{2(194)}{40}$$

$$= 9.700^{**}$$

This value is to be compared with a χ^2-distribution with $a - 1 = 2$ degrees of freedom. Since $Q = 9.700 > \chi^2_{.01[2]} = 9.210$, we reject the null hypothesis that the proportion of plants in good condition remained the same as the season progressed. It is reasonable to conclude, therefore, that when the plants are deprived of their protecting ants, their condition gradually deteriorates. Many of the plants died a few months later, but the control plot showed little change.

case of randomized blocks in which there are only two treatments; hence each individual or block is tested twice. By analogy to this we have for attributes the *McNemar test for significance of changes*. In Box 16.12 we show an example in which thirty ticks were each tested twice for their phototactic response—first in a 1-inch arena and then in a 2-inch arena. The data consist of thirty paired scores in which a plus was recorded if the tick left the arena toward the light and a minus if it moved away from the light. The thirty paired observations can be conveniently recorded in a 2 × 2 table as shown in the box and also repeated below.

		1-inch arena −	*1-inch arena* +	Σ
2-inch arena	−	8	5	13
	+	9	8	17
	Σ	17	13	30

Unlike the situation in Section 16.4, where we were interested in the proportions implied by the cells of the table, we are now more interested in the marginal proportions. For example, it is no longer of particular interest that 8 out of 17 of the ticks that had a negative response in the 1-inch arena also responded negatively in the 2-inch arena. But rather, we are interested in the fact that in the 1-inch arena 17 out of 30 ticks showed a negative reaction, whereas in the 2-inch arena 13 out of 30 ticks showed a negative reaction. We could have rearranged the table as follows:

	−	+	Σ
1-inch arena	17	13	30
2-inch arena	13	17	30
Σ	30	30	60

But the effect of this would be to imply that our sample size was $n = 60$ ticks, twice the actual number of ticks involved. It is for this reason that we cannot use the methods of Section 16.4 on such data. The proportions within this second table are not independent of one another. They are all based upon the same thirty ticks.

Inspecting the earlier table, we note that the difference between the marginal proportions in which we are interested is a function only of the number of ticks that are "changers"—those which responded differently to the second test (the 2-inch arena) than they did to the first test. If there are no changers, or if the two types of changers are equally frequent, the marginal frequencies will be the same for both sets. In our example, 5 ticks changed from a negative to a positive reaction in the 2-inch arena and 9 changed from an earlier positive to a later negative response. We may state as our null hypothesis that the proportion of individuals reacting negatively will be the same in both sizes of arenas. This would imply that we expect the frequencies of the two types of changers to be equal, and we can easily construct a chi-square test of this hypothesis. If we label the four cells of the table in the conventional manner (a, b, c, d), it is possible to construct the following simple formula for X^2 (approximating χ^2 with one degree of freedom):

$$X^2 = \frac{(b - c)^2}{b + c} \tag{16.16}$$

In the present example this is

$$\frac{(5 - 9)^2}{5 + 9} = \frac{(-4)^2}{14} = \frac{16}{14} = 1.429$$

When the quantity $(b + c)$ is small, say less than 200, Yates' correction should be used. The equation for the adjusted chi-square test is as follows:

$$X^2_{\text{adj}} = \frac{(|b - c| - 1)^2}{b + c} \tag{16.16Y}$$

It is also possible to test this hypothesis with the G-statistic, but there is no shortcut equation. One must carry out the test on a null hypothesis that the two frequencies (of changers) are equal according to the standard procedure given before. Yates' correction should also be applied in the G-test when $(b + c)$ is less than 200.

When the sample sizes are very small $[(b + c) < 10]$, an "exact" binomial test is recommended. This is carried out as follows. If we let $k = b + c$, and let Y equal b or c, whichever is greater, then the desired probability is twice the probability of observing Y or more events of one kind out of k from a binomial distribution with $p = q = 0.5$. We can employ the methods of Section 13.11 to evaluate this probability, either computing the probability or inferring it by the use of confidence limits for $\hat{p}$ in Table **W**.

When individuals have been tested three or more times for the presence or absence of some attribute, a test analogous to the randomized blocks design is possible. Such a test is called *Cochran's Q-test*. The equation for Q is

$$Q = \frac{(a-1)\left[a\sum^{a}\left(\sum^{b} Y\right)^2 - \left(\sum^{a}\sum^{b} Y\right)^2\right]}{a\sum^{a}\sum^{b} Y - \sum^{b}\left(\sum^{a} Y\right)^2} \tag{16.17}$$

Q is expected to follow the chi-square distribution with $a - 1$ degrees of freedom (where a is the number of times each of the b individuals has been tested). This expression is based on a different model than those used for the previous chi-square or G-tests. The original observations are replaced by 1's and 0's ($Y = 1$ corresponds to presence of attribute and $Y = 0$ corresponds to the absence of the attribute). The resulting data are entered in a two-way table and analyzed according to the standard equations for a randomized block design. For purposes of tests of significance, however, the usual test statistic is modified so that the results are expected to follow the chi-square rather than the F-distribution. If we let A stand for the columns (or the treatment effects) and B for the rows (or blocks) of the randomized block design (the individuals), the test statistic Q is equal to

$$Q = \frac{SS_A}{(SS_{\text{total}} - SS_B)/(df_{\text{total}} - df_B)}$$

Since the data consist only of ones and zeros, it is possible to do a great deal of simplification so that the resulting equation becomes Expression (16.17).

In Appendix A1.16 it is shown that McNemar's test is a special case of Cochran's Q-test. There does not appear to be a G-test equivalent to the Cochran Q-test.

In the second half of Box 16.12, Cochran's Q-test is applied to data, demonstrating the fact that acacias gradually deteriorate when the ants of a species that normally lives within its stems are removed. The test shows that there is a significant change in the frequency of plants in "good condition" when the plants are observed in successive months after having been deprived of the ant. This effect is caused by the fact that normally the ants prevent herbivores from feeding on the tender growing shoots. The example is explained in sufficient detail in Box 16.12.

Exercises 16

16.1 In an experiment to determine the mode of inheritance of a *green* mutant, 146 wild type and 30 mutant offspring were obtained when F_1 generation houseflies were crossed. Test whether the data agree with the hypothesis that the ratio of wild type to mutants is 3:1. ANS. $X^2 = 5.939$, $G = 6.4624$.

16.2 Locality A has been exhaustively collected for snakes of species S. An examination of the 167 adult males that have been collected reveals that 35 of these have pale-colored bands around their necks. From locality B, 90 miles away, we obtain a sample of 27 adult males of the same species, 6 of which show the bands. What is the chance that both samples are from the same statistical population with respect to frequency of bands?

16.3 Of 445 specimens of the butterfly *Erebia epipsodea* from mountainous areas, 2.5% have light color patches on their wings. Of 65 specimens from the prairie 70.8% have such patches (unpublished data by P. R. Ehrlich). Is this difference significant? ANS. $X^2 = 266.488$, $G_{adj} = 170.998$.

16.4 In a study of polymorphism of chromosomal inversions in the grasshopper *Moraba scurra*, Lewontin and White (1960) gave the following results for the composition of a population at Royalla "B" in 1958.

		Chromosome CD		
		St/St	St/Bl	Bl/Bl
Chromosome EF	Td/Td	22	96	75
	St/Td	8	56	64
	St/St	0	6	6

Are the frequencies of the three different combinations of chromosome EF independent of those of the frequencies of the three combinations of chromosome CD? ANS. $G = 7.396$

16.5 The following data were presented by Verner (1964) on the sex ratios of adult Long-billed Marsh Wrens at Seattle.

	Date	*Males*	*Females*
1961	May 6	14	11
	May 11	14	12
	May 25	14	12
	June 13	13	11
	June 19	13	9
	June 25	12	8
	July 8	9	6
1962	April 18	8	8
	May 22	6	8
	May 31	6	7
	June 2	6	7
	June 20	8	8
	July 23	7	5

Test whether there is a 1:1 sex ratio for each date and whether the observations made on different dates are homogeneous. If there is an indication of heterogeneity, perform an a posteriori test by the *STP* procedure to obtain sets of dates homogeneous for sex ratios. Is there any evidence of heterogeneity between years?

16.6 Test whether the percentage of nymphs of the aphid *Myzus persicae* that developed into winged forms depends on the type of diet provided. Stem mothers had been placed on the diets one day before the birth of the nymphs (data by Mittler and Dadd, 1966).

Type of diet	*% Winged forms*	*n*
Synthetic diet	100	216
Cotyledon "sandwich"	92	230
Free cotyledon	36	75

16.7 Test agreement of observed frequencies to those expected on the basis of a binomial distribution for the data given in Tables 5.1 and 5.2.

16.8 Test agreement of observed frequencies to those expected on the basis of a Poisson distribution for the data given in (a) Table 5.5, (b) Table 5.6, (c) Table 5.8, (d) Table 5.9, (e) Table 5.10, and (f) Table 5.11. ANS. (a) $G = 157.527$.

16.9 For the pigeon data of Exercise 2.4 test the agreement of the observed distribution to a normal distribution.

(a) Use the Komolgorov-Smirnov test on the ungrouped data. ANS. $D = 0.0709$.

(b) Use chi-square and G-tests for goodness of fit to the expected frequencies calculated in Exercise 6.1.

16.10 Repeat Exercise 16.9 with the pigeon data as transformed in Exercise 2.6. Use expected frequencies computed in Exercise 6.2.

16.11 Calculate expected normal frequencies for the butterfat data of Exercise 4.3. Test the goodness of fit of the observed to the expected frequencies by means of the Kolmogorov-Smirnov test.

16.12 If a standard drug will cure disease X 60% of the time, what sample size would be required in an experiment designed to have an 80% chance of detecting (at the 5% level of significance) a new drug that would cure disease X (a) at least 65% of the time, (b) at least 70% of the time, and (c) at least 80% of the time.

16.13 The following table of data is from a study of caste differentiation in the ant *Myrmica rubra* by Brian and Hibble (1964).

Frequencies of larvae of two sizes in three colonies at weekly intervals

	Week	*Small larvae*	*Large larvae*
Colony 1	2	12	47
	3	19	29
	4	9	52
	5	5	42
	6	8	58
	7	2	43
	8	0	35

Frequencies of larvae of two sizes in three colonies at weekly intervals

	Week	Small larvae	Large larvae
Colony 1	9	7	50
	10	9	38
Colony 2	2	0	47
	3	28	36
	4	2	44
	5	1	45
	6	4	33
	7	2	46
	8	3	38
	9	3	33
	10	5	50
Colony 3	2	55	23
	3	71	4
	4	40	21
	5	29	9
	6	42	17
	7	26	8
	8	28	21
	9	29	13
	10	57	7

Are the three factors (larval size, colonies, and weeks) jointly independent? If not, analyze to test if the factors taken two at a time are independent. Is the degree of association between weeks and larval size homogeneous over the three colonies—that is, is there any interaction?

16.14 In studying the effect of 1% indol-acetic acid in lanolin on fruit setting in muskmelon, Burrell and Whitaker (1939) obtained the results presented in the following three-way table.

Number of flowers that did or did not set fruit

		Treatment		Control	
		Fruit set	Fruit not set	Fruit set	Fruit not set
Replication	1	13	10	5	23
	2	2	10	1	7
	3	13	1	6	10
	4	5	6	5	9
	5	13	5	6	14

Is the effect of the treatment on fruit setting independent of the effect of different replications? You could test the independence of treatment and fruit-setting separately for each replicate, and sum the G-values. From this you could subtract the G resulting from a test of independence of the data pooled over all replications. What is the meaning of the resulting quantity? HINT: $G_{\text{Total}} - G_{\text{Pooled}} = G_{\text{Heterogeneity}}$.

16.15 A doctor had 25 patients who suffered from disease X during a recent epidemic. Although antibiotic B appeared to have no curative value against X, only 6 out of 11 patients administered B developed secondary infections, but 11 out of 14 patients not given B developed such infections. Would the doctor be justified in believing that the administration of B was beneficial to his patients? Compare the results given, using X^2, G, X^2_{adj}, G_{adj}, and Fisher's exact test. ANS. $P = 0.3976$ from Fisher's exact text.

16.16 In F_2 crosses between progeny of yellow-green and chlorophyll-deficient strains of lettuce, Whitaker (1944) obtained the following results.

Family number	*Yellow green*	*Chlorophyll deficient*
15487	482	156
15395	92	40
15488	28	12
15396	256	96
15397	158	47

Test agreement to a 3:1 ratio. Are the proportions of the two color types homogeneous over the families?

17 MISCELLANEOUS METHODS

We conclude our introduction to biometry with the presentation of three methods that do not easily fit into any of the previous chapters. All three fall in the general category of being nonparametric tests of significance and are useful in a variety of situations.

Section 17.1 furnishes a method for combining the outcome of several experiments to obtain an overall significance test of a given hypothesis. In Section 17.2 we discuss runs tests, which are tests for randomness in a sequence of events. In Section 17.3 we introduce the notion of a randomization test, which is an extremely versatile method for testing the significance of a variety of statistics about whose distribution we find it impossible to make any assumption. The only requirement is random sampling of the data. This method is generally very tedious, but the wide availability of digital computers has made it the technique of choice in a variety of problems.

The topic of randomization tests leads us naturally to the final section (17.4) in which we discuss the outlook for future developments in biometry and introduce the concept of data analysis, which in conjunction with the computer revolution promises considerable progress in biometric research.

17.1 Combining probabilities from tests of significance

This is a very useful technique that we personally have employed repeatedly. A series of separate significance tests on different sets of data may test the same scientific (but not statistical) hypothesis. Each of these tests furnishes a probability value for the particular outcome, assuming the null hypothesis to be correct. These probabilities may be low enough to be suggestive, but no one of them may be sufficiently low to enable you to establish statistical significance. It stands to reason that some way of combining these probabilities would be desirable. Fisher (1954, section 21.1) has developed such a technique.

Although, as we have said, we have used this method repeatedly, we have chosen to illustrate it with a hypothetical example in Box 17.1 to show the full range of applicability of this technique. A physiologist has been studying oxygen consumption in two species of crayfish. Experiment 1 is a careful study of oxygen consumption of individuals of species A and B, with separate readings obtained for each of 14 individuals of A and 12 of B. In this and all subsequent experiments, species A appears to have the higher oxygen consumption. A t-test results in a value of $t_s = 1.90$ and, since the degrees of freedom are 24, $P = 0.07$ is obtained. The test cannot be considered significant, yet is suggestively low.

The investigator has recourse to four other experiments. In a published study by another author, equal numbers of the two species had been left in separate containers with the same volume of water and the proportion of survivors recorded after a fixed time interval. Fewer of species A survived, and the significance of this was tested by a chi-square test for independence with one degree of freedom, yielding $X^2 = 2.95$, corresponding to a probability of 0.086. The investigator undertakes a similar test; however, he puts only ten individuals of each species into a flask, limits oxygen supply by pouring paraffin over the surface, and studies time to death of each crayfish, assuming that those with the greater oxygen consumption will die first. Though he has fairly exact times of death for all crayfish, in view of the uncertainty of the distribution of this variable he prefers to analyze it by a nonparametric technique involving ranking and employs the Mann-Whitney statistic. Although it is difficult to interpolate in the table furnished by us (Table **CC**), we obtain an approximate probability of 0.10 for $U_s = 161$ for this experiment.

A fourth study involving an analysis of separate batches of four species of crayfish (including species A and B) led to an analysis of variance. A single degree of freedom comparison between A and B yields a suggestive F_s-value with a corresponding P-value near 0.080. Finally, intrigued by all these results, he undertakes yet another experiment involving measurements of individual crayfish. However, he entrusts parts of this experiment to an inaccurate technician, resulting in a much greater variance for the oxygen readings for species A than those for species B, so that a regular t-test could not be applied. Using the interpolated value (see Section 13.4 and Box 13.4) he obtains a value of $t' = 2.040$ at 16.2 degrees of freedom, corresponding to a probability of 0.060. What can he conclude as a result of all these findings?

The actual computation is based on the fact that $\ln P$ is distributed as $-\frac{1}{2}\chi^2_{[2]}$ or $-2 \ln P$ is distributed as $\chi^2_{[2]}$. By evaluating twice the negative natural logarithm of each of the five probabilities, considering each to be a $\chi^2_{[2]}$, and summing these values in the manner of Section 16.3, we obtain a total, which can be looked up under ten degrees of freedom in this example, there being two degrees of freedom for each probability value looked up. We have analyzed this case in Box 17.1 and the resulting $-2\sum \ln P$ when compared with $\chi^2_{[10]}$ yields a probability between 0.01 and 0.001, leading to an overall rejection of the null hypothesis. It appears, therefore, that on the basis of all these experiments, species A clearly has a higher oxygen consumption than species B. Details of the computation are furnished in Box 17.1.

BOX 17.1

Combining probabilities from independent tests of significance.

Five experiments testing oxygen consumption in two species of crayfish. See Section 17.1 for detailed account of each experiment and statistical test.

Test	P	$\ln P$
1. t-test $n_1 = 14$, $n_2 = 12$ 24 df $t_s = 1.90$	0.070	−2.6593
2. Chi-square test for independence 1 df $X^2 = 2.95$	0.086	−2.4534
3. Mann-Whitney U-test $n_1 = 17$, $n_2 = 12$ $U_s = 161$	0.10	−2.3026
4. Anova, single degree of freedom comparison $\nu_1 = 1$, $\nu_2 = 20$ $F_s = 3.522$	0.080	−2.5257
5. t' (t-test with unequal variances) 16.2 df $t_s' = 2.040$	0.060	−2.8134
Total		−12.7544

For P accurate to two decimal places, the $\ln P$ values may be looked up in Table **F**. For P values given to more decimal places, such as $P = 0.086$, Table **C** (common logarithms) may be used and the result multiplied by 2.30259 to convert the value to natural logarithms.

$$-2\sum \ln P = -2(-12.7544) = 25.5088$$

If all the null hypotheses in the experiments were true, this quantity would be distributed as χ^2 with $2k$ degrees of freedom (k = the number of separate tests and probabilities).

Since our value of $-2\sum \ln P$ is greater than $\chi^2_{.01[10]} = 23.209$, we reject the null hypothesis that there was no difference between the oxygen consumption of the two species in the five experiments.

The usefulness of this method should be apparent. It can also be used when the separate significance tests to be combined are all of the same kind, but where for one reason or another a joint overall statistical analysis is not possible. For instance, suppose we measure the differences in production of a metabolite by two groups of mice, each with different diets. This experiment is repeated three times over intervals of a week. This could be set up as a two-way analysis of variance with weeks being blocks and the two diets being the columns of the table. However, it may be that for one reason or another the replication of this experiment is unequal, so that we would be faced with the tedious computational problem of trying to solve a two-way anova with unequal and disproportional subclass numbers. Rather than do this, we might feel it more desirable to carry out separate anovas or t-tests on each week, obtain the probability value for each of the tests, and combine it in the manner just shown (but, of course, one would not be able to test for the presence of interaction by this technique).

A limitation of this method is that the exact probability is required for each significance test. The conventional statistical tables only furnish selected probabilities and either inverse interpolation must be practiced in these tables or more comprehensive tables of the various probability distribution functions must be consulted.

17.2 Tests for randomness: runs tests

In this section, we discuss ways of testing whether events occur in a random sequence or whether the probability of a given event is a function of the outcome of a previous event. These tests, which are of great general usefulness, are known as *runs tests* (for reasons that will become obvious shortly). Runs tests are best introduced by means of an actual example. We shall turn to a hypothetical case familiar to everyone, which has undisputable biological implications. Imagine a line of 20 youngsters in front of a cinema box office, with an equal number of boys and girls. There is a large number of ways in which they could be lined up, based on the permutations of the 10 boys and 10 girls. We shall not work out exactly how many such arrangements are possible, but will concentrate on extreme departures from random arrangement. If these youngsters were eight-year-olds, a most natural arrangement would be 10 boys followed by 10 girls or, shown symbolically,

BBBBBBBBBBGGGGGGGGGG

Clearly, this is not a random arrangement and would be quite unlikely to occur by chance. The other extreme would be an arrangement likely to be found in the same group of youngsters ten years later:

BGBGBGBGBGBGBGBGBGBG

We find here a regular alternation between the sexes. How can we devise a test for these departures from randomness?

Let us call a sequence of one or more like elements preceded and/or followed by unlike elements a *run*. The initial and terminal sequence can, of course, only be followed or preceded, respectively, by unlike elements. In our case the eight-year-olds comprised two runs, one of 10 boys, followed by one of 10 girls. On the other hand, the eighteen-year-olds comprised 20 runs, since each sequence consisted of only a single individual preceded and/or followed by a member of the opposite sex.

Statisticians have worked out the expected mean and distribution of the number of runs in dichotomized samples containing n_1 individuals of one kind and n_2 individuals of the other kind. The critical number of runs at both tails of the distribution of runs have been tabulated for sample sizes between 2 and 20 for n_1 and n_2. They are given in Table **BB**. Any number of runs that is *equal to or less than* the desired critical value at the left half of any row in the table or is *equal to or greater than* the desired critical value in the right half of any row, leads to a rejection of the hypothesis of random arrangement. For larger samples, we use a normal approximation to the expected value and standard deviation of the distribution of runs. We can then test whether the observed number of runs is significantly deviant from the expected value, basing our conclusion on the table of areas of the normal curve. For small sample sizes there are essentially no computations, and even for large sample sizes the computation is quite simple.

Box 17.2 shows a runs test for dichotomized data, such as the sexual composition of the box office line. There are three major ways in which we can apply the runs test for dichotomized data in biometric work.

BOX 17.2

A runs test for dichotomized data.

A wasp produced 18 offspring in the following sequence, where F stands for females (fertilized egg) and M for males (unfertilized eggs): n_1 (females) $= 12$; n_2 (males) $= 6$; r (number of runs) $= 4$ (sequences of one or more like elements preceded and/or followed by unlike elements). The runs have been underlined.

$$\underline{\text{F F F F F}}\quad \underline{\text{M M M}}\quad \underline{\text{F F F F F F F}}\quad \underline{\text{M M M}}$$

Critical values ($P = 0.05$) for r from Table **BB**:

Lower bound (from column 0.025) $= r = 4$

Upper bound (from column 0.975) $= r = 13$

We conclude that the two types of eggs were not laid in random sequence. The eggs are laid in sequences of one sex. A sample value of $r = 13$ would have indicated an alternating sequence of male and female eggs.

When one of the two sample sizes is greater than 20, we use a normal approximation and test

$$t_s = \frac{r - \mu_r}{\sigma_r} = \frac{r - [2n_1n_2/(n_1 + n_2)] + 1}{\sqrt{[2n_1n_2(2n_1n_2 - n_1 - n_2)]/[(n_1 + n_2)^2(n_1 + n_2 - 1)]}}$$

where μ_r is the expected number of runs, and σ_r is its standard deviation. For $n_1 = n_2 = n$ this formula simplifies to

$$t_s = \frac{r - \mu_r}{\sigma_r} = \frac{[r - n - 1]}{\sqrt{(n - 1)/(2n^2 - n)}}$$

We reject the null hypothesis at the 5% level if t_s is greater than 1.96 (using Table **P**).

The first of these is where the dichotomy occurs naturally, as in differences between two sexes, two color phases, two different species, and the like. The example in Box 17.2 is a case in point. Among bees and wasps, fertilized eggs give rise to females, unfertilized eggs to males. Females fertilize some of these eggs by releasing sperm stored in the spermatheca during copulation, while laying others unfertilized, which result in males. It is of interest to know whether fertilized eggs are laid randomly or whether they occur in batches. In Box 17.2 the sequence of male and female eggs is shown with runs underlined. There are 4 runs. Consulting Table **BB,** we find that the probability of getting as few as 4 runs or as many as 13 runs if the sequence is random, is only 0.05. We may therefore conclude that the two types of eggs were not laid in a random sequence, but were laid in sequences of one sex. The boys and girls discussed earlier, who were lined up in front of the box office, would be another instance of naturally occurring dichotomies.

A second application is the study of sequences of dichotomies defined by the investigator. An example might be the plus and minus signs representing deviations from expectation in Table 5.4, the sex ratios in 6115 sibships of 12. You will remember that we fitted expected binomial frequencies to these data, discovering that observed frequencies were higher than expected at the tails and deficient at the center of the distribution. The chi-square test for goodness of fit discussed in Section 16.2 substantiated the departure from expectation by a significant value of X^2. By looking only at the signs of the deviations, we are ignoring their magnitude. Thus it is conceivable (but improbable) that we might have a sequence of positive deviations for the first half of the distribution, followed by a sequence of negative deviations, with all of these deviations being quite minute and the observed data fitting their expectations quite faithfully. However, in cases where the fit to expectation is satisfactory, it is much more likely that the pattern of pluses and minuses among the deviations is random, and that if there is a departure from randomness, especially in the direction of long sequences, this will be matched by large deviations from expectation.

Let us look at the pattern in column 9 of Table 5.4. Notice that there are 9 plus signs and 4 minus signs and that there are only 3 runs in the data. This is significant at $P = 0.025$. We conclude that the sequence of signs of the deviations is nonrandom. Of course, this does not prove that these data do not fit the binomial expectations. This was not the hypothesis being tested. But the test does, by an extremely simple method, lead us to suspect that all is not as expected in this instance and leads us to further tests of various hypotheses.

A special instance of the runs test for dichotomized data is the so-called *runs test above and below the median.* This is especially useful in testing the random sequence of a series of observations. As you will recall, we have stressed that randomness of the observations is a fundamental assumption of the analysis of variance without which the value of the significance test is put into question (Section 13.1). At the time you were referred to a test for randomness to be learned later. This is the runs test above and below the median. We find the median item in a given sample to be used for a t-test or an anova, and label all items above the median item + and all those below − (any variates equal to the median must be ignored). Now, we array these items in their natural order in the sample and undertake a runs test of the pluses and minuses. If individual variates are independent of their predecessors—that is, the data are in random order—the values above and below the median (the plus and minus signs) should be in random sequence. If, on the other hand, the data lack independence, there may be either more or fewer runs than expected.

It is obvious how a sequence over an area as, for example, a transect across a field, might lack independence. Several plots might occur on a particularly rich patch of ground followed by other plots on relatively infertile soil. However, a similar lack of independence might occur in a temporal sequence. A technician weighing specimens for a given analysis might commit various types of errors. He might have a bias for high readings, which he would periodically overcorrect by deliberate low readings, creating an unusual alternation of high and low readings—that is, a greater number of runs than expected. Or the balance may gradually go out of adjustment during the experiment, yielding a trend in which all the later-weighed specimens would tend to be lighter than the earlier-weighed specimens. This would tend to place earlier-weighed specimens above the median, and later-weighed ones below the median. A similar trend would occur if the specimens had been dried in an oven from which they were all removed before weighing, permitting them to absorb water from the atmosphere before being weighed. Those specimens weighed last would have had the greatest opportunity to absorb water and would be likely to be heaviest. In a perfect linear trend, the plus or minus deviations from the median would be in two groups to each side of the median. In such a case the runs test provides a simple, distribution-free test for the significance of regression, although it cannot, of course, substitute for regression analysis, since it does not fit a line or serve to estimate the dependent variable. Also, it is less efficient if all the assumptions of regression analysis are actually met.

We shall briefly consider a second type of runs test especially suited for trend data, called *runs up and down*. If n items are ordered in their natural sequence of occurrence and the sign of the difference from the previous value is recorded for each item but the first, we can carry out a runs test on the resulting sequence of $n - 1$ signs. These signs would all be alike if the data were monotonically increasing or decreasing. Cyclical data would show more than the number of runs expected in a random sequence of values. Although the expected distribution of runs has been tabulated for sample sizes up to $n = 25$, such a table is not furnished here, since the normal approximation shown in Box 16.15 is adequate for samples of $n \geq 10$.

Box 17.3 shows how a test for runs up and down is carried out. The data are from a selection experiment in which survival to the pupal stage was a measure of fitness, possibly correlated with the progress of selection. The simple analysis is adequately illustrated in Box 17.3 and it is obvious from

BOX 17.3

A runs test for trend data (runs up and down).

Percent survival to the pupal stage in 32 generations of a line of *Drosophila melanogaster* selected for central pupation site. The data are graphed in Figure 17.1. From the differences between successive generations the following set of signs is obtained (the runs have been underlined): (+) and (−) indicate an increase or a decrease from the previous generation.

Generation	1	2	3	4	5	6	7	8	9	10	11
Increase or decrease		−	+	−	+	−	+	+	−	−	+

Generation	12	13	14	15	16	17	18	19	20	21	22
Increase or decrease	−	+	−	+	−	−	−	−	−	+	−

Generation	23	24	25	26	27	28	29	30	31	32
Increase or decrease	+	−	+	−	−	+	−	−	+	−

Source: Data from Sokal (1966).

Number of points or elements, $n = 32$. Number of runs (sets of like signs preceded or followed by unlike signs), $r = 23$. We test the latter value by means of a normal approximation:

$$t_s = \frac{r - \mu_r}{\sigma_r} = \frac{r - [(2n - 1)/3]}{\sqrt{(16n - 29)/90}}$$

where μ_r is the expected number of runs and σ_r is its standard deviation. The numbers in this expression are constants independent of the specific problem. In this example,

$$t_s = \frac{23 - [63/3]}{\sqrt{[16(32) - 29]/90}} = \frac{(23 - 21)}{\sqrt{5.3667}} = \frac{2}{2.317} = 0.86$$

Clearly, the number of runs is not significantly different from expectation (t_s being less than 1.96). We cannot show any departure from a random trend in survival over the 32 generations tested.

To obtain 5% significance in this sample would require $\mu_r \pm 1.96\sigma_r$ runs, which works out to $21 \pm 1.96(2.317) = 16.5$ or 25.5 runs. Thus, with 16 or fewer runs we would have concluded that one or more systematic trends were exhibited by the data. With more than 25 runs we would be led to suspect a more or less regular alternation of differences, as might be caused by some cyclical phenomenon.

The normal approximation is satisfactory for n as low as 10.

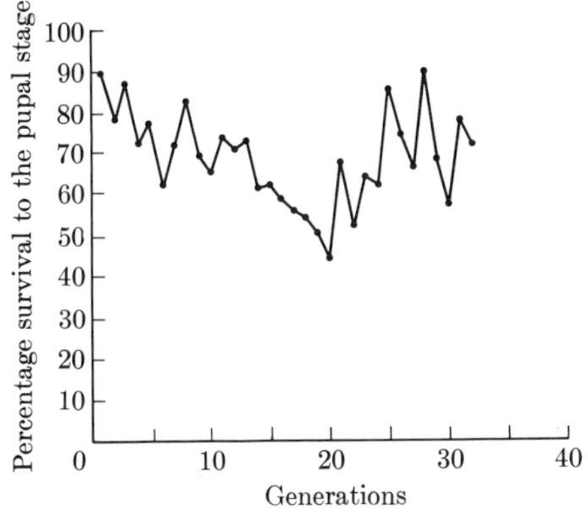

FIGURE 17.1 Percent survival to pupal stage in the CP line of *Drosophila melanogaster* selected for peripheral pupation site. Abscissa: Generations of selection. Ordinate: Percent survival. (Data from Sokal, 1966.)

the results that no significant trend in successive differences can be shown. Yet, when the data are inspected (see Figure 17.1), an apparent downward trend followed by a recovery after generation 20 seems evident. This illustrates two points. First, the need for careful statistical testing of what often appears obvious should again be emphasized. Second, all runs tests apply to attributes only. They work with dichotomized attributes, such as pluses and minuses or males and females, but they take neither absolute magnitude nor order of magnitude into consideration. Thus a small positive deviation will count as much as a big positive deviation. This explains, for instance, why the data of Figure 17.1 can be shown to exhibit a significant downward trend up to generation 20, using ordinary regression analysis, even though this cannot be shown by the runs test, which uses less information from the data. With this caution, the runs tests can be readily recommended as "quick and dirty" methods for a rapid assessment of trends and randomness. They are also of use when the investigator is unwilling to make assumptions necessary for more elaborate tests such as a regression analysis. When ranks or actual measurements are not at hand, runs tests are often the only practical alternatives.

17.3 Randomization tests

As the quantitative analysis of biological data becomes ever more common, we are frequently faced with the necessity of making tests of significance in situations where we know little or nothing about the expected distribution of the variables or statistics being tested; or we may in fact know that the data to be analyzed do not meet the assumptions required for the customary statistical tests. In such instances, the following approach has been found to be a powerful tool, and because of its structural simplicity it has considerable appeal. The basic idea of a randomization test involves three steps. (1) Consider an observed sample of variates or frequencies as one of many possible different outcomes that could have arisen by chance. (2) Enumerate many or all the possible outcomes that could be obtained by randomly rearranging the variates or frequencies. (3) On the basis of the resulting distribution of outcomes, decide whether the single outcome to be tested is deviant (= improbable) enough to warrant rejection of the null hypothesis. That is, can the single outcome be considered significantly different from a random arrangement of the data? Although many randomization tests involve heavy computation, the general availability of computers has made them feasible so that they are quite frequently applied. For a recent comprehensive treatment of the subject, see Bradley (1968).

There are several types of randomization tests and the terminological distinctions among them are not always made. It may not be necessary to carry out the actual enumeration and computation of all possible outcomes in a given problem. On the basis of probability theory we may predict the probability of a given outcome or class of outcomes to permit a decision on the significance of a sample observation. Such a test is an *exact probability test*, examples of which we are already familiar with. In Section 5.2 we worked out the probability of a sibship with 14 females and 3 males, using binomial expectations and the assumption of a sex ratio of 50:50. In Section 16.4 Fisher's exact test was a similar evaluation of the exact probability of a given outcome (and of more deviant outcomes), assuming independence of two criteria of classification. It must be realized that in such tests we could, if we wished, enumerate all possible outcomes and base our decision on a count of how many outcomes deviate more than the observed sample. Should this number be fewer than, say, 5%, we would reject the null hypothesis. If we do not know the relevant probability theory, enumeration of all outcomes is the only possible way to attack such a problem. Before we turn to other types of randomization tests, let us first solve a problem by an exact probability test.

In a selection experiment in the flour beetle, *Tribolium castaneum*, Sokal and Sonleitner (1968) studied average deviations of zygotic frequencies of eggs as compared with equilibrium expectations based on adults of the parent generation. In one series of results, deviations were recorded for generations 2 through 8 as well as for generations 18 through 28, no observations having been taken for generations 9 through 17. Preliminary results had suggested that the eggs showed an increase of mutant homozygotes and a corresponding decrease of wild type homozygotes. It was of interest to show whether this trend was uniform through the experiment or whether it was altered during the later generations. The results showed that the deviation pattern, "loss of wild type—increase of mutants," included the 7 early generations as well as 1 of the late generations. All other patterns of deviations, considered as a single group, contained only late generations. These data suggest that there was a distinctive pattern of deviations of zygotic frequencies during the early generations. We can represent the results as follows:

Pattern of deviations	*Generations*	
Loss of wild type—increase of mutants	2, 3, 4, 5, 6, 7, 8, 20	$n_1 = 8$
Other types of deviations	18, 19, 21, 22, 23, 24, 25, 26, 27, 28	$n_2 = 10$

Note that 7 of the 8 in the first deviation group are early generations, and all 10 in the second deviation group are late generations.

Let us assume that the arrangement of the generations in the two deviation groups of size 8 and 10 is entirely due to chance. In how many different ways can we arrange the total of 18 generations into two groups of 8 and 10 generations, respectively? This is a simple problem if we remember some of the material learned in Section 5.2. We can evaluate the number of different arrangements by computing the number of ways in which 8 items can be selected from 18 to form the first deviation group (of necessity leaving behind the second group of 10 items). There are $C([n_1 + n_2], n_1) = C(18, 8) = 18!/8!10! = 43{,}758$ ways of setting up the groups as specified. Out of this very large number of possible arrangements, how many could have resulted in the distribution of the items as shown in the original sample (that is, 7 early and 1 late generation in the first deviation group)? Since there are 7 early generations and 11 late ones, there are 11 ways in which 1 late generation can combine with the 7 early generations to form the sample of 8 making up the first group of deviations. Thus, the probability of obtaining a result as unusual as the one observed is 11 out of 43,758 outcomes, yielding a probability of only 0.000251. Since this is a two-tailed test (it could have been possible for the late generations to favor deviations of the first type) we double the probability obtained to yield 0.000502. We must clearly reject our null hypothesis that the arrangement of the early and late generations in the two groups of deviations is entirely at random.

Suppose the outcome had not been so clear-cut and there had been 6 early generations and 2 late ones in the sample of eight deviations of the first type. There are $C(7, 6) = 7$ ways of obtaining 6 early generations out of the 7 available, and these have to be matched with 2 late generations, which can be taken in $C(11, 2) = 55$ ways out of the 11 late ones. Therefore, there are $7 \times 55 = 385$ ways of obtaining 6 early and 2 late generations for a sample of eight deviations. Even this outcome would have represented only 385 out of 43,758 possible outcomes, or a probability of 0.008798. Doubling this value, we would still conclude that the arrangement is unlikely to have arisen randomly. Remember that you would usually add the probability of "all worse" outcomes. Adding the value of 0.000251 computed above yields an overall two-tailed probability of 0.018098.

When a group of eight deviations is composed of 5 early and 3 late generations it would have a probability of 0.079186, there being 3,465 ways of obtaining such an arrangement. The cumulative two-tailed probability, including all worse outcomes is 0.17647. We would be unable to reject the null hypothesis of random allocation of the generations to the two types of deviations.

For the actual experiment we therefore conclude that during the early generations eggs were produced representing fewer wild type and more mutant genotypes than their parents, but that during the later generations other types of deviations prevailed.

In the example just discussed, we were able to take advantage of our knowledge of probability theory to forecast the outcome of repeated random selection of eight deviations out of the universe of eighteen. We were not required to list all possible types of outcomes of the sampling procedure, since we could directly calculate the probability of those outcomes that interested us. However, in many randomization tests this is not possible because the statistic studied is some function of the variables, such as the mean, standard deviation, g_1, or even an unconventional statistic.

An example will make this clear. In an experiment studying the effects of selection 25 larval ticks were exposed to cold shock treatment. Only 9 larvae survived. Measurements of the scutum were made for both dead and surviving larvae, the results being as follows (in microns):

Surviving larvae		$n_1 = 9$, $\bar{Y}_1 = 210.63$, $s_1^2 = 16.4025$	
211.3	218.5	211.2	205.1
211.9	204.9	211.4	211.9
209.5			

Killed larvae		$n_2 = 16$, $\bar{Y}_2 = 214.43$, $s_2^2 = 51.0036$	
219.2	211.1	210.4	219.5
205.1	222.8	210.1	218.4
213.4	210.2	213.1	204.6
206.7	212.7	224.4	229.2

It was the hypothesis of the investigator that variation in scutum length reflects general genetic variability and that only ticks from a restricted portion of the genetic spectrum would survive. This might be tested by demonstrating that the variance of scutum length of the sample of ticks surviving the cold shock is lower than that for the killed ticks.

How can we test this hypothesis? We can calculate a variance, s_2^2, of the scutum length of the 16 nonsurviving ticks and divide it by the variance, s_1^2, based on the 9 survivors. In the actually observed data, $s_1^2 = 16.4025$, $s_2^2 = 51.0036$, and $s_2^2/s_1^2 = 3.1095$. There appears to be a suggestion that the hypothesis is justified. However, this ratio does not follow the F-distribution. The two variances are not obtained from independent, normally distributed samples, but from a (nonrandom) sorting out of individuals within a circumscribed universe, the 25 experimental organisms. Also, the scutum lengths on inspection do not appear to be normally distributed. Although an appropriate statistical technique may have been developed for such a test, none was known to the investigator, who proceeded by means of a randomization test.

There are $C([n_1 + n_2], n_1) = C(25, 9) = 25!/9!16! = 2{,}042{,}975$ ways of taking 9 ticks out of the sample of 25. For each of these ways one can compute the two variances and their ratio s_2^2/s_1^2 and then see whether the ratio observed in the sample is sufficiently deviant in terms of the total distribution of the sampled ratios to merit rejection of the null hypothesis of random assortment. But such an undertaking would involve an astronomical number of computations. The investigator therefore obtained 500 random partitions of the data into samples of 9 and 16 from the population of the 25 scutum lengths and computed s_1^2, s_2^2, and s_2^2/s_1^2 for each case. Figure 17.2 illustrates the distribution of ratios obtained by putting the problem on the computer. It is obvious from this figure that the ratio 3.1095 is not in the upper 5% tail of the distribution of 500 empirically obtained ratios. We cannot support our hypothesis with the evidence from this experiment. Of the randomized variance-ratios, 8% are equal to or greater than the observed value 3.1095. Based on a sample of 500, the 99% confidence limits of this percentage are 5.23–11.60 (see Table W). Thus it is improbable that the true level of significance for the entire population of 2,042,975 partitions is less than 5%.

The basic computer program to carry out the computations leading up to Figure 17.2 is given in Appendix A3.16 as an example of a randomization test. For carrying out another type of randomization test, we would have to replace the present function subprograms STAT and COMPR (explained in A3.16) with appropriately modified subprograms for the new problem.

Such a test is a *sampled randomization test.* We take random samples, compute our statistic, empirically find a distribution of this statistic, and decide on the significance of our original observation. Sometimes there is a relatively small number of possible outcomes so that all, rather than only a

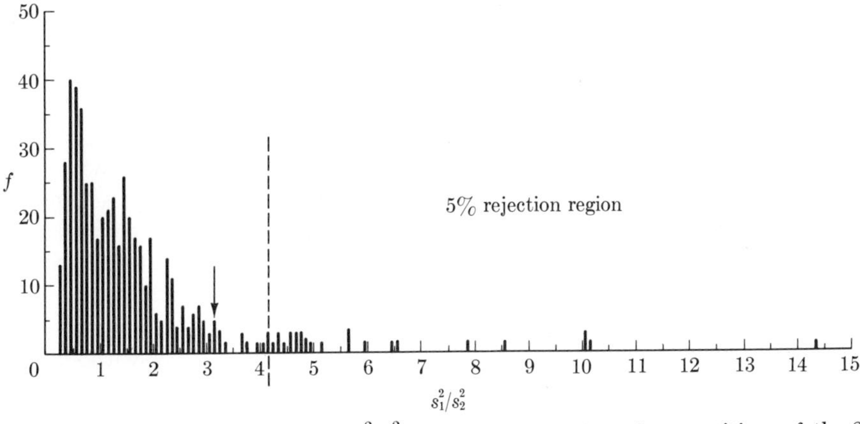

FIGURE 17.2 Distribution of ratios s_2^2/s_1^2 obtained from 500 random partitions of the 25 scutum lengths into two sets of 9 and 16 each. The one-tailed 5% rejection region is marked off by a broken line. The arrow points to the observed ratio being tested.

random sample of them, may be evaluated. Such a test is called a *permutation test.*

We would like to illustrate the latter procedure with a different example taken from numerical taxonomy. In Table 17.1 we show a half matrix of correlation coefficients indicating the degree of similarity among 10 species

TABLE 17.1

Similarity coefficients (correlation coefficients) among 10 species selected from the *Hoplitis*-complex of bees. Species code numbers are shown at the margins of the half-matrix.

Species	4	5	8	26	35	36	40	50	67	68
4	—									
5	.65	—								
8	.70	.84	—							
26	.40	.43	.49	—						
35	.41	.45	.49	.60	—					
36	.41	.42	.48	.52	.94	—				
40	.17	.39	.44	.57	.42	.29	—			
50	.35	.48	.52	.45	.41	.33	.51	—		
67	.36	.37	.41	.20	.25	.29	.20	.49	—	
68	.33	.36	.43	.18	.25	.29	.19	.46	.96	—

Source: Sokal (1958).

of bees whose numerical code names are shown at the margins of the matrix. The correlation coefficients are computed from a set of variates representing the characters of these species. From the matrix it is possible to read off the similarity between any pair of species. Thus, the similarity between species 26 and 36 is 0.52. Conventional taxonomic work had considered species 4, 5, and 8 to be in a single genus, *Proteriades*. We now would like to justify this judgment by developing a criterion for distinctness of a group. By distinctness we mean a measure of homogeneity or cohesion of the members of a group relative to their similarity with other species. A very simple and intuitively obvious criterion would be to take average similarity among the members of the group (that is, the average similarity between species 4 and 5, 4 and 8, and 5 and 8) and subtract from it the average similarity of each of the species 4, 5, and 8 with all the other species that are not in this group. When this is done, we obtain a positive or negative value indicating the distinctness of the original group. The higher the positive value, the more distinct the group; that is, the similarity is high within the group, and low with species outside the group. Conversely, if the distinctness value is negative, the average similarity with species outside of the group is higher than that within the group, and hence there would seem little justification for its formal recognition. When we calculate such a distinctness value for group 4, 5, 8, we obtain $\frac{1}{3}(0.65 + 0.70 + 0.84) - \frac{1}{21}(0.40 + 0.41 + 0.41 + \cdots + 0.43 + 0.45 + \cdots + 0.49 + \cdots + 0.41 + 0.43) = 0.730 - 0.409 = 0.321$, indicating that the group of these three species possesses considerable distinctness.

How can we compute the significance of such a finding? The similarity coefficients (in this case correlation coefficients) in the half-matrix of Table 17.1 have an unknown distribution and this is even more true of the distinctness values. The problem was solved by computing distinctness values for all possible samples of three out of the ten species. There are $C(10, 3) = 120$ different ways of making sets of 3 out of 10 species. The distribution of the 120 distinctness values is shown in Figure 17.3. Five per cent of the distinctness values lie beyond -0.148 and 0.210 at the two tails of the distribution. Thus the distinctness 0.321 of group 4, 5, and 8 is significantly deviant. On the other hand, an artificial group made up of species 4, 26, and 50 (which have been placed in different genera) form a group with little cohesion (distinctness $= -0.039$, not significant by Figure 17.3). Had this been a larger problem (with more possible outcomes), we might have decided to carry out a sampled randomization test by taking several hundred samples and computing distinctness values for these, rather than for all outcomes.

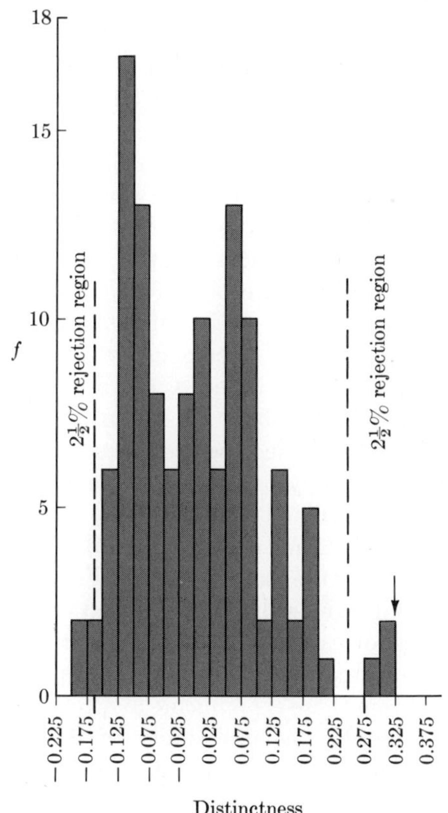

FIGURE 17.3 Frequency distribution of "distinctness" values for all sets of three species from ten species of bees listed in Table 17.1. The broken lines mark off $2\frac{1}{2}$% rejection regions. The arrow points out the distinctness value of group 4, 5, 8 being tested.

In a similarity matrix among 100 species there are 161,700 possible sets of 3, and we clearly would not wish to compute a distribution of distinctness values for all of these.

Permutation and randomization tests are most useful when the statistic to be computed has an unknown distribution or a distribution other than that required for the customary tests of significance. The variance-ratio and distinctness index discussed above fall into this category. With the help of a computer we can apply this randomization test to any statistic of interest, perhaps one generated especially for the problem at hand, which measures some aspect of the data of particular interest to us. When such a new statistic was developed in the past, it often required considerable mathematical skill to find its expected distribution and critical values. In some instances these problems proved intractable even for experienced mathematical statisticians. By randomization tests and with the help of a digital computer, a nonmathematician can empirically obtain a sampling distribution of any statistic he cares to define, and as a result of this empirical distribution decide whether a given observation is significantly deviant from expectation.

It is clear that a randomization test can involve extremely complicated functions requiring the processing of large amounts of data, which would be quite impossible without powerful computational facilities. Thus, Rohlf (1965) carried out randomization tests to establish critical values for a certain type of correlation coefficient (cophenetic correlations) in numerical taxonomy, in which each single randomization is essentially equivalent to carrying out a full numerical taxonomic study involving an appreciable amount of time on even the fastest computer. Yet without such approaches there would be no hope of establishing significance levels for these statistics. We hope that the reader is impressed by the tremendous power and versatility of this approach.

Randomization tests belong to the general category of *Monte Carlo methods*, the solution of complex mathematical and statistical problems by random sampling from a simulated population on a computer. We shall furnish one more example to illustrate the general Monte Carlo approach. Lewontin and Felsenstein (1965) were concerned about chi-square tests of independence in $R \times C$ tables, where $R = 2$ and the expected frequencies in one of the rows would be very small (considerably below the minimum of 5 generally recommended). They set up 54 different types of tables, varying the number of rows and the marginal totals. For each type of table they randomly allocated frequencies limited only by the marginal totals. From the table they computed X^2 in the conventional manner. Each randomization and X^2-computation was repeated 5000 times for each of the 54 types of tables. The resulting distributions of X^2 were compared with chi-square distributions for the appropriate degrees of freedom. From these experiments and comparisons, Lewontin and Felsenstein conclude that the probability of Type I error given by the conventional chi-square test is, in general, conservative for 5 or more degrees of freedom even when expectations are very small in each cell. This important reassurance to the research worker, carrying out a test of independence with low expectations in one row of the table, could not easily have been obtained without computers and the Monte Carlo method. Readers may recognize that the sampling experiments on confidence limits and the X^2- and G-statistics done by computer to illustrate Chapters 7 and 16 fall into this general category.

17.4 The future of biometry: data analysis

Although we indicated in Chapter 1 that biometry is a rapidly changing field and that it would be risky to try to anticipate its future development, we shall nevertheless at least offer a glimpse of one of the developments that seem to loom over the horizon. Our discussion of randomization tests in the last section leads us naturally into the area of greatest promise, which is coming to be known as *data analysis*. By this is meant the systematic search through a set of data to reveal information and relationships of interest. The procedures of data analysis, although essentially numerical, are also experimental. In the randomization tests of the last section, we were, in a manner of speaking, carrying out experiments with a set of data. These include such techniques as *transformations* to see what effect this would have on the distribution of a sample or the relationships between two variables in a sample; *winsorizing*, which is another name for truncating the data (discarding a certain percentage of the extreme observations), followed by a recalculation of statistics such as means and standard deviations; *randomizations;* and so forth. In essence these techniques are nothing new. They have been carried out by perceptive research workers and statisticians for many years. Yet the development of high-speed computation—and especially of on-line conversational mode data processing—permits almost instantaneous replies to any query an investigator wishes to ask of his data. Therefore he can ask many more questions, in the hope that some of them will yield insights through results that can then be followed up with further questions or possibly with further experiments to elaborate and confirm any new point that has been uncovered.

Why should there be this sudden emphasis on the analysis of the actual research data as contrasted with the computation of statistics that are simply abstractions or summaries from these data? Tukey and Wilk (1965) list several obvious reasons. More and more data are being obtained simply through the enormous expansion of scientific activity and through the development of automatic data-gathering devices. These confront research workers in many disciplines with masses of data the like of which had never been reckoned with before. Their digestion presents a serious challenge to the meaningful interpretation of underlying factors and regularities. A second reason is the ever-increasing tendency toward quantification in all fields of human knowledge, as numerical methods enter fields previously untouched by quantification.

Most important, however, is the revolution in speeds and capabilities of computing devices and in visual and graphic display units for the results of computation. By means of cathode ray tubes and light pens it is now quite possible to truncate a sample visually and automatically recompute means and standard deviations or to transform a variable and study its new distribution curve, with all of these computations being carried out essentially instantaneously. Many attempts at data analysis, which previous researchers might wistfully have thought of undertaking but abandoned because of the tremendous amount of labor involved, can now be followed up. Clearly, much of what will be done will be useless and wasted; yet enough new information should emerge to make the overall approach of great promise.

We can thus view data analysis as the systematic and largely automated scrutiny of sets of data to yield both summaries of and fresh insights into the relationships in a given study. Tukey and Wilk place special emphasis on the analysis of residuals when carrying out an anova or a regression analysis. We have generally considered only the sum of the squared residuals and not the individual deviations giving rise to the error estimate. In data analysis as much emphasis is paid to the residual deviations as to the main effects and interactions. An unusually large or small deviation is looked at critically to see whether we can find any reasons for such a deviation. Only with automatic computational devices is it feasible to inspect, routinely and systematically, all residual deviations. For example, the factorial anova program in Appendix A3.5 allows the outputting of all residuals for all of the interaction terms. Since these deviations represent the remaining variability, they are, of course, the object of inquiry for further scientific research. General rules and procedures for data analysis cannot yet be given, inasmuch as the hardware and software necessary to carry out such work is not yet widely available. But it is quite clear that data analysis will be a major aspect of future biometric work.

Exercises 17

The following problems do not require any actual computations, but serve as review problems for the student, who is required to indicate the appropriate method for a solution. The problems range over the entire subject matter of the book.

The following steps should be taken for *each* of the problems.

(a) State statistical method appropriate for providing solution, and why.
(b) Outline the setup of the data.
(c) Give formulas for computations.
(d) State degrees of freedom; if none applicable, state "none."
(e) Draw attention to any requirements to which the data must conform in order to be tested by the method you have chosen, as well as to any corrections that might have to be applied.
(f) Enumerate the specific answers that could emerge as the result of your method.
(g) List any alternative method applicable.

17.1 A normal probability curve has been fitted to certain data. There are 15 classes. Are the deviations such as could be expected from accidents of sampling?

17.2 The amount of inorganic phosphorus in the blood plasma of 10 unrelated men is determined from specimens taken every morning for a fortnight. Are there differences among individuals and/or does the amount vary from morning to morning? Would you expect the interaction to be negligible? If not, how would you design the experiment to demonstrate interaction?

17.3 An ecologist claims that there is a narrow optimum size for a burrowing animal, above or below which the animal is at a disadvantage. How could you investigate this in a frequency distribution of 300 body diameters of this species?

17.4 You are in the possession of data on the physical characteristics of 1221 sets of 4 siblings. Among other characters you know whether each individual's eye color was blue or not. Without studying any pedigrees how could you demonstrate that blue eyes are controlled wholly or in part by genetic factors? (If we can prove that sibs are more like each other than randomly chosen individuals, we shall consider this sufficient evidence, although theoretically sibs could be more alike because of common environment as well.)

17.5 We are measuring tarsal claw length of male lice of species P taken from eight different individuals of host species X. Is there an added variance due to differences between hosts? Remember that we are unlikely to obtain the same number of lice from each host individual.

17.6 Counts are made of number of mites found on blades of grass species G in the sunlight and in the shade. Are there more mites per blade in the shade? Twenty blades of grass were found in the sun and 31 in the shade. The number of mites per blade varied from 3 to 32.

17.7 An animal behaviorist is studying mating behavior in animal species X. Females of equal ages are introduced singly into a mating cage where they are exposed to several males of known and demonstrable aggressiveness. However, in species X the female can accept or reject copulation at will. Thus number of copulations per unit time is an index of the female's submissiveness or sexual appetite. The investigator tests 100 females, and they range from 0 to 11 matings per unit time, with a mean at 3.2 matings. He has a hypothesis that the females can be subdivided into at least two races or groups differing in submissiveness. How can he support his findings with an analysis of the above experiment? Another way of looking at this analysis is to ask what results the scientist would expect if copulations were independent of the genetics of the female, and merely a randomly varying variable.

17.8 A study of respiration under the influence of drugs in salamanders has yielded data on 7 animals for each of five drugs and 7 control animals. The variates are volume of oxygen consumed, and a preliminary investigation has shown that the variances for each drug are anything but homogeneous. Before the investigator attempts to equalize variances by transformation, how could he test for differences among his drugs? If such differences are not present, he might not think it worthwhile to spend time looking for a suitable transformation.

17.9 An investigator studying weight gain in rats on different diets is comparing two groups of 20 rats each. Not only are the variances of the two samples different, but the variates are clearly not normally distributed. No transformation known to the investigator appears able to take care of the difficulty. Suggest a method of testing for differences in weight gain between the two samples.

17.10 Samples of moths were collected from four localities. The percentage of melanistic individuals varied from sample to sample. Can the four samples be considered as coming from a homogeneous population as regards incidence of melanistic forms? (You are in possession of the original data from which the percentages were calculated.) If not, are there any homogeneous subsets of localities?

17.11 An investigator wishes to demonstrate normality in a great number of frequency distributions. None of the frequency distributions has more than 20 items.

17.12 Measurements are taken of the lengths of 50 skulls of a certain species of field mouse from each of two localities. Is there a significant difference in average length? (Variances of the two samples do not appear to differ greatly.)

17.13 You are attempting to find whether variations in body weight and amount of bile pigment in urine are due to common causes. You have data on 1243 male white Americans, ages 45 to 50.

17.14 The number of nests of bird species A per square mile of a certain forested area has been determined on the basis of a study of 60 square miles. This bird is known to be territorial (that is, nonrandomly distributed). How can you show this from your data and would data based on square mile units necessarily show the phenomenon?

17.15 The mean number of dorsocentral bristles in dichaete *Drosophila melanogaster* is determined for flies emerging on each of five 2-day intervals. Separate readings were obtained for each of twelve replicate bottles. Can the whole set of data be treated as from a homogeneous population?

17.16 A biologist collects 357 specimens of a species of lizard in one locality. He measures tail length on all of them. How many must he measure in the future from any other locality so that he can pick up a significant difference equal to 5% of the present mean tail length of his lizards?

17.17 The variation "banded" is found in the dragonfly species L. Both males and females show this variation. During an afternoon's intensive collection in a rice paddy area in the South, students in a field zoology class collected 117 mating pairs (in copula). As each pair was collected the students classified it as belonging to one of the following four categories:

(a) both sexes common (unbanded) type
(b) both sexes banded
(c) ♀♀ common, ♂♂ banded
(d) ♀♀ banded, ♂♂ common

We wish to test whether mating was at random or whether these dragonflies prefer to mate with their own type.

17.18 Sixty-seven field mice and 23 pocket gophers were collected from a certain area. Femur length is measured for both species. Is length of the femur more variable in the mice?

17.19 Seven doses of chemical A (1, 2, 4, 8, 16, 32, and 64 mg) are administered to 3-day-old chicks, 15 chicks per dose. All 105 chicks are caged together and distinguished by headstains denoting the dose. At the end of three weeks the chicks are killed and the glycogen content of the livers determined separately for each chick. Did changes in the dose have effects on glycogen content and can this relation be quantified? Is a linear equation adequate to describe the effect of chemical A dose on glycogen content?

17.20 A botanist collects five samples of herbs of species B from different localities. The sample sizes vary. The correlation between height of plant and number of flowers has been calculated for all the samples and is significant for every one. Could the correlation between height and flower number be the same for the five localities? If so, what estimate can you give of ρ?

17.21 You would like to predict the total length of a cat from the length of its femur. What are the confidence limits of your predicted value for a given femur length?

17.22 An investigator wishes to establish a correlation between number of ovarioles and fecundity in drosophila. He first raises the female under optimal conditions, counting her egg output over a 2-week period. Then he dissects out her ovaries, sections them, and counts the ovarioles. He has counted fecundity on 100 individuals but has sectioned only 25 ovaries to date. Before proceeding with his work he would like to have reassurance that correlation actually occurs in his data. What preliminary test would you suggest on the 25 paired readings amassed to date?

17.23 An ecologist watches a bird nest to record the food brought in by the female for her young. He can distinguish two size classes of insect prey, big and small. Looking over a sequence of thirty consecutive observations taken in the course of one afternoon and containing 20 big and 10 small insects, he wonders whether the sequence of size classes is random.

17.24 To test one aspect of a new theory, an ecologist wished to compare two populations with respect to a rather unconventional statistic. It was unlikely to be normally distributed and he did not know its standard error.

17.25 The progeny of a cross was expected to yield dominants, heterozygotes, and recessives in a 1:2:1 ratio. Separate replicates were reared under five environmental conditions. Do the data fit the expected 1:2:1 ratio? Is there any evidence of differential survivorship in the different environments?

17.26 A pharmacologist wished to compare the effects of two drugs on the longevity of hamsters. He mainly wished to know which drug had the greatest effect in shortening the life span. The variable recorded (length of life in days) was not normally distributed. He wished to complete the experiment in as short a time as was possible but a few animals (of each group) lived a very long time.

17.27 An immunologist compared the strength of two antigens by two separate tests. The first test induced an all-or-nothing skin reaction and was analyzed by a 2×2 test for independence. The second experiment was by means of a precipitin test and the difference in turbidity was analyzed by a t-test (after the variates had been transformed). Both tests were inconclusive. How could an overall test of significance for both tests be constructed?

APPENDIX 1

MATHEMATICAL APPENDIX

A1.1 Demonstration that the sum of the deviations from the mean is equal to zero.

We have to learn two common rules of statistical algebra. We can open a pair of parentheses with a $\sum$ sign in front of them by treating the $\sum$ as though it were a common factor.

We have

$$\sum_{i=1}^{n} (A_i + B_i) = (A_1 + B_1) + (A_2 + B_2) + \cdots + (A_n + B_n)$$
$$= (A_1 + A_2 + \cdots + A_n) + (B_1 + B_2 + \cdots + B_n)$$

Therefore

$$\sum_{i=1}^{n} (A_i + B_i) = \sum_{i=1}^{n} A_i + \sum_{i=1}^{n} B_i$$

Also, when $\sum_{i=1}^{n} C$ is developed during an algebraic operation, where C is a constant, this can be computed as follows:

$$\sum_{i=1}^{n} C = C + C + \cdots + C \qquad (n \text{ terms})$$
$$= nC$$

Since in a given problem a mean is a constant value, $\sum^{n} \bar{Y} = n\bar{Y}$. If desired, you may check these rules, using simple numbers. In the subsequent demonstration and others to follow, whenever all summations are over n items we have simplified the notation by dropping subscripts for variables and superscripts above summation signs.

We wish to prove that $\sum y = 0$. By definition,

$$\begin{aligned}\sum y &= \sum(Y - \bar{Y}) \\ &= \sum Y - n\bar{Y} \\ &= \sum Y - \frac{n\sum Y}{n} \qquad \left(\text{since } \bar{Y} = \frac{\sum Y}{n}\right) \\ &= \sum Y - \sum Y\end{aligned}$$

Therefore $\sum y = 0$.

A1.2 Demonstration of the effects of additive, multiplicative, and combination coding on means, variances, and standard deviations.

For this proof we have to learn one more convention of statistical algebra. We have seen in Appendix A1.1 that $\sum C = nC$. However, when the $\sum$ precedes both a constant and a variable, as in $\sum CY$, the constant can be placed before the $\sum$, since

$$\begin{aligned}\sum_{i=1}^{n} CY_i &= CY_1 + CY_2 + \cdots + CY_n \\ &= C(Y_1 + Y_2 + \cdots + Y_n) \\ &= C\left(\sum_{i=1}^{n} Y_i\right)\end{aligned}$$

Therefore

$$\sum_{i=1}^{n} CY_i = C \sum_{i=1}^{n} Y_i$$

Thus $\sum CY = C\sum Y$, $\sum C^2y^2 = C^2\sum y^2$, and $\sum 2\bar{Y}Y = 2\bar{Y}\sum Y$, because both 2 and $\bar{Y}$ are constants.

Means of coded data

Additive coding.—The variable is coded $Y_c = Y + C$, where C is a constant, the additive code. Therefore

$$\sum Y_c = \sum(Y + C) = \sum Y + nC$$

and

$$\bar{Y}_c = \frac{\sum Y_c}{n} = \frac{\sum Y}{n} + C = \bar{Y} + C$$

To decode $\bar{Y}_c$, subtract C from it and you will obtain $\bar{Y}$; that is, $\bar{Y} = \bar{Y}_c - C$.

Multiplicative coding.—The variable is coded $Y_c = DY$, where D is a constant, the multiplicative code. Therefore

$$\sum Y_c = D\sum Y$$

and

$$\overline{Y}_c = \frac{\sum Y_c}{n} = D\frac{\sum Y}{n} = D\overline{Y}$$

To decode $\overline{Y}_c$ divide it by D and you will obtain $\overline{Y}$; that is, $\overline{Y} = \overline{Y}_c/D$.

Combination coding.—The variable is coded $Y_c = D(Y + C)$, where C and D are constants, the additive and multiplicative codes, respectively. Therefore

$$\sum Y_c = D\sum(Y + C) = D\sum Y + nDC$$

and

$$\overline{Y}_c = \frac{\sum Y_c}{n} = D\frac{\sum Y}{n} + DC = D\overline{Y} + DC$$

To decode $\overline{Y}_c$, divide it by D, then subtract C and you will obtain $\overline{Y}$; that is,

$$\overline{Y} = \frac{\overline{Y}_c}{D} - C.$$

Variances and standard deviations of coded data

Additive coding.—The variable is coded $Y_c = Y + C$, where C is a constant, the additive code. By definition, $y = Y - \overline{Y}$, and

$$\begin{aligned} y_c &= Y_c - \overline{Y}_c \\ &= [(Y + C) - (\overline{Y} + C)] \qquad \text{(as shown for means above)} \\ &= [Y + C - \overline{Y} - C] \\ &= [Y - \overline{Y}] \\ &= y \end{aligned}$$

Therefore $\sum y_c^2 = \sum y^2$, and

$$\frac{\sum y_c^2}{n-1} = \frac{\sum y^2}{n-1}$$

Thus additive coding has no effect on sums of squares, variances, or standard deviations.

Multiplicative coding.—The variable is coded $Y_c = DY$, where D is a constant, the multiplicative code. By definition, $y = Y - \overline{Y}$, and

$$\begin{aligned} y_c &= Y_c - \overline{Y}_c \\ &= DY - D\overline{Y} \qquad \text{(as shown for means above)} \\ &= D(Y - \overline{Y}) \\ &= Dy \end{aligned}$$

Therefore $y_c^2 = D^2y^2$, $\sum y_c^2 = D^2\sum y^2$, and

$$\frac{\sum y_c^2}{n-1} = D^2\frac{\sum y^2}{n-1}$$

$$s_c^2 = D^2s^2$$

Thus, when data have been subjected to multiplicative coding, a sum of squares or variance can be decoded by dividing it by the *square* of the multiplicative code; a standard deviation can be decoded by dividing it by the code itself, that is, $s^2 = s_c^2/D^2$ and $s = s_c/D$.

Combination coding.—The variable is coded $Y_c = D(Y + C)$, where C and D are constants, the additive and multiplicative codes, respectively. By definition $y = Y - \overline{Y}$, and

$$\begin{aligned} y_c &= Y_c - \overline{Y}_c \\ &= [D(Y + C) - (D\overline{Y} + DC)] \qquad \text{(as shown for means above)} \\ &= [DY + DC - D\overline{Y} - DC] \\ &= D[Y - \overline{Y}] \end{aligned}$$

Therefore $y_c = Dy$, as before.

Thus in combination coding only the multiplicative code needs to be considered when decoding sums of squares, variances, or standard deviations.

A1.3 Demonstration that Expression (4.7), the computational formula for the sum of squares, equals Expression (4.6), the expression originally developed for this statistic.

We wish to prove that $\sum(Y - \overline{Y})^2 = \sum Y^2 - ((\sum Y)^2/n)$. We have

$$\begin{aligned} \sum(Y - \overline{Y})^2 &= \sum(Y^2 - 2Y\overline{Y} + \overline{Y}^2) \\ &= \sum Y^2 - 2\overline{Y}\sum Y + n\overline{Y}^2 \\ &= \sum Y^2 - \frac{2(\sum Y)^2}{n} + \frac{n(\sum Y)^2}{n^2} \qquad \left(\text{since } \overline{Y} = \frac{\sum Y}{n}\right) \\ &= \sum Y^2 - \frac{2(\sum Y)^2}{n} + \frac{(\sum Y)^2}{n} \end{aligned}$$

Hence

$$\sum(Y - \overline{Y})^2 = \sum Y^2 - \frac{(\sum Y)^2}{n}$$

A1.4 Derivation of simplified formulas for standard error of the difference between two means.

The standard error squared from Expression (9.2) is

$$\left[\frac{(n_1 - 1)s_1^2 + (n_2 - 1)s_2^2}{n_1 + n_2 - 2}\right]\left(\frac{n_1 + n_2}{n_1 n_2}\right).$$

When $n_1 = n_2 = n$, this simplifies to

$$\left[\frac{(n - 1)s_1^2 + (n - 1)s_2^2}{2n - 2}\right]\left(\frac{2n}{n^2}\right) = \left[\frac{(n - 1)(s_1^2 + s_2^2)(2)}{2(n - 1)(n)}\right] = \frac{1}{n}(s_1^2 + s_2^2)$$

which is the standard error squared of Expression (9.3).

When $n_1 \neq n_2$, but each is large so that $(n_1 - 1) \approx n_1$ and $(n_2 - 1) \approx n_2$, the standard error squared of Expression (9.2) simplifies to

$$\left[\frac{n_1 s_1^2 + n_2 s_2^2}{n_1 + n_2}\right]\left(\frac{n_1 + n_2}{n_1 n_2}\right) = \left[\frac{n_1 s_1^2}{n_1 n_2} + \frac{n_2 s_2^2}{n_1 n_2}\right] = \frac{s_1^2}{n_2} + \frac{s_2^2}{n_1}$$

which is the standard error squared of Expression (9.4).

A1.5 Demonstration that t_s^2 obtained from a test of significance of the difference between two means (as in Box 9.6) is identical to the F_s-value obtained in a single classification anova of two equal-sized groups (as in Box 9.5).

$$t_s \text{ (from Box 9.6)} = \frac{\bar{Y}_1 - \bar{Y}_2}{\sqrt{\dfrac{1}{n(n-1)}\left(\sum^n y_1^2 + \sum^n y_2^2\right)}}$$

$$t_s^2 = \frac{(\bar{Y}_1 - \bar{Y}_2)^2}{\dfrac{1}{n(n-1)}\left(\sum^n y_1^2 + \sum^n y_2^2\right)} = \frac{n(n-1)(\bar{Y}_1 - \bar{Y}_2)^2}{\sum^n y_1^2 + \sum^n y_2^2}$$

In the 2-sample anova,

$$MS_{\text{means}} = \frac{1}{2-1}\sum^2 (\bar{Y}_i - \bar{\bar{Y}})^2$$

$$= (\bar{Y}_1 - \bar{\bar{Y}})^2 + (\bar{Y}_2 - \bar{\bar{Y}})^2$$

$$= \left(\bar{Y}_1 - \frac{\bar{Y}_1 + \bar{Y}_2}{2}\right)^2 + \left(\bar{Y}_2 - \frac{\bar{Y}_1 + \bar{Y}_2}{2}\right)^2$$

(since $\bar{\bar{Y}} = (\bar{Y}_1 + \bar{Y}_2)/2$)

$$= \left(\frac{\bar{Y}_1 - \bar{Y}_2}{2}\right)^2 + \left(\frac{\bar{Y}_2 - \bar{Y}_1}{2}\right)^2$$

$$= \tfrac{1}{2}(\bar{Y}_1 - \bar{Y}_2)^2$$

since the squares of the numerators are identical. Then

$$MS_{\text{groups}} = n \times MS_{\text{means}} = n[\tfrac{1}{2}(\bar{Y}_1 - \bar{Y}_2)^2]$$

$$= \frac{n}{2}(\bar{Y}_1 - \bar{Y}_2)^2$$

$$MS_{\text{within}} = \frac{\sum^n y_1^2 + \sum^n y_2^2}{2(n-1)}$$

$$F_s = \frac{MS_{\text{groups}}}{MS_{\text{within}}}$$

$$= \frac{\dfrac{n}{2}(\bar{Y}_1 - \bar{Y}_2)^2}{\left(\sum^n y_1^2 + \sum^n y_2^2\right) \Big/ [2(n-1)]}$$

$$= \frac{n(n-1)(\bar{Y}_1 - \bar{Y}_2)^2}{\left(\sum^n y_1^2 + \sum^n y_2^2\right)}$$

$$= t_s^2$$

A1.6 Derivation of Expression (11.3) for the expected error mean square of a completely randomized design estimated from the mean squares of a randomized complete blocks design.

Randomized complete blocks design

Source of variation	*df*	*MS*	*Expected MS*
Blocks	$b - 1$	MS_B	$\sigma^2_{(RB)} + a\sigma^2_B$
Treatments	$a - 1$	$MS_{A(RB)}$	$\sigma^2_{(RB)} + \frac{b}{a-1}\Sigma\alpha^2$
Error	$(a-1)(b-1)$	$MS_{E(RB)}$	$\sigma^2_{(RB)}$
Total	$ab - 1$		

Completely randomized design

Source of variation	*df*	*MS*	*Expected MS*
Treatments	$a - 1$	$MS_{A(CR)}$	$\sigma^2_{(CR)} + \frac{b}{a-1}\Sigma\alpha^2$
Error	$a(b-1)$	$MS_{E(CR)}$	$\sigma^2_{(CR)}$
Total	$ab - 1$		

If we assume that the total SS of the two designs is the same, we can write the following identity where each side represents the total SS:

$$(b-1)MS_B + (a-1)MS_{A(RB)} + (a-1)(b-1)MS_{E(RB)} = (a-1)MS_{A(CR)} + a(b-1)MS_{E(CR)}$$

Rewriting the identity in terms of the variance components of the expected mean squares, we obtain

$$(b-1)(\sigma^2_{(RB)} + a\sigma^2_B) + (a-1)\left(\sigma^2_{(RB)} + \frac{b}{a-1}\sum\alpha^2\right) + (a-1)(b-1)\sigma^2_{(RB)}$$

$$= (a-1)\left(\sigma^2_{(CR)} + \frac{b}{a-1}\sum\alpha^2\right) + a(b-1)\sigma^2_{(CR)}$$

$$[(b-1) + (a-1) + (a-1)(b-1)]\sigma^2_{(RB)} + a(b-1)\sigma^2_B + b\sum\alpha^2 = [(a-1) + a(b-1)]\sigma^2_{(CR)} + b\sum\alpha^2$$

$$(ab-1)\sigma^2_{(RB)} + a(b-1)\sigma^2_B = (ab-1)\sigma^2_{(CR)}$$

$$\sigma^2_{(CR)} = \sigma^2_{(RB)} + \frac{a(b-1)}{ab-1}\sigma^2_B$$

We rewrite this formula for $\sigma^2_{(CR)}$ in terms of expected MS. Since $\sigma^2_{(RB)} = MS_{E(RB)}$ and $\sigma^2_B = (MS_B - MS_{E(RB)})/a$, we obtain

$$MS_{E(CR)} = MS_{E(RB)} + a(b-1)\frac{MS_B - MS_{E(RB)}}{a(ab-1)}$$

$$= MS_{E(RB)} + (b-1)\frac{MS_B}{ab-1} - (b-1)\frac{MS_{E(RB)}}{ab-1}$$

$$= [(ab-1)-(b-1)]\frac{MS_{E(RB)}}{ab-1} + (b-1)\frac{MS_B}{ab-1}$$

$$= \frac{b(a-1)MS_{E(RB)} + (b-1)MS_B}{ab-1}$$

This is Expression (11.3).

A1.7 Demonstration that t_s^2 obtained from a paired-comparisons significance test (as in Box 11.3) is identical to the F_s-value for treatments obtained in a two-way anova without replication (same box).

We shall use the simpler $\sum^b (Y_1 - Y_2)$ in place of $\sum_{i=1}^{b} (Y_{i1} - Y_{i2})$. Also remember that in such an anova $a = 2$.

$$t_s = \frac{\bar{D} - (\mu_1 - \mu_2)}{s_{\bar{D}}} \qquad \text{(from Box 11.3)}$$

where $\bar{D} = \bar{Y}_1 - \bar{Y}_2$, $\mu_1 - \mu_2$ is hypothesized to equal zero, and

$$s_{\bar{D}} = \sqrt{\sum^b [(Y_1 - Y_2) - (\bar{Y}_1 - \bar{Y}_2)]^2/(b-1)b}$$

Squared, it can be written, with a little rearrangement,

$$t_s^2 = \frac{b(b-1)(\bar{Y}_1 - \bar{Y}_2)^2}{\sum^b [(Y_1 - Y_2) - (\bar{Y}_1 - \bar{Y}_2)]^2}$$

In a randomized block anova with $a = 2$ groups

$$F_s = \frac{MS_{\text{groups}}}{MS_{AB}}$$

where in our case, since we only have two groups,

$$MS_{\text{groups}} = b\sum^a (\bar{Y} - \bar{\bar{Y}})^2 = \frac{(\sum Y_1 - \sum Y_2)^2}{2b} = \frac{b}{2}(\bar{Y}_1 - \bar{Y}_2)^2$$

$$MS_{AB} = \frac{\sum_i^a \sum_j^b (Y_{ij} - \bar{Y}_i - \bar{Y}_j + \bar{\bar{Y}})^2}{b-1}$$

Here $\bar{\bar{Y}} = (\bar{Y}_1 + \bar{Y}_2)/2$ because of equal sample size, and likewise $\bar{Y}_j = (Y_{1j} + Y_{2j})/2$. Since there are only two groups, we may drop the summation sign over a and write out the two squared terms for each pair of observations:

$$MS_{AB} = \frac{1}{b-1}\sum_j^b \left\{\left(Y_{1j} - \bar{Y}_1 - \frac{Y_{1j} + Y_{2j}}{2} + \frac{\bar{Y}_1 + \bar{Y}_2}{2}\right)^2 + \left(Y_{2j} - \bar{Y}_2 - \frac{Y_{1j} + Y_{2j}}{2} + \frac{\bar{Y}_1 + \bar{Y}_2}{2}\right)^2\right\}$$

With a little rearrangement we find that

$$MS_{AB} = \frac{1}{b-1}\sum_j^b \left\{\left[\left(Y_{1j} - \frac{1}{2}Y_{1j} - \bar{Y}_1 + \frac{1}{2}\bar{Y}_1\right) + \left(-\frac{Y_{2j}}{2} + \frac{\bar{Y}_2}{2}\right)\right]^2 + \left[\left(Y_{2j} - \frac{1}{2}Y_{2j} - \bar{Y}_2 + \frac{\bar{Y}_2}{2}\right) + \left(-\frac{Y_{1j}}{2} + \frac{\bar{Y}_1}{2}\right)\right]^2\right\}$$

$$= \frac{1}{b-1}\sum_j^b \{[\tfrac{1}{2}(Y_{1j} - \bar{Y}_1) - \tfrac{1}{2}(Y_{2j} - \bar{Y}_2)]^2 + [-\tfrac{1}{2}(Y_{1j} - \bar{Y}_1) + \tfrac{1}{2}(Y_{2j} - \bar{Y}_2)]^2\}$$

The above two terms in square brackets are identical except for the sign, which is not important since the terms are squared. The constant $\frac{1}{2}$ may also now be factored out (becoming $\frac{1}{4}$). We can therefore simplify the above expression as follows:

$$MS_{AB} = \frac{1}{b-1}\,\tfrac{1}{4}\sum_j^b 2\,[(Y_{1j} - \bar{Y}_1) - (Y_{2j} - \bar{Y}_2)]^2$$

$$= \frac{1}{2(b-1)}\sum_j^b [(Y_{1j} - Y_{2j}) - (\bar{Y}_1 - \bar{Y}_2)]^2$$

F_s may now be written

$$F_s = \frac{\left[\dfrac{b}{2}(\bar{Y}_1 - \bar{Y}_2)^2\right]}{\dfrac{1}{2(b-1)}\displaystyle\sum_j^b [(Y_{1j} - Y_{2j}) - (\bar{Y}_1 - \bar{Y}_2)]^2}$$

After canceling the two 2's and moving the $(b - 1)$ into the numerator, the expression for F_s is identical to our expression for t_s^2 given above.

A1.8 Demonstration that Expression (14.5), the computational formula for the sum of products, equals $\sum(X - \bar{X})(Y - \bar{Y})$, the expression originally developed for this quantity.

All summations are over n items. We have

$$\begin{aligned}
\sum xy &= \sum(X - \bar{X})(Y - \bar{Y}) \\
&= \sum XY - \bar{X}\sum Y - \bar{Y}\sum X + n\bar{X}\bar{Y} && (\text{since } \sum\bar{X}\bar{Y} = n\bar{X}\bar{Y}) \\
&= \sum XY - \bar{X}n\bar{Y} - \bar{Y}n\bar{X} + n\bar{X}\bar{Y} && (\text{since } \sum Y/n = \bar{Y}, \\
& && \sum Y = n\bar{Y};\text{ similarly, } \sum X = n\bar{X}) \\
&= \sum XY - n\bar{X}\bar{Y} \\
&= \sum XY - n\bar{X}\sum Y/n \\
&= \sum XY - \bar{X}\sum Y
\end{aligned}$$

Similarly

$$\sum xy = \sum XY - \bar{Y}\sum X$$

and

$$\sum xy = \sum XY - \frac{(\sum X)(\sum Y)}{n} \tag{14.5}$$

A1.9 Derivation of computational formula for $\sum d_{Y\cdot X}^2 = \sum y^2 - ((\sum xy)^2/\sum x^2)$.

By definition, $d_{Y\cdot X} = Y - \hat{Y}$. Since $\overline{Y} - \hat{Y}$, we can subtract $\overline{Y}$ from both Y and $\hat{Y}$ to obtain

$$d_{Y\cdot X} = y - \hat{y} = y - bx \qquad \text{(since } \hat{y} = bx\text{)}$$

Therefore

$$\sum d_{Y\cdot X}^2 = \sum (y - bx)^2 = \sum y^2 - 2b \sum xy + b^2 \sum x^2$$

$$= \sum y^2 - 2 \frac{\sum xy}{\sum x^2} \sum xy + \frac{(\sum xy)^2}{(\sum x^2)^2} \sum x^2 = \sum y^2 - 2 \frac{(\sum xy)^2}{\sum x^2} + \frac{(\sum xy)^2}{\sum x^2}$$

or

$$\sum d_{Y\cdot X}^2 = \sum y^2 - \frac{(\sum xy)^2}{\sum x^2} \tag{14.7}$$

A1.10 Demonstration that the sum of squares of the dependent variable in regression can be partitioned exactly into explained and unexplained sums of squares, the cross products canceling out.

By definition (Section 14.5)

$$y = \hat{y} + d_{Y\cdot X}$$
$$\sum y^2 = \sum (\hat{y} + d_{Y\cdot X})^2 = \sum \hat{y}^2 + \sum d_{Y\cdot X}^2 + 2 \sum \hat{y} d_{Y\cdot X}$$

If we can show that $\sum \hat{y} d_{Y\cdot X} = 0$, then we have demonstrated the required identity. We have

$$\sum \hat{y} d_{Y\cdot X} = \sum bx(y - bx) \qquad [\text{since } \hat{y} = bx \text{ from Expression (14.3) and } d_{Y\cdot X} = y - bx \text{ from Appendix A1.9}]$$

$$= b \sum xy - b^2 \sum x^2$$

$$= b \sum xy - b \frac{\sum xy}{\sum x^2} \sum x^2 \qquad \text{(since } b = \sum xy / \sum x^2\text{)}$$

$$= b \sum xy - b \sum xy$$
$$= 0$$

Therefore $\sum y^2 = \sum \hat{y}^2 + \sum d_{Y\cdot X}^2$

or, written out in terms of variates,

$$\sum (Y - \overline{Y})^2 = \sum (\hat{Y} - \overline{Y})^2 + \sum (Y - \hat{Y})^2$$

A1.11 Proof that the variance of the sum of two variables is

$$\sigma_{(Y_1 + Y_2)}^2 = \sigma_1^2 + \sigma_2^2 + 2\rho_{12}\sigma_1\sigma_2$$

where σ_1 and σ_2 are standard deviations of Y_1 and Y_2, respectively, and ρ_{12} is the parametric correlation coefficient between Y_1 and Y_2.

If $Z = Y_1 + Y_2$, then

$$\sigma_Z^2 = \frac{1}{n} \sum (Z - \overline{Z})^2 = \frac{1}{n} \sum \left[(Y_1 + Y_2) - \frac{1}{n} \sum (Y_1 + Y_2) \right]^2$$

$$= \frac{1}{n}\sum\left[(Y_1 + Y_2) - \frac{1}{n}\sum Y_1 - \frac{1}{n}\sum Y_2\right]^2 = \frac{1}{n}\sum\left[(Y_1 + Y_2) - \bar{Y}_1 - \bar{Y}_2\right]^2$$

$$= \frac{1}{n}\sum\left[(Y_1 - \bar{Y}_1) + (Y_2 - \bar{Y}_2)\right]^2 = \frac{1}{n}\sum\left[y_1 + y_2\right]^2$$

$$= \frac{1}{n}\sum\left[y_1^2 + y_2^2 + 2y_1y_2\right] = \frac{1}{n}\sum y_1^2 + \frac{1}{n}\sum y_2^2 + \frac{2}{n}\sum y_1y_2$$

$$= \sigma_1^2 + \sigma_2^2 + 2\sigma_{12}$$

But, since $\rho_{12} = \sigma_{12}/\sigma_1\sigma_2$, we have

$$\sigma_{12} = \rho_{12}\sigma_1\sigma_2$$

Therefore,

$$\sigma_Z^2 = \sigma_{(Y_1+Y_2)}^2 = \sigma_1^2 + \sigma_2^2 + 2\rho_{12}\sigma_1\sigma_2$$

Similarly,

$$\sigma_D^2 = \sigma_{(Y_1-Y_2)}^2 = \sigma_1^2 + \sigma_2^2 - 2\rho_{12}\sigma_1\sigma_2$$

The analogous expressions apply to sample statistics. Thus

$$s_{(Y_1+Y_2)}^2 = s_1^2 + s_2^2 + 2r_{12}s_1s_2 \qquad (15.9)$$
$$s_{(Y_1-Y_2)}^2 = s_1^2 + s_2^2 - 2r_{12}s_1s_2 \qquad (15.9a)$$

A1.12 Proof that in a goodness of fit test with only two classes the deviations are always equal in magnitude but opposite in sign.

Let f_1 and f_2 be the two observed frequencies ($f_1 + f_2 = n$), and let $\hat{p}$ be the expected probability of the event which has an observed frequency of f_1.

The expected frequencies $\hat{f}_1$ and $\hat{f}_2$ would then be

$$\hat{f}_1 = n\hat{p}$$
$$\hat{f}_2 = n(1 - \hat{p}) = n - n\hat{p}$$

The deviations from expectation are

$$f_1 - \hat{f}_1 = f_1 - n\hat{p}$$

and

$$\begin{aligned} f_2 - \hat{f}_2 &= f_2 - n + n\hat{p} \\ &= (n - f_1) - n + n\hat{p} \qquad \text{(since } f_1 + f_2 = n,\ f_2 = n - f_1\text{)} \\ &= -f_1 + n\hat{p} \\ &= -(f_1 - n\hat{p}) \qquad \text{(and since } n\hat{p} = \hat{f}_1\text{)} \\ &= -(f_1 - \hat{f}_1) \end{aligned}$$

A1.13 Proof that Expression (16.1), $\sum^{2}(f_i - \hat{f}_i)^2/\hat{f}_i$ equals Expression (16.2), $(f_1 - \hat{p}n)^2/\hat{p}\hat{q}n$, or Expression (16.2a), $(f_2 - \hat{q}n)^2/\hat{p}\hat{q}n$.

We have

$$X^2 = \frac{\sum^{2}(f_i - \hat{f}_i)^2}{\hat{f}_i} = \frac{(f_1 - \hat{f}_1)^2}{\hat{f}_1} + \frac{(f_2 - \hat{f}_2)^2}{\hat{f}_2}$$

We have shown in Appendix A1.12 that $(f_2 - \hat{f}_2) = -(f_1 - \hat{f}_1)$. The squares of these terms are therefore equal and we can write

$$\frac{(f_1 - \hat{f}_1)^2}{\hat{f}_1} + \frac{(f_2 - \hat{f}_2)^2}{\hat{f}_2} = \frac{(f_1 - \hat{f}_1)^2}{\hat{f}_1} + \frac{(f_1 - \hat{f}_1)^2}{\hat{f}_2}$$

$$= \frac{(f_1 - \hat{f}_1)^2(\hat{f}_1 + \hat{f}_2)}{\hat{f}_1\hat{f}_2}$$

$$= \frac{(f_1 - \hat{f}_1)^2 n}{\hat{f}_1(n - \hat{f}_1)} \qquad (\text{since } \hat{f}_1 + \hat{f}_2 = n, \hat{f}_2 = n - \hat{f}_1)$$

$$= \frac{(f_1 - \hat{p}n)^2 n}{\hat{p}n(n - \hat{p}n)} \qquad (\text{since } \hat{f}_1 = \hat{p}n)$$

$$= \frac{(f_1 - \hat{p}n)^2}{\hat{p}(1 - \hat{p})n} \qquad (\text{by definition, } \hat{q} = 1 - \hat{p})$$

$$= \frac{(f_1 - \hat{p}n)^2}{\hat{p}\hat{q}n} \tag{16.2}$$

By replacing $(f_1 - \hat{f}_1)^2$ with $(f_2 - \hat{f}_2)^2 = (f_2 - \hat{q}n)^2$ in the numerator of the right side of the derivation, we could obtain

$$X^2 = \frac{(f_2 - \hat{q}n)^2}{\hat{p}\hat{q}n} \tag{16.2a}$$

A1.14 Derivation of computational formula for X^2 [Expression (16.4)] from Expression (16.3).

Expanding Expression (16.3),

$$X^2 = \sum^{a} \frac{(f_i - \hat{f}_i)^2}{\hat{f}_i}$$

$$= \sum^{a} \frac{f_i^2}{\hat{f}_i} + \sum^{a} \frac{\hat{f}_i^2}{\hat{f}_i} - 2\sum^{a} \frac{f_i\hat{f}_i}{\hat{f}_i}$$

$$= \sum^{a} \frac{f_i^2}{\hat{f}_i} + \sum^{a} \hat{f}_i - 2\sum^{a} f_i$$

But, since $\sum^{a} f_i = \sum^{a} \hat{f}_i = n$

$$X^2 = \sum^{a} \frac{f_i^2}{\hat{f}_i} - n \tag{16.4}$$

A1.15 Proof that the general expression for the G-test can be simplified to Expressions (16.6) and (16.7).

In general, G is twice the natural logarithm of the ratio of the probability of the sample with all parameters estimated from the data and the probability of the sample assuming the null hypothesis is true. Assuming a multinomial distribution, this ratio is

$$L = \frac{\dfrac{n!}{f_1!f_2!\cdots f_a!}p_1^{f_1}p_2^{f_2}\cdots p_a^{f_a}}{\dfrac{n!}{f_1!f_2!\cdots f_a!}\hat{p}_1^{f_1}\hat{p}_2^{f_2}\cdots \hat{p}_a^{f_a}}$$

$$= \prod_{i=1}^{a}\left(\frac{p_i}{\hat{p}_i}\right)^{f_i}$$

$$G = 2\ln L$$

$$= 2\sum^{a} f_i \ln\left(\frac{p_i}{\hat{p}_i}\right)$$

Since $f_i = np_i$ and $\hat{f}_i = n\hat{p}_i$,

$$G = 2\sum^{a} f_i \ln\left(\frac{f_i}{\hat{f}_i}\right) \tag{16.6}$$

If we now replace $\hat{f}_i$ by $n\hat{p}_i$,

$$G = 2\sum^{a} f_i \ln\left(\frac{f_i}{n\hat{p}_i}\right) = 2\left[\sum^{a} f_i \ln f_i - \sum^{a} f_i \ln \hat{p}_i - \sum^{a} f_i \ln n\right]$$

$$= 2\left[\sum^{a} f_i \ln f_i - \sum^{a} f_i \ln \hat{p}_i - n \ln n\right] \tag{16.7}$$

A1.16 Proof that Expression (16.16) for the McNemar test is a special case of Expression (16.17) for Cochran's Q-test with $a = 2$ columns (single repetition of the test).

$$Q = \frac{(a-1)\left[a\sum^{a}\left(\sum^{b} Y\right)^2 - \left(\sum^{a}\sum^{b} Y\right)^2\right]}{a\sum^{a}\sum^{b} Y - \sum^{b}\left(\sum^{a} Y\right)^2} \tag{16.17}$$

The data to be analyzed in this manner would have the following appearance arranged in a two-way table.

Individuals	*Tests*	
	1	*2*
1	1	1
2	0	0
3	0	1
4	1	0
5	1	1
.		
.		
.		
b	0	1

We let zero stand for one type of response and 1 for the other.

We could summarize these data in a 2×2 table, employing standard notation for the cell frequencies. However, we print $\mathfrak{a}$, $\mathfrak{b}$, $\mathfrak{c}$, and $\mathfrak{d}$, in script letters to avoid confusion with a and b referring to number of columns and rows in Expression (16.17).

		Results of first test	
	Type of response →	0	1
Results of second test	0	$\mathfrak{a}$	$\mathfrak{b}$
	1	$\mathfrak{c}$	$\mathfrak{d}$

The following quantities in Expression (16.17) can be expressed in terms of $\mathfrak{a}$, $\mathfrak{b}$, $\mathfrak{c}$, and $\mathfrak{d}$:

$$\sum^{a}(\sum^{b} Y)^2 = (\mathfrak{b} + \mathfrak{d})^2 + (\mathfrak{c} + \mathfrak{d})^2$$

since $\sum^{b} Y_1$ is the number of "1's" in test 1 which equals $\mathfrak{b} + \mathfrak{d}$, and $\sum^{b} Y_2$ is the number of "1's" in test 2 which equals $\mathfrak{c} + \mathfrak{d}$.

$$\sum^{a}\sum^{b} Y = \mathfrak{b} + \mathfrak{c} + 2\mathfrak{d}$$

This is the total number of "1's" in the table.

$$\sum^{b}(\sum^{a} Y)^2 = \mathfrak{b} + \mathfrak{c} + 4\mathfrak{d}$$

This equation is obtained as follows. $(\sum^{a} Y)^2$ will equal 1 for any individual responding 1, 0. There are $\mathfrak{b}$ such individuals. Similarly, $(\sum^{a} Y)^2$ will equal 1 for individuals responding 0, 1. There are $\mathfrak{c}$ such individuals. For individuals responding 1, 1, quantity $(\sum^{a} Y)^2 = 4$; and there are $\mathfrak{d}$ such individuals.

Using the terms developed above and remembering that $a = 2$, we can rewrite Q as

$$Q = \frac{(2 - 1)\{2[(\mathfrak{b} + \mathfrak{d})^2 + (\mathfrak{c} + \mathfrak{d})^2] - (\mathfrak{b} + \mathfrak{c} + 2\mathfrak{d})^2\}}{2(\mathfrak{b} + \mathfrak{c} + 2\mathfrak{d}) - (\mathfrak{b} + \mathfrak{c} + 4\mathfrak{d})}$$

$$= \frac{[2\mathfrak{b}^2 + 4\mathfrak{b}\mathfrak{d} + 2\mathfrak{d}^2 + 2\mathfrak{c}^2 + 4\mathfrak{c}\mathfrak{d} + 2\mathfrak{d}^2 - \mathfrak{b}^2 - \mathfrak{c}^2 - 4\mathfrak{d}^2 - 2\mathfrak{b}\mathfrak{c} - 4\mathfrak{b}\mathfrak{d} - 4\mathfrak{c}\mathfrak{d}]}{\mathfrak{b} + \mathfrak{c}}$$

$$= \frac{(\mathfrak{b}^2 + \mathfrak{c}^2 - 2\mathfrak{b}\mathfrak{c})}{\mathfrak{b} + \mathfrak{c}}$$

$$= \frac{(\mathfrak{b} - \mathfrak{c})^2}{\mathfrak{b} + \mathfrak{c}}$$

$$= \frac{(b - c)^2}{b + c} \tag{16.16}$$

APPENDIX 2

OPERATION OF DESK CALCULATORS

Facility in machine computation is an achievement that cannot be overemphasized. Two hours spent now, learning efficient and correct ways of operating a calculator, can save you many hours on your exercises, not to mention time saved during research later on. Beginners tend to get into simple but time-consuming machine habits and shy away from learning more intricate, but more efficient, operations. Yet much disappointment over incorrect results and many weary hours could be eliminated by an early acquisition of good computation habits. The exercises following the description of the features of the calculator are designed to give you an opportunity to do this. Study the particular model of desk calculator you will be using and locate the following features on it. You may have to consult the manufacturer's instruction booklet in order to ascertain whether the various features mentioned by us are present and to discover ways of carrying out various operations. This is especially true if you have an unusual model or a printing calculator in which the flow of operations differs appreciably from that of the rotary calculator described here.

A fully automatic calculator consists of two main parts, the body of the machine (containing the keyboard and the operational keys) and the carriage (containing the dials). The main *keyboard* serves to enter numbers into the machine. Some calculators have a second numerical keyboard for entering multipliers.

The *add bar* serves to add a number in the keyboard into the long (accumulating) dials of the carriage, while simultaneously a "1" will appear in the short (counter) dials indicating that a single machine cycle has been carried out. The *subtract bar* subtracts a number in the keyboard from a number in the long dials and turns the short dials in a negative direction. Since the calculator does not have a negative sign associated with either of the dials, negative numbers are shown in complement form (see below).

A *multiply bar* may or may not be present. If it is, this is an evolutionary relic from the semiautomatic machine. Its function is to permit the repeated addition or subtraction of items entered in the keyboard. On some machines a *repeat add key* when engaged changes the add and subtract bars into positive or negative multiply bars, respectively.

Automatic multiplication is carried out by entering the multiplier via the regular keyboard or a multiplier keyboard, entering the multiplicand into the main keyboard, and activating an *automatic multiplication key*. In a typical operation the long and short dials are first cleared and the product will appear in the long dials, the multiplier in the short dials. The multiplicand is lost during the operation, although most machines possess a key that will retain it in the keyboard if desired. Auxiliary keys necessary for automatic multiplication are: a correction key for removing errors in the multiplier before the multiplication is carried out, a locking device to enable employment of a constant multiplier without having to enter the same number time and again, and a squaring device which, by one means or another, employs the same number as a multiplier and multiplicand, the number having been entered in a keyboard only once. The *accumulative multiplier key* is an important one. It will carry out a multiplication as indicated above but will not clear the dials prior to the multiplication. This permits the accumulation of products of numbers in the long dials, and of multipliers in the short dials, a valuable feature for statistical work. The *negative accumulative multiplier key* will subtract the product from the long dials.

The *carriage shift keys* move the carriage in both directions to align numbers for a variety of operations.

Division is customarily accomplished by entering a dividend into the long dials through the main keyboard, following this up with entering the divisor in the same keyboard, and activation of the *division key*. Each machine has its own method of decimal point alignment to insure correct placing of the decimal point in the quotient. A feature built in automatically for most division work is the *ones removal* or *suppression* during division. When the dividend is placed into the long dials by means of the add bar the machine should ordinarily record a "1" in the short dial, one operating cycle having been completed. The quotient would be added to this "1," giving an erroneous result unless the "1" were removed during the division operation. Some machines accomplish this by having a special *dividend entry key* which places the dividend into the long dial but suppresses the "1" in the short dial. Other machines remove the "1" from the short dial upon activation of the division key before the actual division operation commences. A *dividend-divisor line-up* feature is an important improvement available on all of the recent models. When a large number, such as 1000, is to be divided by a small number, such as 2, machines not equipped with this feature would spin through 500 cycles to produce the correct quotient. To avoid this delay the operator has to line up the leading digit of the dividend with that of the divisor manually by means of the carriage shift. On machines equipped with a dividend-divisor line-up feature this process is carried out automatically, removing a common source of operator error and annoyance. The process of division can be interrupted by the *division stop key*, which is useful when the number of significant figures in the quotient already obtained is sufficient or when it is realized that the figures are in error after the division key has been actuated. When through oversight a dividend is divided by zero, the machine would start spinning interminably unless the dividend stop key is depressed. All machines have devices for *"negative" division*, i.e., for subtracting the quotient from the contents of the short dials.

The *clear keys* permit the manual or electric clearing of all dials and the keyboard together or separately. Separate *dial locks* exist for each set of dials to prevent the clearing of a particular one during a general clearing operation. They are extremely useful in accumulating long columns of numbers; without such locks accidental clearing might erase a number representing the result of a substantial amount of work. Long dials generally have a *dial split*, which permits half of the dial to be locked while the other half can be cleared out. All features, such as this one, which give flexibility to machine operation become very useful during a variety of computational procedures. All machines have *keyboard locks* of one type or another that prevent the clearing of keys depressed on the keyboard. The *counter control key* changes the direction of rotation of the short dials in doing a given mathematical operation without affecting the outcome of the operation in the long dials. As will be seen later this is useful in a variety of ways. The *nonentry key*, another useful feature, suppresses rotation of the short dials during addition, subtraction, and multiplication, and thus permits the saving of important intermediary products of a computation in the short dials without contamination from multipliers during subsequent multiplication. *Tabs* are present on most machines and these generally determine the number of decimal places shown as a result of division. They are similar in operation to tabs on a typewriter.

Special features found only on one or a few models include a square root device permitting the automatic computation of square roots. Without such a feature square roots have to be obtained by one of the several methods given below. An automatic square root feature increases the price of a machine materially and is probably worthwhile only for persons having very extensive and continuous statistical computations. A recent development is an added dial or other storage space inside the machine to serve as a memory. Such a feature enables intermediate products of the computation to be stored temporarily while other computations proceed. The number stored in the memory may be used subsequently in later computations. Such a memory is a very convenient feature, taking the place of a note pad at the operator's side and precluding the possibility of transcription error from the machine to the note pad and from the note pad back into the machine. Transfer devices from various dials also facilitate the work of the calculator operator. Transfers from the long dials to the short dials are easily accomplished by means of division by one. Several models have a transfer key from the long dials into the multiplier dials or into the keyboard which enables the operator to carry out chain multiplications of the type $A \times B \times C \times D \cdots$. Machines equipped with a memory also have keys for transferring a number into memory and from memory into the multiplier dials or the long dials.

Some general notes on desk calculator computing are presented below, and special pointers for desk calculator computation will be presented throughout the book as the various statistical methods are taken up.

After you have studied the particular model of desk calculator you will be using and located the features described above, it is strongly recommended that you do the following *machine practice exercises*. Before you start, make sure that the keyboard and all of the dials have been cleared.

Basic Mathematical Operations

Addition: $(a + b)$

Ascertain that the add key is set so that the keyboard clears after each addition. Add the following columns of numbers and check your answers.

1	2.4	837	3.39
10	4.4	537	0.16
32	0.6	404	2.68
94	1.0	51	4.29
99	2.7	755	9.48
44	6.8		
11	4.0	2584	20.00
38	4.3		
8	0.0		
37	4.5		
374	30.7	Grand Total = 3008.7	

Subtraction and conversion of a negative number in complement form into a positive number: $(a - b)$

Whenever possible, problems should be set up to avoid negative numbers, but if this is not convenient the following procedures should be learned. Negative numbers are indicated in complement form in the dials of a rotary calculator. For example, clear the dials, then subtract, say, 27. You should get 99. . .9973. (The exact number of 9's depends upon the number of dials, usually from 18 to 21, but the 9's will extend all the way to the left side of the long dials.) To convert this number back to its noncomplemented form one can do one of three things. (1) Starting at the extreme right mentally subtract the first nonzero digit, 3, from ten, the next and all of the following digits from 9, yielding 000027. (2) A quicker method for longer numbers is to copy the number into the keyboard, set the add-key for repeat add, and press the subtract bar once. The long dial should be zero if you copied correctly. Then press the subtract bar once more to yield the final answer. (3) If the calculator has a transfer multiplication key, the answer may be obtained by simply setting the counter control key to negative and pressing the transfer multiplication key. The answer will appear in the short dials. (Note that on some models a "9" will be left at the extreme left of the short dials, but this is to be ignored.) Printing calculators handle negative numbers directly, preceded by a minus sign or printed in a different color, avoiding work with complements.

Perform the indicated operations and convert the final answer from its complemented form if necessary.

289	6.5	−51	−.8
−4	−5.5	59	8.9
−98	6.6	−73	−5.9
−41	−3.0	52	2.3
−28	5.9	−27	−6.7
118	10.5	−40	−2.2
		(99 . . . 99960)	(99 . . . 9997.8)

Multiplication and decimal point alignment: $(a \times b)$

The exact steps for multiplication vary a bit on different calculators. Some have a separate keyboard for the multiplier, whereas in others one enters both the multiplier and the multiplicand in the main keyboard. On the latter type of calculator the first number entered is the multiplier, which is transferred to the multiplier dials by pressing the enter-multiplier key. On both types the multiplicand is then entered on the keyboard, followed by depression of the multiplication key.

Unless the counter control lever is set to minus or the nonentry key is activated, the multiplier will appear in the short dials after the multiplication is performed. This is a very important feature that is used extensively in statistical calculations as discussed in Sections 4.9 and 14.4. Perform the following operations, entering all numbers at the extreme right of the keyboard.

$17 \times 23 = 391 \qquad 122 \times 427 = 52{,}094 \qquad 69 \times 61 = 4209$

Note the difference in time it takes for the calculator to carry out 9×99999 and $99999 \times 9 = 899991$. (On most calculators you should use the smaller factor as the multiplier, just as you would normally do when multiplying by hand.)

The multiplication of numbers other than integers poses an additional problem—decimal point alignment. Some of the newest calculators possess truly automatic decimal point placement. Set a dial for the desired decimal point and your answer will be automatically expressed correctly. On other calculators one simply presses a tabulating key to indicate the location of the decimal point on the keyboard. The location of the decimal point is then automatically indicated on the long and short dials (as long as both factors are entered lined up identically around the same decimal point). On simpler calculators one must manually set the movable decimal point indicators, using the following rule. The number of dials to the right of the decimal place on the long dial is equal to the number of keyboard columns (or multiplier digits) to the right of the decimal point in the multiplier (hence, also in the short dials after multiplication) *plus* the number of keyboard columns to the right of the decimal in the multiplicand.

Example: $23.8 \times 0.131 = 3.1178$

One may either ignore the decimal points and multiply the terms as if they were whole numbers and then move the decimal point four places to the left in the product or one can set the decimal between, say, the third and fourth keyboard columns in both the multiplier and multiplicand. Then the decimal point in the product will be located between the sixth and seventh dials in the long dial. The advantage of this latter procedure is that if one has a long series of multiplications to perform on numbers of the same order of magnitude, once the decimal is correctly set for the first problem it will remain correct for all of the others (as long as the keyboard decimal position is not changed). This permits us to perform accumulative multiplications (either positive or negative) simply by using the accumulative multiplication key or the negative multiplication keys.

Do the following problems and check your answers.

Single multiplication

$15.1 \times 37 = 558.7$ $\quad$ $0.0136 \times 12.501 = 0.1700136$

$1986.12 \times 0.0134 = 26.614008$

Accumulative multiplication

$(17 \times 23) + (35 \times 15) + (11 \times 2) = 938$

$(15.1 \times 37) + (16.1 \times 17) + (99.0 \times 86) = 9346.4$

$(0.0124 \times 6.382) + (0.0264 \times 36.724) = 1.0486504$

Negative multiplication

$(16 \times 701) - (12 \times 42) = 10712$

$(361.3 \times 12.261) - (17.2 \times 1.111) - (15.1 \times 15.000) = 4184.2901$

$(13.92 \times 63.01) - (98.49 \times 17.69) = -865.1889$

The latter registers as the complement 99 . . . 99134.81110.

A special multiplicative operation used extensively in statistics is squaring. One can, of course, simply enter the number twice (as multiplier and multiplicand) but since squaring needs to be done so often, most calculators have special devices so that the number to be squared need be entered only once. Work the following problems.

$(37)^2 = 1369$ $(16.2)^2 = 262.44$ $(136.771)^2 = 18706.306441$

$(0.0046)^2 = 0.00002116$

Division: (a/b)

Division is performed similarly on all calculators. First, make sure the counter control key is set to positive and that the dials are cleared. Press the tabulating key that corresponds to the number of significant decimal places desired in the quotient (these are usually located in a row along the short dial). You should set it for two more decimal places than you want in the final answer and plan to round the quotient, since the answer given by the calculator is *not* rounded. Then enter the dividend into the keyboard, depress the enter dividend key, enter the divisor around the same decimal point, and depress the divide key. The quotient will now appear in the short dial with the decimal point corresponding to the tab key you pressed. If you divide a large number by a much smaller number on a calculator without an automatic division alignment feature, the division will take a long time. If this happens, or if you accidentally divide by zero, press the division stop key. If it is necessary to divide a large number by a very small number, it is best to ignore the decimal point, line the two numbers up on the keyboard, and then move the decimal in the quotient to the left by the number of places by which the divisor was shifted to line up the numbers.

Perform the following divisions and round to the number of significant decimal places indicated:

2 decimal places

$16/7 = 2.29$ $18{,}000/7 = 2571.43$

4 decimal places

$3.21/176.61 = 0.0182$ $0.00063/0.00725 = 0.0869$

Square roots: $(\sqrt{a})$

Some calculators have this feature built in so that one need only enter the number in the keyboard and depress a single button and the square root appears in the short dials. On most calculators one must either use special tables or calulate the square root directly. The tables and instructions can be obtained from the manufacturers of the calculators. We furnish such square root and cube root tables for calculating machines in Tables **A** and **B**, respectively. Instructions for their use are given with the tables. Lacking such tables, one can with a little practice learn to calculate a single square root by the direct method about as fast as one can search through a bookshelf to find the table. A direct method for calculating a square root on a desk calculator (called the method of successive odd integers) is as follows.

a. Clear the tab stops, set the counter control to negative, then set the add key for repeated addition, and shift the carriage all the way to the right.

b. Enter the number at the extreme left of the keyboard (ignoring decimal points), depress add key, and remove the 9's from the short dials. On some machines this can be accomplished by means of the dividend entry key. Clear the keyboard.

c. If the number has an odd number of digits to the left of the decimal point (or odd number of zeros to its right if the number is less than unity), then start step **d** with the keyboard column at the extreme left. If the number of digits to the left of the decimal point is even (or there are no zeros or an even number of them to the right of the decimal point), start step **d** with the keyboard column next to the leftmost one.

d. Subtract, successively, the odd numbers 1, 3, 5, 7, 9, 11, . . . , 19 *until* the long dial becomes negative (on some machines a bell will ring). For example, to find the square root of 12.3, we would successively subtract 1, 3, 5, and 7 from the units position in 12.3 which would give us 999630 . . . 0.

e. Depress the add bar once (giving us 000330 . . . 0 in the long dial and a 3 in the short dial in the present example).

f. Change the number in the keyboard (7) to the next lowest even number (6).

g. Shift the carriage one position to the left.

h. Now subtract successive odd numbers again, but this time with the units position starting in the column to the right of the previously used one until the long dial again becomes negative (subtract 61, 63, 65, 67, 69, and 71 yielding 9999340 . . . 0 and 0360 . . . 0 in the long and short dials).

i. Go back to steps **e** through **h**. (In the present example the repetition of step **e** would yield 0000050 . . . 0 and 0350 . . . 0 in the long and short dials. In **f** the number in the keyboard would become 070. After shifting in **g**, **h** would yield 99999799 and 0351, and in **e** we would have 0000050 . . . 0 and 03500 . . . 0.)

This process is continued until, after step **e**, the number in the short dial (the square root) is accurate to one-half the number of significant figures that we desire in the final answer. When this point is reached, press the divide key to double the number of significant figures in the answer. (Note this must be a *positive* division, hence on some models the counter control key must now be set to plus.) The final answer is 3.5071356 correct to 8 figures. The answer can be checked by squaring, which should yield the original number (within accuracy of the square root). Practice on the following problems.

$$\sqrt{37} = 6.08276 \qquad \sqrt{370} = 19.2354 \qquad \sqrt{15.98} = 3.99750$$
$$\sqrt{13732} = 117.184 \qquad \sqrt{11.752} = 3.42812 \qquad \sqrt{0.00261} = 0.0510882$$

Other Mathematical Operations Frequently Performed

Division of a sum or difference: $[(a \pm b)/c]$
Enter a into the keyboard and depress enter dividend key (tab stop should have been set for the desired number of significant decimal places), then add (or subtract) b. Enter c into the keyboard and depress divide. Make certain short dials are clear or will clear before division. The answer will appear in the short dials.

Problems

$$\frac{17 + 24}{71} = 0.577 \qquad \frac{13.1 - 16.24}{11.73} = -0.268$$

(Note that in the second problem you must change the signs in the numerator since you cannot divide a complemented number.)

Division by a sum or difference: $[a/(b \pm c)]$
Only on calculators with a "memory" device for transferring numbers from the long dial into a storage register can one perform the operation without the operator having to copy the result $(b \pm c)$ and then enter dividend a followed by the divisor $(b \pm c)$.

Problems

$$\frac{17}{24 + 39} = 0.270 \qquad \frac{171.361}{37.94 + 115.6} = 1.116$$

Division of a product: $[(a \times b) \pm c]$
This can be done on all machines by multiplying a times b and then dividing by c. On some machines the short dials must be cleared to receive the quotients. Note that the location of the decimal place in the quotient is determined by its location on the keyboard, not by the tab keys.

Problems

$$\frac{73 \times 12}{29} = 30.21 \qquad \frac{11.231 \times 15.2697}{17.125} = 10.0142$$

Division by a product: $[a/(b \times c)]$
As with a sum, there is no short cut except for machines that can transfer a number from the long dial $(b \times c)$ to the memory and then enter a dividend a followed by division by the number in the memory $(b \times c)$.

Problems

$$\frac{11}{23 \times 34} = 0.0141 \qquad \frac{23.569}{19.001 \times 16.2} = 0.0766$$

Multiplication by a constant: $[k(a, b, c, \ldots)]$
The basic procedure for this type of problem is to lock in k as a constant multiplier. The other factors are then entered into the keyboard and multiplied. On calculators with a memory unit the constant can be entered into the memory and then each of the factors multiplied by the constant in the memory. For a problem of the type $k(a \pm b \pm c \ldots)$ one would ordinarily first evaluate $(a \pm b \pm c \ldots)$ and then multiply this quantity by k. On calculators with a transfer key one may transfer the sum to the multiplier and then multiply it by k in the keyboard. For only a few terms $(a \pm b \pm c \ldots)$ when using a machine without a transfer key it may be more efficient to enter k as a constant multiplier and the other factors into the keyboard, multiplying each accumulatively (positively or negatively, as indicated).

Problems

$13 \times 23 = 299$ $\qquad 13.216 \times 17.301 = 228.6500160$
$13 \times 64 = 832$ $\qquad 13.216 \times 0.0076 = 0.1004416$
$13 \times 171 = 2223$
$13(23 + 64 + 171) = 3354$ $\qquad 13.216(17.301 + 0.0076) = 228.7504576$

Division by a constant: $[(a, b, c, \ldots)/k]$
In a calculator without a memory the quantity 1/k should be computed and then used as a constant multiplier, as above. In calculators with a memory the divisor could be placed in the memory and then each of the numbers a, b, $c \ldots$ can be divided by it.

Problems

$36/11 = 3.273$ $\qquad 3.9/17.2 = 0.227$
$17/11 = 1.545$ $\qquad 2.5/17.2 = 0.145$
$52/11 = 4.727$ $\qquad 0.32/17.2 = 0.0186$

Multiplication of a product: $(a \times b \times c)$
In calculators with a transfer key it is possible to transfer a number in the long dials (which can be the result of a previous calculation, $a \times b$) into the multiplier and multiply it by a number in the keyboard, c.

Problems:

$17 \times 15 \times 26 = 6630$ $\qquad 2.1 \times 3.6 \times 17.5 = 132.3$

Addition (or subtraction) of a quotient to a constant: $[a \pm (b/c)]$
This is a useful trick that can be done on all machines. Set the tab stops for the same number of decimal places in the short dials that you will use in the

keyboard. Enter a into the multiplier and depress the multiply key. (a should now appear properly lined up in the short dials.) Lock the short dials and enter dividend b and then divide (either positively or negatively) by c. The answer is in the short dials. Note, b could be also the result of some other computation in the long dials, e.g., $\left(a - \frac{b^2}{c}\right)$ or $[a + (b + d + e)/c]$. When calculating b^2, or $b + d + e$, etc., the nonentry key must be activated to avoid destroying a, which had previously been placed in the short dials.

Problems

$$10 - \frac{20}{5} = 6 \qquad 3.8 + \frac{2.6}{0.8} = 7.05 \qquad 7.4 - \frac{(1.2)^2}{5} = 7.112$$

$$4.6 + \frac{3.1 + 2.8}{6.3} = 5.5365$$

Summation of quotients: $[(a/b) + (c/d) + (e/f) + \cdots]$
The above procedure can easily be extended to add (or subtract) the quotients of a series of divisions.

Problems

$$\frac{7}{2} + \frac{1.5}{3.2} + \frac{6}{13} = 4.4303 \qquad \frac{(6)^2}{13} + \frac{(17)^2}{1.8} - \frac{(4.5)^2}{16.2} = 162.0748$$

$$\frac{(10)^2}{11.5} + \frac{(13)^2}{12.6} + \frac{(16)^2}{14.9} - 39 = 0.2896$$

APPENDIX 3 STATISTICAL COMPUTER PROGRAMS

Introduction

These programs are furnished to provide a set of basic statistical routines. The programs have been written in FORTRAN IV in as simple and general a form as possible so that they can be directly implemented at most computation centers without any reprogramming. For this reason certain more sophisticated ways of programming have been avoided, as have been specialized features of our local version of FORTRAN IV.

The first section presented below is a glossary of some commonly used terms in the separate write-ups for each program. These terms must be clearly understood before any one write-up is consulted. Program write-ups for the 16 programs follow in numerical sequence. They describe the statistical computations carried out by each program, the preparation of data for input to the program, and the appearance of the output. Most programs need one or more functions and subroutines. These subprograms are identified by a code word—for example, COMOUT—and are listed for each program at the beginning of the FORTRAN print-up. Since most subprograms are used in more than one program, they are presented in alphabetical sequence at the end of the numbered main programs.

To solve a problem by means of these programs, a punched-out main FORTRAN deck should be followed by all necessary subprograms and these in turn must be followed by all the required input cards. The entire deck must be presented at the computation center. At most centers it will also be necessary to insert systems control cards. These differ for every center, and no attempt therefore has been made to describe such cards here. Instructions for preparing such cards are usually quite simple and are furnished to the user of every computation center.

The limits for numbers of variates or classes specified in the various programs can be increased for given computer installations with sufficient memory. Usually the DIMENSION statements in the programs must be modified. Some of the limits specified in the programs below may have to be reduced to adapt them to a small computer.

All programs have been checked out on the version of FORTRAN IV for the GE 625 system. Users employing these programs on other computers may obtain slightly different numerical answers because of the different word lengths of these other computers. Dr. Rohlf, who wrote the programs, will appreciate hearing from any users who discover errors in the programs.

Persons planning to implement most or all of the 16 programs might wish to obtain a punched deck containing all the programs rather than run the risk of introducing errors during repunching from the print-ups below. Persons interested in purchasing such a deck should write to Mrs. Patricia Rohlf, Route 1, Box 278, Lawrence, Kansas 66044.

The assistance of Mr. Koichi Fujii in the writing of the programs is much appreciated. Edwin H. Bryant, David R. Fisher, Koichi Fujii, John L. Kishpaugh, and David Wool helped to check out the programs.

Glossary of technical terms employed in the program write-ups

b. Columns designated by b must be left blank.

Comment cards. These cards, interspersed in the FORTRAN programs and identified by a C in column 1, are not read by the compiler program. Comment cards are used for labeling programs and inserting explanatory comments about the program. Thus the label CA3.1 identifies program A3.1, and CKSRD identifies subroutine KSRD.

Ignored. Card columns not considered by the program. These can be employed by the user for identifying information if he so wishes.

Integer number. A number not involving a decimal fraction, or a decimal point. Thus 3, 123, or 10372 are integer numbers. They must always be entered right-adjusted (*q.v.*) into a card field.

Real number. A number containing a decimal point. Examples are 12.3, 0.23, 3., .001. Note in the third example that when the program write-up specifies that a whole number be expressed as a real number, this can be done easiest by terminating the number with a decimal point. Real numbers can be placed anywhere within an allocated card field and unused card columns are left blank. Thus .001 can be placed into a seven-digit field as . 0 0 1 b b b, b b . 0 0 1 b, or b b b . 0 0 1.

Right-adjusted. Means that a number with fewer digits than a card field allocated to it should be aligned with the right end of this field. For example, if the number 123 has to be entered in a card field allowing for six digits, it would be punched b b b 1 2 3 to be right-adjusted.

Transformation code. The following set of arbitrary codes (integer numbers, *q.v.*) has been designated to provide the indicated transformations of real numbers.

0 — no transformation
1 — $|Y|$
2 — $\sqrt{Y}$
3 — $\sqrt{(Y + ½)}$
4 — Y^2
5 — $\log_{10} Y$
6 — $\log_{10} (Y + 1)$
7 — $\arcsin (\sqrt{Y})$ expressed in degrees, where Y is assumed to be a proportion between zero and one
8 — $\arcsin (\sqrt{Y/100})$ expressed in degrees, where Y is a percentage between zero and 100 percent
9 — $1/Y$

Note: These transformations are performed by the function subprogram TRANSG. All of the programs in Appendix 3 allow the use of these transformations except programs A3.3, A3.11, A3.14, A3.15, and A3.16. The codes should be entered in the card columns specified for each appropriate program. In programs with the option to read in summary statistics, such as means or variances, the transformation code is ignored.

Variable format card. This card contains a series of codes to specify the format of the input data. The following format specifications are likely to be employed by users of our programs.

F-format code. This code is for real numbers (*q.v.*) which are specified as $nFw.d$, where n is an integer indicating how many times this specification is to be repeated in sequence, w is an integer specifying the width of the entire card field allocated to one number, and d is an integer specifying the number of card columns in the field assumed to be to the right of the decimal point (if the input data lack punched-in decimal points). Thus if we wish to allocate ten card columns to one field and assume the decimal point to be between the second and third columns from the right end of the field, we would specify the code $F10.2$. Three real numbers in sequence, not exceeding five digits and lacking decimal fractions, could be coded $3F5.0$ into three fields of five columns. If some numbers are negative, the width of the field has to include a column for a preceding minus sign. If the input data have decimal points punched-in, the value of d in the F-format code is ignored.

I-format code. This code is for integer numbers (*q.v.*) which are specified as nIw, where n is an integer indicating how many times this specification is to be repeated in sequence, and w is an integer specifying the width of the entire card field allocated to one number. Thus if we wish to allocate seven card columns to one field, we would specify the code $I7$. If some numbers are negative, the width of the field has to include a column for a preceding minus sign. Decimal points must not be punched into fields specified by the I-code.

X-format code. This code specifies that certain card columns are to be ignored. It is stated as nX, where n is the number of columns to be ignored.

The variable format card consists of the format-specification codes enclosed within parentheses. There is no prescribed position for punching the codes on the card. Several examples should familiarize the reader with the correct formulation of a variable format card. The variable format card ($7F10.3$) specifies that each card would have seven fields of ten digits, where the decimal point is assumed to lie between the third and fourth digit from the right. Thus a card punched as $\underline{b}\,\underline{b}\,\underline{b}\,\underline{b}\,\underline{1}\,\underline{2}\,\underline{3}\,\underline{6}\,\underline{7}\,\underline{5}\,|\,\underline{b}\,\underline{b}\,\underline{b}\,\underline{b}\,\underline{b}\,\underline{2}\,\underline{2}\,\underline{1}\,\underline{2}\,\underline{0}\,|\,\underline{b}\,\underline{b}\,\underline{b}\,\underline{b}\,\underline{b}\,\underline{-}\,\underline{0}\,\underline{7}\,\underline{2}\,\underline{3}\,|\,\underline{b}\,\underline{b}\,\underline{b}\,\underline{b}\,\underline{4}\,\underline{2}\,\underline{0}\,\underline{0}\,\underline{0}\,\underline{0}\,|\,\underline{b}\,\underline{b}\,\underline{b}\,\underline{1}\,\underline{6}\,\underline{7}\,\underline{9}\,\underline{2}\,\underline{3}\,\underline{0}\,|\,\underline{b}\,\underline{b}\,\underline{b}\,\underline{b}\,\underline{6}\,\underline{2}\,\underline{4}\,\underline{0}\,\underline{0}\,\underline{2}\,|\,\underline{b}\,\underline{b}\,\underline{b}\,\underline{b}\,\underline{b}\,\underline{b}\,\underline{b}\,\underline{4}\,\underline{6}\,\underline{5}\,|\,\underline{b}\,\underline{b}\,\underline{b}\,\underline{b}\,\underline{b}\,\underline{b}\,\underline{b}\,\underline{b}\,\underline{1}\,\underline{2}\,|$ would be read by the computer program as 123.675, 22.120, −0.723, 420.000, 1679.230, 624.002, 0.465. Note that the last ten columns on the card are ignored by the program. They can be used to identify the card (as shown by the code number 12 in the example). The leading zero in a number such as 0.723 can be punched-in or left out, as preferred.

Many computer users, especially beginners, would prefer to punch decimal points into their data. This enables checking of the input data in familiar form for correct placement of the decimal point. Since the decimal point in the data overrides the value of d in the F-format code, the latter may be specified simply as ($7F10.0$).

A second example is the variable format card ($10X$, $I5$, $F10.4$). This specifies that on each card the first ten columns are ignored (probably used for identification), that columns 11–15 contain an integer number and that columns 16–25 contain a real number with the decimal point between card columns 21 and 22. Thus a card punched as b b b b b b b 2 0 1 | b b 1 2 2 | b b b b 1 2 6 2 3 1 | b b b b b b 2 7 5 6 | b b b b 8 9 . 3 4 2 | b b b . . . would be read by the program as 122, 12.6231. Note that the first ten columns on the card are ignored and that other data on the card past the end of the format specifications (column 25) are also ignored. In this way specified data can be extracted from cards containing a variety of data.

Further refinements of variable format specifications can be found in any FORTRAN manual.

A3.1 Basic statistics for ungrouped data.

This program reads in samples of ungrouped continuous or meristic variates, optionally performs transformations on these data, and outputs various basic statistics for each sample.

INPUT

Three types of cards are used.

Card Type 1. *Input parameters* (one card per sample).

Card columns	
1–5	Ignored.
6–10	Sample size, n (≤ 1000; as an integer number).
11–14	b b b b
15	Transformation code (defined in the glossary).
16–20	$\alpha > 0.0001$ (as a real number) for computing $100\,(1-\alpha)\%$ confidence intervals.
21–25	$t_{\alpha[n-1]}$ (as a real number) from a two-tailed t-table such as Table **Q**, where n is sample size.
26–80	Ignored.

Card Type 2. *Variable format card* (one card per sample) for input data (real numbers only).

Card Type 3. *Data cards* (variable number of cards per sample) punched as specified by the variable format card (card type 2).

OUTPUT

The output consists of a table of the various statistics computed—mean, median, variance, standard deviation, coefficient of variation, g_1, g_2, and the Kolmogorov-Smirnov statistic $D_{\max}$ resulting from a comparison of the observed sample with a normal distribution based on the sample mean and variance—followed by their standard errors and $100(1 - \alpha)\%$ confidence intervals where applicable.

The number of observations, n, and the transformation code are reproduced with the output.

Test data

The aphid femur length data from Box 2.1 can be used as test data for this program. The input and output are shown below.

```
                                  INPUT
      25     0 .05   2.06                   BOX 2.1


(10F5.1)
 3.3  3.5  3.6  3.6  3.6  3.6  3.8  3.8  3.8  3.8
 3.9  3.9  3.9  4.1  4.1  4.2  4.3  4.3  4.3  4.3
 4.4  4.4  4.5  4.7  4.4
```

```
                                  OUTPUT


         BASIC STATISTICS
   N=     25        0 CLASSES   TRANSFORMATION CODE =    0


                  STATISTIC    STAND.ERROR        CONFIDENCE LIMITS
                                                  (95.00,PER CENT)

 MEAN              4.00400        0.07314       3.85333    4.15467
 MEDIAN            3.90000        0.09167       3.71117    4.08883
 VARIANCE          0.13373
 STAND.DEV.        0.36570
 COEFF. VAR        9.13326        1.30237       6.45038   11.81614
 G1               -0.01807        0.46368      -0.92707    0.89093
 G2               -0.94508        0.90172      -2.71281    0.82265
 K-S DMAX          0.13194
```

A3.2 Basic statistics for data grouped into a frequency distribution.

This program is similar to program A3.1, but is intended for data grouped into a frequency distribution.

INPUT

Three types of cards are used.

Card Type 1. *Input parameters* (one card per sample).

Card columns

1–5	Number of classes (≤ 30; as an integer number).
6–10	Ignored.
11–80	Same parameters as in corresponding columns of program A3.1 (see description there).

Card Type 2. *Variable format card* (one card per sample) for input data. Format must be specified for class marks and frequencies in that order (both stated as real numbers).

Card Type 3. *Frequency distribution cards* (variable number of cards per sample). Class marks Y_i and the frequencies of each of the classes f_i punched in that order as specified by the variable format card (card type 2). There will be one card for each class in a frequency distribution.

OUTPUT

The output of program A3.2 is identical to that of A3.1 except that it first outputs the observed frequency distribution, including the upper class limits for each class, and the expected frequency $\hat{f}_i$ (FHAT) for each class assuming a normal distribution with $\mu = \bar{Y}$ and $\sigma = s$.

Test data

The birth weight data from Box 4.3 can be used as test data for this program. The input and output are shown below.

INPUT

```
   15          0   .05 1.96              BOX 4.3
 (F6.1,F5.0)
  59.5    2
  67.5    6
  75.5   39
  83.5  385
  91.5  888
  99.5 1729
 107.5 2240
 115.5 2007
 123.5 1233
 131.5  641
 139.5  201
 147.5   74
 155.5   14
 163.5    5
 171.5    1
```

OUTPUT

Y	UPPER CLASS LIMIT	F	FHAT
59.50000	63.50000	2.	3.0
67.50000	71.50000	6.	19.4
75.50000	79.50000	39.	97.5
83.50000	87.50000	385.	350.5
91.50000	95.50000	888.	899.6
99.50000	103.50000	1729.	1648.5
107.50000	111.50000	2240.	2157.6
115.50000	119.50000	2007.	2017.1
123.50000	127.50000	1233.	1347.0
131.50000	135.50000	641.	642.4
139.50000	143.50000	201.	218.8
147.50000	151.50000	74.	53.2
155.50000	159.50000	14.	9.2
163.50000	167.50000	5.	1.1
171.50000	175.50000	1.	0.1

BASIC STATISTICS

N= 9465 15 CLASSES TRANSFORMATION CODE = 0

	STATISTIC	STAND.ERROR	CONFIDENCE LIMITS (95.00,PER CENT)	
MEAN	109.89958	0.13973	109.62570	110.17345
MEDIAN	109.51250	0.17512	109.16926	109.85574
VARIANCE	184.80101			
STAND.DEV.	13.59415			
COEFF. VAR	12.36961	0.09127	12.19073	12.54850
G1	0.18936	0.02517	0.14001	0.23871
G2	0.08913	0.05034	-0.00956	0.18782
K-S DMAX	0.01194			

A3.3 Goodness of fit to discrete frequency distributions.

Options are provided for the following computations.

1) Compute a binomial or Poisson distribution with specified parameters.

2) Compute the deviations of an observed frequency distribution from a binomial or Poisson distribution of specified parameters or based on appropriate parameters estimated from the observed data. A G-test for goodness of fit is carried out.

3) A series of up to 10 observed frequency distributions may be read in and individually tested for goodness of fit to a specified distribution, followed by a test of homogeneity of the series of observed distributions.

4) A specified expected frequency distribution (other than binomial or Poisson) may be read in and used as the expected distribution. This may be entered in the form of relative expected frequencies or simply as ratios (for example, 1:2:1). The maximum number of classes for all cases is thirty. In the case of binomial and Poisson, the class marks cannot exceed $Y_i = 29$.

INPUT

Four types of cards are used.

Card Type 1. *Input parameters* (one card per problem).

Card columns	
1–3	b b b
4–5	Set zero (00) if no observed frequency distribution is to be read in. Or punch (integer number) number of observed frequency distributions to be read in (up to 10). For example, punch 01 if a single observed frequency distribution is to be read in.
6–9	b b b b
10	0 if specified expected distribution (other than binomial or Poisson) is to be read in for comparison with the observed data. 1 computes expected binomial distribution. 2 computes expected Poisson distribution.
11–14	b b b b
15	0 if the parameters for the expected distribution are to be read in on cards. Also set to 0 if column 10 was punched 0. 1 if the parameters for the expected distribution are to be estimated from the observed frequency distribution.
16–18	b b b
19–20	Number of classes (≤ 30; integer number) in the observed frequency distribution (must be the same for all of the observed frequency distributions to be read in, if there are more than one). In the binomial and Poisson distributions we shall start with class mark $Y_i = 0$ in all cases. The number of classes to be specified may be greater than the highest observed class mark. This is so because we may wish to test deviations between zero observed frequencies against expected frequencies for classes above the highest observed class mark. Given the space limitations of the program as written here, 30 classes (class mark $Y_i = 29$) is the upper limit for this parameter.
21–23	b b b
24–25	k (= sample size in the binomial) entered as an integer number in those cases where expected binomial frequencies are to be computed. In other cases, leave blank.

Card Type 2. *Specified expected frequency distribution* (other than binomial or Poisson). If such a distribution is to be read in (0 in column 10 of card type 1), prepare one or more cards in the following manner.

For relative expected frequencies punch as follows.

Card columns	
1–10	First relative expected frequency, $\hat{f}_{rel}$, punched as a real number.
11–20	Second relative expected frequency in the same format.

Card columns (contd.)

21–30	Third relative expected frequency in the same format.
.	.
.	.
.	.

If more than eight expected classes must be read in, the expected frequencies are continued on a second card (and on succeeding cards, if necessary).

For ratios such as are used in genetics (e.g., 9:3:3:1) these cards are punched as follows.

Card columns

1–10	9 . b b b b b b b b
11–20	3 . b b b b b b b b
21–30	3 . b b b b b b b b
31–40	1 . b b b b b b b b
41–80	Ignore.

Different ratios could be defined following the example given above. If the ratios involve more than eight classes, a second card is added for the remaining classes (and succeeding cards, if necessary).

Card Type 3. *Parameters for binomial and Poisson distributions.* Card to be included if such parameters are to be read in (1 or 2 in column 10, and 0 in column 15 of card type 1).

Card columns

1–9	b b b b b b b b b
10	1
11–20	$\hat{p}$ (parametric proportion of events of the first kind), or μ (parametric mean), respectively, for binomial or Poisson distributions punched as real numbers.
21–80	Ignored.

Card Type 4. *Observed frequency distributions* (if needed, as shown by nonzero integer in columns 4–5, card type 1). There will be one or two cards for each observed frequency distribution prepared as follows.

Card columns

1–5	First observed frequency (integer number).
6–10	Second observed frequency (integer number).
.	.
.	.
.	.
71–75	Last observed frequency.
76–80	Ignored.

There will be a second card per frequency distribution if there are more than fifteen classes.

OUTPUT

The output is largely self-explanatory. The input parameters and frequencies are reproduced in the output. If observed frequencies are present, the results of a *G*-test are printed. If several samples of observed frequencies have been read in to be tested against the same expected frequencies, a pooled observed frequency distribution is also printed, as well as the results of a *G*-test for heterogeneity. FHAT = $\hat{f}$, FHATR = $\hat{f}_{rel}$.

Test data

Owing to the several possible combinations of the various options, three sets of test data are featured for this program. The first example is a goodness of fit to an 18:6:6:2:12:4:12:4 ratio for the data in Box 16.1, the second is a goodness of fit test to a binomial distribution ($\hat{p}$ to be estimated from the observed data in Box 16.2), and the third example is the replicated goodness of fit test to a 9:7 ratio in Box 16.4. The input and output are shown below. Only the beginning and the end of the output for the third example are shown to conserve space.

INPUT

```
 1     0     0     8                                          BOX 16.1
18.         6.           6.          2.          12.          4.          12.          4.
63    31    28    12    39    16    40    12
 1     1     1    13    12                      BOX 16.2
 7    45   181   478   829  1112  1343  1033   670   286   104    24     3
 8     0     0     2                                          BOX 16.4
9.          7.
33    47
77    43
10    96
92    58
51    31
48    61
70    42
35    66
```

OUTPUT

```
NO. OF FREQ. DIST. =  1      DIST. CODE = 0
ESTIMATION CODE = 0 NO. OF CLASSES = 8

ARBITRARY DIST.
3.00000    6.00000    6.00000    2.00000   12.00000    4.00000   12.00000    4.00000


    Y          F         FHAT       DIFF

    0.        63.        67.8       -4.8
    1.        31.        22.6        8.4
    2.        28.        22.6        5.4
    3.        12.         7.5        4.5
    4.        39.        45.2       -6.2
    5.        16.        15.1        0.9
    6.        40.        45.2       -5.2
    7.        12.        15.1       -3.1
```

```
G-TEST STATISTIC =      8.824, DF =   7

NO. OF FREQ. DIST. =  1      DIST. CODE = 1
ESTIMATION CODE = 1 NO. OF CLASSES =13

BINOMIAL DIST. K =  12., P =    0.48078
```

Y	F	FHAT	DIFF
0.	7.	2.3	4.7
1.	45.	26.1	18.9
2.	181.	132.8	48.2
3.	478.	410.0	68.0
4.	829.	854.2	-25.2
5.	1112.	1265.6	-153.6
6.	1343.	1367.3	-24.3
7.	1033.	1085.2	-52.2
8.	670.	628.1	41.9
9.	286.	258.5	27.5
10.	104.	71.8	32.2
11.	24.	12.1	11.9
12.	3.	0.9	2.1

```
G-TEST STATISTIC =     97.008, DF =  11

NO. OF FREQ. DIST. =  8      DIST. CODE = 0
ESTIMATION CODE = 0 NO. OF CLASSES = 2

ARBITRARY DIST.
9.00000   7.00000
```

Y	F	FHAT	DIFF
0.	83.	73.1	9.9
1.	47.	56.9	-9.9

```
G-TEST STATISTIC =      3.101, DF =   1
```

Y	F	FHAT	DIFF
0.	77.	67.5	9.5
1.	43.	52.5	-9.5

```
G-TEST STATISTIC =      3.112, DF =   1
```

⋮

Y	F	FHAT	DIFF
0.	85.	84.9	0.1
1.	66.	66.1	-0.1

```
G-TEST STATISTIC =      0.000, DF =    1

POOLED DISTRIBUTION

     Y        F        FHAT       DIFF

     0.      616.      596.2      19.7
     1.      444.      463.7     -19.7

G-TEST STATISTIC =      1.500, DF =    1

HETEROGENEITY G  =     16.514, DF =    7
```

A3.4 Single classification and nested anova.

This program performs either a single classification or a k-level nested analysis of variance following the techniques presented in Chapters 9 and 10. The basic anova table as well as the variance components are computed. The program allows for unequal sample sizes at any level.

INPUT

Four types of cards are used.

Card Type 1. *Input parameters* (one card per analysis).

Card columns

1–3	b b b
4–5	The number of levels, k (≤ 10; integer number). For a single classification anova $k = 1$.
6–9	b b b b
10	Transformation code (defined in the glossary).
11–80	Ignored.

Card Type 2. *Variable format card* (one card per analysis) for input data (real numbers only). The format must allow for $k + 1$ integer fields followed by a specification sufficient to handle the largest sample (see illustration under card type 3, below).

Card Type 3. *Data cards* (variable number of cards per analysis) punched as specified by the variable format card. For a single classification anova the first integer field gives the number of the group listed on the card. The second integer field is the sample size n of that group. The next n fields (which may be on additional cards if n is large) contain the n variates for this group. In nested anovas, data for each lowest level subgroup must start on a new card. There will be k integer fields for a k-level nested anova. These fields identify numerically the number of the highest level group, number of the group at the next level, . . . until the number of the group at the lowest level. The next integer field [the $(k + 1)$th field] is the size n of this subgroup. The next n fields (which may be on additional cards if n is large) contain the n variates for this subgroup. For example, the data for the second subgroup in the third group in the two-level nested anova in Box 10.4 could be punched as

b 3 b 2 b 4 b 4 5 . b 3 3 . b 4 0 . b 4 6 . b b . . . b

if the variable format was (2*I*2, *I*2, 5*F*4.0). The 5*F*4.0 is needed, since the largest subgroup contains five variates.

Card Type 4. *End of data card.* This is a blank card which indicates to the program that all the data have been read in.

OUTPUT

The input parameters are reproduced in the output and followed by a standard anova table giving *SS*, *df*, *MS*, and F_s. For nested anovas with unequal sample sizes, synthetic mean squares and their approximate degrees of freedom (using Satterthwaite's approximation) are given below each *MS* and *df*. Each F_s is the result of dividing the *MS* on its line by the synthetic *MS* from the level below it. When sample sizes are equal the synthetic mean squares and their degrees of freedom are the same as their ordinary counterparts, but are printed out nevertheless by the program. No pooling is performed. The anova table is followed by a list of the estimated variance components expressed both in the original units and as percentages, which is in turn followed by a table of the coefficients of the expected mean squares.

Test data

The tick data of Box 9.1 (single classification anova) and the blood *p*H data of Box 10.4 (a two-level nested anova) may be used as test data for this program. The input and output are shown below.

INPUT

```
   1     0                                      BOX 9.1
(I2,I3,13F5.0)
1  8   380  376  360  368  372  366  374  382
2 10   350  356  358  376  338  342  366  350  344  364
3 13   354  360  362  352  366  372  362  344  342  358  351  348  348
4  6   376  344  342  372  374  360

   2     0                       BOX 10.4
(1X,2I2,I2,2X,5F3.0)
 1 1 4    48 48 52 54
 1 2 4    48 53 43 39
 2 1 5    45 43 49 40 40
 2 2 4    50 45 43 36
 3 1 4    40 45 42 48
 3 2 4    45 33 40 46
 3 3 5    40 47 40 47 47
 4 1 3    38 48 46
 4 2 4    37 31 45 41
 5 1 5    44 51 49 51 52
 5 2 4    49 49 49 50
```

```
 5 3 3    48 59 59
 6 1 4    54 36 36 40
 6 2 4    44 47 48 48
 6 3 5    43 52 50 46 39
 7 1 4    41 42 36 47
 7 2 5    47 36 43 38 41
 7 3 5    53 40 44 40 45
 8 1 3    52 53 48
 8 2 5    40 48 50 40 51
 9 1 4    40 34 37 45
 9 2 4    42 37 46 40
10 1 5    39 31 30 41 48
10 2 4    50 44 40 45
11 1 5    52 54 52 56 53
11 2 5    56 39 52 49 48
12 1 5    50 45 43 44 49
12 2 4    52 43 38 33
13 1 5    39 37 33 43 42
13 2 3    43 38 44
13 3 4    46 44 37 54
14 1 4    50 53 51 43
14 2 4    44 45 39 52
14 3 5    42 48 45 51 48
15 1 5    47 49 45 43 42
15 2 5    45 42 52 51 32
15 3 5    51 51 53 45 51
```

OUTPUT

SINGLE CLASSIFICATION ANOVA. TRANSFORMATION CODE = 0

ANOVA TABLE

LEVEL	SS	DF	MS	FS
1	1807.687	3.	602.5625	5.2632
0	3778.063	33.	114.4867	

VARIANCE COMPONENTS

LEVEL	VAR. COMP.	PERCENT
1	54.1764	32.1211
0	114.4867	67.8789

TABLE OF COEFFICIENTS

1   9.0

```
2 LEVEL NESTED ANOVA      TRANSFORMATION CODE =  0


                       ANOVA TABLE
LEVEL          SS          DF              MS              FS

  2         1780.164      14.           127.1546         3.4726

  1          800.230      22.            36.3741         1.4705
                          21.4           36.6163

  0         3042.551     123.            24.7362

VARIANCE COMPONENTS
     LEVEL    VAR. COMP.          PERCENT

       2          8.5213         23.6884
       1          2.7148          7.5469
       0         24.7362         68.7646


TABLE OF COEFFICIENTS

  2     4.4    10.6
  1     4.3
```

A3.5 Factorial anova.

This program reads in data for a complete factorial analysis of variance with no replications. Using the techniques described in Section 12.5 it is possible to use this program for single classification anova with equal sample sizes, multiway analysis of variance with equal replications, and other completely balanced designs. The program produces the standard anova table and provides as well an optional output of a table of deviations for all possible one-, two-, three-, four-way (and more) tables.

INPUT

Three types of cards are used.

Card Type 1. *Input parameters* (one card per analysis).

Card columns

1–4	b b b b
5	The number of factors, $k \leq 9$, for k-way anova (integer number).
6–9	b b b b
10	0 if table of deviations is not wanted.
	1 if the deviations are desired.
11–14	b b b b
15	Transformation code (defined in the glossary).
16–60	The number of levels of each of the k factors (integer num-

bers). Five card columns are allowed for each quantity. If, for example, there are $k = 3$ factors with $a = 2$, $b = 3$, and $c = 4$ levels, respectively, then punch a 2 in column 20, a 3 in column 25, and a 4 in column 30. The other columns must be left blank. The product $(a + 1)\ (b + 1)\ (c + 1)\ \ldots$ must not exceed 9000. This limitation will not in fact permit a nine-factor anova since for even two levels per factor $(2 + 1)^9 > 9000$. If for any problem the above product exceeds 9000, DIMENSION of array X should be increased accordingly, if possible.

61–80 Ignored.

Card Type 2. *Variable format card* (one card per analysis) for input data (real numbers only).

Card Type 3. *Data cards* (variable number of cards per analysis) punched as specified by the variable format card. The order of the input data cards is very important. If a variate in a three-way anova is denoted by the symbol Y_{ijk}, then the variates must be entered in an order such that the first subscript (i) changes most rapidly and the last subscript (k) changes least rapidly. The order of the input data for 2, 3, and 4 levels of 3 factors would be as follows: $Y_{111}, Y_{211}, Y_{121}, \ldots, Y_{132}, Y_{232}, Y_{113}, \ldots, Y_{234}$. Each time the level of the second factor (j) changes, one must begin a new input data card. Probably the simplest method for assuring that the data are arranged properly is to punch the variates for all levels of factor A at one combination of levels for factors B, C, . . . on one card and to identify the card by punching the subscripts j, k, . . . for these levels of factors, B, C, . . . on the card. Then sort the cards on a card sorter into the proper order.

OUTPUT

The input parameters are reproduced in the output. If requested, tables of the deviations of the means are included in the output marked by a series of codes to identify the means. There are two types of identifications. First, the level numbers for each of the factors are given (factors not employed in the study are labeled with level 1) and this is followed by a column labeled SOURCE. The numbers in this column correspond to various sources of variation in the anova table presented below. All the deviations with the same identifying number are used to compute the sums of squares for a particular source of variation with the same identifying number (ID NO.) in the anova table given below. This output is especially useful as input to various programs for testing differences among means and can be inspected for homogeneity of interaction terms. If a source of variation is nonsignificant, one would expect all of the deviations making up a sum of squares to be approximately of the same magnitude.

This table of deviations is followed by the anova table itself. The various sources of variation, their sums of squares, degrees of freedom, and mean squares are given.

Test data

The limpet data from Box 11.1 may be used. Since there actually are eight replications per cell, the first factor, A, is a dummy factor and the sums of squares and degrees of freedom for factor A and its interactions with factors B and C must be pooled together to give the error sum of squares and degrees of freedom, respectively. The input and output are shown below. Only the beginning and end of the list of deviations are shown to conserve space.

INPUT

```
   3     1     0     8     2     3                             BOX 11.1
 (8F6.2)
  7.16  6.78 13.60  8.93  8.26 14.00 16.10  9.66       1      1
  6.14  3.86 10.40  5.49  6.14 10.00 11.60  5.80       2      1
  5.20  5.20  7.18  6.37 13.20  8.39 10.40  7.18       1      2
  4.47  9.90  5.75 11.80  4.95  6.49  5.44  9.90       2      2
 11.11  9.74 18.80  9.74 10.50 14.60 11.10 11.80       1      3
  9.63  6.38 13.40 14.50 14.50 10.20 17.70 12.30       2      3
```

OUTPUT

```
3 FACTORS OUTPUT CODE =     1
TRANSFORMATION CODE =  0
NO. LEVELS PER FACTOR =     8    2    3
  DEVIATIONS        A  B  C  D  E  F  G  H  I  SOURCE

 -1.00583           1  1  1  1  1  1  1  1  1    1
  0.21917           2  1  1  1  1  1  1  1  1    1
 -1.04917           3  1  1  1  1  1  1  1  1    1
  1.86750           4  1  1  1  1  1  1  1  1    1
 -0.97917           5  1  1  1  1  1  1  1  1    1
 -0.69417           6  1  1  1  1  1  1  1  1    1
  0.79583           7  1  1  1  1  1  1  1  1    1
  0.84583           8  1  1  1  1  1  1  1  1    1
  0.97750           9  1  1  1  1  1  1  1  1    5
  1.00583           1  2  1  1  1  1  1  1  1    1
 -0.21917           2  2  1  1  1  1  1  1  1    1
  1.04917           3  2  1  1  1  1  1  1  1    1
 -1.86750           4  2  1  1  1  1  1  1  1    1
  0.97917           5  2  1  1  1  1  1  1  1    1
  0.69417           6  2  1  1  1  1  1  1  1    1
 -0.79583           7  2  1  1  1  1  1  1  1    1
 -0.84583           8  2  1  1  1  1  1  1  1    1
 -0.97750           9  2  1  1  1  1  1  1  1    5
 -0.01042           1  3  1  1  1  1  1  1  1    3
 -1.03208           2  3  1  1  1  1  1  1  1    3
  1.10292           3  3  1  1  1  1  1  1  1    3
 -1.63708           4  3  1  1  1  1  1  1  1    3

                                .
                                .
                                .
```

```
   -2.33458           1  3  4  1  1  1  1  1  1     4
   -2.64292           2  3  4  1  1  1  1  1  1     4
    1.90208           3  3  4  1  1  1  1  1  1     4
   -0.14792           4  3  4  1  1  1  1  1  1     4
   -0.02792           5  3  4  1  1  1  1  1  1     4
    0.99375           6  3  4  1  1  1  1  1  1     4
    2.43708           7  3  4  1  1  1  1  1  1     4
   -0.17958           8  3  4  1  1  1  1  1  1     4
    0.                9  3  4  1  1  1  1  1  1     8
```

ANOVA TABLE

SOURCE	SS	DF	MS	ID NO.
ABC	118.4709	14	8.4622	1
AB	35.7534	7	5.1076	2
A C	109.0868	14	7.7919	3
A	138.2102	7	19.7443	4
BC	23.9262	2	11.9631	5
B	16.6381	1	16.6381	6
C	181.3210	2	90.6605	7
	4441.7462	1	4441.7462	8

LAST LINE GIVES CT FOR MEAN

A3.6 Sum of squares *STP*.

This program tests the homogeneity of all subsets of means in anova using the sums of squares simultaneous test procedure (Section 9.7). This program is an extensive modification of an earlier one written by K. R. Gabriel.

INPUT

Five types of cards are used, but only three or four types are employed in any one problem.

Card Type 1. *Input parameters* (one card per problem).

Card columns

1–3	b b b
4–5	a, the number of means to be tested (≤ 60; integer number). As the number of means increases, computer processing time increases very rapidly.
6–9	b b b b
10	0 if the original data are to be read in. 1 if only the means are to be read in.
11–14	b b b b
15	Transformation code (defined in glossary; can be used only if the original data are to be read in).

Card columns (contd.)

16–25	If the means and not the original data are to be read in, then the error mean square must be punched into this card field as a real number. If original data are read in, leave this field blank.
26–30	A critical value of $F_{[a-1,\nu_2]}$ (Table **S**; real number). $\nu_2 = df$ of the error MS which is computed by the program or punched in the preceding field.
31–35	α, the significance level of the F-value punched in the preceding field (real number). This information is used for labeling the output only.
36–80	Ignored.

Card Type 2. *Sample sizes* (one card per problem). This card is to be included *only* if the original data are read in. The sample sizes are punched as integer numbers (≤ 100) with five card columns allowed per input sample size starting with column 1. A maximum of fifteen sample sizes may be punched on a single card, and columns 76–80 are ignored. If more than fifteen samples are included in a single study, additional cards may be prepared following the same format.

Card Type 3. *Variable format card* (one card per problem). If the original data are read in, this card describes the format of the input data (card type 5; real numbers only). Or, if means rather than the original data are to be read in, it describes the format of the mean cards (card type 4).

Card Type 4. *Mean and sample size cards* (one card per mean). These cards are used if the original data are *not* read in. Each card contains the sample size (integer number) followed by the corresponding mean (real number) for each of the a samples. Thus there will be a cards. Follow format specified in card type 3.

Card Type 5. *Original data cards* (one set of cards per sample). The original observations are to be punched following the format specified by the variable format card above (card type 3). The data for each sample must begin on a new card.

OUTPUT

The input parameters are reproduced in the output. These are followed by a list of sample sizes and means numbered in the sequence of input. If original data are used in the computation, sample variances are also printed. These data are followed by a listing of all the maximal nonsignificant subsets. No more than one hundred such subsets can be printed out. If there are more than that number the message LIST OF SUBSETS INCOMPLETE appears. Each list consists of the identification code numbers for those samples that are not significantly different. This is followed by the average of these means and their sums of squares.

Test data

The data from Box 9.10 may be used as test data. The input and output are shown below.

```
                                    INPUT

    8    1     0.02645      2.05.05                     BOX 9.10
(I5,F15.8)
   20        3.5123
   20        3.6940
   20        3.7223
   14        3.7229
   10        3.7376
   15        3.7479
   11        3.7843
   20        3.8451
```

```
                                    OUTPUT

 SS-STP
      8 MEANS, INPUT CODE =  1 TRANSFORMATION CODE =  0


    SAMPLE           N       MEAN
    NUMBER
         1          20     3.5123
         2          20     3.6940
         3          20     3.7223
         4          14     3.7229
         5          10     3.7376
         6          15     3.7479
         7          11     3.7843
         8          20     3.8451

 ERROR MS =     0.0265, FPOINT =      2.05 0.0500 LEVEL
 CRITICAL VALUE =                0.37955750

 MAXIMUM NONSIGNIFICANT SUBSETS
 SUBSET    1 OF MEANS
       2  3  4  5  6  7  8
                AVERAGE =   3.7506   SS =           0.28376595
 SUBSET    2 OF MEANS
       1  2
                AVERAGE =   3.6031   SS =           0.33014901
 SUBSET    3 OF MEANS
       1  4
                AVERAGE =   3.5990   SS =           0.36525520
 SUBSET    4 OF MEANS
       1  5
                AVERAGE =   3.5874   SS =           0.33840147
```

A3.7 Student-Newman-Keuls test.

This program performs a Student-Newman-Keuls a posteriori multiple range test, following the procedures given in Boxes 9.9 and 9.10.

INPUT

Five types of cards are used, but only two or four types are employed in any one problem.

Card Type 1. *Input parameters* (one or more cards per problem).

Card columns	
1–2	b b
3–5	a, the number of means to be tested (≤ 100; integer number).
6–9	b b b b
10	0 if the original data are to be read in. 1 if only the means are to be read in.
11–14	b b b b
15	Transformation code (defined in glossary; can be used only if the original data are to be read in).
16–30	Error MS (real number). If original data are read in (0 in column 10), leave this field blank.
31–70	On the remaining portion of this card and for as many additional cards as are required, punch the critical values of $Q_{\alpha[k,\nu]}$, the Studentized range (Table **U**), for the desired level of significance α, and for ν the degrees of freedom of the error mean square. Enter all values of $Q_{\alpha[k,\nu]}$ from k = 2 to k = a. There should be a total of $a - 1$ such values. Ten card columns are allowed per value of Q (real numbers). Four such numbers may be punched on the first card and maximally seven such numbers on all subsequent cards on which we can start on column 1.
71–80	Ignored (in first and also in subsequent cards).

Card Type 2. *Sample sizes* (one or more cards per problem). Included *only* if the original data are read in. Punch the sample sizes as integer numbers (≤ 100) with five card columns allowed per input sample size starting with column 1. A maximum of fifteen sample sizes per card may be punched on a single card and columns 76–80 are ignored. If more than fifteen samples are included in a single study, additional cards may be prepared following the same format.

Card Type 3. *Variable format card* (one card per problem). If the original data are to be read in, this card should describe the format of the input data (card type 4) which must be real numbers. If the means and sample sizes are to be read in, it describes the format of these cards (card type 5).

Card Type 4. *Original data cards* (one set of cards per sample). If the original data are to be read in, they are included at this point. Punch the data as real numbers, following the format specified on card type 3. Data for each sample must begin on a new card.

Card Type 5. *Mean and sample size cards* (one card per mean). These cards are used if the original data are not read in. Each card contains the sample size (integer number) followed by the corresponding mean (real number). There will be a cards for the a means.

OUTPUT

The input parameters are reproduced in the output and a table of the means and sample sizes sorted by ascending magnitude of the sample means is also produced. The numerical identification of the means is assigned automatically by the program corresponding to the order in which the means are entered into the computer. Following this are given the maximum non-significant ranges of means. These subsets are sequentially numbered and list the identifying numbers of the two means constituting the end points of the range. Thus 2, 3, 5 means subset 2, the range being bounded by means of samples 3 and 5.

Test data

The data from Box 9.10 can be used as test data for this program. The input and output are shown below.

INPUT

```
    8    1    0   .02645        2.8         3.3563     3.685     3.917
4.096     4.241     4.363                              BOX 9.10
(I5,F10.5)
   20     3.5123
   20     3.6940
   15     3.7479
   11     3.7843
   20     3.8451
   20     3.7223
   14     3.7229
   10     3.7376
```

OUTPUT

```
    SNK TEST     8 MEANS, TRANSFORMATION CODE =   0
    ERROR MS =             0.0265

 Q =   2.800   3.356   3.685   3.917   4.096   4.241   4.363

 TABLE OF MEANS (SORTED)
          SAMPLE          N        MEAN
          NUMBER
              1.         20.      3.5123
              2.         20.      3.6940
              6.         20.      3.7223
              7.         14.      3.7229
              8.         10.      3.7376
              3.         15.      3.7479
              4.         11.      3.7843
              5.         20.      3.8451
```

```
MAXIMUM NONSIGNIFICANT RANGES
     SUBSET       SAMPLE NUMBERS
       1             2        5
```

A3.8 Tests of homogeneity of variances.

This program performs Bartlett's test of homogeneity of variances and the F_{max}-test (Box 13.1).

INPUT

Five types of cards are used, but only two or four types are employed in any one problem.

Card Type 1. *Input parameters* (one card per problem).

Card columns

1–2	b̲ b̲
3–5	a, the number of samples (≤ 100; integer number).
6–9	b̲ b̲ b̲ b̲
10	0̲ if the original data are to be read in.
	1̲ if only the variances and their sample sizes are to be read in.
11–14	b̲ b̲ b̲ b̲
15	Transformation code (defined in glossary; can be used only if the original data are to be read in).

Card Type 2. *Sample sizes* (one or more cards per problem). Included *only* if the original data are to be read in. Punch the sample sizes as integer numbers (≤ 100) with five card columns allowed per input sample size starting with column 1. A maximum of fifteen sample sizes may be punched on a single card and columns 76–80 are ignored. If more than fifteen samples are included in a single study, additional cards may be prepared following the same format.

Card Type 3. *Variable format card* (one card per problem). This card should be included only if the original data are to be read in. It should describe the format of the input data (card type 4), which must be real numbers.

Card Type 4. *Original data cards* (one set of cards per sample). If the original data are to be read in, they are included at this point. Punch the original data as real numbers, following the format specified on card type 3. Data for each sample must begin on a new card.

Card Type 5. *Variance and sample size cards* (one card per variance). These cards are used if the original data are not read in. There will be a such cards for the a variances.

Card columns

1–5	Sample size (integer number).
6–15	Ignored.
16–25	Sample variance (real number).
26–80	Ignored.

OUTPUT

The input parameters are reproduced in the output followed by the results of Bartlett's test (yielding an adjusted chi-square value). This should be compared with a value from the chi-square table (Table **R**) with degrees of freedom $a - 1$ as specified in the output. This is followed by the results of the F_{max}-test, which gives a maximum F-ratio to be compared with the values in the F_{max} table (Table **T**). When the original data are read, means are also computed and printed out.

Test data

The data given in Box 13.1 may be used as test data for this program. The input and output are shown below.

```
                                                      INPUT

      8       1      0                                 BOX 13.1
     18     3.88              .0707
     13     4.61              .1447
     17     4.79              .0237
     16     4.73              .0836
      8     4.92              .2189
     11     4.96              .1770
     10     5.20              .0791
     10     6.58              .2331


                                                      OUTPUT

      8 SAMPLES, INPUT CODE = 1 TRANSFORMATION CODE = 0

   ADJUSTED X SQUARE =   20.91379 TO BE COMPARED WITH A
  CHISQUARE DISTRIBUTION WITH     7 DF

   THE F-MAX TEST

   FS=    9.83544 WITH DEGREES OF FREEDOM  8 AND  9
```

A3.9 Test of the equality of means with heterogeneous variances.

This program performs an approximate test of the equality of means when the variances are assumed to be heterogeneous (Box 13.2).

INPUT

Five types of cards are used, but only two or four types are employed in any one problem.

Card Type 1. *Input parameters* (one card per problem).

Card columns

1–2	b̲ b̲
3–5	a, the number of samples ($\leq$100; integer number).
6–9	b̲ b̲ b̲ b̲

10	0 if the original data are to be read in.
	1 if the means, variances, and sample sizes are to be read in.
11–14	b b b b
15	Transformation code (defined in glossary; can be used only if the original data are to be read in).
16–80	Ignored.

Card Type 2. *Sample sizes* (one or more cards per problem). Included *only* if the original data are to be read in. Punch the sample sizes as integer numbers (≤ 100) with five card columns allowed per input sample size starting with column 1. A maximum of fifteen sample sizes may be punched on a single card and columns 76–80 are ignored. If more than fifteen samples are included in a single study, additional cards may be prepared following a similar format.

Card Type 3. *Variable format card* (one card per problem). This card should be included only if the original data are to be read in. It should describe the format of the input data (card type 4), which must be real numbers.

Card Type 4. *Original data cards* (one set of cards per sample). If the original data are to be read in, they are included at this point. Punch the original data as real numbers, following the format specified on card type 3. Data for each sample must begin on a new card.

Card Type 5. *Mean, variance, and sample size cards* (one card per sample). These cards are to be included only if the original data are not read in.

Card columns

1–5	Sample size (integer number).
6–15	Sample mean (real number).
16–25	Sample variance (real number).

OUTPUT

The input parameters are reproduced in the output along with a listing of the means and variances for each sample. This is followed by the sample variance ratio F_s' and the degrees of freedom required for looking up the critical F-value.

Test data

The data given in Box 13.2 may be used as test data for this program. The input and output are shown below.

INPUT

```
    8     1    0                                 BOX 13.2
   18  3.88          .0707
   13  4.61          .1447
   17  4.79          .0237
   16  4.73          .0836
    8  4.92          .2189
   11  4.96          .1770
   10  5.20          .0791
   10  6.58          .2331
```

```
                         OUTPUT


 8 SAMPLES, INPUT CODE = 1 TRANSFORMATION CODE = 0

       N        YBAR        VAR
      18       3.880      0.071
      13       4.610      0.145
      17       4.790      0.024
      16       4.730      0.084
       8       4.920      0.219
      11       4.960      0.177
      10       5.200      0.079
      10       6.580      0.233

FS =         46.95020

DEGREES OF FREEDOM ARE     7.   AND   34.5
```

A3.10 Tukey's test for nonadditivity.

This program performs Tukey's test for nonadditivity (Box 13.4).

INPUT

Four types of cards are used, but only two or four types are employed in any one problem.

Card Type 1. *Input parameters* (one card per problem).

Card columns	
1–3	b b b
4–5	*b*, the number of rows in the two-way table (≤10; integer number).
6–8	b b b
9–10	*a*, the number of columns in the two-way table (≤10; integer number).
11–12	b b
13–15	*n*, the number of replications for each cell (≤100; integer number; must be the same for all cells).
16–19	b b b b
20	0 if the original data are to be read in. 1 if the means are to be read in.
21–24	b b b b
25	Transformation code (defined in the glossary; used only if the original data are to be read in).
26–80	Ignored.

Card Type 2. *Variable format card* (one card per problem). This card is to be included only if the original data are to be read in. It should describe the format of the original data (card type 3), which must be real numbers.

Card Type 3. *Original data cards* (one set of cards per cell of the two-way table). If the original data are to be read in, they are included at this point. Punch the original data as real numbers following the format specified on card type 2. The data for each cell must begin on a new card. The data for the cells are entered row-wise; that is, all cells for row 1 are entered before those for row 2, and so forth. Within each row the cells are ordered by columns. The last cell to be entered would be that for the last row and the last column.

Card Type 4. *Mean cards* (one set of cards per problem). These cards are to be included only if the original data are not read in. The means for each cell are punched on a separate card as a real number in columns 6–15. The cards must be arranged so that the cell means are read in row-wise (see card type 3 above for a definition of this item).

OUTPUT

The input parameters are reproduced in the output, followed by an anova table which partitions the interaction sums of squares into a component for nonadditivity and a residual *SS*.

Test data

The limpet data given in Boxes 11.1 and 13.4 may be used as test data for this program. The input and output are shown below.

INPUT

```
    3     2     8     0     0           BOX 13.4
(8F6.2)
  7.16   6.78 13.60   8.93   8.26 14.00 16.10   9.66
  6.14   3.86 10.40   5.49   6.14 10.00 11.60   5.80
  5.20   5.20   7.18   6.37 13.20   8.39 10.40   7.18
  4.47   9.90   5.75 11.80   4.95   6.49   5.44   9.90
 11.11   9.74 18.80   9.74 10.50 14.60 11.10 11.80
  9.63   6.38 13.40 14.50 14.50 10.20 17.70 12.30
```

OUTPUT

```
    3 ROWS     2 COLUMNS     8 REPLICATIONS PER CELL
INPUT CODE  = 0 TRANSFORMATION CODE = 0
    SAMPLE  N      MEAN    VARIANCE
    NUMBER

       1    8    10.5612       12.3389

       2    8     7.4287        7.9012

       3    8     7.8900        7.5053

       4    8     7.3375        7.6869

       5    8    12.1737        9.5523

       6    8    12.3262       12.3757
```

SOURCE	DF	SS	MS	FS
AXB	2	23.9262		
NONADD.	1	4.2399	4.2399	0.2154
RESID.	1	19.6863	19.6863	

A3.11 Kruskal-Wallis test.

This program performs the Kruskal-Wallis test for equality in the "location" of several samples.

INPUT

Three types of cards are used.

Card Type 1. *Input parameters* (one card per problem).

Card columns

1–2	b̲ b̲
3–5	a, the number of samples (≤ 100; integer number).
6–74	On the remainder of the card, and on a second card if more than fourteen samples are to be analyzed, punch the sample sizes as integer numbers, allowing five card columns per sample size. On additional cards start at column 1.
75–80	Ignored (on first and additional cards).

Card Type 2. *Variable format card* (one card per problem). This card describes the format of the data (card type 3) which follow. They must be real numbers.

Card Type 3. *Data cards* (one set of cards per sample). Punch real numbers as specified by the variable format card (card type 2). The sum of the sample sizes must not exceed one thousand. The data for each sample must begin on a new card. There must not be more than one hundred sets of tied variates in any one problem.

OUTPUT

The input parameters and sample sizes are reproduced in the output, followed by the Kruskal-Wallis statistic H (adjusted, if necessary), which is to be compared with a chi-square distribution for degrees of freedom equal to $a - 1$.

Test data

The data given in Box 13.5 may be used as test data for this program. The input and output are shown on the next page.

INPUT

```
    5   10   10   10   10   10   A3.11/DATA FROM BOX 9.4
(10F5.0)
   75   67   70   75   65   71   67   67   76   68
   57   58   60   59   62   60   60   57   59   61
   58   61   56   58   57   56   61   60   57   58
   58   59   58   61   57   56   58   57   57   59
   62   66   65   63   64   62   65   65   62   67
```

OUTPUT

```
KRUSKAL-WALLIS TEST
       5 GROUPS
    SAMPLE SIZES
   10   10   10   10   10

    THE KRUSKAL-WALLIS STATISTIC H =    38.4368
    IS TO BE COMPARED WITH A CHI SQUARE DISTRIBUTION WITH
   4 DEGREES OF FREEDOM
```

A3.12 Linear regression analysis.

This program performs an analysis of regression with one or more Y-values corresponding to each X-value (Boxes 14.1 and 14.4).

INPUT

Seven types of cards are used, but only three or six types are employed in any one problem.

Card Type 1. *Input parameters* (one card per problem).

Card columns

1–2	b b
3–5	a, the number of samples or X-values to be read in (in Box 14.1 for single Y-values per value of X, this quantity is called n; punch as integer number not to exceed 100).
6–9	b b b b
10	0 if the list of all X-values is to be followed by lists of one or more Y-values for each value of X. Thus first all X-values are entered, then all Y-values for the first value of X, then the Y's for the second value of X, and so on. 1 if each value of X is paired on a card with its corresponding n, $\overline{Y}$, and s_Y^2. Note that if one has a single value of Y for each value of X, then one may enter the data by using either input code. A list of X-values may be followed by the Y-values, each punched on a separate card (code = 0) or a list of X and Y pairs

may be read in with each pair punched on a separate card (code = 1). When preparing data by the latter procedure, the sample size and sample variance in card type 7 may be left blank, although they *must* still be allowed for in the first variable format card (card type 2).

11–14 b b b b
15 Transformation code for the X-values (defined in glossary).
16–19 b b b b
20 Transformation code for the Y-values (defined in the glossary; can be used only if original Y-values are read in).
21–30 $t_{\alpha[a-2]}$ [real number; use significance level α to yield 100 $(1 - \alpha)$% confidence limits].
31–35 α (real number; used to label output).
36–80 Ignored.

Card Type 2. *First variable format card* (one card per problem). If the original Y-values are to be read in (0 in column 10 of card type 1), this card describes the format in which the X-values (card type 3) are punched. If means of Y for each value of X are to be read in (1 in column 10 of card type 1), this card describes the format for the cards containing the X-values, sample sizes, means, and variances (card type 7).

Card Type 3. X-*values.* Punch real numbers as specified by card type 2. Entered only when original list of Y-values is to be read in.

Card Type 4. *Sample sizes.* This card is to be included only if the original Y-values are read in (0 in column 10 of card type 1). The sample sizes for the a samples are punched as integers (≤ 100), allowing five card columns per sample size and maximally 15 sample sizes per card. Start at column 1 and ignore columns 76–80.

Card Type 5. *Second variable format card* (one card per problem). This card is to be included only if the original Y-values are read in (0 in column 10 of card type 1). This card describes the format of the Y-values in original data form (card type 6; real numbers).

Card Type 6. *Original* Y-*values.* Punch real numbers as specified by the second variable format card (card type 5). The Y-values for each sample (X-value) must begin on a new card.

Card Type 7. X, n, $\bar{Y}$, $s_{\bar{Y}}^2$ *cards* (one card for each value of X). These cards are to be included only if the original data are not read in (1 in column 10 of card type 1). Punch real numbers as specified by the first variable format card (card type 2). For each value of X, the card contains the X-value, sample size, the mean of Y and sample variance in that order.

OUTPUT

The input parameters are reproduced in the output followed by the regression equation, and means and variances of X and Y. The confidence limits of β are also shown. This is followed by an analysis of variance table. Note that the code -9999.9999 is used to indicate inappropriate F-ratios in this standard anova table. (These will appear when one has only a single Y-value for each X-value.) In the anova the MS for linear regression is always tested over the MS for deviation. The latter is tested over the error MS. The anova table is followed by a table giving the X and Y-values ordered by sequence of input. The table also features $\hat{Y}$ (YHAT), lower and upper $100(1 - \alpha)\%$ confidence limits to $\hat{Y}$ (L1 and L2), and $d_{Y \cdot X}$ (DEVIATION) for each value of X.

Test data

The data from Boxes 14.1 and 14.4 may be used as test data for the cases of a single value of Y and several Y-values for each X-value, respectively. The input and output for the two examples are shown below.

INPUT

```
    9    1    0    0  2.36    .05                                      BOX 14.1
(F5.0,I1,F5.2,F1.0)
 0.   8.98
12.   8.14
29.5  6.67
43.   6.08
53.   5.90
62.5  5.83
75.5  4.68
85.   4.20
93.   3.72
    4    0    0    0  2.16    .05                                      BOX 14.4
( 4F5.0)
   5.   20.   50. 100.
    5    4    3    3
(5F10.2)
     61.68     58.37     69.30     61.68     69.30
     68.21     66.72     63.44     60.84
     58.69     58.37     58.37
     53.13     49.89     49.82
```

OUTPUT

```
   9 X-VALUES           1 = INPUT CODE, TRANSFORMATION CODES  0,  0
  T = 2.360 AT P =0.0500


THE REGRESSION EQUATION IS Y =    8.70403 +  -0.05322 X
MEAN OF X =  50.38889   MEAN OF Y =    6.02222   TOTAL N=     9.
VARIANCE OF X =      1037.67357   VARIANCE OF Y =          3.01632


 95.0 PERCENT CONFIDENCE LIMITS FOR THE SLOPE ARE   -0.0609 AND   -0.0455
```

ANALYSIS OF VARIANCE

SOURCE	SS	DF	MS	FS
GROUPS	24.1306	8	3.0163	-9999.9999
LINEAR	23.5145	1	23.5145	267.1841
DEV.	0.6161	7	0.0880	-9999.9999
ERROR	0.	0	0.	
TOTAL	24.1306	8		

SAMPLE NUMBER	X	Y	L1	YHAT	L2	DEVIATION
1	0.	8.98000	8.25194	8.70403	9.15612	0.27597
2	12.00000	8.14000	7.68922	8.06536	8.44150	0.07464
3	29.50000	6.67000	6.85073	7.13397	7.41722	-0.46397
4	43.00000	6.08000	6.17529	6.41547	6.65566	-0.33547
5	53.00000	5.90000	5.64902	5.88325	6.11749	0.01675
6	62.50000	5.83000	5.12640	5.37764	5.62889	0. 5236
7	75.50000	4.68000	4.38294	4.68575	4.98857	-0.00575
8	85.00000	4.20000	3.82631	4.18014	4.53398	0.01986
9	93.00000	3.72000	3.35228	3.75437	4.15646	-0.03437

4 X-VALUES 0 = INPUT CODE, TRANSFORMATION CODES 0, 0
T = 2.160 AT P =0.0500

SAMPLE NUMBER	N	MEAN	VARIANCE
1	5	64.0660	24.6550
2	4	64.8025	10.9486
3	3	58.4767	0.0341
4	3	50.9467	3.5764

THE REGRESSION EQUATION IS Y = 65.96004 + -0.14701 X
MEAN OF X = 37.00000 MEAN OF Y = 60.52066 TOTAL N= 15.
VARIANCE OF X = 1335.00000 VARIANCE OF Y = 40.17059

95.0 PERCENT CONFIDENCE LIMITS FOR THE SLOPE ARE -0.1967 AND -0.0973

ANALYSIS OF VARIANCE

SOURCE	SS	DF	MS	FS
GROUPS	423.7015	3	141.2338	11.2020
LINEAR	403.9280	1	403.9280	40.8555
DEV.	19.7735	2	9.8868	0.7842
ERROR	138.6867	11	12.6079	
TOTAL	562.3883	14		

SAMPLE NUMBER	X	Y	L1	YHAT	L2	DEVIATION
1	5.00000	64.06600	62.85804	65.22499	67.59194	-1.15899
2	20.00000	64.80250	61.07345	63.01984	64.96623	1.78266
3	50.00000	58.47667	56.74077	58.60953	60.47829	-0.13287
4	100.00000	50.94667	47.67143	51.25902	54.84662	-0.31236

A3.13 Product-moment correlation coefficient.

This program computes the Pearson product-moment correlation coefficient for a pair of variables (Box 15.1) and its confidence limits (Box 15.3). In addition, the program computes and prints the means, variances, standard deviations, standard errors, and covariance for the variables, as well as the equation of the principal and minor axes. The confidence limits for the slope of the principal axis are also computed and the coordinates of eight points are given for plotting confidence ellipses (Box 15.5) for the bivariate mean.

INPUT

Three types of cards are used.

Card Type 1. *Input parameters* (one card per problem).

Card columns	
1–2	b̲ b̲
3–5	n, sample size of pairs of points (≤ 500; integer number).
6–10	$\chi^2_{\alpha[1]}$ (real number) for the desired level of significance α and 1 degree of freedom. This quantity is used for computing the confidence limits to the slope of the principal axis.
11–15	$F_{\alpha[2,n-2]}$ (real number). This value is used for computing the equal frequency ellipse.
16–20	α (real number). This is the level of significance of the previous two critical values. It is also used to set confidence limits to the correlation coefficient.
21–24	b̲ b̲ b̲ b̲
25	Transformation code for variable Y_1 (defined in the glossary).
26–29	b̲ b̲ b̲ b̲
30	Transformation code for variable Y_2 (defined in the glossary).
31–80	Ignored.

Card Type 2. *Variable format card* (one card per problem). This card describes the format of the input data cards (real numbers) which follow.

Card Type 3. *Data cards* (one card for each pair of variates). Punch real numbers as specified by the variable format card (card type 2).

OUTPUT

The input parameters are reproduced in the output followed by the means, variances, standard deviations, and standard errors for each of the variables, and by the covariance between the two variables. Next, the correlation and its $100(1 - \alpha)\%$ confidence limits are given along with the equations of the principal and minor axes and $100(1-\alpha)\%$ confidence limits to the slope of the principal axis. Finally, the coordinates for eight pairs of points are given to be used for plotting $100(1-\alpha)\%$ equal frequency ellipses. These represent points a through h shown in Box 15.5.

Test data

The data given in Boxes 15.1 and 15.5 may be used as test data for this program. The input and output are shown below.

INPUT

```
 12 3.844.10  .05      0    0                       BOX 15.1
(6X,2F10.2)
   1  159.        14.40
   2  179.        15.20
   3  100.        11.30
   4   45.         2.50
   5  384.        22.70
   6  230.        14.90
   7  100.         1.41
   8  320.        15.81
   9   80.         4.19
  10  220.        15.39
  11  320.        17.25
  12  210.         9.52
```

OUTPUT

```
N =    12, CHI SQUARE =      3.840, F-VALUE =      4.100, ALPHA =.0500
THE TRANSFORMATION CODES ARE  0 AND  0

                                                        Y1          Y2


                          MEAN                       195.583      12.047
                          VARIANCE                 11306.265      42.043
                          STANDARD DEVIATION         106.331       6.484
                          STANDARD ERROR              30.695       1.872
                          COVARIANCE                        596.511
THE PRODUCT-MOMENT CORRELATION COEFFICIENT IS    0.86519


THE  95.0 PER CENT CONFIDENCE LIMITS ARE L1 =  0.5785 AND L2 =  0.9616

THE EQUATION OF THE PRINCIPAL AXIS IS
     Y1= -32.55301 +   18.93641 Y2

THE EQUATION OF THE MINOR AXIS IS
     Y1= 196.21954 +   -0.05281 Y2

THE 95.0 PER CENT CONFIDENCE LIMITS TO THE SLOPE OF THE PRINCIPAL AXIS ARE
     L1=  14.25702      L2=  28.16317
```

```
FOLLOWING PAIRS SHOW THE COORDINATES OF POINTS A TO H
FOR PLOTTING CONFIDENCE ELLIPSE (SEE BOX 15.5).
                   Y1                     Y2

A             241.8112              12.0475
B             149.3555              12.0475
C             195.5833              14.8665
D             195.5833               9.2285
F             287.7707              16.9158
F             103.3959               7.1792
G             195.4349              14.8587
H             195.7318               9.2363
```

A3.14 Fisher's exact test.

This program performs Fisher's exact test for independence in a 2 × 2 contingency table (Box 16.7).

INPUT

One card type is used.

Card Type 1. *Input data* (one card per problem).

Card columns

1–5	Frequency in cell a (integer number).
6–10	Frequency in cell b (integer number).
11–15	Frequency in cell c (integer number).
16–20	Frequency in cell d (integer number).

OUTPUT

The program puts out the observed 2 × 2 table of frequencies and the probability of observing these particular frequencies. This is followed by tables of successively worse cases and their probabilities. The last such value is followed by two times the cumulative probability. If $ad–bc$ is positive, the frequencies in the table will be interchanged to simplify computations. If this quantity is zero, no computations are performed, since all other combinations of frequencies are "worse" cases and the sample data show perfect independence. In this case the message AD = BC is printed out.

Test data

The data from Box 16.7 may be used as test data for this program. The input and output are shown below.

```
                                        INPUT

    2   13   10    3                        BOX 16.7

                                        OUTPUT

  OBSERVED DATA

    2    13

   10     3
```

```
PROBABILITY     OF     OBSERVED     SAMPLE     =  0.000987

NEXT WORSE CASE

    1     14

   11      2

PROBABILITY     OF     THIS     CASE     =  0.000038

NEXT WORSE CASE

    0     15

   12      1

PROBABILITY     OF     THIS     CASE     =  0.000000

2X     CUMULATIVE     PROBABILITY     =  0.002052
```

A3.15 $R \times C$ test of independence in contingency tables.

The program performs a test of independence in an $R \times C$ contingency table by means of the G-test. Optionally it carries out an a posteriori test of all subsets of rows and columns in the $R \times C$ contingency table by the simultaneous test procedure.

INPUT

Two types of cards are used.

Card Type 1. *Input parameters* (one card per problem).

Card columns

1–3	b̲ b̲ b̲
4–5	b, number of rows in the contingency table (≤ 30; integer number).
6–8	b̲ b̲ b̲
9–10	a, number of columns in the contingency table (≤ 30; integer number).
11–20	Critical chi-square value (real number), with degrees of freedom equal to $(r - 1)(c - 1)$ and at significance level α.
21–30	α, the level of significance (real number). This value is used for labeling the output.
31–34	b̲ b̲ b̲ b̲
35	0̲ if all combinations of rows and columns are to be tested using the STP procedure. 1̲ if only the complete set is to be tested (simple test of independence). If the analysis is of a 2 × 2 table, Yates' correction will be performed automatically and the code in column 35 will be ignored.

Card Type 2. *Data cards* (one set of cards per row of the contingency table). The frequencies (integer numbers) in the cells of the $R \times C$ contingency table are to be punched row-wise with five card columns allotted per frequency and a maximum of 15 frequencies to be punched per card. The frequencies for each row must begin on a new card. The frequencies are to be punched as integers.

OUTPUT

The output parameters are reproduced in the output and followed by the input frequencies. A G-value for the entire table is printed. If all combinations of rows and columns tested are significant, then a message stating that the entire set is significant will be printed. Otherwise, the maximal nonsignificant sets of rows and columns will be printed out, preceded by the G-value for that set. The first row of integers is a list of those rows included in the set being tested and the second row of integers indicates those columns in the set being tested.

Test data

The data given in Box 16.8 may be used as test data for this program. The input and output are shown below.

```
                              INPUT


    4    2 7.81        .05                          BOX 16.8
   29   11
  273  191
    8   31
   64   64


                              OUTPUT


    4 ROWS     2 COLUMNS
  CRITICAL CHI SQUARE =      7.810, ALPHA =0.0500
  TEST CODE = 0

  29.   11.
 273.  191.
   8.   31.
  64.   64.

  G =    28.5964 FOR TOTAL SET

 SUBSET    1      G-STATISTIC =     7.0766
           ROWS =  1  2  4
           COLS =  1  2
```

A3.16 Randomization test.

This program may be used to perform randomization tests involving two samples. The user must furnish a function whose name is STAT, which computes the desired statistic for a set of data. As an example, a routine is furnished which computes the variances of two samples and their ratio. The user must also furnish a routine called COMPR, which will compare the observed value of the sample statistic with the value of the statistic computed for a random partition of the data and increments one of two counters which keep track of how many samples were more deviant than the observed value of the sample statistic in each tail of the distribution. A routine is furnished which performs these operations for comparing variances. See Section 17.3 for an account of this method.

INPUT

Three types of cards are used.

Card Type 1. *Input parameters* (one card per problem).

Card columns	
1–2	b̲ b̲
3–5	Size of the first sample (integer number).
6–7	b̲ b̲
8–10	Size of the second sample (integer number).
11–15	Number of random partitionings of the data to be performed (integer number).
16–26	An integer to be used by the pseudorandom number generator function as an initial starting random pattern. It is suggested that one select a series of digits from a random number table for this quantity. The maximum size of this integer number depends upon the word length of the computer. On many computers, a ten-digit integer may be entered.

Card Type 2. *Variable format card* (one card per problem). This card describes the format of the data (real numbers) to be read in.

Card Type 3. *Data cards.* The data for each sample (real numbers) must be punched as specified by the variable format card (card type 2). The data for the second sample must begin on a new card. The total sample size for both samples cannot exceed 500.

OUTPUT

The input parameters are reproduced in the output. This is followed by the observed value of the sample statistic being tested and an account of the number of samples more deviant than the observed statistic (in each tail of the distribution) out of the total number of samples taken.

Test data

The problem in Section 17.3 is used as test data for this program. However, these data represent a different run of the experiment. Hence the answers are not identical to those furnished there (or to those obtained during a possible repetition by the reader). The input and output are given below.

INPUT

```
   9   16  500 1234554321                    TEST DATA FOR A3.16
(13F6.1)
211.3 218.5 211.2 205.1 211.9 204.9 211.4 211.9 209.5                        1
219.2 211.1 210.4 219.5 205.1 222.8 210.1 218.4 213.4 210.2 213.1 204.6 206.7 2
212.7 224.4 229.2                                                             3
```

OUTPUT

```
SAMPLE SIZES ARE     9 AND    16
 500 ITERATIONS WITH THE INITIAL RANDOM PATTERN EQUAL TO  1234554321



OBSERVED VALUE OF THE STATISTIC =    3.10944


   47. SAMPLES WERE MORE DEVIANT THAN THE OBSERVED STATISTIC
   18. WERE IN THE LEFT TAIL AND    29. WERE IN THE RIGHT TAIL
(OUT OF    500 SAMPLES)
IN PROPORTIONS THESE ARE 0.0360 AND 0.0580 RESPECTIVELY

FINAL RANDOM INTEGER =-32075034847
```

BASIC STATISTICAL PROGRAMS IN FORTRAN

```
CA3.1          BASIC STATISTICS FOR UNGROUPED DATA
C.... USES SUBROUTINES  COMOUT, FNORMD, KSRD, PROBN, RSORT, TRANSG.
      DIMENSION Y ( 1000 ), FMT ( 14 )
C.... READ PARAMETERS
  1   READ ( 5, 2 ) N, ITRANS, ALPHA, TALPH
  2   FORMAT (5X,2I5, 2F5.3)
C.... CLEAR CELLS FOR SUMS
      SUMY = 0.
      SUMY2 = 0.
      SUMY3 = 0.
      SUMY4 = 0.
      FN = N
C.... READ VARIABLE FORMAT
      READ ( 5, 50 ) FMT
 50   FORMAT (13A6,A2)
C.... READ IN RAW DATA ACCORDING TO VARIABLE FORMAT
 100  READ ( 5, FMT ) ( Y ( I ), I = 1, N )
C.... TRANSFORM DATA (IF DESIRED) AND COMPUTE MEAN.  THIS SLOWER MODE
C.... OF COMPUTATION IS USED IN ORDER TO MINIMIZE ROUNDING ERRORS.
      DO 106I = 1, N
      IF ( ITRANS ) 106, 106, 105
 105  Y ( I ) = TRANSG ( ITRANS, Y ( I ))
 106  SUMY = SUMY + Y ( I )
      YBAR = SUMY / FN
      DO 115I = 1, N
      YDEV = Y ( I ) - YBAR
      YDEV2 = YDEV * YDEV
      SUMY2 = SUMY2 + YDEV2
      YDEV3 = YDEV2 * YDEV
      SUMY3 = SUMY3 + YDEV3
 115  SUMY4 = SUMY4 + YDEV3 * YDEV
      S2 = SUMY2 /( FN - 1. )
      S = SQRT ( S2 )
C.... SORT Y() AND DO K-S TEST
      CALL KSRD ( N, Y, YBAR, S, DMAX )
C.... COMPUTE MEDIAN (NOTE THE USE OF THE TRUNCATING FEATURES OF INTEGER
C.... ARITHMETIC).
      IND1 = ( N + 1 )/ 2
      IND2 = ( N + 2 )/ 2
      FMEDN = ( Y ( IND1 ) + Y ( IND2 ))/ 2.
C.... CALL OUTPUT PROGRAM TO FINISH UP COMPUTATIONS AND OUTPUT RESULTS
      CALL COMOUT ( N, 0, ITRANS, ALPHA, TALPH, YBAR, S, S2, SUMY3, SUMY
     14, FMEDN, DMAX )
      GO TO 1
      END
```

```
CA3.2  BASIC STATISTICS FOR DATA GROUPED INTO A FREQUENCY DISTRIBUTION
C.... USES SUBROUTINES COMOUT, FNORMD, KSFD, PROBN, TRANSG
      DIMENSION Y ( 30 ), YUPLM ( 30 ), F ( 30 ), FMT ( 14 ), FHAT ( 30
     1)
  1   READ ( 5, 2 ) NCLAS, ITRANS, ALPHA, TALPH
  2   FORMAT (I5, 5X, I5, 2F5.3)
C.... CLEAR CELLS FOR SUMS
      SUMY = 0.
      SUMY2 = 0.
      SUMY3 = 0.
      SUMY4 = 0.
C.... READ VARIABLE FORMAT
      READ ( 5, 50 ) FMT
 50   FORMAT (13A6, A2)
C.... READ IN FREQ. DIST. ACCORDING TO THE VARIABLE FORMAT.  TRANSFORM
```

```
C.... DATA.
 200  FN = 0.
      DO 201I = 1, NCLAS
 201  READ ( 5, FMT ) Y ( I ), F ( I )
      DO 207I = 1, NCLAS
      IF ( ITRANS ) 204, 204, 205
 204  YUPLM ( I ) = ( Y ( I ) + Y ( I + 1 ))/ 2.
      GO TO 206
 205  YUPLM ( I ) = TRANSG ( ITRANS, ( Y ( I ) + Y ( I + 1 ))/ 2. )
      Y ( I ) = TRANSG ( ITRANS, Y ( I ))
 206  SUMY = SUMY + Y ( I )* F ( I )
 207  FN = FN + F ( I )
C.... ESTIMATE LAST UPPER CLASS LIMIT.
      YUPLM ( NCLAS ) = 2. * YUPLM ( NCLAS - 1 ) - YUPLM ( NCLAS - 2 )
      YBAR = SUMY / FN
      N = FN
C.... COMPUTATION OF BASIC STATISTICS (NOTE THAT DEVIATIONS FROM THE
C.... MEAN ARE ACCUMULATED TO AVOID LARGE ROUNDING ERRORS).
      DO 215I = 1, NCLAS
      YDEV = Y ( I ) - YBAR
      YDEV2 = YDEV * YDEV * F ( I )
      SUMY2 = SUMY2 + YDEV2
      YDEV3 = YDEV2 * YDEV
      SUMY3 = SUMY3 + YDEV3
 215  SUMY4 = SUMY4 + YDEV3 * YDEV
      S2 = SUMY2 /( FN - 1. )
      S = SQRT ( S2 )
C.... DO K-S TEST
      CALL KSFD ( N, NCLAS, YUPLM, F, YBAR, S, DMAX, FHAT )
      WRITE ( 6, 220 )
 220  FORMAT(1H1/14X,1HY,10X,5HUPPER,14X,1HF,11X,4HFHAT/25X,5HCLASS/
     125X,5HLIMIT/)
      WRITE ( 6, 225 ) ( Y ( KK ), YUPLM ( KK ), F ( KK ), FHAT ( KK ),
     1KK = 1, NCLAS )
 225  FORMAT(2F15.5,F15.0,F15.1)
      FMDO BS = FN / 2.
      YCUM = 0
      DO 235I = 1, NCLAS
      YCUM = YCUM + F ( I )
      IF ( YCUM - FMDO BS ) 235, 230, 230
 230  K = I
      GO TO 240
 235  CONTINUE
 240  FMEDN = YUPLM ( K ) - (( YCUM - FMDO BS )/ F ( K ))*( YUPLM ( K )
     1- YUPLM ( K - 1 ))
C.... CALL OUTPUT ROUTINES TO FINISH UP COMPUTATIONS AND OUTPUT THE
C.... RESULTS.
      CALL COMOUT ( N, NCLAS, ITRANS, ALPHA, TALPH, YBAR, S, S2, SUMY3,
     1SUMY4, FMEDN, DMAX )
      GO TO 1
      END
```

```
CA3.3         GOODNESS OF FIT TO DISCRETE FREQUENCY DISTRIBUTIONS
C.... USES SUBROUTINES BINOM, POISN
      DIMENSION Y ( 30 ), F ( 30, 11 ), FHAT ( 30 ), PARM ( 4 ), FN ( 11
     1 ), FHATR ( 30 ), DEV ( 30 )
C.... READ PARAMETERS
 1    READ ( 5, 2 )INFD,IDIST, IESTIM, NCLAS, K
  2   FORMAT (5I5)
      WRITE ( 6, 21 ) INFD, IDIST, IESTIM, NCLAS
 21   FORMAT(1H1/ 5X, 21HNO. OF FREQ. DIST. =    ,  I2, 5X,12HDIST. CODE =
```

```
1   , I2/  5X,   17HESTIMATION CODE =    , I2,17H NO. OF CLASSES =   ,
2  I2)
      INDEX = IDIST + 1
      INFDP1 = 1
      IF ( INFD - 1 ) 3, 4, 3
 3    INFDP1 = INFD + 1
C.... WILL PARAMETERS FOR DISTRIBUTIONS BE READ (IESTIM = 0) OR ESTIM
C.... ATED FROM THE DATA (IESTIM POSITIVE).
 4    IF ( IESTIM ) 100, 100, 200
C.... WILL ENTIRE DISTRIBUTION BE READ IN.(IDIST = 0) OR ONLY THE PARA
C.... METERS (IDIST POSITIVE).
 100  IF ( IDIST ) 105, 105, 150
C
 105  READ ( 5, 106 ) ( FHAT ( I ), I = 1, NCLAS )
 106  FORMAT (8F10.5)
C.... NORMALIZE RATIOS, I.E., CONVERT TO PROPORTIONS.
      SUM = 0.
      DO 120I = 1, NCLAS
 120  SUM = SUM + FHAT ( I )
      DO 125I = 1, NCLAS
 125  FHATR ( I ) = FHAT ( I )/ SUM
      WRITE ( 6, 126 ) ( FHAT ( I ), I = 1, NCLAS )
 126  FORMAT(/5X, 16HARBITRARY DIST.    /( 8F10.5))
      NEST = 0
      GO TO 200
C
C.... READ PARAMETERS FOR THE DISTRIBUTIONS.
 150  READ ( 5, 151 ) NP, ( PARM ( I ), I = 1, NP )
 151  FORMAT (I10, 4F10.5)
      GO TO  ( 161, 162 ), IDIST
C.... BINOMIAL
 161  PARM ( 2 ) = PARM ( 1 )
      PARM ( 1 ) = K
      GO TO 180
C.... POISSON O.K.
 162  CONTINUE
 180  CONTINUE
      NEST = 0
C.... (INFD POSITIVE AND EQUAL TO THE NUMBER OF      DISTRIBUTIONS TO BE
C.... READ IN).
C.... CHECK IF OBS. FREQUENCY DIST. ARE TO BE READ IN (IF NOT CLEAR F,
C.... GENERATE Y)
 200  IF ( INFD ) 201, 201, 205
 201  DO 202I = 1, NCLAS
      F ( I, 1 ) = 0.
 202  Y ( I ) = I - 1
      GO TO 300
 205  DO 206J = 1, INFDP1
 206  FN ( J ) = 0.
 210  DO 212J = 1, INFD
 212  READ ( 5, 215 ) ( F ( I, J ), I = 1, NCLAS )
 215  FORMAT (15F5.0)
      DO 230I = 1, NCLAS
      Y ( I ) = I - 1
C.... POOLED DISTRIBUTION.
      SUM = 0.
      DO 221J = 1, INFD
      FN ( J ) = FN ( J ) + F ( I, J )
      IF ( INFD - 1 ) 230, 230, 221
 221  SUM = SUM + F ( I, J )
      F ( I, INFDP1 ) = SUM
      FN ( INFDP1 ) = FN ( INFDP1 ) + SUM
```

```
 230   CONTINUE
C
C.... PERFORM TESTS
       GSUM = 0.
 300   DO 600J = 1, INFDP1
       IF ( INFD - 1 ) 306, 306, 301
 301   IF ( J - INFDP1 ) 306, 302, 302
 302   WRITE ( 6, 303 )
 303   FORMAT(/2X,20HPOOLED DISTRIBUTION  /)
C.... ARE PARAMETERS TO BE ESTIMATED FROM THE OBSERVED DATA.
 306   IF ( IESTIM ) 400, 400, 310
C.... SET UP PARAMETERS FOR DISTRIBUTIONS
 310   SUMFY = 0
       DO 312I = 1, NCLAS
 312   SUMFY = SUMFY + Y ( I )* F ( I, J )
       GO TO  ( 400, 350, 355 ), INDEX
C.... BINOMIAL DIST.
 350   PARM ( 1 ) = K
       PARM ( 2 ) = SUMFY /( PARM ( 1 )* FN ( J ))
       NEST = 1
       GO TO 400
C.... POISSON DIST.
 355   PARM ( 1 ) = SUMFY / FN ( J )
       NEST = 1
C.... COMPUTE EXPECTED FREQUENCIES
 400   GO TO  ( 500, 411, 412 ), INDEX
C.... BINOMIAL DISTRIBUTION
 411   CALL BINOM ( PARM ( 1 ), PARM ( 2 ), Y, FHATR, NCLAS )
       WRITE ( 6, 4111 ) PARM ( 1 ), PARM ( 2 )
 4111  FORMAT(/5X,18HBINOMIAL DIST. K =   ,F5.0,6H, P =    , F10.5/)
       GO TO 500
C.... POISSON DISTRIBUTION
 412   CALL POISN ( PARM ( 1 ), Y, FHATR, NCLAS )
       WRITE ( 6, 4121 ) PARM ( 1 )
 4121  FORMAT(/5X,21HPOISSON DIST. YBAR =    , F10.5/)
C
C.... COMPARE WITH OBSERVED DISTRIBUTION (IF ANY)
 500   IF ( INFD ) 501, 501, 510
 501   WRITE ( 6, 502 ) ( Y ( I ), FHATR ( I ), I = 1, NCLAS )
 502   FORMAT (//9X, 1HY, 5X, 5HFHATR//(F10.0, F10.5))
       GO TO 600
 510   SUM = 0.
       DO 515I = 1, NCLAS
       FHAT ( I ) = FHATR ( I )* FN ( J )
       DEV ( I ) = F ( I, J ) - FHAT ( I )
       R = F ( I, J )/ FHAT ( I )
       IF ( R - 0.000001 ) 515, 515, 514
 514   SUM = SUM + F ( I, J )* ALOG ( R )
 515   CONTINUE
 519   WRITE ( 6, 520 ) ( Y ( I ), F ( I, J ), FHAT ( I ), DEV ( I ), I =
     1 1, NCLAS )
 520   FORMAT (//9X, 1HY, 9X, 1HF, 6X, 4HFHAT,6X,4HDIFF//(2F10.0,2F10.1))
       SUM = 2. * SUM
       IDF = NCLAS - 1 - NEST
       WRITE ( 6, 530 ) SUM, IDF
 530   FORMAT (//5X, 18HG-TEST STATISTIC = , F10.3,6H, DF = ,I4/)
       GSUM = GSUM + SUM
C.... ARE WE NOW WORKING ON THE POOLED DISTRIBUTION. (YES = 551)
       IF ( INFD - 1 ) 600, 600, 550
 550   IF ( J - INFDP1 ) 600, 551, 600
 551   GHET = GSUM - 2. * SUM
       IDF = ( INFD - 1 )*( NCLAS - 1 - NEST )
```

```
      WRITE ( 6, 555 ) GHET, IDF
 555  FORMAT (//5X, 18HHETEROGENEITY G  =,F10.3,6H, DF = ,I4//)
 600  CONTINUE
      GO TO 1
      END


CA3.4        SINGLE CLASSIFICATION AND NESTED ANOVA
C
C.... THIS PROGRAM USES FUNCTION TRANSG
C.... FOR K GREATER THAN 10 THE FOLLOWING DIMENSION STATEMENTS
C.... MUST BE CHANGED
      DIMENSION FMT ( 14 ), LABEL ( 10 ), LABELP ( 10 ), Y ( 100 ), SY (
     1 12 ), SY2 ( 12 ), SN ( 12 ), SN2 ( 12, 12 ), SNQ ( 12, 12 ), MS (
     2 12 ), SS ( 12 ), DF ( 12 ), C ( 12, 12 ), PVC ( 12 ), VC ( 12 ),
     3SMS ( 12 ), W ( 12 ), SDF ( 12 ), FS ( 12 )
      REALMS
C.... READ PARAMETER
    1 READ ( 5, 100 ) K, ITRANS
 100  FORMAT(2I5)
      IF ( K - 1 ) 101, 102, 105
  101 STOP
  102 WRITE (  6, 103 ) ITRANS
  103 FORMAT(1H1/54H SINGLE CLASSIFICATION ANOVA.  TRANSFORMATION CODE
     1= ,I3)
      GO TO 111
  105 WRITE ( 6, 110 ) K, ITRANS
 110  FORMAT(1H1/I5,19H LEVEL NESTED ANOVA ,5X, 21HTRANSFORMATION CODE =
     1, I3)
  111 READ ( 5, 115 ) FMT
 115  FORMAT(13A6,A2)
      LASTCD = 1
      KP1 = K + 1
      KM1 = K - 1
C.... CLFAR ARRAYS
      KP2 = K + 2
      DO 120I = 1, KP2
      SY ( I ) = 0.
      SY2 ( I ) = 0.
      SN ( I ) = 0.
      DF ( I ) = 0.
      LABELP ( I ) = 1
      DO 120J = I, KP2
      SN2 ( I, J ) = 0.
      SNQ ( I, J ) = 0.
  120 CONTINUE
C.... READ A DATA CARD
 122  READ ( 5, FMT ) ( LABEL ( I ), I = 1, K ), N, ( Y ( I ), I = 1, N
     1)
C.... CHECK IF END OF DATA      (N=0)
      IF ( N ) 124, 124, 125
 124  LASTCD = 0
      INDEX = K + 1
      GO TO 210
C.... CHECK LABELS TO FIND PLACE IN HIERARCHY.
 125  DO 130I = 1, K
      LAST = I
      IF ( LABELP ( I ) - LABEL ( I )) 132, 130, 132
 130  CONTINUE
 132  DO 133I = LAST, K
 133  LABELP ( I ) = LABEL ( I )
C.... ACCUMULATE SUMS IN HIERARCHY IF DATA ARE NOT IN THE SAME LOWEST
```

```
C.... LEVEL SUBGROUP
      INDEX = K - LAST + 1
      IF ( INDEX - 1 ) 240, 240, 210
 210  DO 230I = 2, INDEX
      SY ( I + 1 ) = SY ( I + 1 ) + SY ( I )
      SY2 ( I + 1 ) = SY2 ( I + 1 ) + SY ( I )* SY ( I )/ SN ( I )
      SN ( I + 1 ) = SN ( I + 1 ) + SN ( I )
      SN2 ( I, I ) = SN ( I )* SN ( I )
      SN2 ( I, I + 1 ) = SN2 ( I, I + 1 ) + SN2 ( I, I )
      IM1 = I - 1
      DO 215J = 1, IM1
 214  SN2 ( J, I + 1 ) = SN2 ( J, I + 1 ) + SN2 ( J, I )
      SNQ ( J, I ) = SNQ ( J, I ) + SN2 ( J, I )/ SN ( I )
  215 SN2 ( J, I ) = 0.0
      SNQ ( I, I ) = SNQ ( I, I ) + SN ( I )
      DF ( I + 1 ) = DF ( I + 1 ) + 1.
      DF ( I ) = DF ( I ) - 1.
      SY ( I ) = 0.
      SN ( I ) = 0.
 230  CONTINUE
      IF ( LASTCD ) 300, 300, 240
C.... COMPUTE LOWEST LEVEL STATISTICS
 240  SUM = 0.
      SUM2 = 0.
      DO 245J = 1, N
      IF ( ITRANS ) 242, 242, 241
C.... TRANSFORM DATA
 241  Y ( J ) = TRANSG ( ITRANS, Y ( J ))
 242  SUM = SUM + Y ( J )
  245 SUM2 = SUM2 + Y ( J )* Y ( J )
      FN = N
C.... UPDATE SUMS IN BOTTOM OF HIERARCHY.
      SY ( 1 ) = SUM
      SY ( 2 ) = SY ( 2 ) + SUM
      SY2 ( 1 ) = SY2 ( 1 ) + SUM2
      SY2 ( 2 ) = SY2 ( 2 ) + SUM * SUM / FN
      SN ( 1 ) = FN
      SN ( 2 ) = SN ( 2 ) + FN
      DF ( 1 ) = DF ( 1 ) + FN - 1.
      DF ( 2 ) = DF ( 2 ) + 1.
      SN2 ( 1, 1 ) = FN * FN
      SN2 ( 1, 2 ) = SN2 ( 1, 2 ) + SN2 ( 1, 1 )
      SNQ ( 1, 1 ) = SNQ ( 1, 1 ) + FN
C.... GO BACK AND READ NEXT CARD
      GO TO 122
C.... DATA ALL READ IN, NOW COMPLETE ANOVA.
 300  SN2 ( KP2, KP2 ) = SN ( KP2 )* SN ( KP2 )
      DO 320I = 1, KP1
      SS ( I ) = SY2 ( I ) - SY2 ( I + 1 )
 320  MS ( I ) = SS ( I )/ DF ( I )
C.... - - - - - - - - - - - - - - - - - - - - - - - - - - - - - - - -
C.... COMPUTE COEFFICIENTS
      DO 501I = 1, K
      DO 501J = I, K
  501 C ( I, J ) = ( SNQ ( I, J ) - SNQ ( I, J + 1 ))/ DF ( J + 1 )
C.... ESTIMATE VARIANCE COMPONENTS
      VC ( 1 ) = MS ( 1 )
      VCSUM = VC ( 1 )
      DO 520I = 1, K
      VC ( I + 1 ) = MS ( I + 1 ) - VC ( 1 )
      IM1 = I - 1
      IF ( IM1 ) 512, 512, 505
```

```
  505 DO 510J = 1, IM1
  510 VC ( I + 1 ) = VC ( I + 1 ) - VC ( I )* C ( J, I )
  512 VC ( I + 1 ) = VC ( I + 1 )/ C ( I, I )
  520 VCSUM = VCSUM + VC ( I + 1 )
      DO 525I = 1, KP1
  525 PVC ( I ) = VC ( I )* 100. / VCSUM
C.... COMPUTE SYNTHETIC MS
      IF ( K - 1 ) 651, 651, 526
  526 DO 530I = 2, K
      SMS ( I ) = VC ( 1 )
      IM1 = I - 1
      DO 530J = 1, IM1
  530 SMS ( I ) = SMS ( I ) + C ( J, I )* VC ( J + 1 )
C.... COMPUTE WEIGHTS
C.... LOOP FROM MS(2) TO MS(K)
      DO 650I = 2, K
      IM1 = I - 1
C.... FOR EACH MS LOOP TO COMPUTE WEIGHTS  W(1) TO W(I)
      IF ( I.NE.2 ) GO TO 610
      W ( 2 ) = C ( 1, 2 )/ C ( 1, 1 )
      GO TO 620
  610 DO 615J = 1, IM1
      IND = I - J + 1
      W ( IND ) = C ( IND - 1, I )/ C ( IND - 1, IND - 1 )
      IF ( J.EQ.2 ) GO TO 615
      DO 611L = IND, IM1
  611 W ( IND ) = W ( IND ) - W ( IND + 1 )* C ( IND - 1, IND )/ C ( IND
     1 - 1, IND - 1 )
  615 CONTINUE
  620 W ( 1 ) = 1.0
      DO 625J = 2, IM1
  625 W ( 1 ) = W ( 1 ) - W ( J )
C.... COMPUTE DEGREES OF FREEDOM, VARIANCE RATIOS
      SUM = 0.0
      DO 630J = 1, I
  630 SUM = SUM + (( W ( J )* MS ( J ))*( W ( J )* MS ( J )))/ DF ( J )
      SDF ( I ) = SMS ( I )* SMS ( I )/ SUM
      FS ( I + 1 ) = MS ( I + 1 )/ SMS ( I )
  650 CONTINUE
  651 FS ( 2 ) = MS ( 2 )/ MS ( 1 )
C.... PRINT RESULTS
      WRITE ( 43, 652 )
 652  FORMAT(///20X,11HANOVA TABLE  /1X,5HLEVEL,8X,2HSS,10X,2HDF,12X,
     12HMS,12X,2HFS  )
      WRITE ( 6, 655 ) K, SS ( KP1 ), DF ( KP1 ), MS ( KP1 ), FS ( KP1 )
      IF ( K - 1 ) 661, 661, 653
  653 DO 660I = 2, K
      IND = K - I + 2
      L = IND - 1
      WRITE ( 6, 655 ) L, SS ( IND ), DF ( IND ), MS ( IND ), FS ( IND )
     1, SDF ( IND ), SMS ( IND )
  655 FORMAT(//I4, F15.3, F9.0,1X, 2F15.4 / 19X, F10.1, F15.4//)
  660 CONTINUE
  661 L = 0
      WRITE ( 6, 655 ) L, SS ( 1 ), DF ( 1 ), MS ( 1 )
      WRITE ( 6, 690 )
 690   FORMAT(///20H VARIANCE COMPONENTS /7X,5HLEVEL,5X,9HVAR.COMP.,  6X
     1, 8HPERCENT //)
      DO 701I = 1, KP1
      IND = KP1 - I + 1
      L = IND - 1
 701  WRITE ( 6, 705 ) L, VC ( IND ), PVC ( IND )
```

```
 705   FORMAT(I10,2F15.4)
       WRITE ( 6, 725 )
 725   FORMAT(///22H TABLE OF COEFFICIENTS  //     )
       DO 730I = 1, K
       L = K - I + 1
 730   WRITE ( 6, 731 ) L, ( C ( J, L ), J = 1, L )
 731   FORMAT(I3,10F7.1)
C
C..... END   GO BACK TO READ NEW PROBLEM.
       GO TO 1
       END
```

```
CA3.5                          FACTORIAL ANOVA
C
C..... THE LOGIC OF THE PROGRAM IS WRITTEN TO HANDLE A MAXIMUM OF 9
C..... FACTORS, BUT DUE TO STORAGE LIMITATIONS THIS CAN ONLY BE ATTAINED
C.....  ON LARGE SCALE COMPUTERS
C..... SECTIONS 1 AND 2 ARE PATTERNED AFTER METHOD DESCRIBED BY H.O.
C..... HARTLEY IN MATH. METHODS FOR DIGITAL COMPUTERS EDITED BY RALSTON
C..... AND WILF (1962).
C-----------------------------------------------------------------------
       DIMENSION X ( 9000 ), IS ( 10 ), SUMSQ ( 256 ), NLEVEL ( 9 ), LIMD
      1 ( 9 ), TEMP ( 20 ), FLEVEL ( 18, 2 ), ISYMB ( 18, 2 ), FLDF ( 18,
      2 2 ), NPOW ( 18 ), FMT ( 14 )
       EQUIVALENCE  ( IALIM, NLEVEL ( 1 )), ( IBLIM, NLEVEL ( 2 )), ( ICL
      1IM, NLEVEL ( 3 )), ( IDLIM, NLEVEL ( 4 )), ( IELIM, NLEVEL ( 5 )),
      2 ( IFLIM, NLEVEL ( 6 )), ( IGLIM, NLEVEL ( 7 )), ( IHLIM, NLEVEL (
      3 8 )), ( IILIM, NLEVEL ( 9 ))
       COMMON X
       DATA (ISYMB(I,1), I=1,18)/9*1H ,1HA, 1HB, 1HC, 1HD, 1HE, 1HF, 1HG
      1,1HH,1HI/,(ISYMB(I,2),I=10,18)/9*1H  /
       I7 = 43
       MAXDIM = 9000
    23 DO 22I = 1, 9
C..... NPOW IS USED IN SECTIONS 3 TO COMPUTE STORAGE LOCATION OF SS
       NPOW ( I ) = 1
       NPOW ( I + 9 ) = 2 **( I - 1 )
    22 NLEVEL ( I ) = 1
C
C..... SET UP ONES IN FLEVEL AND FTEMP AND IN FTPDF AND FLDF
C..... THESE ARE USED IN SECTION 4
       DO 600I = 1, 9
       FLEVEL ( I, 1 ) = 1.0
       FLEVEL ( I + 9, 1 ) = 1.0
       FLEVEL ( I + 9, 2 ) = 1.
       FLDF ( I, 1 ) = 1.
       FLDF ( I + 9, 1 ) = 1.0
 600   FLDF ( I + 9, 2 ) = 1.
C..... READ THE NO. OF FACTORS AND THE NO. OF LEVELS OF EACH
 1     READ ( 5, 2 ) K, IDEV, ITRANS, ( NLEVEL ( I ), I = 1, K )
 2     FORMAT(4I5, 9I5)
       IF ( K.LE.1 ) STOP
       WRITE ( 6, 200 ) K, IDEV, ITRANS, ( NLEVEL ( I ), I = 1, K )
 200   FORMAT(1H1/I6, 1X,7HFACTORS,1X,13HOUTPUT CODE =I5/4X,         22H
      1 TRANSFORMATION CODE =,I3/5X,23HNO. LEVELS PER FACTOR =,9I5)
C..... READ VARIABLE FORMAT CARD
       READ ( 5, 204 ) FMT
 204   FORMAT (13A6,A2)
 2041 IF ( IDEV ) 201, 201, 202
C..... PRINT TITLE
C
```

```
  202 WRITE ( 6, 203 )
 203  FORMAT( 4X,     11H DEVIATIONS,5X,27 H  A  B  C  D  E  F  G  H  I ,
     27H SOURCE //)
C.... CONVERT NUMBER OF LEVELS OF EACH FACTOR TO FLOATING POINT
  201 DO 61I = 1, K
      FLDF ( I + 9, 1 ) = NLEVEL ( I ) - 1
   61 FLEVEL ( I + 9, 1 ) = NLEVEL ( I )
      KP1 = K + 1
C.... IS( )   CONTAINS THE STEP INCREMENT (=SMALLS( ))
C.... IS(KP1)=MAXIMUM AMOUNT OF STORAGE USED BY DATA
C....  (= S ASTRIX)
      IS ( 1 ) = 1
      DO 3I = 2, KP1
 3    IS ( I ) = ( NLEVEL ( I - 1 ) + 1 )* IS ( I - 1 )
      LIM = IS ( K + 1 )
      IF ( LIM - MAXDIM ) 300, 300, 301
  301 WRITE ( 6, 302 ) LIM
  302 FORMAT(42H JOB TOO LARGE FOR THIS SYSTEM, PRODUCT = , I6, 12H JOB
     1DELETED)
      GO TO 99
C.... CLEAR STORAGE TO ZEROS
 300  DO 30I = 1, LIM
 30   X ( I ) = 0.0
C-----------------------------------------------------------------------
C
C.... SECTION 1 READ AND STORE DATA
C
C.... THE INDEXING IS USED IN ORDER TO STORE THE DATA SO THAT SPACES
C.... ARE LEFT FOR SUBSEQUENT CALCULATIONS (DELTA OPERATIONS).
C.... A DATA ELEMENT IS DENOTED X(I,J,KI,L,M,M6,M7,M8,M9) WHERE THE SUBS
C.... CORRESPOND TO FACTORS A,B,C,D,E,F,G,H,I.
C.... THE DATA MUST BE READ IN  SO THAT THE FIRST SUBSCRIPT VARIES MOST
C.... RAPIDLY AND THE LAST THE LEAST,I.E., COLUMNWISE.
C.... EACH NEW LEVEL OF FACTOR B MUST START ON A NEW CARD.
      DO 20M9 = 1, IILIM
      IND1 = IS ( 9 )*( M9 - 1 )
      DO 20M8 = 1, IHLIM
      IND2 = IND1 + IS ( 8 )*( M8 - 1 )
      DO 20M7 = 1, IGLIM
      IND3 = IND2 + IS ( 7 )*( M7 - 1 )
      DO 20M6 = 1, IFLIM
      IND4 = IND3 + IS ( 6 )*( M6 - 1 )
      DO 20M = 1, IELIM
      IND5 = IND4 + IS ( 5 )*( M - 1 )
      DO 20L = 1, IDLIM
      IND6 = IND5 + IS ( 4 )*( L - 1 )
      DO 20KI = 1, ICLIM
      IND7 = IND6 + IS ( 3 )*( KI - 1 )
      DO 20J = 1, IBLIM
      IND8 = IND7 + IS ( 2 )*( J - 1 )
 33   READ ( 5, FMT ) ( TEMP ( I ), I = 1, IALIM )
 34   DO 20I = 1, IALIM
      IND9 = IND8 + I
   20 X ( IND9 ) = TRANSG ( ITRANS, TEMP ( I ))
C-----------------------------------------------------------------------
C
C.... SECTION 2 SIGMA AND DELTA OPERATIONS, ONE
C.... FOR EACH FACTOR
C
      DO 9IF = 1, K
      FNL = NLEVEL ( IF )
C.... ILOC = BIG S  (USED AS A COUNTER)
```

```
      ILOC = 1
C.... ILOCP = BIG S PRIMED   (USED AS A COUNTER)
      ILOCP = 1
C.... SIGMA OPERATION STARTS HERE
      SUM = 0
   31 NLIM = NLEVEL ( IF )
      DO 4N = 1, NLIM
      SUM = SUM + X ( ILOC )
    4 ILOC = ILOC + IS ( IF )
      X ( ILOC ) = SUM
C.... END OF SIGMA
C.... DELTA OPERATION
      DO 5N = 1, NLIM
      X ( ILOCP ) = FNL * X ( ILOCP ) - SUM
    5 ILOCP = ILOCP + IS ( IF )
C.... END OF DELTA
      SUM = 0
C.... TEST IF END OF LAST PASS (WHEN ILOC = LIM)
      IF ( ILOC - LIM ) 6, 9, 9
    6 IF ( ILOC - LIM + IS ( IF )) 7, 7, 8
    7 ILOC = ILOC + IS ( IF )
      ILOCP = ILOCP + IS ( IF )
C.... GO BACK TO DO THE NEXT LEVEL OF FACTOR NUMBER IF.
      GO TO 31
    8 ILOC = ILOC + IS ( IF ) + 1 - LIM
      ILOCP = ILOCP + IS ( IF ) + 1 - LIM
C.... GO BACK TO START THE FIRST LEVEL OF THE NEXT FACTOR
      GO TO 31
 9    CONTINUE
C------------------------------------------------------------------
C
C.... SECTION 3 FORM RAW DEVIATIONS SQUARED
C
      ITWOK = 2 ** K
      DO 40I = 1, ITWOK
   40 SUMSQ ( I ) = 0.0
      ITOT = 1
      DO 81I = 1, K
   81 ITOT = ITOT * NLEVEL ( I )
      ATOT = ITOT
      GMEAN = X ( LIM )/ ATOT
C.... FIX UP DO LOOP RANGES AND DIVISORS
      DO 42I = 1, 9
      LIMD ( I ) = NLEVEL ( I ) + 1
      IF ( NLEVEL ( I ) - 1 ) 99, 42, 41
   41 NLEVEL ( I ) = LIMD ( I )
   42 CONTINUE
   60 LOC = 1
C.... NOTE THAT THE DIVISORS ARE SUCH THAT THE DIVISION  RESULTS IN A
C.... ZERO EXCEPT FOR THE LAST TIME THROUGH EACH LOOP.
C.... THE PURPOSE OF THIS IS TO CALCULATE THE ADDRESS OF THE PROPER SS
C.... FOR EACH DEVIATION.
      DO 80M9 = 1, IILIM
      IND1 = 1 + NPOW ( K + 1 )*( M9 / LIMD ( 9 ))
      DO 80M8 = 1, IHLIM
      IND2 = IND1 + NPOW ( K + 2 )*( M8 / LIMD ( 8 ))
      DO 80M7 = 1, IGLIM
      IND3 = IND2 + NPOW ( K + 3 )*( M7 / LIMD ( 7 ))
      DO 80M6 = 1, IFLIM
      IND4 = IND3 + NPOW ( K + 4 )*( M6 / LIMD ( 6 ))
      DO 80M = 1, IELIM
      IND5 = IND4 + NPOW ( K + 5 )*( M / LIMD ( 5 ))
```

```
      DO 80L = 1, IDLIM
      IND6 = IND5 + NPOW ( K + 6 )*( L / LIMD ( 4 ))
      DO 80KI = 1, ICLIM
      IND7 = IND6 + NPOW ( K + 7 )*( KI / LIMD ( 3 ))
      DO 80J = 1, IBLIM
      IND8 = IND7 + NPOW ( K + 8 )*( J / LIMD ( 2 ))
      DO 80I = 1, IALIM
      IND9 = IND8 + NPOW ( K + 9 )*( I / LIMD ( 1 ))
      IF ( IDEV ) 79, 79, 790
  790 DEV = X ( LOC )/ ATOT
      IF ( LOC.EQ.LIM ) DEV = 0.
C.... PRINT DEVIATIONS
      WRITE ( 6, 791 )      DEV, I, J, KI, L, M, M6, M7, M8, M9, IND9
 791  FORMAT ( F14.5, 6X, 9I3, I5)
   79 SUMSQ ( IND9 ) = X ( LOC )* X ( LOC ) + SUMSQ ( IND9 )
   80 LOC = LOC + 1
C-----------------------------------------------------------------------
C
C.... SECTION 4
C.... CONVERT RAW RESULTS OF SUM OF SQUARED DEVIATIONS
C.... INTO SUMS OF SQUARES AND MS AND PREPARE OUTPUT
C.... PUNCH TITLE
      WRITE ( 6, 65 )
 65   FORMAT (1H1/50X, 20H  ANOVA  TABLE          //
     1  4X,        7H SOURCE, 15X, 2HSS,11X, 2HDF,16X,3HMS ,5X,7H ID NO.)
C
      LOC = 1
      DO 95M9 = 1, 2
      DO 95M8 = 1, 2
      DO 95M7 = 1, 2
      DO 95M6 = 1, 2
      DO 95M = 1, 2
      DO 95L = 1, 2
      DO 95KI = 1, 2
      DO 95J = 1, 2
      DO 95I = 1, 2
C.... COMPUTE SS
      SUMSQ ( LOC ) = SUMSQ ( LOC )/( ATOT * FLEVEL ( K + 1, M9 )* FLEVE
     1L ( K + 2, M8 )* FLEVEL ( K + 3, M7 )* FLEVEL ( K + 4, M6 )* FLEVE
     2L ( K + 5, M )* FLEVEL ( K + 6, L )* FLEVEL ( K + 7, KI )* FLEVEL
     3( K + 8, J )* FLEVEL ( K + 9, I ))
C.... COMPUTE DEGREES OF FREEDOM
      DF = FLDF ( K + 1, M9 )* FLDF ( K + 2, M8 )* FLDF ( K + 3, M7 )* F
     1LDF ( K + 4, M6 )* FLDF ( K + 5, M )* FLDF ( K + 6, L )* FLDF ( K
     2+ 7, KI )* FLDF ( K + 8, J )* FLDF ( K + 9, I )
 94   IDF = DF
C.... COMPUTE MEAN SQUARE
      FMSQ = SUMSQ ( LOC )/ DF
C.... PUNCH RESULTS
      WRITE ( 6, 63 ) ISYMB ( K + 1, M9 ), ISYMB ( K + 2, M8 ), ISYMB (
     1K + 3, M7 ), ISYMB ( K + 4, M6 ), ISYMB ( K + 5, M ), ISYMB ( K +
     26, L ), ISYMB ( K + 7, KI ), ISYMB ( K + 8, J ), ISYMB ( K + 9, I
     3), SUMSQ ( LOC ), IDF, FMSQ, LOC
 63   FORMAT(/2X,9A1, F20.4, I10, F20.4,I9)
C.... TEST IF LAST SS HAS BEEN CALCULATED (WHEN LOC = ITWOK).
      IF ( LOC - ITWOK ) 95, 96, 96
C.... BOTTOM OF LOOP
   95 LOC = LOC + 1
   96 WRITE ( 6, 97 )
   97 FORMAT (28H LAST LINE GIVES CT FOR MEAN )
      GO TO 23
C-------------------- ----------------------------------------------------
 99   CALL EXIT
      END
```

```
CA3.6       SUM OF SQUARES STP
C      USES SUBROUTINES INPUT, TRANSG
       DIMENSION OB ( 60 ), YBAR ( 60 ), YBARN ( 60 ), Y2N ( 60 ), NST (
      1100, 60 ), FMT (12 ), NUMBER ( 60 ), MK ( 60 ), IOB ( 60 )
       EQUIVALENCE  ( OB ( 1 ), IOB ( 1 ))
C.... K IS NUMBER OF MEANS IN ANOVA AND MUST NOT EXCEED 60
C.... ERRVAR IS UNBIASED ERROR VARIANCE ESTIMATE, HYPDF IS HYPOTHESIS
C.... DEGREES OF FREEDOM(K-1), EF(M) AND SIGLV(M) THE TABULATED F
C.... VALUES AND THEIR SIGNIFICANCE LEVELS. THE D.F. OF F ARE (K-1) OVER
C.... SUM OF (OB(J)-1)
    1  READ ( 5, 41 ) K, ICODE, ITRANS, ERRVAR, EF, SIGLV
   41  FORMAT (3I5, F10.4, 2F5.2)
       IF ( K.EQ.0 ) STOP
       WRITE ( 6, 2 ) K, ICODE, ITRANS
    2  FORMAT(1H1/7H SS-STP/5X,I3,20H MEANS, INPUT CODE =,I3,22H TRANSFOR
      1MATION CODE =   , I3)
       HYPDF = K - 1
       IF ( ICODE ) 20, 20, 30
   20  CALL INPUT ( K, IOB, YBAR, Y2N, ERRVAR, ITRANS )
       GO TO 35
   30  READ ( 5, 31 ) FMT
   31  FORMAT(12A6)
C.... IOB(J) IS NUMBER OF OBSERVATIONS FOR MEAN X(J)
       READ ( 5, FMT ) ( IOB ( J ), YBAR ( J ), J = 1, K )
       WRITE ( 6, 33 ) ( J, IOB ( J ), YBAR ( J ), J = 1, K )
   33  FORMAT(//4X,6HSAMPLE,9X,1HN,6X,4HMEAN/4X,6HNUMBER/(I10,I10,F10.4
      1))
   35  DO 36J = 1, K
   36  OB ( J ) = IOB ( J )
C
C.... F= (( GROUP SS/(GP.D.F.))/(ERROR VAR)
C.... THEREFORE CRITICAL VALUE FOR GROUP SS= ERROR VAR*D.F.*F
   40  CRIT = ERRVAR * HYPDF * EF
       WRITE ( 6, 51 ) ERRVAR, EF, SIGLV
   51  FORMAT(/11H ERROR MS =,F9.4,10H, FPOINT =, F9.2, F7.4, 6H LEVEL)
       WRITE ( 6, 53 ) CRIT
   53  FORMAT (17H CRITICAL VALUE =, 3X, F20.10)
       WRITE ( 6, 592 )
  592  FORMAT (/33H MAXIMUM NONSIGNIFICANT SUBSETS       )
       DO 62J = 1, K
       YBARN ( J ) = YBAR ( J )* OB ( J )
   62  Y2N ( J ) = YBAR ( J )* YBARN ( J )
       NNS = 0
       JJ = K + 1
  100  JJ = JJ - 1
       IF ( JJ - 1 ) 1, 1, 101
  101  DO 102I = 1, JJ
  102  MK ( I ) = I
  105  ENUM = 0.
       SUM1 = 0.
       SUM2 = 0.
C.... COMPUTE YBAR AND SS FOR THOSE GROUPS IN THE SUBSET
       DO 107I = 1, JJ
       J = MK ( I )
       ENUM = ENUM + OB ( J )
       SUM1 = SUM1 + YBARN ( J )
       SUM2 = SUM2 + Y2N ( J )
  107  CONTINUE
       AVE = SUM1 / ENUM
       SOS = SUM2 - SUM1 * AVE
C
C - - - - - - - - - - - - - - - - - - -- - - - - - - - - - - - - - - - - - -
C.... CHECK IF NON-SIGNIFICANT
```

```
      IF ( SOS - CRIT ) 130, 400, 400
 130  IF ( NNS ) 210, 210, 200
C.... START SEARCH
  200 DO 250IC = 1, NNS
      DO 225J = 1, JJ
      IND = MK ( J )
      IF ( NST ( IC, IND )) 250, 250, 225
  225 CONTINUE
      GO TO 400
  250 CONTINUE
C.... NOT A SUBSET, ADD TO LIST
 210  NNS = NNS + 1
      DO 213J = 1, K
 213  NST ( NNS, J ) = 0
      DO 214I = 1, JJ
      J = MK ( I )
 214  NST ( NNS, J ) = 1
      WRITE ( 6, 609 ) NNS, ( MK ( I ), I = 1, JJ )
      WRITE ( 6, 610 ) AVE, SOS
      IF ( NNS - 100 ) 400, 498, 498
 400  MK ( JJ ) = MK ( JJ ) + 1
 412  IF ( MK ( JJ ) .LE.K ) GO TO 105
      JI = 0
 450  JI = JI + 1
      J = JJ - JI
      IF ( J.LE.0 ) GO TO 100
      MK ( J ) = MK ( J ) + 1
      IF ( MK ( J ) .GT.K - JI ) GO TO 450
      DO 460LL = 1, JI
      NN = LL + J
 460  MK ( NN ) = MK ( NN - 1 ) + 1
      GO TO 105
 498  WRITE ( 6, 499 )
  499 FORMAT (27H LIST OF SUBSETS INCOMPLETE   //)
  609 FORMAT (8H SUBSET ,I3,9H OF MEANS/(5X,30I3))
  610 FORMAT (15X,10H AVERAGE =,F9.4,6H  SS =,F20.10)
 999  GO TO 1
      END
```

```
CA3.7                 STUDENT-NEWMAN-KEULS MULTIPLE RANGE TEST
C.... USES SUBROUTINES INPUT, RSORT, TRANSG
      DIMENSION Q ( 100 ), YBAR ( 100 ), FREQ ( 100 ), VAR ( 100 ), IFRE
     1Q ( 100 ), TABLE ( 100, 3 ), NST ( 100, 2 ), FMT ( 14 )
      EQUIVALENCE  ( YBAR ( 1 ), TABLE ( 1, 2 )), ( FREQ ( 1 ), TABLE (
     11, 3 ), IFREQ ( 1 ))
  1   READ ( 5, 2 ) IA, ICODE, ITRANS, WAVVAR, ( Q ( I ), I = 2, IA )
  2   FORMAT(3I5,F15.5,4F10.3/(7F10.3))
      IF ( IA.LE.0 ) STOP
      WRITE ( 6, 3 ) IA, ITRANS, WAVVAR, ( Q ( I ), I = 2, IA )
  3   FORMAT(1H1/5X,8HSNK TEST,I5,1X,28HMEANS, TRANSFORMATION CODE =,
     1I3/5X,10HERROR MS =,F15.4//4H Q =,10F7.3/(4X,10F7.3))
      IF ( ICODE ) 10, 10, 12
 10   CALL INPUT ( IA, IFREQ, YBAR, VAR, WAVVAR, ITRANS )
      GO TO 20
 12   READ ( 5, 13 ) FMT
 13   FORMAT(13A6,A2)
 15   READ ( 5, FMT ) ( IFREQ ( I ), YBAR ( I ), I = 1, IA )
 20   STD = SQRT ( WAVVAR )
      DO 21I = 1, IA
      Q ( I ) = Q ( I )* STD
      FREQ ( I ) = IFREQ ( I )
```

```
 21   TABLE ( I, 1 ) = I
      CALL RSORT ( TABLE, IA, 3, 100, 2 )
      WRITE ( 6, 30 ) ( TABLE ( I, 1 ), TABLE ( I, 3 ), TABLE ( I, 2 ),
     1I = 1, IA )
 30   FORMAT (/25H TABLE OF MEANS (SORTED) /9X,6HSAMPLE,9X,1HN,6X,
     14HMEAN/9X,6HNUMBER/(5X,F10.0,F10.0,F10.4))
      NNS = 0
      IAM1 = IA - 1
      DO 50I = 1, IAM1
      IAMI = IA - I
      DO 50IFIRST = 1, I
      LAST = IFIRST + IAMI
      IF ( NNS ) 40, 40, 35
 35   DO 38J = 1, NNS
 36   IF ( IFIRST - NST ( J, 1 )) 38, 37, 37
 37   IF ( LAST - NST ( J, 2 )) 50, 50, 38
 38   CONTINUE
C.... TEST
 40   CRIT = Q ( IAMI + 1 )* SQRT ( 0.5 *( 1. / FREQ ( IFIRST ) + 1. / F
     1REQ ( LAST )))
      IF ( ABS ( YBAR ( LAST ) - YBAR ( IFIRST )) - CRIT ) 41, 50, 50
C.... NOT SIGNIFICANT
  41  NNS = NNS + 1
      NST ( NNS, 1 ) = IFIRST
      NST ( NNS, 2 ) = LAST
 50   CONTINUE
C.... PRINT RESULTS
      IF ( NNS ) 60, 60, 62
 60   WRITE ( 6, 61 )
 61   FORMAT(//35H  ALL MEANS SIGNIFICANTLY DIFFERENT)
      GO TO 1
 62   WRITE ( 6, 63 )
 63   FORMAT(///31H MAXIMUM NONSIGNIFICANT RANGES /4X,6HSUBSET,5X,
     114HSAMPLE NUMBERS)
      DO 65I = 1, NNS
      IF = NST ( I, 1 )
      IL = NST ( I, 2 )
      IFIRST = TABLE ( IF, 1 )
      LAST = TABLE ( IL, 1 )
 65   WRITE ( 6, 66 ) I, IFIRST, LAST
 66   FORMAT(2X,I5,I12,5X,I3)
      GO TO 1
      END
```

```
CA3.8       TESTS OF HOMOGENEITY OF VARIANCES
C.... USES SUBROUTINES INPUT AND TRANSG
C
C.... BARTLETT-S TEST OF HOMOGENEITY OF VARIANCES
      DIMENSION IFRE ( 100 ), VAR ( 100 ), YBAR ( 100 )
C.... READ THE NUMBER OF VARIANCES TO BE TESTED AND THE INPUT CODE.
 1    READ ( 5, 2 ) IA, ICODE, ITRANS
 2    FORMAT (3I5)
      IF ( IA.LE.0 ) STOP
      WRITE ( 6, 3 ) IA, ICODE, ITRANS
 3    FORMAT (1H1/I5,22H SAMPLES, INPUT CODE =, I2, 22H TRANSFORMATION C
     1ODE =, I2//)
      A = IA
      IF ( ICODE ) 10, 10, 13
   10 CALL INPUT ( IA, IFRE, YBAR, VAR, WAVVAR, ITRANS )
      GO TO 15
C     READ IN ONLY FREQUENCIES AND VARIANCES
```

```
   13 READ ( 5, 14 ) ( IFRE ( I ), VAR ( I ), I = 1, IA )
   14 FORMAT (I5, 10X, F10.5)
      WAVVAR = 0.0
      SDF = 0.0
      DO 145I = 1, IA
      DF = IFRE ( I ) - 1
      SDF = SDF + DF
  145 WAVVAR = WAVVAR + VAR ( I )* DF
      WAVVAR = WAVVAR / SDF
   15 SLOGS = 0.
      SRECDF = 0.
      SUMDF = 0.
      DO 16I = 1, IA
      DF = IFRE ( I ) - 1
      SUMDF = SUMDF + DF
      SLOGS = SLOGS + DF * ALOG ( VAR ( I ))
   16 SRECDF = SRECDF + 1. / DF
      XLWAV = ALOG ( WAVVAR )
C.... CALCULATION OF X SQUARE
      XSQ = SUMDF * XLWAV - SLOGS
C.... CALCULATION OF THE CORRECTION FACTOR
      C = 1. + 1. /( 3. *( A - 1. ))*( SRECDF - 1. / SUMDF )
      ADJXSQ = XSQ / C
      IDF = IA - 1
      WRITE ( 6, 20 ) ADJXSQ, IDF
   20 FORMAT (20H0ADJUSTED X SQUARE =F10.5,22H TO BE COMPARED WITH A /
     129H CHI SQUARE DISTRIBUTION WITH, I5, 3H DF)
C
C.... F-MAX TEST
C.... FIND THE GREATEST AND SMALLEST VARIANCES AND THEIR DEGREES OF
C.... FREEDOM.
      GRVAR = VAR ( 1 )
      SMVAR = VAR ( 1 )
      IDFGR = IFRE ( 1 ) - 1
      IDFSM = IFRE ( 1 ) - 1
   50 DO 62I = 2, IA
      IF ( VAR ( I ) - GRVAR ) 52, 52, 51
   51 GRVAR = VAR ( I )
      IDFGR = IFRE ( I ) - 1
      GO TO 62
   52 IF ( VAR ( I ) - SMVAR ) 61, 62, 62
   61 SMVAR = VAR ( I )
      IDFSM = IFRE ( I ) - 1
   62 CONTINUE
C.... FIND THE LESSER OF THE DEGREES OF FREEDOM
      IF ( IDFGR - IDFSM ) 70, 70, 71
   70 IDF = IDFGR
      GO TO 80
   71 IDF = IDFSM
   80 IDFU = IDF - 1
C.... CALCULATION OF MAXIMUM VALUE OF FS
      FS = GRVAR / SMVAR
      WRITE ( 6, 89 )
   89 FORMAT (15H0THE F-MAX TEST)
      WRITE ( 6, 90 ) FS, IDFU, IDF
   90 FORMAT (/4H FS=F10.5,24H WITH DEGREES OF FREEDOM,I3,4H AND,I3)
      GO TO 1
      END
```

```
CA3.9      TEST OF THE EQUALITY OF MEANS WITH HETEROGENEOUS VARIANCES
C.... USES SUBROUTINES INPUT, TRANSG
```

```
      DIMENSION IFRE ( 100 ), W ( 100 ), VAR ( 100 ), YBAR ( 100 )
  1   READ ( 5, 2 ) IA, ICODE, ITRANS
    2 FORMAT (3I5)
      IF ( IA.LE.0 ) STOP
      WRITE ( 6, 12 ) IA, ICODE, ITRANS
   12 FORMAT(1H1, I5, 22H SAMPLES, INPUT CODE =, I2, 22H TRANSFORMATION
     1CODE =, I2//)
      IF ( ICODE ) 20, 20, 30
   20 CALL INPUT ( IA, IFRE, YBAR, VAR, WAVVAR, ITRANS )
      GO TO 50
   30 READ ( 5, 40 ) ( IFRE ( I ), YBAR ( I ), VAR ( I ), I = 1, IA )
   40 FORMAT (I5, 2F10.5)
      WRITE ( 6, 41 ) ( IFRE ( I ), YBAR ( I ), VAR ( I ), I = 1, IA )
   41 FORMAT(9X,1HN,6X,4HYBAR,7X,3HVAR/(I10,2F10.3))
C.... CALCULATION OF W(I), SIGMA W(I), W(I)Y(I), SIGMA W(I)Y(I), AND
C.... SIGMA SQUARE
  50  SIGYWY = 0.
      SIGW = 0.
      SIGWY = 0.
      DO 70I = 1, IA
      FREQ = IFRE ( I )
      W ( I ) = FREQ / VAR ( I )
      WY = W ( I )* YBAR ( I )
      SIGW = SIGW + W ( I )
      SIGWY = SIGWY + WY
      SIGYWY = SIGYWY + YBAR ( I )* WY
   70 CONTINUE
C.... CALCULATION OF THE WEIGHTED GRAND MEAN
      WGMY = SIGWY / SIGW
C.... CALCULATION OF THE WEIGHTED CORRECTION TERM
      CTW = SIGWY ** 2 / SIGW
C.... CALCULATION OF THE WEIGHTED SUM OF SQUARE OF GROUPS
      SSGR = SIGYWY - CTW
      SIGC7 = 0.
      DO 90I = 1, IA
      FRE = IFRE ( I )
      SIGC7 = SIGC7 + ( 1. - W ( I )/ SIGW )** 2 /( FRE - 1. )
   90 CONTINUE
      A = IA
C.... CALCULATION OF FS
      FS = ( SSGR /( A - 1. ))/( 1. + ( 2. *( A - 2. )/( A ** 2 - 1. )*
     1SIGC7 ))
C.... CALCULATION OF DEGREES OF FREEDOM
      V1 = A - 1.
      V2 = ( A ** 2 - 1. )/( 3. * SIGC7 )
      WRITE ( 6, 100 ) FS
  100 FORMAT (5H0FS =F15.5)
      WRITE ( 6, 110 ) V1, V2
  110 FORMAT (24H0DEGREES OF FREEDOM ARE F5.0,1X,4H AND, F6.1)
      GO TO 1
      END
```

```
CA3.10       TUKEY-S TEST FOR NONADDITIVITY
C...., USES SUBROUTINES INPUT, TRANSG
      DIMENSION RC ( 10, 10 ), SIGR ( 10 ), SIGC ( 10 ), DEVR ( 10 ), DE
     1VC ( 10 ), Q ( 10 ), ICL ( 10 ), YBAR ( 100 ), VAR ( 100 )
      READ ( 5, 2 ) I, J, N, ICODE, ITRANS
    2 FORMAT (5I5)
      WRITE ( 6, 3 ) I, J, N, ICODE, ITRANS
    3 FORMAT(1H1/I5,5H ROWS,I5,8H COLUMNS,I5,22H REPLICATIONS PER CELL/
     1 13H INPUT CODE =,I2,22H TRANSFORMATION CODE =,I2/)
```

```
      IF ( I * J.LE.0 ) STOP
      XI = I
      XJ = J
      XN = N
      IF ( ICODE ) 31, 31, 35
 31   CALL INPUT2 ( I * J, N, YBAR, VAR, WAVVAR, ITRANS )
      GO TO 39
   35 LS = 1
      DO 37II = 1, I
      LEND = LS + J + 1
      READ ( 5, 36 ) ( YBAR ( L ), L = LS, LEND )
   36 FORMAT(5X,F10.5)
   37 LS = LEND + 1
C.... CALCULATION OF CELL TOTALS
C.... CALCULATION OF ROW TOTALS DIVIDED BY J, SIGR(I)
   39 SSBN = 0.
      L = 0
      DO 5IV = 1, I
      SIGRN = 0.
      DO 4JV = 1, J
      L = L + 1
      RC ( IV, JV ) = YBAR ( L )* XN
      SIGRN = RC ( IV, JV ) + SIGRN
    4 CONTINUE
      SSBN = SSBN + SIGRN * SIGRN
      SIGR ( IV ) = SIGRN / XJ
    5 CONTINUE
      SSB = SSBN /( XN * XJ )
C.... CALCULATION OF COLUMN TOTALS DIVIDED BY I, SIGC(J)
C.... CALCULATION OF GRAND TOTAL DIVIDED BY IXJ, SIGT
      SSAN = 0.
      SSSBGN = 0.
      SIGTN = 0.
      DO 7JV = 1, J
      SIGCN = 0.
      DO 6IV = 1, I
      SIGCN = SIGCN + RC ( IV, JV )
      SSSBGN = SSSBGN + RC ( IV, JV )* RC ( IV, JV )
    6 CONTINUE
      SIGTN = SIGTN + SIGCN
      SSAN = SSAN + SIGCN * SIGCN
      SIGC ( JV ) = SIGCN / XI
    7 CONTINUE
      SSA = SSAN /( XN * XI )
      SSSBG = SSSBGN / XN
      SIGT = SIGTN /( XI * XJ )
C.... CALCULATION OF DEVIATION OF SIGR(I)-SIGT
      DO 9IV = 1, I
      DEVR ( IV ) = SIGR ( IV ) - SIGT
    9 CONTINUE
C.... CALCULATION OF Q(I)=SIGMA (CELL TOTAL) X (SIGR(I)-SIGT), I=1,I
      DO 11JV = 1, J
      Q ( JV ) = 0.00
      DO 11IV = 1, I
      Q ( JV ) = RC ( IV, JV )* DEVR ( IV ) + Q ( JV )
 11   CONTINUE
C.... CALCULATION OF QQ=SIGMA(Q(I)) X (SIGC(J)-SIGT), J=1,J
C.... CALCULATION OF DEVIATION SIGC(J)-SIGT
C.... CALCULATION OF K
      QQ = 0.00
      SSDEVC = 0.00
      DO 12JV = 1, J
      DEVC ( JV ) = SIGC ( JV ) - SIGT
```

```
      QQ = Q ( JV )* DEVC ( JV ) + QQ
      SSDEVC = DEVC ( JV )** 2 + SSDEVC
   12 CONTINUE
      SSDEVR = 0.00
      DO 14IV = 1, I
      SSDEVR = DEVR ( IV )** 2 + SSDEVR
   14 CONTINUE
      XK = SSDEVC * SSDEVR
      SSNAD = QQ ** 2 /( XK * XN )
C.... CALCULATION OF SS(AXB)
C.... CALCULATION OF CORRECTION TERM
      L = ( I - 1 )*( J - 1 )
      M = L - 1
      XM = M
      CT = SIGTN ** 2 /( XI * XJ * XN )
      SSAXB = SSSBG - SSA - SSB + CT
      SSRES = SSAXB - SSNAD
      XMSNAD = SSNAD
      XMSRES = SSRES / XM
      FS = XMSNAD / XMSRES
      DO 98JV = 1, J
      ICL ( JV ) = JV
   98 CONTINUE
      WRITE ( 6, 104 )
  104 FORMAT (/ 2X, 6HSOURCE,4X,3H DF,11X,3H SS,12X,3H MS,12X,3H FS)
      WRITE ( 6, 105 ) L, SSAXB
  105 FORMAT (/ 3X, 3HAXB,6X,I3,1X,F15.4)
      WRITE ( 6, 106 ) SSNAD, XMSNAD, FS
  106 FORMAT (/ 3X, 7HNONADD.,3X,2H 1,1X,3F15.4)
      WRITE ( 6, 107 ) M, SSRES, XMSRES
  107 FORMAT (/ 3X, 6HRESID.,3X,I3,1X,2F15.4)
      STOP
      END
```

```
CA3.11               KRUSKAL AND WALLIS TEST
C.... USES SUBROUTINE RSORT
      DIMENSION Y ( 1000, 2 ), T ( 100 ), IFREQ ( 100 ), FMT ( 14 )
      COMMON Y, T, IFREQ, FMT
    1 READ ( 5, 2 ) IA, ( IFREQ ( I ), I = 1, IA )
    2 FORMAT(15I5)
      READ ( 5, 5 ) FMT
    5 FORMAT(13A6, A2)
      NTOT = 0
      WRITE ( 6, 7 ) IA, ( IFREQ ( I ), I = 1, IA )
    7 FORMAT(/20H KRUSKAL-WALLIS TEST/5X, I5, 7H GROUPS/5X,13H SAMPLE SI
     1ZES    /(20I5))
      ILAST = 0
      DO 10I = 1, IA
      IFIRST = ILAST + 1
      ILAST = IFIRST + IFREQ ( I ) - 1
      READ ( 5, FMT ) ( Y ( L, 1 ), L = IFIRST, ILAST )
      NTOT = NTOT + IFREQ ( I )
      DO 10L = IFIRST, ILAST
   10 Y ( L, 2 ) = I
C.... SORT DATA
      CALL RSORT ( Y, NTOT, 2, 1000, 1 )
C.... LOOK FOR TIES
      IT = 1
      YI = Y ( 1, 1 )
      L = 0
      NTOTP1 = NTOT + 1
```

```
      Y ( NTOTP1, 1 ) = - 99999.
      DO 30I = 2, NTOTP1
      IF ( Y ( I, 1 ) - YI ) 25, 21, 25
21    IT = IT + 1
      GO TO 30
25    YI = Y ( I, 1 )
      IF ( IT - 1 ) 26, 26, 28
C.... NO TIE
26    Y ( I - 1, 1 ) = I - 1
      IT = 1
      GO TO 30
C.... AT LEAST ONE TIED RANK
28    AVTIE = FLOAT ( I - IT ) + FLOAT ( IT - 1 )/ 2.
      L = L + 1
      IF ( L.GT.100 ) GO TO 999
      T ( L ) = IT
      DO 29K = 1, IT
      IND = I - IT - 1 + K
29    Y ( IND, 1 ) = AVTIE
      IT = 1
30    CONTINUE
C.... SORT BACK INTO ORIGINAL GROUPS
      CALL RSORT ( Y, NTOT, 2, 1000, 2 )
      R2SUM = 0.
      ILAST = 0.
      DO 50I = 1, IA
      RSUM = 0
      FN = IFREQ ( I )
      IFIRST = ILAST + 1
      ILAST = IFIRST + IFREQ ( I ) - 1
      DO 45J = IFIRST, ILAST
45    RSUM = RSUM + Y ( J, 1 )
50    R2SUM = R2SUM + RSUM * RSUM / FN
      FNTOT = NTOT
      H = 12. * R2SUM /( FNTOT *( FNTOT + 1. )) - 3. *( FNTOT + 1. )
      IF ( L ) 70, 70, 60
60    SUMT = 0
      DO 62I = 1, L
62    SUMT = SUMT + T ( I )*( T ( I ) - 1. )*( T ( I ) + 1. )
      D = 1. - SUMT /(( FNTOT - 1. )* FNTOT *( FNTOT + 1. ))
      H = H / D
C.... PRINT RESULTS
70    IAM1 = IA - 1
      WRITE ( 6, 75 ) H, IAM1
75    FORMAT(/5X, 32HTHE KRUSKAL-WALLIS STATISTIC H = ,F10.4/5X,  53HIS
     1TO BE COMPARED WITH A CHI SQUARE DISTRIBUTION WITH/ I5 ,9H DEGREES
     2  ,10HOF FREEDOM  /1H1 )
      GO TO 1
999   WRITE ( 6, 80 )
80    FORMAT (///14H TOO MANY TIES//)
      GO TO 1
      END


CA3.12                  LINEAR REGRESSION ANALYSIS PROGRAM
C.... USES SUBROUTINES INPUT, TRANSG
      DIMENSION IFREQ ( 100 ), YBAR ( 100 ), VAR ( 100 ), FMT ( 14 ), X
     1( 100 ), FREQ ( 100 )
      EQUIVALENCE  ( FREQ ( 1 ), IFREQ ( 1 ))
C.... READ INPUT PARAMETERS,IA = NO. OF X VALUES, ICODE = INPUT CODE,
C.... TRANX  AND TRANY ARE TRANSFORMATION CODES, TALPH = T VALUE AT
C.... P = ALPHA
```

```
      I1 = 1
  1   READ ( 5, 2 ) IA, ICODE, ITRANX, ITRANY, TALPH, ALPHA
  2   FORMAT (4I5,F10.3,F5.4)
      IF ( IA.LE.1 ) STOP
      WRITE ( 6, 3 ) IA, ICODE, ITRANX, ITRANY, TALPH, ALPHA
  3   FORMAT(1H1/I5, 9H X-VALUES  ,I10,13H = INPUT CODE ,22H, TRANSFORMA
     1TION CODES  , I3,1H,, I3/3X,3HT =,F6.3,7H AT P =,F6.4//)
      READ ( 5, 4 ) FMT
  4   FORMAT (13A6, A2)
      DF = 0.
      IF ( ICODE ) 5, 5, 7
  5   READ ( 5, FMT ) ( X ( I ), I = 1, IA )
  6   CALL INPUT ( IA, IFREQ, YBAR, VAR, WAVVAR, ITRANY )
      GO TO 20
  7   WAVVAR = 0.
      DO 12I = 1, IA
      READ ( 5, FMT ) X ( I ), IFREQ ( I ), YBAR ( I ), VAR ( I )
      IF ( IFREQ ( I ) - 1 ) 10, 10, 11
 10   IFREQ ( I ) = 1
      YBAR ( I ) = TRANSG ( ITRANY, YBAR ( I ))
 11   FNM1 = IFREQ ( I ) - 1
      WAVVAR = WAVVAR + FNM1 * VAR ( I )
 12   DF = DF + FNM1
      IF ( DF ) 20, 20, 13
 13   WAVVAR = WAVVAR / DF
C.... COMPUTE REGRESSION STATISTICS
 20   SUMX = 0.
      SUMY = 0.
      TOTN = 0.
      DO 25I = 1, IA
      X ( I ) = TRANSG ( ITRANX, X ( I ))
      FREQ ( I ) = IFREQ ( I )
      SUMX = SUMX + FREQ ( I )* X ( I )
      SUMY = SUMY + FREQ ( I )* YBAR ( I )
 25   TOTN = TOTN + FREQ ( I )
      NTOTM1 = TOTN - 1.
      IAM1 = IA - 1
      IAM2 = IA - 2
      FIAM1 = IAM1
      FIA = IA
      DF = TOTN - FIA
      NDF = DF
      XBAR = SUMX / TOTN
      YBARB = SUMY / TOTN
      SUMX2 = 0.
      SUMY2 = 0.
      SUMXY = 0.
      DO 30I = 1, IA
      XD = X ( I ) - XBAR
      YD = YBAR ( I ) - YBARB
      SUMX2 = SUMX2 + XD * XD * FREQ ( I )
      SUMY2 = SUMY2 + YD * YD * FREQ ( I )
 30   SUMXY = SUMXY + XD * YD * FREQ ( I )
C....
      VARX = SUMX2 /( TOTN - 1. )
      SSERR = DF * WAVVAR
      SSTOT = SUMY2 + SSERR
      VARY = SSTOT /( TOTN - 1. )
      B = SUMXY / SUMX2
      A = YBARB - B * XBAR
      WRITE ( 6, 41 ) A, B, XBAR, YBARB, TOTN, VARX, VARY
 41   FORMAT(//31H THE REGRESSION EQUATION IS Y =,F10.5,2H +,F10.5,2H X
     1/,12H MEAN OF X =,F10.5,3X,11HMEAN OF Y =,F10.5,3X, 9HTOTAL N= ,
```

```
     2F5.0 /16H VARIANCE OF X =,F15.5,3X,15HVARIANCE OF Y =, F15.5 / )
      VARG = SUMY2 / FIAM1
      SSEXP = SUMXY * SUMXY / SUMX2
      SSUNEX = SUMY2 - SSEXP
      VUNEX = SSUNEX / FLOAT ( IA - 2 )
C.... CONFIDENCE LIMITS TO SLOPE
      SBXT = SQRT ( VUNEX / SUMX2 )* TALPH
      B1 = B - SBXT
      B2 = B + SBXT
      PCENT = ( 1. - ALPHA )* 100.
      WRITE ( 6, 42 ) PCENT, B1, B2
 42   FORMAT(//F6.1,44H PERCENT CONFIDENCE LIMITS FOR THE SLOPE ARE  ,
     1F10.4, 4H AND, F10.4)
      FSGP = - 9999.9999
      FSDEV = - 9999.9999
      IF ( NDF ) 45, 45, 43
 43   FSGP = VARG / WAVVAR
      FSDEV = VUNEX / WAVVAR
 45   FSEXP = SSEXP / VUNEX
C.... PRINT ANOVA TABLE
      WRITE ( 6, 50 ) SUMY2, IAM1, VARG, FSGP, SSEXP, I1, SSEXP, FSEXP,
     1SSUNEX, IAM2, VUNEX, FSDEV, SSERR, NDF, WAVVAR, SSTOT, NTOTM1
 50   FORMAT(//25X,20HANALYSIS OF VARIANCE//7H SOURCE, 7X, 2HSS, 20X,
     12H DF, 14X, 2HMS, 18X, 2HFS // 7H GROUPS , F13.4, I17,2F20.4/
     22X, 6HLINEAR, F15.4, I17, 2F20.4/ 2X,4HDEV., F17.4, I17, 2F20.4//
     3 6H ERROR, F14.4, I17,  F20.4// 6H TOTAL, F14.4, I17)
C.... PREPARE TABLE OF YHAT,CONFIDENCE LIMITS, AND DEVIATIONS
      WRITE ( 6, 60 )
 60   FORMAT(///2X,6HSAMPLE,5X,1HX,11X,1HY,10X,2HL1,10X,4HYHAT,8X,
     12HL2,9X,9HDEVIATION /2X,6HNUMBER  //)
      DO 70I = 1, IA
      YHAT = A + B * X ( I )
      DEV = YBAR ( I ) - YHAT
      STDRXT = SQRT ( VUNEX *( 1. / TOTN + ( X ( I ) - XBAR )*( X ( I )
     1- XBAR )/ SUMX2 ))* TALPH
      Y1 = YHAT - STDRXT
      Y2 = YHAT + STDRXT
      WRITE ( 6, 65 ) I, X ( I ), YBAR ( I ), Y1, YHAT, Y2, DEV
 65   FORMAT(I8,6F12.5)
 70   CONTINUE
      GO TO 1
      END


CA3.13        PRODUCT MOMENT CORRELATION COEFFICIENT
C.... CALLS SUBROUTINES TRANSG, FNDRMD
      DIMENSION Y1 ( 500 ), Y2 ( 500 ), PY1 ( 8 ), PY2 ( 8 ), FMT ( 14
C     ARITH STATEMENT FUNCTIONS
      Z ( A ) = 0.5 * ALOG (( 1. + A )/( 1. - A ))
      R ( A ) = ( EXP ( A ) - EXP ( - A ))/( EXP ( A ) + EXP ( - A ))
 1    READ ( 5, 2 ) N, CHISQ, FDIS, ALPHA, ITRAN1, ITRAN2
    2 FORMAT(I5,3F5.3,2I5)
      IF ( N.LE.1 ) STOP
      WRITE ( 6, 3 ) N, CHISQ, FDIS, ALPHA, ITRAN1, ITRAN2
    3 FORMAT(1H1/4H N =,I5,14H, CHI SQUARE =,F10.3,11H, F-VALUE =, F10.3
     1,9H, ALPHA =,F5.4/29H THE TRANSFORMATION CODES ARE,I3,4H AND,I3)
      READ ( 5, 5 ) FMT
    5 FORMAT(13A6,A2)
      WRITE ( 6, 6 )
    6 FORMAT(//52X,2HY1,8X,2HY2//)
      DO 10K = 1, N
      READ ( 5, FMT ) Y1 ( K ), Y2 ( K )
```

```
   10 CONTINUE
C.... CALCULATION OF THE SUMS AND SUMS OF SQUARES FOR EACH VARIABLE
      FN = N
      SUMY1 = 0.
      SUMY2 = 0.
   20 DO 30I = 1, N
      IF ( ITRAN1 ) 25, 25, 21
 21   Y1 ( I ) = TRANSG ( ITRAN1, Y1 ( I ))
 25   SUMY1 = SUMY1 + Y1 ( I )
      IF ( ITRAN2 ) 30, 30, 26
 26   Y2 ( I ) = TRANSG ( ITRAN2, Y2 ( I ))
   30 SUMY2 = SUMY2 + Y2 ( I )
      Y1MEAN = SUMY1 / FN
      Y2MEAN = SUMY2 / FN
      SSQY1 = 0.
      SSQY2 = 0.
      SPROD = 0.
      DO 39I = 1, N
      SSQY1 = SSQY1 + ( Y1 ( I ) - Y1MEAN )** 2
      SSQY2 = SSQY2 + ( Y2 ( I ) - Y2MEAN )** 2
   39 SPROD = SPROD + ( Y1 ( I ) - Y1MEAN )*( Y2 ( I ) - Y2MEAN )
C.... CALCULATION OF SUM OF SQUARES OF Y1 AND Y2 AND SUM OF PRODUCTS
C.... CALCULATION OF THE PRODUCT-MOMENT CORRELATION COEFFICIENT
   70 COCOEF = SPROD / SQRT ( SSQY1 * SSQY2 )
C.... CALCULATION OF VARIANCE, COVARIANCE, AND MEAN
      DF = FN - 1.
      VARY1 = SSQY1 / DF
      VARY2 = SSQY2 / DF
      COV = SPROD / DF
      SDEV1 = SQRT ( VARY1 )
      SDEV2 = SQRT ( VARY2 )
      SERR1 = SDEV1 / SQRT ( FN )
      SERR2 = SDEV2 / SQRT ( FN )
      WRITE ( 6, 80 ) Y1MEAN, Y2MEAN, VARY1, VARY2, SDEV1, SDEV2, SERR1,
     1 SERR2, COV, COCOEF
   80 FORMAT(///25X,4HMEAN,17X,2F10.3/
     11H ,24X,8HVARIANCE,13X,2F10.3/
     21H ,24X,18HSTANDARD DEVIATION,3X,2F10.3/
     31H ,24X,14HSTANDARD ERROR,7X,2F10.3/
     41H ,24X,10HCOVARIANCE,16X,F10.3/
     548H THE PRODUCT-MOMENT CORRELATION COEFFICIENT IS    ,F8.5)
      ZOB = Z ( COCOEF )
      SZ = 1. / SQRT ( FN - 3. )
      TALPH = FNORMD ( ALPHA / 2. )
      Z1 = ZOB - TALPH * SZ
      Z2 = ZOB + TALPH * SZ
      R1 = R ( Z1 )
      R2 = R ( Z2 )
      PCENT = ( 1. - ALPHA )* 100.
      WRITE ( 6, 81 ), PCENT, R1, R2
 81   FORMAT(//4H0THE F6.1,36H PER CENT CONFIDENCE LIMITS ARE L1 =
     1F8.4, 9H AND L2 =  ,F8.4)
C.... CALCULATION OF THE EIGENVALUES
      IF ( COV ) 84, 82, 84
   82 XLMDA1 = VARY1
      XLMDA2 = VARY2
      B1 = 1.0E - 10
      GO TO 1
   84 D = SQRT (( VARY1 + VARY2 )** 2 - 4. *( VARY1 * VARY2 - COV ** 2 )
     1)
      XLMDA1 = ( VARY1 + VARY2 + D )/ 2.
      XLMDA2 = VARY1 + VARY2 - XLMDA1
C.... CALCULATION OF THE SLOPE OF THE PRINCIPAL AXIS
```

```
      B1 = COV /( XLMDA1 - VARY1 )
C.... WRITE THE EQUATION OF THIS AXIS
      WRITE ( 6, 85 )
   85 FORMAT(38H0THE EQUATION OF THE PRINCIPAL AXIS IS)
      A = Y1MEAN - B1 * Y2MEAN
      WRITE ( 6, 89 ) A, B1
C.... CALCULATION OF THE SLOPE OF THE MINOR AXIS
      B2 = - 1. / B1
C.... WRITE THE EQUATION OF THE MINOR AXIS
      WRITE ( 6, 87 )
   87 FORMAT(34H0THE EQUATION OF THE MINOR AXIS IS)
      A = Y1MEAN - B2 * Y2MEAN
      WRITE ( 6, 89 ) A, B2
   89 FORMAT(1H ,5X,3HY1=,F10.5,2H +,F10.5,3H Y2)
C.... CALCULATION OF CONFIDENCE LIMITS FOR B1
      H = CHISQ /( XLMDA1 / XLMDA2 + XLMDA2 / XLMDA1 - 2. )/ FN
      A = SQRT ( H /( 1. - H ))
      B1LIM1 = ( B1 - A )/( 1. + B1 * A )
      B1LIM2 = ( B1 + A )/( 1. - B1 * A )
      PCENT = ( 1. - ALPHA )* 100.
      WRITE ( 6, 95 ) PCENT, B1LIM1, B1LIM2
   95 FORMAT(4H0THE,F5.1, 66H PER CENT CONFIDENCE LIMITS TO THE SLOPE OF
     1 THE PRINCIPAL AXIS ARE//1H ,5X,3HL1=,F10.5,5X,3HL2=,F10.5)
C.... CALCULATION OF CONFIDENCE ELLIPSES FOR THE BIVARIATE MEAN
      C05 = XLMDA1 * XLMDA2 *( FN - 1. )* 2. * FDIS / FN /( FN - 2. )
C.... CALCULATION OF THE COORDINATES OF PARTICULAR POINTS FOR WHICH
C.... MOST OF THE EQUATION REDUCES TO ZERO
      WRITE ( 6, 97 )
   97 FORMAT (54H1FOLLOWING PAIRS SHOW THE COORDINATES OF POINTS A TO H/
     148H FOR PLOTTING CONFIDENCE ELLIPSE (SEE BOX 15.5).
      WRITE ( 6, 100 )
  100 FORMAT(1H ,15X,2HY1,20X,2HY2)
      DO 690I = 1, 8
      IF (( I / 2 * 2 + 1 ) .EQ.I ) GO TO 602
      SIGN = - 1.
      GO TO 603
  602 SIGN = 1.
  603 IF ( I - 2 ) 610, 610, 611
  610 PY2 ( I ) = Y2MEAN
      PY1 ( I ) = Y1MEAN + SIGN * SQRT ( C05 / VARY2 )
      GO TO 690
  611 IF ( I - 4 ) 620, 620, 621
  620 PY1 ( I ) = Y1MEAN
      PY2 ( I ) = Y2MEAN + SIGN * SQRT ( C05 / VARY1 )
      GO TO 690
  621 IF ( I - 6 ) 630, 630, 640
  630 PY2 ( I ) = Y2MEAN + SIGN * SQRT ( C05 /( XLMDA2 *( 1. + B1 ** 2 )
     1))
      PY1 ( I ) = Y1MEAN + B1 *( PY2 ( I ) - Y2MEAN )
      GO TO 690
  640 PY2 ( I ) = Y2MEAN + SIGN * SQRT ( C05 /( XLMDA1 *( 1. + B2 ** 2 )
     1))
      PY1 ( I ) = Y1MEAN + B2 *( PY2 ( I ) - Y2MEAN )
  690 CONTINUE
      WRITE ( 6, 700 ) ( PY1 ( I ), PY2 ( I ), I = 1, 8 )
  700 FORMAT(/2H A, 2F20.4/2H B, 2F20.4/2H C, 2F 20.4/2H D, 2F20.4/
     12H F, 2F20.4/2H F, 2F20.4/2H G, 2F20.4/2H H, 2F20.4)
      GO TO 1
      END
```

```
CA3.14                  FISHER-S EXACT TEST
C.... USES SUBROUTINE FACT
 150  READ ( 5, 1 ) A, B, C, D
   1  FORMAT (4F5.0)
      XRSRT = 0.000000
      IA = A
      IB = B
      IC = C
      ID = D
      WRITE ( 6, 2 ) IA, IB, IC, ID
   2  FORMAT (14H1OBSERVED DATA//2I6//2I6//)
C.... CALCULATION OF FACTORIALS OF MARGINAL TOTALS AND TOTAL FREQUENCY
      RC13 = FACT ( IA + IB )
      RC23 = FACT ( IC + ID )
      RC31 = FACT ( IA + IC )
      RC32 = FACT ( IB + ID )
      RC33 = FACT ( IA + IB + IC + ID )
C.... SUM OF LOG, OF MARGINAL TABLES MINUS LOG, OF TOTAL FREQUENCY
      RCUP = RC13 + RC23 + RC31 + RC32 - RC33
C.... TEST OF A*D-B*C
C.... IF POSITIVE, GO TO 50 AND INTERCHANGE CELLS AND GO BACK TO 14
C.... IF ZERO, NO CALCULATION
      IF ( A * D - B * C ) 14, 100, 50
C.... COMPUTE NUMBER OF WORSE CASES
  14  I = MIN1 ( A, D )
      ISTOU = I + 1
C.... CALCULATION OF SUM OF LOGS OF THE OBSERVED OR WORSE CASE(S)
  15  DO 33J = 1, ISTOU
      ISTOV = IA - J + 1
      IF ( ISTOV ) 17, 17, 16
  16  RC11 = FACT ( ISTOV )
      GO TO 18
  17  RC11 = 0.000000
  18  ISTOW = IB + J - 1
      RC12 = FACT ( ISTOW )
      ISTOX = IC + J - 1
      RC21 = FACT ( ISTOX )
      ISTOY = ID - J + 1
      IF ( ISTOY ) 20, 20, 19
  19  RC22 = FACT ( ISTOY )
      GO TO 21
  20  RC22 = 0.0
  21  RCLP = RC11 + RC12 + RC21 + RC22
      XRS = RCUP - RCLP
C.... TRANSFORM THE PROBABILITY INTO REAL VALUE FROM LOG
      XRSR = EXP ( XRS )
      IF ( J - 1 ) 22, 22, 26
  22  XRSRO = XRSR
C.... WRITE THE PROBABILITY OF OBSERVED CASES
      WRITE ( 6, 23 ) XRSRO
  23  FORMAT (44HOPROBABILITY    OF    OBSERVED    SAMPLE    =F10.6)
      GO TO 33
  26  WRITE ( 6, 27 ) ISTOV, ISTOW, ISTOX, ISTOY
  27  FORMAT (16HONEXT WORSE CASE//2I6//2I6//)
C.... WRITE THE PROBABILITY OF WORSE CASES
      WRITE ( 6, 28 ) XRSR
  28  FORMAT (44H PROBABILITY     OF     THIS      CASE      =F10.11)
      XRSRT = XRSRT + XRSR
  33  CONTINUE
  35  XRSRV = XRSRT + XRSRO
      GO TO 99
C.... INTERCHANGE THE CELLS TO PUT TABLE IN STANDARD FORM
```

```
 50   ISAVE = IA
      SAVE = A
      IA = IB
      A = B
      IB = ISAVE
      B = SAVE
      ISAVE = IC
      SAVE = C
      IC = ID
      C = D
      ID = ISAVE
      D = SAVE
      WRITE ( 6, 51 )
 51   FORMAT (19H0CELLS INTERCHANGED)
C.... GO BACK TO 14 FOR CALCULATION
      GO TO 14
C.... MULTIPLY THE CUMULATIVE PROBABILITY BY 2 FOR 2-TAILED TEST
 99   XRSRU = 2. * XRSRV
      WRITE ( 6, 101 ) XRSRU
101   FORMAT (44H02X        CUMULATIVE        PROBABILITY      =F10.6)
      GO TO 150
100   WRITE ( 6, 103 )
103   FORMAT (8H AD = BC)
      GO TO 150
      END
```

```
CA3.15               R X C TEST OF INDEPENDENCE
C.... USES FUNCTIONS  G, TEST, AND SUBROUTINE CHECK
      COMMON STAT, SIGLV, MR ( 30 ), MC ( 30 ), CHISQ, KR, KC, FNCOL ( 3
     10 ), FN ( 30, 30 ), NNSIG, IRSIG, ICSIG, NR, NC
 1    READ ( 5, 4 ) NR, NC, CHISQ, SIGLV, ICODE
 4    FORMAT (2I5, 2F10.5,I5)
      IF ( NR * NC ) 2, 2, 3
 2    STOP
 3    WRITE ( 6, 5 ) NR, NC, CHISQ, SIGLV, ICODE
 5    FORMAT(1H1/I5,5H ROWS,I5,8H COLUMNS/22H CRITICAL CHI SQUARE =,
     1F10.3,9H, ALPHA =,F6.4/12H TEST CODE =,I2//)
      NNSIG = 0
      JJEND = 1
      IIEND = 1
      IF ( ICODE ) 9, 9, 6
  6   JJEND = NR - 1
      IIEND = NC - 1
  9   DO 15I = 1, NR
      READ ( 5, 10 ) ( FN ( I, J ), J = 1, NC )
 10   FORMAT(15F5.0)
 15   WRITE ( 6, 11 ) ( FN ( I, J ), J = 1, NC )
 11   FORMAT ( 1X, 15F5.0 )
      IF (( NR.EQ.2 ) .AND. ( NC.EQ.2 )) GO TO 70
C.... START LOOPS
 16   JJ = NR + 1
 20   JJ = JJ - 1
      KR = JJ
      IF ( JJ - JJEND ) 50, 50, 21
 21   DO 22I = 1, JJ
 22   MR ( I ) = I
      IRSIG = 0
C.... CALCULATE COLUMN SUMS FOR ROWS BEING TESTED.
 25   DO 27K = 1, NC
      FNSUM = 0
      DO 26K2 = 1, KR
```

```
      K3 = MR ( K2 )
 26   FNSUM = FNSUM + FN ( K3, K )
 27   FNCOL ( K ) = FNSUM
C.... INNER LOOP
 30   II = NC + 1
 31   II = II - 1
      KC = II
      IF ( II - IIEND ) 40, 40, 32
 32   DO 33I = 1, II
 33   MC ( I ) = I
      ICSIG = 0
 35   IF ( TEST ( II ) .EQ.0. ) CALL CHECK
      IF ( NNSIG - 100 ) 36, 998, 998
C.... COMPUTE NEXT COMBINATION OF COLUMNS.
 36   MC ( II ) = MC ( II ) + 1
      IF ( MC ( II ) .LE.NC ) GO TO 35
      IJ = 0
 38   IJ = IJ + 1
      K = II - IJ
      IF ( K ) 37, 37, 371
C.... IF ALL N.S. GO TO BOTTOM OF COLUMN LOOP
  37  IF ( ICSIG ) 40, 40, 31
 371  MC ( K ) = MC ( K ) + 1
      IF ( MC ( K ) .GT.NC - IJ ) GO TO 38
      DO 39L = 1, IJ
      NN = K + L
 39   MC ( NN ) = MC ( NN - 1 ) + 1
      GO TO 35
C.... COMPUTE NEXT COMBINATION OF ROWS.
 40   MR ( JJ ) = MR ( JJ ) + 1
      IF ( MR ( JJ ) .LE.NR ) GO TO 25
      JI = 0
 45   JI = JI + 1
      K = JJ - JI
      IF ( K ) 451, 451, 452
C.... IF ALL N.S. THEN GO TO END.
 451  IF ( IRSIG ) 50, 50, 20
 452  MR ( K ) = MR ( K ) + 1
      IF ( MR ( K ) .GT.NR - JI ) GO TO 45
      DO 46L = 1, JI
      NN = K + L
 46   MR ( NN ) = MR ( NN - 1 ) + 1
      GO TO 25
C.... BOTTOM OF LOOPS
C.... END, CHECK IF TOTAL SET SIGNIFICANT
  50  IF ( NNSIG ) 51, 51, 1
  51  WRITE ( 6, 52 )
  52  FORMAT(//23H ENTIRE SET SIGNIFICANT)
      GO TO 1
 70   CORR = 0.5
C.... 2X2 USE YATES CORRECTION
      IF ( FN ( 1, 1 )* FN ( 2, 2 ) - FN ( 1, 2 )* FN ( 2, 1 )) 71, 16,
     175
 71   CORR = - 0.5
 75   FN ( 1, 1 ) = FN ( 1, 1 ) - CORR
      FN ( 2, 2 ) = FN ( 2, 2 ) - CORR
      FN ( 1, 2 ) = FN ( 1, 2 ) + CORR
      FN ( 2, 1 ) = FN ( 2, 1 ) + CORR
      WRITE ( 6, 76 ) (( FN ( I, J ), J = 1, 2 ), I = 1, 2 )
 76   FORMAT (/5X,33HYATES CORRECTION HAS BEEN APPLIED /(2F6.1))
      GO TO 16
 998  WRITE ( 6, 999 )
```

```
 999  FORMAT ( // 30H NO MORE SUBSETS CAN BE TESTED // )
      GO TO 1
      END
```

```
CA3.16                              RANDOMIZATION TEST
C.... USES SUBROUTINES STAT, COMPR, RAND, AND RSORT.
      INTEGER IFREQ ( 65 ), COUNT
      COMMON LOT ( 10 ), X ( 500, 2 ), FMT ( 14 )
C.... READ SAMPLE SIZES FOR THE TWO GROUPS, THE NUMBER OF ITERATIONS TO
C.....BE CARRIED OUT, AND THE INITIAL RANDOM NUMBER.
    1 READ ( 5, 10 ) N1, N2, ITER, IRPAT
   10 FORMAT(3I5,I11)
      IF ( N1 * N2.EQ.0 ) STOP
      WRITE ( 6, 20 ) N1, N2, ITER, IRPAT
   20 FORMAT( 1H1 /  17H SAMPLE SIZES ARE, I5, 4H AND, I5/I5,   52H ITER
     1ATIONS WITH THE INITIAL RANDOM PATTERN EQUAL TO, I12//)
      READ ( 5, 30 ) FMT
   30 FORMAT(13A6,A2)
C.... READ DATA FOR THE FIRST GROUP.
      READ ( 5, FMT ) ( X ( I, 1 ), I = 1, N1 )
      N1P1 = N1 + 1
      N = N1 + N2
C.... READ DATA FOR THE SECOND GROUP.
      READ ( 5, FMT ) ( X ( I, 1 ), I = N1P1, N )
      COUNTL = 0
      COUNTU = 0
C.... COMPUTE THE STATISTIC FOR THE INITIAL TWO GROUPS.
      OBSVAL = STAT ( X, N1 + N2, N1, N2 )
      WRITE ( 6, 50 ) OBSVAL
   50 FORMAT(//1X,33HOBSERVED VALUE OF THE STATISTIC =  ,F10.5)
C.... PARTITION THE DATA INTO TWO GROUPS AT RANDOM AND RECOMPUTE THE
C.... STATISTICS.
      DO 70I = 1, ITER
      DO 60J = 1, N
 60   X ( J, 2 ) = RAND ( IRPAT )
      CALL RSORT ( X, N, 2, 500, 2 )
      VALU = STAT ( X, N1 + N2, N1, N2 )
   70 CALL COMPR ( VALU, OBSVAL, COUNTL, COUNTU )
C.... PRINT THE RESULTS.
      FITER = ITER
      PCNTL = COUNTL / FITER
      PCNTU = COUNTU / FITER
      CSUM = COUNTL + COUNTU
      WRITE ( 6, 80 ) CSUM, COUNTL, COUNTU, ITER, PCNTL, PCNTU
   80 FORMAT(//    F7.0, 55H SAMPLES WERE MORE DEVIANT THAN THE OBSERVED
     1 STATISTIC  /F7.0, 26H WERE IN THE LEFT TAIL AND, F6.0,23H WERE IN
     2 THE RIGHT TAIL/8H (OUT OF  , I6, 9H SAMPLES) , / 1X,       24HIN
     3PROPORTIONS THESE ARE  , F7.4, 4H AND , F7.4, 14H RESPECTIVELY )
      WRITE ( 6, 81 ) IRPAT
  81  FORMAT(/23H FINAL RANDOM INTEGER =,I12)
      GO TO 1
      END
```

```
CBINOM SUBROUTINE TO COMPUTE THE FIRST NCLAS TERMS OF THE BINOMIAL
C.... DISTRIBUTION WITH K=FK, P=P
      SUBROUTINE BINOM ( FK, P, Y, FHATR, NCLAS )
      DIMENSION FHATR ( 30 )
      Q = 1. - P
      FHATR ( 1 ) = Q ** FK
```

```
      FACTOR = P / Q
      DO 10I = 2, NCLAS
      FI = I
   10 FHATR ( I ) = FHATR ( I - 1 )* FACTOR *( FK - FI + 2. )/( FI - 1.
     1)
      RETURN
      END


CCHECK       SUBROUTINE TO CHECK FOR MAXIMUM NONSIGNIFICANT SUBSETS
C     FOR PROGRAM A3.15
      SUBROUTINE CHECK
      COMMON STAT, SIGLV, MR ( 30 ), MC ( 30 ), CHISQ, KR, KC, FNCOL ( 3
     10 ), FN ( 30, 30 ), NNSIG
      DIMENSION LIST ( 100, 60 )
      IF ( NNSIG ) 51, 51, 2
    2 DO 50I = 1, NNSIG
      DO 10II = 1, KR
      IND = MR ( II )
      IF ( LIST ( I, IND )) 50, 50, 10
C.... IF = ZERO THEN PRINT, SINCE THIS SET CANNOT BE A SUBSET
   10 CONTINUE
C.... ROW SUBSET, NOW CHECK COLUMNS
      DO 20JJ = 1, KC
      IND = MC ( JJ ) + 30
      IF ( LIST ( I, IND )) 50, 50, 20
   20 CONTINUE
C.... ALSO A COLUMN SUBSET
      RETURN
   50 CONTINUE
C.... NOT A SUBSET, ALL TO LIST
   51 NNSIG = NNSIG + 1
      DO 55I = 1, 60
   55 LIST ( NNSIG, I ) = 0
      DO 60II = 1, KR
      IND = MR ( II )
   60 LIST ( NNSIG, IND ) = 1
      DO 70JJ = 1, KC
      IND = MC ( JJ ) + 30
   70 LIST ( NNSIG, IND ) = 1
      WRITE ( 6, 80 ) NNSIG, STAT, ( MR ( I ), I = 1, KR )
   80 FORMAT(/7H SUBSET,I4,5X,13HG-STATISTIC =,F10.4/10X,6HROWS =,30I3)
      WRITE ( 6, 81 ) ( MC ( J ), J = 1, KC )
   81 FORMAT(10X,6HCOLS =,30I3)
      RETURN
      END


CCOMOUT              COMPUTE BASIC STATISTICS AND OUTPUT THEM
      SUBROUTINE COMOUT ( N, NCLAS, ITRANS, ALPHA, TALPH, YBAR, S, S2, S
     1UMY3, SUMY4, FMEDN, DMAX )
      FN = N
  300 CV = ( S / YBAR )* 100.
      G1 = ( FN /( FN - 1. ))* SUMY3 /(( FN - 2. )* S2 * S )
      G2 = (( FN + 1. )/( FN - 1. ))*( FN /( FN - 2. ))*( SUMY4 /(( FN -
     1 3. )* S2 * S2 )) - 3. *( FN - 1. )*( FN - 1. )/(( FN - 2. )*( FN
     2- 3. ))
C.... STANDARD ERRORS, CONFIDENCE LIMITS
      SYBAR = S / SQRT ( FN )
      YBARL1 = YBAR - TALPH * SYBAR
      YBARL2 = YBAR + TALPH * SYBAR
```

```
      SMED = 1.2533 * SYBAR
      YMEDL1 = FMEDN - TALPH * SMED
      YMEDL2 = FMEDN + TALPH * SMED
      SCV = CV * SQRT (( 1. + 2. *( CV / 100. )*( CV / 100. ))/( 2. * FN
     1 ))
      CVL1 = CV - TALPH * SCV
      CVL2 = CV + TALPH * SCV
      SG1 = SQRT ( 6. * FN *( FN - 1. )/(( FN - 2. )*( FN + 1. )*( FN +
     13. )))
      FND = FNORMD ( ALPHA / 2. )
      G1L1 = G1 - FND * SG1
      G1L2 = G1 + FND * SG1
      SG2 = SQRT ( 24. * FN *( FN - 1. )*( FN - 1. )/(( FN - 3. )*( FN -
     1 2. )*( FN + 3. )*( FN + 5. )))
      G2L1 = G2 - FND * SG2
      G2L2 = G2 + FND * SG2
C.... BASIC STAT. OUTPUT
C
      WRITE ( 6, 401 ) N, NCLAS, ITRANS
 401  FORMAT(1H /10X,16HBASIC STATISTICS  /5X, 3HN= , I5, I9, 8H CLASSES
     1,2X,21HTRANSFORMATION CODE =   , I4///)
      CONF = 100. *( 1. - ALPHA )
      WRITE ( 6, 403 ) CONF
 403  FORMAT(16X,9HSTATISTIC,4X,11HSTAND.ERROR,8X,17HCONFIDENCE LIMITS /
     148X,  1H( , F5.2,  1H, 9HPER CENT) //)
      WRITE ( 6, 405 ) YBAR, SYBAR, YBARL1, YBARL2
 405  FORMAT (2X, 4HMEAN,4X,2F15.5,          3X, 2F11.5 /)
      WRITE ( 6, 407 ) FMEDN, SMED, YMEDL1, YMEDL2
 407  FORMAT (2X, 6HMEDIAN,2X,2F15.5  ,3X,2F11.5/)
      WRITE ( 6, 409 ) S2
 409  FORMAT ( 2X 8HVARIANCE,  F15.5/)
      WRITE ( 6, 411 ) S
 411  FORMAT( 2X,10HSTAND.DEV.   , F13.5   /)
      WRITE ( 6, 412 ) CV, SCV, CVL1, CVL2
 412  FORMAT(2X,10HCOEFF. VAR   , F13.5 ,F15.5,3X,2F11.5 /)
      WRITE ( 6, 413 ) G1, SG1, G1L1, G1L2
 413  FORMAT (2X,  2HG1,6X, 2F15.5, 3X,2F11.5/)
      WRITE ( 6, 415 ) G2, SG2, G2L1, G2L2
 415  FORMAT ( 2X, 2HG2, 6X, 2F15.5, 3X, 2F11.5/)
      WRITE ( 6, 417 ) DMAX
 417  FORMAT (2X, 8HK-S DMAX, F15.5 /)
      RETURN
      END
```

```
CCOMPR                VARIANCE RATIO TEST FOR PROGRAM A3.16
C.... VALU IS THE CURRENT VALUE OF THE STATISTIC FOR A RANDOM PARTITION
C.... OF THE DATA.  OBSVAL IS THE ORIGINAL OBSERVED VALUE OF THE
C.... STATISTIC. COUNTL IS A COUNTER USED TO RECORD THE NUMBER OF
C.... PARTITIONS WHICH RESULT IN VALU BEING MORE DEVIANT THAN THE
C.... OBSERVED VALUE IN THE LEFT (LOWER) TAIL AND COUNTU IS A SIMILAR
C.... COUNTER FOR THE RIGHT (UPPER) TAIL
C.... (NOTE-VARIANCES MUST BE COMPARED IN TERMS OF RATIOS).
      SUBROUTINE COMPR ( VALU, OBSVAL, COUNTL, COUNTU )
      IF (( VALU.GE.OBSVAL ) .AND. ( VALU.GE.1. / OBSVAL )) COUNTU = COU
     1NTU + 1.
      IF (( VALU.LE.OBSVAL ) .AND. ( VALU.LE.1. / OBSVAL )) COUNTL = COU
     1NTL + 1.
      RETURN
      END
```

```
CFACT                  FUNCTION TO COMPUTE THE LOG OF N FACTORIAL.
C.... THIS FUNCTION IS USED BY PROGRAM A3.14
      FUNCTION FACT ( N )
      FACT = 0.000000
      M = N
      DO 1I = 1, M
      FACT = FACT + ALOG ( FLOAT ( I ))
    1 CONTINUE
      RETURN
      END
```

```
CFNORMD  COMPUTES NORMAL DEVIATE CORRESPONDING TO A GIVEN PROBABILITY
      FUNCTION FNORMD ( PROB )
      T = SQRT ( ALOG ( 1. /( PROB * PROB )))
      FNORMD = T - ( 2.515517 + T *( 0.802853 + T * 0.010328 ))/( 1. + T
     1 *( 1.432788 + T *( 0.189269 + T * 0.001308 )))
      RETURN
      END
```

```
CG                 F LOG F FOR A3.15
      DOUBLE PRECISION FUNCTION G ( FI )
      DOUBLE PRECISION X, DLOG
      G = 0D0
      IF ( FI.EQ.0. ) RETURN
      X = FI
      G = 2D0 * X * DLOG ( X )
      RETURN
      END
```

```
CINPUT           INPUT SUBROUTINE
      SUBROUTINE INPUT ( IA, IFRE, YBAR, VAR, WAVVAR, ITRANS )
      DIMENSION IFRE ( 20 ), YBAR ( 20 ), VAR ( 20 ), DATA ( 100 ), FMT
     1( 14 )
      READ ( 5, 2 ) ( IFRE ( I ), I = 1, IA )
    2 FORMAT (15I5)
      WRITE ( 6, 30 )
 30   FORMAT (1H 3X,6HSAMPLE,2X,1HN,5X,4HMEAN,3X,8HVARIANCE/4X,6HNUMBER)
      SUMDF = 0.
C.... CALCULATION OF MEANS, VARIANCES, SUM OF DF, AND WEIGHTED AVERAGE
      WAVVAN = 0.0
      READ ( 5, 20 ) FMT
 20   FORMAT(13A6,A2)
      DO 70I = 1, IA
      IFREQ = IFRE ( I )
      FREQ = IFREQ
      DF = FREQ - 1.
      READ ( 5, FMT ) ( DATA ( J ), J = 1, IFREQ )
      SUM = 0.
      DO 40J = 1, IFREQ
      IF ( ITRANS ) 35, 35, 31
 31   DATA(J)=TRANSG(ITRANS,DATA(J))
 35   SUM = SUM + DATA ( J )
 40   CONTINUE
      YBART = SUM / FREQ
      YBAR ( I ) = YBART
      VARN = 0.
      IF ( IFREQ - 1 ) 45, 45, 49
```

```
45    VART = 0.
      GO TO 51
49    DO 50J = 1, IFREQ
      VARN = VARN + ( DATA ( J ) - YBAR ( I ))** 2
50    CONTINUE
      VART = VARN / DF
51    VAR ( I ) = VART
      SUMDF = SUMDF + DF
      WRITE ( 6, 60 ) I, IFREQ, YBART, VART
60    FORMAT(/ I7,I5,F10.4,F12.4)
      WAVVAN = WAVVAN + VARN
70    CONTINUE
      WAVVAR = WAVVAN / SUMDF
      RETURN
      END
```

```
CINPUT2          INPUT2 SUBROUTINE        USED ONLY BY A3.10
      SUBROUTINE INPUT2 ( IA, IFREQ, YBAR, VAR, WAVVAR, ITRANS )
      DIMENSION YBAR ( 20 ), VAR ( 20 ), DATA ( 100 ), FMT ( 14 )
      WRITE ( 6, 30 )
 30   FORMAT (1H 3X,6HSAMPLE,2X,1HN,5X,4HMEAN,3X,8HVARIANCE/4X,6HNUMBER)
      SUMDF = 0.
C...CALCULATION OF MEANS, VARIANCES, SUM OF DF, AND  A WEIGHTED AVERAGE
      WAVVAN = 0.0
      READ ( 5, 20 ) FMT
 20   FORMAT(13A6,A2)
      DO 70I = 1, IA
      FREQ = IFREQ
      DF = FREQ - 1.
      READ ( 5, FMT ) ( DATA ( J ), J = 1, IFREQ )
      SUM = 0.
      DO 40J = 1, IFREQ
      IF ( ITRANS ) 35, 35, 31
 31   DATA(J)=TRANSG(ITRANS,DATA(J))
 35   SUM = SUM + DATA ( J )
 40   CONTINUE
      YBART = SUM / FREQ
      YBAR ( I ) = YBART
      VARN = 0.
      IF ( IFREQ - 1 ) 45, 45, 49
 45   VART = 0.
      GO TO 51
 49   DO 50J = 1, IFREQ
      VARN = VARN + ( DATA ( J ) - YBAR ( I ))** 2
 50   CONTINUE
      VART = VARN / DF
 51   VAR ( I ) = VART
      SUMDF = SUMDF + DF
      WRITE ( 6, 60 ) I, IFREQ, YBART, VART
 60   FORMAT(/ I7,I5,F10.4,F12.4)
      WAVVAN = WAVVAN + VARN
 70   CONTINUE
      WAVVAR = WAVVAN / SUMDF
      RETURN
      END
```

```
CKSFD            K - S TEST FOR DATA IN A FREQUENCY DISTRIBUTION
      SUBROUTINE KSFD ( N, NCLAS, YUPLM, F, YBAR, S, DMAX, FHAT )
      DIMENSION Y ( 1000 ), F ( 30 ), FHAT ( 30 ), YUPLM ( 30 )
```

```
C.... Y IS A LIST OF UPPER CLASS LIMITS FOR EACH CLASS.  NOTE THAT THE
C.... FIRST AND LAST CLASSES ARE CONSIDERED TO EXTEND TO - AND +
C.... INFINITY.
      YHCUM1 = 0.
      YCUM = 0.
      FN = N
      DMAX = 0.
      Y ( NCLAS ) = YBAR + 100. * S
      DO 10I = 1, NCLAS
      YCUM = YCUM + F ( I )/ FN
      YHCUM2 = PROBN ( YUPLM ( I ), YBAR, S )
      ABDIFF = ABS ( YHCUM2 - YCUM )
      FHAT ( I ) = ( YHCUM2 - YHCUM1 )* FN
      YHCUM1 = YHCUM2
      IF ( ABDIFF - DMAX ) 10, 10, 5
  5   DMAX = ABDIFF
 10   CONTINUE
      RETURN
      END
```

```
CKSRD                     K - S TEST ON RAW UNORDERED DATA
      SUBROUTINE KSRD ( N, Y, YBAR, S, DMAX )
C.... CALLS SUBROUTINE RSORT,REQUIRED BY PROGRAM A3.1
      DIMENSION Y ( 1000 )
      FN = N
      DELTA = 1. / FN
      CALL RSORT ( Y, N, 1, N, 1 )
      DMAX = 0
      YCUM = 0
      DO 10I = 1, N
      YCUM = YCUM + DELTA
      YHCUM = PROBN ( Y ( I ), YBAR, S )
      ABDIFF = ABS ( YCUM - YHCUM )
      IF ( ABDIFF - DMAX ) 10, 10, 5
 5    DMAX = ABDIFF
10    CONTINUE
      RETURN
      END
```

```
CPOISN SUBROUTINE TO COMPUTE THE FIRST NCLAS TERMS OF THE POISSON
C.... DISTRIBUTION WITH MEAN EQUAL TO YBAR.
      SUBROUTINE POISN ( YBAR, Y, FHATR, NCLAS )
      DIMENSION F ( 30 ), FHATR ( 30 )
      FHATR ( 1 ) = EXP ( - YBAR )
      DO 10I = 2, NCLAS
      FI = I - 1
  10  FHATR ( I ) = FHATR ( I - 1 )* YBAR / FI
      RETURN
      END
```

```
CPROBN   COMPUTES THE PROBABILITY OF OBTAINING A VARIATE LESS THAN OR
C.... EQUAL TO Y FROM A NORMAL DISTRIBUTION WITH MEAN YBAR AND STANDARD
C.... DEVIATION S.
      FUNCTION PROBN ( Y, YBAR, S )
      Z = ( Y - YBAR )/ S
      X = ABS ( Z )
      T = 1. /( 1. + 0.2316419 * X )
      PROBN = 1.0 - 0.3989423 * EXP ( - X * X / 2. )* T *( 0.31938153 +
```

```
     1T *( - .35656378 + T *( 1.78147794 + T *( - 1.82125598 + T * 1.330
     227443 ))))
      IF ( Z ) 5, 10, 10
  5   PROBN = 1. - PROBN
 10   RETURN
      END


CRAND       PSEUDO RANDOM NUMBER GENERATOR FOR PROGRAM A3.16
C.... THIS TYPE OF ROUTINE IS QUITE COMPUTER DEPENDENT.  OPTIONAL METHOD
C.... DEPENDS UPON WORD LENGTHS OF COMPUTER AND WHETHER BINARY OR
C.... DECIMAL ARITHMETIC IS USED.  IT IS SUGGESTED THAT THE USER SHOULD
C.... CHECK HIS LOCAL COMPUTATION CENTER FOR AN APPROPRIATE ROUTINE AND
C.... SUBSTITUTE IT FOR THE  CALL TO -RCM- GIVEN BELOW.
      FUNCTION RAND ( IPAT )
      RAND = RCM ( IPAT )
      RETURN
      END


CRSORT             SUBROUTINE TO SORT AN ARRAY (  IROW BY ICOL  )
C.... BY ROWS ACCORDING TO THE VALUES IN COLUMN  ICOLST
      SUBROUTINE RSORT ( NAME, IROW, ICOL, IROWD, ICOLST )
      REALNAME ( IROWD, ICOL )
  1   NSWIT = 1
      IROWM1 = IROW - 1
      DO 12I = 1, IROWM1
      IF ( NAME ( I, ICOLST ) - NAME ( I + 1, ICOLST )) 12, 12, 8
  8   NSWIT = 2
      DO 10K = 1, ICOL
      TEMP = NAME ( I, K )
      NAME ( I, K ) = NAME ( I + 1, K )
 10   NAME ( I + 1, K ) = TEMP
 12   CONTINUE
      GO TO  ( 13, 1 ), NSWIT
 13   RETURN
      END


CSTAT    TEST VARIANCES FOR PROGRAM A3.16
      FUNCTION STAT ( X, NTOT, N1, N2 )
C.... X IS AN ARRAY  THE FIRT N1 ENTRIES ARE FOR SAMPLE NO. 1 AND THE
C.... NEXT N2 ENTRIES ARE FOR THE SECOND SAMPLE.
C.... NTOT=N1+N2
C.... THE FUNCTION RETURNS AS ITS VALUE THE DESIRED STATISTIC
      DIMENSION X ( NTOT )
C.... ARITH STATEMENT FUNCTION
      VAR ( SUM, SUM2, FN ) = ( SUM2 - SUM * SUM / FN )/( FN - 1. )
      FN1 = N1
      SUM = 0.
      SUM2 = 0.
      DO 10I = 1, N1
      SUM = SUM + X ( I )
 10   SUM2 = SUM2 + X ( I )* X ( I )
      V1 = VAR ( SUM, SUM2, FN1 )
      N1P1 = N1 + 1
      FN2 = N2
      SUM = 0.
      SUM2 = 0.
      DO 20I = N1P1, NTOT
      SUM = SUM + X ( I )
```

```
 20    SUM2 = SUM2 + X ( I )* X ( I )
       V2 = VAR ( SUM, SUM2, FN2 )
       STAT = V2 / V1
       RETURN
       END


CTEST              TEST FUNCTION  FOR A3.15
       FUNCTION TEST ( IDUM )
C..... NOTE ARGUMENT IDUM IS NOT ACTUALLY USED.
       COMMON STAT, SIGLV, MR ( 30 ), MC ( 30 ), CHISQ, KR, KC, FNCOL ( 3
      10 ), FN ( 30, 30 ), NNSIG, IRSIG, ICSIG, NR, NC
       DOUBLE PRECISION G, GSTAT
       GSTAT = 0
       DO 3II = 1, KR
       FNSUM = 0.0
       I = MR ( II )
       DO 2JJ = 1, KC
       J = MC ( JJ )
       GSTAT = GSTAT + G ( FN ( I, J ))
    2  FNSUM = FNSUM + FN ( I, J )
    3  GSTAT = GSTAT - G ( FNSUM )
       FNTOT = 0
       DO 4JJ = 1, KC
       JI = MC ( JJ )
       GSTAT = GSTAT - G ( FNCOL ( JI ))
 4     FNTOT = FNTOT + FNCOL ( JI )
       STAT = GSTAT + G ( FNTOT )
       IF (( NR.EQ.KR ) .AND. ( NC.EQ.KC )) WRITE ( 6, 99 ) STAT
       TEST = 0
       IF ( STAT - CHISQ ) 12, 12, 10
  10   TEST = 1
       IRSIG = 1
       ICSIG = 1
  12   RETURN
 99    FORMAT (/ 2X, 3HG =, F10.4,1X, 13 HFOR TOTAL SET /)
       END


CTRANSG       THIS SUBROUTINE PERFORMS VARIOUS COMMON TRANSFORMATIONS
       FUNCTION TRANSG ( ICODE, DATA )
C..... THIS FUNCTION IS USED BY PROGRAMS A3.1,A3.2,A3.4,A3.5,A3.6,A3.7,
C..... A3.8,A3.9,A3.10,A3.12,AND A3.13.
C..... THE FOLLOWING IS AN ARITHEMATIC STATEMENT FUNCTION
       ASIN ( X ) = ATAN ( X / SQRT ( 1. - X * X ))
       IF ( ICODE ) 99, 99, 100
 99    TRANSG = DATA
       RETURN
 100   GO TO  ( 1, 2, 3, 4, 5, 6, 7, 8, 9 ), ICODE
 1     TRANSG = ABS ( DATA )
       RETURN
 2     TRANSG = SQRT ( DATA )
       RETURN
 3     TRANSG = SQRT ( DATA + 0.5 )
       RETURN
 4     TRANSG = DATA * DATA
       RETURN
 5     TRANSG = ALOG10 ( DATA )
       RETURN
 6     TRANSG = ALOG10 ( DATA + 1.0 )
       RETURN
```

```
7     IF ( DATA - 1. ) 75, 71, 71
71    TRANSG = 90.
      RETURN
75    TRANSG = ASIN ( SQRT ( DATA ))* 57.29578
      RETURN
8     IF ( DATA - 100. ) 85, 71, 71
85    TRANSG = ASIN ( SQRT ( DATA / 100. ))* 57.29578
      RETURN
9     TRANSG = 1.0 / DATA
      RETURN
      END
```

APPENDIX

4 TABULAR GUIDE TO STATISTICAL METHODS

The following tables provide a guide to the statistical methods featured in the text. Research problems are classified by their main purpose (e.g., obtaining a statistic of location or of association), by whether they are based on one or more samples, by whether the data are measurement variables (continuous and meristic variables) or attributes, and by similar considerations. References are given to the boxes or other places in the book in which a method is featured.

Preliminary Key

1. The statistical analysis is for a single variable only 2
   The statistical analysis is for two variables 3
2. The main purpose of the analysis is to
   - Estimate a statistic of location . . . . Table A4.1, page 744
   - Estimate a statistic of dispersion . . . . Table A4.2, page 746
   - Estimate a statistic describing shape or distribution of a sample or compute a theoretical frequency distribution . . . . Table A4.3, page 747
   - Miscellaneous procedures other than the above Table A4.6, page 749
3. The main purpose of the analysis is to
   - Establish a functional relationship between two variables . . . . Table A4.4, page 748
   - Estimate association between two variables . . Table A4.5, page 748

TABLE A4.1
Methods Aimed at Statistics of Location.

Type of analysis	*Arrangement of samples*	*Data are expressed as measurement variables*	*Data are expressed as ranks*	*Data are frequencies of attributes*
Analysis of a single sample		Computing median of frequency distribution: Box 4.1 Computing arithmetic mean from unordered sample: Box 4.2; from frequency distribution: Box 4.3 Setting confidence limits for the mean: Box 7.2 Comparison of a sample statistic with an expected value: Box 7.4		Confidence limits for a percentage: Section 16.1
Analysis of two or more samples	Samples are logically arranged by a single classification	Single classification anova with unequal sample sizes: Box 9.1; with equal sample sizes: Box 9.4 A priori comparison of means in anova: Box 9.8; single degree of freedom comparisons of means: Box 14.10 A posteriori comparison of means; *SNK* test for means based on equal sample sizes: Box 9.9; on unequal sample sizes: Box 9.10; *STP* test, refer to Section 9.7 Test of equality of means when variances are heterogeneous: Box 13.3	Kruskal and Wallis test: Box 13.5 A posteriori comparison of means by a nonparametric *STP*: Box 13.7	*G*-test for homogeneity of percentages: Boxes 16.5 and 16.8
	Samples are logically arranged in a nested classification	Two-level nested anova based on equal sample sizes: Box 10.1; on unequal sample sizes: Box 10.4 Three-level nested anova based on equal sample sizes: Box 10.3; on unequal sample sizes: Box 10.5		

TABLE A4.1 (*continued*)

Type of analysis	*Arrangement of samples*	*Data are expressed as measurement variables*	*Data are expressed as ranks*	*Data are frequencies of attributes*
	Samples are logically arranged by a two-way or multi-way classification	Two-way anova with replication: Box 11.1; without replication: Box 11.2; with unequal but proportional subclass sizes: Box 11.4; with a single missing observation: Box 11.5 Three-way anova: Box 12.1 More-than-three-way classification: see Section 12.3 and Box 12.2 Test for nonadditivity in a two-way anova: Box 13.1	Friedman's method for randomized blocks: Box 13.8	Randomized blocks for frequency data (repeated testing of the same individuals): Box 16.12
Special methods for two samples	Individual variates in the two samples are independent	Single classification anova with unequal sample sizes: Box 9.1; with equal sample sizes: Box 9.5 *t*-test comparing two means or other statistics of location (sample size equal or unequal): Box 9.6 *t*-test comparing a single observation with a sample mean: Box 9.7 Testing difference between two means with unequal variances: Box 13.4	Mann-Whitney *U*-test: Box 13.6	Testing difference between two percentages: Box 16.10
	Individual variates in the two samples are paired	Paired comparisons test: Box 11.3	Wilcoxon's signed ranks test: Box 13.9	Repeated testing of the same individuals: Box 16.12

TABLE A4.2
Methods Aimed at Statistics of Dispersion.

Type of analysis	*Data are expressed as measurement variables*
Analysis of a single sample	Computing standard deviation from unordered sample: Box 4.2; from frequency distribution: Box 4.3 Setting confidence limits to the variance: Box 7.3
Analysis of two or more samples*	Estimate variance components based on unequal sample sizes: Box 9.2; on equal sample sizes: Box 9.3 Setting confidence limits to a variance component (based on equal sample sizes): Box 9.3 Test of homogeneity of variances: Box 13.2
Special method for two samples	Test of equality of two variances: Box 8.1

* All Model II anovas are fundamentally concerned with dispersion. Rather than duplicate here the table to the structure of anovas, we refer the reader to the analogous portion of Table A4.1 for statistics of location (rows for two or more samples), where most of the methods are applicable to Model II anova as well.

TABLE A4.3

Methods Aimed at Describing Shape or Distribution of a Sample, Including Computation of a Theoretical Frequency Distribution.

Type of analysis	*Data are expressed as measurement variables or as frequencies of attributes*	*Data are expressed as measurement variables only*	*Data are frequencies of attributes*
Computing expected frequencies for theoretical frequency distributions		Normal expected frequencies: Box 6.1	Binomial expected frequencies: Box 5.1 Poisson expected frequencies: Box 5.2
Analysis of a single sample		Procedure for grouping a frequency distribution: Box 2.1 Computing g_1 and g_2: Box 6.2	
Comparison of a single sample with an expected frequency distribution	Goodness of fit tests with parameters from an extrinsic hypothesis: Box 16.1; from an intrinsic hypothesis: Box 16.2; chi-square test for goodness of fit with more than two classes: Table 16.2	Kolmogorov-Smirnov test of goodness of fit: Box 16.3 Graphic "tests" for normality; large sample sizes: Box 6.3; small sample sizes (rankit test): Box 6.4 Test of sample statistic of shape against expected value: Box 7.4	
Comparison of several samples with an expected frequency distribution	For these tests, see: Box 16.4 A posteriori analysis of replicated tests of goodness of fit by *STP*: Box 16.5		

TABLE A4.4

Methods Aimed at Describing a Functional Relationship Between Two Variables.

Type of analysis	*Data are expressed as measurement variables*
Analysis of a single sample	Regression statistics; computation: Box 14.1; standard errors: Box 14.2; confidence limits: Box 14.3 Estimation of X from Y: Box 14.7 Test for curvilinear regression by orthogonal polynomials: Table 14.3
Three or more samples, each sample representing a different value of X	Regression with more than one value of Y per value of X; unequal sample sizes: Box 14.4; equal sample sizes: Box 14.6; standard errors and confidence limits for these regression statistics: Box 14.5 Test for curvilinear regression by orthogonal polynomials: Table 14.3
Two or more regression equations obtained from separate samples	Comparison of regression coefficients: Box 14.8; a posteriori comparison of these by *STP*: Box 14.9

TABLE A4.5

Methods Aimed at Statistics of Association.

Type of analysis	*Data are expressed as measurement variables*	*Data are expressed as ranks*	*Data are frequencies of attributes*
Analysis of a single sample	Computation of correlation coefficient from unordered sample: Box 15.1; from two-way frequency distribution: Box 15.2 Significance tests and confidence limits for correlation coefficients: Box 15.3 Principal axes and confidence regions for bivariate scattergram: Box 16.5 Olmstead and Tukey's corner test for association: Box 15.7	Kendall coefficient of rank correlation: Box 15.6	2×2 test for independence by G-test: Box 16.6; by chi-square test: Section 16.4; by Fisher's exact test: Box 16.7 $R \times C$ test for independence: Box 16.8 Three-way classification test of independence: Box 16.9
Analysis of two or more samples	Test of homogeneity of correlation coefficients: Box 15.4		Test for interaction in a three-way table: Box 16.9

TABLE A4.6

Miscellaneous Procedures Other Than the Above.

Type of analysis	*Reference*	
Combining probabilities of separate significance tests	For this test, see: Box 17.1	
Tests for randomness of measurement variables or attributes	Runs test for dichotomized data: Box 17.2 Runs tests for trends: Box 17.3	
	Data are expressed as measurement variables	*Data are frequencies of attributes*
Planning experiments	Finding number of replicates for tests on means: Box 9.11 Optimum allocation of resources: Box 11.6	Sample size required for detecting a given true difference between two percentages: Box 16.11
Transformations of continuous or meristic variables	Logarithmic: Table 13.2; square root: Table 13.3; arcsine: Table 13.4	

BIBLIOGRAPHY

Alder, H. L., and E. B. Roessler. 1968. *Introduction to Probability and Statistics.* 4th ed. W. H. Freeman and Company, San Francisco and London. 333 pp.

Allee, W. C., and E. Bowen. 1932. Studies in animal aggregations: Mass protection against colloidal silver among goldfishes. *J. Exp. Zool.*, **61**:185–207.

Anderson, R. L., and T. A. Bancroft. 1952. *Statistical Theory in Research.* McGraw-Hill, New York. 399 pp.

Archibald, E. E. A. 1950. Plant populations. II. The estimation of the number of individuals per unit area of species in heterogeneous plant populations. *Ann. Bot. N.S.*, **14**:7–21.

Ball, G. H. 1965. Data analysis in the social sciences: What about the details? *Proc. Fall Joint Computer Conference*, pp. 553–559.

Bancroft, T. A. 1964. Analysis and inference for incompletely specified models involving the use of preliminary test(s) of significance. *Biometrics*, **20**:427–442.

Band, H. T., and P. T. Ives. 1963. Genetic structure of populations. I. On the nature of the genetic load in the South Amherst population of *Drosophila melanogaster. Evolution*, **17**:198–215.

Banta, A. M. 1939. Studies on the physiology, genetics, and evolution of some Cladocera. Carnegie Institution of Washington, Dept. Genetics, Paper 39. 285 pp.

Barnes, H., and F. A. Stanbury. 1951. A statistical study of plant distribution during the colonization and early development of vegetation on china clay residues. *J. Ecol.*, **39**:171–181.

Blakeslee, A. F. 1921. The globe mutant in the Jimson Weed (*Datura stramonium*). *Genetics*, **6**:241–264.

Bliss, C. I., and D. W. Calhoun. 1954. *An Outline of Biometry.* Yale Co-op. Corp., New Haven, Conn. 272 pp.

Bliss, C. I., and R. A. Fisher. 1953. Fitting the negative binomial distribution to biological data and note on the efficient fitting of the negative binomial. *Biometrics*, **9**:176–200.

Block, B. C. 1966. The relation of temperature to the chirp-rate of male snowy tree crickets, *Oecanthus fultoni* (Orthoptera: Gryllidae). *Ann. Entomol. Soc. Amer.*, **59**:56–59.

Bortkiewicz, L. von. 1898. *Das Gesetz der Kleinen Zahlen.* Teubner, Leipzig. 52 pp.

Bradley, J. V. 1968. *Distribution-free Statistical Tests.* Prentice-Hall, Englewood Cliffs, N.J. 388 pp.

Brian, M. V., and J. Hibble. 1964. Studies of caste differentiation in *Myrmica rubra* L. 7. Caste bias, queen age and influence. *Insectes Sociaux*, **11**:223–228.

Brower, L. P. 1959. Speciation in butterflies of the *Papilio glaucus* group. I. Morphological relationships and hybridization. *Evolution*, **13**:40–63.

Brown, B. E., and A. W. A. Brown. 1956. The effects of insecticidal poisoning on the level of cytochrome oxidase in the American cockroach. *J. Econ. Entomol.*, **49**:675–679.

Brown, F. M., and W. P. Comstock. 1952. Some biometrics of *Heliconius charitonius* (Linnaeus) (Lepidoptera, Nymphalidae). *Amer. Mus. Novitates*, **1574**. 53 pp.

Burr, E. J. 1960. The distribution of Kendall's score S for a pair of tied rankings. *Biometrika*, **47**:151–171.

Burrell, P. C., and T. W. Whitaker. 1939. The effect of indol-acetic acid on fruit-setting in muskmelons. *Amer. Soc. Hort. Science*, **37**:829–830.

Carter, G. R., and C. A. Mitchell. 1958. Methods for adapting the virus of Rinderpest to rabbits. *Science*, **128**:252–253.

Clark, F. H. 1941. Correlation and body proportion in mature mice of the genus *Peromyscus*. *Genetics*, **26**:283–300.

Cochran, W. G., and G. M. Cox. 1957. *Experimental Designs.* 2nd ed. Wiley, New York. 611 pp.

Cohen, A. C., Jr. 1960. Estimating the parameter in a conditional Poisson distribution. *Biometrics*, **16**:203–211.

Cohen, A. I. 1954. Studies on glycosis during early development of the *Rana pipiens* embryo. *Physiol. Zool.* **27**:128–140.

Cooley, W. W., and P. R. Lohnes. 1962. *Multivariate Procedures for the Behavioral Sciences.* Wiley, New York. 211 pp.

Cowan, I. M., and P. A. Johnston. 1962. Blood serum protein variations at the species and subspecies level in deer of the genus *Odocoileus*. *Syst. Zool.*, **11**:131–138.

Croxton, F. E., D. J. Cowden, and S. Klein. 1967. *Applied General Statistics.* 3rd ed. Prentice-Hall, New York. 754 pp.

Davis, E. A., Jr. 1955. Seasonal changes in the energy balance of the English sparrow. *Auk*, **72**(4):385–411.

Dixon, W. J. (ed.). 1965. *Biomedical Computer Programs.* Health Sciences Computing Facility, Univ. California, Los Angeles. 620 pp.

Dixon, W. J., and F. J. Massey, Jr. 1957. *Introduction to Statistical Analysis.* 2nd ed. McGraw-Hill, New York. 488 pp.

Dupraw, E. J. 1965. Non-Linnean taxonomy and the systematics of honeybees. *Syst. Zool.*, **14**:1–24.

Dwass, M. 1960. Some k-sample rank-order tests, In *Contributions to Probability and Statistics, Essays in Honor of Harold Hotelling.* I. Olkin et al. (eds.). Stanford University Press, Stanford. pp. 198-202.

Ehrlich, P. R. 1955. The distribution and subspeciation of *Erebia epipsodea* Butler (Lepidoptera: Satyridae). *Univ. Kansas Sci. Bull.*, **37**:175–194.

Eisenhart, C. 1947. The assumptions underlying the analysis of variance. *Biometrics*, **3**:1–21.

Falconer, D. S. 1960. *Introduction to Quantitative Genetics*. Oliver & Boyd, Edinburgh. 365 pp.

Farina, M. V. 1966. *FORTRAN IV Self-Taught*. Prentice-Hall, Englewood Cliffs, N.J. 426 pp.

Federer, W. T. 1955. *Experimental Design*. Macmillan, New York. 544(+47) pp.

Finney, D. J. 1963. *Probit Analysis*. Rev. ed. Cambridge University Press, London. 318 pp.

Fisher, R. A. 1954. *Statistical Methods for Research Workers*. 12th ed. Oliver & Boyd, Edinburgh. 356 pp.

Fisher, R. A., and F. Yates. 1963. *Statistical Tables for Biological, Agricultural and Medical Research*. 6th ed. Oliver & Boyd, Edinburgh. 138 pp.

Fröhlich, F. W. 1921. *Grundzüge einer Lehre vom Licht- und Farbensinn. Ein Beitrag zur allgemeinen Physiologie der Sinne*. Fischer, Jena. 86 pp.

Fruchter, B. 1954. *Introduction to Factor Analysis*. Van Nostrand, New York. 280 pp.

Gabriel, K. R. 1964. A procedure for testing the homogeneity of all sets of means in analysis of variance. *Biometrics*, **20**:459–477.

Gartler, S. M., I. L. Firschein, and T. Dobzhansky. 1956. Chromatographic investigation of urinary amino-acids in the great apes. *Am. J. Phys. Anthropol.*, **14**:41–57.

Geissler, A. 1889. Beiträge zur Frage des Geschlechtsverhältnisses der Geborenen. *Z. K. Sächs. Stat. Bur.*, **35**:1–24.

Gnedenko, B. V., and A. YA. Khinchin. 1961. *An Elementary Introduction to the Theory of Proability. W. H. Freeman and Company, San Francisco and London.* 139 pp.

Golden, J. T. 1965. *FORTRAN IV Programming and Computing*. Prentice-Hall, Englewood Cliffs, N.J. 270 pp.

Goldstein, A. 1964. *Biostatistics*. Macmillan, New York. 272 pp.

Goulden, C. H. 1952. *Methods of Statistical Analysis*. 2nd ed. Wiley, New York. 467 pp.

Greenwood, M., and G. U. Yule. 1920. An inquiry into the nature of frequency-distributions of multiple happenings. *J. Roy. Stat. Soc.*, **83**:255–279.

Greig-Smith, P. 1964. *Quantitative Plant Ecology*. 2nd ed. Butterworths, Washington, D.C. 256 pp.

Grizzle, J. E. 1967. Continuity correction in the x^2-test for 2 x 2 tables. *Amer. Statist.*, Oct. **1967**:28–32.

Gumbel, E. J. 1954. Statistic theory of extreme values and some practical applications. U.S. Dept. of Commerce, National Bureau of Standards, Appl. Math. Ser. 33, 51 pp.

Haltenorth, T. 1937. Die verwandtschaftliche Stellung der Grosskatzen zueinander. *Z. Säugetierk.*, **12**:97–240.

Hanna, B. L. 1953. On the relative importance of genetic and environmental factors in the determination of human hair pigment concentration. *Am. J. Human Genetics*, **5**:293–321.

Harman, H. H. 1960. *Modern Factor Analysis*. University of Chicago Press, Chicago. 469 pp.

Hartley, H. O. 1962. Analysis of variance, In *Mathematical Methods for Digital Com-*

puters, Vol. 1. A. Ralston and H. S. Wilf (eds.). Wiley, New York. pp. 221–230.

Hasel, A. A. 1938. Sampling error in timber surveys. *J. Agr. Res.*, **57**:713–737.

Hotelling, H. 1953. New light on the correlation coefficient and its transforms. *J. Roy. Stat. Soc.*, Ser B, **15**:193–232.

Hubbs, C. L., and C. Hubbs. 1953. An improved graphical analysis and comparison of series of samples. *Syst. Zool.*, **8**:50–56.

Hunter, P. E. 1959. Selection of *Drosophila melanogaster* for length of larval period. *Z. Vererbungsl.*, **90**:7–28.

Johnson, N. K. 1966. Bill size and the question of competition in allopatric and sympatric populations of Dusky and Gray Flycatchers. *Syst. Zool.*, **15**:70–87.

Karten, I. 1965. Genetic differences and conditioning in *Tribolium castaneum*. *Physiol. Zoology*, **38**:69–79.

Kendall, M. G., and W. R. Buckland. 1960. *A Dictionary of Statistical Terms*. 2nd ed. Hafner, New York. 575 pp.

Kendall, M. G., and A. Stuart. 1961. *The Advanced Theory of Statistics*. Vol. 2. Hafner, New York. 676 pp.

King, J. A., D. Maas, and R. G. Weisman. 1964. Geographic variation in nest size among species of *Peromyscus*. *Evolution*, **18**:230–234.

Kouskolekas, C. A., and G. C. Decer. 1966. The effect of temperature on the rate of development of the potato leafhopper, *Empoasca fabae* (Homoptera: Cicadellidae). *Ann. Entomol. Soc. Amer.*, **59**:292–298.

Kruskal, W. H., and W. A. Wallis. 1952. Use of ranks in one-criterion variance analysis. *J. Am. Stat. Assoc.*, **47**:583–621.

Kullback, S. 1959. *Information Theory and Statistics*. Wiley, New York. 395 pp.

Leggatt, C. W. 1935. Contributions to the study of the statistics of seed testing. *Proc. Internat. Seed Testing Assoc.*, **5**:27–37.

Lewontin, R. C., and J. Felsenstein. 1965. The robustness of homogeneity tests in 2 x n tables. *Biometrics*, **21**:19–33.

Lewontin, R. C., and M. J. D. White. 1960. Interaction between inversion polymorphisms of two chromosome pairs in the grasshopper, *Moraba scurra*. *Evolution*, **14**:116–129.

Li, J. C. R. 1964. *Statistical Inference*. Edwards Bros., Ann Arbor, Mich. Vol. I, 658 pp.; Vol. II, 575 pp.

Littlejohn, M. J. 1965. Premating isolation in the *Hyla ewingi* complex. *Evolution*, **19**:234–243.

MacArthur, J. W. 1931. Linkage studies with the tomato. III. Fifteen factors in six groups. *Trans. R. Canad. Inst.*, **18**:1–19.

Mandel, J. 1964. *The Statistical Analysis of Experimental Data*. Wiley, New York. 410 pp.

McCracken, D. D. 1965. *A Guide to FORTRAN IV Programming*. Wiley, New York. 151 pp.

Miller, J. C. P. 1954. *Tables of Binomial Coefficients*. Royal Society Mathematical Tables. Vol. 3.

Millis, J., and Y. P. Seng. 1954. The effect of age and parity of the mother on birth weight of the offspring. *Ann. Human Genetics*, **19**:58–73.

Mittler, T. E., and R. H. Dadd. 1966. Food and wing determination in *Myzus persicae* (Homoptera: Aphidae). *Ann. Entomol. Soc. Amer.*, **59**:1162–1166.

Molina, E. C. 1942. *Poisson's Exponential Binomial Limit*. Van Nostrand, New York. 47 pp.

Morrison, D. F. 1967. *Multivariate Statistical Methods*. McGraw-Hill, New York. 338 pp.

Nelson, V. E. 1964. The effects of starvation and humidity on water content in *Tribolium confusum* Duval (Coleoptera). Unpublished Ph.D. thesis, University of Colorado. 111 pp.

Newman, K. J., and H. V. Meredith. 1956. Individual growth in skeletal bigonial diameter during the childhood period from 5 to 11 years of age. *Am. J. Anatomy*, **99**:157–187.

O'Connell, C. P. 1953. The life history of the Cabezon *Scorpaenichthys marmoratus* (Ayres). *Calif. Div. Fish and Game, Fish Bull.*, **93**. 76 pp.

Olson, E. C., and R. L. Miller. 1958. *Morphological Integration*. University of Chicago. Chicago Press. 317 pp.

Ostle, B. 1963. *Statistics in Research*. 2nd ed. Iowa State University Press, Ames. 585 pp.

Park, W. H., A. W. Williams, and C. Krumwiede. 1924. *Pathogenic Microörganisms*. Lea & Febiger, Philadelphia and New York. 811 pp.

Pearce, S. C. 1965. *Biological Statistics*. McGraw-Hill, New York. 212 pp.

Pearson, E. S., and H. O. Hartley. 1958. *Biometrika Tables for Statisticians*. Vol. I. 2nd ed. Cambridge University Press, London. 240 pp.

Phillips, E. F. 1929. Variation and correlation in the appendages of the honeybee. *Cornell Univ. Agric. Exp. Sta. Mem.* **121**. 52 pp.

Phillips, J. R., and L. D. Newsom. 1966. Diapause in *Heliothis zea* and *Heliothis virescens* (Lepidoptera: Noctuidae). *Ann. Entomol. Soc. Amer.*, **59**:154–159.

Pimentel, R. A. 1958. Taxonomic methods, their bearing on sub-speciation. *Syst. Zool.*, **7**:139–159.

Powick, W. C. 1925. Inactivation of vitamin A by rancid fat. *J. Agric. Res.*, **31**:1017–1027.

Price, R. D. 1954. The survival of *Bacterium tularense* in lice and louse feces. *Am. J. Trop. Med. Hyg.*, **3**:179–186.

Radi, M. H., and D. C. Warren. 1938. Studies on the physiology and inheritance of feathering in the growing chick. *J. Agr. Res.*, **56**:679–707.

Rao, C. R. 1952. *Advanced Statistical Methods in Biometric Research*. Wiley, New York. 390 pp.

Rohlf, F. J. 1965. A randomization test of the non-specificity hypothesis in numerical taxonomy. *Taxon*, **14**:262–267.

Scheffé, H. 1959. *The Analysis of Variance*. Wiley, New York. 477 pp.

Seal, H. 1964. *Multivariate Statistical Analysis for Biologists*. Wiley, New York. 207 pp.

Shapovalov, L., and H. C. Taft. 1954. The life histories of the Steelhead Rainbow Trout (*Salmo gairdneri gairdneri*) and Silver Salmon (*Oncorhynchus kisutch*) with special reference to Waddell Creek, California, and recommendations regarding their management. *Calif. Div. Fish Game, Fish Bull.*, **98**. 375 pp.

Siegel, S. 1956. *Nonparametric Statistics for the Behavioral Sciences*. McGraw-Hill, New York, Toronto, and London. 312 pp.

Simpson, G. G. 1961. *Principles of Animal Taxonomy*. Columbia University Press, New York. 440 pp.

Simpson, G. G., A. Roe, and R. C. Lewontin. 1960. *Quantitative Zoology*. Rev. ed. Harcourt, Brace and Co., New York. 440 pp.

Sinnott, E. W. and D. Hammond. 1935. Factorial balance in the determination of fruit shape in *Cucurbita*. *Amer. Nat.*, **64**:509–524.

Skinner, J. J., and F. E. Allison. 1923. Influence of fertilizers containing borax on growth and fruiting of cotton. *J. Agric. Res.*, **23**:433–445.

Smith, F. L. 1939. A genetic analysis of red seed-coat color in *Phaseolus vulgaris*. *Hilgardia*, **12**:553–621.

Snedecor, G. W. 1956. *Statistical Methods*. 5th ed. Iowa State College Press, Ames. 534 pp.

Sokal, R. R. 1958. Quantification of systematic relationships and of phylogenetic trends. *Proc. Tenth Internat. Congr. Entomol.*, **1**:409–415.

Sokal, R. R. 1962. Variation and covariation of characters of alate *Pemphigus populi-transversus* in eastern North America. *Evolution*, **16**:227–245.

Sokal, R. R. 1965. Statistical methods in systematics. *Biol. Rev.* (Cambridge), **40**:337–391.

Sokal, R. R. 1966. Pupation site differences in *Drosophila melanogaster*. *Univ. Kansas Sci. Bull.*, **46**:697–715.

Sokal, R. R. 1967. A comparison of fitness characters and their responses to density in stock and selected cultures of wild type and black *Tribolium castaneum*. *Tribolium Inf. Bull.*, **10**:142–147.

Sokal, R. R., and P. E. Hunter. 1955. A morphometric analysis of DDT-resistant and non-resistant housefly strains. *Ann. Entomol. Soc. Amer.* **48**:499–507.

Sokal, R. R., and I. Karten. 1964. Competition among genotypes in *Tribolium castaneum* at varying densities and gene frequencies (the black locus). *Genetics*, **49**:195–211.

Sokal, R. R., and R. C. Rinkel. 1963. Geographic variation of a late *Pemphigus populi-transversus* in Eastern North America. *Univ. Kansas Sci. Bull.*, **44**:467-507.

Sokal, R. R., and P. H. A. Sneath. 1963. *Principles of Numerical Taxonomy*. W. H. Freeman and Co., San Francisco and London. 359 pp.

Sokal, R. R., and F. J. Sonleitner. 1969. The ecology of selection in hybrid populations of *Tribolium castaneum*. *Ecol. Monogr.*, **38**. (In press)

Sokal, R. R., and P. A. Thomas. 1965. Geographic variation of *Pemphigus populi-transversus* in Eastern North America: Stem mothers and new data on alates. *Univ. Kansas Sci. Bull.*, **46**:201–252.

Sokoloff, A. 1955. Competition between sibling species of the *Pseudoobscura* subgroup of *Drosophila*. *Ecol. Monogr.*, **25**:387–409.

Sokoloff, A. 1966. Morphological variation in natural and experimental populations of *Drosophila pseudoobscura* and *Drosophila persimilis*. *Evolution*, **20**:49–71.

Steel, R. G. D., and J. H. Torrie. 1960. *Principles and Procedures in Statistics*. McGraw-Hill, New York. 481 pp.

Student (W. S. Gossett). 1907. On the error of counting with a haemacytometer. *Biometrika*, **5**:351–360.

Sullivan, R. L., and R. R. Sokal. 1965. Further experiments on competition between strains of houseflies. *Ecology*, **46**:172–182.

Swanson, C. O., W. L. Latshaw, and E. L. Tague. 1921. Relation of the calcium content of some Kansas soils to soil reaction by the electrometric titration. *J. Agr. Res.*, **20**:855–868.

Tate, R. F., and G. W. Klett. 1959. Optimal confidence intervals for the variance of a normal distribution. *J. Am. Stat. Assoc.*, **54**:674–682.

Teissier, G. 1960. Relative Growth, In *The Physiology of Crustacea*. Volume 1. T. H. Waterman (ed.). Academic Press, New York. pp. 537–560.

Thomson, G. H. 1951. *The Factorial Analysis of Human Ability*. 5th ed. Houghton Mifflin, New York. 383 pp.

Tukey, J. W. 1951. Quick and dirty methods in statistics, part II, Simple analyses for standard designs. *Proc. 5th Ann. Convention, Amer. Soc. for Quality Control.* pp. 189–197.

Tukey, J. W., and M. B. Wilk. 1965. Data analysis and statistics: techniques and approaches. Proceedings of the Symposium of Information Processing in Sight Sensory Systems, 1-3 November 1965, California Institute of Technology, Pasadena.

Utida, S. 1943. Studies on experimental population of the Azuki bean weevil, *Callosobruchus chinensis* (L.). VIII. Statistical analysis of the frequency distribution of the emerging weevils on beans. *Mem. Coll. Agr. Kyoto Imp. Univ.*, **54**:1–22.

Verner, J. 1964. Evolution of polygamy in the Long-billed Marsh Wren. *Evolution*, **18**:252–261

Vollenweider, R. A., and M. Frei. 1953. Vertikale und zeitliche Verteilung der Leitfähigkeit in einem eutrophen Gewässer während der Sommerstagnation. *Schweiz. Z. Hydrol.* **15**:158–167.

Von Frisch, K. 1950. *Bees. Their Vision, Chemical Senses, and Language.* Cornell University Press, Ithaca, N.Y. 199 pp.

Webber, L. G. 1955. The relationship between larval and adult size of the Australian sheep blowfly *Lucilia cuprina* (Wied.). *Austral. J. Zool.*, **3**:346–353.

Weir, J. A. 1949. Blood *p*H as a factor in genetic resistance to mouse typhoid. *J. Inf. Dis.*, **84**:252–274.

Whitaker, T. W. 1944. The inheritance of chlorophyll deficiencies in cultivated lettuce. *J. Heredity*, **35**:317–320.

Whittaker, R. H. 1952. A study of summer foliage insect communities in the Great Smoky Mountains. *Ecol. Monogr.*, **22**:1–44.

Williams, C. B. 1964. *Patterns in the Balance of Nature.* Academic Press, London and New York. 324 pp.

Willis, E. R., and N. Lewis. 1957. The longevity of starved cockroaches. *J. Econ. Entomol.*, **50**:438-440.

Woodson, R. E., Jr. 1964. The geography of flower color in butterflyweed. *Evolution*, **18**:143–163.

Wuhrmann, K., and H. Woker. 1953. Über die Giftwirkungen von Ammoniak und Zyanidlösungen mit verschiedener Sauerstoffspannung und Temperatur auf Fische. *Schweiz. Z. Hydrol.* **15**:235–260.

Yule, G. U., and M. G. Kendall. 1950. *An Introduction to the Theory of Statistics.* 14th ed. Hafner, New York. 701 pp.

INDEX